Electrical Systems

based on the

2011 NEC®

National Electrical Code

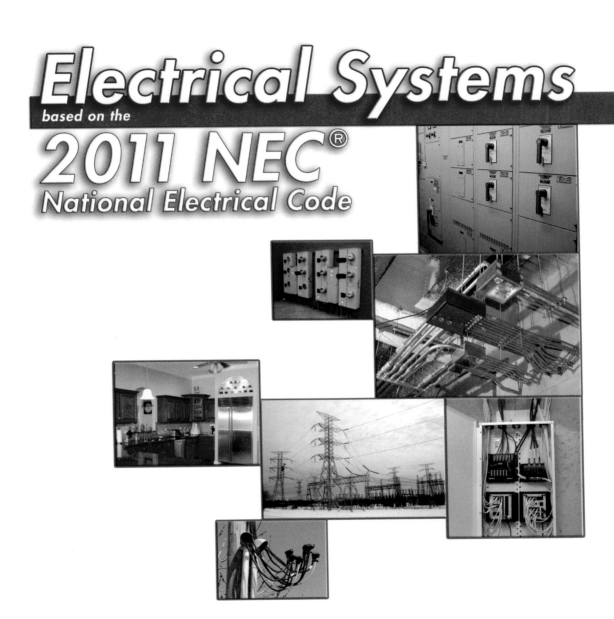

AMERICAN TECHNICAL PUBLISHERS
ORLAND PARK, ILLINOIS 60467-5756

Michael I. Callanan
Bill Wusinich

American Technical Publishers, Inc., Editorial Staff

Editor in Chief:
Jonathan F. Gosse
Vice President—Production:
Peter A. Zurlis
Art Manager:
James M. Clarke
Multimedia Manager:
Carl R. Hansen
Technical Editor:
Scott C. Bloom
Copy Editor:
Catherine A. Mini
Cover Design:
Jennifer M. Hines
Illustration/Layout:
Jennifer M. Hines
Samuel T. Tucker
Melanie G. Doornbos
CD-ROM Development:
Gretje Dahl
Adam T. Schuldt
Robert E. Stickley
Nicole S. Polak
Daniel Kundrat

1 2 3 4 5 6 7 8 9 – 11 – 9 8 7 6 5

Printed in the United States of America

ISBN 978-0-8269-1644-0

This book is printed on recycled paper.

Acknowledgments

The authors and publisher are grateful to the following companies and organizations for providing technical information and assistance.

ABB Power T&D Company, Inc.
Atlas Technologies, Inc.
Baldor Electric Co.
Carrier Corporation
Cooper Bussmann
Cooper Crouse-Hinds
Cutler-Hammer
Delta Star, Inc.
Federal Pacific
Federal Signal Corporation
Greenlee Textron, Inc.
Henny Penny Corporation
Hubbell Incorporated (Delaware), Wiring Devices-Kellems
Ideal Industries, Inc.
Maytag
Ruud Lighting, Inc.
Saftronics, Inc
Solar World Industries America
Square D Company
Steel Tube Institute
Wacker Corporation
W. W. Grainger, Inc.

Contents

Interactive CD-ROM Contents

- Using This Interactive CD-ROM
- Quick Quizzes®
- Illustrated Glossary
- Flash Cards
- Interactive Load Calculation Forms
- Media Clips
- ATPeResources.com

Introduction

Electrical Systems Based on the 2011 NEC® is designed for use by journeyman and master electricians, inspectors, contractors, electrical designers, and others who use the National Electrical Code® (NEC®). The textbook includes an overview of the NEC®, wiring methods and materials, conductors and overcurrent protective devices, branch circuits and feeders, grounding, transformers, services, special locations, and calculations. Included procedures cover all aspects of applying the Code when designing electrical systems.

Each chapter begins with a listing of important changes in the new 2011 NEC®. Electrical system topics and related Code references are thoroughly covered. Located after the chapter opener, Learning the Code is designed to be a conversational sharing of insights and hands-on trade experience with learners. Topics include the relevance of specific parts of the Code and how the Code is typically applied in the field. Calculation forms have been designed specifically for this book. Blank copies of these forms are located at the back of the book and may be reproduced for instructional use.

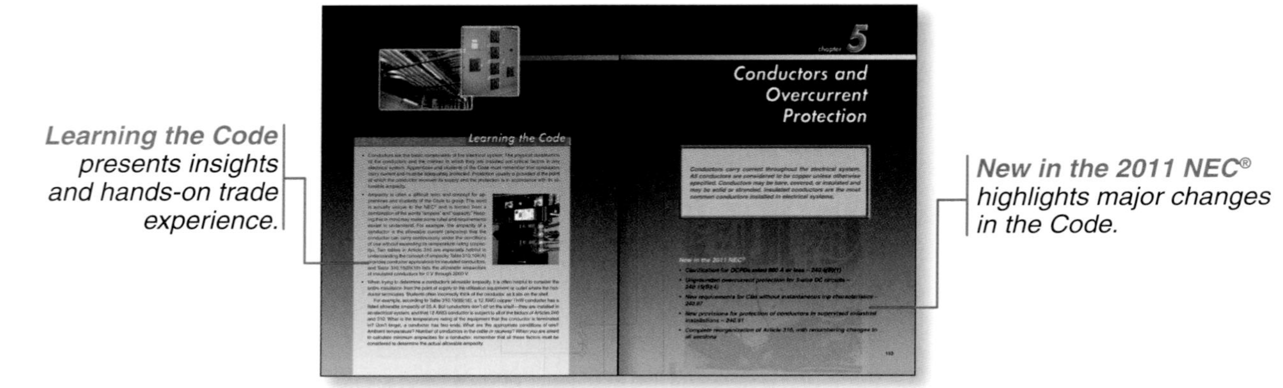

Learning the Code presents insights and hands-on trade experience.
chapter 5
Conductors and Overcurrent Protection
Learning the Code
New in the 2011 NEC® highlights major changes in the Code.

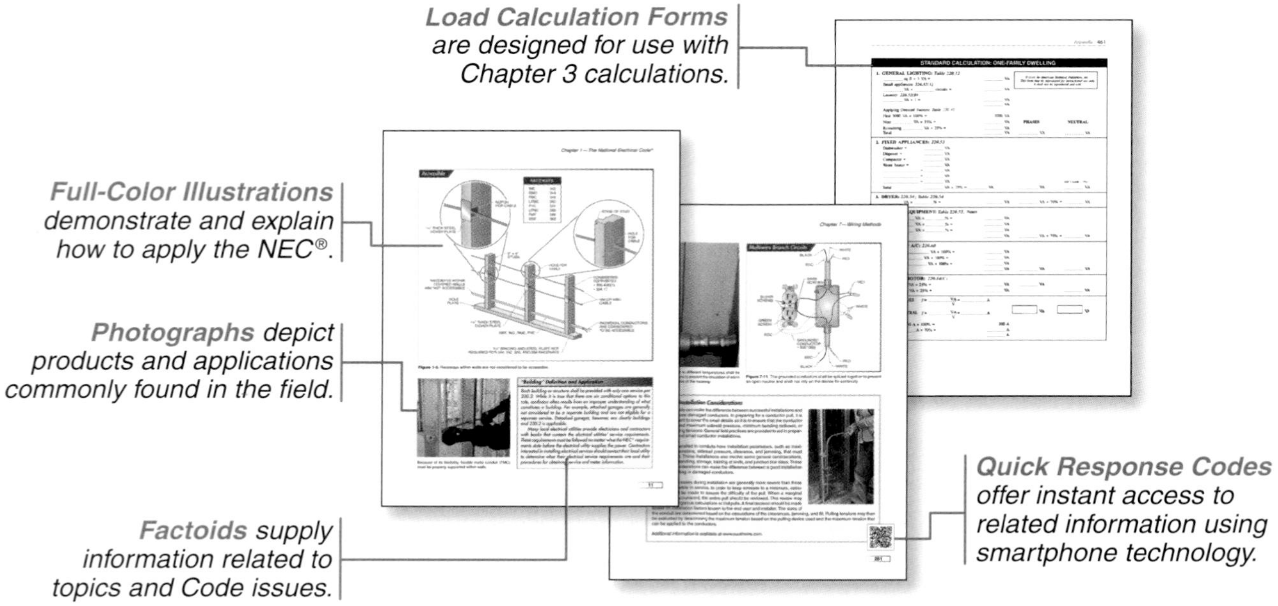

Load Calculation Forms are designed for use with Chapter 3 calculations.
Full-Color Illustrations demonstrate and explain how to apply the NEC®.
Photographs depict products and applications commonly found in the field.
Factoids supply information related to topics and Code issues.
Quick Response Codes offer instant access to related information using smartphone technology.

REVIEW QUESTIONS AND TRADE COMPETENCY TESTS

Electrical Systems Based on the 2011 NEC® contains Review Questions and Trade Competency Tests at the end of Chapters 1–13. These tests should be completed after studying the corresponding chapter. Chapter 14 contains the Final Exam. This exam includes questions drawn from the content of the entire text. The Final Exam should be taken after completing the text, Review Questions, and Trade Competency Tests.

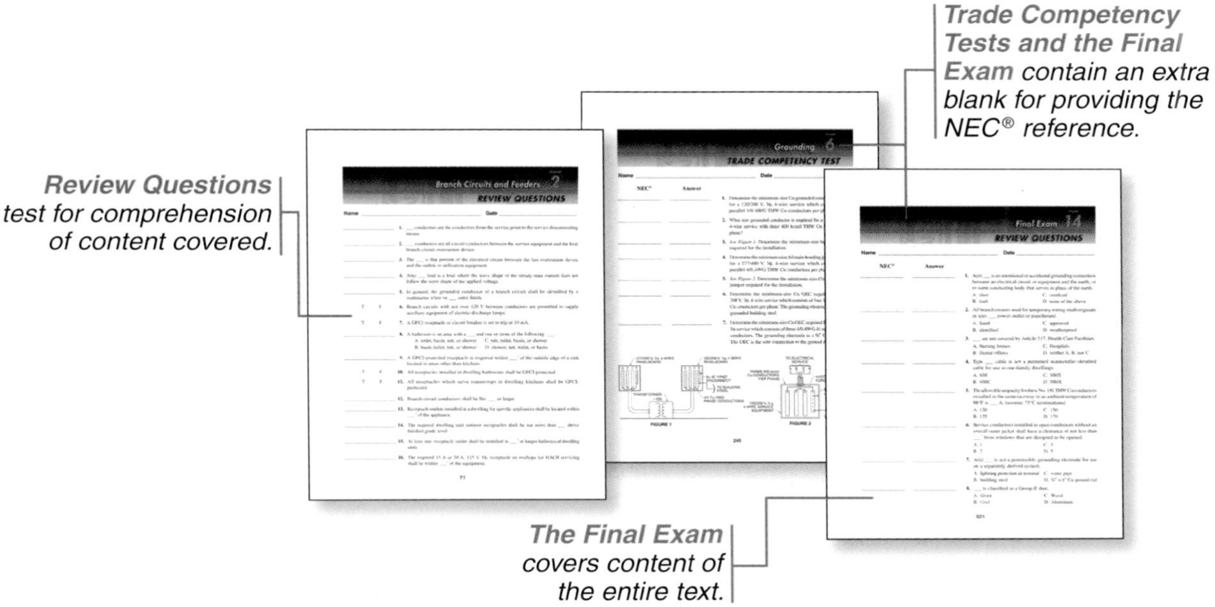

Trade Competency Tests and the Final Exam contain an extra blank for providing the NEC® reference.
Review Questions test for comprehension of content covered.
The Final Exam covers content of the entire text.
Branch Circuits and Feeders 2
REVIEW QUESTIONS
Grounding 6
TRADE COMPETENCY TEST
Final Exam 14
REVIEW QUESTIONS

Review question formats include true-false, multiple choice, and completion. Always record the answer in the space provided or circle either T or F as appropriate. Trade Competency Tests and the Final Exam contain an additional blank for providing the NEC® reference, if required by your instructor. Answers and NEC® references for Review Questions and Trade Competency Tests are given in the *Answer Key.*

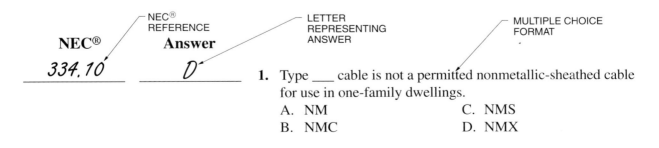

NEC® REFERENCE — LETTER REPRESENTING ANSWER — MULTIPLE CHOICE FORMAT

NEC®	Answer	
334.10	D	**1.** Type ___ cable is not a permitted nonmetallic-sheathed cable for use in one-family dwellings.

A. NM   C. NMS
B. NMC   D. NMX

Interactive CD-ROM Features

The *Electrical Systems Based on the 2011 NEC®* CD-ROM in the back of the book is designed as a study aid to enhance book content.

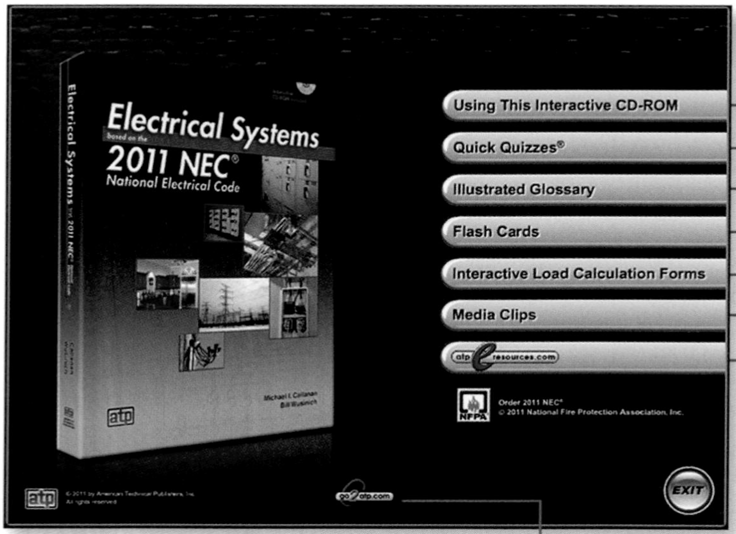

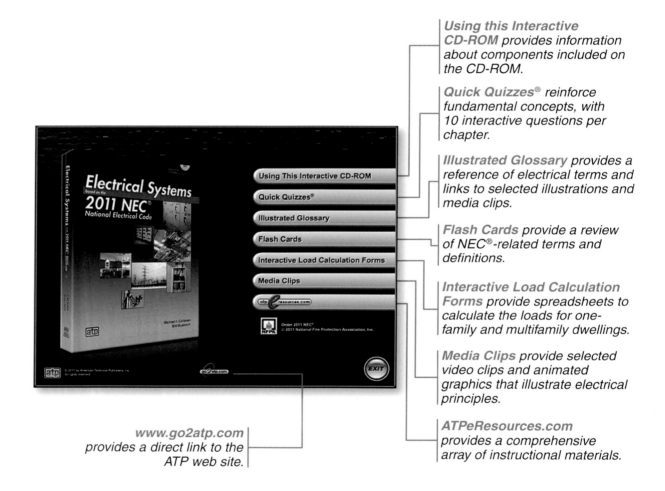

Using this Interactive CD-ROM provides information about components included on the CD-ROM.

Quick Quizzes® reinforce fundamental concepts, with 10 interactive questions per chapter.

Illustrated Glossary provides a reference of electrical terms and links to selected illustrations and media clips.

Flash Cards provide a review of NEC®-related terms and definitions.

Interactive Load Calculation Forms provide spreadsheets to calculate the loads for one-family and multifamily dwellings.

Media Clips provide selected video clips and animated graphics that illustrate electrical principles.

ATPeResources.com provides a comprehensive array of instructional materials.

www.go2atp.com provides a direct link to the ATP web site.

Information about using the *Electrical Systems Based on the 2011 NEC®* CD-ROM is included on the last page of the book. To obtain information about related training products, visit the American Technical Publishers web site at www.go2atp.com.

The Publisher

About the NEC®

NATIONAL ELECTRICAL CODE®

The National Electrical Code® is sponsored and controlled by the National Fire Protection Association, Inc. (NFPA). The primary function of the NEC® is to safeguard people and property against electrical hazards. It is mandatory that *Electrical Systems Based on the 2011 NEC®* be used only in conjunction with the current NEC®. Copies of the current NEC® (NFPA No. 70) may be ordered directly from its publisher:

National Fire Protection Association, Inc.
1 Batterymarch Park
Quincy, MA 02169

WATTS AND VOLT-AMPERES

In general, within the NEC®, the term watts (W) has been superseded by the term volt-amperes (VA) for the computation of loads. However, references to nameplate ratings still reflect the term watts on certain loads.

CALCULATIONS

When total wattage or VA is to be divided by phase-to-phase (3ϕ) voltage times 1.732, the following values may be substituted:

for 208 V × 1.732, use 360
for 230 V × 1.732, use 398
for 240 V × 1.732, use 416
for 440 V × 1.732, use 796
for 460 V × 1.732, use 797
for 480 V × 1.732, use 831
for 2400 V × 1.732, use 4157
for 4160 V × 1.732, use 7205

MANDATORY USE OF SHALL

Section 90.5(A) states that mandatory rules use the word *shall*. Always refer to the NEC® for mandatory rules.

COPPER CONDUCTORS

Unless otherwise specified, copper conductors are sized on THW per Table 310.15(B)(16).

Learning the Code

- Studying the National Electrical Code® (NEC®) is a rewarding but often challenging task. The NEC® contains the rules and requirements for installing electrical systems and equipment but does not address the reasons for such rules. Apprentices and students are often left with the difficult task of understanding and applying the Code provisions. Making the task even more challenging, the NEC® has its own language and format. To aid apprentices and students of the Code, a "Learning the Code" sidebar is included at the beginning of each chapter. The points contained in the sidebar are designed and intended to assist apprentices and students in navigating their study of the NEC®.

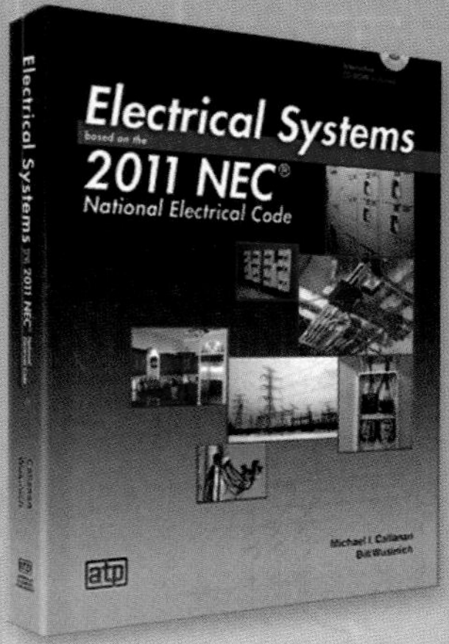

- Chapter 1 of *Electrical Systems* describes the Code-making process and the structure and format of the NEC®. In time, students of the Code will probably want to get involved with the Code development process. For now, aim to have a basic understanding of the process. Most importantly, it is helpful to understand the organizations and companies that are active in the Code-making process.
- Article 100 is the dictionary of the NEC®. Make certain to understand these definitions, as they are critical to understanding many Code rules and requirements. As discussed in Chapter 1, remember that not all terms are defined in Article 100. Only those terms that appear in two or more articles are included.
- Article 110 is a "bread and butter" article of the Code. It is too often misunderstood or neglected. Make sure you spend enough time studying the provisions of this article. These are general provisions that apply throughout the Code.

The National Electrical Code®

The defined purpose of the Code is the practical safeguarding of persons and property from hazards arising from the use of electricity. The NEC® was first published in 1897, and today is published every three years. The NEC® is published by the National Fire Protection Association, Inc. (NFPA).

New in the 2011 NEC®

- *Clarification of the NEC® scope – 90.2(B)(5)(c)(d)*
- *The addition of Informative Annexes – 90.5(D)*
- *New definition of arc-fault circuit interrupter – Article 100*
- *Revised definition of bathroom – Article 100*
- *Revised definition of ground fault – Article 100*
- *Field marking requirements for available fault currents – 110.24*
- *Revised requirements and exceptions for working space 600 V and less – 110.26(A)*
- *New requirements for electrical vaults over 600 V – 110.31(A)*

THE NATIONAL ELECTRICAL CODE®

The National Electrical Code® (NEC®) is one of the most widely used and recognized consensus standards in the world today. It is a true consensus standard because members from throughout the electrical industry contribute to its development. The NEC® is revised and updated every three years to reflect current trends in the electrical industry. The purpose of the NEC® is the practical safeguarding of persons and property from hazards arising from the use of electricity.

NEC® Process

The National Fire Protection Association, Inc. (NFPA) sponsors the development of the NEC®. The NFPA publishes guidelines in the Regulations Governing Committee Projects for the procedures for all of the standards it publishes. For the NEC®, the procedures call for essentially a five-step process:

1. The Call for Proposals
2. Report on Proposals (ROP)
3. Report on Comments (ROC)
4. The Technical Report Session of the NFPA Annual Meeting
5. Standards Council Consideration and Issuance

The Call for Proposals. The first step in the Code process involves issuing a public notice that the NEC® revision process has begun and asking for interested parties to submit proposals for revising the Document. The notice is placed in appropriate publications such as the *NFPA News*, the *U.S. Federal Register*, and the American National Standards Institute's *Standards Action*. This step provides approximately twenty weeks for respondents to submit their proposals.

Report on Proposals. The second step in the code process is the receipt of proposals. Anyone can submit a proposal to change the NEC® provided it contains the required information. **See Figure 1-1.**

The key to successful code proposals is proper substantiation for the proposed changes. The time window to submit proposals is very small. Usually, the closing date for receipt of proposals is in the fall of the same year a new Code is issued. For example, the last date to submit changes for the 2011 NEC® was November 2008. After the last date to receive proposals, the Code-Making Panels meet to discuss and vote on each of the proposed changes. These votes are recorded and published in the Report on Proposals.

NEC® Proposals

NEC® proposals shall contain:

1. Identification of the submitter (name and organization or company affiliation).
2. Identification of the specific Code section and type of revision that is proposed.
3. A statement of the problem and substantiation for why the change is necessary.
4. The actual wording of the revised text or the wording to be deleted.

Figure 1-1. NEC® proposals shall contain the required information in the order to be acted on.

Receipt of Comments. Once the Report on Proposals is published, everyone has the opportunity to submit a comment on each of the proposed changes whether the Code-Making Panel voted to accept or reject them. A closing date for comments is published, and after the closing date the Code-Making Panel meets to vote on each of the comments.

The Report on Comments contains all comments and the votes of the respective Code-Making Panels. If a proposal or comment is not accepted by the Code-Making Panel, the statement must state the reason the proposal or comment was not accepted. New proposals cannot be submitted during this stage of the Code cycle. The Code-Making Panel can only take action on proposals that have received adequate public review during the proposal stage.

NFPA Annual Meeting. Once the Code-Making Panels have reviewed the proposals and comments, the next step is to present the changes at the NFPA Annual Meeting. Floor action on individual proposals can occur during this meeting. Beginning with the 2008 NEC®, the NFPA modified the method by which floor actions can be implemented at the NFPA Annual Meeting. The revised process permits votes only in cases where a Notice of Intent to Make a Motion (NITMAM) has been properly filed.

Voting on floor actions at the NFPA Annual Meeting is limited to NFPA members only. A simple majority vote is required for a floor action to pass. Actions that occur on the floor of the NFPA Annual Meeting are still subject to review by the NFPA Standards Council.

Standards Council Issuance. The Standards Council has the responsibility for overseeing all of the codes and standards developed for the NFPA. The NEC® Correlating Committee works directly under the Standards Council. The Correlating Committee steers the Code panels through the process, ensuring that each proposal and comment received is handled according to an established operating procedure. Once the process is complete, the Standards Council reviews the entire process and actually issues the document for publication.

The NEC® is a legal document designed to be adopted by local and/or state governmental bodies. Local jurisdictions may choose to adopt the Code in its entirety, with specific additions or exceptions, or they may choose not to adopt the Code at all.

Code-Panel Membership. Each NEC® proposal or comment is reviewed by the representatives of various segments of the electrical industry. There are 20 Code-Making Panels to cover the articles in the NEC®. The members of each Code panel represent labor, manufacturing, electrical utilities, electrical inspectors, contractor associations, and testing laboratories.

NFPA Operating Procedures for the National Electrical Code® Committee state that no interest group shall comprise more than one-third of the total voting panel membership. Membership is designated as either principal or alternate. Alternate members assume the participation and voting rights only when the principal member is not present.

The official vote on panel proposals and comments occurs on a written ballot after the meetings conclude. Although each proposal is voted on during the Code-Making Panel discussions, the official vote on each proposal occurs when the Code-Making Panel member returns a written ballot which was mailed after the conclusion of the Code-Making Panel meetings. All members who vote against a panel action must state the reasons for doing so, and their comments are recirculated to each panel member.

Using the NEC®

The NEC® is available in soft-cover, loose-leaf, and CD-ROM versions. While the NEC® is not difficult to use, it does have its own method of organization that must be understood for ease of use.

Revisions and Extracted Text. Beginning with the 2008 NEC®, vertical bars are no longer the only means to identify code changes. Shading behind the revised text is now used to identify changes. **See Figure 1-2.** The Code also contains extracted text that has been taken from other NFPA documents identified with brackets. The number, title and edition of the NFPA document from which the extract is taken appears at the beginning of the article in which the extract has been used. For example, in Article 517, extracted text from NFPA 99-2002 *Standard for Health Care Facilities* appears throughout the article. These editorial markings are important for helping to trace the development of Code requirements. Rules for formatting the NEC® are found in Chapter 2 of the NFPA *National Electrical Code®* Style Manual.

Inspections of work in progress by building officials certify that all applicable codes are met before occupancy occurs.

NEC® Revisions and Extracted Text	
SHADED TEXT	Shading behind text is used to indicate revisions from the previous issue of the Code. For example, see Article 100 where the definition of bathroom had been revised from the previous edition of the NEC®.
[EXTRACTED TEXT]	Brackets are used to denote extracted text from another document that appears in the Code. For example, see 517.35(A) which was extracted from the *Standard for Health Care Facilities*, NPFA 99-2002, Section 4.4.1.1.4.

Figure 1-2. NEC® editorial markings are useful tools for studying the latest NEC® changes.

Outline Format. The NEC® is arranged in a simple outline format. Section numbers are designated by a period following the article number. For example, 210.8 indicates that the NEC® section is taken from Article 210, Section 8.

Generally articles and sections are numbered consecutively but there are occasions where gaps are left in the numbering sequence. This occurs when sections have been deleted from the NEC® or perhaps if the Code-Making Panel wanted to leave room for future Code sections.

Subsections are part of the sections and are designated by lowercase letters set off in parentheses. For example, 210.8(A) is a subsection of 210.8. Subsections are further broken down by numbers to indicate parts of subsections. For example, 210.8(A)(4) is a part of subsection 210.8(A). In a few instances, the NEC® has further subsections of subsections. In most cases, the format remains in this number-letter-number sequence. The only time the NEC® deviates from this fashion is when sections do not contain any subsections, but do contain numbered items. For example, 230.43(4) is a numbered item of Section 230.43.

Because the NEC® is organized in an outline format, exceptions are applied only to the section they follow or as noted. In some cases, where lists are utilized, the exception will specifically list the number of the list to which the exception applies. See for example, the exceptions following 210.8(A)(3) and 210.8(B)(4).

Exceptions. Exceptions permit alternate methods to the general (main) rule. Exceptions are permitted in lieu of the main rule when all of the provisions of the exception are met. For example, 210.8(A)(5) requires that, in general, all receptacles in dwelling unit unfinished basements shall be ground-fault circuit interrupter (GFCI) protected. There is an exception to this section, however, that permits receptacles that supply only a permanently installed fire alarm or burglar alarm system to be installed without GFCI protection. The reason for the exception is to ensure that these life-safety and security systems remain in operation and are not inadvertently deenergized or rendered inoperable in the event of a GFCI tripping.

Similarly, in 210.8(B)(3) there is an exception from the requirement for GFCI protection for outdoor locations for snow-melting or deicing equipment. Note that where exceptions contain qualifying provisions or conditions, all of the conditions of the exception must be met to apply the exception. In this case, the snow-melting or deicing equipment must be 1) not readily accessible, and 2) supplied from a dedicated branch circuit. Section 210.52(E), *Outdoor Outlets*, provides a little guidance in determining if the outlet is readily accessible by providing a maximum elevation of 6½′ above grade for an outlet to be considered accessible from grade level.

Informational Notes. Prior to the 2011 NEC®, Fine Print Notes, or FPNs, were used throughout the Code to provide additional explanatory information on certain Code requirements. Frequently, the FPNs were used to cross-reference related Code provisions. Sometimes they were used to provide explanatory information or to reference other appropriate industry standards.

In any case, FPNs were never intended to include mandatory requirements. In the 2011 NEC, the term FPN was dropped, and a new term, "Informational Note," was used to replace all existing references to FPNs. Section 90.5(C) covers Informational Notes. Informational Notes, like FPNs, contain only explanatory information and do not contain any Code requirements, so the information remains an unenforceable part of the Code.

For example, Section 110.11 prohibits the exposure of electrical conductors and equipment to conditions or agents that could have a deteriorating effect on the conductors or equipment. Informational Note No. 2 reminds users of the Code that some cleaning and lubricating agents could potentially damage conductor insulation. This forewarns the users that when using cable preparation agents, they should avoid using compounds that could have a deteriorating effect on cable insulation.

Layout. The arrangement of the NEC® is specified in 90.3. There is an Introduction and nine chapters in the NEC®. The first four chapters apply generally, unless they are modified by the latter chapters. For example, 300.14 requires that, in general, a minimum of 6″ of free conductor be provided at each outlet for splices or terminations. This applies throughout the NEC® unless modified.

Since 1937 Chapter 8 of the Code had been granted "independent" status. This had been widely interpreted as to the exact applicability of other Code provisions to the articles within Chapter 8. During the 2002 NEC® cycle, several revisions were made to sections 90.2, 90.3 and 90.4 to clarify that, while Chapter 8 is not subject to the requirements of Chapter 1 through 7 in general, specifically referenced provisions do apply and in all cases, communication systems are within the scope of the Code and ought to be inspected and installed in a manner consistent with other requirements in the Code.

Chapter 9 contains tables which are referenced throughout the entire NEC® and are applicable as referenced. The Informative Annexes, which appear at the end of the Code, do not contain any mandatory provisions, but they do contain information that may be helpful in designing or installing electrical installations. For example, Informative Annex C contains many useful tables for determining the maximum number of conductors permitted in a raceway.

Applying the NEC®

The National Electrical Code® is adopted for use by local (village, town, city, etc.), county or parish, and state authorities. These authorities are generally represented by building officials who issue permits for jobs and make periodic inspections of work in progress. Additionally, these authorities certify that the completed building meets all applicable codes before occupancy occurs.

Cutler-Hammer

The two-fold purpose of the NEC® is to protect people and property from the dangers associated with the use of electricity.

Administration and Enforcement

Informative Annex H of the Code is intended to provide a set of administrative and enforcement provisions to supplement the provisions of the Code. Informative Annex H is not a mandatory part of the NEC® and is included for informational purposes only. These provisions cover 1) inspection of electrical installations, 2) investigation of electrical fires, 3) electrical plan review, 4) electrical system design and use provisions, and 5) regulatory provisions for special events. The placement of these rules in Informative Annex H is intended to make them only applicable where local jurisdictions that adopt the NEC® specifically adopt Informative Annex H. See Informative Annex H, Administration and Enforcement.

The NEC® contains requirements that are necessary for safety and is not intended to be an instruction manual.

Purpose and Intent – 90.1. The purpose of the Code is the practical safeguarding of persons and property from hazards arising from the use of electricity. This two-fold purpose should be kept in mind whether applying the NEC® or submitting new proposals to modify the NEC®.

It is important to note that the NEC® only contains requirements that are necessary for safety. Section 90.1(B) clearly states that following the provisions of the NEC® will only result in an installation that is "essentially free from hazard." The NEC® rules and requirements are a minimum standard. Following the NEC® provisions does not ensure an electrical installation that will be suitable for future use or expansion. This point is especially important for those charged with designing electrical systems and installations.

The NEC® is not intended to be used as an instruction manual per 90.1(C). As such, reasons for the inclusion of the rules are not provided in the Code text. In order to understand the reason for a rule, one would need to know when the rule was placed in the NEC®. The next step would be to find a copy of the Report on Proposals for that year's version of the Code and read the substantiation for the proposal that was accepted by the Code panel. For example, Article 353 covers the use, installation, and construction specifications for high-density polyethylene (HDPE) conduit and its associated fittings. The article was new to the 2005 NEC®. The Report on Proposals for the 2005 Code cycle, in the proposal to add the article, indicates in the substantiation that HDPE is a listed product that is restricted in its uses and that the new article is needed to provide the installation requirements and construction specifications for HDPE.

The NEC® is also not a design specification per 90.1(C). It contains the rules and necessary provisions for electrical systems, but leaves the design and layout of the electrical systems to others.

Scope – 90.2. Before applying the NEC®, the scope should be reviewed to determine if the electrical installation falls within the jurisdiction of the Code. Installations which are not covered are given in 90.2(B). Often, the distinction between what is covered and what is not covered is very fine and can lead to some difficult interpretations. For example, installations for electrical utilities are not covered if they involve the metering, generation, transmission, or distribution of electric energy. Those installations, however, that are used by an electric utility but are not an integral part of the electrical energy transmission or distribution are covered.

Over the past several Code cycles, there has been considerable discussion and debate as to the definition of an electrical utility and the scope of a utility's work that is subject to the provisions of the Code. An Informational Note to 90.2(B)(4)(5) is intended to clarify that utilities are entities designated or recognized by law or regulation that install, maintain, and operate the electric supply. The Informational Note also states that the utilities may be subject to Code provisions and are not intended to receive a blanket exemption from all codes and standards compliance.

Attempts to provide additional clarification of the scope of the NEC® as it relates to installations under the exclusive control of an electrical utility continue in the 2011 NEC®. Sections 90.2(B)(5)(c) and (d) were both revised to more clearly state how an electrical utility property or a structure may be exempt from NEC® provisions if it is located in a legally established easement or right-of-way or if other written agreements are designated by public service or utility commissions. These questions need to be clearly and precisely defined and agreed upon by the local parties and the AHJ before any projects are begun.

Enforcement – 90.4. All applications of the NEC® are ultimately subject to approval by the authority having jurisdiction (AHJ). The NEC® is designed to be used as a legal document. Any interpretations and approval of equipment rest with the AHJ. In many cases, the AHJ is a local or municipal building official.

Some localities permit the use of independent third-party electrical inspection agencies. In other cases, the AHJ may be a governmental official, local fire marshal, or an insurance official with jurisdiction over the installation. In any case, whoever the AHJ is, they can permit alternate methods or even waive specific requirements if they believe the installation is equally effective at meeting the intended objective. **See Figure 1-3.**

Authority Having Jurisdiction (AHJ)

Duties and responsibilities:

1. Make interpretations of NEC® rules.
2. Approve equipment and materials.
3. Grant special permission. (see Art. 100 definition)
4. Waive specific requirements. (must be in writing)
5. Permit alternate methods. (must be in writing)

Figure 1-3. The authority having jurisdiction (AHJ) is responsible for enforcement of the NEC® per 90.4.

Requirements and Explanatory Provisions – 90.5

The Code contains two types of rules. The first are mandatory rules and the second are permissive. Mandatory rules are designated by the use of the word "shall" or "shall not." For example, 110.12 requires that electrical equipment be installed in a "neat and workmanlike manner." This provision is mandatory.

Permissive rules are rules that are not required but are permitted. These are designated by the use of terms like "shall be permitted" or "shall not be required." For example, see 250.21(A) which lists AC systems of 50 volts to 1000 volts that shall be permitted to be grounded, but shall not be required to be grounded. This is an example of permissive rules.

Equipment Examination – 90.7

In general, equipment and materials referred to in the Code are not required to be examined prior to use. It is intended that factory-installed internal wiring is sufficient and not within the scope of the installer's responsibility. The only exception to this rule is where such equipment or materials have been damaged or altered to a degree that requires examination by the installer. However, installers of electrical equipment need to ensure that the equipment is listed, identified, or otherwise labeled as being suitable for the intended use. They should also be aware that many state and local authorities require product listing for equipment used within their jurisdiction. In addition, OSHA includes requirements for listing and listing laboratories in its code. Article 100 provides definitions of "listed," "labeled," and "identified."

Future Use – 90.8. Sections 90.1(B) and 90.8, taken together, indicate that installations completed in accordance with the NEC® merely meet a minimum established standard. The installations may not be adequate for future use, and while it is not required, allowance for future expansion is generally in the best interest of all concerned. Essentially, the NEC® is a minimum standard. The designer or installer can choose to provide for future use if desired. Installations in accordance with the NEC® are not necessarily efficient and may not meet the needs and demands for electricity.

Metric and Standard Units – 90.9

The NEC® utilizes a dual system of measurement. Metric units of measurement are shown utilizing the International System of Units (SI). The US customary units are shown using standard inch-pound units. Section 90.9(B) requires that the SI units appear first and the inch-pound units follow immediately and are shown in parentheses. Section 90.9(D) states that compliance with either unit of measurement is permitted. Paragraph (C) designates the rules for conversion between units and where such conversion is required and where it is optional.

Definitions – Article 100

Applying and interpreting the NEC® incorrectly can often be traced to a failure to properly understand the terms used. To assist in the proper application of the NEC®, terms that are used in two or more articles are defined in Article 100. In addition, terms may be defined in the particular article to which they apply. Many articles contain a separate section with the specific definitions for that article only.

Other definitions may be included within the particular section to which they apply. For example, 210.8(A)(5) defines an unfinished basement for the purposes of applying the general rule which requires all receptacles in unfinished basements of dwellings to be GFCI-protected. **See Figure 1-4.**

Accessible – Article 100. There are two definitions of "accessible" in Article 100. One is for use with equipment and the other is for use with wiring methods. Accessible equipment admits close approach and is not guarded by locked doors, elevation, etc. Many NEC® references require that the wiring methods used for electrical installations be accessible. For example, 314.29 requires that conduit bodies, junction boxes, and handhole enclosures be installed so that the wiring within is accessible. Such wiring is capable of being removed without damage to the building or the finish. Boxes with accessible wiring methods should not be installed in such a method that the wall would be taken down or damaged to gain access to the boxes. **See Figure 1-5.**

Readily Accessible – Article 100. *Readily accessible* equipment is equipment capable of being reached quickly. Equipment is often required to be installed in a readily accessible location. Such equipment cannot be located behind locked doors. Readily accessible equipment cannot be located so that a ladder is necessary to reach it, and it cannot be located behind obstacles. **See Figure 1-6.**

Ampacity – Article 100. *Ampacity* is the maximum current that a conductor can carry continuously, under the conditions of use. Ampacity is derived from the words ampere and capacity.

The type of insulation is a major factor in determining a conductor's allowable ampacity. For example, Table 310.15(B)(16), which lists the allowable ampacities of insulated conductors in a raceway, cable, or earth classifies conductor insulation according to the temperature rating of the conductor. There are three ratings for copper (Cu) and aluminum (Al) conductors. The allowable ampacity of a 8 AWG Cu conductor changes from 40 A for 60°C insulation, to 50 A for 75°C insulation, and to 55 A for 90°C insulation.

NFPA Membership

Electricians are encouraged to join and participate in the NFPA and the NEC® development process. Such membership helps individuals maintain an awareness of the latest technology and changes in the industry. Participation in professional organizations provides opportunities for and aids in gaining knowledge about electrical wiring processes.

NEC® Definitions

100	Generally, only those terms that are used in two or more NEC® articles are defined in Article 100.
OTHER ARTICLES	Some articles contain a section that has definitions of terms used in that particular article only. For example, see 517.2, 550.2, 551.2, 680.2, and 690.2. Generally, article definitions will be located in the ".2" Sections.
INDIVIDUAL SECTIONS	Sometimes the definition is included in the specific NEC® section to which the definition is applicable. For example, see the definition of unfinished basement in 210.8(A)(5).

Figure 1-4. Definitions in the NEC® appear in Article 100 or in the particular article or section in which they are used.

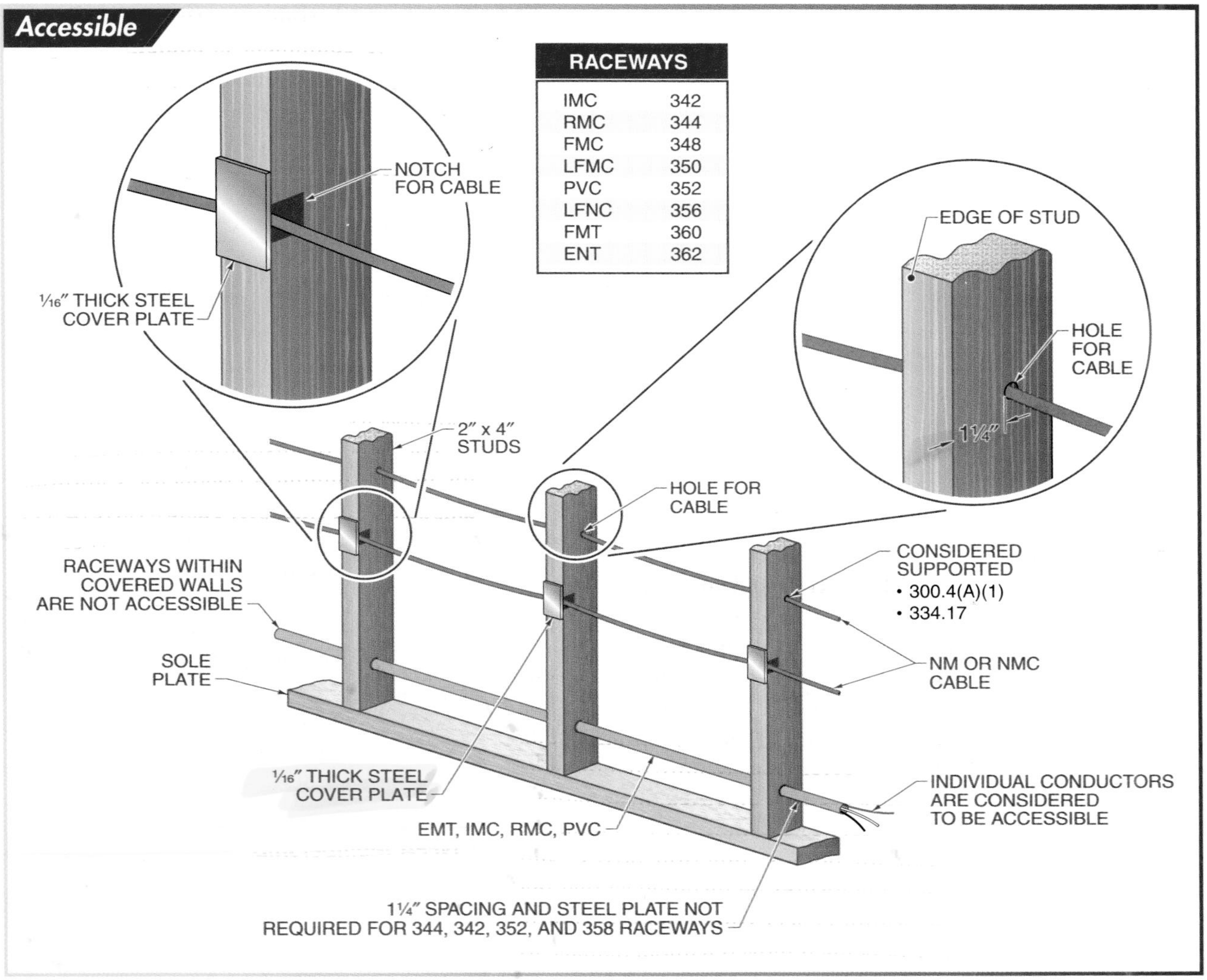

Figure 1-5. Raceways within walls are not considered to be accessible.

Because of its flexibility, flexible metal conduit (FMC) must be properly supported within walls.

"Building" Definition and Application

Each building or structure shall be provided with only one service per 230.2. While it is true that there are six conditional options to this rule, confusion often results from an improper understanding of what constitutes a building. For example, attached garages are generally not considered to be a separate building and are not eligible for a separate service. Detached garages, however, are clearly buildings and 230.2 is applicable.

Many local electrical utilities provide electricians and contractors with books that contain the electrical utilities' service requirements. These requirements must be followed no matter what the NEC® requirements state before the electrical utility supplies the power. Contractors interested in installing electrical services should contact their local utility to determine what their electrical service requirements are and their procedures for obtaining service and meter information.

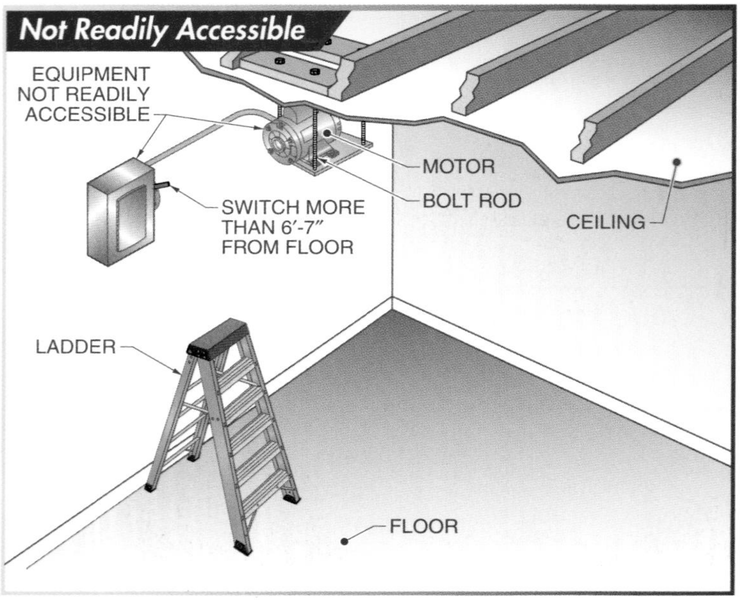

Figure 1-6. If portable ladders are necessary to reach equipment, the equipment is not readily accessible.

Other factors that affect a conductor's allowable ampacity are the ambient temperature surrounding the conductor, and the number of current-carrying conductors in the raceway or cable. *Ambient temperature* is the temperature of air around a conductor or a piece of equipment. Informational Note No. 1 to 310.15(B)(16) states that the temperature rating of any conductor is determined in part by the maximum temperature at any point along its length that the conductor can withstand over a long period of time without sustaining damage to the conductor's insulation. If a conductor passes through an area with a high ambient temperature, the conductor's ampacity is based on the highest ambient temperature to which the conductor is exposed.

Table 310.15(B)(2)(a) has correction factors for making ambient temperature corrections. Likewise, the allowable ampacities listed in Table 310.15(B)(16) are based on not more than three current-carrying conductors in a raceway, cable, or earth. If the number of conductors is more than three, then the conductor's allowable ampacity shall be derated. This is because, as the number of conductors increases, the ability of the conductor to dissipate heat away from the raceway or cable decreases. 310.15(B)(3)(a) lists adjustment factors for more than three current-carrying conductors in a raceway or cable. See Chapter 5 for a full discussion of conductor ampacity.

Approved – Article 100. *Approved* is acceptable to the authority having jurisdiction. Conductors and equipment that are either required or permitted by the NEC® must be approved to be acceptable. See 110.2. The Informational Note to 110.2 references 90.7, 110.3, and the definitions of identified, labeled, and listed to provide further guidance to the AHJ in the approval process.

Building – Article 100. A *building* is a standalone structure or is a structure separated from adjoining structures by fire walls. One of two conditions must be present in order to meet the definition of a building for the purposes of the NEC®. First, the structure must stand alone. Additions to an existing building do not constitute a building. For example, a detached garage is a building while an attached garage may not be.

The second condition involves a structure that is attached to or part of an existing structure, but is separated by fire walls and approved fire doors. In this case, a single structure can be classified as two buildings. **See Figure 1-7.**

Conductors – Article 100. A *conductor* is a slender rod or wire that is used to control the flow of electrons in an electrical circuit. The NEC® defines three types of conductors: bare, covered, and insulated.

A *bare conductor* is a conductor with no insulation or covering of any type. Bare conductors are often permitted to be used for the equipment grounding conductor (EGC), bonding jumpers, and sometimes for the grounded conductor.

A *covered conductor* is a conductor not encased in a material recognized by the NEC® as an insulating material. Conductors installed in free air may not require a specific insulation and can merely be covered with a suitable material. An example of a material that is not classified as an insulation is the rubber covering used on conductors installed in free air. The rubber covering protects the actual conductor, but the covering has not been evaluated for its insulating properties.

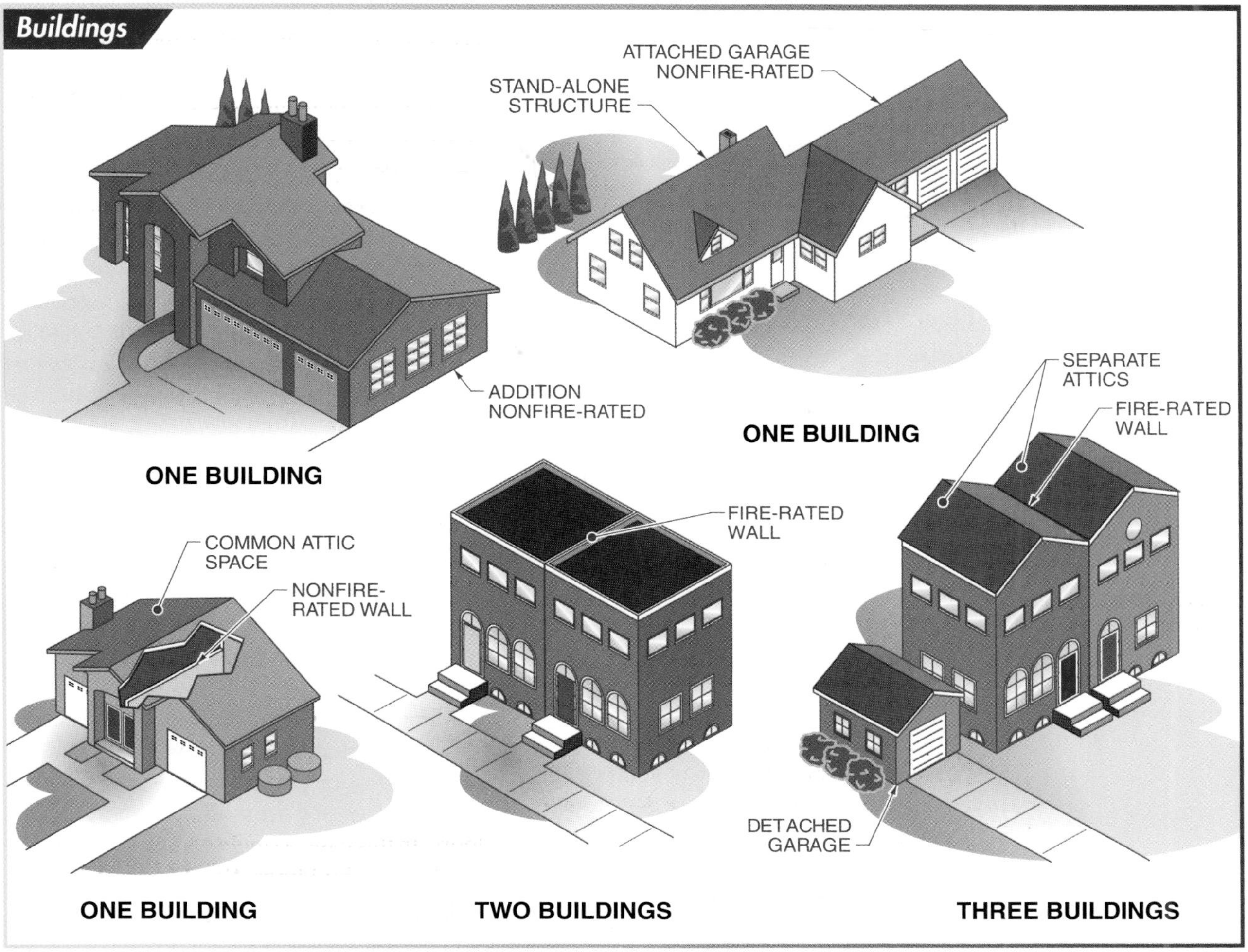

Figure 1-7. A structure that stands alone may or may not be more than one building depending on the use of fire walls.

An *insulated conductor* is a conductor that is encased using a material of composition and thickness that is recognized by the NEC® as an insulating material. Table 310.104(A) contains conductor insulation trade names and characteristics. The table includes provisions regarding the construction and application of recognized insulation types.

Continuous Load – Article 100. A *continuous load* is a load in which the maximum current is expected to continue for three hours or more. Much of the equipment installed and serviced under the NEC® is intermittent in its use. An *intermittent load* is a load in which the maximum current does not continue for three hours. Equipment and the conductors feeding it have varying periods of load and no-load conditions. Other equipment stays on under load conditions for longer periods. Such equipment operating for more than three hours constitutes a continuous load.

Coordination (Selective) – Article 100. The concept of selective coordination is an important one as it impacts the reliability and performance of the electrical distribution system. The purpose of selective coordination is to minimize power outages to unaffected loads by isolating a fault condition to the nearest upstream overcurrent device. The degree to which an electrical system is "coordinated" depends upon the selection and design of the overcurrent protective devices. Isolation is critical because it ensures that the operation of an overcurrent device does not result in outages to other loads. See 240.12, 517.17 and 620.62 where the Code addresses selective coordination for orderly shutdown, health care facilities, and multiple elevator circuits.

Device – Article 100. A *device* is any unit of an electrical system that carries or controls electricity. Receptacles and switches are two types of devices. The principal function of devices is to control or carry electrical energy. **See Figure 1-8.**

Dwelling – Article 100. Four types of occupancies are included under the definition of dwelling. These are dwelling unit, multifamily dwelling, one-family dwelling, and two-family dwelling. A *dwelling* is a structure that contains eating, living, and sleeping space, and permanent provisions for cooking and sanitation. **See Figure 1-9.**

A *dwelling unit* is a single unit for one or more persons that includes permanent provisions for living, sleeping, cooking, and sanitation. A dwelling unit provides complete and independent living facilities. A *multifamily dwelling* is a dwelling with three or more dwelling units. A *one-family dwelling* is a dwelling with one dwelling unit. A *two-family dwelling* is a dwelling with two dwelling units.

Section 210.8(B) refers to occupancies that are classified as other than dwelling units. The NEC® does not define commonly used classifications such as residential, commercial, or industrial even though the terms are used throughout the NEC®.

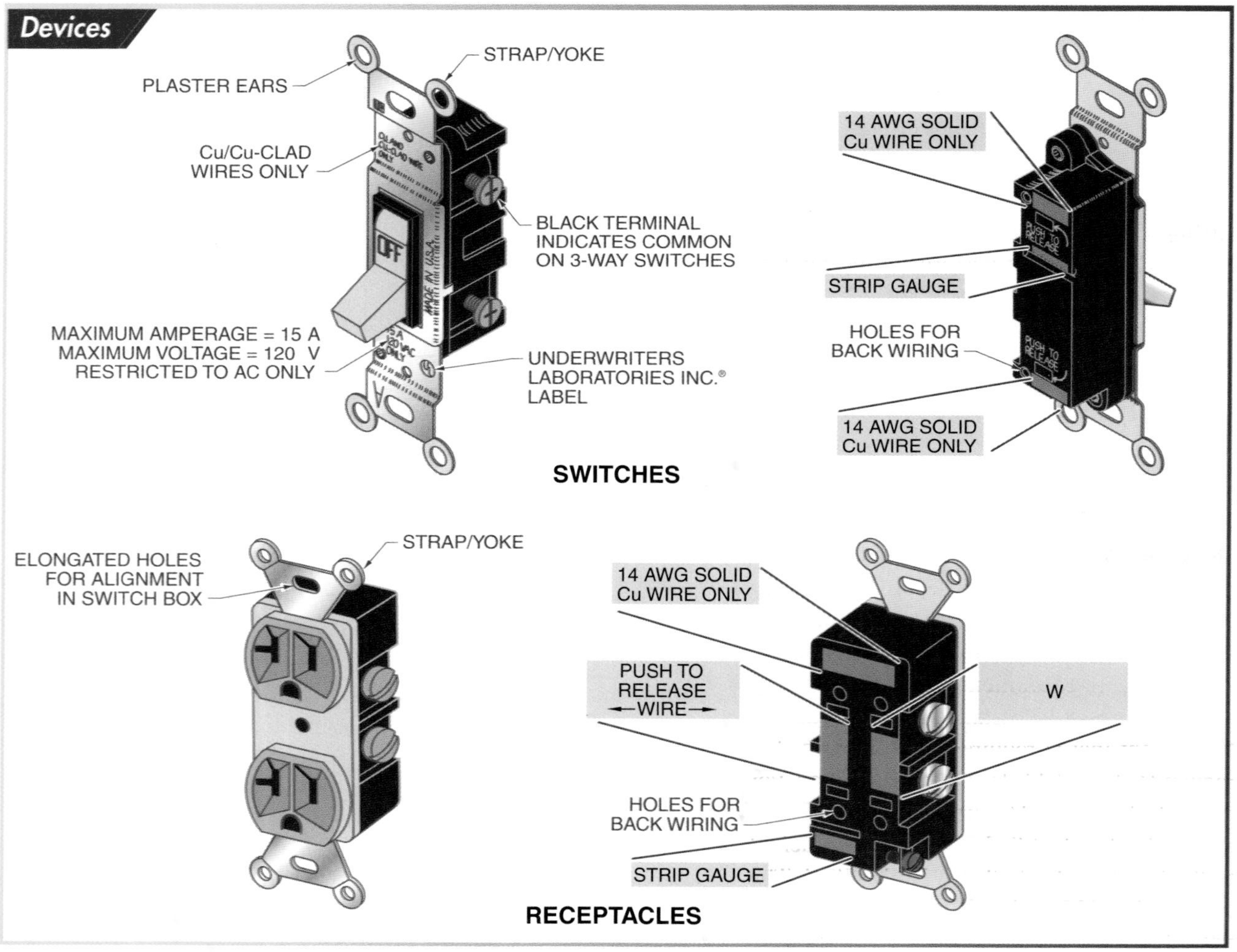

Figure 1-8. Switches and receptacles are devices because they carry or control electricity but do not use it.

Energized – Article 100. The definition for the term "energized" establishes important safety considerations for personnel that are referenced throughout the entire Code. *Energized* is being electrically connected to voltage or being a source of voltage. This is a necessary distinction because some energized components, such as capacitors, batteries, and conductors with induced voltages, are not literally connected to a source of voltage but are in fact sources of voltage themselves. Many of the Code provisions are designed to protect personnel against energized electrical equipment and components.

Exposed – Article 100. *Exposed*, as applied to wiring methods, means on a surface or behind panels which allow access. For example, wiring methods such as AC cable and NM cable have requirements which must be adhered to when installed in exposed locations. Exposed locations are not limited to those locations that are visible. Installations that are behind panels, such as drop ceiling tiles which are designed to be removed, are considered to be exposed. **See Figure 1-10.**

Fitting – Article 100. A *fitting* is an electrical system accessory that performs a mechanical function. Connectors, couplings, locknuts, bushings, etc., are components of the electrical system that serve a mechanical function. Although not designed to serve an electrical function, they can be an important part of the equipment grounding path when the raceway serves as an equipment grounding conductor. **See Figure 1-11.**

Identified – Article 100. *Identified* is recognized as suitable for the use, purpose, etc. Misapplication of the NEC® often results when equipment which is not suitable for the purpose is used. For example, equipment suitable for damp locations cannot be installed in wet locations. Equipment that is required to be identified is required to be suitable for its use or function. For example, boxes used to support ceiling fans are required to be identified and a label is attached to the box to indicate that it is identified. The Informational Note that follows this definition provides some guidance on determining the suitability of equipment.

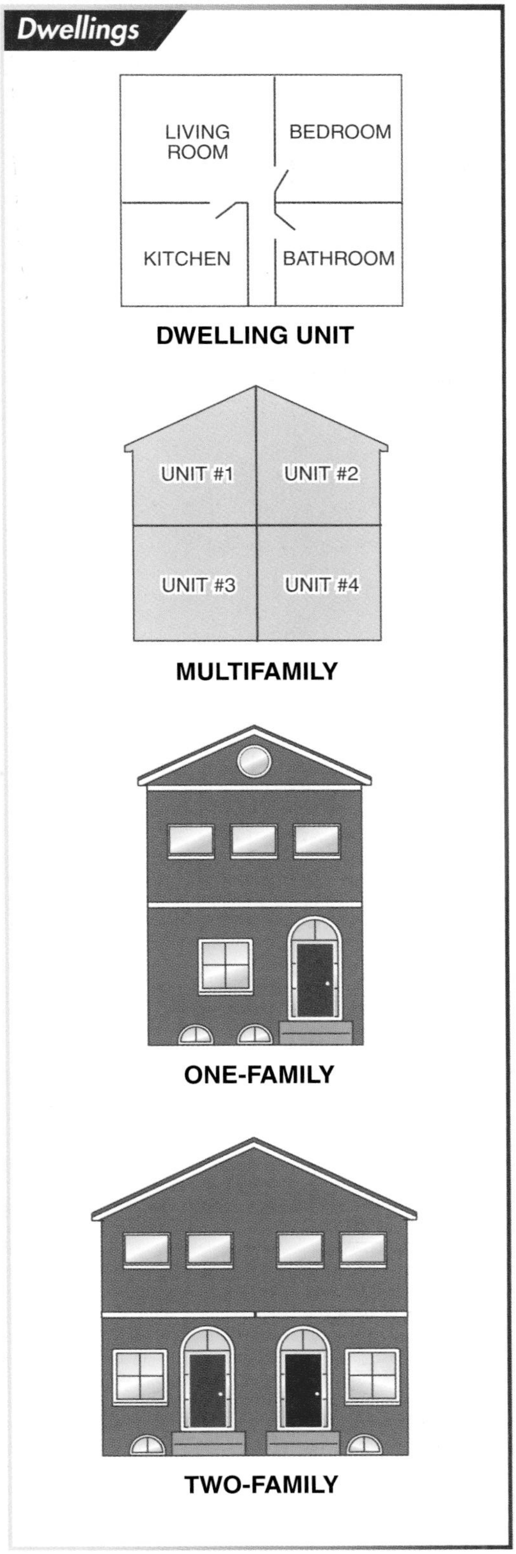

Figure 1-9. Four types of occupancies are included in the definition of dwelling.

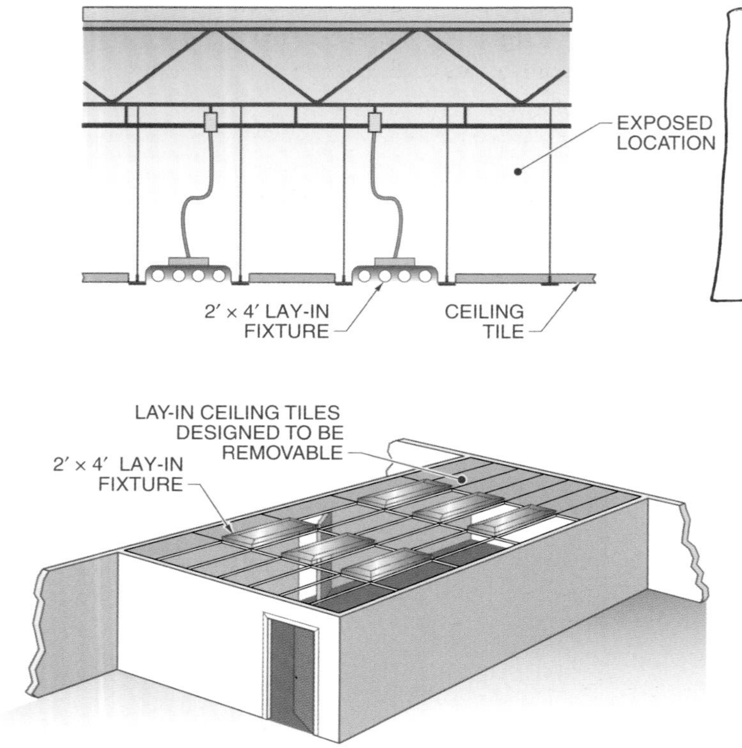

Figure 1-10. Wiring methods installed above lay-in ceilings are considered to be exposed, even if they are not visible.

In Sight From – Article 100. *In sight from* is visible and not more than 50′ away. Some NEC® installation provisions require certain equipment to be located in sight from, within sight, or within sight from other equipment. Anytime this provision is stated, the equipment shall be visible and not more than 50′ from the other equipment. Visible means an unobstructed view. **See Figure 1-12.**

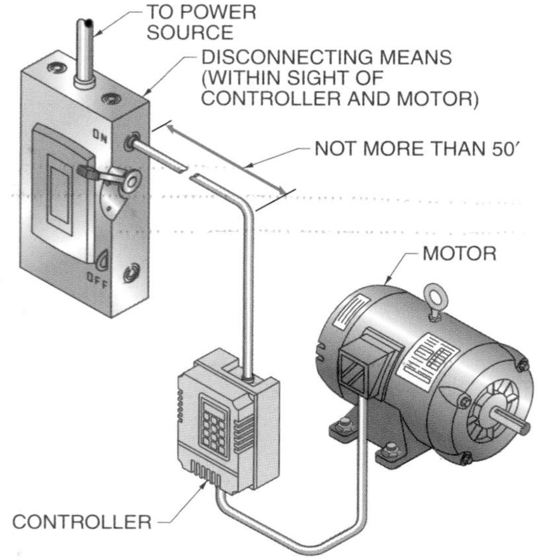

Figure 1-12. The disconnecting means shall be located within sight from the equipment.

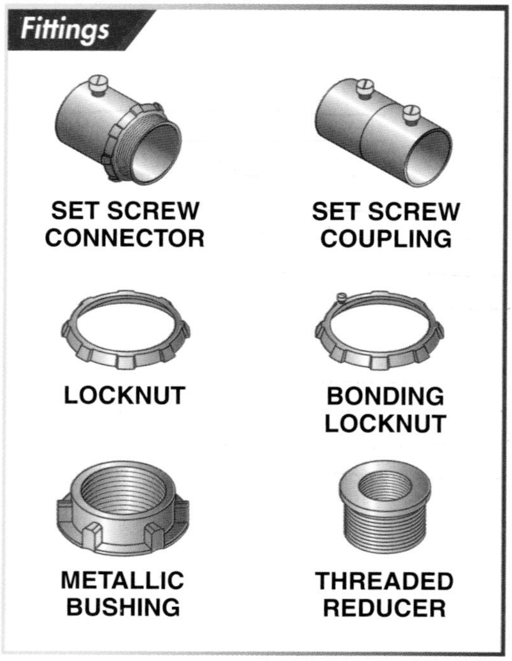

Figure 1-11. Fittings provide a mechanical function in constructing electrical systems.

Labeled – Article 100. *Labeled* is equipment acceptable to the AHJ and to which a label has been attached. Labeling identifies equipment or material which has been evaluated by a suitable testing laboratory and found to be acceptable and in compliance with established performance and/or construction standards. Some equipment, connectors, couplings, fittings, etc., may be difficult to label because of the physical makeup or size of the equipment. Often, this type of equipment contains labeling information on the carton or box in which it is packaged.

Listed – Article 100. *Listed* is equipment, materials, or services that are included in a list published by an organization acceptable to the AHJ. Like labeling, listing of a product certifies that the equipment has been evaluated by a testing laboratory suitable to the AHJ. Equipment which has been listed is published in a directory by the testing laboratory certifying that it has met established standards and is periodically reviewed to ensure standards compliance.

Location – Article 100. Electrical equipment can be installed in various types of locations subjecting it to all types of weather. Locations are classified as dry, damp, or wet to help determine suitability of equipment.

A *dry location* is a location which is not normally damp or wet. Most electrical equipment is installed in dry locations. Dry locations are out of the direct effects of moisture. If a location is normally dry, but temporarily subject to moisture, it may still be classified as a dry location.

A *damp location* is a partially protected area subject to some moisture. Damp locations are subject to moderate degrees of moisture, but are generally out of the weather. Outdoor locations which are partially covered and interior locations such as basements are classified as damp locations.

A *wet location* is any location in which a conductor is subject to saturation from any type of liquid or water. Locations that are subject to more than moderate degrees of moisture are classified as wet locations. Equipment to be installed in wet locations should be designed so that it is suitable for locations exposed to weather without any protection. Underground installations and those in direct contact with the earth are classified as wet locations. An example of rules for damp and wet locations is found in 314.15(A). Equipment in the scope of Article 314 must be placed or equipped to prevent moisture from entering or accumulating within the box, conduit body, or fitting. Boxes, conduit bodies, or fittings installed in wet locations are required to be listed for wet locations. There is no requirement for listing boxes, conduit bodies, or fittings installed in damp locations for either damp or wet locations. **See Figure 1-13.**

Safety in the Electrical Industry

There is a great deal of activity in the electrical industry concerning electrical safety. The focus is on the two greatest electrical hazards to workers: shock and arc flash. In recent years significant knowledge has been gained through testing and analysis concerning arc-flash hazards and how to contend with this type of hazard. This hazard exists when a worker is working on or near exposed, energized electric conductors or circuit parts that have not been placed in an electrically safe work condition. If an arc fault occurs, the tremendous energy released in a fraction of a second can result in serious injury or death. However, there is a great challenge in getting the message to the populace of the electrical industry so that safer system designs and safer work procedures and behaviors result. Workers continue to sustain life altering injuries or death.

NFPA 70E, *Standard for Electrical Safety in the Workplace,* 2009 Edition, is the foremost consensus standard on electrical safety. Why is there an NFPA 70E? In 1976 a new electrical standards development committee was formed to assist the Occupational Safety and Health Administration (OSHA) in preparing electrical safety standards. This committee on Electrical Safety Requirements for Employee Workplaces, NFPA 70E, was needed for a number of reasons, including (1) the NEC® is an installation standard while OSHA also addresses employee safety in the workplace, (2) not all sections in the NEC® relate to worker safety and these are therefore of little value to OSHA's focus and needs, (3) many safety-related work and maintenance practices are not covered, or not adequately covered, in the NEC®, and (4) a national consensus standard on electrical safety for workers did not exist, but was needed—an easy to understand document that addressed worker electrical safety. The first edition was published in 1979.

Additional information is available at www.cooperbussmann.com.

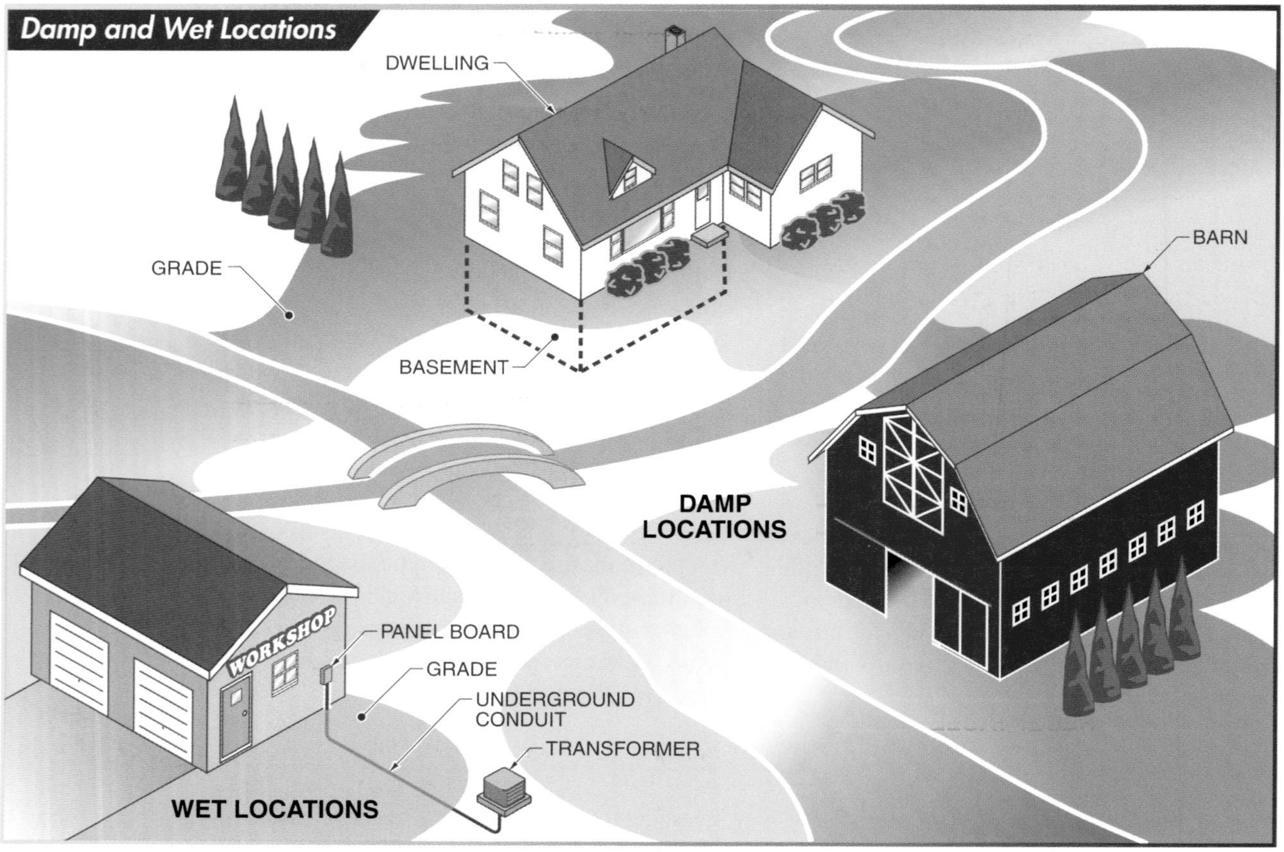

Figure 1-13. Equipment and materials shall be suitable for the location in which they are installed.

Neutral Conductor – Article 100. A *neutral conductor* is a conductor connected to the neutral point of a system that is intended to carry current under normal conditions. A *neutral point* is the common point on a wye-connection in a polyphase system or midpoint on a single phase, 3-wire system, or midpoint on a single-phase portion of a 3-phase delta system, or midpoint of a 3-wire, direct current system. The Informational Note states that at the neutral point of the system, the vectorial sum of the nominal voltages from all other phases within the system that utilizes the neutral, with respect to the neutral point, is zero.

Outlet – Article 100. An *outlet* is any point in the electrical system where current supplies utilization equipment. The purpose of electrical systems is to deliver electrical power at convenient points in order to operate electrical equipment. The points at which the electrical system is accessed are outlets. A *receptacle outlet* is an outlet that provides power for cord-and-plug-connected equipment. A *lighting outlet* is an outlet intended for the direct connection of a lampholder, luminaire (lighting fixture), or pendant cord terminating in a lampholder. **See Figure 1-14.**

Overcurrent – Article 100. *Overcurrent* is any current in excess of that for which the conductor or equipment is rated. All equipment and conductors are designed to operate at a specified current rating. When the rated current is exceeded, the result is an overcurrent. Overcurrents are caused by overloads, ground faults, or short circuits.

A *ground fault* is an unintentional connection between an ungrounded conductor and any grounded raceway, box, enclosure, fitting, etc. Ground faults can result in a large-magnitude current flow in the ground-fault path. The amount of current that flows depends upon the resistance or impedance of the ground path. For example, if a 120 V, ungrounded conductor in a metal raceway is nicked during installation, a ground fault can occur. The amount of current that flows depends upon the impedance of the metal conduit back to the source of power. Section 250.2, in its definition of effective ground-fault current path, provides some guidance on what is required in the design of an effective ground-fault path.

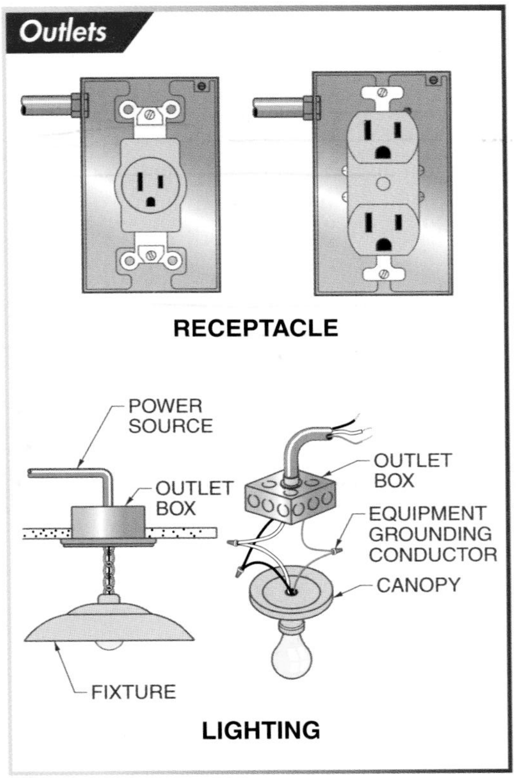

Figure 1-14. Outlets do not use electricity, but provide a means to access it.

A *short circuit* is the unintentional connection of two ungrounded conductors that have a potential difference between them. A short circuit also occurs when an ungrounded conductor is connected to a grounded conductor. The largest overcurrent results when a short circuit occurs. A short circuit occurs when two ungrounded conductors or an ungrounded and grounded conductor come in contact with each other. A short circuit is not an overload. In general, all equipment should be suitable for the available fault current. See 110.9 and 110.10.

Overload – Article 100. An *overload* is the operation of equipment in excess of normal, full-load rating, or a conductor that has an excess amount of current. Where equipment or conductors are subjected to the excess current for sufficient periods, damage or dangerous overheating may occur. An overload is a specific type of overcurrent condition that involves small amounts of excess current. Most overloads will not cause the overcurrent device protecting the circuit to trip unless the device is equipped with overload protection. Faults such as short circuits and ground faults are not overloads. Overloads are typically found in motor applications where bearings need lubrication, a drive belt is too tight, the motor is subjected to a load larger than it was designed for, or an ambient temperature is in excess of the rated temperature.

Qualified Person – Article 100. A *qualified person* is a person who has knowledge skills related to the construction and operation of electrical equipment and has received appropriate safety training. The training must focus on the recognition and avoidance of hazards.

Raceway – Article 100. A *raceway* is a metal or nonmetallic enclosed channel for conductors. The primary purpose of raceways is to support the electrical conductors and protect them from physical damage. Raceways may have other functions such as equipment grounding conductors. See 250.118.

Installation requirements for specific raceways are found in Article 300 and within the particular raceway article. For example, 300.11(A) requires that all raceways be securely fastened in place while 358.30(A) requires that electrical metallic tubing be securely fastened in place at least every 10′.

Voltage, Nominal – Article 100. *Nominal voltage* is a nominal value assigned to a circuit or system for the purpose of designating its class. Most voltages referenced in the NEC® are nominal voltages. For example, 210.8(A) lists the requirements for GFCI protection of 125 V receptacles in dwelling units. The 125 V is a nominal voltage. Receptacles of 110 V, 115 V, and 120 V are also covered; these are actual voltages that vary from the nominal. See 220.5(A). The actual voltages at which circuits operate may fall in a range around the nominal circuit voltage that will allow the satisfactory operation of the equipment. Section 220.5(A) lists 120 V as one of the nominal system voltages that should be used for load calculations, where the actual voltage is unknown. In many cases the voltage supplied by the utility is not the nominal voltage of 120 V but will fall within a range between 110 V and 130 V. Equipment rated at 120 V will, in most cases, operate properly where the actual voltage is within this range.

Voltage-to-Ground – Article 100. *Voltage-to-ground* is the difference of potential between a given conductor and ground. However, for ungrounded systems, the voltage-to-ground is the maximum voltage between any two conductors of the circuit. In some instances, the NEC® specifies voltage between conductors. In these cases, voltage shall be considered to be either line-to-line or line-to-neutral voltage, depending on the type of circuit.

All electrical equipment and conductors installed under the requirements of the NEC® shall be approved to be deemed acceptable by the AHJ.

Conduit Trade Sizes

Electricians working in the United States typically reference conduit sizes as trade size values such as ½, ¾, and 1 and so on. With the SI system, values are expressed as 16, 21, 27 and so on. Per the NEC®, there is standardization between the SI and English systems, as each trade size has a corresponding metric designator. Most electricians working with conduit use trade size values when selecting and installing conduit.

ELECTRICAL INSTALLATIONS – Article 110

The basic building block for electrical installations throughout the NEC® is Article 110. Electrical installation requirements concerning equipment, conductors, and terminations are included. Additionally, work space clearance requirements which are critical for protection of workers are also included.

Approval – 110.2

Section 110.2 requires that all electrical equipment and conductors installed under the requirements of the NEC® shall be approved to be deemed acceptable. The AHJ is charged with the responsibility for approving both equipment and installations. The AHJ, in deciding whether to grant approval, should determine if the equipment is identified, listed, or labeled for the intended use.

Usage of Equipment – 110.3

Equipment shall be suitable for its intended use. Factors such as mechanical strength, conductor termination space, conductor insulation, thermal effects, and arcing effects should be considered along with the primary factor, which is to ensure that equipment will not in any way place people at risk during operation.

Misapplication of equipment is one of the most frequent factors contributing to poor electrical installations. Section 110.3(B) requires that all listed or labeled equipment shall be installed, used, or both in a manner consistent with any instructions included with the listing.

The Informational Note that follows 110.3(A)(1) indicates the importance of ensuring the suitability of all electrical equipment for the environment or conditions under which it will be installed and used. Users of the Code should rely upon the listing and labeling of the equipment to ensure proper suitability. In addition, manufacturers of electrical equipment and conductors are providing more markings directly on the equipment and conductors to indicate any special conditions of use or limitations of the product.

Listed Equipment

Contractors or installers frequently interpret 110.3(B) to mean that the use of nonlisted material and equipment is prohibited. This section does not require that only listed equipment be installed. Such confusion may result from job specifications that require only listed equipment or from Code rules that may require some particular material or equipment to be listed. For example, 110.14(B) requires that all splicing devices for direct earth burial be listed.

Section 110.3(B) requires that if listed or labeled equipment is used or installed, then it must be done in accordance with any instructions included in the listing or labeling. For example, ceiling fans are listed with specific instructions detailing the minimum clearance from the floor for the blades. If the clearances are not met, the installation would be a Code violation despite the fact that there are no specific minimum height requirements for ceiling fans provided in the NEC®.

Conductors – 110.5

In general, the NEC® recognizes three types of conductors: copper, aluminum, and copper-clad aluminum. Unless specified, all conductors in the NEC® are considered to be copper. There are better conductors than copper. For example, silver is a better conductor of electricity than copper; however, copper offers the most ampacity at the least cost per foot.

Conductor sizes are expressed in American Wire Gauge (AWG) or circular mils (CM) per 110.6. Table 8, Chapter 9, contains some useful information on conductor properties and can also be used to show the relationship between AWG sizes and circular mils. Basically, the AWG is a system for comparing the relative area of a conductor. For the purposes of the NEC®, sizes start at 18 AWG (small) and run through 1 AWG (larger).

After 1 AWG, the AWG uses the aught sizes. Four sizes are listed using this method: 1/0, AWG 2/0 AWG, 3/0 AWG, and 4/0 AWG. For sizes other than aught sizes, as the number value increases, the size of the conductor decreases. Thus, a 4 AWG conductor is larger than a 8 AWG conductor.

After the 4/0 AWG size, the Table switches to the use of kcmils. A *mil* is .001″. A *circular mil* is a measurement used to determine the cross-sectional area of a conductor. The prefix "k" stands for thousands, thus kcmil is equal to thousands of circular mils. As the number of kcmils increases, the size of the conductor increases.

Equipment – 110.9

In addition to the factors listed in 110.3(A), other factors shall be considered before making a final determination as to the suitability of equipment. Section 110.9 requires that all equipment used to interrupt current at fault levels shall have an interrupting rating not less than the available nominal voltage and circuit current. Proper selection of fuses and circuit breakers requires that the installer know what the available fault current is at the line terminals of the equipment.

A fuse must have an interrupting rating suitable for the current it is designed to interrupt.

Many manufacturers of these devices publish information designed to assist the installer in calculating the available fault current at the line terminals of the equipment. Equipment that is not designed to open the circuit under fault conditions, such as snap switches and contactors etc., shall have an interrupting rating suitable for the current it is designed to interrupt.

Mechanical Installation – 110.12

All electrical equipment shall be installed in a neat and workmanlike manner. There is no definition of what constitutes neat and workmanlike, but equipment that is installed without any consideration given to it being level, plumb, or adequately supported would violate this section.

Unused openings in equipment shall be effectively closed per 110.12(A). Unless they are provided for the normal operation of the equipment. For the purposes of this section, an effective seal is one that is equivalent to the material from which the equipment is constructed.

Section 110.12(B) requires that electrical equipment, especially internal components, shall be protected against foreign materials. Often, equipment may be left unprotected from materials like paint, plaster, abrasives, etc., that are used in construction of the job. Problems may arise where internal parts become contaminated and are cleaned. Section 110.11, Informational Note 2 warns that some cleaning and lubricating compounds can cause severe deterioration of many plastic materials that are used for insulating and structural applications in equipment. Suitable protection shall be given to the equipment to ensure that nothing will adversely affect its safe operation.

Mounting and Cooling – 110.13

Aside from the requirement to install all equipment in a neat and workmanlike manner, all equipment shall be securely fastened on the surface to which it is mounted per 110.13(A). Proper installation requires that the correct fastener for the specific installation be used. For example, wood plugs are not suitable for use in holes in masonry, concrete, or plaster.

One of the primary considerations that must be taken into account when designing electrical installations is the effect of temperature on the equipment. Section 110.13(B) requires that proper ventilation be provided for electrical equipment which depends on the natural circulation of air.

Some equipment, such as transformers, are constructed with ventilation openings which shall not be obstructed by adjacent walls or equipment. The transformer nameplate specifies the required distance between the wall and the transformer that the transformer must be installed to provide for proper air circulation and cooling. See 450.9.

Electrical connections require selection of the termination device designed for the specific conductor size and material.

Electrical Connections – 110.14

At the core of any electrical system is the electrical connection and termination. Just like equipment, electrical connections shall be properly designed and installed to avoid problems associated with excessive heat. Improper electrical terminations lead to high resistance connections which in turn lead to electrical system failures.

The first consideration in making good electrical connections is the type of conductor material involved. The NEC® recognizes copper, aluminum, and copper-clad aluminum, each of which has different physical characteristics. The basic rule when selecting termination devices is to always use a termination device identified for the type of conductor material. For example, when using lugs with aluminum conductors, make sure the lug is identified for use with aluminum conductors. Similarly, if conductors of dissimilar materials are to be used, the termination device shall be identified for use with both types of conductor material.

Splicing devices and terminations utilizing mechanical pressure are common in the electrical industry. Most commonly, screw-on wire connectors are used for splicing conductors. Many manufacturers offer low-cost wire nuts that accomplish this task by use of a mechanical insert within an insulating cover. It is important that a screw-on connector be selected that matches the allowable wire combinations specified by the manufacturer. Generally, wire nuts are identified by color for different wire sizes. For example, a connector may be designed to accommodate up to three 12 AWG conductors.

Another commonly used mechanical connector is the split-bolt connector. This connector utilizes mechanical pressure to establish a splice that maintains excellent mechanical and electrical properties. The split-bolt connector must then be covered with insulation equivalent to that of the conductors being spliced. Table 310.13 provides the insulation thickness values based on insulation type and wire size. An insulating device identified for the purpose may be used as an alternative to the use of other insulating materials. This type of splicing device is commonly used when tapping one conductor from another. Section 240.21 permits branch-circuit taps, feeder taps, and transformer secondary taps under specified conditions.

Recently, some manufacturers have been listing cable splicing devices that incorporate a mechanical device within an insulated protective enclosure. These insulated multiple connectors offer many combinations of wire inputs in many different wire size configurations. It is important to match the application to the correct product to ensure the reliability of the connection. Factors such as wire material (copper, aluminum, bronze), wire size, and splicing device construction must be properly considered.

Another popular splicing device is the bolted terminal connector or lug. Lugs are commonly used for both low- and medium-voltage applications. They are easy to use and are frequently found where power distribution conductors terminate in transformers and power panelboards. The design of the lug can accommodate a range of conductor sizes and material types. One disadvantage with this type of connector is that the termination is subject to influences that may impact the overall reliability of the termination. For example, vibration or heat cycling may result in a loose connection that raises the resistance of the connection, thus subjecting the termination to an increased temperature rise. Failure to properly maintain the connection, beginning with proper torquing of the termination, could result in a premature failure of the termination and a resulting power outage.

Compression connectors are also widely used within the electrical industry. The range of applications is extremely broad and they can be utilized for both low-voltage and medium-voltage applications. The principle behind the connector is the application of irreversible pressure, either pneumatically or by hand, to ensure the connection. Many manufacturers have pneumatic tools that use interchangeable dies that permit a wide range of conductor sizes and applications.

Terminal connectors can be designed to accommodate a varying range of conductor sizes.

When the termination is a terminal block, care must be taken to ensure a good electrical connection without damaging the conductor. Terminals should not be used for more than one conductor, unless they are identified for such use. When it is necessary to splice conductors, identified splicing devices shall be used. If the splice is underground, the device shall be listed for direct-burial use. **See Figure 1-15.** Although not common, splices are permitted to be made by brazing, welding, or soldering with a fusible metal or alloy, provided the conductors are mechanically and electrically secured first.

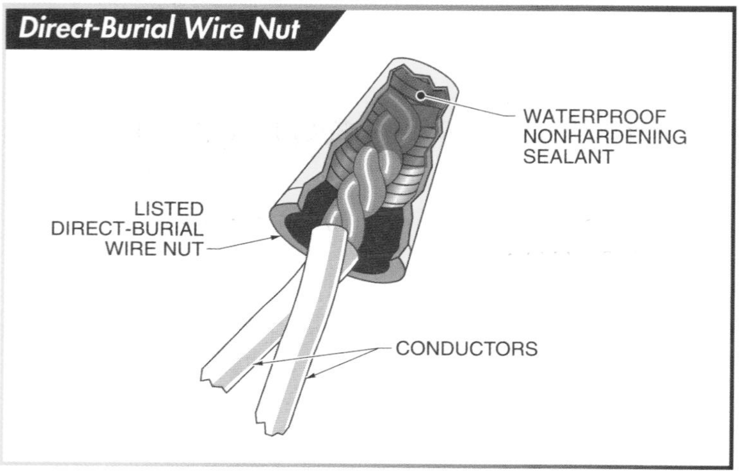

Figure 1-15. Listed direct-burial wire nuts contain a waterproof, nonhardening sealant.

Another consideration when making electrical terminations is the temperature rating of the termination device. Equipment, such as conductors, are rated according to how much heat rise can occur at the conductor termination. For example, if a panelboard is labeled with a 75°C temperature rating, the temperature at the conductor termination can rise up to 75°C. Conductors, in this case, would require insulation ratings of at least 75°C. If 60°C conductors were used, the insulation on the conductor would not be sufficient for the termination heat rise, and insulation failure could occur.

Section 110.14(C) discusses the temperature limitations of electrical connections. In general, it requires the ampacity of conductors to be selected and coordinated so that it will not exceed the lowest temperature rating of any termination, conductor, or device found in the circuit. Conductors with temperature ratings higher than the selected temperature rating are permitted to be used for ampacity adjustment or correction, or both. For example, a 12 AWG THHN copper conductor rated at 90°C may be terminated on a 20 A circuit breaker rated at 60° C as long as the load on the circuit does not exceed the allowable ampacity found in the 60°C column of Table 310.15(B)(16) for a 12 AWG conductor. Section 110.14(C)(1) references 110.14(C)(1)(a) and 110.14(C)(1)(b) for the determination of termination provisions of equipment. The two subsections break the provisions into two basic categories. Circuits rated 100 A or less, or marked for 14 AWG through 1 AWG are rated for 60°C. Circuits rated over 100 A, or marked larger than 1 AWG are rated for 75°C. These temperatures may be increased where the equipment is listed and identified for use at the higher ratings. For example 75°C conductors can be terminated on a circuit breaker that is listed and labeled for 60/75°C and use the allowable ampacity found in the 75°C column of Table 310.15(B)(16), as long as all other terminations, equipment, and conductors found in the circuit are rated for at least 75°C. Where any portion of the circuit is rated at 60°C, then the allowable ampacity for the load on the circuit may not exceed the allowable ampacity found in the 60°C column of Table 310.15(B)(16) for the rated circuit amperage.

Both the conductor and equipment temperature ratings shall be considered when designing or installing electrical systems. The ampacity of the conductor shall be selected so that the lowest temperature rating of the conductor or any equipment in the circuit is never exceeded.

Identifying High-Leg in Delta 4-Wire Systems – 110.15

The circuit conductor with the higher voltage-to-ground in a delta 4-wire system shall be identified. The means of identification shall be by orange finish, tagging, or other effective means. This identification shall occur wherever the high-leg conductor and the neutral are accessible, such as junction boxes, pull boxes, troughs, or other accessible locations. **See Figure 1-16.**

Arc Flash Protection – 110.16

In addition to the risk of electrical shock, other hazards are present when work is performed on or near energized electrical equipment or conductors. Two serious hazards are electrical arc blasts and electrical arc flashes. Section 110.16 contains a field marking requirement that applies to switchboards, panelboards, MCCs, meter socket enclosures, and industrial control panels that are located in other than dwelling units and which are likely to require examination or servicing while in an energized condition. The marking must be clearly visible to personnel and warn them of potential electrical arc flash hazards.

Informational Note No. 1 refers to NFPA 70E®-2008, *Standard for Electrical Safety in the Workplace®* and follows this section. This Standard is intended to provide protection for employees in the workplace from the hazards of working on or near energized electrical equipment and conductors. NFPA 70E® requires that all energized equipment and circuits be de-energized and an electrically safe working condition be established before anyone may work on or near a circuit or equipment. Establishing an electrically safe working condition requires, among other things, de-energizing the circuit or equipment and locking and tagging the circuit or equipment in the de-energized position. Only qualified persons are permitted to work on or near energized equipment of 50 V or greater. Such work is permitted only where it is demonstrated that it is not feasible to de-energize or it would result in a greater hazard to de-energize the circuit or equipment.

Fault Current Field Markings – 110.24

In the 2011 NEC®, new provisions were added to Article 110 that should greatly enhance worker safety for those charged with installing, servicing and maintaining electrical equipment and installations. 110.24(A) requires that for other than dwelling units, service equipment shall be field marked with the maximum available fault current. These field markings must be legibly marked using a method that ensures that the marking will be sufficiently durable for the environment.

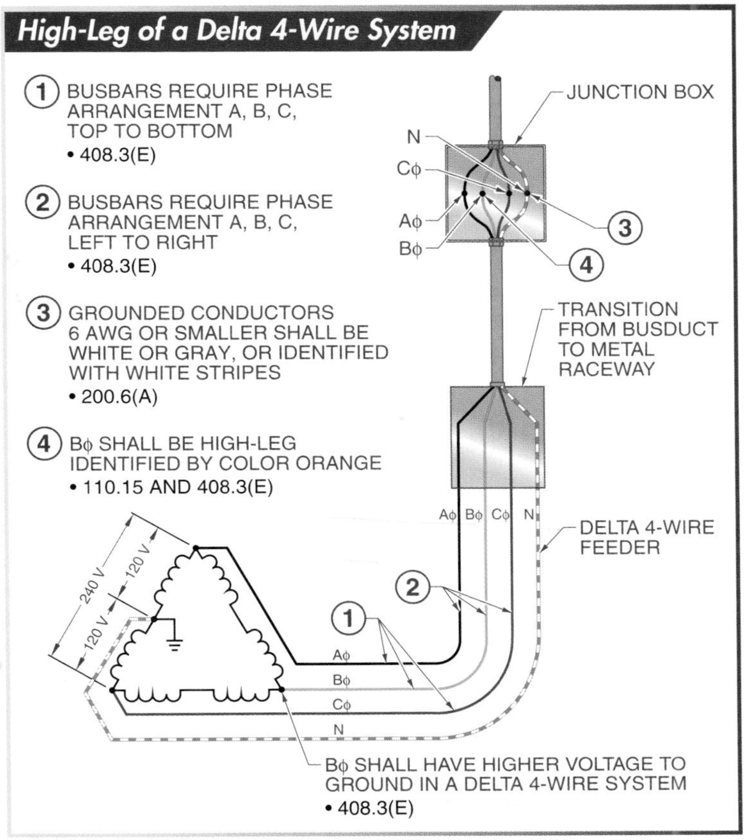

Figure 1-16. The circuit conductor with the higher voltage-to-ground in a delta 4-wire system shall be identified.

The field markings must indicate the date the fault calculation study was performed. 110.24(B) adds the additional requirement that when modifications are made to the electrical distribution system that could affect the maximum available fault current at the service, the fault calculation must be performed again or otherwise recalculated to determine the impact of the modification on the marked fault-current ratings.

There is an exception to this provision that excludes installations in industrial establishments provided the conditions of maintenance and supervision are such that only qualified persons will service the equipment. Users of the Code will see that this type of exception is used throughout the Code to exempt "industrial installations" from the intended Code provision Granted the assurance that only qualified persons will service the installation is helpful, but the fact remains that these types of exceptions from the rule speak more to the nature of the code-making process than the safety of the installation and the worker.

Spaces About Electrical Equipment – 110.26

Section 110.26 includes installation and design considerations to provide for safe and efficient operation and maintenance of electrical equipment rated at 600 V, nominal, or less. Working space for equipment rated over 600 V is found in 110.32. Requirements regarding work space clearance limitations, lighting, headroom, and means of entrance and exit are specified to help ensure that workers can perform repairs and maintenance operations safely.

The working space around panelboards shall be sufficiently illuminated.

Depth of Working Space – 110.26(A)(1). Equipment which may need to be repaired, serviced, maintained, etc., while energized, shall be provided with minimum working clearances. Table 110.26(A)(1) lists the minimum working clearances.

Two factors affect the minimum working clearance dimensions. The first factor is the nominal voltage-to-ground at which the equipment operates. In most cases, higher voltage-to-ground equipment requires more work space clearance. The second factor is the condition of use. For voltages-to-ground above 150 V and up to 600 V, the minimum working clearances increase depending on the relationship of exposed live parts to ground and other exposed live parts. **See Figure 1-17.**

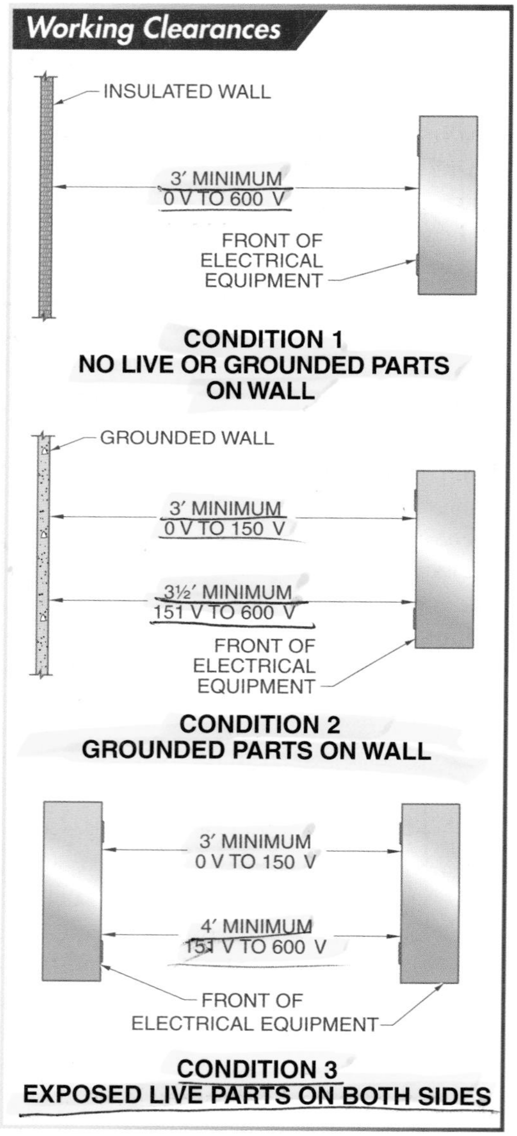

Figure 1-17. Three conditions determine working clearances of electrical equipment less than 600 V.

In general, these minimum work space clearances only apply to front working space, unless there are renewable components located in the rear of the equipment or all connections are not accessible from either the front or the sides. In these cases, the minimum clearances apply to the rear work space as well. See 110.26(A)(1)(a).

Other conditions in which these work space clearances may be reduced are specialized conditions with strict limitations. See 110.26(A)(1)(b) and (c). In any of these cases, the distances required shall be measured from the live parts, if exposed, or from the face of the enclosure, if one is provided. Concrete, brick, and tile walls are considered to be grounded when determining distances to ground.

Width of Working Space – 110.26(A)(2). In addition to the front and rear minimum working clearances, 110.26 requires that the work space clearance width in front of the equipment shall be a minimum of 30″. If the equipment is wider than 30″, then the width of the workspace shall be at least the width of the equipment. In all cases, equipment with doors or hinged panels shall be installed so that the doors and panels can be opened at least 90°.

Height of Working Space – 110.26(A)(3). To ensure that all electrical equipment and installations can be worked on and maintained in a safe manner, the Code requires that the work space around electrical equipment be clear and extend from the floor, grade, or platform that any equipment may be mounted on to a height of 6½′ or the height of the equipment, whichever is greater. If the electrical installation results in electrical equipment that is "stacked" or mounted directly above or below other equipment, the height requirements of 110.26(A)(3) permit a 6″ infringement of the space, provided the equipment is associated with the electrical installation. This would permit, for example an auxiliary gutter with a depth of 6″ to be installed above a service equipment panelboard with a depth of only 4″.

There are two general exceptions to this provision. Exception No. 1 permits service equipment or panelboards, in existing dwelling units only and not rated in excess of 200 A to be installed in spaces that do not have a minimum of 6½′ of headroom. Exception No. 2 is intended to permit meters that are installed in meter sockets to extend beyond the front of other electrical equipment.

Clear Spaces – 110.26(B). Electrical equipment shall be installed and maintained in areas which are free from stored items which could limit free access to the electrical equipment. Section 110.26(B) requires that any working space provided for electrical equipment shall not be used for storage. This section also requires suitable guarding if the normally enclosed live parts are exposed. For example, when electrical troubleshooting is being performed on a panelboard that is normally enclosed and the cover is removed, exposing live parts, the working space, if in a passageway or general open space, must be suitably guarded.

Entrance to and Egress from Working Space – 110.26(C). At least one entrance of sufficient area is required to provide access to and egress from the required work space around electrical equipment. Equipment rated at 1200 A or more and over 6′ wide that contains overcurrent devices, switching devices, or control devices requires an entrance at each end of the working space. These doors must be at least 24″ wide and 6½′ high.

Personnel doors for the required access and egress, located less than 6′ from the working space, must have doors that open in the direction of egress. The doors must be equipped with panic bars, pressure plates, or other devices that are normally latched but open under simple pressure.

Sections 110.26(C)(2)(a) and 110.26(C)(2)(b) provide two conditions under which the requirement for two doors could be changed to allow for the use of one door. The first condition is where the location provides a continuous, unobstructed egress path. Under the section path, one door may be installed where the depth of the working space required by 110.26(A)(1) is doubled. In this case, the entrance must be installed so that the distance from the equipment to the nearest edge of entrance is not less that the minimum clear distance provided in Table 110.26(A)(1). **See Figure 1-18.**

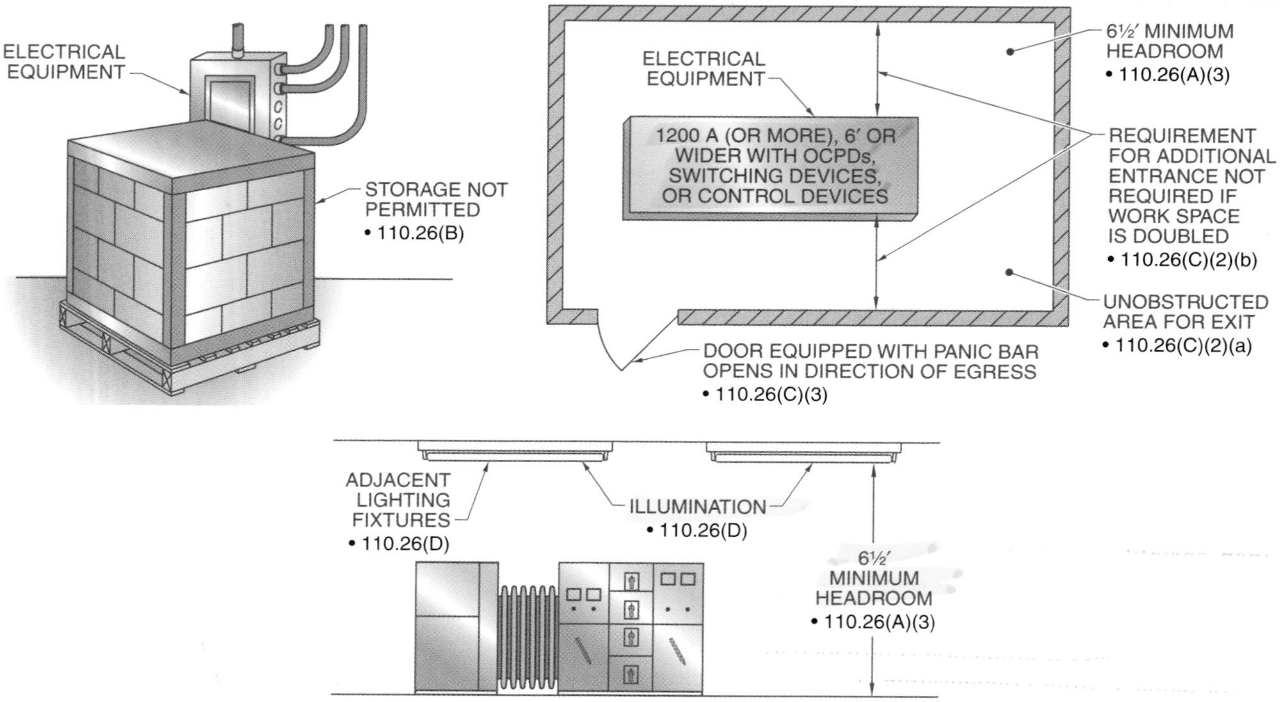

Figure 1-18. Persons working on live electrical equipment shall have suitable headroom, illumination, and means of access and entrance.

If personnel doors are provided in rooms that contain electrical equipment that includes overcurrent devices, switching devices or control devices, the doors must be equipped with panic type hardware and must open in the direction of egress where such equipment is rated at 1200 A or more and over 6′ in width. This requirement only applies where a personnel door is located less than 25′ from the nearest edge of the working space. These requirements are consistent with those found in 450.43(C) for transformer vaults and are intended to ensure that workers can quickly exit the space in the event of an electrical incident.

The requirement for an additional entrance is not necessary if there is a clear and direct means of exit provided or if the working space clearance specified in Table 110.26(A)(1) is doubled. See 110.26(C)(2)(b).

Illumination – 110.26(D). Proper illumination is required about electric equipment. The work space around service equipment, panelboards, switchboards, main control centers (MCCs), etc., shall be sufficiently illuminated. It is not the intent of this section to require a dedicated luminaire (lighting fixture) to meet the illumination requirement. However, in electrical equipment rooms, the illumination shall not be controlled by automatic means only.

Electrical equipment must be installed with work space clearance for personnel performing maintenance and troubleshooting tasks.

Dedicated Equipment Space – 110.26(E). Dedicated equipment space is intended to make alterations and additions to the electrical system easier for the installer. These requirements were moved out of Article 384 in the 1996 NEC® and into Article 110 in the 1999 NEC®. Electrical equipment, such as panelboards, switchboards, MCCs, distribution boards, etc., shall be installed in indoor spaces dedicated to the equipment. **See Figure 1-19.** The space shall be equal to the depth and width of the equipment and shall extend from the floor to 6′ above the equipment, unless there is a structural ceiling at a lower height. An exception is provided for control equipment that must be located within sight of the equipment. See 110.26(E) Ex.

In general, all foreign systems other than sprinkler protection shall not be located in dedicated equipment space. Section 110.26(E)(1)(b) does however, permit equipment in this space if protection is provided to avoid damage from condensation, leaks, or broken pipes. For the purposes of this section, a drop or suspended ceiling is not considered the structural ceiling if it is not designed to add strength to the building structure. See 110.26(E)(1)(d).

Because of the potential risks associated with electrical equipment, electrical equipment rooms and enclosures must provide adequate protection to assure there is no inadvertent access or contact by unqualified personnel. At the same time, it is often critical that qualified persons have quick and immediate access to this equipment. See, for example, 430.107, which requires that one of the required motor disconnecting means be readily accessible. To provide for both types of requirements, 110.26(F) states that electrical equipment rooms and enclosures that are controlled by lock and key are considered accessible to qualified personnel.

Electrical Hazard Signs

Rooms and spaces that house electrical equipment must be locked and only accessed by qualified persons. Such rooms and spaces shall be marked as "DANGEROUS—HIGH VOLTAGE—AUTHORIZED PERSONNEL ONLY" with a red and black electrical warning sign.

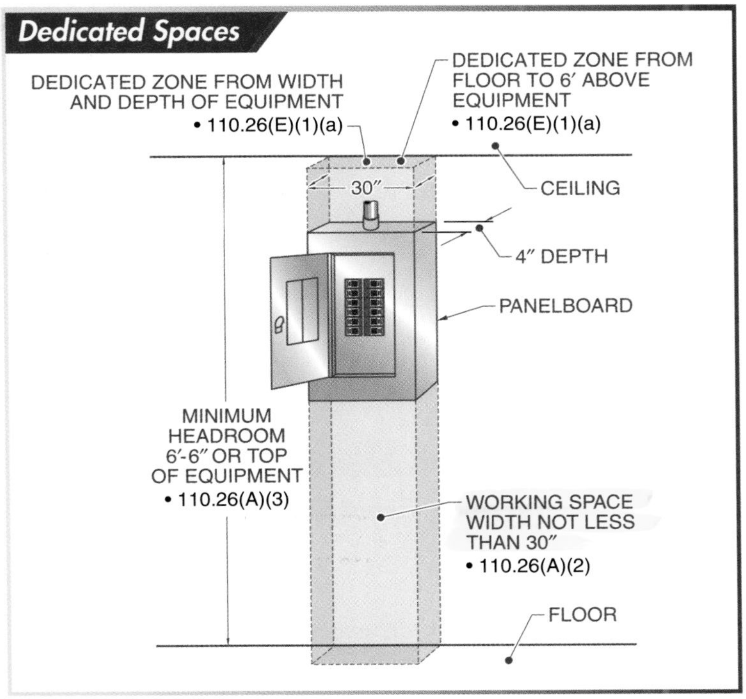

Figure 1-19. No barriers or other electrical equipment shall be located in dedicated spaces of switchboards, panelboards, etc.

Over 600 V – Part III

The layout of the Code provides for applications of over 600 V to be arranged as separate parts within each article. Part III of Article 110 contains special requirements for electrical installations rated over 600 V.

General – 110.30. Generally, the 600 V and less provisions in Part I of Article 110 will apply. The over 600 V provisions in Part III of Article 110 serve only to modify or amend Part I. Because of the increased voltage, special provisions must be made for electrical enclosures, working space, and circuit conductors.

Over 600 V Electrical Enclosures – 110.31. Because of the increased voltage levels, over 600 V installations generally require special enclosure considerations. Typically, dedicated vaults, rooms, or closets are designed for over 600 V equipment and conductors. Access to such rooms or spaces must be limited to only qualified persons. Guarding can be provided by means of walls, screens, fences, locks, or other approved means.

Dedicated equipment space allows access for alterations and additions by the installer.

With the exception of some specially designed enclosed equipment, electrical equipment rated 600 V or over and installed outdoors shall be guarded by the use of a suitable wall, screen, or fence to limit access to the equipment. If a fence is used, it must be at least 7′ in height. However, a 6′ fence with a minimum 1′ extension of three or more strands of barbed wire shall be permitted. Table 110.31 provides the minimum spacing distance requirements for electrical equipment guarded by a fence. The spacing requirements are to ensure that any ground fault in the equipment does not result in a step potential voltage being imposed upon the fence. The minimum distances increase from 10′ for nominal voltage less than 13.8 kV to 18′ for nominal voltage in excess of 230 kV.

Vaults that house electrical equipment over 600 V are required to be constructed of walls, roofs, and doorways that have a minimum fire rating of 3 hr. Typically, concrete is used for vault construction. The Code requires that the floors of vaults in contact with the earth be constructed at least 4″ thick. If the vault is constructed above a vacant space, the vault must be designed with a minimum fire resistance rating of 3 hr. As with transformer vaults, electrical vaults for equipment rated over 600 V are not permitted to be constructed with standard studs-and-wallboard. See Part III of Article 450 for related provisions for transformer vaults.

The location of electrical equipment is an important consideration when designing electrical distribution systems over 600 V. If the equipment is installed indoors in areas that are accessible to unqualified persons, Section 110.31(B) requires the use of metal-enclosed equipment. Caution signs warning of the potential electrical hazards must be installed on the equipment to warn unqualified persons to avoid contact. Manufacturers of such equipment are required to design the equipment with ventilation openings and any similar types of openings so that the openings deflect inserted objects away from any energized electrical components.

Electrical equipment rated over 600 V that is installed outdoors, where subject to vehicular traffic, must be protected by the use of suitable guards. Typically, concrete-filled pipes are installed to provide this protection. In addition, the hardware for such equipment must be designed so that no exposed nuts or bolts can be easily removed. Any doors or covers for equipment such as pull boxes or junction boxes must be locked, bolted, or screwed on to minimize the ease of their removal. Doors must be locked if they permit access from grade level at a height of 8′ or less from grade. Where such equipment is installed in an underground vault, the manhole cover shall be considered as providing this protection provided that weight of the manhole cover is over 100 lb.

Working Space for Electrical Equipment over 600 V–110.32. The requirements for maintaining working space about electrical equipment rated over 600 V are generally similar to those found in Section 110.26 for equipment rated 600 V or less.

For example, the working space must in all cases permit at least a 90° opening of doors or hinged panels where so provided. The minimum clear working space is 6½′ high and 3′ wide in front of the equipment. The height measurements are established vertically from the floor or platform that the equipment is installed on, and the minimum width measurement is measured parallel to the equipment.

These are important working spaces that ensure that when maintenance, troubleshooting, or inspection work is performed on such equipment, it can be done in a manner safe for personnel.

Access to electrical equipment over 600 V must be provided by at least one entrance that provides a 6½′ high and 2′ wide opening. The access is not required to be a door, but where a personnel door is provided, the door must be equipped with panic-type hardware and must swing out in the direction of egress. This important provision is intended to ensure that workers can quickly exit the space in the event of an accident or incident.

Section 110.33(A)(1) contains a special provision for entrances for electrical spaces that contain switchboards and control panels rated over 600 V and exceeding 6′ in width. In these cases, an entrance is required to be installed at each end of the space. This will permit an unobstructed exit from the space regardless of where an electrical incident might occur. There are several conditions that will permit a single entrance, such as doubling the working space required by Table 110.34(A), but these conditions should be used sparingly. The addition of a second entrance is the safest means to provide protection for workers who may need to exit the space quickly.

As for under 600 V installations in Section 110.26(C)(3), personnel doors for access and egress from working space for over 600 V installations shall be provided with panic bars or similar devices if such doors are located less than 25′ from the nearest edge of working space.

Front Working Space – 110.34. Working space about electrical systems must be maintained three-dimensionally. In addition to the height and width requirements, the Code establishes minimum front working space distances as well for electrical equipment rated over 600 V. Table 110.34(A) relates the minimum distance to the nominal voltage to ground and three conditions. Condition 1 occurs where exposed live parts are on one side and no live or grounded parts are on the other side of the working space. This condition also includes installations where exposed live parts are present on both sides of the equipment but are effectively guarded by some insulating material. Condition 2 covers those installations where exposed live parts are on one side and grounded parts are on the other. The note to the table clarifies that concrete, brick, or tile walls are considered to be grounded surfaces. Condition 3 covers installations where exposed live parts are on both sides of the working space and requires the greatest minimum distances. In no case is the minimum depth of clear working space permitted to be less than 3′ in the front of the equipment.

For electrical equipment that contains rear access panels, working space is not required if the panels do not provide access to renewable or adjustable parts and all connections can be made accessible from other locations on the equipment. Where rear access is required to work on de-energized components, the minimum rear working space shall be 30″ measured horizontally from the equipment.

Per 110.34(C), a high-voltage label is required for spaces that contain electrical equipment over 600 V.

Special warning provisions are also required for entrances to rooms, spaces, and enclosures that contain live parts or exposed conductors that operate at over 600 V. Generally, the entrance must be kept locked at all times unless a qualified person monitors access to the space at all times. The Code specifies the text of the warnings in Section 110.34(C). The signs must read "DANGER – HIGH VOLTAGE – KEEP OUT" and must be located so that they can be seen easily.

QUICK QUIZ®

Refer to Chapter 1 Quick Quiz® on CD-ROM.

Additional information is available at ATPeResources.com

One last consideration must be given to spaces that contain electrical equipment rated over 600 V. Other system equipment, such as plumbing, ventilation ducts, and sprinkler piping, shall generally not be installed in these spaces if their malfunction could result in a fault in the operation of the electrical equipment. For example, a leaking water pipe over the equipment could result in an explosion or malfunction. The Code does permit supplemental protection, such as drip pans in this case, but a more practical method would be to avoid the installation of these types of systems. A good rule of thumb is that the foreign system piping, ventilation ducts, etc., should be limited to only those apparatus necessary for the space where the electrical equipment is located.

Over 600 V Circuit Conductors – 110.36. Specific installation requirements for over 600 V circuit conductors are covered by Section 300.37. This section lists the permissible wiring methods for aboveground installations. Section 110.36 provides general guidelines that permit their installation in raceways and in cable trays. This section also permits the use of metal-clad cable, bare wire, and busbars for over 600 V installations. Unlike most 600 V or less installations, Section 110.40 permits conductors to be terminated utilizing the 90°F temperature rating. This provision permits the ampacity calculations to be based on the 90°F columns from Tables 310.60(C)(67) through 310.60(C)(86).

chapter 1

The National Electrical Code®

REVIEW QUESTIONS

Name Bradly Lowry **Date** 22/Jun/2014

Modified 1. The first four chapters of the NEC® apply generally unless they are ___ by the latter chapters.

8 2. Chapter ___ of the NEC® is independent of the other chapters, and none of the provisions apply unless they are directly referenced.

(T) F 3. Equipment designed to open circuits under fault conditions shall have an interrupting rating not less than the available voltage and circuit current at the line side of the equipment.

T (F) 4. Weatherproof equipment is constructed so that water will not enter it.

(T) F 5. Installations underground in mines are not covered by the NEC®.

quickly 6. Equipment that is required to be readily accessible shall be capable of being reached ___.

Ampacity 7. A conductor's ___ is the amount of amperes it can carry continuously without exceeding its insulation rating.

Accessible 8. Boxes installed within walls that are required to be taken down to gain access are not considered ___.

shall 9. The word ___ indicates that an NEC® rule is mandatory.

T (F) 10. The NEC® is revised and updated every six years.

T (F) 11. At least two entrances to electrical rooms shall be provided to give access to the working space about all electrical equipment rated less than 1200 A.

T (F) 12. All terminations and equipment are marked with tightening torques.

Person, Property 13. The two-fold purpose of the NEC® is to protect ___ and ___ from the dangers associated with the use of electricity.

Scope 14. The ___ of the NEC® contains a list of the electrical installations that are covered by the NEC®.

AHJ 15. Interpretations and approval of equipment are granted by the ___.

methods 16. The AHJ can permit alternate ___ for installation, if they believe the installation is equally effective at meeting the intended objective.

Future 17. Although installed in conformance with the NEC®, all electrical installations may not be adequate for ___ use.

Design 18. The NEC® is not intended to be used as a(n) ___ specification.

(T) F 19. Explanatory material appears in the NEC® in Informational Notes.

T (F) 20. The minimum headroom of working spaces about electrical equipment in commercial buildings is 6¼′.

(T) F 21. Electric equipment with ventilating openings shall be installed so that openings are not blocked.

T (F) 22. The NEC® may be used as a design specification by qualified persons.

Approved 23. If the AHJ finds the equipment or installation to be acceptable, it is ___.

Covered 24. ___ conductors are conductors not encased in a material recognized by the NEC® as an insulating material.

fittings 25. Connectors, couplings, locknuts, and bushings are electrical ___.

identified 26. Equipment that is ___ has been found to be suitable for a specific use or function.

50 27. Equipment that is required to be located in sight from other equipment shall be visible and not more than ___′ from the other equipment.

listed 28. Equipment that has been ___ is published in a directory by the testing lab certifying that it has met established standards.

Damage 29. One of the primary purposes of raceways is to protect conductors from physical ___.

approved 30. The NEC® requires that all electric equipment and conductors be ___ to be deemed acceptable.

Copper 31. When not specified, all conductors in the NEC® are considered to be ___.

closed 32. All unused openings other than those intended for the operation of equipment are required to be effectively ___ to afford equivalent protection.

identified for such 33. Terminals for more than one conductor shall be ___.

listed for direct burial 34. Wire connectors for use with conductors in direct-burial applications shall be ___.

disconnecting Means 35. Each ___ required by the NEC® shall be legibly marked to indicate its purpose.

The National Electrical Code®

TRADE COMPETENCY TEST

Name Bradly Lowry Date 01/24/2014

See Electrical Equipment for problems 1 through 4.

NEC®	Answer		
Article 100	T	(F)	**1.** The electrical equipment at A is readily accessible.
	T	F	**2.** The electrical equipment at B is readily accessible.
	T	F	**3.** The electrical equipment at C is readily accessible.
	T	F	**4.** The electrical equipment at D is readily accessible.

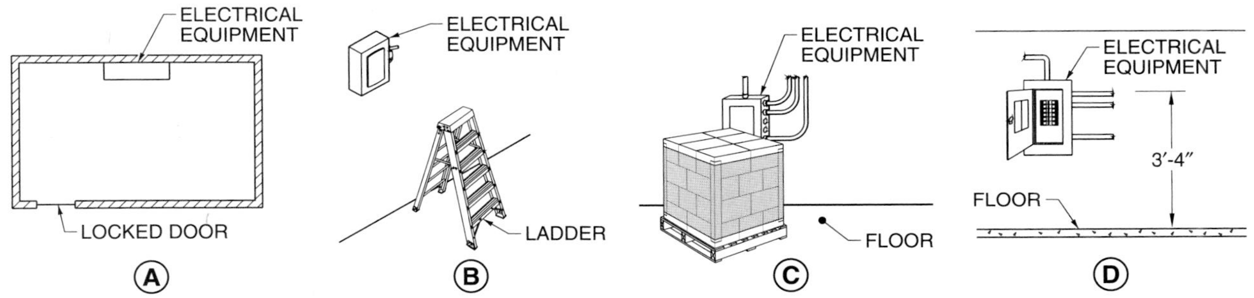

ELECTRICAL EQUIPMENT

See Devices for problems 5 through 8.

NEC®	Answer		
Article 100	(T)	F	**5.** The electrical system unit at A is a device.
	T	F	**6.** The electrical system unit at B is a device.
	T	F	**7.** The electrical system unit at C is a device.
	T	F	**8.** The electrical system unit at D is a device.

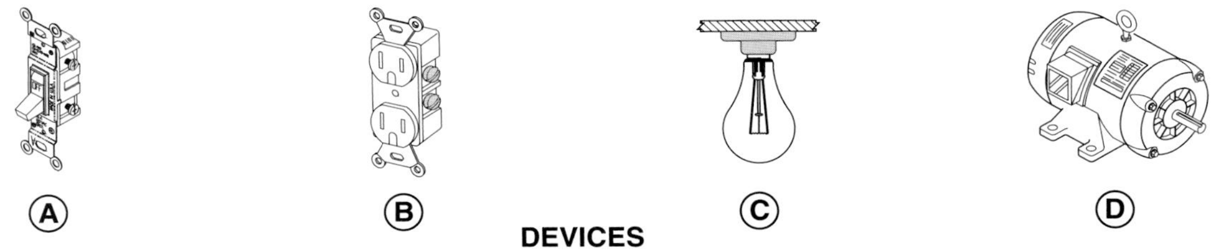

DEVICES

See Loads for problems 9 through 11.

__________ __________ **9.** The load at A is a(n) ___ load.

__________ intermittent **10.** The load at B is a(n) ___ load.

__________ __________ **11.** The load at C is a(n) ___ load.

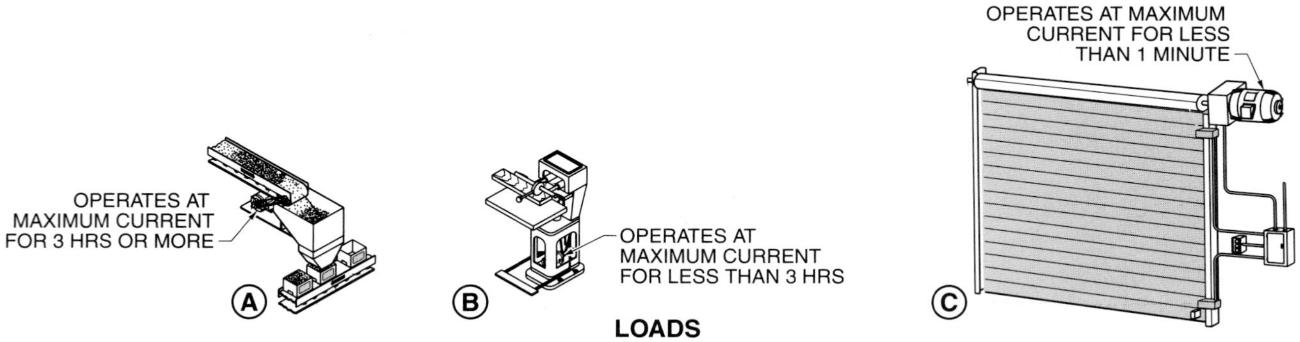

LOADS

See Within Sight for problems 12 through 15.

__________ T F **12.** The equipment at A is within sight.

__________ T F **13.** The equipment at B is within sight.

__________ T F **14.** The equipment at C is within sight.

__________ T F **15.** The equipment at D is within sight.

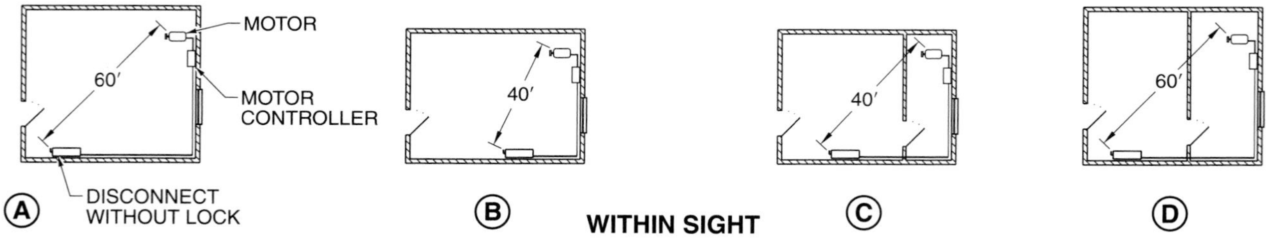

WITHIN SIGHT

See Working Clearances for problems 16 through 20.

table 110.26 (A)(1)

table 110.26 (A)(1) | 3 | **16.** The minimum working clearance at A is ___′.

110.26(A)(1) | ___ | **17.** The minimum working clearance at B is ___′.

___ | ___ | **18.** The minimum working clearance at C is ___′.

___ | ___ | **19.** The minimum working clearance at D is ___′.

___ | ___ | **20.** The minimum working clearance at E is ___′.

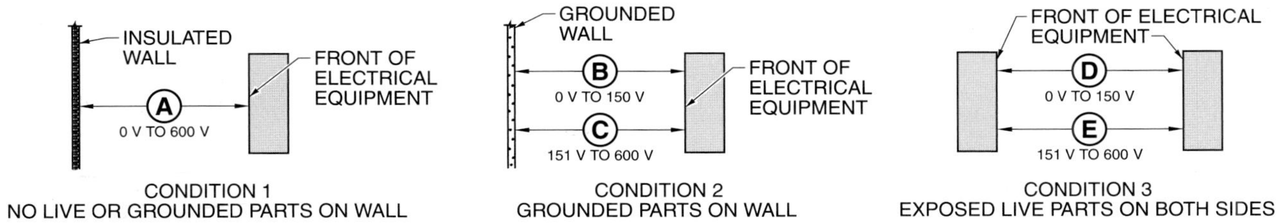

WORKING CLEARANCES

Learning the Code

- The vast majority of electrical work in a building or structure is going to be in the installation of branch circuits and feeders. Chapter 2 covers the requirements for both branch circuits in Article 210 and feeders in Article 215. Pay close attention to the requirements for branch circuits because these provisions are critical to most electrical systems. They are the circuits from the overcurrent device (fuse or circuit breaker) to the outlet. Remember, an outlet is any point on the wiring system where current is taken. It can be a receptacle outlet or can be for lighting.
- One of the keys to understanding Article 210 and correctly applying the provisions is to understand that the rules are often separated between those for dwelling units and those for other than dwelling units. This can be confusing at first, but reviewing the various definitions in Article 100 utilizing the root word "dwelling" can help in sorting it out. A good example is the requirements for GFCI protection in 210.8. GFCI requirements for dwelling units are contained in 210.8(A), and those for other than dwelling units are in 210.8(B).

- Apprentices and students of the Code who are especially interested in residential electrical work need to pay particular attention to 210.52. This important section contains all of the requirements for installing receptacle outlets in dwelling units, including small appliance, kitchen, and countertop provisions.
- The NEC® does not define "residential," "commercial," and "industrial" locations. Apprentices and students will hear these terms in the field, but for the most part the terms remain undefined in the NEC®. It is more important to understand the Code rule or requirement in terms of the Article 100 definitions and the specific applications.

chapter 2

Branch Circuits and Feeders

Feeder circuits are located between the service equipment and the final branch-circuit OCPD. Branch circuits are located between the final OCPD and the outlets or utilization equipment. The four basic types of branch circuits are appliance branch circuits, general-purpose branch circuits, individual branch circuits, and multiwire branch circuits.

New in the 2011 NEC®

- Clarification of the means for identifying grounded conductors – 200.6(A)
- Revision of the requirements for identification of ungrounded branch circuit conductors – 210.5(C)
- Revised requirements for the disconnection of multiple branch circuits on the same yoke – 210.7
- Revised exceptions for AFCI protection in dwelling units – 210.12(A) Ex. 1–2
- Clarification of countertop receptacle spacing requirements – 210.52(C)(4)
- New foyer receptacle outlet requirements for dwelling units – 210.52(C)(I)

BRANCH CIRCUITS – ARTICLE 210

There are basically three types of conductors in any electrical installation. These are the service-entrance conductors, feeder conductors, and branch-circuit conductors. **See Figure 2-1.** Article 100 defines a *service conductor* as a conductor from the service point to the service disconnecting means. Service conductors supply power to the service equipment. Service conductors can be either overhead or underground. See Article 230. Article 100 defines *service equipment* as the necessary equipment, usually consisting of a circuit breaker or switch and fuses and their accessories, connected to the load end of service conductors to a building or other structure, or an otherwise designated area, and intended to constitute the main control and cut-off of the supply. The wires leaving the service equipment are feeder conductors. Article 100 defines *feeder* as all circuit conductors between the service equipment, the source of a separately derived system, or other supply source, and the final branch-circuit overcurrent device.

The *branch circuit* is that portion of the electrical circuit between the last overcurrent device (fuse or circuit breaker) and the outlets or utilization equipment. *Utilization equipment* is equipment that utilizes electric energy for electronic, electromechanical, chemical, heating, lighting, or similar purposes.

Section 210.1 limits the scope of Article 210 to all branch circuits except branch circuits that supply only motor loads that are covered in Article 430. Branch circuits shall comply with Article 210 and with applicable provisions of other articles of the NEC® per 210.2. The provisions in the more specific articles amend the provisions of Article 210. See Table 210.2 for a branch-circuit cross-reference list.

The four types of branch circuits defined in Article 100 are appliance branch circuits, general-purpose branch circuits, individual branch circuits, and multiwire branch circuits. **See Figure 2-2.** An *appliance branch circuit* is a branch circuit that supplies energy to one or more outlets to which appliances are to be connected. Appliance branch circuits shall have no permanently connected luminaires (lighting fixtures) which are not a part of an appliance.

A *general-purpose branch circuit* is a branch circuit that supplies two or more outlets for lighting and appliances. An *individual branch circuit* is a branch circuit that supplies only one piece of utilization equipment. A *multiwire branch circuit* is a branch circuit with two or more ungrounded conductors having a voltage between them, and a grounded conductor having equal voltage between it and each ungrounded conductor, and is connected to the neutral or grounded conductor of the system. The most popular forms of multiwire branch circuits occur in 1ϕ, 3-wire and 3ϕ, 4-wire systems. The main concern in multiwire branch circuits is that the neutral shall never be opened while the circuit is energized.

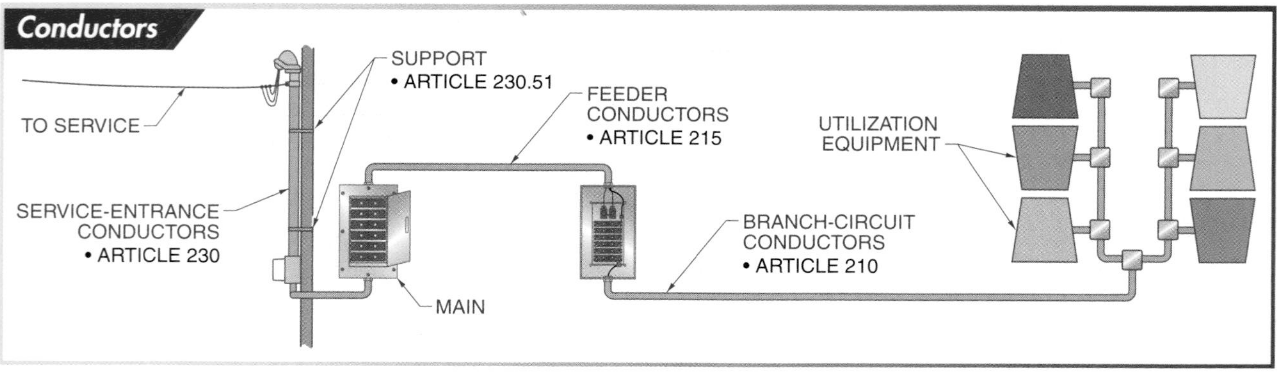

Figure 2-1. Three types of conductors are service-entrance conductors, feeder conductors, and branch-circuit conductors.

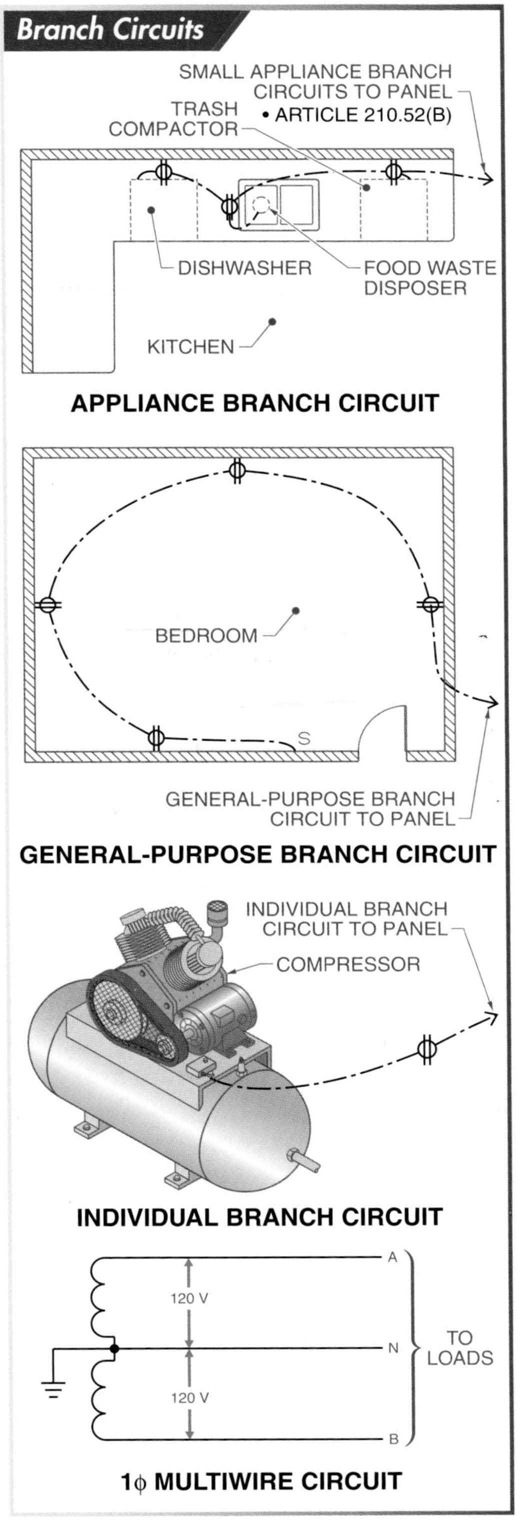

Figure 2-2. Four types of branch circuits are appliance branch circuits, general-purpose branch circuits, individual branch circuits, and multiwire branch circuits.

Underground Feeders

In large industrial facilities, it is common to tap off of an overhead feeder and run the feeder conductors underground to another point in the distribution system. Underground feeders usually operate at voltages over 600 V. When working with conductors that have voltages over 600 V, special training is required for both splicing and terminating conductors, as well as meeting NEC® requirements. Only qualified persons can work on these types of feeder conductors.

Branch-Circuit Ratings – 210.3

Branch circuits shall be rated in accordance with the maximum ampere rating or setting of the overcurrent device. These ratings shall be 15 A, 20 A, 30 A, 40 A, and 50 A in other than individual branch circuits. Where conductors of higher ampacity are used, the rating or setting of the overcurrent device shall determine the circuit rating. For example, a 20 A branch circuit with a load rated at 16 A may have 8 AWG THW Cu conductors due to an ambient temperature that exceeds 86°F. and more than three current-carrying conductors in the raceway. Although, per Table 310.15(B)(16), the conductors have an allowable ampacity of 50 A, the overcurrent protective device is required to be rated at 20 A. Since 210.3 requires that the branch-circuit rating be based on the rating of the overcurrent device, the branch-circuit rating would be 20 A.

Multiwire Branch Circuits – 210.4

Section 210.4 permits multiwire branch circuits to be considered as multiple circuits. All of the conductors in a multiwire branch circuit shall originate from the same panelboard or similar distribution equipment. Per 210.4(A), Informational Note, the neutral conductor in multiwire branch circuits supplying power to nonlinear loads may carry more current than the ungrounded conductors. This is due to the high harmonic currents. A *nonlinear load* is a load where the waveform of the steady-state current does not follow the wave shape of the applied voltage. **See Figure 2-3.** Examples of nonlinear loads include electronic equipment, electronic/electric-discharge lighting, adjustable speed drive systems, and similar equipment.

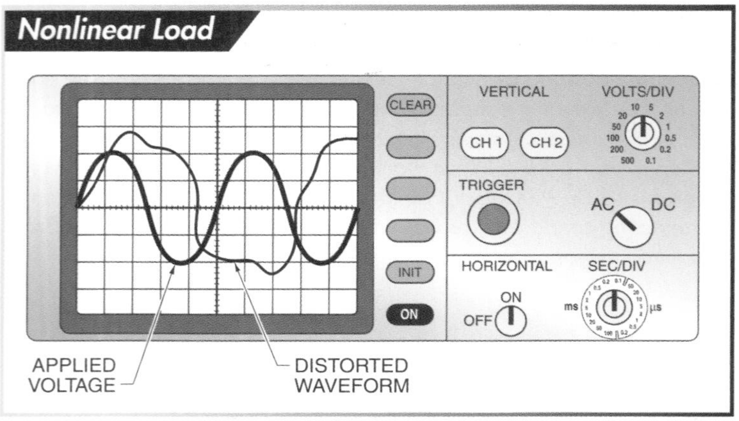

Figure 2-3. A nonlinear load is a load where the waveform of the steady-state current does not follow the waveform of the applied voltage.

Studies have been made of 3ϕ conductors that are perfectly balanced and it was found that the neutral conductor carried more current. For example, if each phase conductor carried 100 A, it was found that the neutral conductor could carry as much as 130 A to 140 A. Engineering manuals suggest that the neutral conductor be increased in multiwire circuits that supply current to nonlinear loads. This is a circuit design issue. However, care should be taken to ensure that the system design allows for the possibility of high harmonic currents on the neutral conductor.

Disconnecting Means. Because of the potential shock hazard, multiwire branch circuits have special disconnecting provisions. Section 210.4(B) requires that a disconnecting means be provided for each multiwire branch circuit. The disconnecting means must disconnect all ungrounded conductors simultaneously and at the point where the branch circuit conductors originate. Article 100 defines a *disconnecting means* as a device or group of devices that separate or isolate the conductors of a circuit from their source of supply.

In the 2008 NEC®, Section 210.4(D) was added to Article 210. The purpose of this requirement is to ensure that where multiwire branch circuits are installed, the grounded conductor is grouped with the associated ungrounded conductors. This is an important means of identification for those who maintain electrical systems. The grouping must occur in at least one location within the panelboard or point of origin, unless the circuit enters from a cable or raceway unique to the circuit and makes the grouping apparent.

Installers and persons performing maintenance on multiwire branch circuits should be aware that the neutral is being shared and that all of the ungrounded conductors that share the neutral must be disconnected from the supply before attempting to work on any portion of the circuit. Section 300.13(B) recognizes the dangers involved and mandates that the grounded conductor of a multiwire branch circuit be spliced and pigtailed rather than being spliced by back-wiring into a device.

Branch-Circuit Identification – 210.5. Section 200.6 specifies identification requirements for the grounded conductor of a branch circuit. The means of identification for grounded conductors is dependent upon the size of the conductor. Section 200.6(A) contains the provisions for grounded conductors sizes 6 AWG or smaller. Section 200.6(B) contains the permissible means of identification for sizes larger than 6 AWG. Generally, the grounded conductor of a branch circuit is identified by the color white or gray on the outer surface or by three continuous white stripes on other than green insulation along the conductors' entire length.

There may be installations where grounded conductors of different systems are installed within the same raceway, cable, or box. In these cases it is imperative that the system grounded conductors remain separated and are not mixed. For this reason, 200.6(D) contains provisions for marking the different system grounded conductors to ensure that each system conductor is adequately distinguished. There are three means that are permitted to establish the identification of the two system grounded conductors. Regardless of which means is selected, the method chosen shall be permanently posted at each branch-circuit panelboard. **See Figure 2-4.**

Color Code for Branch-Circuit Equipment Grounding Conductors (EGCs) – 210.5(B). Section 250.119 specifies identification requirements for the EGC of a branch circuit.

Branch-circuit EGCs shall have a continuous outer finish that is either green or green with one or more yellow stripes unless it is bare. Insulated conductors larger than 6 AWG are permitted to be permanently identified as an EGC at each end and at every point where the conductor is accessible per 250.119(A). Identification shall be accomplished in one of three ways:

- Stripping the insulation or covering from the full exposed length.
- Coloring the exposed insulation or covering green.
- Marking the exposed insulation or covering with green-colored tape or green adhesive labels.

See 250.119(B) for identification requirements for EGCs that are part of a multiconductor cable.

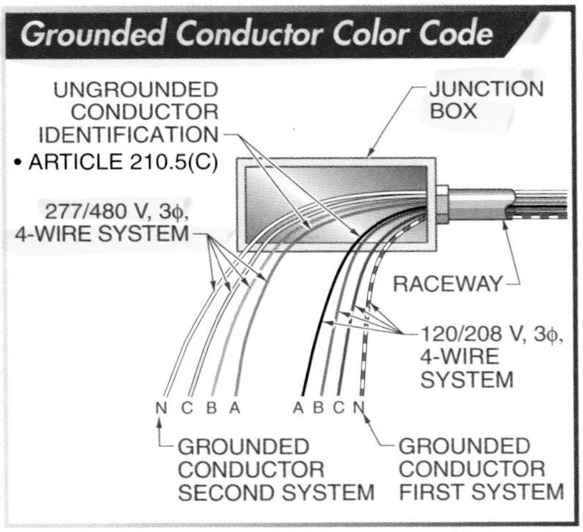

Figure 2-4. The grounded conductors of different systems shall be identified in accordance with 200.6(D).

Identification of Ungrounded Conductors – 210.5(C). In buildings where more than one voltage system exists, each ungrounded branch-circuit conductor shall be identified by phase or line and system. The identification must take place at all termination, connection, and splice points. Conductors may be identified by a separate color code, marked with tape, tagged, or identified by other means acceptable to the AHJ. Whatever method is used, the means of identification must be documented. The documentation must be readily available or permanently posted in each branch-circuit panelboard. **See Figure 2-5.**

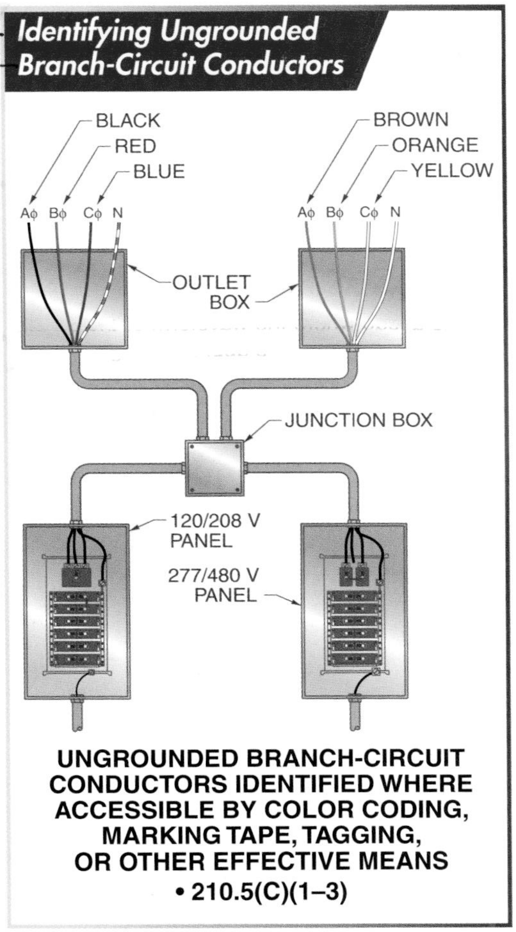

Figure 2-5. Each ungrounded branch circuit shall be identified by phase and system.

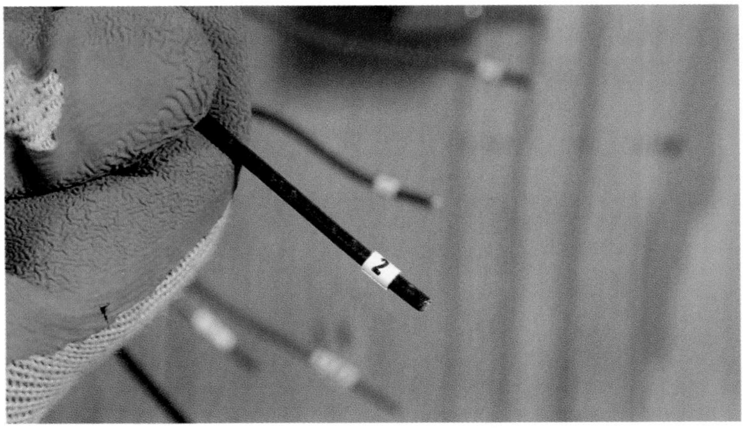

If acceptable by the AHJ, wire markers may be used to identify conductors.

BRANCH-CIRCUIT VOLTAGE LIMITATIONS – 210.6

Voltage limitations for branch circuits are covered by 210.6. The section provides limitations based on occupancy and voltage for lighting loads and utilization equipment.

Occupancy Limitation – 210.6(A)

This section pertains specifically to dwelling units, motels, hotels, and other occupancies, such as dormitories, nursing homes, and similar residential occupancies. In these types of dwellings, any luminaire (lighting fixture) or receptacle for plug-connected loads rated up to 1440 VA, or less than ¼ HP, shall be supplied at not more than 120 V.

120 V between Conductors – 210.6(B)

Branch circuits with voltages not over 120 V between conductors are permitted to supply three types of loads. **See Figure 2-6.** These loads are:

(1) Terminals of lampholders applied within their voltage rating.

(2) Auxiliary equipment of electric-discharge lamps.

(3) Cord-and-plug-connected or permanently connected utilization equipment.

277 V to Ground – 210.6(C)

Circuits exceeding 120 V between conductors, and not exceeding 277 V to ground, are permitted to supply six types of loads. **See Figure 2-7.** These loads are:

(1) Listed electric-discharge or listed light-emitting-diode-type luminaires (lighting fixtures) used within their rating.

(2) Listed incandescent luminaires (lighting fixtures) where supplied at 120 V or less from the output of a step-down autotransformer that is an integral component of the luminaire (lighting fixture) and the outer shell terminal is electrically connected to a grounded conductor of the branch circuit.

(3) Luminaires (lighting fixtures) equipped with mogul-base, screw-shell lampholders.

(4) Other than screw-shell type lampholders applied within their voltage ratings.

(5) Auxiliary equipment of electric-discharge lamps.

(6) Cord-and-plug-connected or permanently connected utilization equipment. Examples include air-conditioning and heating equipment in commercial locations and electric cooking equipment.

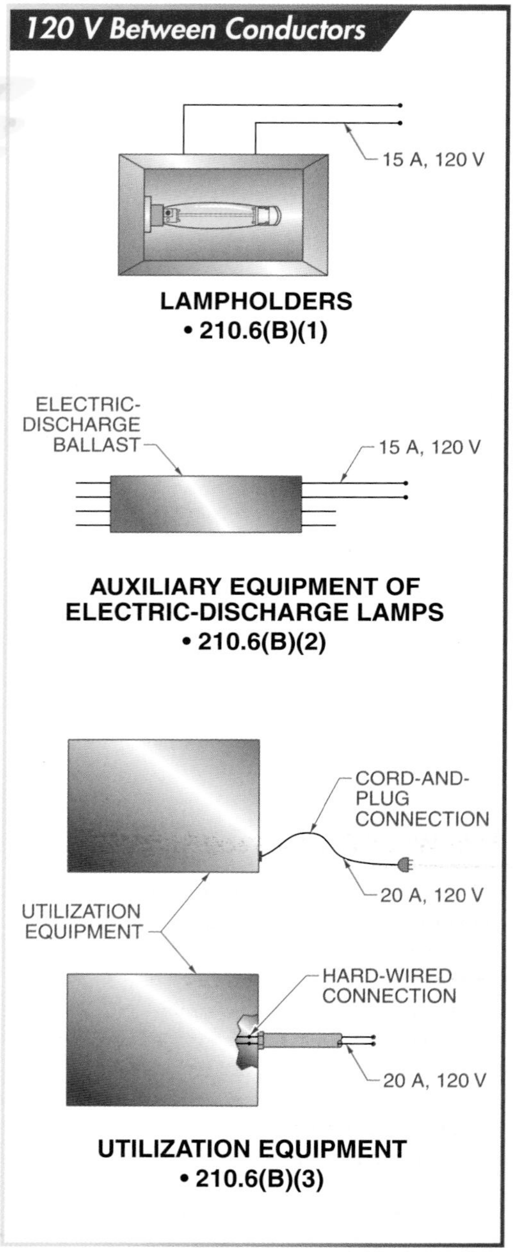

Figure 2-6. Branch circuits with voltages not over 120 V between conductors are permitted to supply three types of loads.

+1-20A
11-15A

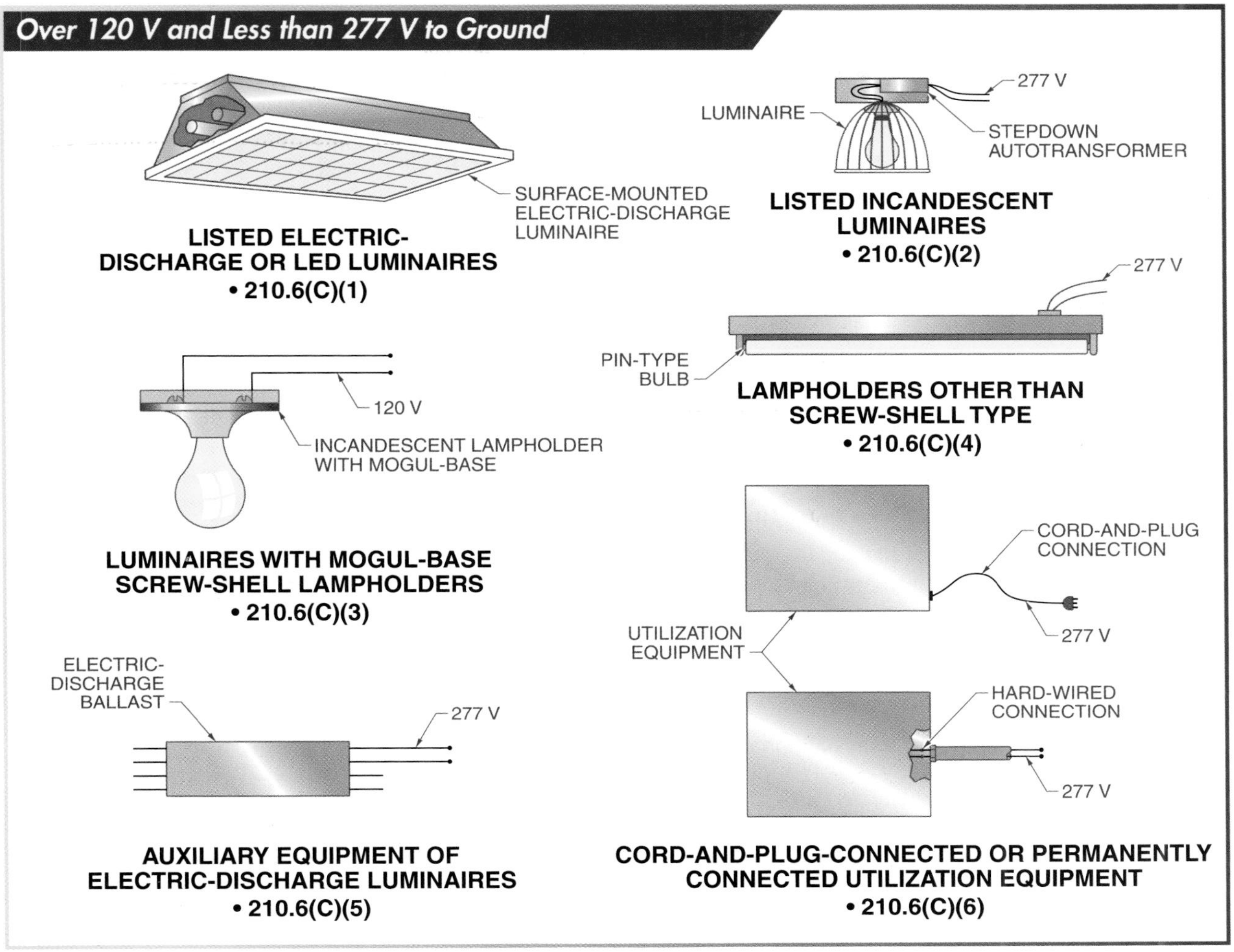

Figure 2-7. Branch circuits with voltages over 120 V, but not more than 277 V to ground, are permitted to supply six types of loads.

Split-Wired Duplex Receptacles

While trying to replace a defective, split-wired duplex receptacle, the homeowner may plug a lamp into one of the receptacle outlets and then open the circuit breakers (CBs) until the lamp turns OFF in order to determine the circuit on which the receptacle is located. The homeowner may then think that both outlets on the duplex receptacle are de-energized and begin to replace the receptacle. On a split-wired receptacle, each outlet of a duplex receptacle could be on a separate circuit. The NEC® requires that the CB be a double-pole type or two single-pole circuit breakers (SPCBs) with a listed tie handle or other means to ensure that they both operate simultaneously.

Plug-connected loads rated up to 1440 VA shall be supplied at not more than 120 V.

Electronic Ballasts

Electronic ballasts offer enhanced lighting performance and energy savings, and are available with different lamp ignition methods. Instant-start electronic ballasts use a high initial voltage to start the lamp without delay or flashing, offer maximum energy savings, and are the most popular type of electronic ballast. Rapid-start ballasts have a separate set of windings to provide a low voltage to electrodes prior to lamp ignition, and then a starting voltage to start the lamp. Programmed-start ballasts heat lamp cathodes to a specific temperature prior to lamp ignition to further reduce thermal stress for a longer lamp life.

600 V between Conductors – 210.6(D)

Branch circuits rated over 277 V to ground and not over 600 V between conductors are permitted to supply three types of loads. **See Figure 2-8.** These loads are:

(1) Electric-discharge auxiliary equipment. Examples include fluorescent, mercury-vapor, and sodium fixtures. Installation is limited to outdoor areas such as roads, bridges, athletic fields, and parking lots. The lamps shall be mounted in permanently installed fixtures not less than 22′ in height or not less than 18′ in height on other structures, such as tunnels.

(2) Cord-and-plug-connected or permanently connected utilization equipment other than luminaires (lighting fixtures).

(3) Luminaires for special direct-current applications. These luminaires must contain a listed, DC-rated ballast that contains isolation provisions to protect against electrical shock during relamping.

GROUND-FAULT CIRCUIT-INTERRUPTER (GFCI) PROTECTION FOR PERSONNEL

A Class A GFCI is a device that trips when the current to ground has a value in the range of 4–6 mA. A *milliampere (mA)* is 1/1000 of an ampere (1000 mA = 1 A). In general, the effects of current flow on the human body are:

- 1 mA = Threshold of sensation
- 2 mA = Mild shock
- 5 mA = GFCI will trip
- 10 mA = Cannot let go
- 20 mA = Muscles contract; breathing difficulty begins
- 50 mA = Breathing difficult; suffocation possible
- 100 mA = Heart stops pumping
- 300 mA = Severe burns; breathing stops

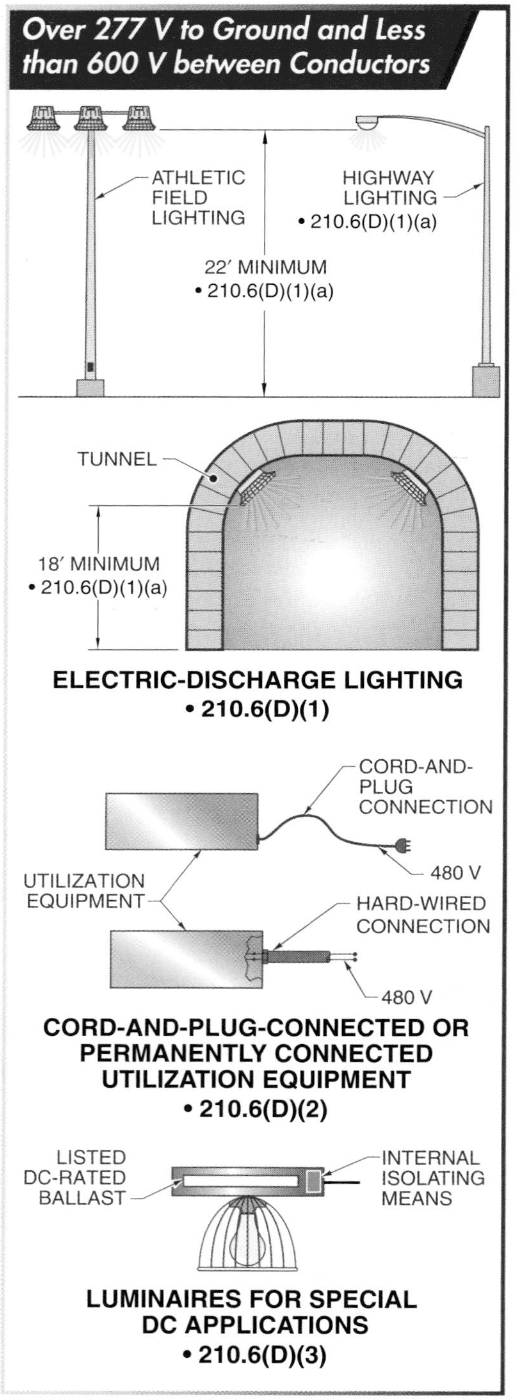

Figure 2-8. Branch circuits with voltages over 277 V, but less than 600 V, between conductors are permitted to supply three types of loads.

GFCI devices monitor for any imbalance in the currents on the ungrounded and grounded conductors of a circuit. The Class A GFCI device will trip the switching contacts when an imbalance exceeds 5 mA. At one time, Class B devices were also available. These types of devices tripped when a ground-fault current exceeded 20 mA. Class B GFCI protectors may only be used to protect underwater swimming pool lighting fixtures installed before the 1965 NEC® was adopted. Pool lighting designed before 1965 often has leakage currents in excess of 5 mA, which causes nuisance tripping when protected by a Class A GFCI. The swimming pool lighting circuit must be disconnected during servicing and relamping operations when a Class B GFCI is used as permitted.

The GFCI does not monitor current flow on equipment grounding conductors. This is normally not a problem since equipment grounding conductors are only meant to carry current under fault conditions and, where the equipment grounding path is designed properly, only for the time it takes to trip the overcurrent protective device protecting the circuit. This design means that the GFCI will not protect personnel from a fault condition where the equipment grounding path has too much impedance, resulting in the fault current being too small to trip the overcurrent device.

It is important for designers and installers to ensure that equipment grounding paths have low impedance that will facilitate the tripping of the overcurrent protective device. Installers, maintenance personnel, and users of GFCI protection devices should also be aware that the sensor may not sense an imbalance where contact is made with the ungrounded conductor and the grounded conductor at the same time. Where an imbalance is not detected, the device will not operate.

The first requirements for GFCI protection were included in the 1971 NEC®. Protection was required in dwelling unit bathrooms and in outdoor receptacles of dwelling units. **See Figure 2-9.** Subsequent editions of the NEC® have continued to expand these requirements. Studies continue to indicate a decreasing trend in the number of electrocutions in the United States since the GFCI was introduced in the 1971 NEC®.

Figure 2-9. The first requirements for ground-fault circuit-interrupter (GFCI) protection were included in the 1971 NEC®.

Dwelling Units – 210.8(A)

All 125 V, 15 A and 20 A, 1ϕ receptacles installed in dwelling units at any of eight locations shall have GFCI protection. These eight locations are bathrooms; garages and accessory buildings; outdoors; crawl spaces; unfinished basements; kitchens; utility, laundry; and boathouses.

Bathrooms – 210.8(A)(1). All bathroom receptacles shall be GFCI-protected. **See Figure 2-10.** A *bathroom* is an area with one or more of the following: a toilet, tub, urinal, shower, bidet, or similar type of plumbing fixture. See Article 100. For an area to qualify as a bathroom, a basin shall be present. Note there are no exceptions to this rule. Any 15 A or 20 A, 125 V receptacle installed in a dwelling unit bathroom must have GFCI protection.

In Article 100, "bathroom" is defined as an area that includes a basin with a toilet, urinal, tub, shower, bidet, or similar plumbing fixture.

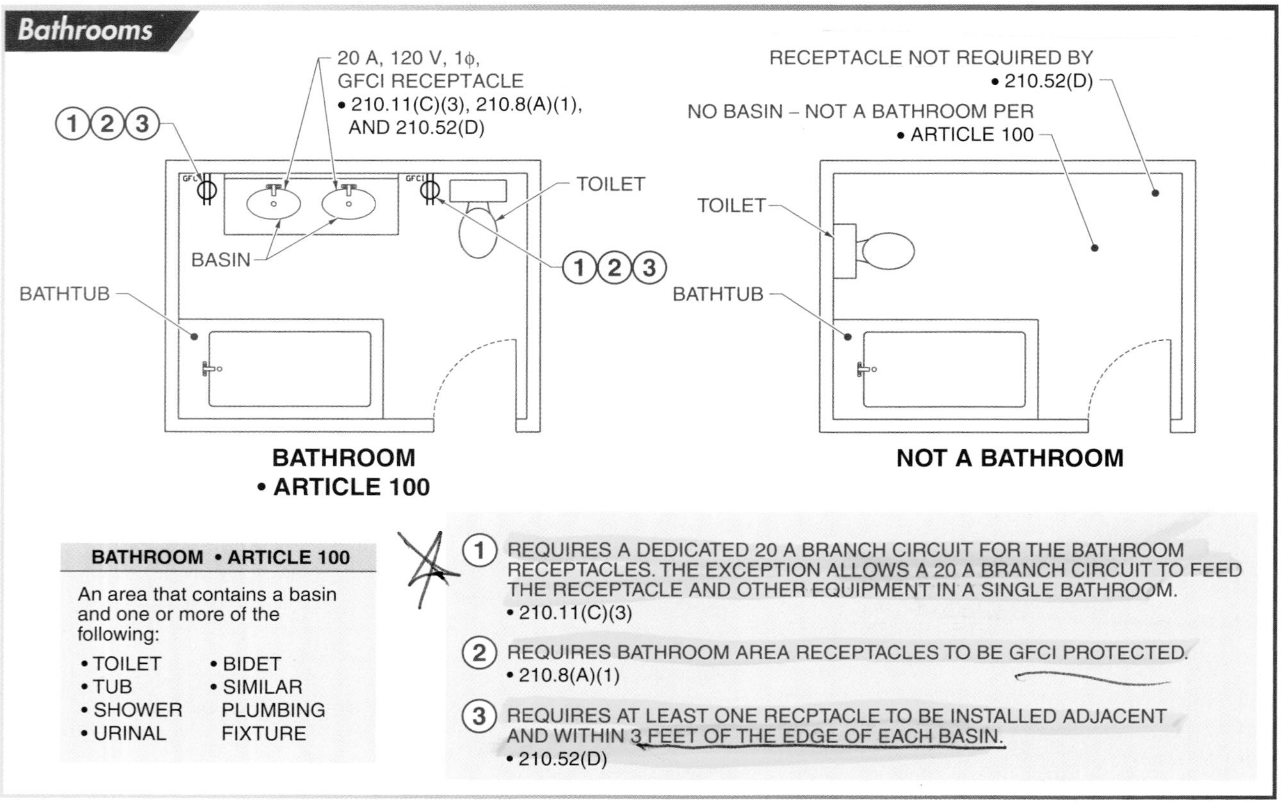
Bathrooms
20 A, 120 V, 1ϕ,
GFCI RECEPTACLE
• 210.11(C)(3), 210.8(A)(1),
AND 210.52(D)
①②③
TOILET
BASIN
BATHTUB
①②③
BATHROOM
• ARTICLE 100
RECEPTACLE NOT REQUIRED BY
• 210.52(D)
NO BASIN – NOT A BATHROOM PER
• ARTICLE 100
TOILET
BATHTUB
NOT A BATHROOM
BATHROOM • ARTICLE 100
An area that contains a basin and one or more of the following:
• TOILET • BIDET
• TUB • SIMILAR
• SHOWER PLUMBING
• URINAL FIXTURE
① REQUIRES A DEDICATED 20 A BRANCH CIRCUIT FOR THE BATHROOM RECEPTACLES. THE EXCEPTION ALLOWS A 20 A BRANCH CIRCUIT TO FEED THE RECEPTACLE AND OTHER EQUIPMENT IN A SINGLE BATHROOM.
• 210.11(C)(3)
② REQUIRES BATHROOM AREA RECEPTACLES TO BE GFCI PROTECTED.
• 210.8(A)(1)
③ REQUIRES AT LEAST ONE RECPTACLE TO BE INSTALLED ADJACENT AND WITHIN 3 FEET OF THE EDGE OF EACH BASIN.
• 210.52(D)

Figure 2-10. All bathroom receptacles shall be GCFI-protected.

All bathroom receptacles must be GFCI-protected.

Garages and Accessory Buildings – 210.8(A)(2). GFCI-protected receptacles shall be installed in garages and grade-level portions of unfinished accessory buildings used for storage or work areas. **See Figure 2-11.** Prior to the 2008 NEC®, there were two exceptions to the GFCI requirements for garages and accessory buildings in dwelling units. The first exception was for receptacles that were not readily accessible. The second exception was for single or duplex receptacles that supplied appliances located in dedicated spaces. Both exceptions were removed to increase the level of personnel protection from shock. Code-Making Panel 2 accepted the argument that the fact that the receptacle is not readily accessible has very little to do with protection against shock hazards. Protection by GFCIs is not related to the location of the receptacle. Beginning with the 2008 NEC®, all 125 V, 1ϕ, 15 A and 20 A receptacles installed in garages and accessory buildings that have a floor located at or below grade level and that are not intended as habitable rooms shall be provided with GFCI protection.

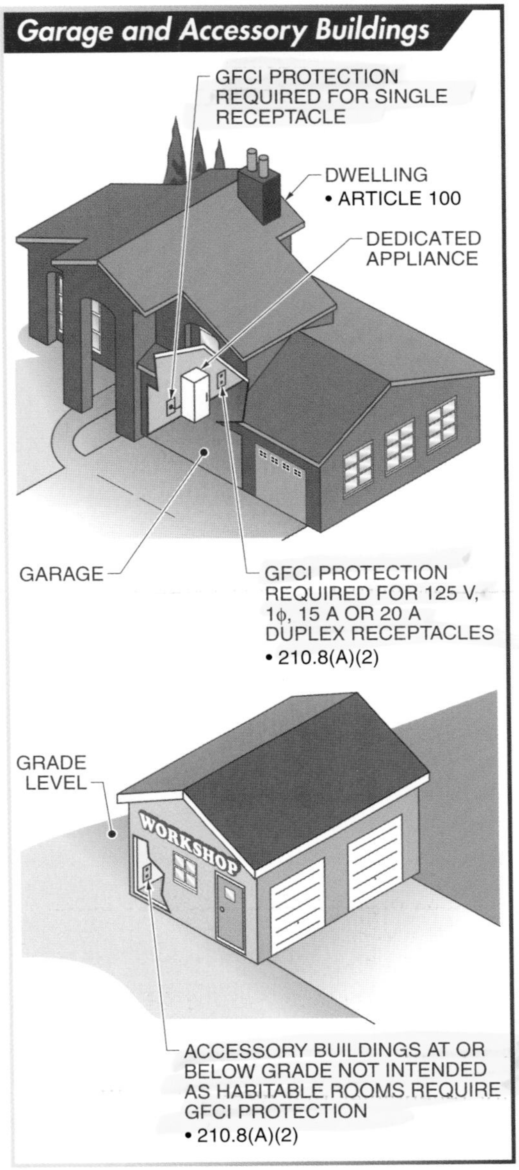
Garage and Accessory Buildings
GFCI PROTECTION REQUIRED FOR SINGLE RECEPTACLE
DWELLING
• ARTICLE 100
DEDICATED APPLIANCE
GARAGE
GFCI PROTECTION REQUIRED FOR 125 V, 1ϕ, 15 A OR 20 A DUPLEX RECEPTACLES
• 210.8(A)(2)
GRADE LEVEL
WORKSHOP
ACCESSORY BUILDINGS AT OR BELOW GRADE NOT INTENDED AS HABITABLE ROOMS REQUIRE GFCI PROTECTION
• 210.8(A)(2)

Figure 2-11. GFCI-protected receptacles shall be installed in garages and accessory buildings having a floor located at or below grade level not intended as habitable rooms and limited to storage areas, work areas, and areas of similar use.

Outdoors – 210.8(A)(3). All outdoor receptacles shall be GFCI-protected. **See Figure 2-12.** The one exception to this requirement is receptacles which are not readily accessible and are supplied from a dedicated branch circuit to serve electric snow-melting, deicing, or pipeline and vessel heating equipment in accordance with 427.22.

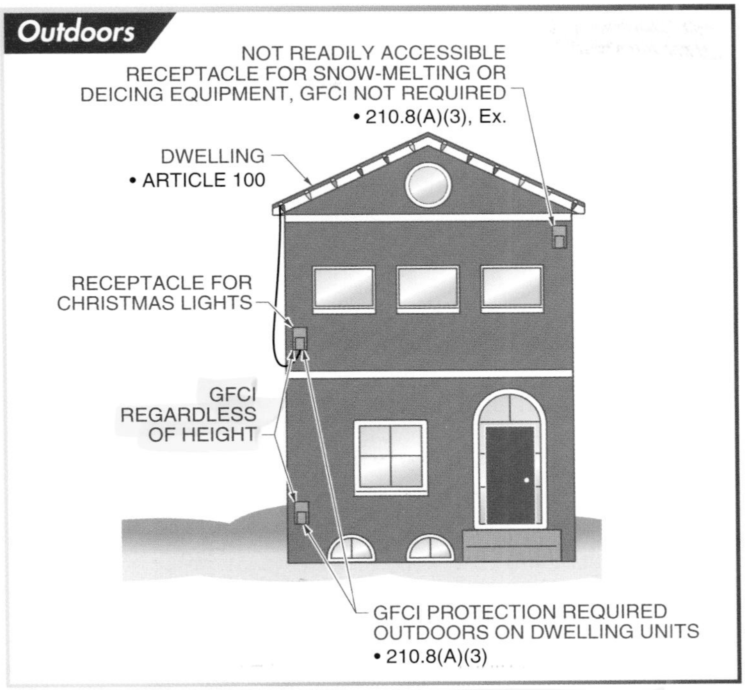
Outdoors
NOT READILY ACCESSIBLE RECEPTACLE FOR SNOW-MELTING OR DEICING EQUIPMENT, GFCI NOT REQUIRED
• 210.8(A)(3), Ex.
DWELLING
• ARTICLE 100
RECEPTACLE FOR CHRISTMAS LIGHTS
GFCI REGARDLESS OF HEIGHT
GFCI PROTECTION REQUIRED OUTDOORS ON DWELLING UNITS
• 210.8(A)(3)

Figure 2-12. All outdoor receptacles shall be GFCI-protected.

These receptacles are permitted to be installed without GFCI protection for personnel. However, 426.28 requires that ground-fault protection of equipment (GFPE) be provided for branch circuits supplying fixed outdoor electric deicing and snow-melting equipment. GFPEs protect the equipment, not personnel. GFPEs are set to trip in the range of 20–30 mA, not the range of 4–6 mA that is necessary to protect personnel.

Outdoor receptacles in dwelling units in most cases must be GFCI-protected.

Crawl Spaces – 210.8(A)(4). In general, all receptacles in crawl spaces where the crawl space is at or below grade level shall be GFCI-protected. **See Figure 2-13.** *Note:* Crawl spaces at or below grade level may be wet or damp locations.

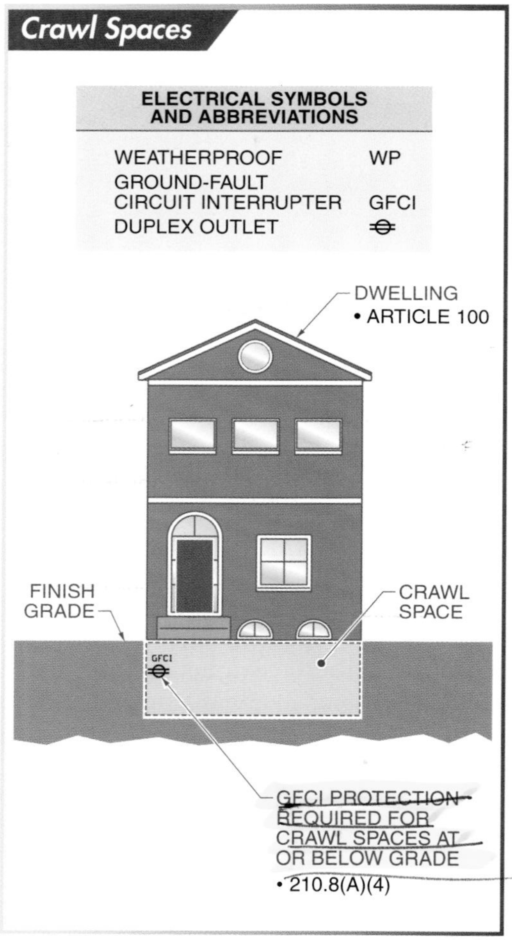

Figure 2-13. All receptacles in crawl spaces at or below grade level shall be GFCI-protected.

Unfinished Basements – 210.8(A)(5). In general, all receptacles in unfinished basements shall be GFCI-protected. **See Figure 2-14.** An *unfinished basement* is the portion or area of a basement which is not intended as a habitable room, but is limited to storage areas, work areas, etc. *Note:* GFCI protection is not generally required in finished basements.

All 125 V, 15 A and 20 A, 1ϕ receptacles installed in kitchens shall have GFCI protection.

sump pump

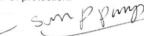

As with the requirements for dwelling unit garages and accessory buildings, the exceptions for GFCI requirements for unfinished basements in dwelling units were deleted in the 2008 NEC® edition. These exceptions excluded those receptacles that were not readily accessible and those that supplied appliances that occupied dedicated spaces. The only exception for receptacles in unfinished basements of dwelling units is for those receptacles that supply permanently installed fire alarm systems or burglar alarm systems. This exception is warranted because of the concern for the inadvertent de-energizing of these two important life safety and security systems in dwelling units. Receptacles installed under this exception shall not be considered as meeting the requirements of 210.52(G).

Kitchens – 210.8(A)(6). All receptacles which serve countertop surfaces installed in kitchens shall be GFCI-protected. **See Figure 2-15.** Most appliances intended for countertop use are not equipped with an equipment grounding conductor (EGC). Since water and many grounded surfaces are present in kitchens, a shock hazard exists. Receptacles installed for dedicated appliances, such as food waste disposers, dishwashers, and trash compactors are not required to have GFCI protection.

See pg 385

Sinks Other Than in Kitchen – 210.8(A)(7). All receptacles installed within 6′ of the outside edge of a sink that is located in an area other than a kitchen shall be GFCI-protected. Per 406.5(E), receptacles shall not be installed in the face-up position of the counter or working surface. **See Figure 2-16.** Such receptacles may be used to operate blenders, ice crushers, etc.

Boathouses – 210.8(A)(8). All 125 V, 15 A and 20 A, 1ϕ receptacles installed in boathouses shall be provided with ground-fault circuit-interrupter protection for personnel. Article 555 covers the installation of wiring and equipment in marinas and similar locations, such as wharves and docking facilities. However, the scope of 555.1 specifically excludes private, noncommercial docking facilities used by private owners or residents of associated single-family dwellings. The requirement in 210.8(A)(8) ensures that GFCI protection will still be afforded in all boatyards associated with dwelling units.

Battery-Based Vessel Electrical Systems

A vessel such as a boat or ship has a DC battery-based power system to supply power to all major vessel loads including engine starting, anchor windlass, pumps, lighting, electronics, communications, navigation equipment, refrigeration, desalination, and entertainment systems. Energy to supply these systems is stored in a battery bank.

Other Than Dwelling Units – 210.8(B)

Section 210.8(B) requires all 15 A and 20 A, 125 V, 1ϕ receptacles installed in other than dwelling units at any of eight locations to have GFCI protection. **See Figure 2-17.** Three locations are bathrooms, kitchens, and rooftops in other than dwelling units.

Per 210.8(B)(1), all receptacles installed in bathrooms shall be GFCI-protected. However, 517.21 does not require GFCI protection if the toilet and basin in health care facilities are installed within critical care areas of the patient's room. Per 210.8(B)(3), all receptacles installed on rooftops shall be GFCI-protected. There is an exception to this requirement for rooftop receptacles that are not readily accessible and are supplied by a dedicated branch circuit to serve electric snow-melt or deicing equipment.

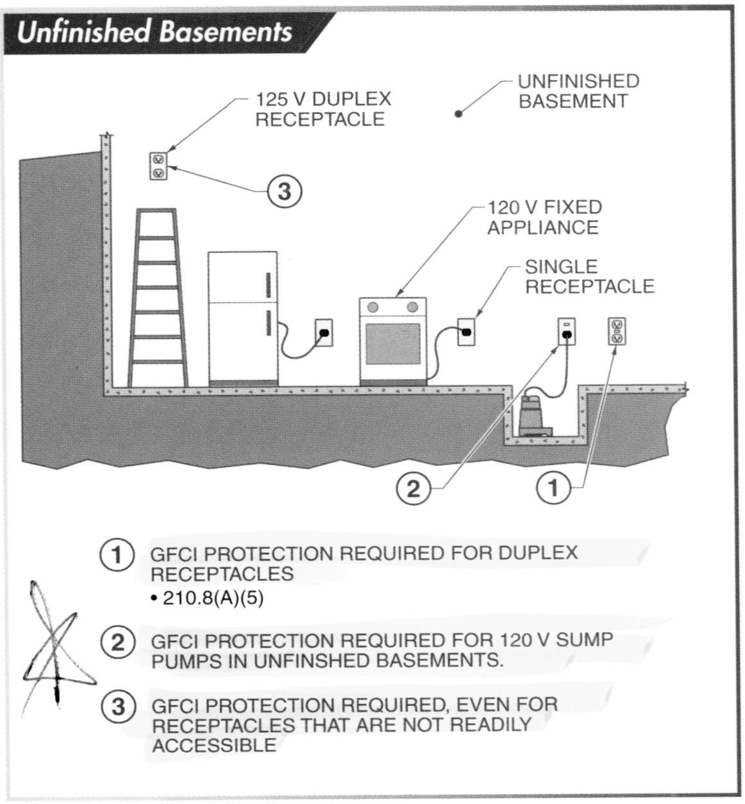
Unfinished Basements
125 V DUPLEX RECEPTACLE
UNFINISHED BASEMENT
120 V FIXED APPLIANCE
SINGLE RECEPTACLE
1 GFCI PROTECTION REQUIRED FOR DUPLEX RECEPTACLES • 210.8(A)(5)
2 GFCI PROTECTION REQUIRED FOR 120 V SUMP PUMPS IN UNFINSHED BASEMENTS.
3 GFCI PROTECTION REQUIRED, EVEN FOR RECEPTACLES THAT ARE NOT READILY ACCESSIBLE

Figure 2-14. All receptacles in unfinished basements shall be GFCI-protected, unless they supply permanently installed fire alarm or burglar alarm systems.

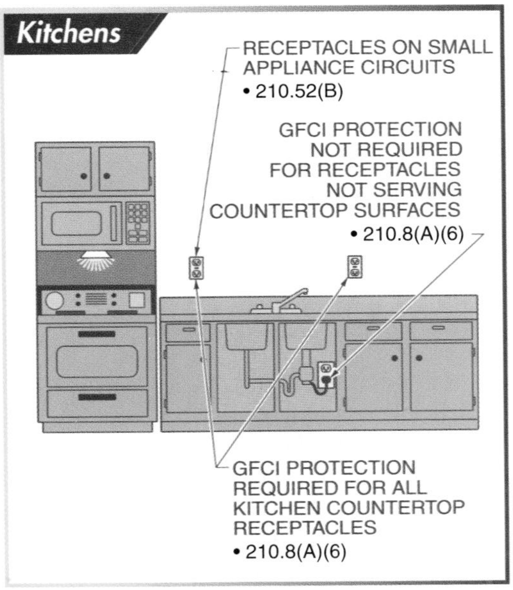
Kitchens
RECEPTACLES ON SMALL APPLIANCE CIRCUITS • 210.52(B)
GFCI PROTECTION NOT REQUIRED FOR RECEPTACLES NOT SERVING COUNTERTOP SURFACES • 210.8(A)(6)
GFCI PROTECTION REQUIRED FOR ALL KITCHEN COUNTERTOP RECEPTACLES • 210.8(A)(6)

Figure 2-15. All receptacles that serve countertop surfaces installed in kitchens shall be GFCI-protected.

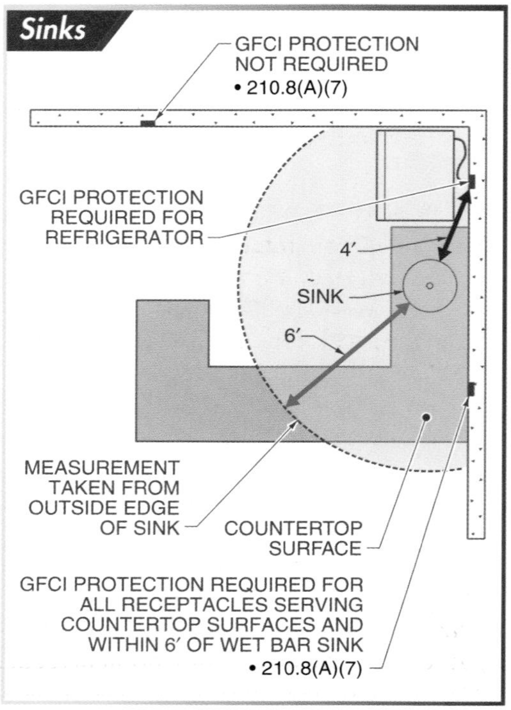

Figure 2-16. All receptacles within 6′ of the outside edge of a sink shall be GFCI-protected.

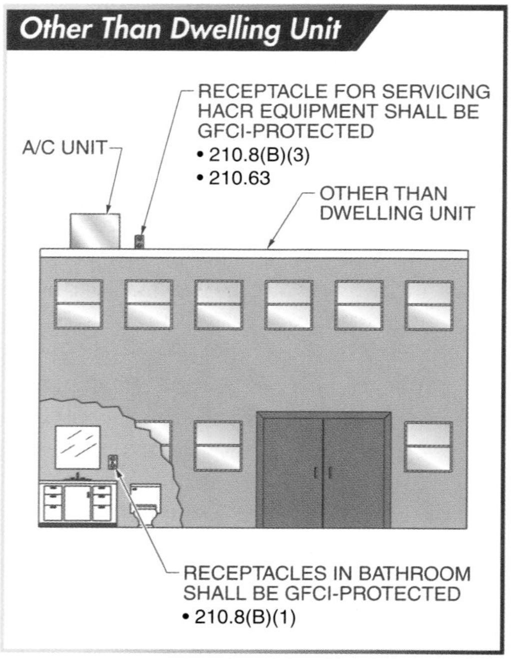

Figure 2-17. All 15 A and 20 A, 125 V, 1ϕ receptacles in bathrooms, kitchens, and on rooftops in other than dwelling units shall be GFCI-protected.

Over the past two Code cycles, the list of GFCI protected locations in other than dwelling units has been expanded. Section 210.8(B)(2) requires that all 125 V, 15 A, and 20 A, single-phase receptacles installed in kitchens have GFCI protection for personnel. A kitchen is defined as an area with a sink and permanent facilities for cooking. This includes all kitchens in commercial and industrial locations. Section 210.8(B)(4) covers outdoor spaces in other than dwelling units. All 125 V, 15 A and 20 A, 1ϕ receptacles installed in public space (any space that the public can access) must be provided with GFCI protection for personnel.

In addition to the requirements listed in 210.8 for GFCI protection for personnel, the installations requiring GFCI/shock protection include:

Section Location

- 210.8(C) – Boat Hoists
- 215.9 – Feeders
- 406.4(D) – Replacement
- 511.12 – Commercial Garages
- 517.21 – Health Care Facilities
- 550.13(B) – Mobile and Manufactured Homes
- 550.32(E) – Mobile and Manufactured Homes
- 551.41(C) – Recreational Vehicles
- 551.71 – Recreational Vehicle Parks
- 555.19(B) – Marinas
- 590.6 – Temporary Wiring
- 600.10(C)(2) – Signs (Mobile or Portable)
- 620.85 – Elevators, Escalators, and Moving Walkways
- 625.22 – Electric Vehicle Charging Systems
- 647.7 – Sensitive Electronic Equipment
- 680.21(C) – Pool Pump Motors
- 680.22(A) – Pools (Permanently Installed)
- 680.32 – Pools (Storable)
- 680.51(A) and 680.56(A) – Fountains
- 680.71 – Hydromassage Bathtubs

Required Branch Circuits – 210.11

Branch circuits shall be provided for all lighting and appliances along with other loads according to the total calculated load as determined by 220.10. In addition to this number of branch circuits, the specific dwelling unit loads detailed in 210.11(C) shall require dedicated branch circuits.

20A circuits only

Section 210.11(A) establishes a minimum number of branch circuits that must be provided in dwellings to serve the total calculated load. The size or rating of the branch circuits is used in the determination of the number of circuits required. For example, a dwelling with a load of 10,000 VA would require a minimum of six 15 A branch circuits (10,000 VA ÷ (15 A × 120 V) = 5.5 circuits). The calculation shows 5.5 circuits, but an installation cannot have half a circuit, so 6 circuits are required. When using 20 A circuits, the same dwelling would require a minimum of 5 circuits (10,000 VA ÷ (20 A × 120 V) = 4.2, or 5 circuits). In all installations, the number of circuits provided by the calculation must be sufficient to supply the load served. The loads placed on circuits must not exceed the values provided for in 220.18.

Section 210.11(B) addresses load calculations that are made on the basis of volt-amperes per square foot. It requires the wiring system, up to and including the branch-circuit panelboard(s), to be sized to serve the calculated load. In addition, the load must be evenly distributed among the multiwire branch circuits in the panelboard(s). In other words, designers and installers are required to balance the loads in branch-circuit panelboards.

Section 210.11(C) provides a list of branch circuits that must be provided in dwelling units, in addition to the branch circuits required by 210.11(A). The required branch circuits are as follows.

- 210.11(C)(1) Small Appliance Branch Circuits. Two or more 20 A small appliance circuits are required to supply the receptacles specified by 210.52(B).
- 210.11(C)(2) Laundry Branch Circuits. One 20 A branch circuit is required for the laundry receptacle outlet(s) required by 210.52(F).
- 210.11(C)(3) Bathroom Branch Circuits. At least one 20 A branch circuit must be installed to supply bathroom receptacle outlet(s). The general rule limits the circuit(s) to the bathroom receptacles only. The exception allows a dedicated 20 A branch circuit to supply a single bathroom and outlets for other equipment located in the same bathroom, in accordance with 210.23(A)(1) and (A)(2).

> ***Fire Alarm Systems***
>
> *A fire alarm system can control or power additional systems, including elevator capture, elevator shutdown, door release, smoke doors and damper control, fire doors and damper control, and fan shutdown.*

AFCI Protection – 210.12

Arc-fault circuit interrupters (AFCIs) are special devices that are intended to provide protection against intermittent arcing faults. These devices recognize the specific signature of the arc and are able to determine whether they are normal arcs, such as occur when a switch is opened, or potential damaging arcs that must be removed.

Section 210.12(B) was significantly changed in the 2008 NEC® cycle to require most 120 V, 15 A and 20 A receptacle outlets in dwelling units to be protected by listed AFCI combination-type devices. By definition, an AFCI combination-type device provides both branch/feeder protection and outlet protection. The branch/feeder-type AFCI provides protection at the origin of a branch circuit or a feeder, typically at the panelboard. The outlet-type AFCI provides protection at the branch circuit outlet point. **See Figure 2-18.**

There are three exceptions to the general rule for AFCI protection for dwelling units in 210.12(A). The first exception states that where RMC, IMC, EMT, Type MC cable, or steel-armored Type AC cable meeting the requirements of 250.118 are used as the wiring method for the portion of the branch circuit between the overcurrent protective device and the first outlet, an outlet branch-circuit-type of AFCI is permitted to be installed at the first outlet to provide protection for the rest of the branch circuit. While this portion of the branch circuit is not technically protected, the fact that metal raceways and cable assemblies are used minimizes the potential for arc fault due to damage to the branch-circuit conductors.

The second exception permits a listed metal or nonmetallic conduit or tubing to be used for the same portion of the circuit, from the branch circuit overcurrent device to the first outlet, provided the wiring method is encased in not less than 2″ of concrete and an outlet branch-circuit-type AFCI is installed at the first outlet to provide protection for the rest of the branch circuit.

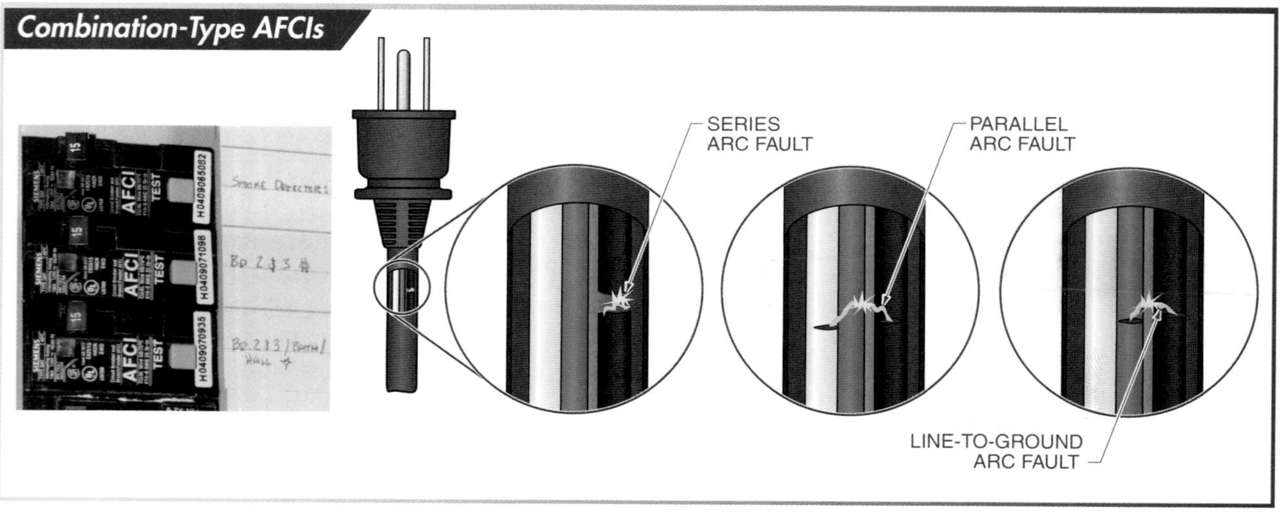

Figure 2-18. Combination-type AFCIs provide protection against series arcs, parallel arcs, and line-to-ground arcs.

The third exception permits the omission of AFCI protection for branch circuits that supply a fire alarm system installed in accordance with 760.41(B) and 760.121(B). The branch circuit must be installed in RMC, IMC, EMT, or steel armor cable, type AC, and metal outlet boxes and junction boxes must be employed.

Section 210.12(B) requires that in dwelling units where branch-circuit wiring modifications are made to any of the areas required by 210.12(A) to have AFCI protection, such protection must be added by one of two methods. First, a listed combination-type AFCI device located at the branch-circuit panelboard may be installed. Second, a listed outlet branch-circuit-type of AFCI device can be installed at the first outlet of the existing branch circuit.

Lampholder Ratings

Lampholders shall have a rating sufficient for the load being served. The intent is to limit branch circuits that supply fluorescent lighting to 20 A. Most lampholders made for use with fluorescent lamps are not of the heavy-duty type and are rated at 250 W or 660 W.

Lampholders connected to branch circuits rated over 20 A are required to be of the heavy-duty type. A medium type lampholder are rated at least 660 W. All other types of heavy-duty lampholder are rated at not less than 750 W. Fluorescent fixture lampholders are not of the heavy-duty type. This prohibits the use of these fixtures on 30 A, 40 A, and 50 A branch circuits.

BRANCH-CIRCUIT RATINGS – ARTICLE 210, PART II

Part II deals with the branch-circuit ratings of conductors, overcurrent devices, and outlet devices. It also deals with maximum and permissible loads. The basic rule is that the branch circuit shall have an ampacity not less than the maximum load to be served. The rating of a branch circuit is determined by the ampere rating or setting of the overcurrent device per 210.3.

Minimum Size Conductors – 210.19(A)(B)

Branch-circuit conductors shall have an ampacity not less than the maximum load to be served. The minimum branch circuit size before applying any adjustment or correction factors shall be equal to or greater than the noncontinuous load plus 125% of the continuous load. **See Figure 2-19.** In addition, conductors of multioutlet branch circuits supplying receptacles for cord-and-plug-connected portable loads shall have an ampacity of not less than the rating of the branch circuit. This is the rating of the overcurrent device. Multioutlet branch circuits for cord-and-plug-connected portable loads have random unpredictable loads; therefore, the conductors shall have an ampacity equal to the rating of the branch circuit overcurrent protective device.

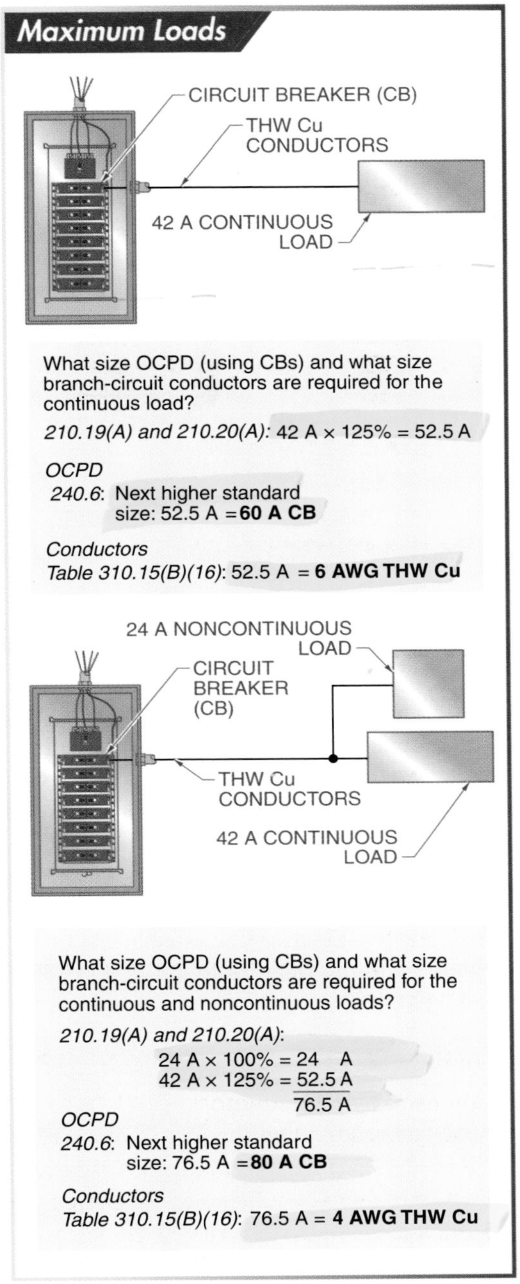

Maximum Loads
CIRCUIT BREAKER (CB)
THW Cu CONDUCTORS
42 A CONTINUOUS LOAD
What size OCPD (using CBs) and what size branch-circuit conductors are required for the continuous load?
210.19(A) and 210.20(A): 42 A × 125% = 52.5 A
OCPD
240.6: Next higher standard size: 52.5 A = 60 A CB
Conductors
Table 310.15(B)(16): 52.5 A = 6 AWG THW Cu
24 A NONCONTINUOUS LOAD
CIRCUIT BREAKER (CB)
THW Cu CONDUCTORS
42 A CONTINUOUS LOAD
What size OCPD (using CBs) and what size branch-circuit conductors are required for the continuous and noncontinuous loads?
210.19(A) and 210.20(A):
24 A × 100% = 24 A
42 A × 125% = 52.5 A
76.5 A
OCPD
240.6: Next higher standard size: 76.5 A = 80 A CB
Conductors
Table 310.15(B)(16): 76.5 A = 4 AWG THW Cu

Figure 2-19. The minimum branch-circuit conductor size shall be equal to or greater than the noncontinuous load plus 125% of the continuous load.

There is an exception to the general rule. The exception allows the size of the branch-circuit conductors to be based on the continuous load plus the noncontinuous load where the assembly, including the overcurrent devices protecting the circuit, is listed for operation at 100% of its rating. In general, if the branch circuit supplies more than one receptacle for cord-and-plug-connected loads, the conductors shall have an ampacity not less than the rating of the branch circuit.

Household Ranges and Cooking Appliances – 210.19(A)(3). Branch-circuit conductors supplying household ranges and other cooking appliances shall have an ampacity not less than the rating of the branch circuit, and not less than the maximum load to be served. For ranges rated 8¾ kW or more, the minimum branch-circuit rating shall be 40 A.

Per 210.19(A)(3), Ex. 1, tap conductors supplying electric ranges and other cooking units from a 50 A branch circuit are permitted to have an ampacity of not less than 20 A. The taps shall not be longer than necessary for servicing the appliance and shall include any leads that are part of the appliance and are smaller than the branch-circuit conductor. **See Figure 2-20.**

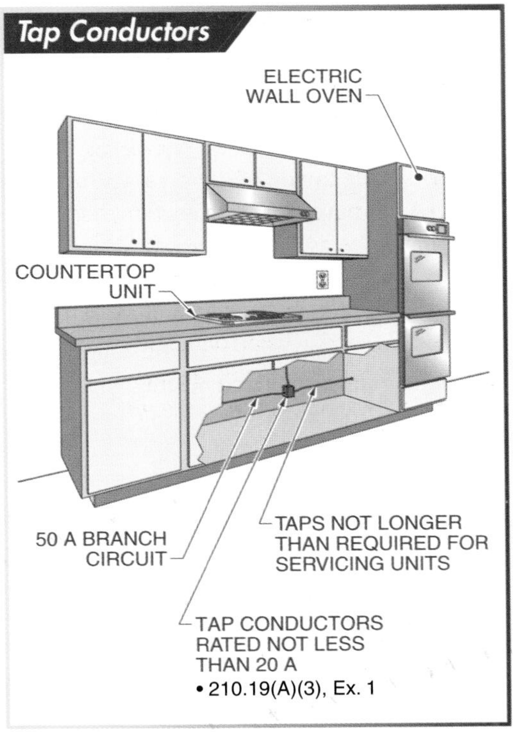

Figure 2-20. Tap conductors supplying electric ranges and other cooking units from a 50 A branch circuit are permitted to have an ampacity of not less than 20 A.

Conductor Sizing

Conductors are sized based on the American Wire Gauge (AWG) system. The AWG system assigns the number 36 to a small conductor with a 0.0050″ diameter. The largest size of a single strand conductor, number 6/0, has a diameter of 0.5800″. As an AWG value decreases, the cross-sectional area of a conductor increases.

Per 210.19(A)(3), Ex. 2, the neutral conductor of a 3-wire branch circuit supplying a household electric range or other cooking units is permitted to be smaller than the ungrounded conductors. This neutral shall have an ampacity of not less than 70% of the branch-circuit rating and shall not be smaller than 10 AWG.

Other Loads – 210.19(A)(4). Branch-circuit conductors shall have an ampacity sufficient for the loads served and shall be 14 AWG or larger. There are two exceptions to this requirement. Per 210.19(A)(4), Ex. 1, tap conductors serving one of five types of loads shall have an ampacity not less than 15 A for circuits rated less than 40 A and not less than 20 A for circuits rated at 40 A or 50 A. The five types of loads are:

- Individual lampholders or fixtures with taps extending no more than 18″ beyond the lampholder or fixture.
- A luminaire with tap conductors per 410.117.
- Individual outlets, other than receptacle outlets, with taps no more than 18″ long.
- Infrared lamp industrial heating appliances.
- Nonheating leads of deicing and snow-melting cables and mats.

Ex. 2 is for fixture wires and cords as permitted in Section 240.5.

Overcurrent Protection – 210.20

Branch-circuit conductors and equipment shall be protected by overcurrent protective devices with a rating or setting which complies with one of the following:

- The rating of the overcurrent device shall not be less than the noncontinuous load plus 125% of the continuous load.
- Conductors shall be protected per 240.4. Flexible cords and fixture wires shall be protected per 240.5.
- Overcurrent protection of equipment shall not exceed that required in applicable Articles as listed in Table 240.3.
- Outlet devices per 210.21.

Outlet Devices – 210.21

Outlet devices shall have an ampere rating not less than the load to be served. Lampholders shall be of the heavy-duty type when connected to a branch circuit in excess of 20 A per 210.21(A). Heavy-duty, and medium lampholders shall have a rating of not less than 660 W. All other lampholders shall have a rating of not less than 750 W.

A single receptacle installed on an individual branch circuit shall have an ampere rating not less than the branch circuit per 210.21(B)(1). **See Figure 2-21.** The important wording here is "single receptacle," not a duplex receptacle. A *single receptacle* is a single contact device with no other contact device on the same yoke. A *multiple receptacle* is a single device with two or more receptacles. As an example, if a single receptacle was installed on a 20 A branch circuit, the single receptacle would have to be rated at 20 A. Two exceptions to the rule are found in 430.81(B) and 630.11(A).

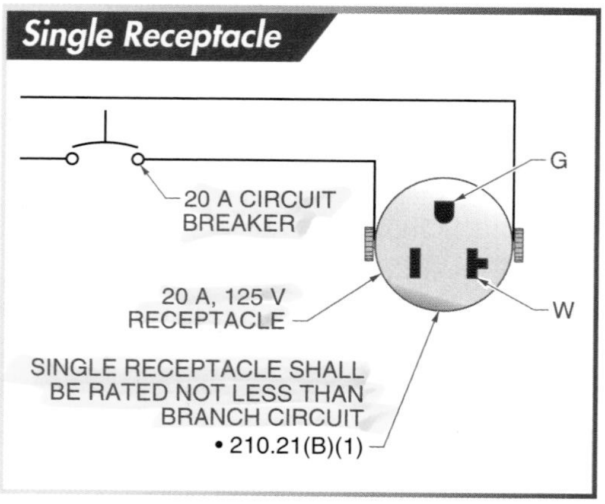

Figure 2-21. A single receptacle on an individual branch circuit shall have an ampere rating not less than the branch circuit.

Table 210.21(B)(2)

There is some confusion as to the intent of Table 210.21(B)(2). Under the column Circuit Rating Amperes, 15 A or 20 A is listed. Then, in the next column, Receptacle Rating Amperes, 15 A is listed. The question that arises is whether a 15 A rated receptacle is permitted to be installed on a circuit that is protected for 20 A.

The answer is NO, because Table 210.21(B)(2) is applied only when there are two or more receptacles, such as a duplex receptacle. The configuration of a 15 A rated receptacle does not accept a 20 A rated plug. Each receptacle of a duplex is rated 15 A.

Per 210.21(B)(2), if a branch circuit supplies two or more receptacles or outlets, the receptacle shall not supply a total cord-and-plug-connected load in excess of the amount specified in Table 210.21(B)(2). The Table provides the following permissible branch-circuit ratings and maximum loads permitted for 15 A, 20 A, and 30 A receptacles for cord-and-plug-connected loads:

- 15 A receptacles may be used on 15 A or 20 A circuits with a maximum load of 12 A.
- 20 A receptacles may be used on 20 A circuits with a maximum load of 16 A.
- 30 A receptacles may be used on 30 A circuits with a maximum load of 24 A.

Permissible Loads – 210.23

The load of a branch circuit shall not exceed the branch-circuit ampere rating. An individual branch circuit may serve any load for which it is rated. Branch circuits supplying two or more outlets or receptacles shall be subject to four limitations:

Section 210.23(A) allows 15 A or 20 A branch circuits to supply lighting units or other utilization equipment, or a combination of both. These circuits shall comply with 210.23(A)(1) and 210.23(A)(2). The exception requires small appliance circuits, laundry circuits, and bathroom circuits addressed by 210.11(C)(1), (C)(2), and (C)(3) to only supply the receptacles specified in the section.

Section 210.23(A)(1) addresses cord-and-plug-connected equipment that is fastened in place. It limits the rating of cord-and-plug-connected utilization equipment that is not fastened in place to a maximum of 80% of the branch-circuit ampere rating. For example, if a microwave oven was connected to a 20 A branch circuit that also supplies other loads, the ampere rating of the microwave oven could not exceed 16 A.

Section 210.23(A)(2) requires that the total rating of utilization equipment fastened in place, other than lighting fixtures, be limited to a maximum of 50% of the branch-circuit ampere rating. This limitation is applied to branch circuits that supply lighting units, cord-and-plug-connected utilization equipment not fastened in place, or both in addition to the utilization equipment fastened in place. For example, a wall-mounted air conditioning unit connected to a receptacle that is part of a 15 A general-purpose branch circuit that also supplies lighting fixtures and other equipment in a living room of a dwelling would be permitted to have a maximum rating of 7.5 A. Equipment that is fastened in place can be hard wired or cord-and-plug-connected. Examples of equipment that is fastened in place would include dishwashers, disposals, and ranges.

Per 210.23(B), 30 A branch circuits are permitted to supply fixed luminaires (lighting fixtures) with heavy-duty lampholders in other than dwelling units. They can also supply utilization equipment in any occupancy. The rating of any one cord-and-plug-connected piece of equipment shall not exceed 80% of the ampere rating of the branch circuit.

Branch circuits may supply lighting loads in addition to receptacles for cord-and-plug-connected loads.

Per 210.23(C), branch circuits rated 40 A and 50 A are permitted to supply fixed lighting units equipped with heavy-duty lampholders or infrared heating units in other than dwelling units. They can also serve cooking appliances that are fastened in place in any occupancy. Nonlighting loads only are permitted, per 210.23(D), to be supplied from branch circuits larger than 50 A.

REQUIRED OUTLETS – 210.50(B)

Receptacle outlets must be installed as required in 210.52 through 210.63. Cord connectors that are supplied by permanent pendant cords are considered to be receptacle outlets. Section 210.50(C) requires that appliance receptacle outlets installed in dwelling units for specific appliances, such as laundry equipment, be located within 6′ of the intended appliance location. The 6′ measurement is a maximum. Other sections of the code provide more restrictive requirements. For example, section 422.16(B)(2) limits the length of the cord for waste disposers to a length of from 18″ to 36″. The section also limits the cord length of trash compactors and dishwashers to a length of from 3′ to 4′. Range hoods, where cord-and-plug connected, shall have a cord length of from 18″ to 36″. **See Figure 2-22.**

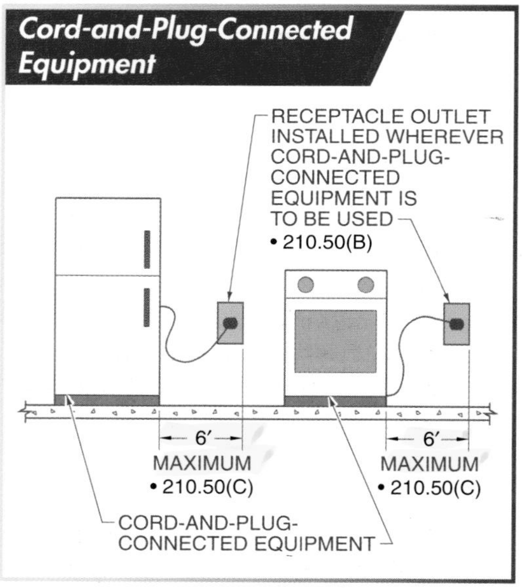

Figure 2-22. Receptacle outlets shall be installed wherever cord-and-plug-connected equipment is used.

Dwelling Unit Receptacle Outlet Provisions – General

Section 210.52 is an important section that establishes the principal receptacle outlet provisions for dwelling units. The requirements are very specific and address most of the types of rooms and spaces found within most dwelling units, such as the kitchen, bathroom, hallways, laundry room, outdoors, and hallways.

Four basic provisions are established in 210.52 that apply throughout these requirements that specify that the receptacle outlet requirements of 210.52 are in addition to any outlets that are 1) part of a luminaire, 2) controlled by a wall switch, 3) located within a cupboard or cabinet, or 4) located more than 5′6″ above the floor.

For example, 210.52(C)(1) requires a receptacle outlet to be installed at each wall countertop space that is 12″ or wider. If such a space had a luminaire located on the wall behind this space and the luminaire contained a receptacle outlet, that outlet could not be considered to meet the requirements of 210.52(C)(1).

Dwelling Unit Receptacle Outlets – 210.52(A)

Receptacle outlets shall be installed so that no point along the floor line in any wall space is more than 6′ from an outlet in kitchens, family rooms, dining rooms, living rooms, parlors, libraries, dens, sun rooms, bedrooms, recreation rooms, or similar rooms or areas of dwelling units. **See Figure 2-23.** Wall space provided by freestanding bar-type counters or railings shall be included in this 6′ measurement. The space occupied by fixed panels in exterior walls, excluding sliding panels, is considered wall space and is included in the six-foot rule. For example, a fixed panel of glass that is part of a sliding door assembly would be considered wall space, while the sliding panel would not be included in the six-foot rule. A dwelling unit is a dwelling with one or more rooms used by one or more persons as a housekeeping unit with eating, living, and sleeping space, and permanent cooking and sanitation provisions.

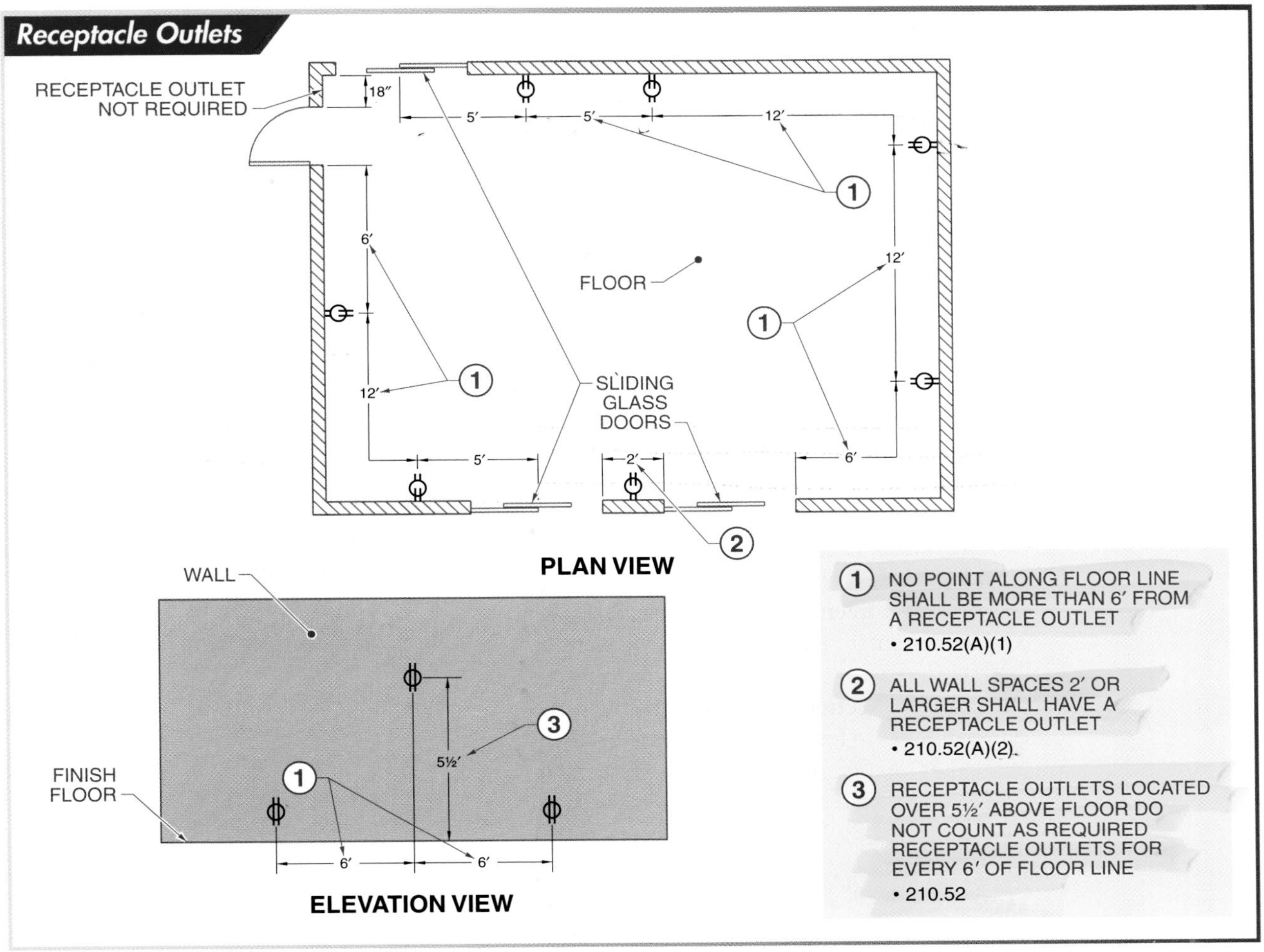
Receptacle Outlets
RECEPTACLE OUTLET NOT REQUIRED
18″
5′
5′
12′
6′
12′
12′
5′
2′
6′
FLOOR
SLIDING GLASS DOORS
PLAN VIEW
WALL
FINISH FLOOR
5½′
6′
6′
ELEVATION VIEW
1 NO POINT ALONG FLOOR LINE SHALL BE MORE THAN 6′ FROM A RECEPTACLE OUTLET
• 210.52(A)(1)
2 ALL WALL SPACES 2′ OR LARGER SHALL HAVE A RECEPTACLE OUTLET
• 210.52(A)(2)
3 RECEPTACLE OUTLETS LOCATED OVER 5½′ ABOVE FLOOR DO NOT COUNT AS REQUIRED RECEPTACLE OUTLETS FOR EVERY 6′ OF FLOOR LINE
• 210.52

Figure 2-23. No point along the floor line in any wall space shall be more than 6′ from a receptacle outlet.

SINGLE RECEPTACLE
DUPLEX RECEPTACLES

A receptacle is a contact device installed at outlets for the connection of cord-connected electrical equipment.

Receptacles are required for all floor line dimensions 2′ or greater and unbroken by doorways, windows, fireplaces, etc. A wall space may include two or more walls of a room (around corners) if it is unbroken at the floor line. Receptacle outlets in the floor shall not be counted as part of the required number of outlets unless they are located within 18″ of the wall.

Dwelling Unit Small Appliances – 210.52(B)

Per 210.52(B)(1), the two or more 20 A small appliance branch circuits required by 210.11(C)(1) shall supply all of the receptacle outlets in the kitchen, pantry, dining room, breakfast room, or similar areas of a dwelling unit. The small appliance branch circuits shall serve the following equipment in these areas.

- All of the wall and floor receptacle outlets required by 210.52(A)
- All of the countertop outlets required by 210.52(C)
- Receptacle outlets for refrigeration equipment

The number of small appliance branch circuits installed ("two or more") is often determined by the load ratings of the appliances that will utilize the receptacle outlets. Section 210.23(A)(1) limits the load of any one cord-and-plug-connected appliance to a maximum of 80% of the branch-circuit ampere rating where other outlets are served by the branch circuit. For example, a small appliance branch circuit serving a refrigerator with a load of 17 A would not be permitted to serve other loads. A refrigerator with a load of 16 A, on the other hand, would be permitted to serve other loads as long as the total load on the branch circuit did not exceed 20 A as determined by 210.19(A)(1). Exception 2 to 210.52(B)(1) allows the use of an individual 15 A or 20 A branch circuit for refrigeration equipment. Exception 1 allows switched receptacles installed on general-purpose branch circuits as defined in 210.70(A)(1) to be installed in addition to the small appliance branch circuits in the areas listed in 210.52(B).

The small appliance branch circuits specified in 210.52(B)(1) shall have no other outlets.

The two small appliance branch circuits required by 210.52(B)(1) shall have no other outlets per 210.52(B)(2). This requirement has two exceptions. Per 210.52(B)(2), Ex. 1, a receptacle outlet is permitted to be installed for an electric clock. Per 210.52(B)(2), Ex. 2, receptacles are permitted to be installed to provide power for lighting and supplemental equipment of gas-fired ranges, ovens, or counter-mounted cooking units.

Section 210.52(B)(3) requires that at least two of the 20 A small appliance branch circuits shall supply the receptacle outlets that are required to serve the countertop surfaces in the kitchen. Additional small appliance branch circuits, if necessary, may be located in the kitchen and other rooms specified in 210.52(B)(1).

Dwelling Unit Countertops – 210.52(C)

The location of receptacle outlets for counter spaces in kitchens, pantries, breakfast rooms, dining rooms, and similar areas of dwelling units is determined by the wall counter space, island counter space, peninsular counter space, and separate spaces in these rooms. A 2008 NEC® change to 210.52(C) addresses the installation of a range, a counter-mounted cooking unit, or a sink in an island or peninsula. If the width of the counter behind the appliance is less than 12″, the range, counter-mounted cooking unit, or sink is considered to divide the countertop space into two separate spaces as defined in 210.2(C)(4). Each of the separate spaces must comply with the applicable requirements of 210.52(C). **See Figure 2-24.**

Wall Counter Space – 210.52(C)(1). Receptacle outlets shall be installed at each wall counter space 12″ or wider. No point along the wall line shall be more than 24″ measured horizontally from a receptacle outlet in that space.

Island Counter Space – 210.52(C)(2). At least one receptacle outlet shall be installed to serve each island countertop space that is 24″ × 12″ or larger. The long dimension is 24″ or larger and the small dimension is 12″ or larger.

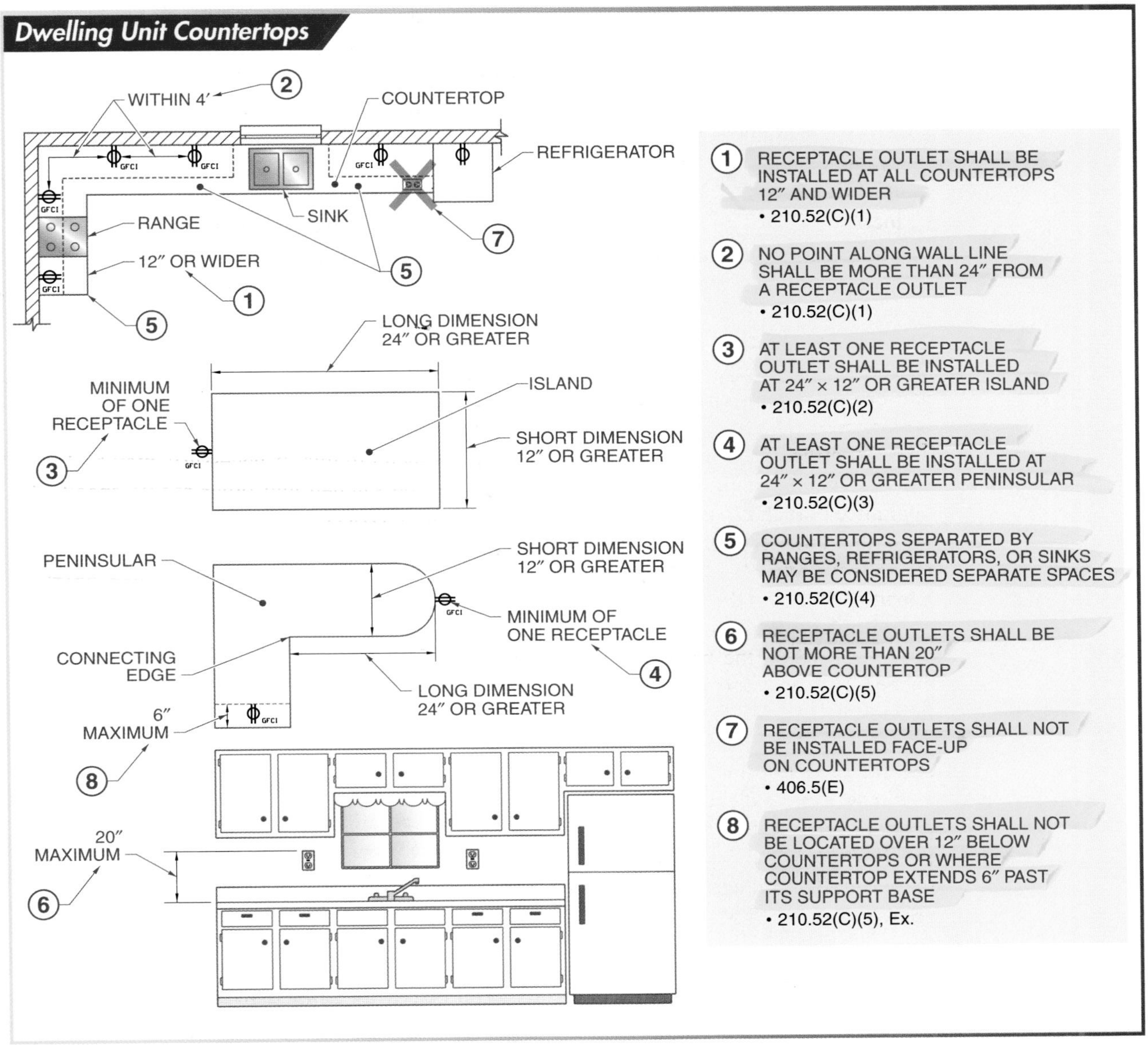
Dwelling Unit Countertops
WITHIN 4′
COUNTERTOP
REFRIGERATOR
GFCI
RANGE
SINK
12″ OR WIDER
LONG DIMENSION 24″ OR GREATER
MINIMUM OF ONE RECEPTACLE
ISLAND
SHORT DIMENSION 12″ OR GREATER
PENINSULAR
SHORT DIMENSION 12″ OR GREATER
MINIMUM OF ONE RECEPTACLE
CONNECTING EDGE
LONG DIMENSION 24″ OR GREATER
6″ MAXIMUM
20″ MAXIMUM
1 RECEPTACLE OUTLET SHALL BE INSTALLED AT ALL COUNTERTOPS 12″ AND WIDER
• 210.52(C)(1)
2 NO POINT ALONG WALL LINE SHALL BE MORE THAN 24″ FROM A RECEPTACLE OUTLET
• 210.52(C)(1)
3 AT LEAST ONE RECEPTACLE OUTLET SHALL BE INSTALLED AT 24″ × 12″ OR GREATER ISLAND
• 210.52(C)(2)
4 AT LEAST ONE RECEPTACLE OUTLET SHALL BE INSTALLED AT 24″ × 12″ OR GREATER PENINSULAR
• 210.52(C)(3)
5 COUNTERTOPS SEPARATED BY RANGES, REFRIGERATORS, OR SINKS MAY BE CONSIDERED SEPARATE SPACES
• 210.52(C)(4)
6 RECEPTACLE OUTLETS SHALL BE NOT MORE THAN 20″ ABOVE COUNTERTOP
• 210.52(C)(5)
7 RECEPTACLE OUTLETS SHALL NOT BE INSTALLED FACE-UP ON COUNTERTOPS
• 406.5(E)
8 RECEPTACLE OUTLETS SHALL NOT BE LOCATED OVER 12″ BELOW COUNTERTOPS OR WHERE COUNTERTOP EXTENDS 6″ PAST ITS SUPPORT BASE
• 210.52(C)(5), Ex.

Figure 2-24. The location of receptacle outlets for counter spaces in kitchen and dining rooms of dwelling units is determined by the wall counter space, island counter space, peninsular counter space, and separate spaces in these rooms.

At least one receptacle outlet must be installed in a kitchen island countertop that is 24″ × 12″ or greater.

Peninsular Counter Space – 210.52(C)(3). At least one receptacle outlet shall be installed to serve each peninsular countertop space, provided the peninsula has a long dimension of 24″ or greater and a short dimension of 12″ or greater. A peninsular countertop is measured from where it connects to the countertop.

Separate Spaces – 210.52(C)(4). Countertop spaces which are separated by range tops, refrigerators, or sinks shall be considered as separate countertop spaces in applying the requirements for placing receptacles on wall counter space, island counter space, or peninsular counter space provided the depth of the countertop behind the range, cooking unit, or sink is less than 12″.

Receptacle Outlet Location – 210.52(C)(5). The location of receptacle outlets shall not be more than 20″ above the countertop surface. Wall cabinets are generally 20″ above the countertops of the base cabinets. Receptacle outlets shall not be installed in a face-up position in the work surfaces of countertops. Receptacle outlet assemblies listed for the application are, however, permitted to be installed in countertops. Receptacle outlets which are not readily accessible, due to fastened-in-place appliances, appliance garages, sinks, or oven tops which prevent ready access, shall not be counted as required receptacle outlets.

The exception to 210.52(C)(5) allows receptacle outlets mounted not more than 12″ below a countertop that does not extend more than 6″ beyond its support base to be considered as the countertop receptacles needed to comply with 210.52(C)(5) under one of two conditions. The first condition is for countertop receptacles installed in a dwelling constructed for the physically impaired. The second condition is for island and peninsular countertops where the countertop is flat across the entire surface (no backsplashes, dividers, etc.) and where there is no means to mount a receptacle within 20″ of the countertop.

Dwelling Unit Bathrooms – 210.52(D)

At least one receptacle outlet shall be installed within 36″ of the outside edge of each basin in all dwelling unit bathrooms. The outlet shall be located on a wall or partition that is adjacent to the location of the basin or basin countertop. **See Figure 2-25.** This clearly eliminates receptacle outlets on a wall opposite the basin location as satisfying the requirements of 210.52(D). Receptacle outlet assemblies listed for the application are, however, permitted to be installed in countertops.

Receptacle outlets are permitted to be installed on the side or face of the bathroom base cabinet not more than 12″ below the countertop to meet the requirements of this section. These receptacle outlets shall be supplied by at least one 20 A branch circuit which will serve no other outlets unless it serves a single bathroom per 210.11(C)(3), Ex. This 20 A bathroom branch circuit is permitted to serve receptacle outlets in more than one bathroom. In this situation the bathroom branch circuit shall serve only receptacle outlets. Receptacle outlets shall not be installed in a face-up position in the countertop in a bathroom basin location. See 406.4(E).

Dwelling Unit Outdoor Outlets – 210.52(E)

At least one receptacle outlet shall be installed outdoors at grade level at the front and the back of each one-family dwelling and each unit of a two-family dwelling. These receptacle outlets shall be accessible at grade level and not more than 6′-6″ above grade. **See Figure 2-26.**

For multifamily dwellings, at least one receptacle outlet, accessible at grade level and not more than 6′-6″ above grade, shall be provided for each dwelling unit that is located at grade level and that has an individual entrance and egress.

Section 210.52(E)(3) was revised in the 2008 Code cycle to require balconies, decks, and porches that are attached to the dwelling and are accessible from the inside of the dwelling unit to have at least one receptacle outlet installed in a location that is accessible from the balcony, deck, or porch. The outlet must be located within the perimeter of the balcony, deck, or porch and shall be no more than 6′-6″ above the surface of the balcony, deck, or porch.

Outdoor receptacle outlets shall be accessible at grade level and not more than 6′-6″ above grade.

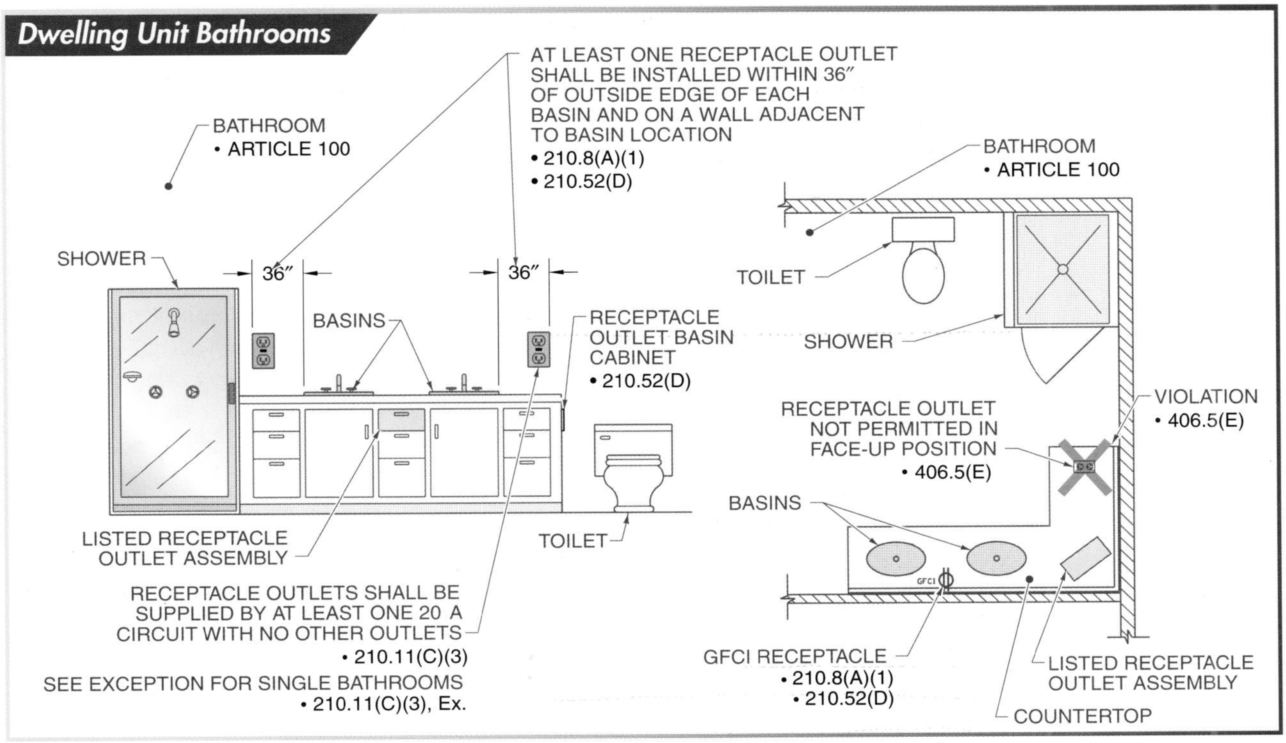

Dwelling Unit Bathrooms
AT LEAST ONE RECEPTACLE OUTLET SHALL BE INSTALLED WITHIN 36″ OF OUTSIDE EDGE OF EACH BASIN AND ON A WALL ADJACENT TO BASIN LOCATION
• 210.8(A)(1)
• 210.52(D)
BATHROOM
• ARTICLE 100
BATHROOM
• ARTICLE 100
SHOWER
36″
36″
BASINS
RECEPTACLE OUTLET BASIN CABINET
• 210.52(D)
TOILET
SHOWER
RECEPTACLE OUTLET NOT PERMITTED IN FACE-UP POSITION
• 406.5(E)
VIOLATION
• 406.5(E)
BASINS
LISTED RECEPTACLE OUTLET ASSEMBLY
TOILET
RECEPTACLE OUTLETS SHALL BE SUPPLIED BY AT LEAST ONE 20 A CIRCUIT WITH NO OTHER OUTLETS
• 210.11(C)(3)
SEE EXCEPTION FOR SINGLE BATHROOMS
• 210.11(C)(3), Ex.
GFCI
GFCI RECEPTACLE
• 210.8(A)(1)
• 210.52(D)
LISTED RECEPTACLE OUTLET ASSEMBLY
COUNTERTOP

Figure 2-25. At least one wall receptacle outlet shall be installed within 36″ of the outside edge of each basin and on a wall adjacent to the basin location.

Dwelling Unit Laundry Areas – 210.52(F)

At least one receptacle outlet shall be installed in dwelling units for the laundry. Per 210.52(F), Ex. 1, the laundry receptacle is not required in each unit of an apartment or living area in a multifamily building where laundry facilities are provided on the premises that are available to the building occupants. Per 210.52(F), Ex. 2, the laundry receptacle is not required in other than one-family dwellings where laundry facilities are not to be installed or permitted.

Dwelling Unit Basements and Garages – 210.52(G)

At least one receptacle outlet shall be installed in each of the following locations in one-family dwellings.

- Each basement
- Each attached garage
- Each detached garage or accessory building that has electric power

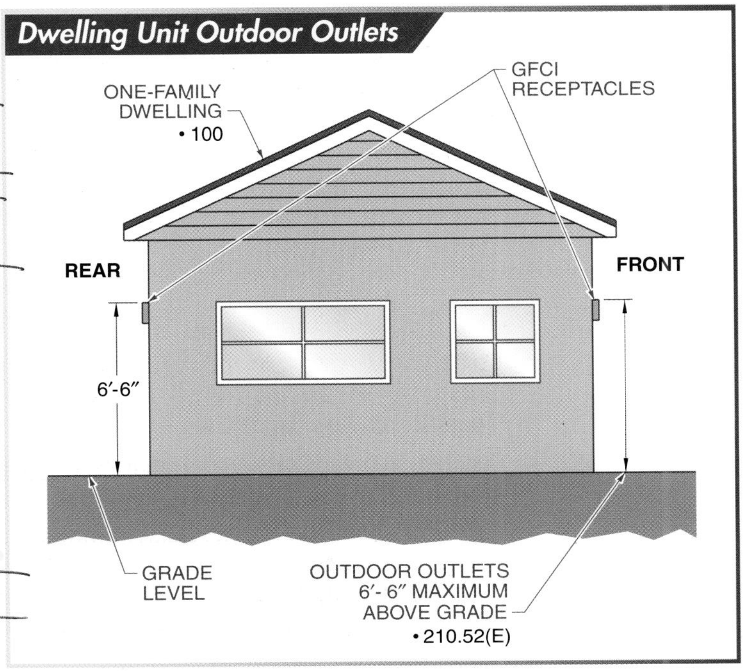

Dwelling Unit Outdoor Outlets
ONE-FAMILY DWELLING
• 100
GFCI RECEPTACLES
REAR
FRONT
6′-6″
GRADE LEVEL
OUTDOOR OUTLETS 6′- 6″ MAXIMUM ABOVE GRADE
• 210.52(E)

Figure 2-26. At least one receptacle shall be installed outdoors at grade level at the front and back of each one-family dwelling and each unit of a two-family dwelling.

This outlet is in addition to any outlets provided for laundry equipment or other specific equipment. Where a portion of the basement is finished into one or more habitable rooms, each separate unfinished portion shall have a receptacle outlet installed.

Dwelling Unit Hallways – 210.52(H)

At least one receptacle outlet shall be installed in 10′ or longer hallways in dwelling units. The hallway length is considered the length along the centerline of the hall without passing through a doorway. This outlet prevents the need to use extension cords for hallway table lamps, vacuum cleaners, etc.

Dwelling Unit Foyers – 210.52(I)

Foyers or other areas of entrance rooms that are not part of hallways and that have an area greater than 60 sq ft shall have a receptacle outlet located in each wall that is 3′ or more in width, provided the space is unbroken by doorways, windows, or similar openings. The requirement was added to the 2011 NEC® because of a growing trend to include very large foyer areas in many new dwelling units. These large areas make it very difficult to ensure that there are sufficient general-purpose outlets for vacuum cleaners and other household appliances in these areas.

Receptacle Outlets on a Single Branch Circuit

It is not uncommon in industrial applications to supply several single receptacle outlets rated 50 A on a single branch circuit. This allows for the relocation of heavy-duty equipment which is used in the production process or for maintenance equipment.

Guest Rooms or Guest Suites – 210.60

Receptacle outlets installed in guest rooms or guest suites of hotels, motels, dormitories, and similar locations must comply with 210.52(A) and 210.52(D). Where permanent provisions for cooking are provided in guest rooms or guest suites, the receptacle outlets must be installed in accordance with all of the applicable rules found in 210.52. The total number of receptacle outlets that must be installed shall not be less than the minimum number required to comply with the provisions of 210.52(A). The Code section allows the required receptacles to be located conveniently for the permanent furniture layout. For example, a guest room in a hotel that would require six receptacle outlets, based on the six-foot rule in 210.52(A)(1), could not have less than six receptacle outlets. These outlets could, however, be relocated behind the furniture even though the spacing between two or more receptacles might be over the 12′ maximum found in 210.52(A)(1).

The section places two conditions on the relocation of the receptacle outlets. One condition is that at least two of the receptacle outlets must be readily accessible. The other condition addresses receptacles installed behind beds. When this occurs, the receptacle must be installed to prevent the bed from contacting any attachment cord that may be plugged into the receptacle or the receptacle must be protected by a suitable guard.

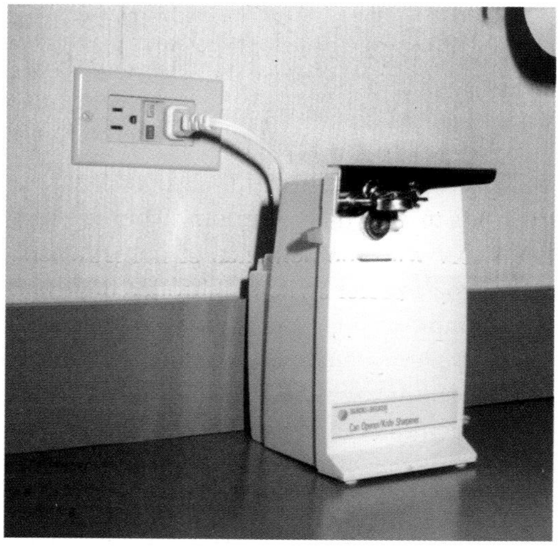

All receptacles intended to serve countertops in dwelling unit kitchens shall be GFCI-protected.

Heating, Air-Conditioning, and Refrigeration Equipment Outlets – 210.63

A 125 V, 1ϕ, 15 A or 20 A receptacle outlet must be located within 25′ of heating, air-conditioning, and refrigeration equipment for servicing the equipment. The receptacle outlet must be accessible and on the same level as the equipment. The receptacle outlet shall not be connected to the load side of the equipment disconnect.

An exception is provided for servicing evaporator coolers in one- and two-family dwellings. See 210.63 Ex.

LIGHTING OUTLETS REQUIRED – 210.70

A lighting outlet is an outlet intended for the direct connection of a lampholder, luminaire (lighting fixture), or pendant cord terminating in a lampholder. Section 210.70 is the section that requires lighting outlets for both dwelling units and other than dwelling units. Lighting outlets are required for safety and convenience in dwelling units, guest rooms, and other locations.

Dwelling Units – 210.70(A)

At least one wall switch-controlled lighting outlet shall be installed in dwelling units as specified in 210.70(A)(1), (2), and (3).

Habitable Rooms – 210.70(A)(1). At least one wall switch-controlled lighting outlet must be installed in every bathroom and habitable room. Exception 1 allows one or more receptacles that are controlled by a wall switch to be used in place of the lighting outlet in all rooms except the kitchen and the bathroom. The intent is to comply with the demands of homeowners. Most people prefer switch-controlled table and floor lamps as opposed to switch-controlled ceiling lighting outlets. Exception 2 permits occupancy sensors to be used in addition to the wall switches required by this section or to replace the wall switch at the customary location of the switch as long as the occupancy sensor is equipped with a manual override that allows the sensor to function as a wall switch.

Additional Locations – 210.70(A)(2). Additional lighting outlets must be installed as follows.

- At least one wall switch-controlled lighting outlet must be installed in hallways, stairways, attached garages, and detached garages with electric power.
- At least one wall switch-controlled lighting outlet must be installed to provide illumination on the exterior side of outdoor entrances and exits with grade-level access. This rule applies to the dwelling units, attached garages, and detached garages with electric power. Vehicle doors in a garage are not considered to be outdoor entrances or exits.
- Where one or more lighting outlets are installed for interior stairways, a wall switch must be installed at each floor level and each landing that has an entryway. This requirement applies to all interior stairways that have six or more risers between floors.

The exception to 210.70(A)(2) allows remote, central, or automatic control of lighting in hallways, stairways, and at outdoor entrances.

Storage or Equipment Spaces – 210.70(A)(3). At least one lighting outlet containing a switch or controlled by a wall switch shall be installed in attics, underfloor spaces, utility rooms, and basements that are used for storage or that contain equipment which requires servicing. At least one control point is required to be installed at the point of entry to these spaces. The lighting outlet must be placed at or near the equipment that requires servicing.

Guest Rooms or Guest Suites – 210.70(B)

At least one wall switch-controlled lighting outlet or wall switch-controlled receptacle shall be installed in guest rooms or guest suites in hotels, motels, or similar occupancies. This is generally located on the inside wall near the entrance door for the convenience of guests.

There are two exceptions to this general rule. The first exception permits the use of one or more receptacle outlets that are controlled by wall switch to meet this requirement, provided the room is not a kitchen or a bathroom. The second exception permits the use of occupancy sensors to control lighting outlets, provided the sensors are in addition to wall switches or located at the normal wall switch location and equipped with a manual override feature that functions like a wall switch.

Other Locations – 210.70(C)

In other than dwellings, at least one lighting outlet containing a switch or controlled by a wall switch shall be installed at or near equipment requiring servicing in attics or under-floor spaces. At least one point of control shall be located at the usual point of entry to the attic or under-floor space. The lighting outlet shall be provided at or near the equipment requiring servicing. This rule is applicable only to buildings that do not meet the definitions of a dwelling unit in Article 100.

FEEDERS – ARTICLE 215

A feeder is all of the circuit conductors between the service equipment, the source of a separately derived system, or other supply source, and the final branch-circuit overcurrent device. The actual load on a feeder depends on the total load connected to the feeder and the demand factor. *Demand factor* is the ratio of the maximum demand of a system, or part of a system, to the total connected load of a system or the part of the system under consideration. See Article 100.

Minimum Rating and Size – 215.2

Feeder conductors rated not more than 600 V shall have an ampacity not less than that required to supply the load. Feeder conductors for dwelling units and mobile homes are not required to be larger than the service-entrance conductors. Section 310.15(B)(7) is used to size service and feeder conductors supplying dwellings with systems rated at 120/240 V, 1ϕ, 3-wire. **See Figure 2-27.**

Feeders installed in metal raceways and sharing a common neutral must comply with 300.20.

Feeders with Common Neutral – 215.4(A)

Feeders containing a common neutral are permitted to supply two or three sets of 3-wire feeders, or two sets of 4-wire or 5-wire feeders. All conductors of all feeders using a common neutral conductor shall be installed in the same raceway or enclosure. The rule applies when the conductors are installed in a metal raceway or metal enclosure. This avoids inductive heating of the surrounding metal. Parallel conductors, which run through multiple raceways, shall contain the conductors from each phase plus the neutral. Feeders installed in metal raceways and sharing a common neutral must comply with section 300.20.

The 2008 Code process added the definitions of neutral conductor and neutral point to Article 100. A *neutral conductor* is a conductor connected to the neutral point of a system that is intended to carry current under normal conditions. A *neutral point* is the common point on a wye connection in a polyphase system; the midpoint on a 1ϕ, 3-wire system; the midpoint of a 1ϕ portion of a 3ϕ delta system; or the midpoint of a 3-wire DC system.

These definitions are necessary to differentiate a neutral conductor from a grounded conductor, which is defined by Article 100 as a system or circuit conductor that is intentionally grounded. Electricians often misapply rules in the Code concerning these two types of conductors. The substantiation in the Report on Proposals for the added definitions indicated that the misapplication was mostly due to the fact that there was not a definition of neutral conductor in Article 100.

Neutral conductors installed in grounded systems are always grounded, but grounded conductors are not always neutral conductors. One example of a grounded conductor that is not a neutral conductor would be a 3ϕ, 3-wire, corner-grounded delta transformer. In this configuration, one of the phase conductors is grounded. Prior to the 2008 change, the NEC® did not provide a section to make it clear that the grounded conductor in this example was not a neutral conductor. The new definitions in the 2008 Code make this perfectly clear since a 3ϕ, 3-wire, corner-grounded delta transformer is not included in the definition of neutral point.

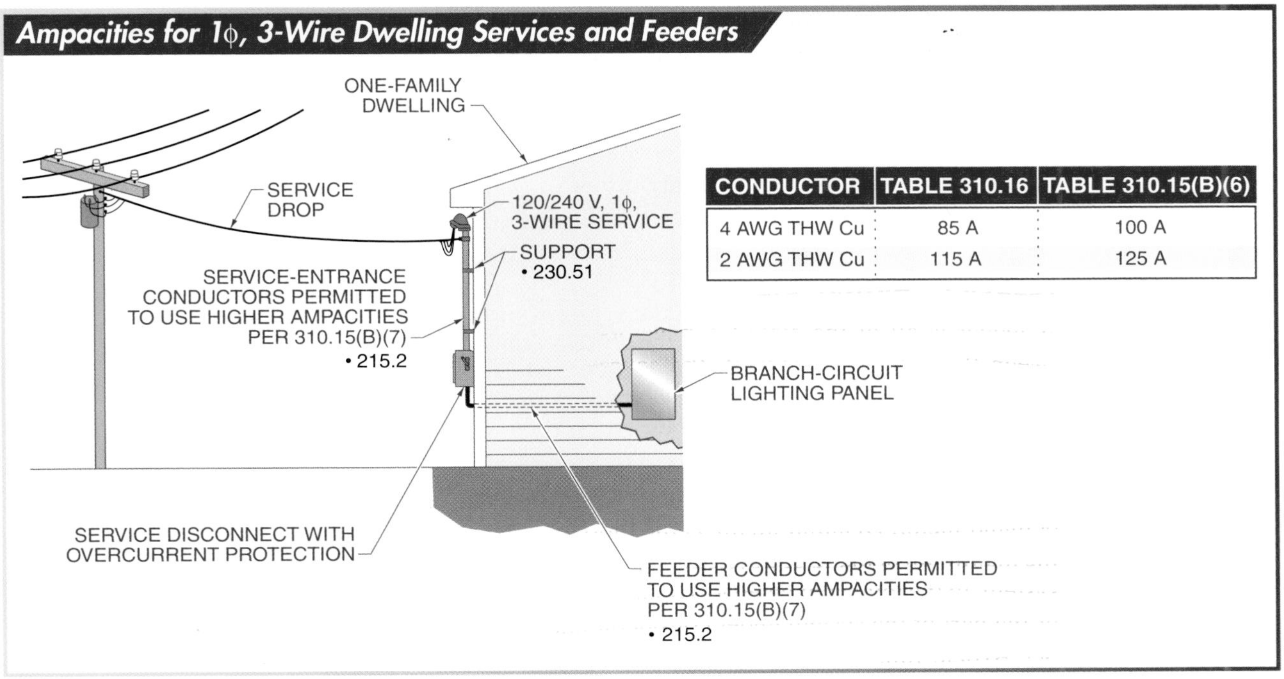

CONDUCTOR	TABLE 310.16	TABLE 310.15(B)(6)
4 AWG THW Cu	85 A	100 A
2 AWG THW Cu	115 A	125 A

Figure 2-27. Section 310.15(B)(7) is used to size service and feeder conductors supplying dwellings with systems rated at 120/240 V, 1ϕ, 3-wire.

The transformer configuration that is most often used in 3ϕ, 4-wire applications is the 3ϕ, 4-wire wye. This configuration has a neutral point at the point where the 3ϕ windings are joined. The neutral conductor is connected to the neutral point and grounded. The two most common voltage systems provided by this configuration are the 208Y/120 V system and the 480Y/277 V system. Line-to-neutral voltages provided by these systems are 120 V and 277 V, respectively. The 120 V, line to neutral, is used to supply receptacle outlets, small motors, lighting fixtures, and other 120 V loads. The 277 V supplies lighting loads, motors, and other similar 277 V loads.

Section 220.61(B)(2) allows the load on the neutral conductor for a feeder or service to be reduced using a 70% demand factor. This demand is permitted to be applied to a neutral conductor that has a load in excess of 200 A. For example, a feeder neutral with a load of 400 A would be permitted to have the demand factor applied to 200 A of the load. This first 200 A would be calculated at 100% and the remaining 200 A at 70%, or 140 A. The 140 A would be added to the 200 A, providing a demand load on the neutral of 340 A. The section does not permit the demand factor to be applied to that portion of the load that is supplied by nonlinear loads. Electric discharge lighting, personal computers, and variable-speed drive controllers are examples of nonlinear loads.

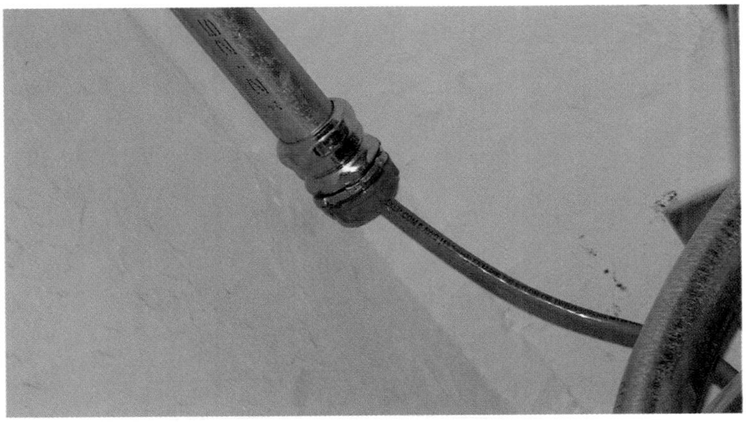

Metal enclosures for GECs that are not electrically continuous violate 250.64(E).

Selective Coordination

Selective coordination is the selection and setting of protective devices in an electrical system in a manner that causes only the smallest part of the electrical system to be de-energized during an overcurrent. The idea is to trip the smallest overcurrent protective device in an electrical system before a larger main overcurrent protective device is tripped. When this happens, the circuit with the overcurrent will be de-energized while the rest of the electrical system remains energized.

Each overcurrent protective device in an electrical system has its own time-current characteristic, which can be coordinated with the time-current characteristics of other overcurrent protective devices. Time-current characteristics of overcurrent protective devices are compared up circuit and down circuit to ensure that there are no overlaps in overcurrent protection.

A short-circuit current study should be conducted to determine the short-circuit currents at all points in an electrical system. An overcurrent coordination study should be conducted to compare the time it takes for each overcurrent protective device to operate when certain levels of current pass through it. The overcurrent coordination study determines the characteristics, ratings, and settings of overcurrent protection to ensure that only the smallest part of the electrical system is interrupted. Also, short-circuit coordination tables should be consulted to provide advanced information on the interaction of overcurrent protective devices to enable the proper selection of these devices.

Additional information is available at www.schneider-electric.com.

Overheated Neutral Conductors

High neutral current is dangerous because it causes overheating in the neutral conductor. Because there is no circuit breaker in the neutral conductor to limit current as in the phase conductors (A, B, and C), overheating of the neutral conductor can become a fire hazard. Excessive current in the neutral conductor can also cause higher than normal voltage drops between the neutral conductor and ground conductor at 120 V outlets.

Section 220.61(B)(1) allows a 70% demand factor to be applied to the portion of the load on a neutral conductor that is supplied by clothes dryers, household ranges, wall-mounted ovens, counter-mounted cooking units, and other household cooking appliances rated over 1750 W.

A 4-wire delta system is popular where there is a large 3ϕ load, a large 1ϕ line-to-line load, and a small 1ϕ line-to-neutral load. This transformer configuration is one that is listed in the definition of neutral point. The neutral point is located at the midpoint of one of the phases. The neutral conductor is connected to the neutral point and grounded. It provides 120 V, line to neutral, for 120 V, 1ϕ loads. This configuration also provides 240 V, 1ϕ, line to line, and 240 V, 3ϕ. Care must be used when these systems are used. The voltage measured between the Bϕ conductor and the neutral conductor is 208 V (120 V × 1.73 = 208 V). This Bϕ conductor is called the high-leg. **See Figure 2-28.**

When tapping into an existing delta 4-wire system in order to take off a 120/240 V, 1ϕ, 3-wire system, it is imperative to know which phase conductor is the high-leg. This is the B phase, which is the conductor with a voltage-to-ground of 208 V. This is the reason that 110.15 requires the phase conductor with the higher voltage-to-ground to be identified by the color orange.

If a circuit breaker for a 120 V load was mistakenly installed on B phase, 208 V would be applied to the load when it was energized. This mistake would destroy the load and could cause injury to anyone who was near the load when the voltage was applied. The misapplication of a higher voltage than a load was rated for could also cause a fire. Electricians must always be aware of the system voltages they are working on.

Feeder Diagrams – 215.5

The authority having jurisdiction may require a feeder diagram that shows all pertinent feeder details. The diagram must indicate the calculated load, demand factors, conductor type and size and area served.

Ground-Fault Protection of Equipment – 215.10

Ground-fault protection of equipment (GFPE) is required for feeder disconnects rated at 1000 A or more. This protection is only required in solidly grounded wye systems where the voltage-to-ground is more than 150 V and the phase-to-phase voltage does not exceed 600 V. For example, protection is required for 480 Y/277 V, 3ϕ, 4-wire systems. A similar requirement is found in 230.95 for services.

The intent of this requirement is to help prevent burndowns on feeders and services in this voltage range. Note that this GFPE required by 215.10 is not the same protection as a GFCI. GFPE provides protection for equipment. A GFCI provides protection for personnel. GFPE is not required on a feeder meeting the requirements of one of the two exceptions to the general rule. Per 215.10, Ex. 1, GFPE is not required in an industrial application with continuous process(es) where a nonorderly shutdown would introduce a new or greater hazard. Per 215.10, Ex. 2, GFPE is not required where ground fault protection is provided on the supply side of the feeder.

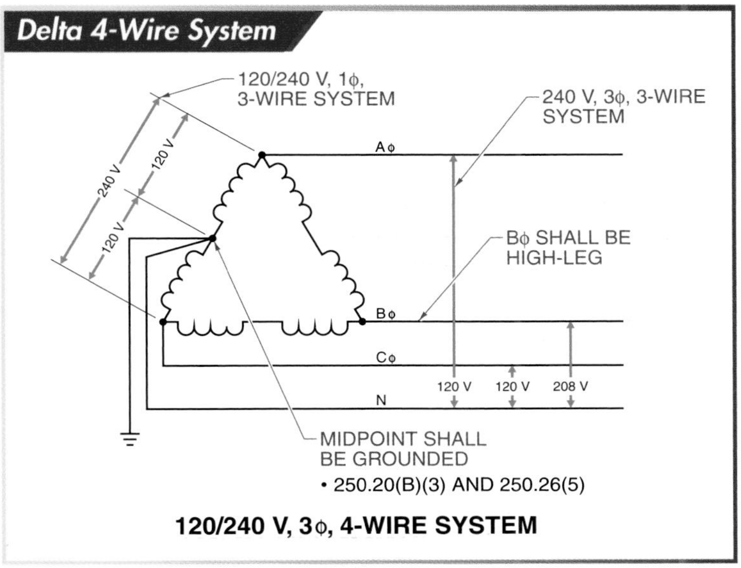

Figure 2-28. A delta 4-wire system is popular where there is a large 3ϕ load and small 1ϕ load.

Feeder Identification – 215.12

As with branch-circuit conductors in 210.5, feeders are required to be properly identified. For feeder grounded conductors, identification must be in accordance with 200.6. For equipment grounding conductors in feeders, the identification must comply with 250.119. Section 215.12(C) requires identification of ungrounded feeder conductors where the wiring system in the building or structure has more than one nominal voltage system. Identification shall be by separate color, marking tape, tagging, or other approved means. Whatever means is selected, the method must be posted at each feeder panelboard or similar distribution equipment and must be readily available for inspection and review.

QUICK QUIZ®

Refer to Chapter 2 Quick Quiz® on CD-ROM.

Large-Size Conductor Marking

Large-size conductors, such as power conductors, are often identified with colored tape at each end before they are routed through conduit. Colored tape is affixed to a conductor prior to the point where insulation is stripped and provides a unique identifier for that specific conductor. Color-coding of insulation also can be used for conductor identification, or wire markers must be attached.

Additional information is available at ATPeResources.com

chapter 2

Branch Circuits and Feeders

REVIEW QUESTIONS

Name ______________________ **Date** ______________________

______ **1.** ___ conductors are the conductors from the service point to the service disconnecting means.

______ **2.** ___ conductors are all circuit conductors between the service equipment and the final branch-circuit overcurrent device.

______ **3.** The ___ is that portion of the electrical circuit between the last overcurrent device and the outlets or utilization equipment.

______ **4.** A(n) ___ load is a load where the wave shape of the steady-state current does not follow the wave shape of the applied voltage.

______ **5.** In general, the grounded conductor of a branch circuit shall be identified by a continuous white or ___ outer finish.

T F **6.** Branch circuits with not over 120 V between conductors are permitted to supply auxiliary equipment of electric-discharge lamps.

T F **7.** A GFCI receptacle or circuit breaker is set to trip at 10 mA.

______ **8.** A bathroom is an area with a ___ and one or more of the following: ___.

A. toilet; basin, tub, or shower
B. basin; toilet, tub, or shower
C. tub; toilet, basin, or shower
D. shower; tub, toilet, or basin

______ **9.** A GFCI-protected receptacle is required within ___′ of the outside edge of a sink located in areas other than kitchens.

T F **10.** All receptacles installed in dwelling bathrooms shall be GFCI-protected.

T F **11.** All receptacles which serve countertops in dwelling kitchens shall be GFCI-protected.

______ **12.** Branch-circuit conductors shall be ___ AWG or larger.

______ **13.** Receptacle outlets installed in a dwelling for specific appliances shall be located within ___′ of the appliance.

______ **14.** The required dwelling unit outdoor receptacles shall be not more than ___ above finished grade level.

______ **15.** At least one receptacle outlet shall be installed in ___′ or longer hallways of dwelling units.

______ **16.** The required 15 A or 20 A, 125 V, 1ϕ, receptacle on rooftops for HACR servicing shall be within ___′ of the equipment.

______ **17.** A wall switch shall be provided at each floor level for a lighting outlet installed in a dwelling stairway with ___ or more steps between floor levels.

______ **18.** ___ is the ratio of the maximum demand of a system to the total connected load.

______ **19.** Branch circuits are rated in accordance with the maximum ampere rating or setting of the ___.

______ **20.** Branch-circuit EGCs shall have a continuous outer finish that is either green or green with one or more ___ stripes unless it is bare.

______ **21.** All conductors of a(n) ___ branch circuit shall originate from the same panelboard.

______ **22.** A current flow of ___ mA will most likely cause the heart to stop pumping.

T F **23.** All receptacles in dwelling unit crawl spaces at or below grade shall be GFCI-protected.

T F **24.** No point along the wall line for a dwelling unit countertop shall be more than 48″ from a receptacle outlet.

T F **25.** At least one wall receptacle outlet shall be installed within 36″ of the outside edge of each bathroom basin in all dwelling units.

Wall Receptacles

______ **1.** The maximum dimension at A is ___′.

______ **2.** The maximum dimension at B is ___′.

______ **3.** The maximum dimension at C is ___′.

______ **4.** The maximum dimension at D is ___′.

T F **5.** A receptacle is required at E.

______ **6.** The maximum dimension at F is ___′.

______ **7.** The maximum dimension at G is ___′.

T F **8.** A receptacle is required at H.

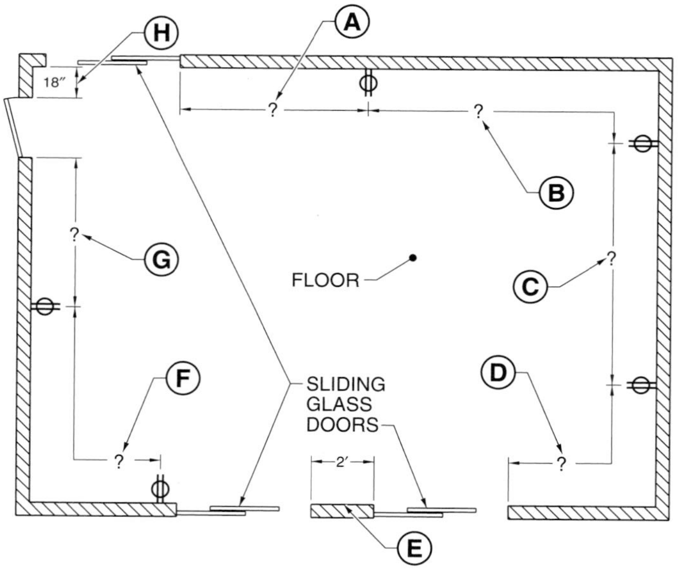

DWELLING PLAN VIEW

Branch Circuits and Feeders

TRADE COMPETENCY TEST

Name ______________________ **Date** ______________

NEC®	**Answer**	
________	________	**1.** A parking lot lighting installation requires the use of 10 AWG THW conductors with a listed ampacity of 35 A to be installed to limit the voltage drop on the circuit. In the panelboard, the 10 AWG conductors are terminated on a 20 A OCPD. Determine the branch-circuit rating.
________	________	**2.** *See Figure 1.* Does the multiwire branch circuit run to supply lighting in a commercial installation violate Article 210 of the NEC®?

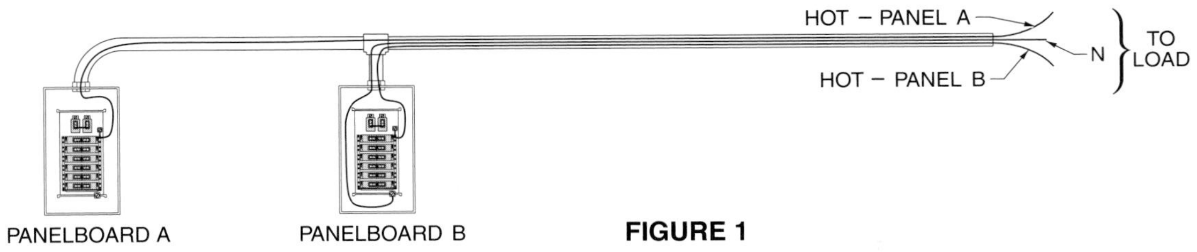

FIGURE 1

________	________	**3.** Determine the minimum distance from ground permitted for auxiliary equipment of electric-discharge luminaires (lighting fixtures) used to illuminate a parking lot. The luminaires (lighting fixtures) operate at 480 V to ground.
________	________	**4.** *See Figure 2.* Does the installation violate the provisions of Article 210 of the NEC®?

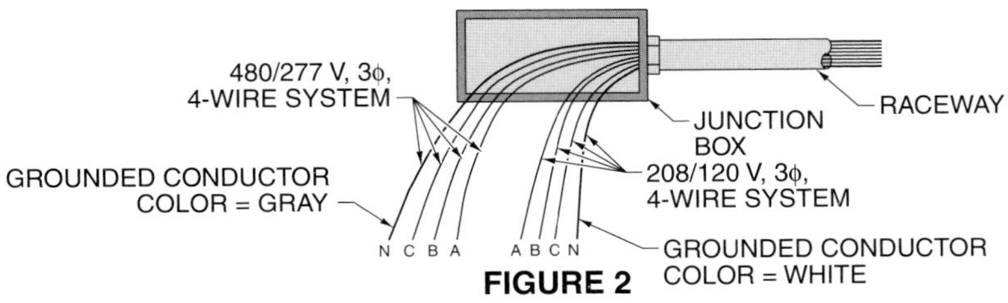

FIGURE 2

______________ ______________ **5.** A contractor installs a 15 A, 125 V, 1ϕ receptacle in the bathroom of a one-family dwelling. The receptacle is not GFCI-protected, but it is installed in a location which is not readily accessible. Does this installation violate Article 210 of the NEC®?

______________ ______________ **6.** A 240 V, 9 kW household electric range is installed in the basement of a one-family dwelling. Determine the minimum branch-circuit rating for the circuit which supplies the range.

______________ ______________ **7.** *See Figure 3.* What size OCPD (using circuit breakers) and what size branch-circuit conductors are required for the continuous load?

______________ ______________ **8.** *See Figure 4.* What size OCPD (using circuit breakers) and what size branch-circuit conductors are required for the continuous and noncontinuous loads?

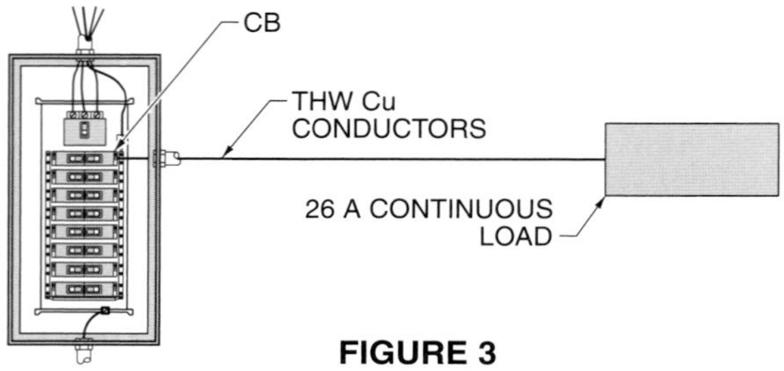

FIGURE 3

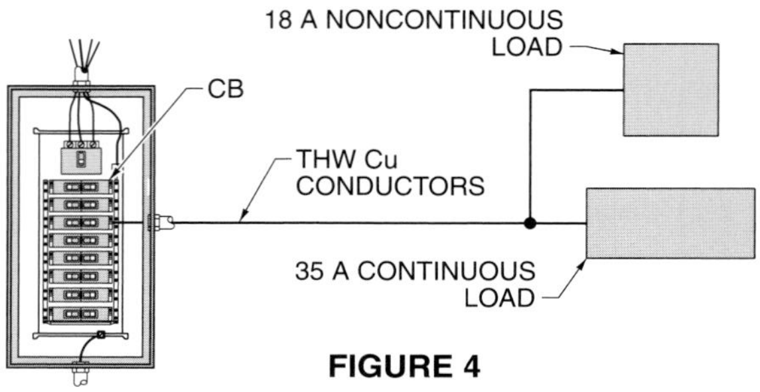

FIGURE 4

______________ ______________ **9.** Determine the maximum rating for a single cord-and-plug- connected piece of utilization equipment which is installed on a 30 A branch circuit.

______________ ______________ **10.** Determine the maximum rating for a single cord-and-plug-connected piece of utilization equipment which is fastened-in-place and installed on a 20 A branch circuit. The branch circuit also supplies lighting units in addition to the utilization equipment.

______________ ______________ **11.** A 40 A branch circuit is installed in a dwelling unit to supply luminaires (lighting fixtures) equipped with heavy-duty lampholders. Does this installation violate Article 210 of the NEC®?

______________ ______________ **12.** *See Figure 5.* Determine the minimum number of receptacles required to be installed in the dwelling unit den.

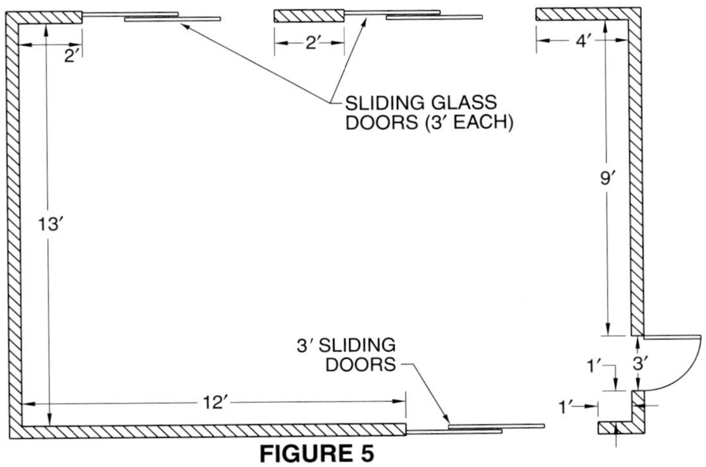

FIGURE 5

Learning the Code

- Nothing can strike more fear into the hearts of apprentices and students of the Code than the thought of doing Code calculations. There is no need to get unnecessarily nervous about Code calculations. They are part of the Code and, most importantly, part of the work that electrical workers do every day on the job site.
- "Calculations" is a broad term that can apply to anything from selecting the correct wire size to determining the size of a service to a building or structure. Chapter 3 takes a step-by-step approach for apprentices and students of the Code. When performing service calculations, it's helpful to think in terms of several individual calculations that, taken together, add up to the total load or demand for the building or structure.

- Over the past few Code cycles, many of the calculations in the NEC® have been relocated to Article 220. This article covers all of the branch-circuit, feeder, and service calculations. Do not forget about Table 220.3, which is a great list of other Code calculations.
- As difficult as these calculations may seem, stay with them and work through the various steps necessary to feel comfortable in performing them. Most journeyman examinations and many local and state licensing examinations contain a minimum of 25% Code calculations questions. If necessary, separate textbooks and workbooks are available that focus specifically on the calculations necessary to be successful when taking Code examinations.

chapter 3

Dwelling Load Calculations

Calculations are used to determine the total load for a building in order to size the service-entrance conductors. The Standard Calculation and the Optional Calculation are the two methods to calculate the load for one-family and multifamily dwellings. The Optional Calculation is generally used with larger loads.

New in the 2011 NEC®

- *Clarification of fraction rounding for amperes – 220.5(B)*
- *Scope clarification for electric dryer and electric cooking appliance load calculations – 220.14(B)*
- *Track-lighting load calculation exception revision – 220.43(B) Exception*
- *Revised table for farm load calculations for other than dwelling units – 220.102*

CALCULATIONS

"Calculation" is a widely used but undefined term in the NEC®. Calculation is defined in the dictionary as the process, act, or result of calculating. The term calculating means to determine by mathematical processes. Designers and electricians use calculations to properly size equipment and conductors for electrical installations. Many of the calculation methods are found in the NEC® but others are found in electrical theory books. For example, guidelines for voltage drop calculations are found in Informational Note No. 4 to 210.19(A)(1) and Informational Note No. 2 to 215.2(A)(3). The actual formulas for calculating voltage drop are not found in the Code but can be located in electrical formula books.

Section 220.1 (Scope) indicates that Article 220 provides the requirements for calculating branch-circuit, feeder, and service loads. The Article has five parts. Part I provides the general rules for calculation methods. Part II covers load calculations for branch circuits. Parts III and IV provide calculation methods for feeders and services. Part V provides rules for farm load calculations.

Parts III and IV are divided into two types of calculations, the Standard Calculation (Part III) and the Optional Calculation (Part IV). The Optional Calculation is generally easier to use and will result in smaller calculated loads than the Standard Calculation. Rules found in Part III can only be applied to the Standard Calculation found in that part. They cannot be used to determine the loads calculated by the Optional Calculation found in Part IV.

Many of the calculations made using both the Standard and the Optional methods will include demands. A *demand* is the amount of electricity required at a given time. The concept behind the use of the term is that the total calculated load is rarely placed on the electrical system. This is because not all of the electrical loads are on at the same time. For example, a household electric range may have a nameplate rating of 14 kW. The range consists of four burners, one broiler, one heating element, and accessories such as fans, lights, and timers. Rarely, if ever, would all of these be used at the same time. Not only do they perform separate functions but they are controlled by thermostats and have different heat settings that vary the amount of load placed on the circuit conductors. The Code recognizes this and permits a demand to be applied to the range load, reducing the calculated loads on the branch-circuit, feeder, and service conductors.

Calculations – Other Articles – 220.3

Article 220 is not the only location in the Code where calculations are found. Section 220.3 refers to Table 220.3 for information on calculations for specialized locations. The rules found in the referenced Code sections are in addition to or modifications to the rules found in Article 220. For example, Table 220.3 references Article 440, Part IV for sizing branch-circuit conductors for air-conditioning and refrigerating equipment. The Table provides an excellent cross-reference for many of the most important calculation requirements found in the NEC®. It should always be reviewed to ensure that specific calculation requirements are not overlooked.

Heating or A/C loads often represent the largest loads in one-family dwellings.

Calculation Parameters – 220.5

Section 220.5 sets some very basic, yet essential, ground rules for performing all calculations. It mandates the use of the following nominal system voltages for calculating branch-circuit, feeder, and service loads where other voltages are not provided:

- 120 V
- 120/240 V
- 208Y/120 V

- 240 V
- 347 V
- 480Y/277 V
- 480 V
- 600Y/347 V
- 600 V

Designers and installers should always use the voltage values provided by the utility, when they are available. Otherwise the values for the applicable voltage system listed in 220.5 should be used. Section 220.5(B) allows fractions of an ampere to be rounded to the nearest whole ampere where the fraction is less than 0.5 A.

BRANCH-CIRCUIT CALCULATIONS – 220.10

Section 220.10 requires branch circuits to be calculated using the rules found Part II of Article 220. This part is not very large, including just three sections, but the information contained is critical to the design of branch-circuit wiring. Branch-circuit ratings for continuous and noncontinuous loads are covered by 210.19 and 210.20. Section 210.3 defines branch-circuit rating as the ampere rating or setting of the overcurrent device protecting the conductors. Section 210.20(A) states that branch-circuit calculations must be based on 100% of the noncontinuous load plus 125% of the continuous load. Additionally, 210.19(A)(1) requires the minimum branch-circuit conductor size, without applying any derating provisions, to have an ampacity that is at least equal to or greater than the noncontinuous load plus 125% of the continuous load. It is easy to see that Code rules found in one article often apply to other articles and that a thorough understanding of the entire NEC® is needed for Code-compliant installations.

General Lighting Loads – 220.12

Section 220.12 and Table 220.12 provide the information needed to calculate the general lighting load for specified occupancies. The general lighting load is based on a volt-ampere per square foot basis. The method of calculation required by the section provides the minimum lighting load. Where the actual loads used in a building or structure exceed the volt-ampere per square foot value provided in Table 220.12 for the applicable occupancy, the larger value shall be used.

There are two steps used in calculating the general lighting load. The first step is to calculate the area of the building or structure. This is done by using the outside dimensions of the building, dwelling unit, or other area involved. Where calculating the total area for a dwelling unit, the calculated floor area shall not include open porches, garages, or unused or unfinished spaces not adaptable for future use.

The second step is to find the volt-ampere per square foot value from Table 220.12 for the applicable occupancy and to multiply that value by the calculated floor area. As an example, a dwelling unit with outside dimensions of 32′ × 40′ has a calculated floor area of 1280 sq ft. Table 220.12 lists the unit lighting load for dwellings as 3 VA per square foot. The general lighting load for the dwelling is 1280 sq ft multiplied by 3 VA per square foot, which is equal to 3840 VA.

Two notes below Table 220.12 provide extra calculation information. These two notes refer to 220.14(J) and 220.14(K). An Informational Note to 220.12 warns that the calculations required by the section are based on 100% power factor and may not provide sufficient capacity for every installation.

Other Loads – 220.14

Section 220.14 provides twelve rules used to determine the minimum load for general-use receptacles and outlets not used for general illumination. The loads must be calculated based on nominal circuit voltages.

1. 220.14(A). Outlets for specific appliances or loads, not included in the eleven other rules must be calculated based on the ampere rating of the appliance or load.
2. 220.14(B). Electric dryers may be calculated per 220.54 and electric ranges and cooking appliances may have their loads determined by 220.55.
3. 220.14(C). Motor loads are calculated per the rules found in 430.22, 430.24, and 440.6.

4. 220.14(D). The calculation of the load for lighting fixtures must use the maximum rating of the equipment and lamps. For example, when calculating the load on a branch circuit supplying five recessed fixtures that are listed for use with lamps rated at 40, 60, 75, and 100 W, the 100 W value must be used even if the plan is to use 40 W lamps for accent lighting. The load calculation must also consider fixture use and multiply the total connected load based on this rule by 125% if the load is expected to operate for 3 hr or more, continuously.
5. 220.14(E). Heavy-duty lampholder outlets must have a minimum of 600 VA included in the calculation. Consideration must also be made for continuous loads.
6. 220.14(F). Outlets for signs and outline lighting must be calculated at a minimum of 1200 VA per branch circuit. This load must be multiplied by 125% where the load on the outlet is considered continuous. Section 600.5(A) requires at least one outlet supplied by a 20 A branch circuit to be installed in each commercial building and commercial occupancy that is accessible to pedestrians. Service hallways or corridors are not considered accessible to pedestrians.
7. 220.14(G). The seventh rule provides two methods for calculating the minimum branch circuit for show window lighting. The first method allows the use of the applicable unit load per outlet required by the eleven other rules found in this section. The second choice is to calculate the minimum load based on 200 VA per foot of show window, multiplying this load by 125% where the load is continuous.
8. 220.14(H). A *multioutlet assembly* is a surface, flush, or freestanding raceway which contains conductors and receptacles. These assemblies can be constructed in the field or they can be factory assembled. Section 220.14(H) provides two methods of calculation. The first calculation is for multioutlet assemblies that are unlikely to have simultaneous loads. In this case, each 5 ft or fraction thereof is calculated at 180 VA. Where appliances are likely to be used simultaneously, the load changes to 180 VA per 1 ft or fraction thereof.
9. 220.14(I). Receptacle outlets, other than those listed in 220.14(I) and (J), are calculated at 180 VA for each single and each multiple receptacle on a strap. A single piece of equipment consisting of four or more receptacles must be calculated at not less than 90 VA per receptacle. This rule does not apply to the small appliance and laundry circuits. **See Figure 3-1.**

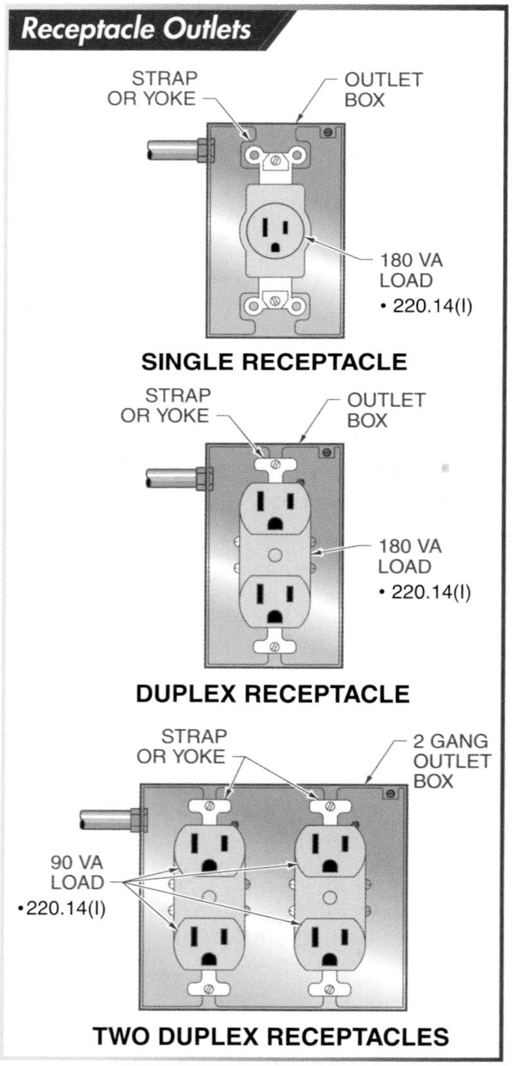

Figure 3-1. Each single or multiple receptacle on a single strap is calculated at 180 VA.

10. 220.14(J). This rule states that the following outlets, where installed in dwelling occupancies are included in the general lighting load calculations and that no other load calculation is required. The outlets included are all general-use outlets of 20 A rating or less, the bathroom receptacle(s) required by 210.11(C)(3), outlets required by 210.52(E) and (G), and lighting outlets required by 210.70(A) and (B). This section is referenced under Table 220.12 and applies to occupancies with a superscript "a".
11. 220.14(K). Receptacle loads in banks and office buildings must be calculated by two methods and the larger load used. The first method is to calculate them at 180 VA per the requirements of 220.14(I). The second calculation is to calculate the receptacle load at 1 VA per square foot. The second calculation is generally used where the number of receptacles is unknown. This situation may exist when plans are submitted for the shell of a building and the interior layout, other than building facilities, will be provided at a future date. This section is referenced under Table 220.12 and applies to occupancies with a superscript "b".
12. 220.14(L). Any outlet loads not covered by the other rules found in 220.14 must be calculated at 180 VA per outlet.

Branch Circuits Required

The minimum number of branch circuits required for lighting and appliances shall be determined from the values obtained from the 220.10 calculations. For specific loads not covered by 220.10, additional branch circuits shall also be provided. Section 210.11 contains provisions for determining the minimum number of branch circuits, including the requirements for small appliance branch circuits, laundry branch circuits, and bathroom branch circuits.

Per 210.11(B), the required branch circuits shall be evenly divided across the branch-circuit panelboard. In addition, branch-circuit overcurrent devices and circuits need only be provided for connected loads. Any future loads need not be considered. See 210.11(A-C).

Conductor Identification

A wire marker is a preprinted peel-off sticker designed to adhere to insulation when wrapped around a conductor. Wire markers resist moisture, dirt, and oil and are used to identify conductors that have the same color but different uses. Wire markers can be used even when different-colored conductors are used. Using wire markers in addition to color-coding clarifies the uses of all conductors.

PART III – STANDARD CALCULATIONS

Interactive Load Calculation Forms

Part III of Article 220 provides the requirements for calculating loads on feeder and service conductors for dwelling units and other than dwelling units (commercial). Sections 220.40, 220.42, 220.50, 220.51, 220.52, 220.53, 220.54, 220.55, 220.60, and 220.61 apply to dwelling load calculations. Sections 220.40, 220.42, 220.43, 220.44, 220.50, 220.51, 220.56, 220.60, and 220.61 apply to other than dwelling load calculations.

The calculated load of a service or a feeder shall not be less than the sum of the loads on the branch circuits supplied, as determined by Part II of Article 220. These loads are permitted to have the demand factors found in Part III applied. Continuous loads shall be applied at 125% per 210.19(A)(1). The Standard Calculation can be used for one- and two-family dwelling units, multifamily dwellings, hotels and motels, and all commercial and industrial occupancies.

Separate load calculations are made to accommodate single-family dwellings.

ONE-FAMILY DWELLINGS – STANDARD CALCULATION

The Standard Calculation for one-family dwellings contains six individual calculations that are performed before the minimum size service or feeder conductors required for the calculated load can be determined. Demand factors may be applied to four of these loads. These six loads, their NEC® references, and those to which demand factors may be applied are:

1. General Lighting – *Table 220.12* and *Table 220.42* (Demand Factors)
2. Fixed Appliances – *220.53* (Demand Factors)
3. Dryer – *220.54; Table 220.54* (Demand Factors)
4. Cooking Equipment – *Table 220.55* (Demand Factors)
5. Heating or A/C – *220.60*
6. Largest Motor – *220.50*

One-Family Dwelling General Lighting – 220.12, 220.52

The general lighting load for one-family dwellings consists of three separate loads which are calculated individually, added together, and then a demand factor may be applied. The first portion of the general lighting load is calculated from the total square footage of the dwelling. **See Figure 3-2.**

Where a one-family dwelling is a multistory structure, the total square footage is obtained by calculating the square footage for each floor using the outside dimensions and adding those values together to obtain a total square footage for the dwelling. Section 220.14(J) allows general-use receptacles installed in dwelling units to be considered as a part of the general lighting load found using the 3 VA per square foot calculation, and no other load calculation is required.

Small Appliances – 220.52(A). The second portion of the general lighting load consists of the calculation for small appliance branch circuits. Section 210.11(C)(1) requires a minimum of two 20 A small appliance branch circuits. Section 210.52(B)(1) states that these branch circuits shall serve all wall and floor receptacle outlets required by 210.52(A), all countertop receptacles required by 210.52(C), and the receptacles for refrigeration equipment installed in a dwelling kitchen, pantry, breakfast room, dining room, or any similar location.

Per 220.52(A), each small appliance branch circuit shall be calculated at 1500 VA when determining the feeder load. If the dwelling has more than two small appliance branch circuits, each circuit shall be calculated at 1500 VA. The section allows the total small appliance load to be added to the general lighting load. The demand factors from Table 220.42 are then applied to the total general lighting load, which includes the small appliance load. **See Figure 3-3.**

Laundry – 220.52(B). The last portion of the general lighting load consists of the laundry load. Sections 210.52(F) and 210.11(C)(2) require that each dwelling be provided with a 20 A branch circuit to supply laundry receptacle outlet(s). Section 220.52(B) requires that, for the purposes of calculating feeder load, each laundry circuit shall be calculated at 1500 VA. A laundry circuit shall always be provided for one-family dwellings. The laundry load is included with the general lighting load and is permitted to be derated by the demand factors of Table 220.42. **See Figure 3-4.**

General Lighting Load – Demand Factors – Table 220.42. Section 220.42 permits a reduction in the general lighting load for dwelling units, hospitals, hotels and motels, and storage warehouses by allowing the application of the demand factors in Table 220.42 to branch circuits that provide general illumination. The note to the table references general lighting loads for hospitals, hotels, and motels and warns that the table cannot be applied to general lighting loads in those occupancies where the loads are likely to be used at one time, such as lighting loads for operating rooms, ballrooms, or dining rooms. Section 220.42 does not permit the demand factors of Table 220.42 to be applied when determining the number of branch circuits that are required for general illumination.

General Lighting—Table 220.12

What is the first portion of the general lighting load for the one-family dwelling?

Area: 30′ × 60′ = 1800 sq ft
Table 220.12: 1800 sq ft × 3 VA = 5400 VA
Lighting Load = **5400 VA**

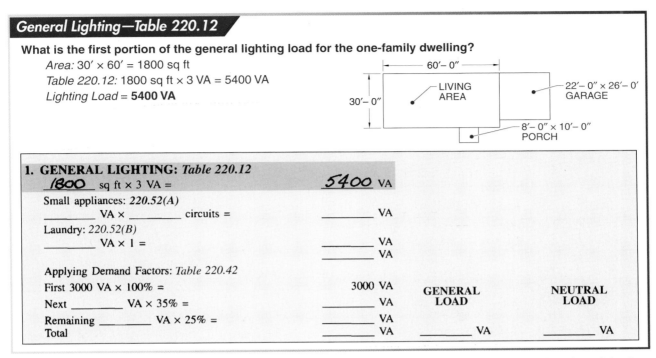

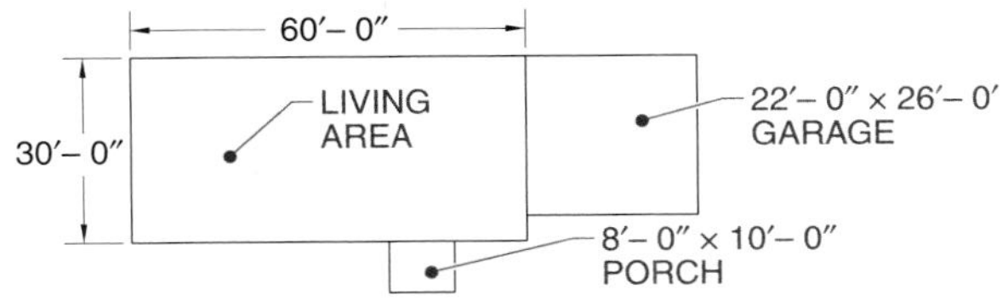

1. GENERAL LIGHTING: ***Table 220.12***
1800 sq ft × 3 VA = 5400 VA
Small appliances: *220.52(A)*
_______ VA × _______ circuits = _______ VA
Laundry: *220.52(B)*
_______ VA × 1 = _______ VA
_______ VA
Applying Demand Factors: *Table 220.42*
First 3000 VA × 100% = 3000 VA
Next _______ VA × 35% = _______ VA
Remaining _______ VA × 25% = _______ VA
Total _______ VA

GENERAL LOAD _______ VA
NEUTRAL LOAD _______ VA

Figure 3-2. The general lighting load for an occupancy is calculated by multiplying the area (in square feet) by the unit load per square foot.

Small Appliances—Table 220.52(A)

What is the small appliance load for a one-family dwelling with three 20 A, 120 V small appliance branch circuits?

220.52(A): 1500 VA × 3 = 4500 VA
Small Appliance Load = **4500 VA**

1. GENERAL LIGHTING: ***Table 220.12***
_______ sq ft × 3 VA = _______ VA
Small appliances: *220.52(A)*
1500 VA × 3 circuits = 4500 VA
Laundry: *220.52(B)*
_______ VA × 1 = _______ VA
_______ VA
Applying Demand Factors: *Table 220.42*
First 3000 VA × 100% = 3000 VA
Next _______ VA × 35% = _______ VA
Remaining _______ VA × 25% = _______ VA
Total _______ VA

GENERAL LOAD _______ VA
NEUTRAL LOAD _______ VA

Figure 3-3. Each small appliance branch circuit is calculated at 1500 VA.

For all occupancies, other than dwelling units, hospitals, hotels, motels, and storage warehouses, the minimum general lighting load must be calculated at 100%. Note that this load must be increased by 125% for any portion of the load that is continuous.

Laundry—Table 220.52(B)

What is the laundry load for a one-family dwelling with one 20 A, 120 V laundry branch circuit?

220.52(B): 1500 VA × 1 = 1500 VA

Laundry Load = **1500 VA**

1. GENERAL LIGHTING: ***Table 220.12***

_______ sq ft × 3 VA = _______ VA

Small appliances: *220.52(A)*

_______ VA × _______ circuits = _______ VA

Laundry: *220.52(B)*

1500 VA × 1 = 1500 VA

_______ VA

Applying Demand Factors: *Table 220.42*

First 3000 VA × 100% = 3000 VA

Next _______ VA × 35% = _______ VA

Remaining _______ VA × 25% = _______ VA

		GENERAL LOAD	NEUTRAL LOAD
Total	_______ VA	_______ VA	_______ VA

Figure 3-4. Each one-family dwelling shall have a 1500 VA laundry branch circuit.

When calculating the general lighting demand for dwelling units, Table 220.42 requires the first 3000 VA of the general lighting load to be calculated at 100%. The next 117,000 VA (3001 VA to 120,000 VA) is calculated at 35%, and the remainder over 120,000 VA is calculated at 25%. **See Figure 3-5.** The neutral load for the general lighting load is the same as the load on the ungrounded conductors.

One-Family Dwelling Fixed Appliances – 220.53

Four or more appliances which are fastened-in-place in a one-family dwelling are calculated at 75% of the total load for all four appliances. This demand factor applies to the nameplate rating of the appliance but does not include electric ranges, clothes dryers, space-heating equipment, or A/C equipment which are served by the same feeder or service in a one-family, two-family, or multifamily dwelling. **See Figure 3-6.**

One-Family Dwelling Dryer – 220.54

The load for household clothes dryers in dwelling units is required by 220.54 to be 5000 VA or the nameplate rating, whichever is greater. Table 220.54 provides demand factors for household electric clothes dryers. The Table does not allow a reduction in the load for one to four dryers supplied by a feeder or service. Where the number of dryers exceeds four, the load is reduced by the applicable demand factor found in the Table. A calculation is required to find the demand for 12 through 22 dryers and another calculation is required for 24 through 42 dryers.

The demand factor for the first four clothes dryers is 100%.

General Lighting—Demand Factors—Table 220.42

What is the general lighting and receptacle load for a 3000 sq ft dwelling with three small appliance branch circuits and one laundry branch circuit?

Table 220.12: 3000 sq ft × 3 VA = 9000 VA
220.52(A): 1500 VA × 3 = 4500 VA
220.52(B): 1500 VA × 1 = 1500 VA
Total VA: 9000 VA + 4500 VA + 1500 VA = 15,000 VA
Table 220.42: First 3000 VA × 100% = 3000 VA
Next 12,000 VA × 35% = 4200 VA
7200 VA

General Lighting and Receptacle Load = **7200 VA**

1. GENERAL LIGHTING: ***Table 220.12***

3000 sq ft × 3 VA =	9000 VA		
Small appliances: *220.52(A)* 1500 VA × 3 circuits =	4500 VA		
Laundry: *220.52(B)* 1500 VA × 1 =	1500 VA		
	15,000 VA		
Applying Demand Factors: *Table 220.42*		**GENERAL LOAD**	**NEUTRAL LOAD**
First 3000 VA × 100% =	3000 VA		
Next 12,000 VA × 35% =	4200 VA		
Remaining ______ VA × 25% =	VA		
Total	7200 VA	7200 VA	7200 VA

Figure 3-5. Demand factors are applied to all general lighting loads of 3001 VA and greater for dwelling units.

Fixed Appliances—Table 220.53

What is the fixed appliance load (line and neutral) for a one-family dwelling with the following 120 V appliances that are fastened in place and supplied by the same feeder?

Water heater – 7500 VA
Dishwasher – 1650 VA
Foodwaste disposer – 950 VA
Trash compactor – 700 VA
Attic fan – 1400 VA

220.53: 7500 VA + 1650 VA + 950 VA + 700 VA + 1400 VA = 12,200 VA
12,200 VA × 75% = 9150 VA

Fixed Appliance Load: Line = **9150 VA**
Neutral = **9150 VA**

2. FIXED APPLIANCES: ***220.53***

Dishwasher =	1650 VA			
Disposer =	950 VA			
Compactor =	700 VA			
Water heater =	7500 VA		**GENERAL LOAD**	**NEUTRAL LOAD**
Attic fan =	1400 VA			
______ =	______ VA			
______ =	______ VA			(120 V Loads × 75%)
Total	12,200 VA × 75% =	9150 VA	9150 VA	9150 VA

Figure 3-6. Four or more fixed appliances in a one-family dwelling are calculated at 75% of the total load for the four appliances.

Energy Star Ratings

Energy Star is a joint program of the EPA and the Department of Energy that promotes energy-efficient products and practices to help save money and protect the environment. Energy Star-qualified household products, whole home retrofits, and new homes contribute to energy efficiency and savings. Businesses can form partnerships with Energy Star, which offers an energy management strategy to help measure current energy performance, set goals, track savings, and reward improvements. Also, the EPA has developed an energy-performance rating system used to rate and recognize top-performing buildings that use Energy Star.

Requirements for sizing services, sizing and selecting distribution equipment, and sizing feeder and branch-circuit conductors are based on the demand for electricity by various loads. An energy-efficient building requires less energy for the operation of loads found within the building. The less energy a load uses means the less demand for electricity when computing load calculations.

Additional information is available at www.energystar.gov.

The calculation for the demand percentage for 12 through 22 dryers requires two steps. Step one involves subtracting 11 from the number of dryers. The result of this subtraction is then subtracted from the number 47 to obtain the demand percentage. For example, the demand percentage for 12 dryers would be 47 – (12 – 11) = 46%. The calculation for the demand percentage for 24 through 42 dryers requires three steps. Step 1 requires the subtraction of the number 23 from the number of dryers. This number is then multiplied by 0.5 and the result of that operation is subtracted from the number 35. For example, the demand percentage for 25 dryers would be 35 – [0.5 × (25 – 23)] = 35 – [0.5 × 2] = 34%. Section 220.61(B)(1) allows the neutral load to be based on 70% of the ungrounded conductor load.

Where two or more 1ϕ dryers are supplied by a 3ϕ, four-wire feeder or service, the total load must be calculated on the basis of twice the maximum number connected between any two phases. This calculation is generally required for multifamily dwellings with 3ϕ electrical service. Single-phase dryers would be connected to a 3ϕ feeder or service as follows:

- A phase to B phase
- B phase to C phase
- A phase to C phase

For example, a multifamily dwelling with 9 dryers rated at 4500 VA each would be calculated by first increasing the load for each dryer to 5000 VA as required by 220.54. The dryers would be connected as follows:

- 3 dryers A phase to B phase
- 3 dryers B phase to C phase
- 3 dryers A phase to C phase

The maximum number of dryers connected between two phases would be 3. The Code section requires the calculation to be based on twice the maximum number of dryers connected between two phases, which in this case would be 6. Multiplying the 6 by 5000 VA would provide a load of 30,000 VA. Table 220.54 provides a demand factor of 75% for 6 dryers. The demand load on the 3ϕ service for the dryers would be 30,000 VA × 75% = 22,500 VA.

One-Family Clothes Dryer Calculations. Table 220.54 requires a demand factor of 100%

for 1–4 dryers. In most cases, one-family dwellings have one dryer. The dryer demand load would be the nameplate rating of the dryer or 5000 VA, whichever is greater. The neutral load may be based on 70% of the ungrounded load. **See Figure 3-7.**

One-Family Dwelling Cooking Equipment – Table 220.55; Notes

Section 220.55 allows the use of the demand factors found in Table 220.55 for calculating electric ranges and cooking appliances that are rated in excess of 1¾ kW. Kilovolt-amperes (kVA) are considered equivalent to kW for the loads calculated under this section. Where two or more 1ϕ ranges are supplied by a 3ϕ, 4-wire feeder or service, the total load shall be calculated on the basis of twice the maximum number connected between any two phases. The connections for the ranges are the same as they are for dryers as discussed in the previous section. The calculation method is the same as the method use for 1ϕ dryers installed on a 3ϕ, 4-wire feeder or service.

Maytag

An electric range demand load can be calculated using the demand factor.

Table 220.55 provides the demand factors and loads for household electric ranges, wall-mounted ovens, counter-mounted cooking units, and other household cooking appliances that are rated over 1¾ kW. The table has three columns of demand factors and five notes that explain the rules for calculating the loads. The table heading requires the use of Column C in all cases except as permitted in Note 3. The Column C heading provides two pieces of information. It indicates that Column C is the maximum demand, in kW, that is required for the Table 220.55 load calculation. The heading also limits the values found in column C to cooking appliances that do not exceed 12 kW. **See Figure 3-8.** The neutral for the range may be reduced to 70% of the ungrounded conductor demand per 220.61(B).

Electrician's Calculators

Special electrician's handheld calculators are available to help electrical tradesworkers perform calculations when performing electrical design or installation work. Electrician's calculators can perform calculations that factor in variables such as conduit type, conductor size, conductor type, and requirements for the 2008 NEC® and the 2011 NEC®, depending on the application.

Dryer—220.54

A one-family dwelling has one dryer with a nameplate rating of 5500 VA. What is the demand load on the line and the neutral?

220.54: The nameplate rating is larger than 5000 VA. Use 5500 VA.
Table 220.54: 5500 VA × 100% = 5500 VA
220.61(B)(1): 5500 VA × 70% = 3850 VA
Dryer Demand Load: Line = **5500 VA**
Neutral = **3850 VA**

3. DRYER: ***220.54 ; Table 220.54***
5500 VA × 100 % = 5500 VA 5500 VA × 70% = 3850 VA

Figure 3-7. For household dryers, the total load is calculated at 5000 VA or the nameplate rating, whichever is greater.

Cooking Equipment—Table 220.55, Col C

What is the range demand load for three household electric ranges rated at 9.5 kW each?

Table 220.55, Col C, Notes: three ranges under 12 kW each demand = 14,000 VA

220.61(B)(1): 14,000 VA × 70% = 9800 VA

Range Demand Load: Line = **14,000 VA**

Neutral = **9800 VA**

4. COOKING EQUIPMENT: *Table 220.55; Notes*

Col A ______ VA ×	______ % =	______ VA			
Col B ______ VA ×	______ % =	______ VA	GENERAL LOAD		NEUTRAL LOAD
Col C 28,500 VA ×	______ % =	14,000 VA			
Total		14,000 VA	14,000 VA × 70% =		9800 VA

Figure 3-8. Household ranges are calculated per the values of Table 220.55.

Where the number of appliances is greater than 25, the Column C value must be calculated. There are two rules for calculating the Column C value for more than 25 appliances; one is for 26–40 appliances and the second for 41 appliances and over. The calculation for 26–40 ranges requires the number of ranges supplied by a feeder or service to be multiplied by 1 kW. The result will then be added to 15 kW. For example, the Column C maximum demand for 30 appliances would be 45 kW (step 1: 30 × 1 kW = 30 kW; step 2: 30 kW + 15 kW = 45 kW). Where the number of appliances exceeds 40, the Column C maximum demand is based on ¾ kW for each range plus 25 kW. For example, the Column C maximum demand for 60 appliances would be 70 kW (step 1: 60 × 0.75 kW = 45 kW; step 2: 45 kW + 25 kW = 70 kW).

Table 220.55, Note 1. Note 1 provides a calculation method for finding the Column C maximum demand for appliances that exceed 12 kW. The largest rating permitted by this note is 27 kW. Appliances rated larger than 27 kW are considered to be commercial cooking appliances and are not subject to the demand factors found in Table 220.55.

Note 1 is applied where one appliance is rated 12 kW through 27 kW. It also applies where two or more appliances rated 12 kW through 27 kW are supplied by the same feeder or service conductors. All of the appliances must have the same rating to apply Note 1. The rule states that the Column C maximum demand must be increased by 5% for every kW or major fraction of a kW that the appliance kW rating exceeds 12 kW. A major fraction is 0.5 or greater. **See Figure 3-9.**

Table 220.55, Note 2. This note applies to household cooking appliances over 8¾ kW through 27 kW where the appliance ratings are not of equal values. This note simply requires a calculation to find the average rating of the appliances supplied by the same feeder or service conductors. Before calculating the average value, all appliances rated less than 12 kW must be increased to 12 kW. Once the average kW rating is determined, the rules are the same as those found in Note 1. The Column C maximum demand is increased by 5% for each kW or major fraction of a kW that the average rating exceeds 12 kW. **See Figure 3-10.**

Maytag

An over-the-range microwave oven is considered a fixed appliance.

Cooking Equipment—Table 220.55, Col C, Note 1

What is the range demand load for the line and neutral for one 15 kW household electric range installed in a one-family dwelling unit?

Table 220.55, Note 1: 15 kW > 12 kW by 3 kW
Table 220.55, Col C: 1 appliance = 8000 VA
3 kW × 5% = 15%
8000 VA × 115% = 9200 VA
220.61(B)(1): 9200 VA × 70% = 6440 VA
Range Demand Load: Line = **9200 VA**
Neutral = **6440 VA**

4. COOKING EQUIPMENT: *Table 220.55; Notes*

Col A	______ VA ×	______ % =	______ VA		
Col B	______ VA ×	______ % =	______ VA	GENERAL LOAD	NEUTRAL LOAD
Col C	8000 VA ×	115 % =	9200 VA		
Total			9200 VA	9200 VA × 70% =	6440 VA

Figure 3-9. Per Note 1, the maximum demand for individual ranges in Table 220.55, Column C is increased by 5% for each additional kW, or major fraction thereof, by which 12 kW is exceeded through 27 kW.

Cooking Equipment—Table 220.55, Col C, Note 2

What is the range demand load for the line and neutral for three household electric ranges with nameplate ratings of 12 kW, 16 kW, and 20 kW?

Table 220.55, Note 2: 12 kW + 16 kW + 20 kW = $\frac{48 \text{ kW}}{3}$ = 16 kW
16 kW > 12 kW by 4 kW
Table 220.55, Col C: 4 kW × 5% = 20%
Demand = 14 kW
14,000 VA × 120% = 16,800
220.61(B)(1): 16,800 VA × 70% = 11,760 VA
Range Demand Load: Line = **16,800 VA**
Neutral = **11,760 VA**

4. COOKING EQUIPMENT: *Table 220.55; Notes*

Col A	______ VA ×	______ % =	______ VA		
Col B	______ VA ×	______ % =	______ VA	GENERAL LOAD	NEUTRAL LOAD
Col C	14,000 VA ×	120 % =	16,800 VA		
Total			16,800 VA	16,800 VA × 70% =	11,760 VA

Figure 3-10. Per Note 2, the maximum demand for ranges of unequal values is increased by 5% of the average kW exceeding 12 kW through 27 kW, or major fraction thereof.

Table 220.55, Note 3. This note provides alternate methods for calculating appliances rated over 1¾ kW through 8¾ kW rating. The alternate demands are found in two columns in Table 220.55, Column A and Column B. To use note 3, the appliances must be separated into groups based on the nameplate rating. The first group applies to any appliances rated over 1¾ kW but less than 3½ kW and uses the demands found in Column A for the number of appliances that fall within the group. The second group applies to any appliances rated 3½ kW through 8¾ kW rating and uses the demands found in Column B for the number of appliances that fall within the group. Any appliances with a rating that exceeds 8¾ kW are calculated using Column C and Notes 1 and 2 where applicable. **See Figure 3-11.**

Cooking Equipment—Table 220.55, Col A and B, Note 3

What is the range demand load for the line and neutral for four household electric ranges with nameplate ratings of 4 kW, 4 kW, 6 kW, and 6 kW?

Table 220.55, Note 3: 4 kW + 4 kW + 6 kW + 6 kW = 20 kW

Table 220.55, Col B: 4 appliances = 50%

20 kW × 50% = 10 kW = 10,000 VA

220.61(B)(1): 10,000 kW × 70% = 7000 VA

Range Demand Load: Line = **10,000 VA**

Neutral = **7000 VA**

4. COOKING EQUIPMENT: *Table 220.55; Notes*

Col A ______ VA ×	______ % =	______ VA			
Col B 20,000 VA ×	50 % =	10,000 VA	GENERAL LOAD		NEUTRAL LOAD
Col C ______ VA ×	______ % =	______ VA			
Total		10,000 VA	10,000 VA × 70% =		7000 VA

Figure 3-11. Per Note 3, nameplate ratings between 1¾ kW and 8¾ kW are added and then multiplied by the appropriate demand from Column A or B.

After the demand is calculated using Note 3, the resulting load must be compared to the demand load found in Column C and the smaller demand load used. For example, the demand load for five 3 kW counter-mounted cooking units using Column A would be 5 × 3 kW = 15 kW × 62% = 9.3 kW. The Column C maximum demand for five appliances is 20 kW. In this case, the lower value of 9.3 kW calculated using Column A is permitted to be used for the feeder or service-conductor load. However, this is not always the case.

For example, the demand load for a 1ϕ feeder supplying ten 8 kW ranges could be determined using Column B. The demand would be 27.2 kW (8 kW × 10 = 80 kW; 80 kW × 34% = 27.2 kW). The maximum demand found in Column C for 10 appliances is 25 kW. The phrase "Maximum Demand" found in the heading for Column C means that the values determined using Column C are the maximum that must be used. In the example of the ten 8 kW ranges, the 25 kW value found in column C is a lower value than the 27.2 kW value calculated using Column B. In this case, 25 kW is the maximum demand required.

Table 220.55, Note 4. The first three words of this note are very important. Note 4 is used to calculate the demand loads for branch-circuit conductors only. **See Figure 3-12.** The note cannot be used to determine the loads for feeder and service conductors. Note 4 has three rules, as follows:

- Rule 1 allows the demand factors of Table 220.55 to be used to determine the load on the branch-circuit conductors supplying a single range.
- Rule 2 applies to a branch circuit that supplies a single counter-mounted cooking unit or a single oven. In this case the load on the branch-circuit conductors must be based on the nameplate rating of the appliance.
- Rule 3 allows the nameplate ratings of a counter-mounted cooking unit and not more than two wall-mounted ovens to be added together and treated as one range for the purposes of calculating the branch-circuit load. The appliances must be supplied by the same branch circuit and must be located in the same room. The demand factors found in Table 220.55 may be applied to the total rating to determine the load on the branch-circuit conductors.

Section 220.61(B)(C) does not allow the use of the 70% demand for the neutral conductor but 210.19(A)(3), Ex. 2, does allow a reduction of not less than 70% of the branch-circuit rating. The derated neutral conductor cannot be smaller than 10 AWG copper.

Cooking Equipment—Table 220.55, Note 4

What is the branch circuit line and neutral load for one 6 kW oven and one 8 kW cooktop installed on the same branch circuit? The oven and cooktop are located in the same room.

Table 220.55, Note 4: 6 kW + 8 kW = 14 kW
Treat as one range

Table 220.55, Note 1: 14 kW – 12 kW = 2 kW
2 kW × 5% = 10%

Table 220.55, Col C: One range = 8 kW
8 kW × 110% = 8800 VA

220.61(B)(1): 8800 VA × 70% = 6160 VA

Branch-Circuit Load: Line = **8800 VA**
Neutral = **6160 VA**

4. COOKING EQUIPMENT: *Table 220.55; Notes*

					GENERAL LOAD	NEUTRAL LOAD
Col A	____ VA ×	____ % =	____ VA			
Col B	____ VA ×	____ % =	____ VA			
Col C	8000 VA ×	110 % =	8800 VA			
Total			8800 VA		8800 VA × 70% =	6160 VA

Figure 3-12. Per Note 4, cooking appliance ratings are added together and treated like a single range when located in the same room.

Table 220.55, Note 5. This note allows the demand factors found in Table 220.55 to be applied to household cooking appliances rated over 1¾ kW where they are used in an instructional program. An example is a high school classroom used to teach cooking to students. If the ranges used are typical household cooking appliances, the demand factors found in Table 220.55 may be applied. The demand factors of Table 220.55 would not, however, be applied to that portion of the load that consists of commercial electric cooking appliances used in the cafeteria kitchen.

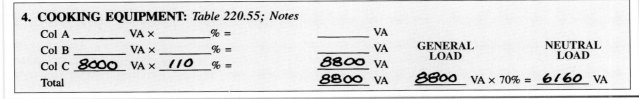

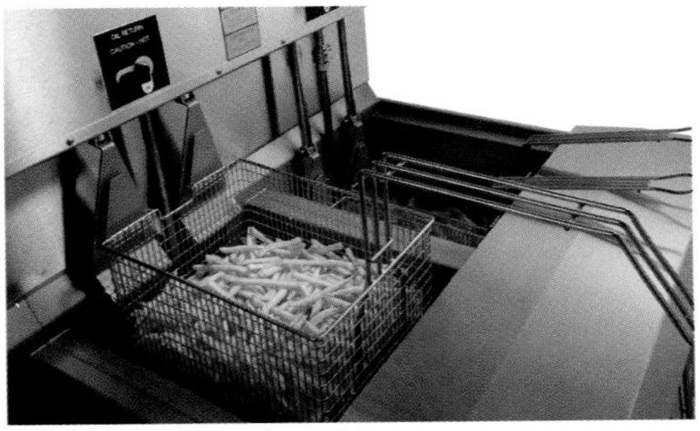

Henny Penny Corporation

The load for kitchen equipment, in other than dwelling units, shall be calculated in accordance with 220.56.

One-Family Dwelling Heating or A/C – 220.60

Heating or A/C loads often represent the largest loads in one-family dwellings. Fortunately, these are noncoincidental loads. A *noncoincidental load* is a load that is not on at the same time as another load. Section 220.60 permits the smaller of the two noncoincidental loads to be omitted from the load calculation, provided it is unlikely that the loads will be in use simultaneously. Heating and A/C loads are both calculated at 100% of the nameplate rating, and the one with the largest load is used in determining the total feeder or service demand. **See Figure 3-13.**

One-Family Dwelling Largest Motor – 220.50

Section 220.50 refers to the following four code sections for determining motor loads:

- 430.24, Several Motors or a Motor(s) and Other Loads
- 430.25, Multimotor and Combination-Load Equipment
- 430.26, Feeder Demand Factor
- 440.6, Hermetic Refrigerant Motor Compressors

Heating or A/C—220.60

What is the total heating and A/C load (line and neutral) for a one-family dwelling unit with a 240 V central heating load of 16,000 VA and an A/C load of 12,000 VA?

220.60: Largest load = 16,000 VA

Heating and A/C Load: Line = **16,000 VA**

Neutral = **0 VA**

5. HEATING or A/C: ***220.60***

			GENERAL LOAD	NEUTRAL LOAD
Heating unit = 16,000 VA × 100% =	16,000 VA			
A/C unit = 12,000 VA × 100% =	— VA			
Heat pump = ______ VA × 100% =	______ VA			
Largest Load	16,000 VA		16,000 VA	______ VA

Figure 3-13. The largest of the heating or A/C load is used.

Section 430.24 requires conductors supplying several motors to have an ampacity based on the full load table currents of all motors supplied plus 25% of the highest-rated motor. This must include 100% of the noncontinuous non-motor load plus 125% of the continuous non-motor load. Sections 430.25 and 430.26 apply to special motor applications that are not generally found in dwelling-unit applications. Section 440.6 covers hermetic refrigerant motor compressors. It does not cover the blower motors found in air handling units of split air conditioning units, which are covered by Article 430.

Motor loads used in dwelling units, other than air conditioning compressors are included in the fixed appliance calculation covered by 220.53. The 25% of the largest motor is a separate calculation that is not included in the fixed appliance calculation. **See Figure 3-14.**

One-Family Dwelling Demand Factors – Neutral – 220.61

The *feeder neutral load* is the maximum unbalance between any of the ungrounded conductors and the grounded conductor. Section 220.60(B) permits an additional demand factor of 70% for a feeder or a service supplying household electric ranges, wall-mounted ovens, counter-mounted cooking units, and electric dryers. A further demand factor of 70% is permitted for that portion of the unbalanced load on a feeder or service in excess of 200 A. The use of this demand is restricted to feeders and services supplied from one of the following systems:

- 3-wire DC or 1ϕ AC system
- 4-wire, 3ϕ system
- 3-wire, 2ϕ system
- 5-wire, 2ϕ system

Determining Largest Motor Load

It is good practice to check with the AHJ prior to sizing service-entrance or feeder conductors for a new dwelling. Different areas may interpret differently or require calculations to be performed in a specific manner. This is often the case in determining the largest motor load. Some areas use all of the connected loads in determining the largest motor load. In these cases, the A/C compressor motor is frequently the largest motor load. In other areas, the A/C load is not considered a motor load. And still in other areas, the A/C load only counts if it is not used when determining the largest of the heating or A/C loads per 220.60. In any event, it is good practice to know how the AHJ determines the largest motor load.

Another type of motor load that often requires the help of the AHJ is one that is included in an appliance such as a dishwasher. The nameplate rating of the dishwasher includes the motor, the heater, and the controls. In most cases, the dishwasher is not considered to be a motor load, but it may be a good idea to verify this.

Additional Calculation for Largest Motor—220.50, 430.24

What is the largest motor load (line and neutral) for a one-family dwelling with the following motor loads:
Blower motor – 660 VA, 120 V
Attic fan – 1300 VA, 120 V
Vent motor – 400 VA, 120 V

220.50: 1300 VA × 25% = 325 VA

Largest Motor Load: Line = **325 VA**
Neutral = **325 VA**

6. LARGEST MOTOR: *220.50*		LINE	NEUTRAL
ϕ 1300 VA × 25% =	325 VA	325 VA	
N 1300 VA × 25% =	325 VA		325 VA

Figure 3-14. The largest motor load is multiplied by 25%.

This 70% reduction does not apply to portions of the load that consist of nonlinear loads that are supplied by a 3ϕ, 4-wire, wye-connected system or any portion of a 3-wire circuit consisting of two phase conductors and the neutral conductor from a 3ϕ, 4-wire, wye-connected system. A nonlinear load is a load where the current wave shape does not follow the applied voltage wave shape. **See Figure 3-15.** Demand factors may be applied to the ungrounded conductors supplying nonlinear loads, but the load for the neutral conductors must be calculated at a minimum of 100% and demand factors are not permitted to be used. Designers and installers must be careful when derating the load on any grounded or neutral conductor. These conductors do not normally have an overcurrent protective device that opens the circuit in the case of an overload.

One-Family Dwelling Conductor Ampacity

After the six loads (General Lighting, Fixed Appliances, Dryer, Cooking Equipment, Heating or A/C, and Largest Motor) are individually calculated, the results of the individual calculations are added to determine the maximum load, in VA, for both the ungrounded and grounded conductors. This total VA value is used to determine the minimum ampacity for the ungrounded and grounded conductors.

For 1ϕ systems, the formula $I = \frac{VA}{V}$ is used to determine minimum ampacity. In this formula, I = amperes (A), VA = volt amperes, and V = voltage. After the minimum ampacity has been determined, Table 310.15(B)(16) and the applicable notes are used to calculate the minimum conductor size for the feeder or service-entrance conductors.

Ungrounded Conductors. The ungrounded phase conductors are not subject to any additional demand factors. Once the phase amperes are calculated by using the formula, the size of the phase conductors is determined. Table 310.15(B)(16) is used to size the ungrounded conductors. In addition, 310.15(B)(7) is applicable to 1ϕ service conductors and feeders for dwelling units that are supplied from 120/240 V, 1ϕ, 3-wire systems. Table 310.15(B)(7) permits reduced conductor sizes for service or feeder ratings from 100 A to 400 A. **See Figure 3-16.**

Grounded Conductors – 220.61. An additional demand of 70% may be applied to that portion of feeder or service neutral conductor loads that exceed 200 A. This only applies to 3-wire DC, 1ϕ AC, 3ϕ 4-wire, and 2ϕ systems that supply other than nonlinear loads. Once the total neutral load is calculated using $I = \frac{VA}{V}$, the demand factor is applied. The first 200 A is taken at 100% and the remainder is taken at 70%.

Demand Factors—Neutral—220.61

What is the neutral load for a one-family dwelling unit with total connected line load of 63.5 kVA of which the neutral load is 52kVA? The service is 120/240 V, 1 ϕ, 3-wire.

Phases: $I = \frac{VA}{V} = \frac{63{,}500}{240} = 264.5\text{ A}$

Neutral: $I = \frac{VA}{V} = \frac{52{,}000}{240} = 216.67\text{ A}$

220.61(B)(2): First 200 A × 100% = 200 A
Remaining 16.67 A × 70% = 11.67 A
200 A + 11.67 A = 211.67 A

Neutral Load = **211.67 A**

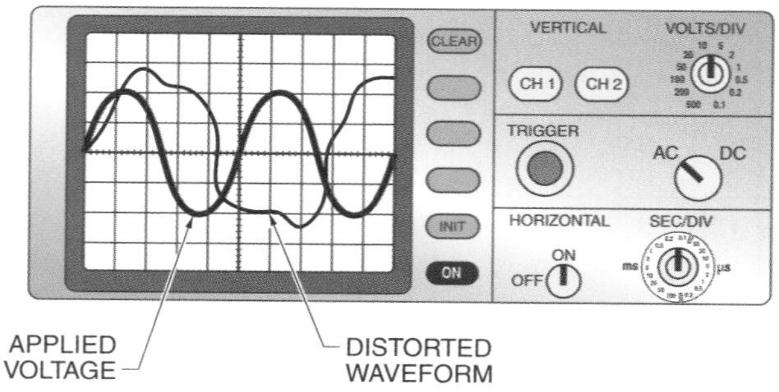

NONLINEAR LOAD • ARTICLE 100

A LOAD WHERE THE CURRENT WAVE SHAPE DOES NOT FOLLOW THE APPLIED VOLTAGE SHAPE

1ϕ service: PHASES $I = \frac{63{,}500\text{ VA}}{240\text{ V}} = 264.5\text{ A}$ — 63,500 VA

NEUTRAL $I = \frac{52{,}000\text{ VA}}{240\text{ V}} = 216.67\text{ A}$ — 52,000 VA

220.61(B)(2); First 200 A × 100% = 200 A
Remaining 16.67 A × 70% = 11.67 A
Total 211.67 A

Figure 3-15. Demand factors are applied to the neutral load.

Sizing Ungrounded Conductors—310.15(B)(7); Table 310.15(B)(7)

What is the minimum size THW, Cu conductors required for a one-family dwelling with a calculated load of 200 A per phase? The service is 120/240 V, 1ϕ, 3-wire.

Table 310.15(B)(16): 200 A = 3/0 AWG
Table 310.15(B)(7): 200 A = 2/0 AWG
Conductors: **2/0 AWG THW Cu**

Figure 3-16. Section 310.15(B)(6) permits reduced conductor sizes for service or feeder ratings from 100 A to 400 A.

The minimum size of the neutral conductor is now selected from Table 310.15(B)(16). Section 310.15(B)(7) permits the grounded conductor (neutral) to be smaller than the ungrounded conductors provided the requirements of 215.2, 220.61, and 230.42 are met. Section 250.24(C)(1) does not permit the grounded conductor to be smaller than the grounding electrode conductor required by Table 250.66. **See Figure 3-17.**

Sizing Grounded Conductors—220.61; Table 310.15(B)(7)

What is the minimum size THW, Cu conductor required for the grounded conductor for a one-family dwelling with a total calculated neutral load of 225 A? The service is 120/240 V, 1ϕ, 3-wire.

220.61: 200 A × 100% = 200 A
25 A × 70% = 17.5 A
200 A + 17.5 A = 217.5 A

Table 310.15(B)(16): 217.5 A = 4/0 AWG [Per Table 310.15(B)(16)]
Grounded Conductor: **3/0 AWG THW Cu**

Figure 3-17. An additional demand factor of 70% is applied to the grounded conductor.

ONE-FAMILY DWELLING – OPTIONAL CALCULATION – ARTICLE 220, PART IV

Part IV of Article 220 provides five Optional Calculations. These calculations provide an alternate method to calculate the loads for one-family dwellings, multifamily dwellings, two-dwelling units, schools, and new restaurants. Optional calculations are, as a rule, easier than calculations made using the Standard Calculation. The Optional Calculation for one-family dwellings is restricted to dwelling units that have the total connected load served by a single 120/240 V or 208Y/120 V set of 3-wire service or feeder conductors that have an ampacity of 100 A or more.

The grounded conductor calculation for the Optional Calculation is permitted by 220.82(A) to be determined by 220.61. There are two methods to accomplish this. The neutral load could be the same as the load on the ungrounded conductors and the 70% demand (per 220.61) applied to that portion of the load in excess of 200 A. This generally does not provide the smallest neutral permitted by Article 220. The second method would be to calculate the neutral load using the rules found in Part III, the Standard Calculation, and then apply the applicable demands found in 220.61. The choice is up to the designer and in many cases the designer chooses to make the grounded conductor the same size as the ungrounded conductors rather than provide a second calculation.

The Optional Calculation for dwelling units consists of two groups of loads. These two groups of loads and their NEC® references are:

1. General Loads – *220.82(B)(1–4)*
2. Heating and A/C Loads – *220.82(C)(1–6)*

One-Family Dwelling Heating and A/C – 220.82(C)(1–6)

Section 220.82(C) lists six heating and A/C load selections which are considered in the Optional Calculation for one-family dwellings. The largest of these selections shall be used in determining the total feeder or service load. **See Figure 3-18.** These six loads are:

(1) A/C and cooling equipment at 100% of nameplate ratings
(2) A/C equipment including heat pump compressor (at 100% of nameplate ratings, no supplemental electric heating)
(3) Central electric space-heating equipment, including integral supplemental heat for heat pumps (at 100% of heat pump compressor and 65% of supplemental electric heating)
(4) Less than four separately controlled electric space-heating units (at 65% of nameplate ratings)
(5) Four or more separately controlled electric space-heating units (at 40% of nameplate ratings) considered
(6) Electric thermal storage loads (at 100% of nameplate ratings) where the usual load is expected to be continuous at the full nameplate value

Heating and A/C—220.82(C)(1–6)

What is the heating and A/C load for a one-family dwelling with a 14 kW central electric space-heating load and a 10 kW A/C load? The service is 120/240 V, 1ϕ, 3-wire with a calculated ampacity of 150 A.

220.82(C): 10,000 VA × 100% = 10,000 VA

14,000 VA × 65% = 9100 VA

Heating or A/C Load = **10,000 VA**

1. HEATING and A/C: ***220.82(C)(1–6)***

Heating units (3 or less) = 14,000 VA × 65% =	9100 VA		
Heating units (4 or more) = ______ VA × 40% =	______ VA		
A/C unit = 10,000 VA × 100% =	10,000 VA		
Heat pump = ______ VA × 100% =	______ VA	**PHASES**	
Largest Load	10,000 VA		
Total	10,000 VA	10,000	VA ÷ 240 = 42 A per phase

Figure 3-18. The largest of six load selections is the "Heating and A/C Load" for the Optional Calculation.

One-Family Dwelling General Loads – 220.82(B)(1–4)

Per Section 220.82(B), General Calculations shall be 100% of the first 10 kVA of all other loads, plus 40% of the remainder of all other loads shall be included. Four loads shall be added together to constitute the "general loads" portion of the Optional Calculation. The other loads include: (1) 3 VA per square foot for the general lighting and general-use receptacle load (as with the Standard Calculation, the floor area is calculated using the outside dimensions of the dwelling, not including open porches, garages, or unfinished space), (2) 1500 VA for each 2-wire, 20 A small appliance and laundry branch circuit, (3) the nameplate rating of all fastened-in-place, permanently connected appliances (ranges, wall-mounted ovens, water heaters, clothes dryers, counter-mounted cooking units), and (4) the nameplate rating, in either amperes or kVA, of all motors not otherwise included in the other loads of (3). **See Figure 3-19.**

Short-Circuit Current Calculations

Miscalculation or ignorance of the magnitude of the possible short-circuit current in a system can result in the installation of breakers or fuses with an inadequate interrupting rating. An inadequate interrupting rating on a breaker or fuse can cause it to explode when subjected to a short circuit, which can cause injury to personnel or damage to property. For additional safety information on the importance of the proper design and operation of electrical equipment and systems, see NFPA 70E®—Electrical Safety in the Workplace 2009 Edition.

MULTIFAMILY DWELLINGS – STANDARD CALCULATION

The Standard Calculation for multifamily dwellings uses the same six loads as used by the Standard Calculation for one-family dwellings. The only difference is that for each of the individual loads, the calculated load is multiplied by the total number of units in the multifamily dwelling to arrive at the total connected load. As with the one-family dwelling calculation, demand factors can be applied.

Multifamily Dwelling General Lighting Load – Table 220.12

The general lighting load for multifamily dwellings consists of the same three loads which were used in the one-family dwelling standard calculation. The first portion of the general lighting load is calculated from the total square footage of the dwelling. Table 220.12 lists the unit load per square foot for dwelling units at 3 VA. For this portion of the lighting load, the total square footage of the dwelling, excluding open porches, garages, and unused or unfinished spaces, is multiplied by 3 VA to determine the calculated load for one dwelling. This value is then multiplied by the number of dwelling units in the multifamily dwelling to determine the first portion of the general lighting load.

Optional Calculation: General Loads—220.82(B)(1–4)

What is the total "general loads" for a 2500 sq ft one-family dwelling with the following loads:

2 – Small appliance 20 A branch circuits and 1 laundry 20 A branch circuit
1 – 8 kW electric range
1 – 2 kW dishwasher
1 – 5 kW clothes dryer
1 – 1500 VA blower motor

220.82(B):	(1) General lighting load		
	2500 sq ft × 3 VA	=	7500 VA
	(2) Small appliance		
	and laundry load	=	4500 VA
	(3) Fixed appliance load	=	15,000 VA
	(4) Motor load	=	1500 VA
	Total load	=	28,500 VA
220.82(B):	10,000 VA × 100%	=	10,000 VA
	18,500 VA × 40%	=	7400 VA
	Total	=	17,400 VA

General Loads = **17,400 VA**

2. GENERAL LOADS: *220.82(B)*

General lighting: *220.82(B)(1)*
2500 sq ft × 3 VA — 7500 VA

Small appliance and laundry loads: *220.82(B)(2)*
1500 VA × 3 circuits = — 4500 VA

Special loads: *220.82(B)(3)(4)*

Dishwasher =	2000	VA
Disposer =		VA
Compactor =		VA
Water heater =		VA
Range =	8000	VA
Dryer =	5000	VA
Blower motor =	1500	VA
=		VA
=		VA
	16,500	VA

16,500 VA
Total 28,500 VA

Applying Demand Factors: *220.82(B)*

First 10,000 VA × 100% =	10,000 VA	
Remaining 18,500 VA × 40% =	7400 VA	
Total	17,400 VA	17,400 VA

Figure 3-19. Four loads constitute the "General Loads" for the Optional Calculation.

For multifamily dwellings in which the square footage of the individual units is different, the square footage of each unit is calculated separately and added together to get the total square footage of the multifamily dwelling. This value multiplied by 3 VA provides the first portion of the total general lighting load. Section 220.14(J) indicates that general-use receptacles in multifamily as well as one-family dwellings are considered outlets for general illumination and are included in the 3 VA per square foot calculation. **See Figure 3-20.**

Optional Calculation: General Lighting—Table 220.12

What is the first portion of the general lighting load for a multifamily dwelling consisting of three individual dwelling units with the following total areas:

Unit No. 1 – 1050 sq ft
Unit No. 2 – 950 sq ft
Unit No. 3 – 1075 sq ft

Table 220.82(B)(1):	1050 sq ft × 3 VA	=	3150 VA
	950 sq ft × 3 VA	=	2850 VA
	1075 sq ft × 3 VA	=	3225 VA
	Total	=	9225 VA

Lighting Load = **9225 VA**

1. GENERAL LIGHTING: ***Table 220.12***

1050 sq ft × 3 VA × 1 units = 3150 VA
950 sq ft × 3 VA × 1 units = 2850 VA
1075 sq ft × 3 VA × 1 units = 3225 VA

Small appliances: *220.52(A)*
_____ VA × _____ circuits × _____ units = _____ VA

Laundry: *220.52(B)*
_____ VA × 1 × _____ units = _____ VA
_____ VA

Applying Demand Factors: *Table 220.42*
First 3000 VA × 100% = 3000 VA
Next _____ VA × 35% = _____ VA
Remaining _____ VA × 25% = _____ VA
Total _____ VA

PHASES	NEUTRAL
_____ VA	_____ VA

Figure 3-20. The general lighting load for an occupancy is calculated by multiplying the area (in square feet) by the unit load per square foot.

Small Appliances – 220.52. As with one-family dwellings, a minimum of two small appliance branch circuits are required by 210.52(B) and 210.11(C)(1) for each multifamily dwelling unit. Per 220.52(A), each small appliance branch circuit shall be calculated at 1500 VA when determining the feeder load. If the dwelling has more than two small appliance branch circuits, each circuit is calculated at 1500 VA.

For multifamily dwellings, the small appliance load is calculated by multiplying 1500 VA times the number of small appliance branch circuits times the number of dwelling units in the multifamily dwelling. The small appliance load is included with the general lighting load and is subject to the demand factors of Table 220.42.

Laundry Load – 220.52(B). The last portion of the general lighting load for multifamily dwellings consists of the laundry circuit load. Per 210.52(F) and 210.11(C)(2), each dwelling shall be provided with a 20 A branch circuit to supply laundry receptacle outlet(s). There are, however, two exceptions which are applicable to multifamily dwellings. Per 210.52(F), Ex. 1, a laundry receptacle may be omitted from dwellings which are part of a multifamily dwelling with separate laundry facilities accessible to all building occupants. Per 210.52(F), Ex. 2, the laundry receptacle may be omitted in other than one-family dwellings, provided laundry facilities are not installed or permitted. In either of these cases, the 1500 VA for the laundry circuit need not be included in the multifamily standard calculation. If, however, each dwelling unit is to have laundry facility provisions, the load shall be added to the total connected load of the multifamily dwelling. This is accomplished by computing 1500 VA for each dwelling unit multiplied by the number of dwelling units in the multifamily dwelling. As with the one-family Standard Calculation, the laundry load is included with the general lighting load and is subject to the demand factors of Table 220.42.

Lighting Load – Demand Factors – Table 220.42. The demand factors of Table 220.42 are applied to the general lighting load of multifamily dwelling units. The first 3000 VA of the lighting load is calculated at 100%. The portion of the general lighting load from 3001 VA to 120,000 VA is calculated at 35%, and the remainder over 120,000 VA is calculated at 25%. These demand factors cannot be applied when determining the number of branch circuits for general illumination. **See Figure 3-21.**

The general lighting load for an office building is calculated using a unit load of 3½ VA per square foot. See Table 220.12.

General Lighting—Table 220.42

What is the general lighting load for the line and neutral of a five-unit multifamily dwelling with one 20 A laundry branch circuit and two 20 A small appliance circuits in each unit, with 1500 sq ft per unit?

Table 220.12: 1500 sq ft × 3 VA x 5 units = 22,500 VA
220.52(A): 1500 VA × 2 circuits × 5 units = 15,000 VA
220.52(B): 1500 VA × 1 circuit × 5 units = 7500 VA
Total connected load = 45,000 VA
Table 220.42: 3000 VA × 100% = 3000 VA
42,000 VA × 35% = 14,700 VA
Total = 17,700 VA
General Lighting Load = **17,700 VA**

1. GENERAL LIGHTING: ***Table 220.12***
1500 sq ft × 3 VA × 5 units = 22,500 VA
____ sq ft × 3 VA × ____ units = ____ VA
____ sq ft × 3 VA × ____ units = ____ VA
Small appliances: *220.52(A)*
1500 VA × 2 circuits × 5 units = 15,000 VA
Laundry: *220.52(B)*
1500 VA × 1 × 5 units = 7500 VA
45,000 VA

Applying Demand Factors: *Table 220.42*

		CALCULATED LOAD	CALCULATED NEUTRAL LOAD
First 3000 VA × 100% =	3000 VA		
Next 42,000 VA × 35% =	14,700 VA		
Remaining ____ VA × 25% =	____ VA		
Total	17,700 VA	17,700 VA	17,700 VA

Figure 3-21. Demand factors of Table 220.42 are applied to the general lighting load of multifamily dwellings.

Multifamily Dwelling Fixed Appliances – 220.53

Each fixed appliance in multifamily dwellings is calculated using the nameplate rating. For four or more appliances, the nameplate ratings are added together and multiplied by the 75% demand factor specified in 220.53. As with one-family dwellings, this demand factor applies to the nameplate rating of the appliance but does not include electric ranges, clothes dryers, space-heating equipment, or A/C equipment served by the same feeder or service. **See Figure 3-22.**

Multifamily Dwelling Dryer – 220.54; Table 220.54

The total load for household electric clothes dryers is calculated at 5000 VA or the nameplate rating, whichever results in the greater value. Table 220.54 lists demand factors that are applied to clothes dryers. The first four clothes dryers are calculated at 100%. Subsequent numbers of dryers are calculated with decreasing demand factor percentages. **See Figure 3-23.**

Multifamily Dwelling Cooking Equipment – Table 220.55, Notes

Household electric ranges rated in excess of $1\frac{3}{4}$ kW shall be calculated per Table 220.55. Column C values are used less frequently for multifamily dwellings since the size of the individual ranges is generally smaller in these occupancies than in one-family dwellings.

Note 3 applies when individual ranges rated over $1\frac{3}{4}$ kW through $8\frac{3}{4}$ kW are used. While Column C could be used for these ranges, Column A or B provides smaller kW demands. Note 3 permits the nameplate ratings of all household cooking appliances rated between $1\frac{3}{4}$ kW and $8\frac{3}{4}$ kW to be added together and multiplied by the appropriate demand from either Column A or B. Note 3 is commonly used in multifamily dwellings with the smaller individual cooking units. As with the one-family calculations, if the cooking appliances are of different ratings that fall under both Columns A and B, the demand factors for each column shall be applied to the appliance for that column and then all of the results added together. **See Figure 3-24.**

Fixed Appliances—220.53

What is the fixed appliance load for an eight-unit multifamily dwelling with the following loads in each of the 1200 sq ft units?

Trash compactor – 1000 VA
Dishwasher – 900 VA
240 V water heater – 4500 VA
Blower motor – 750 VA

220.53: 1000 VA + 900 VA + 4500 VA + 750 VA = 7150 VA
7150 VA × 8 units × 75% = 42,900 VA
1000 VA + 900 VA + 750 VA = 2650 VA
2650 VA × 8 units × 75% = 15,900 VA

Fixed Appliance Load: Line = **42,900 VA**
Neutral = **15,900 VA**

2. FIXED APPLIANCES: *220.53*

Dishwasher =	900 VA			
Disposer =	____ VA			
Compactor =	1000 VA			
Water heater =	4500 VA			
Blower motor =	750 VA			
____ =	____ VA			
____ =	____ VA			(120 V Loads × units × 75%)
Total	7150 VA × 8 units × 75% =	42,900 VA	42,900 VA	15,900 VA

Figure 3-22. A 75% demand factor is applied to four or more fixed appliance loads in multifamily dwellings.

Dryer—220.54

What is the household electric clothes dryer load for a 12-unit multifamily dwelling with a 5.5 kW dryer in each unit?

220.54: 5500 VA × 12 units = 66,000 VA

Table 220.54: 12 dryers = 47% – [number of dryers – 11]

220.61(B)(1): 66,000 VA × 47% – (12 – 11) = 47% – 1 = 46% = 30,360

30,360 VA × 70% = 21,252 VA

Clothes Dryer Load: Line = 30,360 VA

Neutral = 21,252 VA

3. DRYER: *220.54; Table 220.54*

5500 VA × 12 units × 46 % = 30,360 VA   30,360 VA × 70% = 21,252 VA

Figure 3-23. Demand factors of Table 220.54 are applied to clothes dryers in multifamily dwellings.

Cooking Equipment—220.55

What is the cooking equipment load for a six-unit multifamily dwelling with a 3.5 kW household range in each of the dwelling units?

Table 220.55, Col B: 6 appliances = 43%

3.5 kW × 6 = 21 kW

220.61(B)(1): 21,000 VA × 43% = 9030 VA

9030 VA × 70% = 6321 VA

Cooking Equipment Load: Line = **9030 VA**

Neutral = **6321 VA**

4. COOKING EQUIPMENT: *Table 220.55; Notes*

Col A ____ units =	____ VA ×	____ % =	____ VA		**GENERAL LOAD**	**NEUTRAL LOAD**
Col B 3500 VA ×	6 units ×	43 % =	9030 VA			
Col C ____ VA ×	____ units ×	____ % =	____ VA			
Total			9030 VA		9030 VA × 70% =	6321 VA

Figure 3-24. Household ranges over 1¾ kW are calculated per Table 220.55.

Note 4, which applies to branch-circuit loads, can also be used frequently with multifamily dwelling units. Note 4 permits the load to be calculated per Table 220.55. For single wall-mounted ovens or single counter-mounted cooking units, the branch-circuit load shall be the nameplate rating of the appliance.

For installations from a common branch circuit in which a single counter-mounted cooking unit is installed with not more than two wall-mounted ovens, Note 4 permits all of the cooking appliances to be added together and treated like a single range, provided all of the cooking appliances are located in the same room.

Multifamily Dwelling Heating or A/C – 220.60

As with one-family dwellings, the heating or A/C load for multifamily dwellings is often the largest load in the multifamily dwelling calculation. Section 220.60 permits the smaller of the two noncoincidental load calculations to be omitted from the multifamily dwelling load calculation, provided it is unlikely that the loads will be in use simultaneously.

Heating and A/C loads are both calculated at 100% of the nameplate rating. These loads are multiplied by the number of units that contain the loads. The calculation with the largest load is used in determining the total feeder or service demand for the multifamily dwelling. If the individual units contain heating or A/C loads of different values, they are added together and the total loads are compared. The largest load is used in the multifamily dwelling calculation. **See Figure 3-25.**

Carrier Corporation

Heating and A/C loads are often the largest loads in multifamily dwellings.

Multifamily Dwelling Largest Motor – 220.50

Despite the fact that each unit in a multifamily dwelling may have the same motor load, 220.50 requires, by its reference to 430.24, that only 25% of the largest motor in the entire multifamily dwelling be added to the standard calculation for multifamily dwellings. Feeders for several motors are calculated per 430.24 using the sum of the full-load table currents of all motors plus 25% of the largest motor. A motor's full-load current rating is calculated under the fixed appliance portion of the calculation at 100% per 220.53. The largest motor portion of the calculation allows for the 25% that is not covered under 220.53.

Multifamily Dwelling Demand Factors – Neutral – 220.61

Section 220.61, which permits several appliances to have the neutral demand calculated at 70% of the ungrounded conductors, is applicable to multifamily dwellings as well as to one-family dwellings. Likewise, for multifamily dwellings, the deduction of 70% is permitted for that portion of the neutral conductor demand in excess of 200 A. This reduction only applies to 3-wire DC, 1ϕ AC, 3ϕ 4-wire, and 2ϕ systems. The 70% reduction does not apply to portions of the load that consist of nonlinear loads when the supply system is 3ϕ, 4-wire, wye-connected or 3ϕ, 4-wire, wye-connected utilizing two phase wires and the neutral.

Heating or A/C—220.60

What is the heating or A/C load for a nine-unit multifamily dwelling which contains a 6 kW, 240 V A/C and an 8 kW, 240 V heating unit in each of the dwelling units?

Total Heating Load: 8 kW × 9 units = 72 kW

Total A/C Load = 6 kW × 9 units = 54 kW

220.60: Largest load = 72 kW

Heating or A/C Load: = **72,000 VA**

5. HEATING or A/C: ***220.60***

		GENERAL LOAD	NEUTRAL LOAD
Heating unit = 8000 VA × 100% × 9 units =	72,000 VA		
A/C unit = 6000 VA × 100% × 9 units =	54,000 VA		
Heat pump = _____ VA × 100% × _____ units =	_____ VA		
Largest Load	72,000 VA	72,000 VA	– VA

Figure 3-25. The largest of the heating or A/C load is used per 220.60.

Multifamily Dwelling Conductor Ampacity

After the six loads (General Lighting, Cooking Equipment, Dryer, Fixed Appliances, Heating or A/C, and Largest Motor) are individually calculated, the results of the individual calculations are added to determine the maximum load, in VA, for both the ungrounded and grounded conductors. This total VA value is used to determine the minimum ampacity for the ungrounded and grounded conductors.

For 1ϕ systems, the formula $I = \frac{VA}{V}$ is used to determine minimum ampacity. In this formula, I = amperes (A), VA = volt amperes, and V = voltage. After the minimum ampacity has been determined, 310.15 and Table 310.15(B)(16) are used to calculate the minimum conductor size for the feeder or service-entrance conductors.

The ungrounded phase conductors are not subject to any additional demand factors. Once the phase amperes are calculated by using the formula, the size of the phase conductors is determined. Table 310.15(B)(16) can always be used to size the ungrounded conductors. In addition, 310.15(B)(7) is applicable to the feeders or service-entrance conductors for the individual dwelling units of the multifamily dwelling, but not to the service-entrance conductors that supply the entire building. If the individual feeders or service-entrance conductors are 120/240 V, 1ϕ, 3-wire, they may be sized per Table 310.15(B)(7). The service-entrance conductors for the entire building are sized per Table 310.15(B)(16) because 310.15(B)(7) only applies to dwelling units.

Grounded Conductors. Section 220.61 permits an additional demand factor of 70% to be applied to the feeder or service-entrance grounded conductor for that portion in excess of 200 A for multifamily dwellings. This only applies to 3-wire DC, 1ϕ AC, 3ϕ 4-wire, and two-phase systems that supply other than nonlinear loads. Once the total neutral load is calculated using $I = \frac{VA}{V}$, the demand factor can be applied. The first 200 A is taken at 100% and the remainder is taken at 70%. The minimum size of the neutral conductor is selected from Table 310.15(B)(16). Section 310.15(B)(7) permits the grounded conductor (neutral) to be smaller than the ungrounded conductors provided the requirements of 215.2, 220.61, and 230.42 are met. The grounded conductor is not permitted by 250.30(A)(3)(a) to be smaller than the grounding electrode conductor required by Table 250.66.

MULTIFAMILY DWELLING – OPTIONAL CALCULATION – 220.84

In addition to the Standard Calculation, there is a multifamily dwelling Optional Calculation that can be performed if three conditions are present. First, 220.84(A)(1) requires that no dwelling within the multifamily structure be supplied by more than one feeder. The second condition is that each dwelling shall be equipped with electric cooking equipment. If the individual dwelling units have natural gas cooking equipment, the Optional Calculation can not be used. The last condition is that each of the individual dwelling units shall have either electric space-heating equipment, A/C, or both loads present. See 220.84(A)(2)(3).

If all three of these conditions are present, the Optional Calculation for multifamily dwellings permits the demand factors of Table 220.84 to be applied to the total calculated load. The neutral load for the dwelling unit feeders or the service-entrance conductors is permitted to be reduced using the applicable demands of 220.61. In order to apply the appliance demand factors, the neutral load must be recalculated using the Standard Calculation method of Part III.

For dwelling units that qualify for the Optional Calculation, two loads are considered before applying the demand factors of Table 220.84. These are the house loads and the connected loads. A *house load* is an electrical load which is metered separately and supplies common usage areas. For example, typical house loads are hallway and perimeter lighting. In addition, 210.25(B) requires that branch circuits for central alarm, signal, communications, or other needs for public or common areas of multifamily types of occupancies be supplied from equipment that does not supply individual dwelling units or other tenant space. This is to ensure that a single tenant of a multifamily occupancy does not disconnect or de-energize one of these common area branch circuits and create a life safety concern for all of the occupants. The house load is considered to be a commercial load and not a dwelling load. House lighting loads are generally considered to be continuous since they will operate for 3 hr or more continuously. As a result, they will be included in the load calculation at 125%.

The second consideration is calculated loads. Section 220.84(C) lists the five calculated loads that are added together for calculating the total connected load to which the demand factors of Table 220.84 are applied.

Multifamily dwellings typically include individual electrical meters for each unit.

QUICK QUIZ®

Refer to Chapter 3 Quick Quiz® on CD-ROM.

Additional information is available at ATPeResources.com

The calculated loads include (1) 3 VA per square foot for the general lighting and general-use receptacle load, (2) 1500 VA for each 2-wire, 20 A small appliance branch circuit and each laundry circuit, (3) the nameplate rating of all fastened-in-place, permanently connected appliances (ranges, wall-mounted ovens, water heaters, clothes dryers, and counter-mounted cooking units), (4) the nameplate rating, in either amperes or VA, of all motors not included in (3), and (5) the larger of the electric space-heating load or the A/C load.

Once the connected loads are totaled, the demand factors in Table 220.84 are applied, providing a demand load for the feeder or service-entrance conductors. The total load is then calculated by adding the house load at 100% to the demand load. Section 220.84(B) indicates that house loads are required to be calculated by Part III of Article 220 and that they are in addition to the loads calculated using Table 220.84. House loads are included in the "All Others" section of Table 220.42 (Part III). The demand for all other loads is 100%.

For example, the total connected load for a 16-unit multifamily dwelling after adding the five loads listed in 220.84 is 200,000 VA, and the total house load is 10,000 VA. The connected demand load would be 78,000 VA (200,000 VA × 39% = 78,000 VA). The total load on the feeder or service-entrance conductors would be 78,000 VA + 10,000 VA = 88,000 VA. This value is then used to calculate the size of the ungrounded conductors to the multifamily dwelling. Section 220.61 may be used to reduce the load on the neutral conductor by using the Standard Calculation. **See Figure 3-26.**

A multifamily dwelling contains three or more dwelling units.

Optional Calculation: Multifamily Dwelling

What is the line and neutral demand load for a seven-unit multifamily dwelling with the following loads? Each unit is 975 sq ft in area and contains;

- **2 – Small appliance 20 A branch circuits**
- **1 – 12 kW electric range**
- **1 – 950 VA dishwasher**
- **1 – 6.5 kW A/C**
- **1 – laundry branch circuit**
- **1 – 8 kW central heating unit**
- **1 – 5000 VA water heater, 240 V**

1. HEATING or A/C: *220.84(C)(5)*

Heating unit = 8000 VA × 100% × 7 units = 56,000 VA

A/C unit = 6500 VA × 100% × 7 units = 45,000 VA

Heat pump = ____ VA × 100% × ____ units = ____ VA

Largest Load 56,000 VA

Total — **CALCULATED LOAD** 56,000 VA

2. GENERAL LOADS: *220.84(C)*

General lighting: *220.84(C)(1)*

975 sq ft × 3 VA × 7 units = 20,475 VA — 20,475 VA

____ sq ft × 3 VA × ____ units = ____ VA

____ sq ft × 3 VA × ____ units = ____ VA

____ sq ft × 3 VA × ____ units = ____ VA

Small appliance and laundry loads: *220.84(C)(2)*

1500 VA × 3 circuits × 7 units = 31,500 VA — 31,500 VA

Special loads: *220.84(C)(3)*

Dishwasher = 950 VA

Disposer = ____ VA

Compactor = ____ VA

Water heater = 5000 VA

Range = 12,000 VA

____ = ____ VA

____ = ____ VA

____ = ____ VA

____ = ____ VA

Total 17,950 VA × 7 units = 125,650 VA — 125,650 VA

Total Connected Load 233,625 VA

Applying Demand Factors: *Table 220.84*

233,625 VA × 44 % = 102,795 VA — 102,795 VA

NEUTRAL (Loads from Standard Calculation)

1. General lighting = 20,141 VA
2. Fixed appliances = 6650 VA
3. Dryer = — VA
4. Cooking equipment = 15,400 VA
5. Heating or A/C = — VA
6. Largest motor = — VA

Total 42,191 VA

1ϕ service: PHASES $I = \frac{VA}{V} =$ ____ A

3ϕ service: $I = \frac{VA}{V \times \sqrt{3}} =$ ____ A

NEUTRAL $I = \frac{VA}{V} =$ ____ A

$I = \frac{VA}{V \times \sqrt{3}} =$ ____ A

220.61; First 200 A × 100% = 200 A

Remaining ____ A × 70% = ____ A

____ A

Figure 3-26. Demand factors of Table 220.84 are used for multifamily dwellings in the Optional Calculation.

chapter 3

Dwelling Load Calculations

REVIEW QUESTIONS

Name ______________________________ Date ____________________

_______________ **1.** ___ is the amount of electricity required at a given time.

_______________ **2.** General lighting loads for dwelling units are calculated at ___ VA per square foot.

A. 3
B. 3½
C. 4
D. none of the above

_______________ **3.** Each single or multiple receptacle on a single strap is calculated at ___ VA.

_______________ **4.** Demand factors may be applied to ___ of the six loads of the Standard Calculation.

_______________ **5.** Four or more appliances that are fastened-in-place in a one-family dwelling are calculated at ___ of the total load for all four appliances (Standard Calculation).

_______________ **6.** The total load for household clothes dryers is calculated at ___ VA or the nameplate rating, whichever is higher (Standard Calculation).

_______________ **7.** ___ loads are loads that are not on at the same time.

_______________ **8.** The ___ load is the maximum unbalance between any of the ungrounded conductors and the grounded conductor.

_______________ **9.** For 1ϕ systems, the formula ___ is used to determine minimum ampacity of the ungrounded conductors.

A. $I = \frac{V}{VA}$
B. $I = V \times VA$
C. $I = \frac{VA}{V}$
D. none of the above

_______________ **10.** Each dwelling shall be supplied with a(n) ___ A branch circuit to supply laundry receptacle(s).

_______________ **11.** Heating or A/C loads are calculated at ___ of the nameplate rating and the largest load is used (Standard Calculation).

_______________ **12.** After the minimum ampacity of the service has been calculated, Table ___ and 310.15(B) are used to determine the size of feeder or service-entrance conductors.

_______________ **13.** The Optional Calculation for dwelling units consists of heating and A/C loads and ___ loads.

A. lighting
B. small appliance
C. largest motor
D. general

_______________ **14.** The demand load for household electric ranges rated in excess of ___ kW is calculated per Table 220.55 (Standard Calculation).

_______________ **15.** The ___ rating is the ampere rating or setting of the overcurrent device protecting the conductors.

_______________ **16.** The total area of an occupancy is determined from the ___ dimensions of the building or structure.

A. inside
B. outside
C. either A or B
D. none of the above

_______________ **17.** ___ branch circuits supply all receptacle outlets located in the kitchen, pantry, breakfast room, dining room, or similar areas of dwelling units (Standard Calculation).

_______________ **18.** Feeders for several motors are calculated using the sum of the FLC tables of all motors plus ___% of the largest motor (Standard Calculation).

A. 3
B. 15
C. 25
D. none of the above

_______________ **19.** Less than four separately controlled electric space-heating units are calculated at ___ % of their nameplate ratings (Optional Calculation).

T F **20.** The Optional Calculation can be used to find the total ampacity of a multifamily dwelling with natural gas cooking equipment.

chapter 3

Dwelling Load Calculations

TRADE COMPETENCY TEST

Name ______________________ **Date** ______________________

Calculation forms in the Appendix may be copied and used to solve the following problems.

______________ **1.** *Standard Calculation: One-Family Dwelling.* What is the general lighting and receptacle load (line and neutral) for a 2800 sq ft dwelling with two small appliance branch circuits and one laundry branch circuit?

______________ **2.** *Standard Calculation: One-Family Dwelling.* What is the fixed appliance load (line and neutral) for a dwelling with the following 120 V loads supplied from the same feeder?

Water heater – 5000 VA

Dishwasher – 1200 VA

Food waste disposer – 1000 VA

Trash compactor – 900 VA

Attic fan – 1200 VA

______________ **3.** *Standard Calculation: Multifamily Dwelling.* What is the total dryer load (line and neutral) for seven household dryers rated at 6.0 kW each?

______________ **4.** *Standard Calculation: One-Family Dwelling.* What is the range demand load (line and neutral) for one 17 kW household electric range?

______________ **5.** *Standard Calculation: One-Family Dwelling.* What is the range demand load (line and neutral) for three household electric ranges with nameplate ratings of 12 kW, 19 kW, and 20 kW?

______________ **6.** *Optional Calculation: One-Family Dwelling.* What is the total of the "general loads" for a 3000 sq ft dwelling with the following loads?

2 – Small appliance 20 A branch circuits

1 – Laundry 20 A branch circuit

1 – 8 kW electric range

1 – 1.2 kW dishwasher

1 – 5 kW clothes dryer

1 – 1800 VA blower motor

______________ **7.** *Standard Calculation: Multifamily Dwelling.* What is the general lighting load (line and neutral) for an eight-unit multifamily dwelling with one 20 A laundry branch circuit and three 20 A small appliance circuits in each 2000 sq ft unit?

______________ **8.** *Standard Calculation: Multifamily Dwelling.* What is the household electric clothes dryer load (line and neutral) for a 20-unit multifamily dwelling with a 6.0 kW dryer in each unit?

_______________ **9.** *Standard Calculation: One-Family Dwelling.* What is the total connected load (line and neutral), in VA, and the minimum service-entrance conductor ampacity (line and neutral) of a 120/240 V, 1ϕ, 3-wire service for a 2900 sq ft dwelling with the following loads?

General lighting and receptacle loads

Three small appliance loads

One laundry load

Range – 10,500 VA

Dryer – 4800 VA

A/C – 9500 VA

Central heating – 12,000 VA

Water heater (240 V) – 5000 VA

Dishwasher – 1200 VA

Trash compactor – 1000 VA

Garbage disposer – 900 VA

Blower motor – 1600 VA

_______________ **10.** *Optional Calculation: One-Family Dwelling.* Determine the total connected load (line and neutral), in VA, and the minimum service-entrance conductor ampacity (line and neutral) for the dwelling in Problem 9.

_______________ **11.** *Standard Calculation: Multifamily Dwelling.* What is the total connected load (line and neutral), in VA, and the minimum service-entrance conductor ampacity (line and neutral) of a 120/208 V, 3ϕ, 4-wire service which supplies six 1200 sq ft dwellings with the following loads?

General lighting and receptacle loads

Two small appliance loads

One laundry load

Range – 8000 VA

Dryer – 5000 VA

A/C – 6000 VA

Electric space heating – 8000 VA

Water heater (120 V) – 4000 VA

Dishwasher – 800 VA

Garbage disposer – 900 VA

Blower motor – 1000 VA

_______________ **12.** *Optional Calculation: Multifamily Dwelling.* Determine the total connected load (line and neutral), in VA, and the minimum service-entrance conductor ampacity (line and neutral) for the dwelling in Problem 11.

Learning the Code

- The heart of the electrical system is the service. The service consists of the conductors and equipment that are used to deliver the electrical energy from the electrical utility to the wiring system of the building or structure served. Chapter 4 covers the provisions for safely and correctly installing electrical services.

- A service to a building or a structure can range from a relatively simple 100 A service for a one-family dwelling, to a 3000 A service for a large-capacity industrial building. One helpful thought to remember is that a 3000 A service is made up of the basic elements as those for a 100 A service. Concentrate on applying and understanding the provisions for a 100 A service and the transition to larger and more complicated services will be much easier.
- Remember that, in addition to the Article 230 provisions, many local electrical utilities have their own specific requirements for installing electrical services. Students of the Code and apprentices need to obtain a copy of these provisions and become familiar with them before sitting for local Code tests or licensing examinations.
- The most difficult aspect of Article 230 is the arrangement of the article. Apprentices and students of the Code should become familiar with Figure 230.1, which pictorially represents the arrangement of Article 230. The most important thing to remember is that each part of the article applies to a specific aspect of the service. For example, Part II is for overhead service conductors, and Part III is for underground service conductors. As with all Code articles, make sure you are in the correct part of the article when referencing the Code.

Services

The service is the electrical supply, in the form of conductors and equipment, that provides electrical power to the building or structure. The service conductors for a building or structure are either overhead (drop) or underground (lateral). The local utility company brings electricity to the service point. All electrical connections, equipment, conductors, etc., beyond that point are controlled by the NEC®.

New in the 2011 NEC®

- *Additional condition for conductors considered outside of building – 230.6(5)*
- *New exception for overhead service conductor clearances above roofs – 230.24(A), Ex. 5*
- *Revised exception for number of sets of service-entrance conductors from each service drop or lateral – 230.40, Ex. 1*
- *New marking provisions for the use of cable trays with service-entrance conductors – 230.44*
- *New identification marking requirements for grouped service disconnects – 230.72 (A), Ex.*
- *Revised provisions for equipment connected to the supply side of a service disconnect – 230.82(9)*

SERVICE

The heart of the electrical distribution system is the service. Article 230 of the NEC® contains requirements for installing service conductors and service equipment in all types of occupancies. Parts I through VII contain the requirements for services 600 V and less and Part VIII contains the requirements for services exceeding 600 V, nominal.

Designers and installers of electrical services may have other requirements to consider in addition to those contained in Article 230. Most services are supplied by the local power utility company. The local power utility may have additional, and sometimes different, requirements for the installation of electrical services. For example, the NEC® does not address the location of meter sockets. Section 230.66 indicates that individual meter socket enclosures shall not be considered as service equipment. Yet most electrical utilities specify where the meter is located and the height at which it is mounted. Designers and installers, therefore, shall comply with the NEC® and the local power utility company. Contact the local utility to obtain a copy of their service requirements before planning or installing electrical services.

Definitions – Article 100

Article 100 contains several important definitions that relate to electrical service installations. For example, to determine the minimum size of the service conductors, several NEC® sections should be considered. Section 230.23(B) provides the minimum size for overhead service conductors, 230.31(B) provides that information for underground service conductors, and 230.42 is used for service-entrance conductors. The minimum-size service conductor, therefore, depends on the type of service that is used. The requirements for underground service conductors contained in Part III cannot be applied to the installation of overhead service conductors, which are contained in Part II.

Service. The *service* is the electrical supply, in the form of conductors and equipment, that provides electrical power to the building or structure. Note that the source of supply for a service must be from a utility. **See Figure 4-1.** Service is the root word for eight other terms that are defined in Article 100. It is a general term and usually includes the underground service or overhead service conductors, service equipment, and any associated metering equipment.

Service Conductors. Service conductors are the conductors from the service point to the service disconnecting means. This is a general term. Service conductors can either be part of an overhead supply system or part of an underground supply system.

From the design and installation point of view, service conductors are the most important part of the service. This is because, while other components of the service may be installed or designed by the utility, designers and installers are almost always responsible for the installation of these conductors.

Service Conductors—Overhead. A *service conductor—overhead* is a conductor that connects the service equipment for the building or structure with the electrical utility supply conductors at the service point. This connection usually occurs outside of the building or structure. **See Figure 4-2.**

Electrical power is produced and converted for use at a specific voltage and current.

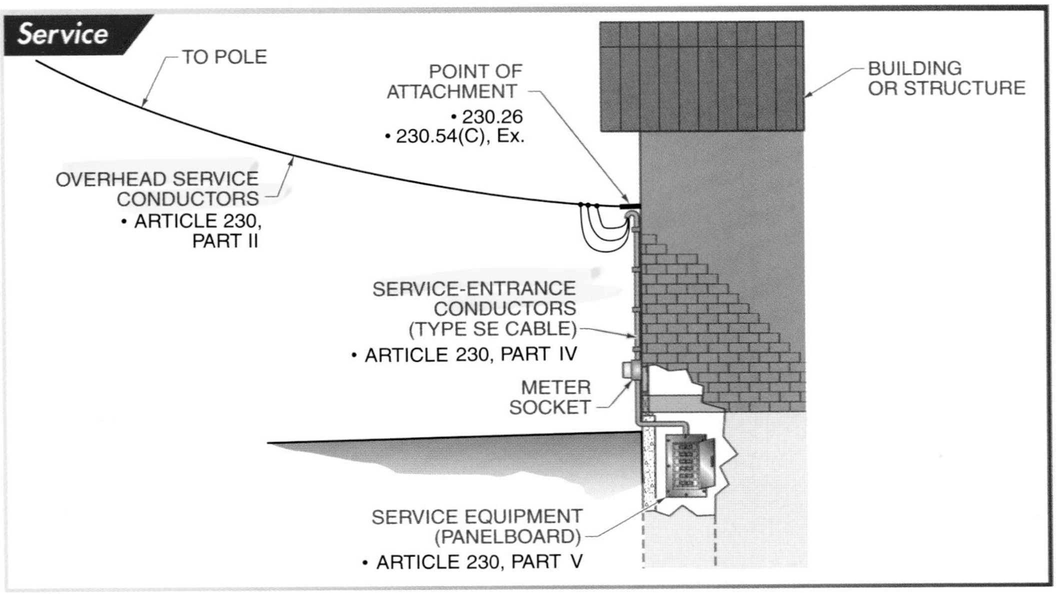

Figure 4-1. The service is the electrical supply, in the form of conductors and equipment, that provides electrical power to a building or structure.

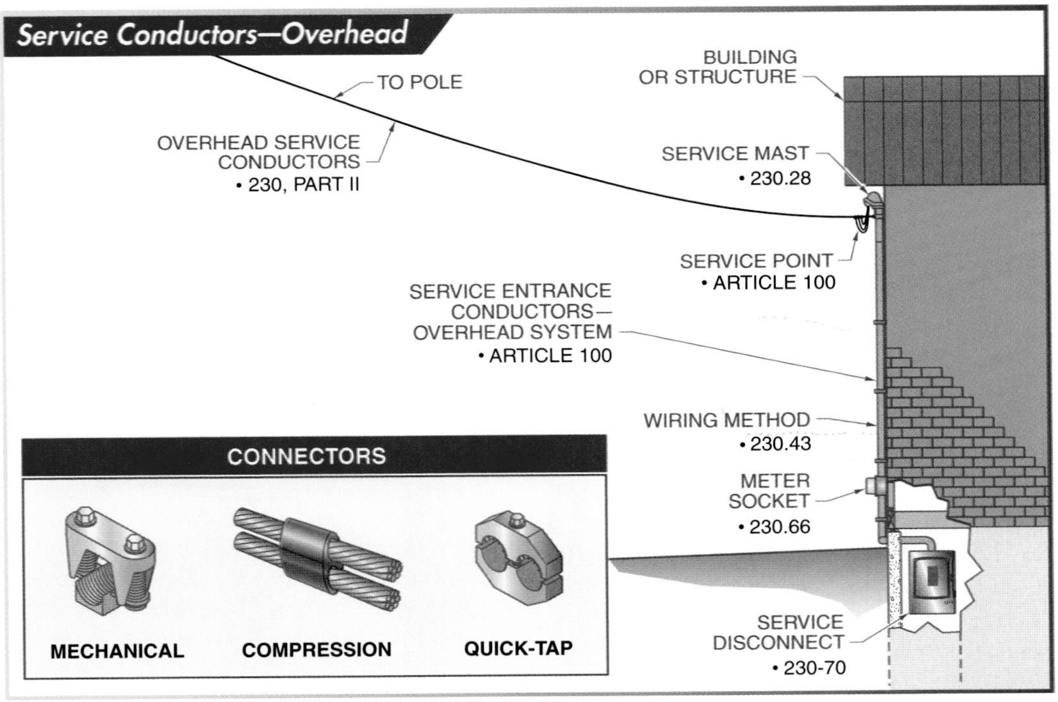

Figure 4-2. Service conductors—overhead are conductors that connect the service equipment for the building or structure with the electrical utility supply conductors at the service point.

The method of connection is usually by splice or tap. Devices such as irreversible compression connectors, service lugs, or quick-taps are often used. Overhead systems are the most common system today. However, utility companies are moving away from overhead systems and prefer to install underground electrical distribution systems whenever possible.

Service-Entrance Conductors—Underground Systems. A *service-entrance conductor—underground system* is a conductor that connects the service equipment with the service lateral. **See Figure 4-3.** The Informational Note, following the definition of underground service-entrance conductors in Article 100, points out that in some electrical service installations there may not be any service-entrance conductors. In other installations, the service-entrance conductors may not even enter the building. This depends on the configuration of the service and the location of the service disconnecting means.

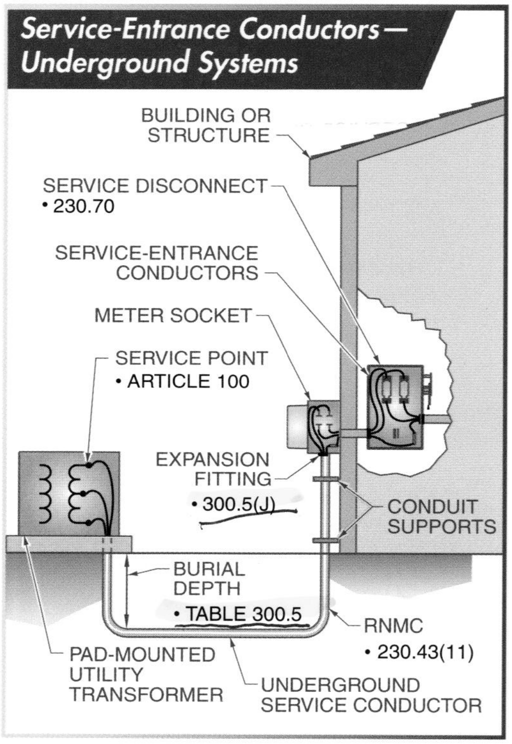

Figure 4-3. Service-entrance conductors—underground systems are conductors that connect the service equipment with the underground service conductors.

Service Drop. The *service drop* is the overhead conductors between the service point and the utility electric supply system. **See Figure 4-4.** Usually, the service-drop conductors originate at the last utility pole or aerial support and continue to their point of attachment on the building or structure. Utility companies frequently use triplex or messenger cable assemblies as the service-drop conductors to the building. In many of these systems, the steel messenger cable is used as the grounded or neutral conductor for the premises wiring system.

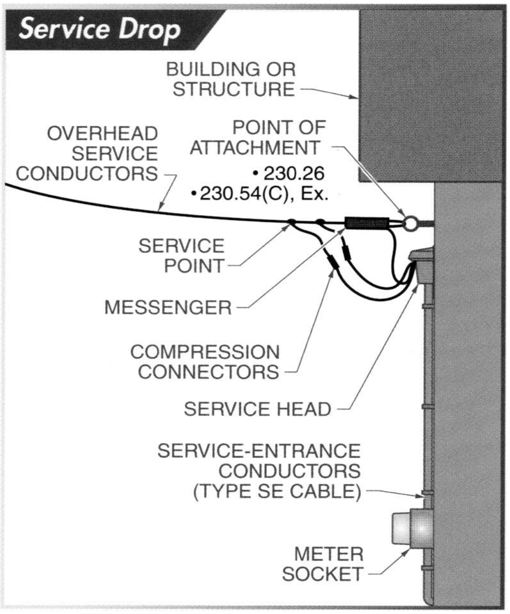

Figure 4-4. The service drop is the overhead conductors between the service point and the utility electric supply system.

Authority Having Jurisdiction (AHJ)

The AHJ has the final say regarding electrical installations no matter what the NEC® states. Due to the possibility of local jurisdiction requiring installation practices that are different from the NEC®, designers and installers should ensure, at the outset of the project, that their intended installation practices meet local as well as NEC® requirements.

For example, many local AHJs require that all electrical services be installed in metal conduit despite the list of possible wiring methods that the NEC® recognizes in 230.43. Other AHJs may prohibit any connections within the metering equipment. Always check with the AHJ to ensure that there are no costly surprises at the end of the job.

Service Lateral. The *service lateral* is the underground service conductors that connect the utility's electrical distribution system and the service point. **See Figure 4-5.** The utility supply system can originate from a pad-mounted transformer or an overhead supply system with the conductors emerging from the ground and continuing up the pole.

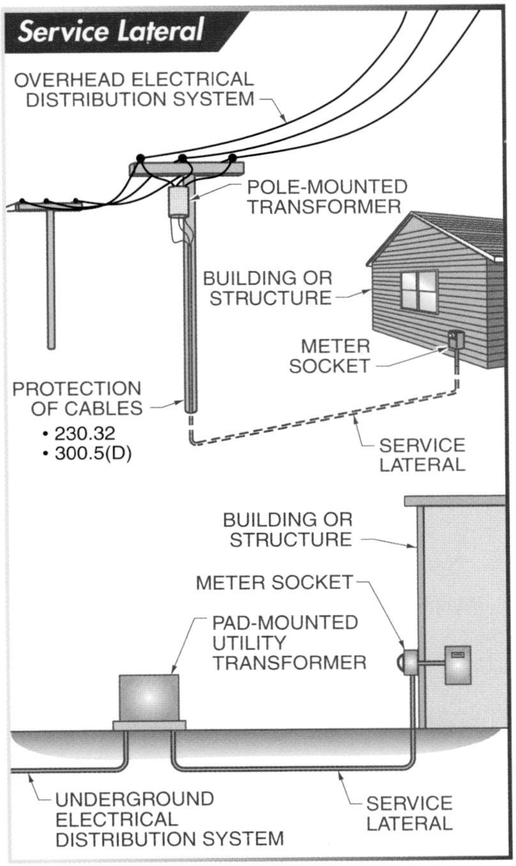

Figure 4-5. The service lateral is the underground service conductors that connect the utility's electrical distribution system with the service-entrance conductors.

The point of connection is between the underground service conductors and the service-entrance conductors. It can occur at a terminal box, meter enclosure, or other type of enclosure, provided there is adequate space for the connection. The connection point may occur inside or outside the building or structure. In the event there is no terminal box or meter enclosure, the point of connection is considered to be the point at which the service conductors enter the building or structure.

Service Point. The *service point* is the point of connection between the local electrical utility company and the premises wiring of the building or structure. **See Figure 4-6.** An important design consideration in any electrical distribution system is the point at which the utility supply ends and the customer-owned or premises wiring begins. The NEC® typically is only applied to the premises wiring.

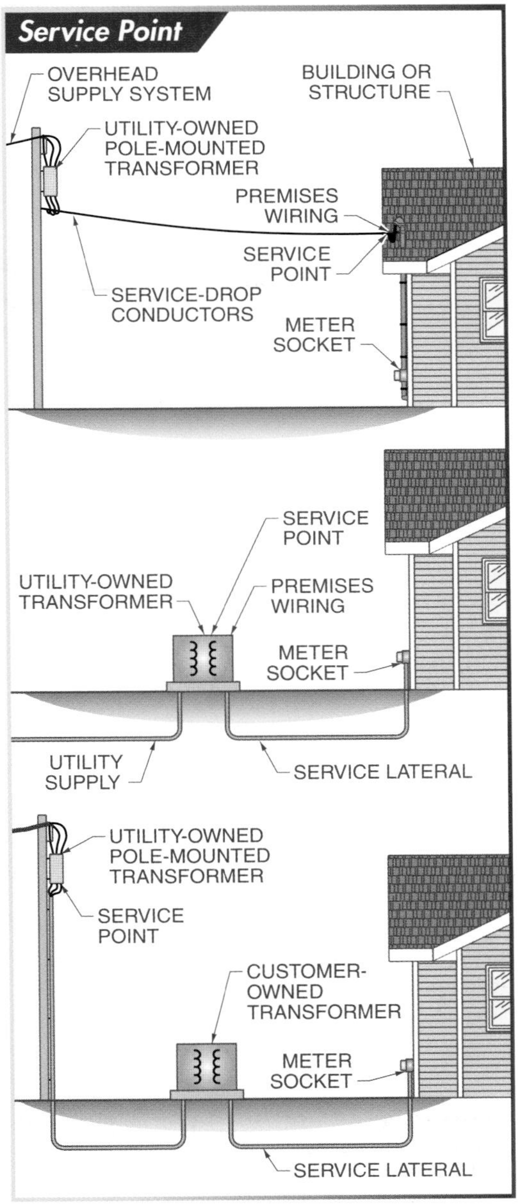

Figure 4-6. The service point is the point of connection between the local electrical utility company and the premises wiring of the building or structure.

In addition to the obvious design concerns, such as who must purchase and maintain the equipment, the service point is important for several legal determinations. The utility company may use its own standards for installation or it may operate under the guidelines established in the *National Electrical Safety Code*. Wiring and equipment connected to the load side of the service point is covered by the NEC®, while conductors located on the line or supply side of the service point are not covered by the NEC®. Unfortunately, as with many of these service-related terms, the location of the service point can change depending upon the design and configuration of the distribution system. Prior to the installation, installers and designers should determine the exact location of the service point and the applicable installation requirements.

Service Equipment. *Service equipment* is all of the equipment necessary to control the supply of electrical power to a building or a structure. It includes circuit breakers, fuses, disconnecting means, panelboards, etc. **See Figure 4-7.** All equipment used as service equipment shall be identified as being suitable for use as service equipment and shall be listed per 230.66. Listing and labeling equipment as suitable for use as service equipment is one method to determine that the equipment is identified. Individual meter socket enclosures are not considered service equipment and need not be identified.

SERVICE LIMITATIONS – Article 230, PART I

Part I contains all of the general requirements applicable to services rated under 600 V, nominal. The general provisions of Part I ensure that adequate protection is provided for the unprotected service conductors. This can be accomplished by maintaining minimum clearances from window and door openings, stairs, balconies, etc.; by limiting service raceways to service conductors only; by protecting service conductors which are installed within a building or structure; and by limiting each building or structure to a single service.

Number of Services – 230.2

The general rule of 230.2 limits each building or structure to a single service. Because the definition of service includes both overhead and underground service conductors, the basic rule permits only a single service drop or service lateral to be supplied to each building or structure. The general rule found in 230.40 limits the number of service-entrance conductor sets for each service drop or lateral to one set. There are five exceptions to the rule that allow more than one set, under certain conditions.

Subsections 230.2(A), (B), (C), and (D) are intended to provide some latitude for the designer and installer for a particular type of occupancy or for a particular type of system. Subsection (E) contains important marking and identification requirements at the service disconnecting means where a building or structure is supplied by more than one service. **See Figure 4-8.**

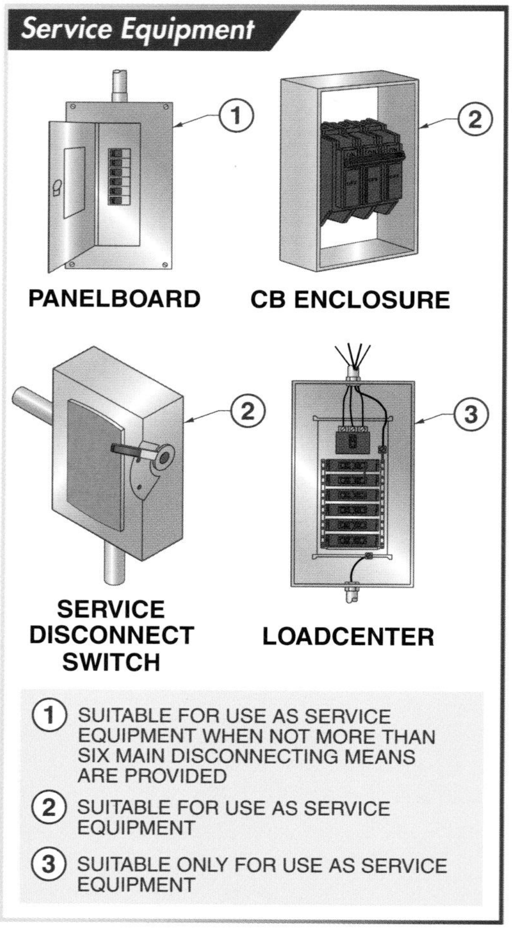
Service Equipment
1
2
PANELBOARD
CB ENCLOSURE
2
3
SERVICE DISCONNECT SWITCH
LOADCENTER
1 SUITABLE FOR USE AS SERVICE EQUIPMENT WHEN NOT MORE THAN SIX MAIN DISCONNECTING MEANS ARE PROVIDED
2 SUITABLE FOR USE AS SERVICE EQUIPMENT
3 SUITABLE ONLY FOR USE AS SERVICE EQUIPMENT

Figure 4-7. Service equipment is all of the equipment necessary to control the supply of electrical power to a building or structure. All service equipment shall be listed.

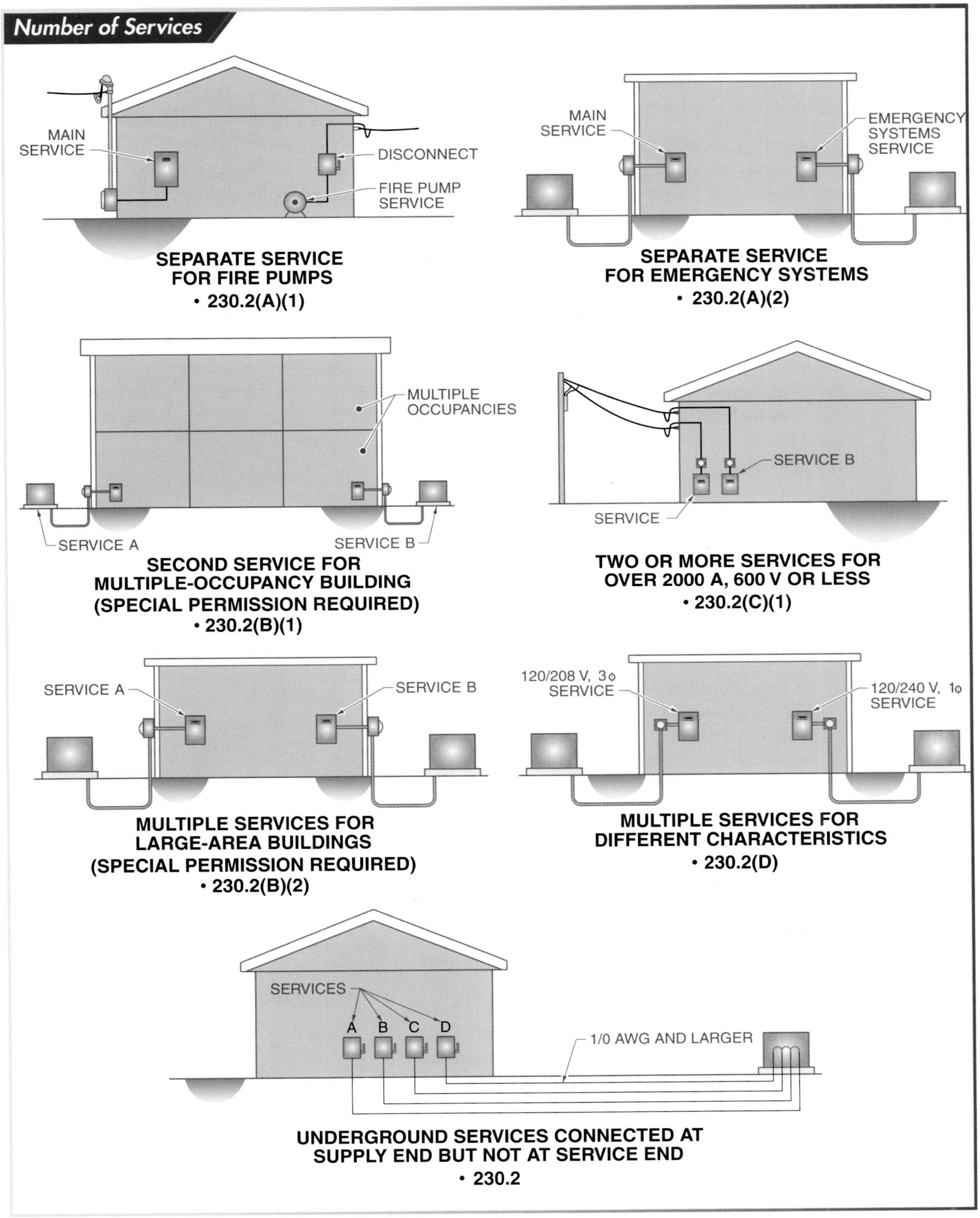

Figure 4-8. Additional services to a common building or structure are permitted in accordance with 230.2(A–E).

Suitable for Use

Installers of electrical service equipment should verify that the equipment is suitable for use as service equipment. All service equipment is required to be listed. There are many types of identification used to mark the suitability of service equipment. Pay particular attention to equipment that is marked "suitable only for use as service equipment." This marking is associated with panelboards which are supplied with the neutral factory-bonded to the enclosure. These panelboards are not intended for and not suitable for use as subpanels. Use of such a panelboard with this marking violates 250.142(B) and could introduce the possibility of a fire and/or shock hazard.

Per 230.2(A)(1), occupancies that utilize a fire pump for life safety protection are permitted to have a separate service provided solely for the fire pump. This helps to ensure that a failure of the primary building service will not remove the life safety protection in the event of a fire.

Emergency or standby systems are permitted to be supplied by additional services by 230.2(A)(2)(3)(4). Standby systems can be either optional or legally required systems. See Articles 701 and 702. These systems provide power to essential building systems, such as exit lights, emergency lighting, HVAC equipment, etc., in the event of a primary electrical service failure. Designers and installers should always review applicable local codes to verify where emergency and standby systems are permitted or required by law.

Designers of electrical systems may also encounter buildings in which there is not adequate space for service equipment that is accessible to all occupants. Multiple services are permitted to these multiple-occupancy buildings per 230.2(B)(1) if special permission is granted. *Special permission* is the written approval of the AHJ. See Article 100.

Another frequent design problem occurs when the electrical load requirements of the occupancy are so large that it makes the installation of a single service impracticable. Per 230.2(C)(1), additional services are permitted to be installed when the capacity requirements of the building or structure exceed 2000 A at 600 V or less. This section also applies when the 1ϕ load requirements exceed that for which the local utility normally supplies a single service. For example, if the occupancy requires a 1ϕ service of 1500 A and the utility company's largest service is 1200 A, a second service can be supplied. This section also permits additional services for capacity when special permission is obtained.

Similarly, there may be large-area buildings that make the utilization of a single service impracticable from the point of view of the distribution system. Per 230.2(B)(2), additional services for these large-area single buildings or structures are permitted if special permission is first obtained. Note there is no definition of what would constitute a building or structure being "sufficiently large;" the AHJ should be consulted before a determination is made.

There also may be design considerations regarding the voltage and frequency requirements of the services or different utility rate schedules for the services. Per 230.2(D), additional services are permitted for these reasons. For example, a local utility company may offer different usage rates for their customers for off-peak services.

Per 230.2, underground sets of conductors are permitted to be run to the same building or structure in accordance with the provisions of 230.40, Ex. 2. The conductors must be sized at least 1/0 AWG and may be connected at their supply end but not at their load end. Essentially, this provision permits multiple service laterals to be considered a single service lateral.

In addition to setting the provisions for the number of services permitted for each building or structure, 230.2 contains important identification and marking requirements. Section 230.2(E) requires that where a building or a structure is permitted to be supplied with more than one service, a permanent plaque or directory must be installed. The location of the plaque or directory must be at each service disconnecting means and must indicate all services, branch circuits, or feeders that supply the structure or building.

Conductors – Outside of Buildings – 230.6

Although service conductors are not provided with overload or short-circuit protection, there may be installations in which there is a need to have the service conductors enter the building. Many high-rise distribution systems, with large power needs, utilize 230.2(C) to bring a second service into the building. Often, the second service is run to an upper-floor electrical room for distribution to the upper levels of the building. In these types of installations, the service conductors enter the building without overcurrent protection. This is accomplished by meeting one of five provisions of 230.6. **See Figure 4-9.**

Roughing in Service Entrances

To make service installation easier, most manufacturers provide service-entrance installation kits, which provide templates to make service-entrance installation quick, safe, and professional.

U7
U7

A utility-owned transformer provides electrical service through underground service conductors, through the meter, to the customer-owned service equipment.

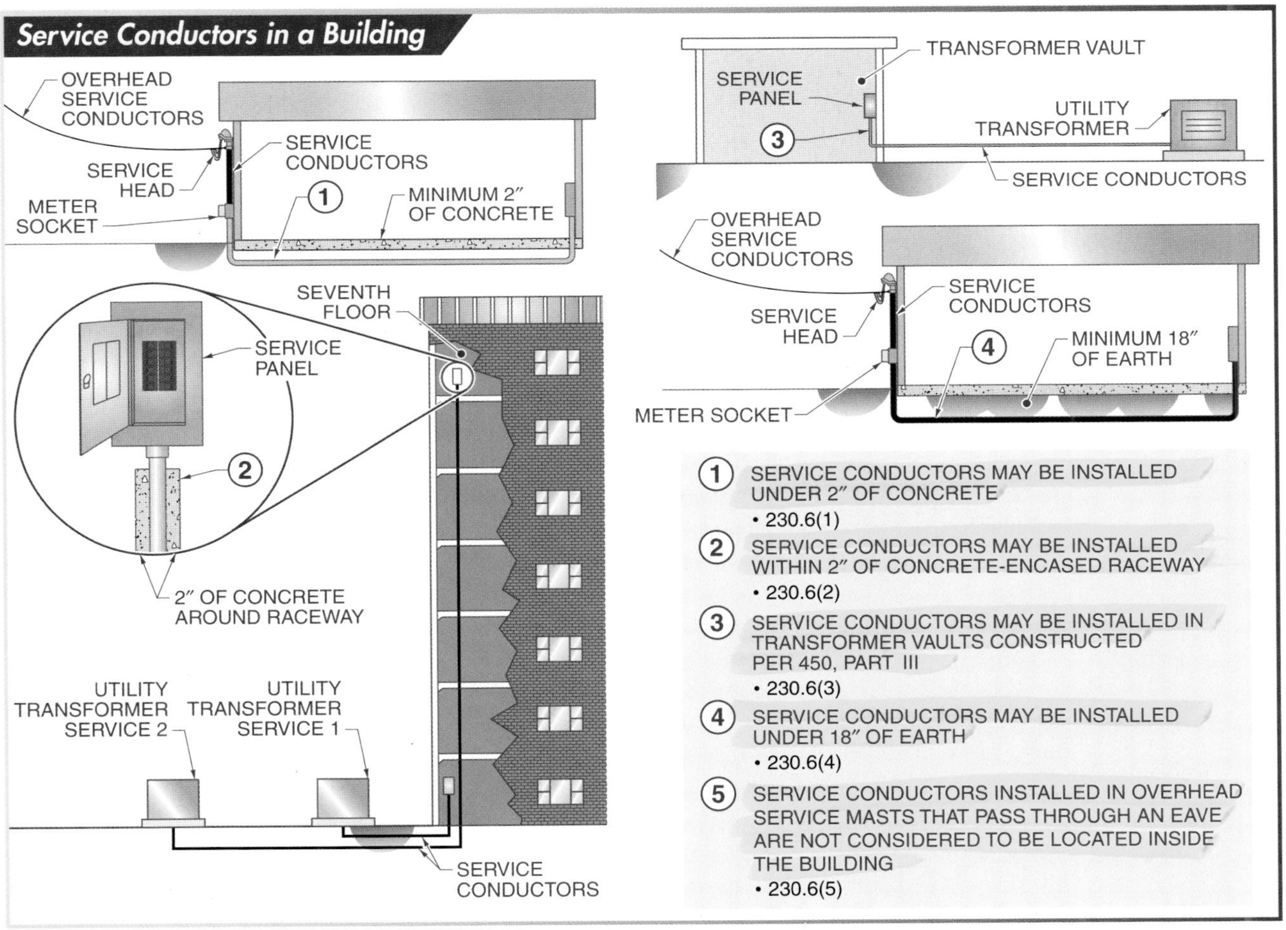
Service Conductors in a Building
OVERHEAD SERVICE CONDUCTORS
SERVICE CONDUCTORS
SERVICE HEAD
METER SOCKET
1
MINIMUM 2″ OF CONCRETE
SEVENTH FLOOR
SERVICE PANEL
2
2″ OF CONCRETE AROUND RACEWAY
UTILITY TRANSFORMER SERVICE 2
UTILITY TRANSFORMER SERVICE 1
SERVICE CONDUCTORS
TRANSFORMER VAULT
SERVICE PANEL
3
UTILITY TRANSFORMER
SERVICE CONDUCTORS
OVERHEAD SERVICE CONDUCTORS
SERVICE HEAD
SERVICE CONDUCTORS
4
MINIMUM 18″ OF EARTH
METER SOCKET
1 SERVICE CONDUCTORS MAY BE INSTALLED UNDER 2″ OF CONCRETE
• 230.6(1)
2 SERVICE CONDUCTORS MAY BE INSTALLED WITHIN 2″ OF CONCRETE-ENCASED RACEWAY
• 230.6(2)
3 SERVICE CONDUCTORS MAY BE INSTALLED IN TRANSFORMER VAULTS CONSTRUCTED PER 450, PART III
• 230.6(3)
4 SERVICE CONDUCTORS MAY BE INSTALLED UNDER 18″ OF EARTH
• 230.6(4)
5 SERVICE CONDUCTORS INSTALLED IN OVERHEAD SERVICE MASTS THAT PASS THROUGH AN EAVE ARE NOT CONSIDERED TO BE LOCATED INSIDE THE BUILDING
• 230.6(5)

Figure 4-9. Service conductors may enter a building.

Section 230.6 establishes five conditions under which conductors can be physically located within a building or structure but can be considered to be outside for the purposes of applying the Code requirements. Subsection 230.6(1) permits service conductors to be routed within the building if they are installed under not less than 2″ of concrete beneath the building or structure. The second condition permits the service conductors to be enclosed in a raceway that is encased within 2″ or more of concrete or brick. The third condition allows the service conductors to be treated as if they were outside the building when they are installed within a transformer vault constructed in accordance with Part III of Article 450.

The 2002 NEC® added a fourth condition that permits the service conductors to be considered as being located outside the building or structure where they are installed in conduit and buried under not less than 18″ of earth beneath a building or structure. The fifth condition was added to the 2011 Code and permits conductors to be considered outside of the building provided they are installed in overhead service masts that are outside of a building and that pass through the building eave.

In each of these cases, the conductors are considered to be outside of the building even though they are physically located within the building. This is permitted by 230.6 because the conductors are suitably protected from physical damage or are inaccessible to unqualified personnel.

Building Services Installation – 230.3; 230.6

Service conductors by definition are the conductors from the service point to the service disconnecting means. Separate and very specific details are included in Article 230 that cover the installation of these conductors. Many of these provisions are included because of the limited overcurrent protection provided by the utility for these conductors. They do not have the same level of overcurrent protection that conductors installed within the building or structure are typically provided with. For this reason, 230.3 prohibits service conductors that supply one building from passing through the interior of another building or structure.

There are, however, instances where a specific installation may require the service conductors to pass through another building or structure. In these cases, 230.6 provides five means to protect the conductors and essentially treat the conductors as if they had not entered the building:

(1) Installing the conductors under 2″ of concrete beneath a building or structure

(2) Encasing the conductors within concrete or brick not less than 2″ thick

(3) Installing the conductors in a vault that conforms with Part III of Article 450

(4) Installing the conductors in conduit at least 18″ below grade and beneath the building or structure

(5) Installing the conductor in an overhead service mast on the outside of a building

Service Raceways and Seals – 230.7; 230.8

Only service conductors shall be installed in service raceways or service cables. Service raceway is not defined in Article 100, but is considered to be any raceway that contains service conductors. Allowing only service conductors to be installed in service raceways ensures that a fault of any type in the service raceway will not affect other conductors. Per 230.7, Ex. 1, this does not apply to grounding or bonding conductors which are permitted within the service raceway. Per 230.7, Ex. 2, load management control conductors are also exempt from the general rule. These control conductors are permitted within the service raceway if they are provided with overcurrent protection. **See Figure 4-10.**

Conductors other than service conductors shall not be installed in the same raceway or service cable.

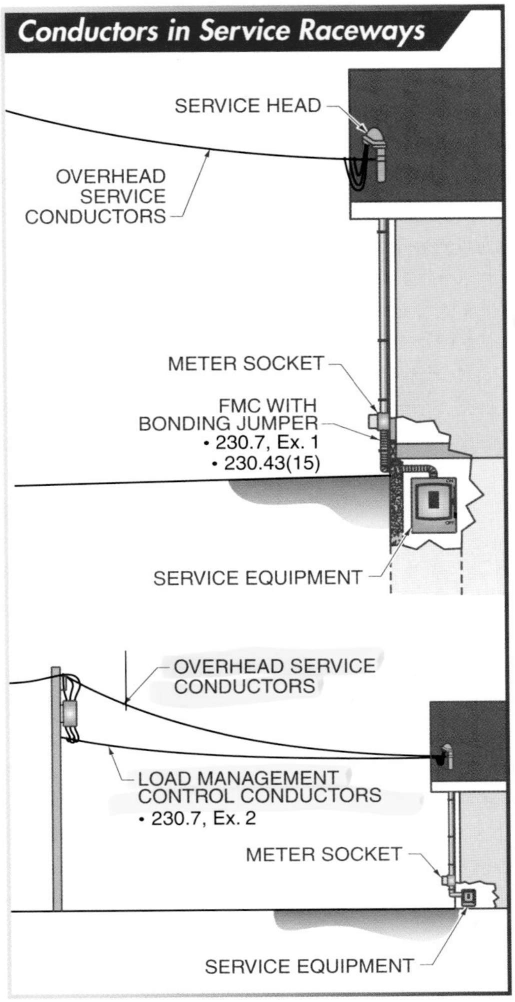

Figure 4-10. Only service conductors shall be installed in service raceways.

Clearance from Openings – 230.9

Because service conductors are not protected by overcurrent protective devices, additional care is taken to ensure that the conductors are not accessible to the general public. Per 230.9(A), whenever service conductors are installed as open conductors or as multiconductor cable without an overall outer jacket, a clearance of 3′ shall be provided from windows that can be opened, doors, porches, balconies, ladders, stairs, fire escapes, or similar locations. **See Figure 4-11.** The intent of this is to protect people from possible contact with these conductors by maintaining a 3′ zone around the conductors. Vertical clearance for overhead spans above or within 3′ measured horizontally from platforms or other surfaces from which the conductors could be reached shall meet the requirements of 230.24(B).

Per 230.9(A), Ex., conductors are permitted to be run above windows within the 3′ zone. The likelihood that the conductors can be reached is reduced by installing them above the window. This only applies to multiconductor cables without an overall outer jacket. It does not apply to Type SE cable with an overall outer jacket. Type SE cable can be installed within the 3′ zone.

The last consideration when installing service conductors near building openings is to ensure that they are not installed beneath openings through which materials may pass. If such openings exist, the service conductors cannot be installed in a manner that obstructs such openings. These types of openings are typical in farm buildings and some commercial buildings.

Section 230.10 prohibits the use of trees and other types of vegetation for the support of overhead service conductors. The NEC® generally does not permit the use of vegetation for the support of overhead conductor spans but it does in some cases permit vegetation to be used for the support of equipment. For example, 590.4(J) in general prohibits trees and vegetation from being used to support overhead spans of temporary branch circuits and feeders. However, 410.36(G) permits outdoor fixtures and lighting equipment to be supported by trees.

Service Lateral Installations

When service originates in a utility hole, work must be arranged with the utility company for joint installation of all conduit and wires. Utility-hole requirements are necessary because utility holes are potentially dangerous spaces and are restricted to utility workmen. When service originates in a sidewalk handhole, electricians are permitted to run conduit and wires into the handhole. The utility company makes the connections in the handhole. When a pole riser is used, the homeowner and power company share the service installation. Electricians can install conduit to a point at least 8′ above grade and pull enough wire to reach the crossarm at the top of the pole. The utility company extends the protection for the drop cables with approved molding to the crossarm upon final connection.

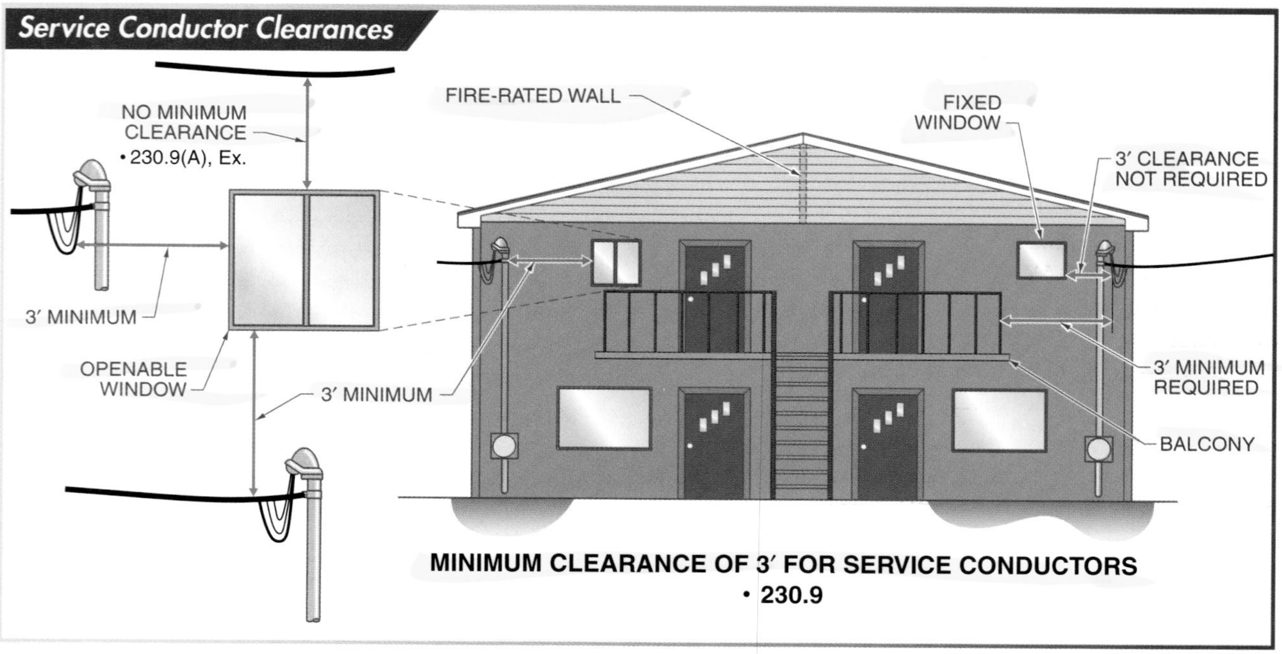

Figure 4-11. Service conductors installed as open conductors or multiconductor cable without an overall outer jacket shall maintain 3′ clearance.

OVERHEAD SERVICE CONDUCTORS – ARTICLE 230, PART II

The overhead service conductors are the overhead service conductors from the utility system to a building, structure, or service point. An example of a structure that could be supplied by an overhead system is a pole. In this case, the overhead service conductors terminate at a meter enclosure or disconnecting means which is mounted on the pole. **See Figure 4-12.**

Overhead service conductors are required to be properly sized for the intended load. In addition, minimum-size conductors are specified. Several other safety concerns are associated with overhead service conductors. Overhead service conductors, in most cases, are required to be covered or insulated to protect personnel who might inadvertently come in contact with them.

Per 230.22, individual overhead service conductors are required to be either insulated or covered. Article 100 defines an insulated conductor as one encased within material of composition and thickness that is recognized as electrical insulation. Covered conductors have an overall outer covering but consist of a material that is not recognized as electrical insulation. For this reason, it is important that unqualified persons stay at least 10′ away from any overhead service conductor rated 600 V, nominal, or less.

The neutral of overhead service conductors, when installed as multiconductor cable, is permitted to be a bare conductor per 230.22, Ex. For example, a weatherproof triplex cable which is commonly used for overhead service conductors for 1ϕ, 120/240 V, 3-wire services is permitted to have a bare neutral.

To avoid contact with personnel, the NEC® requires minimum clearances for overhead service conductors from grade, rooftops, building openings, and swimming pools. Overhead service conductors are required to be adequately supported to minimize lines falling in poor weather conditions, such as severe wind or snow/ice storms.

Size and Rating – 230.23

Overhead service conductors shall have an ampacity that is adequate for the intended load. The load is determined based on the calculations required by Article 220. Another general consideration for overhead service conductors is that they have adequate mechanical strength. Overhead service conductors are often subjected to severe weather conditions which can put a great strain on the conductors due to the additional weight. Likewise, overhead service conductors are sometimes subject to long spans which also increase the stress on the conductors. For this reason, 230.23(B) sets a minimum size of 8 AWG Cu or 6 AWG Al or copper-clad aluminum for most overhead service conductors. While some installations for limited loads of a single branch circuit allow the size of the overhead service conductors to be reduced, they are not permitted to be smaller than 12 AWG hard-drawn copper or the equivalent per 230.23(B), Ex. **See Figure 4-13.**

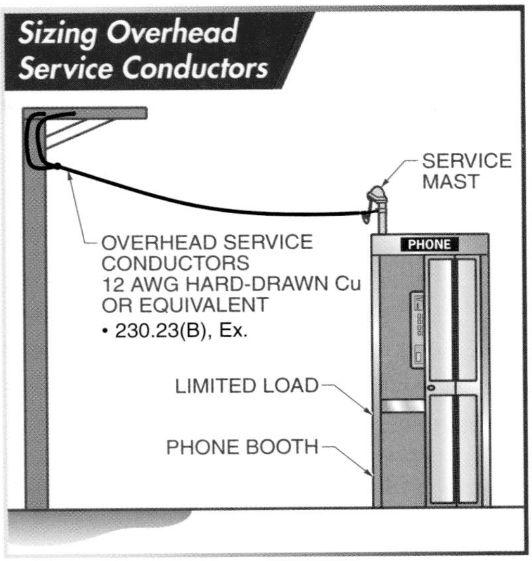

Figure 4-13. Overhead service conductors shall not be smaller than 12 AWG hard-drawn Cu or its equivalent.

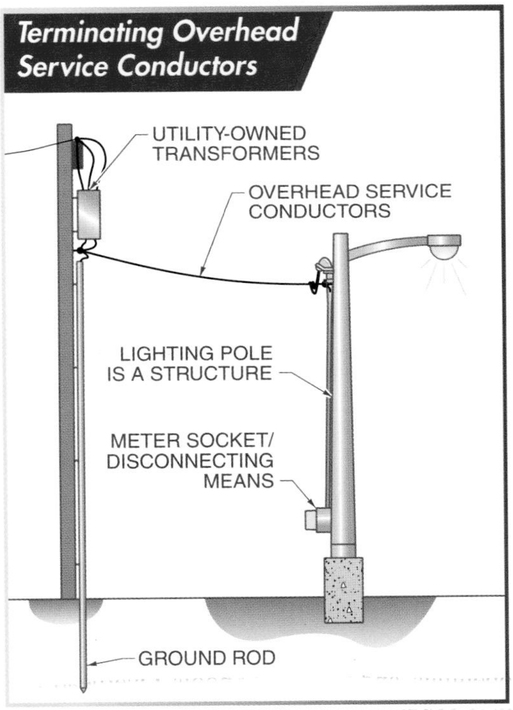

Figure 4-12. Overhead service conductors terminate at a building or structure.

Meter sockets are installed as part of building service installation.

The grounded conductor is permitted by 230.23(C) to be sized per 250.24(C). Essentially, the grounded conductor is sized per Table 250.66, based on the size of the largest service-entrance conductor. In no case shall the grounded conductor be smaller than that required by 250.24(C).

Clearances – 230.24; 230.26

The clearance requirements of Article 230 are provided to protect the overhead service conductors from physical damage and to protect personnel from contact with the conductors. The clearances listed in 230.24 are based on a set of prescribed conditions. These include a no-wind condition and a conductor temperature of 60°F (15°C). Designers and installers of overhead service conductors in conditions other than these shall calculate the additional stress on the conductors as a result of the varying conditions.

The general rule for overhead service conductors requires that they be installed so that they are not readily accessible. Overhead service conductors shall not be easily reached. They shall not be installed so that they can be reached without the use of ladders or other portable means. This requirement does not pose a problem for most installations. One area of concern, however, is when the overhead service conductors pass over roofs. Per 230.24(A), overhead service conductors shall have a minimum vertical clearance of not less than 8′ above rooftops. **See Figure 4-14.**

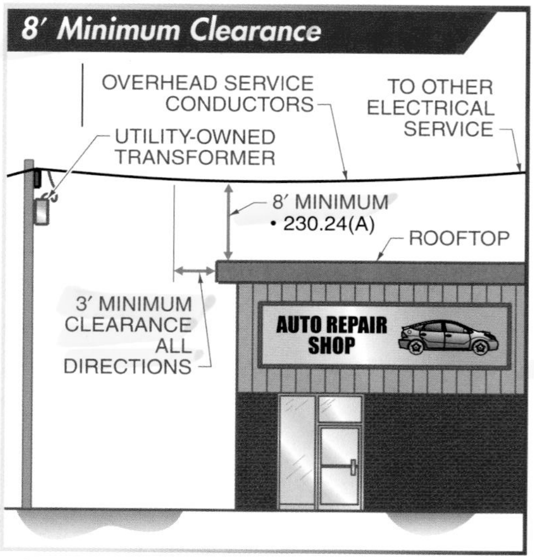

Figure 4-14. Overhead service conductors shall have a minimum vertical clearance of 8′ above rooftops.

The 8′ clearance above the roof shall be maintained in all directions for a minimum distance of 3′ from the edge of the roof. This safety provision ensures that the conductors can not be reached by anyone who happens to be on the roof.

If a roof is subject to pedestrian or vehicular traffic, such as in the case of rooftop parking garages, 230.24(A), Ex. 1 requires that the minimum clearance from the roof surface conform to the requirements of 230.24(B)(4). Therefore, in this installation the minimum clearance above the rooftop is 18′. For services of less than 300 V between conductors, the required clearance is permitted to be reduced to 3′ per 230.24(A), Ex. 2, provided the slope of the roof is not less than 4″ in 12″. **See Figure 4-15.**

Per 230.24(A), Ex. 3, a further clearance reduction is permitted for services not exceeding 300 V between conductors when not more than 6′ of overhead service conductor passes 4′ or less horizontally above a roof overhang and the conductors are terminated in a through-the-roof raceway or approved support. If both of these conditions are met, the minimum clearance can be reduced to 18″ above the overhanging portion of the roof only. **See Figure 4-16.**

Per 230.24(A), Ex. 4, the 3′ vertical clearance from the edge of the roof does not apply to the final span of overhead service conductors when the conductors are attached to the side of the building. This exception is necessary for installations in which the overhead service conductors attach to the side of the building at an angle which does not permit the 3′ clearance in all directions to be maintained. **See Figure 4-17.**

In installations where the voltage between overhead service conductors does not exceed 300 V and the installation is such that the roof is isolated or otherwise guarded, a reduction of the clearance above the roof to 3′ shall be permitted. See Article 100 for the definitions of isolated and guarded.

Service raceways shall be equipped with a service head.

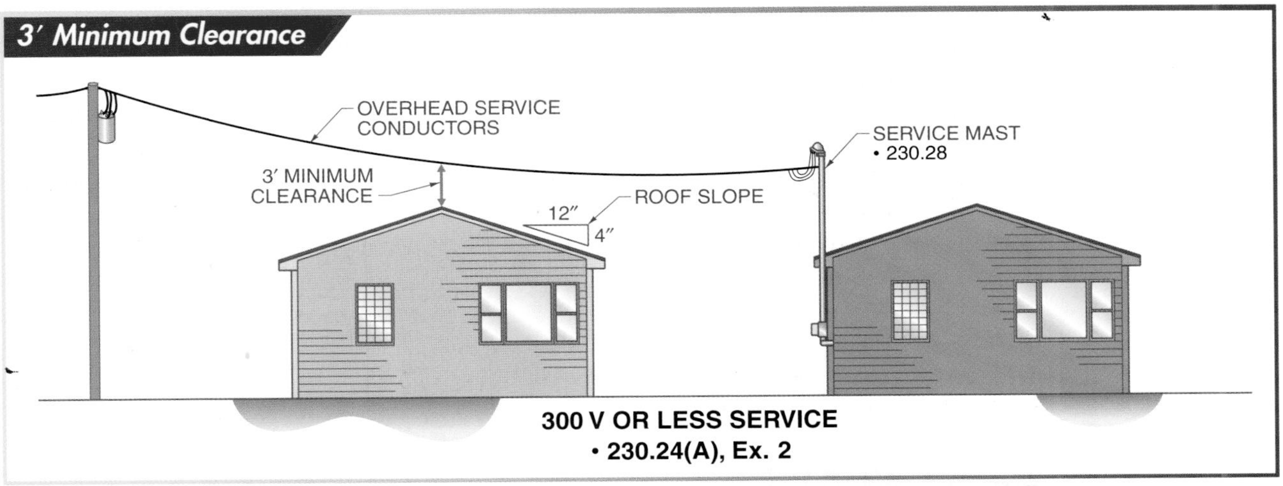

Figure 4-15. The minimum clearance above rooftops is 3′ provided the roof slope is not less than 4″ in 12″ and the voltage does not exceed 300 V.

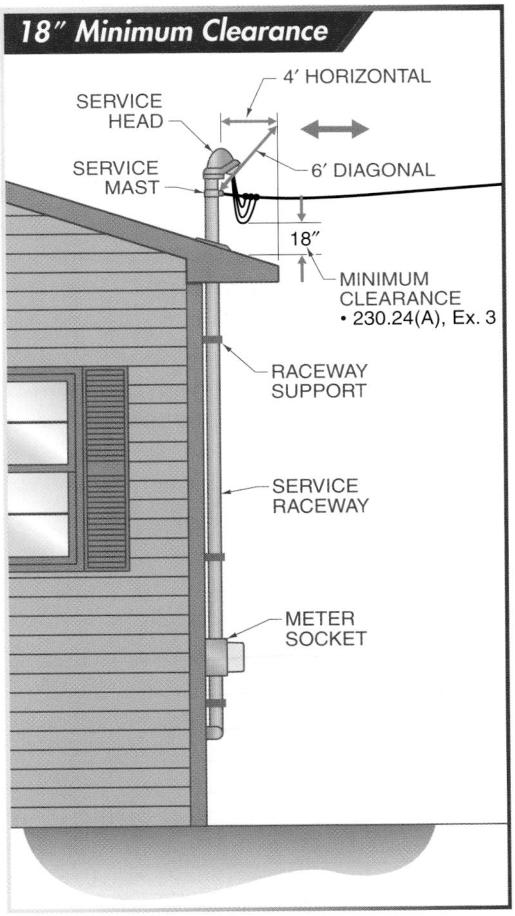

Figure 4-16. The minimum clearance for 300 V or less overhead service conductors terminating in a through-the-roof raceway is 18″ if not more than 6′ of overhead service conductors passes 4′ or less horizontally above the overhang.

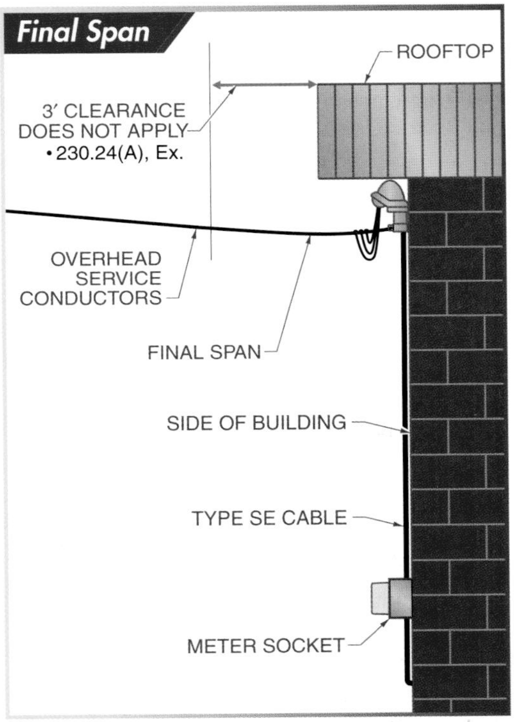

Figure 4-17. The 3′ clearance does not apply to the final span of overhead service conductors attached to the side of a building.

The second type of clearance that shall be maintained for overhead service conductors is vertical clearance from ground. The requirements of 230.24(B) establish four different clearances based on the voltage of the overhead service conductors and the type of traffic area over which the conductors are installed. **See Figure 4-18.**

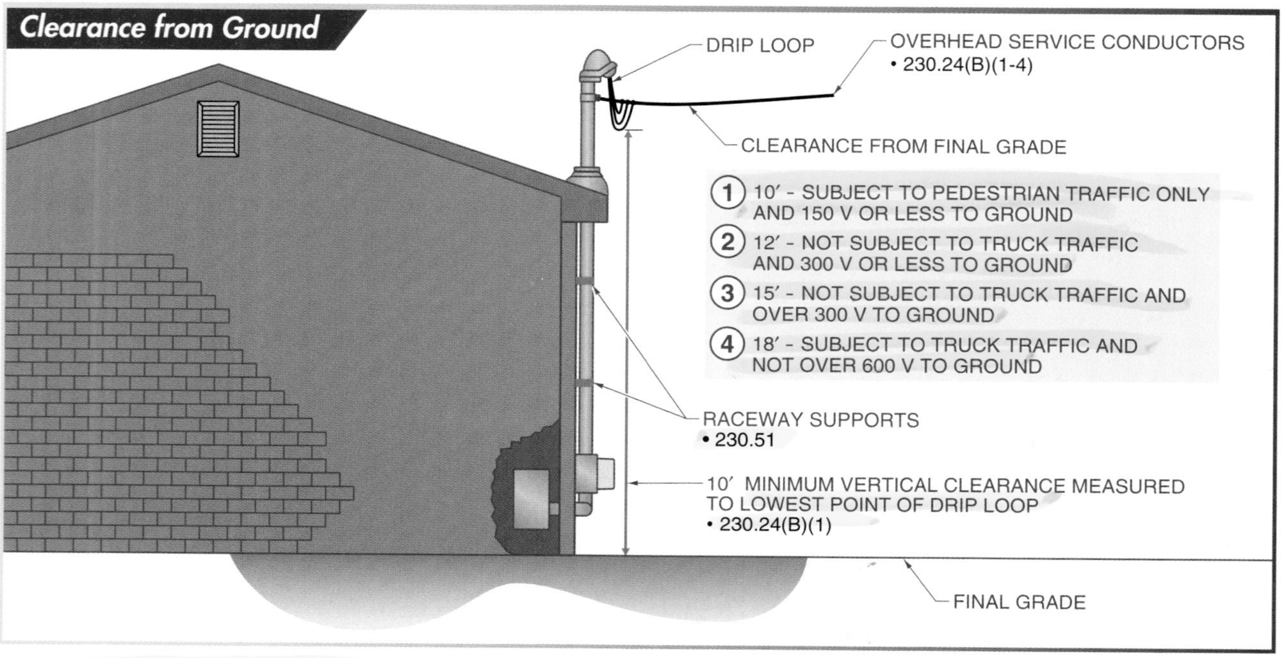

Clearance from Ground
DRIP LOOP
OVERHEAD SERVICE CONDUCTORS
• 230.24(B)(1-4)
CLEARANCE FROM FINAL GRADE
1 10′ - SUBJECT TO PEDESTRIAN TRAFFIC ONLY AND 150 V OR LESS TO GROUND
2 12′ - NOT SUBJECT TO TRUCK TRAFFIC AND 300 V OR LESS TO GROUND
3 15′ - NOT SUBJECT TO TRUCK TRAFFIC AND OVER 300 V TO GROUND
4 18′ - SUBJECT TO TRUCK TRAFFIC AND NOT OVER 600 V TO GROUND
RACEWAY SUPPORTS
• 230.51
10′ MINIMUM VERTICAL CLEARANCE MEASURED TO LOWEST POINT OF DRIP LOOP
• 230.24(B)(1)
FINAL GRADE

Figure 4-18. The clearance of overhead service conductors from a final grade is based on their voltage and type of traffic.

Selecting Service Equipment

Selecting the proper service equipment is an important part of installing an electrical system. Service equipment is the equipment needed to control the supply of electrical power to a building or structure. Considerations such as voltage, number of branch circuits/poles, location, and main device type are used to select service equipment. Service equipment may include panelboards and switches/disconnects.

A panelboard is a single panel or group of assembled panels with buses and overcurrent devices. The panelboard may have switches to control light, heat, or power circuits. To select the proper panelboard, the following should be taken into consideration:

- voltage
- busbar rating
- number of branch circuits/poles
- NEMA enclosure type
- trim type (surface or flush mount)
- cable entry (top or bottom)
- main device (main lug or main circuit breaker)

A switch is a device used to open or close an electrical circuit. To select the proper switch/disconnect, the following should be taken into consideration:

- switch type (heavy-duty, general-duty, or double-throw)
- current rating
- fused, non-fused, or fusible with neutral
- number of poles
- voltage rating
- NEMA enclosure type

Additional information is available at www.cutlerhammer.com.

As a general rule, there shall be a 10′ minimum clearance from final grade for all overhead service conductors. The measurement shall be taken from the lowest point of the drip loop of the overhead service conductors. The 10′ clearance applies to systems with a grounded bare messenger wire where the voltage to ground does not exceed 150 V and the conductors extend over sidewalks and areas accessible to pedestrians only.

When the conductors extend over residential properties and driveways or commercial areas not subject to truck traffic, and the voltage to ground does not exceed 300 V, the minimum clearance from ground shall be 12′. If a similar installation to that of the 12′ condition exists, but the system voltage exceeds 300 V to ground, the clearance from ground shall be increased to 15′. For those installations in which the overhead service conductors extend over public streets, alleys, roads, parking lots subject to truck traffic, etc., the minimum clearance shall be increased to 18′.

Service Mast – 230.28

Overhead service conductors can terminate at the building or structure in several different ways. Often, eyebolts or rafter attachments are used to attach the overhead service conductors. In some installations, however, a through-the-roof assembly is used. In these installations, the overhead service conductors are attached directly to a service mast.

A *service mast* is an assembly consisting of a service raceway, guy wires or braces, service head, and any fittings necessary for the support of overhead service conductors. **See Figure 4-19.** Service masts must be designed to support the weight of the overhead service conductors under varying conditions including those when snow and ice may add additional weight to the conductor span.

Fittings for use with service masts shall be identified for the use. Designers and installers of service masts should check with the local utility company to see if they have additional requirements for the installation of service masts. Finally, service masts shall be used only for the termination and support of power overhead service conductors. Other system conductors, such as CATV or communication systems, are not permitted to be attached to service masts.

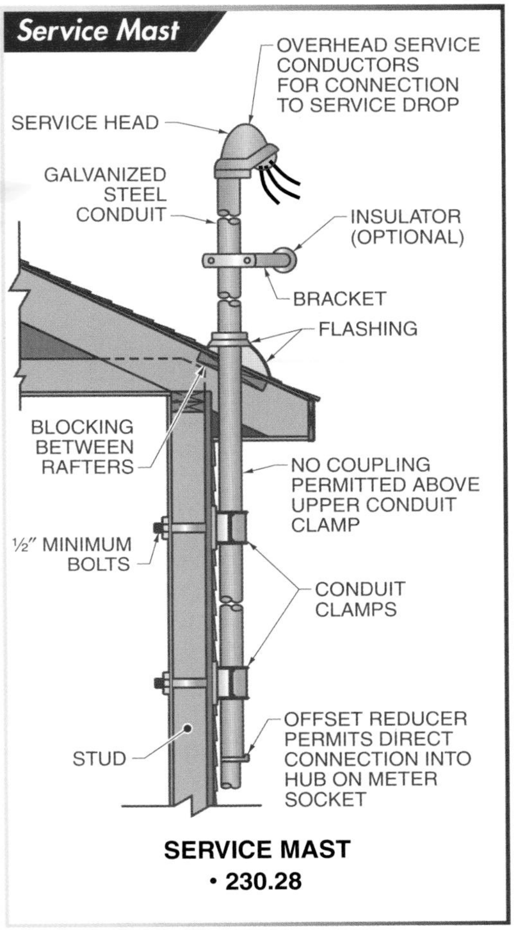

Figure 4-19. A service mast is an assembly consisting of a service raceway, guy wires or braces, service head, and any fittings necessary for the support of overhead service conductors.

> **Residential Service-Entrance Considerations**
>
> *The location of residential service equipment should be carefully considered based on a number of factors. First and foremost are the serving utility service requirements. A careful plan review should also note any special electrical utility requirements, including service characteristics, service equipment locations, and any special installation provisions. Additionally, the designer and installer of the electrical system may want to consider the service equipment location to reduce the overall feeder and branch-circuit wiring costs to minimize the overall length of the circuits.*

UNDERGROUND SERVICE CONDUCTORS – ARTICLE 230, PART III

In general, underground service conductors shall be insulated per 230.30 for the voltage that is applied. Underground service conductors shall be suitable for and capable of withstanding any atmospheric conditions, without insulation degradation which could lead to current leakage. All service conductors, for this reason, shall be insulated for the applied system voltage. Covered conductors are not permitted. Bare conductors are permitted to be used for the grounded conductor per 230.30, Ex., provided they are (1) installed in a raceway; (2) judged suitable for direct burial; (3) part of a cable assembly which is identified for use underground; and (4) aluminum or copper-clad aluminum and part of a cable assembly which is identified for use underground either in a raceway or for direct burial.

Like overhead service conductors, underground service conductors shall be protected against physical damage. Instead of accomplishing this by maintaining clearances, 230.32 requires that the underground service conductors be protected against damage just as any other underground conductors are per 300.5. Per 300.5(D), direct-buried conductors emerging from the ground shall be protected against physical damage from the minimum cover distance required by 300.5(A), up to a point which is at least 8′ above finished grade. Per 300.5(D)(3), for location purposes warning ribbons shall be placed 12″ above all underground service conductors that are not encased in concrete and are buried 18″ or more below grade level. Service-entrance conductors are required to be protected against physical damage by installing them in one of the service raceway types listed in 240.43 or by encasement in concrete. See 230.6.

Section 230.33 contains the permission for the splicing and tapping of underground service conductors. Underground service conductors are permitted to be spliced or tapped but this must be accomplished in accordance with the provisions of 110.14, 300.5(E), 300.13, and 300.15. In general, all splices must be made with devices identified for the purpose. If the location of the splice is underground, the splicing devices must be listed for direct burial.

Size and Rating – 230.31

Underground service conductors shall be sized to handle the calculated load per 230.31(A). The load shall be computed per Article 220. In addition, the underground service conductors shall have adequate mechanical strength for the application. To ensure that the conductors have adequate mechanical strength, 230.31(B) requires that the minimum size for underground service conductors be 8 AWG Cu or 6 AWG Al or copper-clad aluminum. Per 230.31(B), Ex., the minimum-size conductor is permitted to be reduced to 12 AWG Cu or 10 AWG Al or copper-clad aluminum when the load served consists of a single branch circuit only. **See Figure 4-20.**

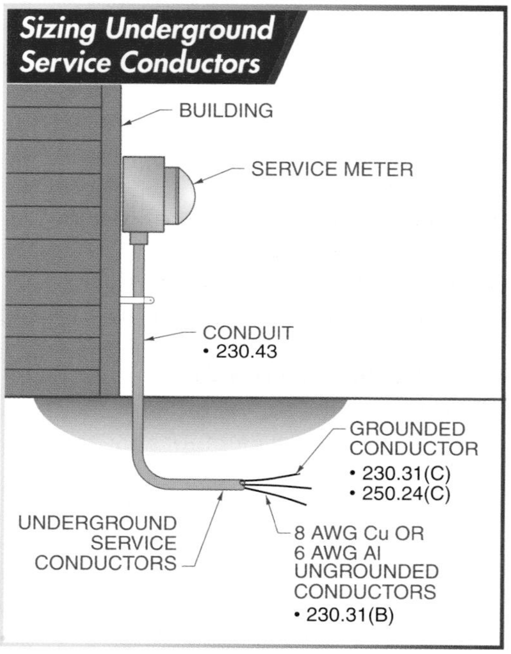

Figure 4-20. The grounded conductor shall be no smaller than that required by 250.24(C).

The size of the grounded conductor must be determined by applying the load calculations of Article 220 and applying the demand factors found in 220.61, where applicable. Section 230.31(C) references 250.24(C) for the minimum-size grounded conductor. That section states that the grounded conductor shall not be smaller than the grounding electrode conductor specified by Table 250.66.

SERVICE-ENTRANCE CONDUCTORS – ARTICLE 230, PART IV

Unlike underground and overhead service conductors, service-entrance conductors are almost always the responsibility of the electrical designer and installer. The utility service and responsibility usually ends at the termination or connection to the underground or overhead service conductors. Part IV of Article 230 contains the requirements for installing service-entrance conductors. An important consideration for any electrical system designer is determining the size and rating of the service-entrance conductors, whether they are supplied from an underground and overhead service. For the installer of the electrical system, an important consideration is the selection of the wiring method for the service-entrance conductors. Requirements for both of these are contained in Article 230, Part IV.

Size and Rating – 230.42

The general rule for sizing service-entrance conductors is that the conductors shall have an ampacity per 310.15 that is suitable for the computed load. The load for the service-entrance conductors shall be calculated per Article 220. **See Figure 4-21.**

The minimum size for ungrounded service-entrance conductors is based upon 230.42 and 230.79.

Per 230.79 and 230.42(B), the service disconnect and the service-entrance conductors in general shall have a rating not less than the load served according to Article 220.

Per 230.79(A) and 230.42(B), the service disconnect and the service-entrance conductors on an installation of a single branch circuit serving limited loads shall have an ampacity of not less than 15 A.

Per 230.79(B) and 230.42(B), the service disconnect and the service-entrance conductors on an installation of not more than two 2-wire branch circuits shall have an ampacity of not less than 30 A.

Per 230.79(C) and 230.42(B), the service disconnect and the service-entrance conductors on an installation of a service serving a one-family dwelling shall have an ampacity of not less than 100 A.

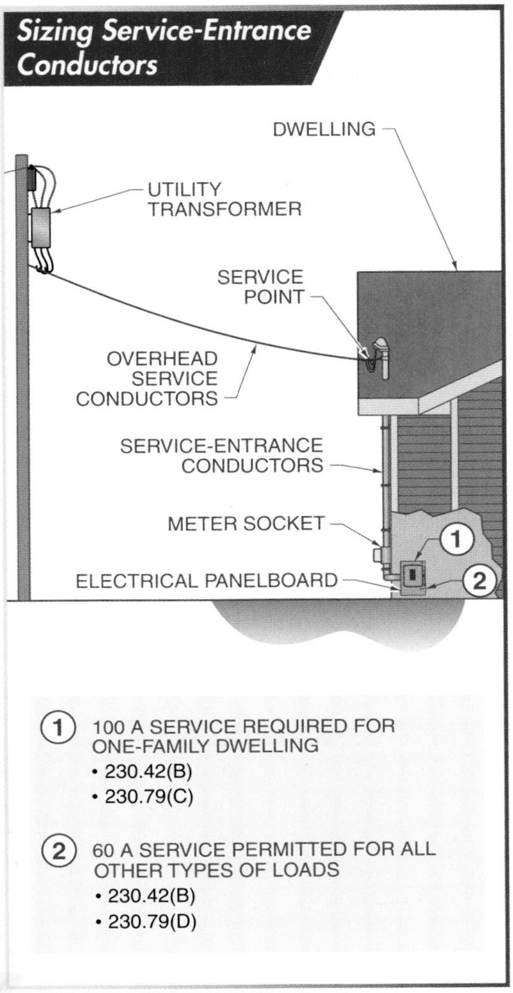

Figure 4-21. The minimum size for service-entrance conductors is based upon 230.79(C) and 230.79(D).

Per 230.79(D) and 230.42(B), the service disconnect and the service-entrance conductors on all other installations shall have an ampacity of not less than 60 A.

The size of the grounded conductor must be determined by applying the load calculations of Article 220 and applying the demand factors found in 220.61, where applicable. Section 230.42(C) references 250.24(C) for the minimum-size grounded conductor. That section states that the grounded conductor shall not be smaller than the grounding electrode conductor specified by Table 250.66.

Service-entrance conductors—overhead systems are conductors that connect the service equipment for the building with the electrical utility supply conductors.

Sets of Service-Entrance Conductors – 230.40

Just as 230.2 limits the maximum number of services for a building or structure to one, the general rule of 230.40 limits the maximum number of sets of service-entrance conductors that can be connected to a single service drop, a set of overhead service conductors, a set of underground service conductors, or a service lateral to one. There may be, however, installations where it is desirable to supply additional sets of service-entrance conductors from a single set of service conductors. **See Figure 4-22.**

Multi-occupancy buildings, for example, are permitted by 230.40, Ex. 1, to have each occupancy or each group of occupancies supplied by a separate set of service-entrance conductors. Per 230.40, Ex. 2, a separate set of service-entrance conductors is permitted to be run to each of the two to six service disconnects that may be used, provided the service disconnects in separate enclosures are grouped at one location. Per 230.40, Ex. 3, an additional set of service-entrance conductors is permitted to be connected to a single service-drop, a set of overhead service conductors, a set of underground service conductors, or service lateral conductors when used to supply a one-family dwelling and a separate structure.

There are two other exceptions to the general rule that each underground service or overhead service is permitted to supply only a single set of service-entrance conductors. Section 230.40, Ex. 4, permits an additional set for common area branch circuits in two-family or multifamily dwellings. Section 210.25 prohibits branch circuits for central lighting, alarms signaling, and communications equipment in these types of dwellings from being supplied by equipment that supplies and individual dwelling unit. Section 230.40, Ex. 5, permits an additional set of service-entrance conductors to supply equipment for load management systems, fire pumps, instrument transformers, etc., that are covered by 230.82 (5) or (6) and permitted to be installed on the supply side of the service disconnect.

Section 230.41 contains specific requirements for the insulation of service-entrance conductors. **See Figure 4-23.** Per 230.41, all ungrounded conductors that are used as service-entrance conductors shall be insulated. The service-entrance conductors shall be suitable for the atmosphere in which they are installed. Service-entrance conductors are not permitted to be covered conductors.

Bare grounded conductors, however, are permitted if they are (a) copper and installed in a raceway or part of a service cable assembly; (b) copper and judged suitable for direct burial; (c) copper and part of a cable assembly which is identified for use underground; or (d) aluminum or copper-clad aluminum and part of a cable assembly which is identified for use underground either in a raceway or for direct burial; or (e) bare grounded conductors used in an auxiliary gutter.

Wiring Methods – 230.43, 230.44

Nineteen listed wiring methods are permitted for installation of service-entrance conductors in 230.43. **See Figure 4-24.** Wiring methods that do not appear in 230.43, such as Type NM cable, are not permitted to be used for service-entrance conductors. These 19 wiring methods consist of various types of raceways, cable assemblies, open wiring, and auxiliary gutters. No matter what wiring method is selected, the service-entrance conductors shall be installed per the specific NEC® requirements for that particular wiring method. For example, if Type MC cable is selected as the wiring method, the service-entrance conductors shall be installed according to all of the provisions of Article 330, *Metal-Clad Cable*.

> **Ungrounded Service Conductors**
>
> *Section 230.42(B) requires the ampacity of the ungrounded service-entrance conductors to be not less than the rating of the disconnecting means per 230.79. The minimum ampacity of conductors for a one-family dwelling is 100 A. Section 310.15 and Table 310.15(B)(7) allow a smaller wire size for service and feeder conductors serving a dwelling with a 120/240 V, 1ϕ, 3-wire system.*

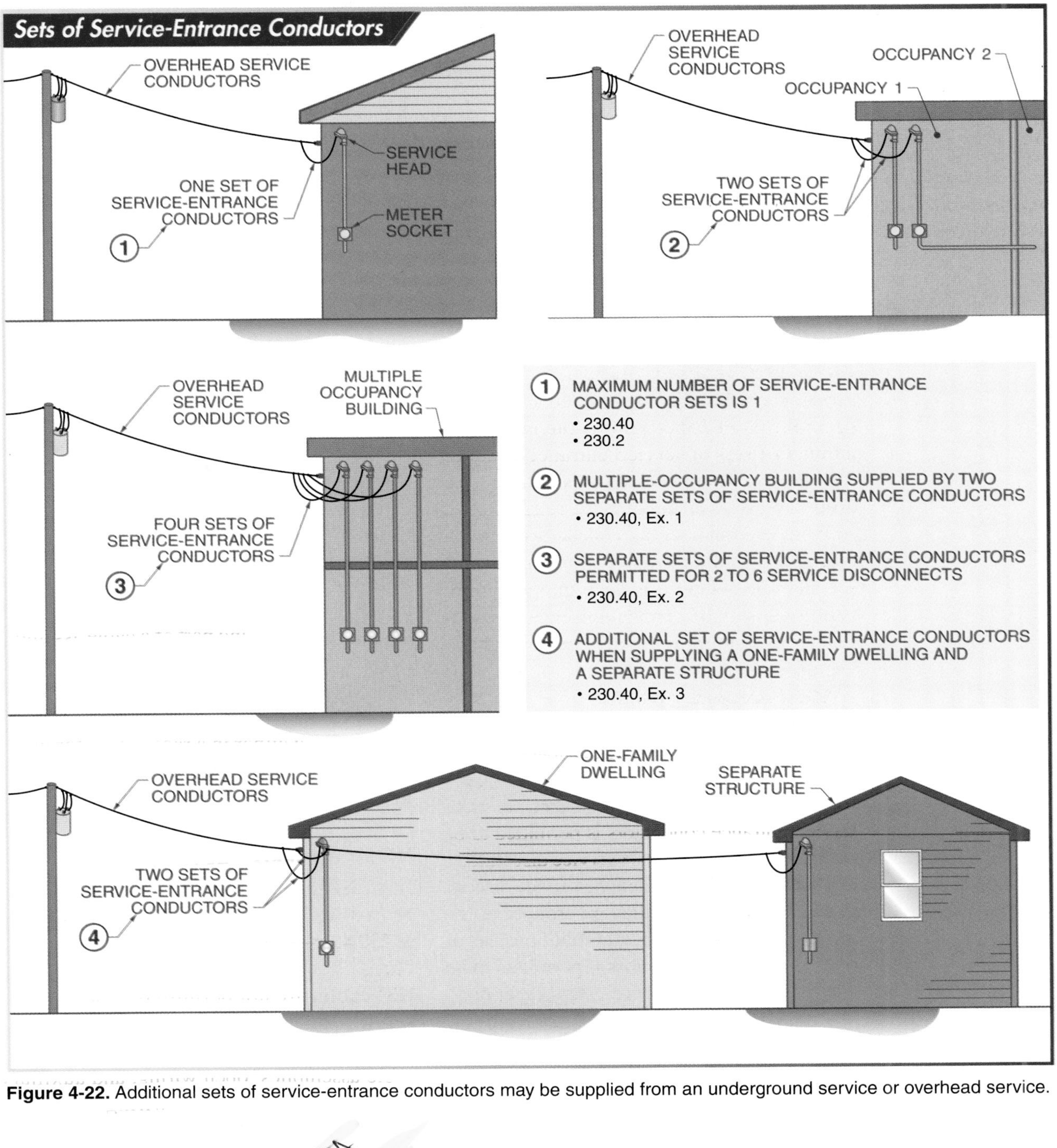
Sets of Service-Entrance Conductors
OVERHEAD SERVICE CONDUCTORS
SERVICE HEAD
ONE SET OF SERVICE-ENTRANCE CONDUCTORS
METER SOCKET
1
OVERHEAD SERVICE CONDUCTORS
OCCUPANCY 2
OCCUPANCY 1
TWO SETS OF SERVICE-ENTRANCE CONDUCTORS
2
OVERHEAD SERVICE CONDUCTORS
MULTIPLE OCCUPANCY BUILDING
FOUR SETS OF SERVICE-ENTRANCE CONDUCTORS
3
1 MAXIMUM NUMBER OF SERVICE-ENTRANCE CONDUCTOR SETS IS 1
• 230.40
• 230.2
2 MULTIPLE-OCCUPANCY BUILDING SUPPLIED BY TWO SEPARATE SETS OF SERVICE-ENTRANCE CONDUCTORS
• 230.40, Ex. 1
3 SEPARATE SETS OF SERVICE-ENTRANCE CONDUCTORS PERMITTED FOR 2 TO 6 SERVICE DISCONNECTS
• 230.40, Ex. 2
4 ADDITIONAL SET OF SERVICE-ENTRANCE CONDUCTORS WHEN SUPPLYING A ONE-FAMILY DWELLING AND A SEPARATE STRUCTURE
• 230.40, Ex. 3
OVERHEAD SERVICE CONDUCTORS
ONE-FAMILY DWELLING
SEPARATE STRUCTURE
TWO SETS OF SERVICE-ENTRANCE CONDUCTORS
4

Figure 4-22. Additional sets of service-entrance conductors may be supplied from an underground service or overhead service.

min 100A Service

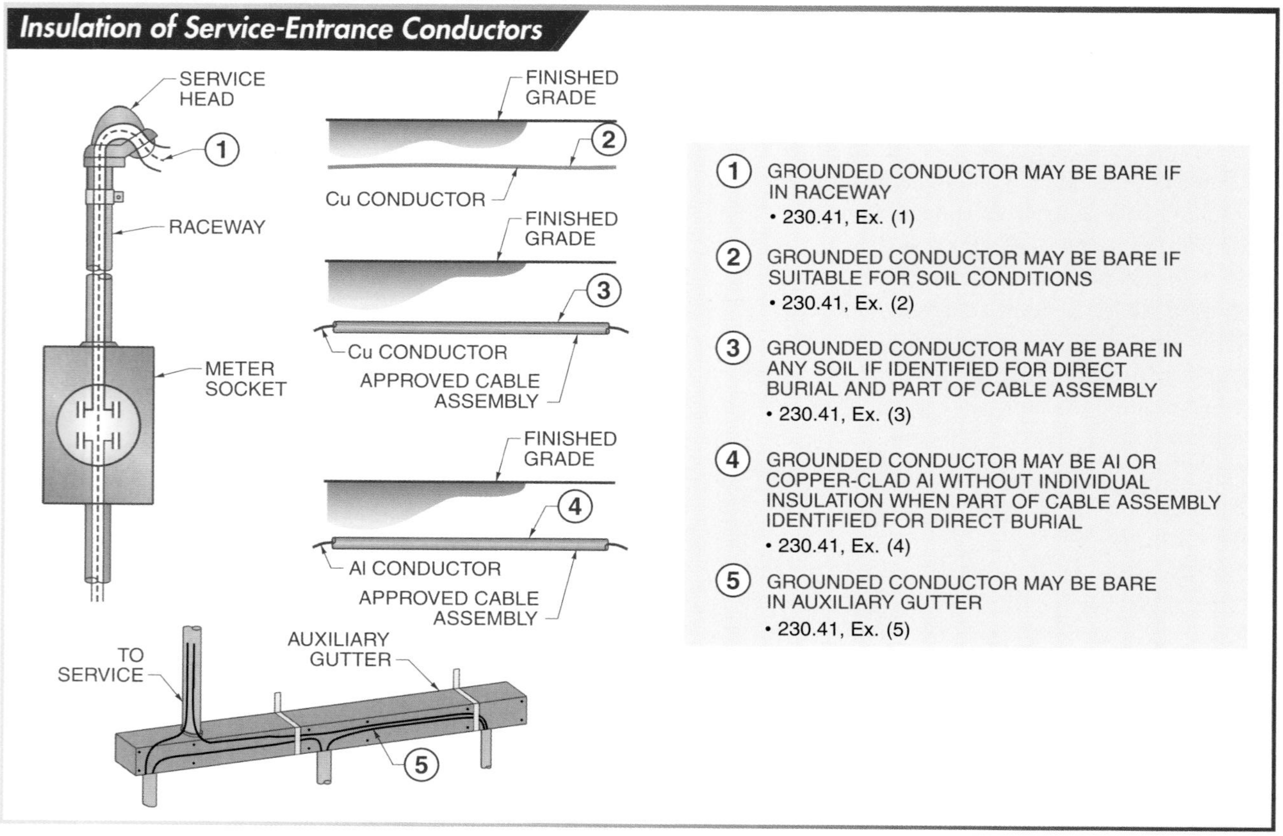

Figure 4-23. Bare, grounded conductors are permitted for service-entrance conductors with conditions.

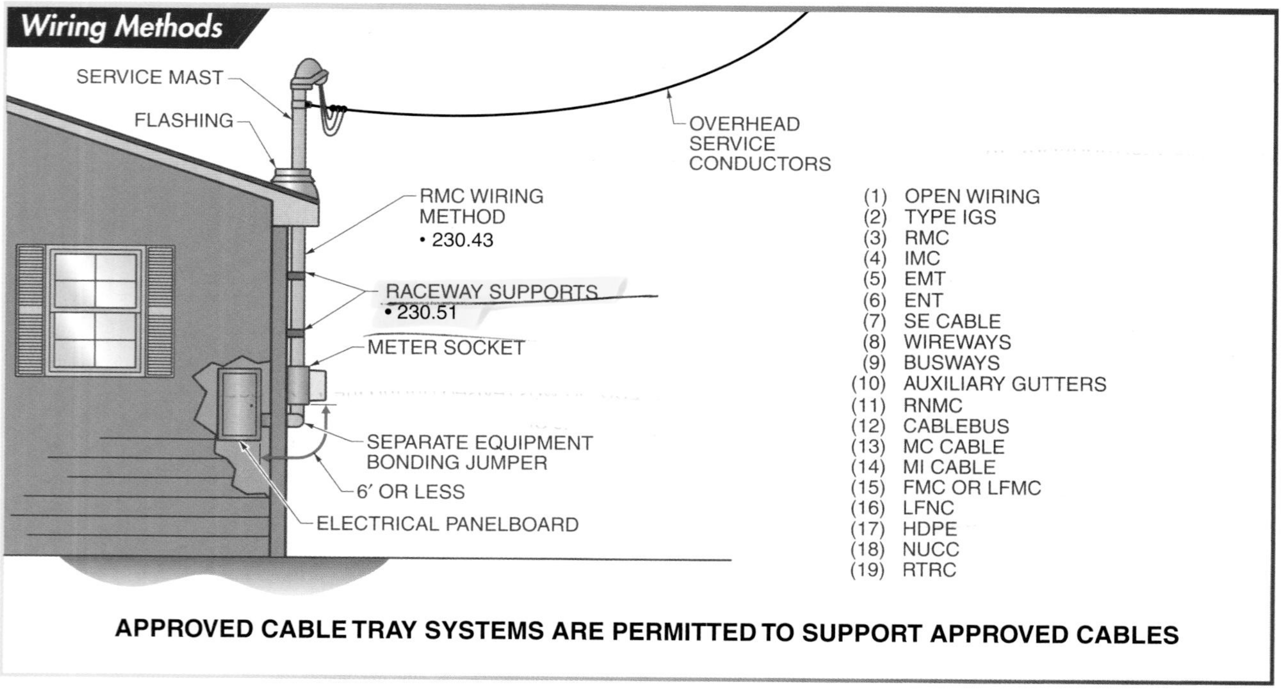

Figure 4-24. Nineteen wiring methods are permitted for use as service-entrance conductors.

Some of the wiring methods have additional limitations when they are used for service-entrance conductors. For example, FMC and LFMC are permitted to be used for service-entrance conductors, but only in lengths not exceeding 6′. In addition, FMC and LFMC shall be installed with an equipment bonding jumper (EBJ) per 250.102(A), (B), (C), and (E).

Section 230.44 permits cable tray systems to be used for the support of service-entrance conductors. Cable tray systems are a unit or an assembly of units or sections, along with the associated fittings, that form a structural system to fasten or support cables, conductors, or raceways. If a cable tray system is used for the support of service-entrance conductors, the system may not contain other types of conductors.

The exception to this rule is 230.44, Ex., which permits other than service-entrance conductors to be installed in the same cable tray system provided a fixed barrier is installed in the cable tray. Where cable trays contain service-entrance conductors, the tray shall be identified and marked to indicate "Service-Entrance Conductors." The marking must be made by labels that are visible after installation and placed such that the service-entrance conductors can be traced through the entire length of the cable tray. There are five wiring methods for service-entrance conductors that are permitted to be installed in cable trays: Type SE cable, Type MC cable, Type MI cable, Type IGS cable and single thermoplastic-insulated conductors 1/0 and larger with a CT (cable tray) rating.

Protection – 230.50

Type SE, service-entrance cable, is the most commonly used wiring method for 120/240 V, 1ϕ, 3-wire services in one-family and multi-family dwellings. The primary reasons for this are the ease of installation and the relatively low cost of the wiring method. However, there may be local codes and standards or local utility requirements that require all service-entrance conductors to be installed in metal conduits.

Type SE cable is a factory-assembled cable assembly, consisting of one or more conductors with a flame-retardant, moisture-resistant covering, used primarily for services. There are other types of SE cables available, such as those without an overall covering and those for use underground, but Type SE cable is by far the most prevalent. See Article 338 for other requirements for installing service-entrance cables.

Type SE cable is constructed with a flame-retardant and moisture-resistant covering to protect the conductors within the cable assembly. This covering, however, does not provide a great deal of protection against physical damage. All service-entrance conductors are required to be protected against physical damage per 230.50. Specific protection is provided by RMC, IMC, RNC, or equivalent. **See Figure 4-25.**

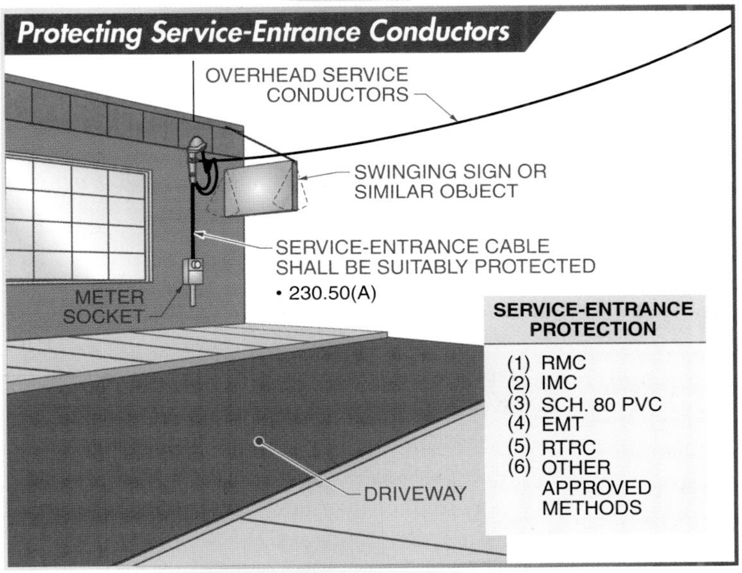

Figure 4-25. Service-entrance cables shall be suitably protected.

Type SE Cable

Section 230.50 was revised during the 1999 Code cycle to clarify that the requirements of this section apply to all types of service cables, not just Type SE cable. The requirement was also revised by removing the laundry list of types of installations that would require protection from physical damage. Previously, protection had to be provided where SE cable was subject to physical damage, such as when installed near sidewalks, driveways, etc. This language has been removed because it was too vague. Installers should check with their AHJ and local electrical utility before settling on a wiring method for service-entrance conductors to ensure that any concerns about potential damage to the conductors are addressed.

When service-entrance cables, such as Type SE cable, are installed in locations subject to physical damage they shall be suitably protected. Suitable protection can be accomplished by use of RMC, IMC, Schedule 80 PVC, EMT, RTRC, or other approved methods. Per 230.50(A), installations where the cables are subject to physical damage shall be protected. Per 230.50(B)(2), open wiring and cables other than service-entrance cables shall not be installed within 10′ of grade level or in locations where they would be subject to physical damage.

Supports – 230.51

Specific support requirements shall be followed when cable assemblies or open wiring are used for service-entrance conductors. Often, these wiring methods are subjected to severe weather conditions which may cause damage to the electrical components and the threat of a power outage if the system is not securely fastened to and supported by the building or structure to which it is mounted. Per 230.51(A), service-entrance cables shall be supported at intervals not exceeding 30″. Check with the AHJ for specific interval requirements. In addition, the cables shall be supported within 12″ of every service head, gooseneck, or point of connection to a raceway or enclosure. The means of support shall be by a cable strap or other approved means. **See Figure 4-26.**

When other types of cables that are not designed for mounting in direct contact with the building are used, 230.51(B) requires that the cables shall be mounted on insulating brackets installed at intervals that do not exceed 15′. The cables shall be installed in a manner that maintains a 2″ minimum clearance from the surface over which they pass.

If the service-entrance conductors consist of individual open conductors, 230.51(C) requires that the conductors be installed per Table 230.51(C). This table establishes maximum support distances and minimum clearances between the conductors, and from the surfaces, depending on the maximum voltage of the conductors.

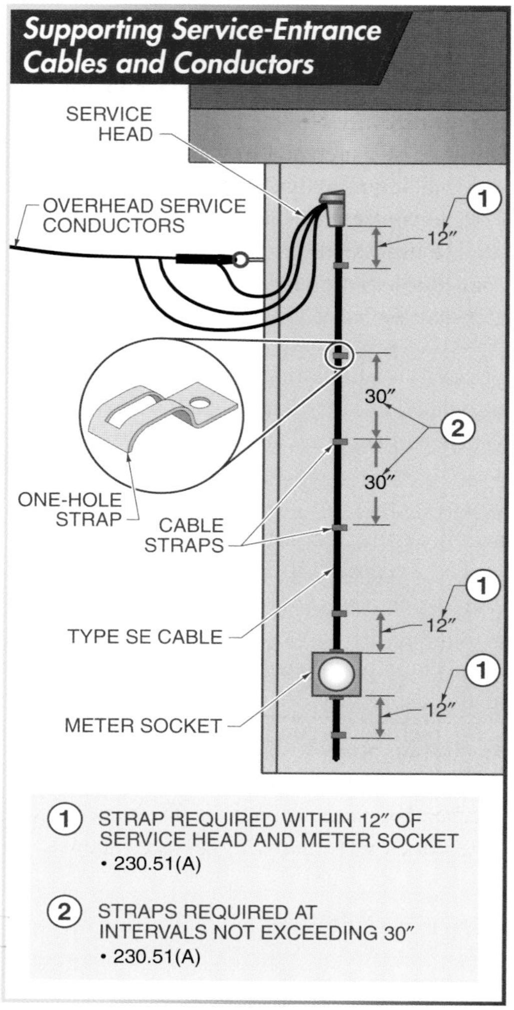

Figure 4-26. Service-entrance cables shall be supported by cable straps or other approved means.

Raceways – Outdoors 230.53

Raceways are frequently installed in locations that are subject to weather. Where these raceways contain service-entrance conductors, 230.53 requires that the service raceway be suitable for use in wet locations and be arranged to drain. The intent is to minimize the collection of water or moisture within the service raceway and the potential harm to the insulation of the service-entrance conductors. Where the service raceway is embedded in masonry, the raceways must also be installed to drain properly.

Overhead Service Locations – 230.54

The overhead service location shall be selected to minimize the possibility of water entering the service raceway. **See Figure 4-27.** To avoid the entrance of water and moisture, service raceways are required by 230.54(A) to be provided with a raintight service head. The raintight service head shall be installed at the point of connection to the overhead service conductors and shall be listed for use in wet locations.

When service-entrance cables are used in lieu of service raceways, similar protection against the entrance of water and moisture shall be provided. Two options are permitted to provide protection from water and moisture for service-entrance cables. The first is to use a listed service head per 230.54(B). The service head consists of a cap and openings for the conductors. The exception to 230.54(B) permits SE cable to be formed into a gooseneck and taped with a self-sealing, weather-resistant thermoplastic.

Point of Attachment – 230.54(C). The service head or gooseneck shall be installed at a point that is above the point of attachment for the overhead service conductors. This ensures that water does not enter the service raceway or service-entrance cable. However, there may be installations where it is impracticable to locate the service head or gooseneck above the point of attachment. In these cases, 230.54(C), Ex., permits the point of attachment to be located above the service head or gooseneck. In no case, however, is the service head or gooseneck permitted to be located more than 24″ from the point of attachment. **See Figure 4-28.**

Service raceways are required to be suitable for use in wet locations and must be installed to drain properly.

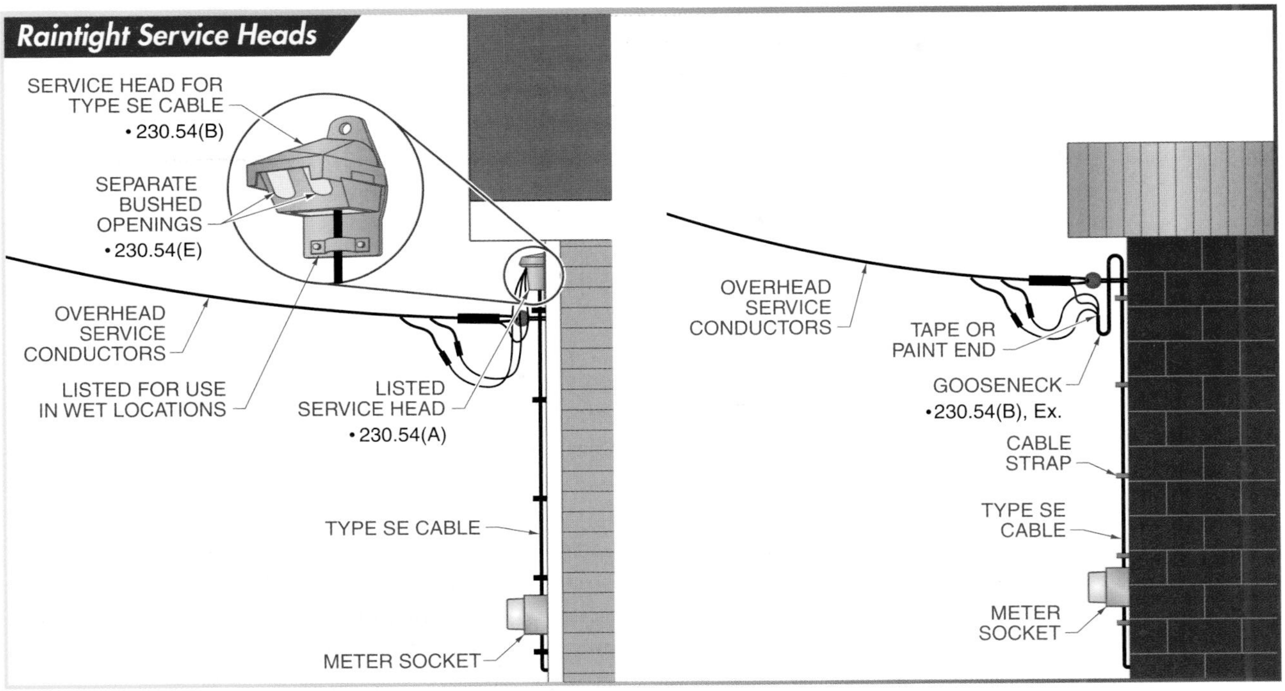

Figure 4-27. The overhead service location shall be raintight.

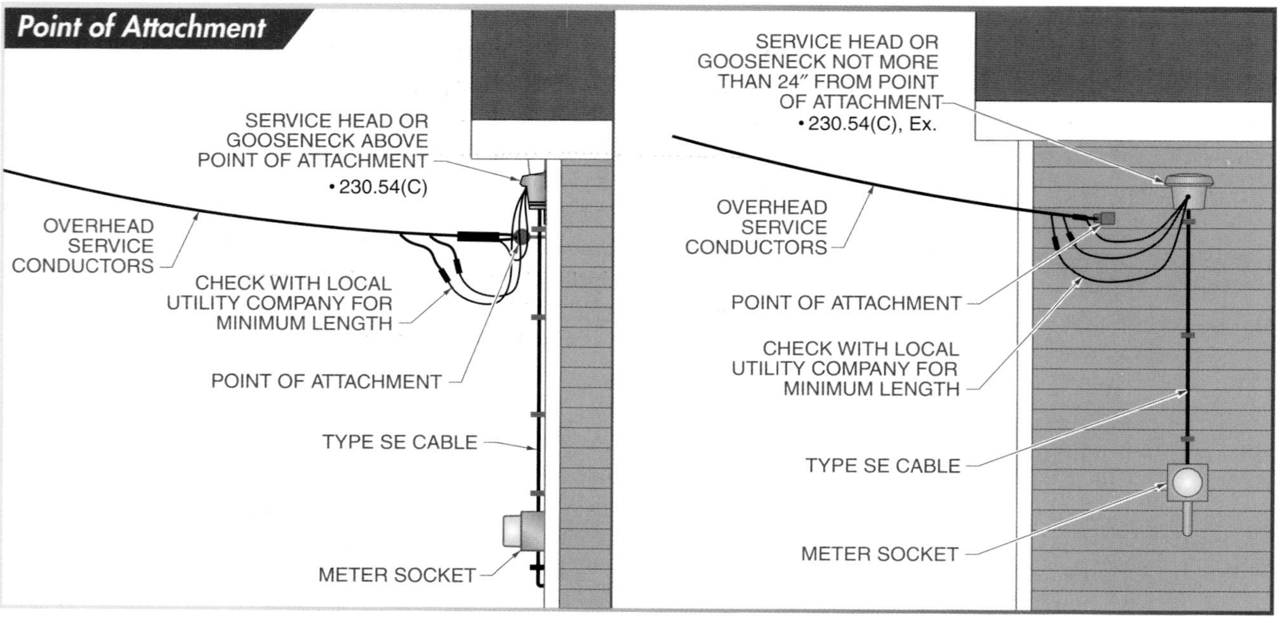

Figure 4-28. Where practicable, the service head or gooseneck shall be installed above the point of attachment of the service-drop conductors.

Drip Loops 230.54(F). The final requirement to prevent the entrance of water into the service raceway or service-entrance cable is the formation of drip loops on all of the individual service-entrance conductors. In addition, 230.54(F) requires that the point of connection between the overhead service conductors and the service-entrance conductors be either below the level of the service head or below the level of the termination point of the cable sheath.

While the Code does not contain any specific length requirement for the drip loops, installers of services should check the installation requirements for their particular electrical utility company. Many utility companies require a minimum of 3′ for the drip loop and for connection to the overhead service conductors.

SERVICE EQUIPMENT – ARTICLE 230, PARTS V AND VI

Designers and installers of electrical systems are required to ensure that the service equipment selected for the electrical system is suitable for the intended use. Article 100 defines service equipment as all of the necessary equipment for the control of electrical power to a building or a structure. Service equipment typically includes the main circuit breaker, or switch and fuses, and any accessories.

No matter what equipment is selected, 230.62 requires that the energized parts shall be either enclosed or guarded. Enclosing or guarding the energized parts protects them from accidental contact by personnel and from physical damage. Proper work clearances shall be provided about service equipment. Work space clearances shall be in accordance with those specified in Section 110.26 for installations 600 V, nominal, or less.

Location of Service Equipment

The local utility company determines the point of attachment of the overhead service conductors to the building or structure. Always check with the local utility prior to locating service equipment.

AIR Ratings – 110.9

The service equipment shall be suitable for the available system short-circuit current. Section 110.9 requires that all equipment intended to break current at fault levels shall have an ampere interrupting rating (AIR) sufficient for the available short-circuit current.

Any service equipment selected for use in an electrical distribution system shall have an AIR sufficient for the available fault current at its supply terminals. For example, a 100 A main circuit breaker with an AIR of 16,000 A could not be installed in service equipment where the available fault current is 23,000 A.

The UL's *Panelboard Marking Guide* requires that each panelboard be marked with the phrase "Short-Circuit-Current-Rating" and the rating in rms symmetrical amperes. Such markings list the maximum fault current to which the panelboard could be subjected and still withstand the magnetic forces generated by the fault current and ensure that the OCPDs are capable of clearing the circuit under fault conditions.

The available short-circuit current depends on several factors. Typically, as the size of the electrical system increases, the available fault current also increases. Conversely, the further away from the main power supply the equipment is located, the lower the available fault current. The effect of the requirement of 110.9 is that designers and installers of electrical services should contact their local utility company to determine how much available fault current is present at the location in which the service equipment is being installed.

Identification – 230.66

In addition to the AIR rating, installers and designers of electrical distribution systems shall ensure that all service equipment is identified as being suitable for use as service equipment. All service equipment rated at 600 V or less shall be marked to show such suitability per 230.66. Marking is typically in the form of a label which is attached to the equipment. All service equipment is required to be listed.

Some equipment may also be marked to indicate that the equipment is "suitable only for use as service equipment." Equipment with this type of marking usually has a factory connection or bonding between the neutral terminal bar and the enclosure. This equipment is not suitable for use in sub-panel applications. Section 230.66 also states that individual meter socket enclosures are not considered service equipment and therefore need not be marked as suitable for service equipment.

Disconnecting Means – 230.70

A means shall be provided to disconnect all service-entrance conductors from all of the other conductors in the building. This isolates the electrical power from the building in the event of a fire or life-safety hazard. The service disconnect is used to accomplish this objective.

> **_Service-Entrance Conductors_**
>
> *For several NEC® cycles, attempts have been made to place a limitation in 230.70(A)(1) on the distance that service-entrance conductors can run in a building or structure before entering the service disconnecting means. CMP-4 has continually rejected such proposals on the basis that each installation presents unique requirements.*
>
> *For example, if a large oil heating tank is installed directly inside the point of entrance of the service conductors, enough of the service-entrance conductors could be installed to place the service disconnecting means at a readily accessible location. This does not, however, permit installers to run the service-entrance conductors to the other side of the basement merely for convenience. If there are any doubts about the location of the service-disconnecting means, check with the AHJ before beginning the installation.*

Location – 230.70(A)(1–2). Service disconnecting means are not permitted to be installed in bathrooms. In addition, the service disconnecting means shall be installed in a readily accessible location. Article 100 defines a readily accessible location as one that is not obstructed and is therefore capable of being reached quickly, without requiring the use of ladders or other portable means.

The service disconnecting means is permitted to be installed either outside or inside of the building or structure. When the service disconnecting means is installed inside of the building or structure, it shall be installed nearest the point of entrance of the service conductors. There is no maximum distance established that the unprotected service conductors are permitted inside the building or structure. The location shall be at the first readily accessible point nearest the entrance of the service conductors.

Marking – 230.70(B). The service disconnecting means shall be clearly and permanently marked to distinguish its purpose. As with other identification requirements in the NEC®, such as 110.22(A), this is designed to aid any personnel who might be working on the service equipment or emergency operations personnel who may need to quickly identify the service disconnecting means and disconnect power from the building.

Suitable for Use – 230.70(C). Service equipment shall be suitable for the location in which it is installed. For example, if the service disconnect is located outside the building, as permitted by 230.70(A)(1), then it shall be suitable for or identified for wet locations. Likewise, if service equipment is installed in a hazardous location, then it shall be suitable for the particular Class and Division for which it is installed and shall comply with all of the applicable requirements of Articles 500 through 517.

Means shall be provided to disconnect all conductors in a building from the service-entrance conductors.

Maximum Number of Disconnects – 230.71. The service disconnecting means shall consist of not more than six switches or circuit breakers. The six devices are permitted to be mounted in a single enclosure, installed in separate enclosures provided they are grouped, or installed in or on a switchboard. The two to six service disconnects are required to be grouped and clearly marked to indicate the load which they serve per 230.72(A).

The purpose of limiting the maximum number of service disconnects and requiring that they be grouped together is to ensure that the entire service, or services, as permitted by 230.2, can be shut down at a single location with no more than six operations of the hand. Permission to install the two to six service disconnects gives latitude to the electrical designer, yet provides for the quick interruption of power in the event of an emergency. **See Figure 4-29.**

Grouping – 230.72. The two to six disconnects shall be grouped per 230.72(A). Each disconnect shall be marked to indicate its load. Grouping facilitates installation and maintenance of the disconnects in addition to providing a central location to turn the disconnects OFF.

Section 230.72(B) permits the one or more service disconnects installed for a fire pump, emergency systems, legally required standby services, or optional standby services to be installed at a location that is sufficiently remote from the other service disconnecting means. This helps to ensure that the emergency power or fire pump service is not inadvertently disconnected by emergency personnel trying to interrupt normal power to the building.

Rating – 230.79. As a general rule, the service disconnecting means is required to have a rating which is equal to or greater than the load to be carried. The service load shall be calculated per Article 220. Subparts (A), (B), (C), and (D) to 230.79 establish minimum service disconnecting ratings.

For limited service loads that consist of a single branch circuit, 230.79(A) requires a minimum service disconnect rating of 15 A. For service loads that consist of not more than two branch circuits, 230.79(B) requires a minimum service disconnect rating of 30 A. Service loads to a one-family dwelling require a minimum service disconnect rating of 100 A.

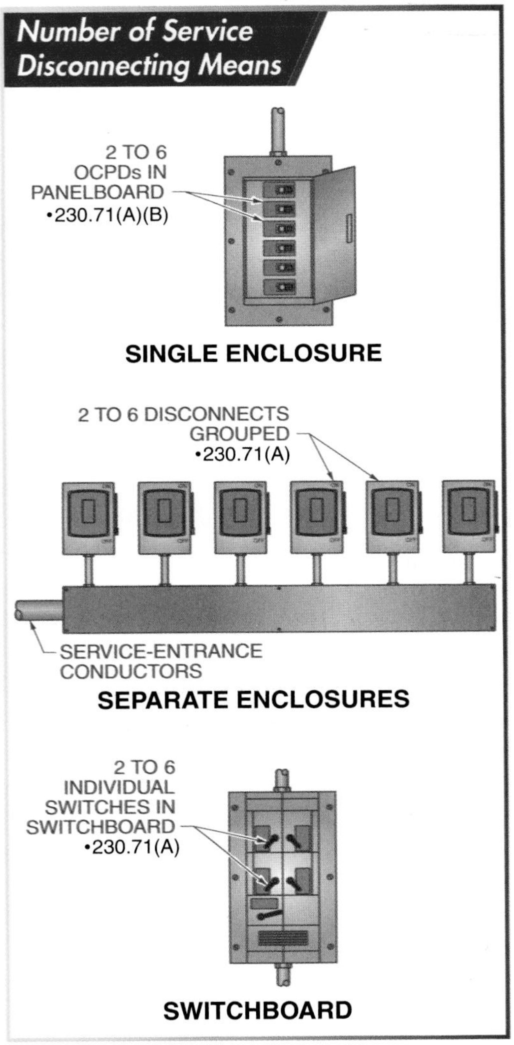

Figure 4-29. The service disconnecting means shall consist of not more than six switches or circuit breakers.

For all other service loads, 230.79(D) requires a minimum service disconnect rating of 60 A. If two to six disconnects are installed per 230.71, then 230.80 requires that the total of all the service disconnecting switches or circuit breakers be equal to or greater than the ratings required by 230.79. For example, if a one-family dwelling is served by two disconnecting means, the combined rating of the disconnects is required to be at least 100 A.

Line-Side Connections – 230.82. In general, there shall be no connections on the line or supply side of the service disconnecting means. Conductors connected to the line side of the service disconnect are not provided with any overcurrent protection and are not permitted for most installations. Often, however, line-side taps or connections are necessary for some applications. Section 230.82 lists nine specific applications where such connections are permitted.

The most common application on the line side of the service disconnecting means is for the connection of metering equipment. Meters are permitted to be connected to the line side of the service disconnecting means per 230.82(2). The meters shall be rated not above 600 V, nominal, and the meter sockets or enclosures shall be grounded and bonded per Article 250.

Another application in which line-side taps are permitted is for cable limiters or other current-limiting devices. These devices are permitted to be connected to the line side of the service disconnecting means per 230.82(1). Such devices are intended to protect downstream conductors and equipment from dangerous let-through currents that might result from short circuits or ground faults.

Taps on the line side are also commonly used to supply power for fire pumps. Article 695 contains the provisions under which this connection can be made and 230.82(5) permits such a connection to be made. See 695.3(A)(1).

Overcurrent Protection – 230.90

In general, service-entrance conductors shall be provided with overcurrent protection. Overcurrent protection for service equipment consists mainly of overload protection for the conductors. Service overcurrent protection, in most cases, does not provide protection against short circuits or ground faults that may occur on the line side of the service overcurrent protection. Section 230.90(A) specifies that each ungrounded conductor shall be provided with this protection by placing an OCPD in series with the conductor. The maximum rating or setting of the OCPD shall not exceed the allowable ampacity of the conductors.

There are particular installations where application of this rule can be difficult. Motor-starting current, for example, can increase up to six times the full-load running current. In these cases, the OCPD shall be sized to permit the motor to start. A rating higher than the allowable ampacity is permitted by 230.90(A), Ex. 1, when necessary to handle motor-starting currents. **See Figure 4-30.**

The requirement for overload protection does not include the grounded conductor. Opening the grounded conductor without opening the ungrounded conductors could result in dangerous voltages at the equipment. Section 230.90(B) prohibits OCPDs from being placed in series with a grounded conductor unless the device is a circuit breaker, which simultaneously opens all conductors of the circuit.

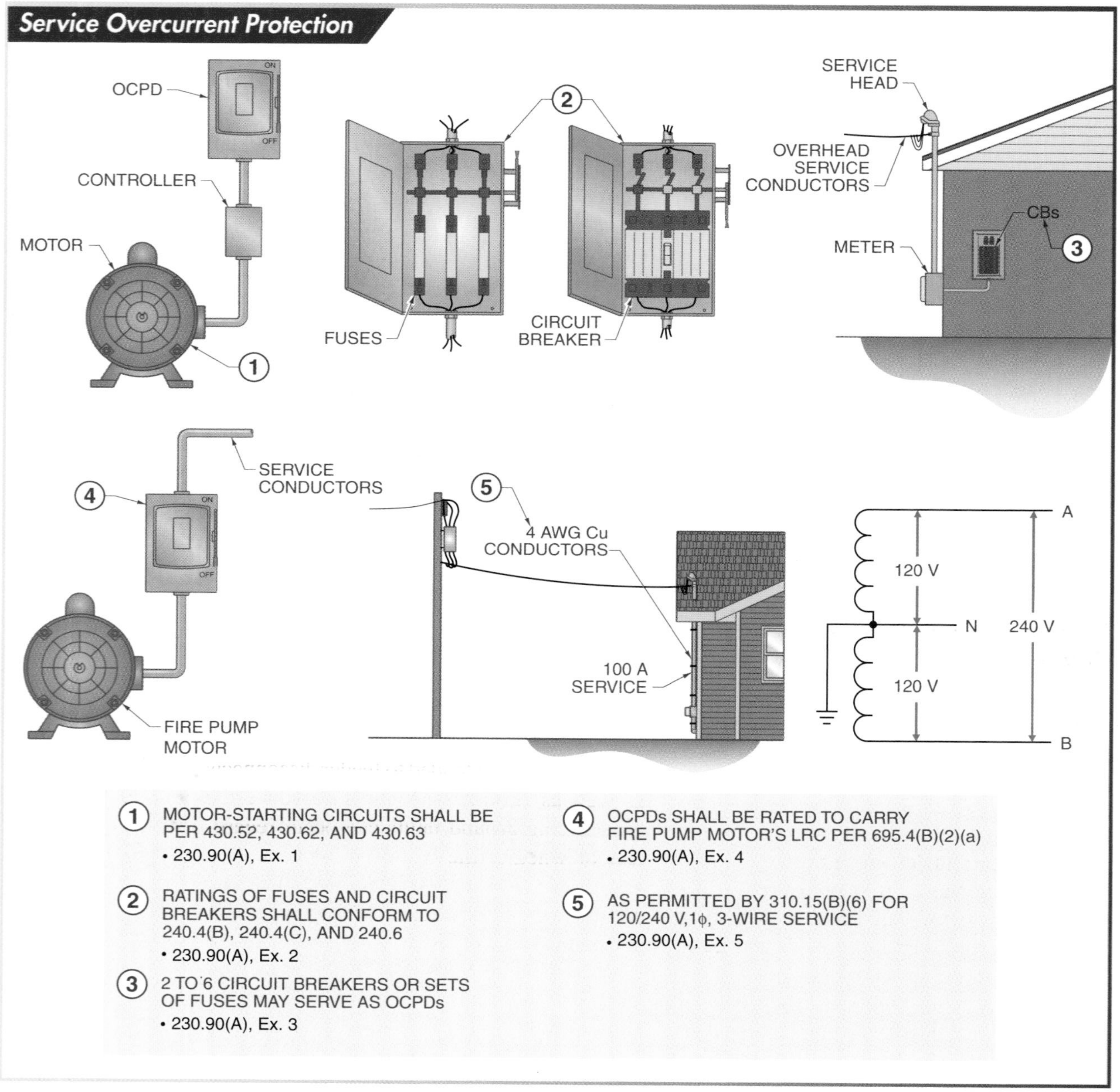
Service Overcurrent Protection
OCPD
ON
OFF
CONTROLLER
MOTOR
1
2
FUSES
CIRCUIT BREAKER
SERVICE HEAD
OVERHEAD SERVICE CONDUCTORS
METER
CBs
3
4
SERVICE CONDUCTORS
FIRE PUMP MOTOR
5
4 AWG Cu CONDUCTORS
100 A SERVICE
A
120 V
N
240 V
120 V
B
1 MOTOR-STARTING CIRCUITS SHALL BE PER 430.52, 430.62, AND 430.63
• 230.90(A), Ex. 1
2 RATINGS OF FUSES AND CIRCUIT BREAKERS SHALL CONFORM TO 240.4(B), 240.4(C), AND 240.6
• 230.90(A), Ex. 2
3 2 TO 6 CIRCUIT BREAKERS OR SETS OF FUSES MAY SERVE AS OCPDs
• 230.90(A), Ex. 3
4 OCPDs SHALL BE RATED TO CARRY FIRE PUMP MOTOR'S LRC PER 695.4(B)(2)(a)
• 230.90(A), Ex. 4
5 AS PERMITTED BY 310.15(B)(6) FOR 120/240 V,1ϕ, 3-WIRE SERVICE
• 230.90(A), Ex. 5

Figure 4-30. Service-entrance conductors shall be provided with overcurrent protection.

Clearance requirements are provided to protect the overhead service conductors from physical damage and to protect personnel from contact with the conductors.

In order to protect the service equipment, 230.91 requires that the location of the service OCPD be within the service disconnecting means or immediately adjacent to it. In addition, each occupant of a multi-occupancy building is required by 240.24(B) to have access to the service overcurrent devices. Note that such access is not required where the building management has continuous building supervision. In these occupancies, the service OCPDs need only be accessible to authorized management personnel. See 240.24(B)(1),(2). A similar requirement for access to the service disconnecting means in multi-occupancy buildings is included in 230.72(C), Ex.

Ground-Fault Protection of Equipment (GFPE) – 230.95

There is another type of protection required by the NEC® that is sometimes confused with, but is in fact very different from, GFCI protection. GFCI protection provides protection for personnel by disconnecting the circuit in about 1/40 of a second in the event of a 5 mA discrepancy between the current flowing on the ungrounded conductor and the current returning on the grounded conductor. GFCI protection is most commonly found on 125 V, 15 A and 20 A receptacles. Ground-fault protection of equipment (GFPE), on the other hand, provides protection for equipment. GFPE is designed to operate at settings not greater than 1200 A and with a maximum delay of 1 second for fault currents equal to or greater than 3000 A.

GFPE protection is designed to ensure that arcing faults of wye-connected systems operating at over 150 V to ground, but less than 600 V phase-to-phase, do not occur on each service disconnect rated 1000 A or more. Studies have shown that these systems, particularly the 277/480 V, 3ϕ, 4-wire, solidly grounded, wye-connected systems have a tendency to permit arcing ground faults to occur, which can severely damage the service equipment.

The GFPE requirements are tightly drawn to cover these types of systems when the service disconnect is rated at 1000 A or more. For example, if a 1000 A rated service disconnecting means is installed with an 800 A OCPD and conductors are rated at 800 A, GFPE is still required. It is the rating of the service disconnect, not the ampacity of the conductors or the setting of the OCPD, that is the determining factor.

The setting of the GFPE shall allow the device to open the system in 1 second for ground faults equal to or greater than 3000 A per 230.95(A). The maximum setting permitted for the GFPE is 1200 A. Section 230.95(C) is one of the few instances in the NEC® where a specific performance test of a system is required. Upon completion of the GFPE system, 230.95(C) requires that a test of the system be done and a written record of the test be made and be available to the AHJ.

In the 1990 NEC®, the provisions for GFPE protection were extended to feeder disconnects as well as service disconnects because the same arcing ground faults can occur regardless of whether the circuit is a feeder or a service. Section 215.10 refers to 230.95 and incorporates all of the provisions for service disconnects to feeders.

QUICK QUIZ®

Refer to Chapter 4 Quick Quiz® on CD-ROM.

Additional information is available at ATPeResources.com

chapter 4

Services

REVIEW QUESTIONS

Name Bradly Lowry Date ______________

______________ **1.** The ___ is the electrical supply, in the form of conductors and equipment, that provides electrical power to the building or structure.

______________ **2.** The ___ is the overhead conductors that extend from the utility supply system to the service-entrance conductors at the building or structure.

______________ **3.** The ___ is the underground conductors that connect the utility's electrical distribution system with the service-entrance conductors.

Service point **4.** The ___ is the point of connection between the local electrical utility company and the premises wiring of the building or structure.

T F **5.** Occupancies which utilize a fire pump for life safety protection are permitted to have a separate service provided solely for the fire pump.

(T) F **6.** Fittings for use with service masts shall be identified for the use.

T F **7.** GFPE is designed to operate at settings not greater than 200 A.

service **8.** ___ conductors are the conductors that extend from the service disconnecting means to the service point.

special permission **9.** ___ is the written approval of the AHJ.

10 **10.** As a general rule, there shall be a(n) ___′ minimum clearance from final grade for all overhead service conductors.

(T) F **11.** As a general rule, only service conductors shall be installed in service raceways.

(T) F **12.** For a panelboard to be suitable for use as service equipment, not more than six main disconnecting means may be provided.

D **13.** Where it is impractical to mount service heads above the point of attachment, the service head or gooseneck shall be permitted to be located not more than ___″ from the point of attachment, of the overhead service conductors.

A. 6
B. 12
C. 18
D. 24

______________ **14.** Individual meter socket enclosures ___ service equipment.

A. are not considered
B. need not be marked as suitable for
C. all of the above
D. none of the above

C **15.** The service disconnecting means shall consist of not more than ___ switches or circuit breakers.

A. two
B. four
C. six
D. none of the above

C **16.** As a general rule, the service disconnecting means is required to have a rating which is ___ the load to be carried.

A. 75% of
B. equal to or less than
C. equal to or greater than
D. 125% of

T F **17.** Conductors located on the line or supply side of the service point are covered by the NEC®.

T F **18.** Service conductors are provided overcurrent protection by overcurrent protective devices that have a rating or setting not higher than the allowable ampacity of the conductor.

T F **19.** Overhead service conductors are required to be insulated or covered.

3 **20.** The 8′ minimum vertical clearance for overhead service conductors shall be maintained in all directions for a minimum distance of ___′.

Service mast **21.** A(n) ___ is an assembly consisting of a service raceway, guy wires or braces, service head, and any fittings necessary for the support of overhead service conductors.

8 **22.** Direct-buried conductors emerging from the ground shall be protected from physical damage up to a point which is at least ___′ above finished grade.

1 1/4 conduit

(T) F **23.** A 100 A service is the minimum rating required for any one-family dwelling.

24. Type ___ cable is the most commonly used wiring method for 120/240 V, 1ϕ, 3-wire services in one-family dwellings.

T (F) **25.** Service disconnecting means are permitted to be installed in bathrooms.

T F **26.** A minimum service disconnect rating of 30 A is required for service loads that consist of not more than two branch circuits.

27. ___ is all of the necessary equipment to control the supply of electrical power to a building or a structure.

2 **28.** Service conductors are permitted to be routed within a building if they are installed under not less than ___″ of concrete beneath the building or structure.

C **29.** The minimum size for most overhead service conductors is ___.

A. 8 AWG Cu
B. 6 AWG Al
C. all of the above
D. none of the above

30. SE cable shall be supported at intervals not exceeding ___″ and within ___″ of every service head, gooseneck, or point of connection to a raceway or enclosure.

A. 12; 12
B. 12; 30
C. 30; 12
D. 30; 30

Clearances

A 1. The clearance from final grade at A is ___′.

A. 10 C. 15
B. 12 D. 18

B 2. The clearance from final grade at B is ___′.

A. 10 C. 15
B. 12 D. 18

C 3. The clearance from final grade at C is ___′.

A. 10 C. 15
B. 12 D. 18

D 4. The clearance from final grade at D is ___′.

A. 10 C. 15
B. 12 D. 18

A 5. The clearance from final grade at E is ___′.

A. 10 C. 15
B. 12 D. 18

D 6. The minimum clearance at F is ___″.

A. 10 C. 15
B. 12 D. 18

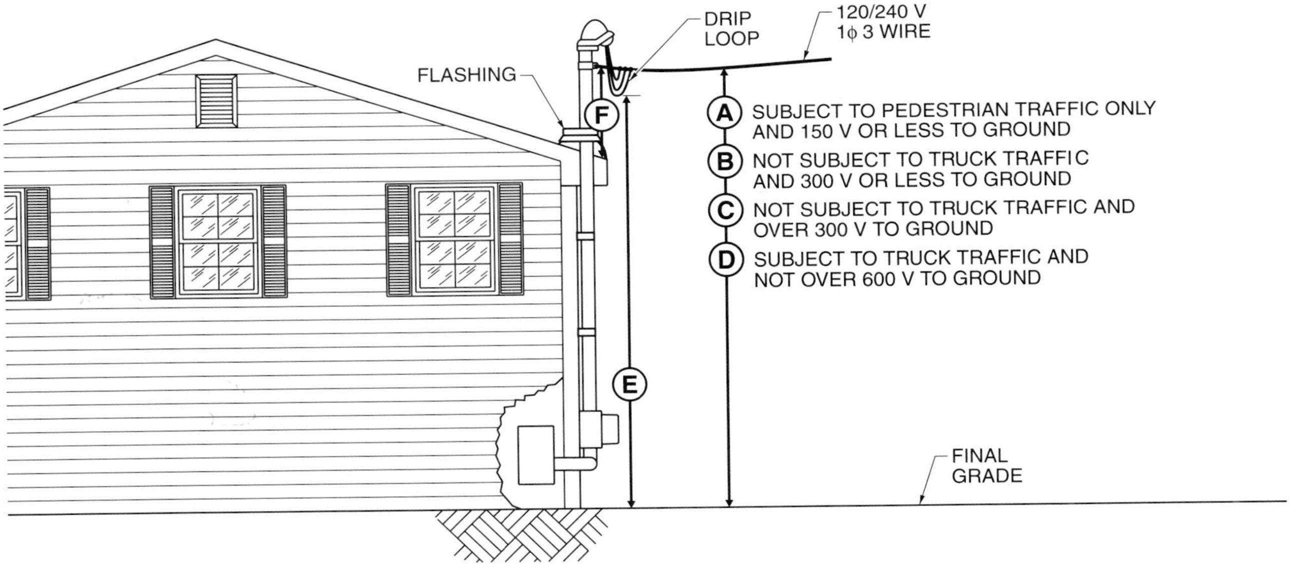

Connectors

______ 1. Compression

______ 2. Mechanical

______ 3. Quick-tap

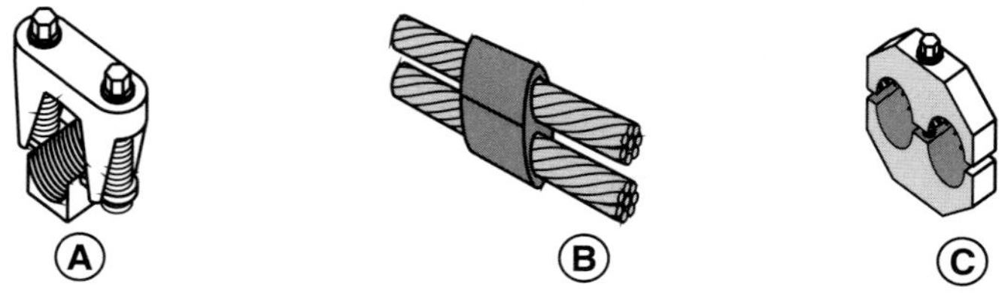

Service Conductor Clearances

___A___ 1. The minimum clearance at A is ___.
- A. no minimum
- B. 18′
- C. 3′
- D. 6′

___C___ 2. The minimum clearance at B is ___.
- A. no minimum
- B. 18′
- C. 3′
- D. 6′

___C___ 3. The minimum clearance at C is ___.
- A. no minimum
- B. 18′
- C. 3′
- D. 6′

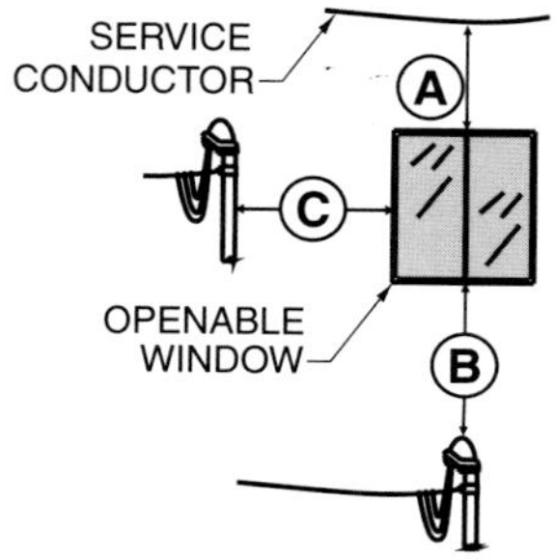

Service Equipment

______ 1. Loadcenter

______ 2. Circuit breaker enclosure

______ 3. Panelboard

______ 4. Service disconnect switch

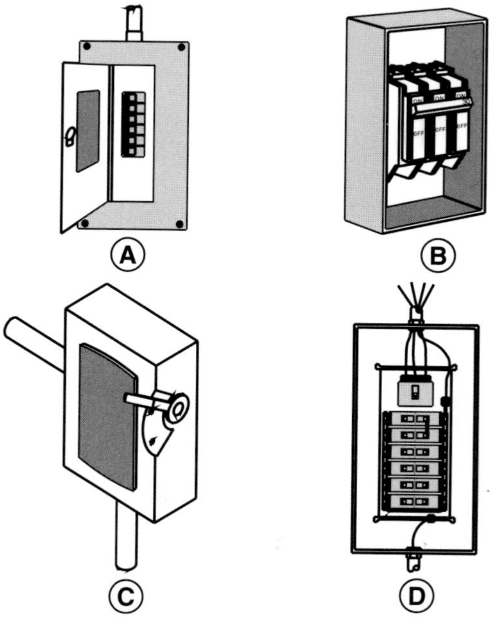

KCC

Kankakee Community
100 Coll
Kankakee,
p. 815.8
f. 815.8
w. ww

Payment Deadline: 5pm-May 2, 2014

Bradly L. Lowry
1200 Stratford Dr West
Bourbonnais IL 60914

Dear Bradly:

Make College Easier to Pay For with a Tuition Payment Plan.

Your school partners with Nelnet Business Solutions to let you pay tuition and fees over time, making college more affordable.

PAYMENT PLAN BENEFITS:

1. **Easy online enrollment**
2. **Flexible payment options**
3. **No interest**

SEE REVERSE SIDE TO LEARN MORE AND ENROLL TODAY!

chapter 4

Services

TRADE COMPETENCY TEST

Name ______________________________ Date ____________________

NEC®	Answer	
230.24(A) 230.24(B)	18 feet	**1.** Determine the minimum clearance for 277/480 V service conductors which are installed above a rooftop and extend over an alley subject to truck traffic.
230.26	10 feet	**2.** What is the minimum clearance from ground permitted for the point of attachment of the overhead service conductors to a building or structure?
		3. A one-family dwelling has an initial net computed load of 9.5 kVA and five branch circuits. Determine the minimum ampacity for the service ungrounded conductors.
		4. What is the maximum number of sets of service-entrance conductors that are permitted to be connected to a single overhead service for one building with four separate occupancies?
		5. Determine the minimum size of the Al ungrounded service-entrance conductors for a structure with a limited load of two, 2-wire branch circuits.
		6. A 14′ vertical run of Type SE cable is installed on the outside of a building between the service head and the meter socket. What is the minimum number of cable supports needed for this installation?
		7. Two separate service disconnecting means are grouped together to supply the service for a one-family dwelling. The ratings of the service disconnects are 30 A and 60 A. The initial computed net load of the building is 10.75 kVA. Does this installation meet the requirements of Article 230, Part F?
		8. Determine the minimum size permitted for Al service-lateral ungrounded conductors which supply a limited load consisting of a single branch circuit.
230.24(A) exception #2	3 feet	**9.** *See Figure 1.* Determine the minimum clearance from the roof required for the overhead service conductors at A.

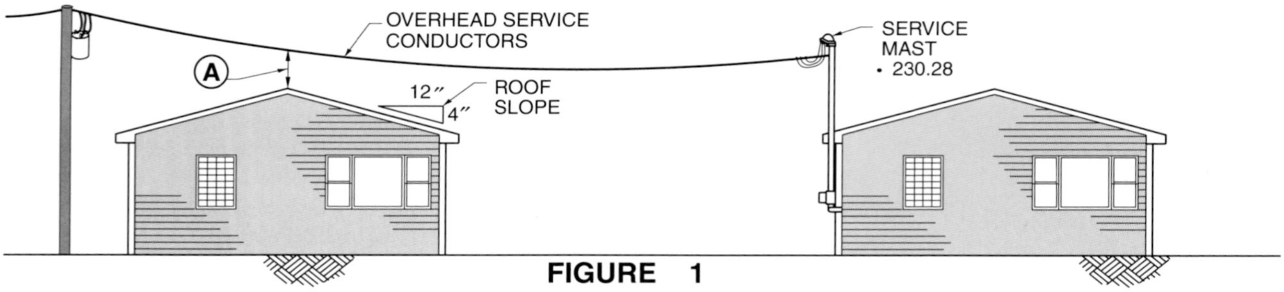

FIGURE 1

230.9(A) — 3 feet

10. Determine the minimum clearance from a building fire escape for service conductors installed as multiconductor cable without an overall outer jacket.

230.24(B)(4) — 18 feet

11. A 120/240 V, 1ϕ overhead service is installed above a public street which is subject to truck traffic. Determine the minimum vertical clearance from ground for the overhead service conductors.

______ ______

12. *See Figure 2.* Determine the minimum size Cu overhead service conductors required for the limited load, single branch-circuit service.

230.24(A) — 8 feet

13. *See Figure 3.* Determine the minimum clearance from the roof for the overhead service conductors at A.

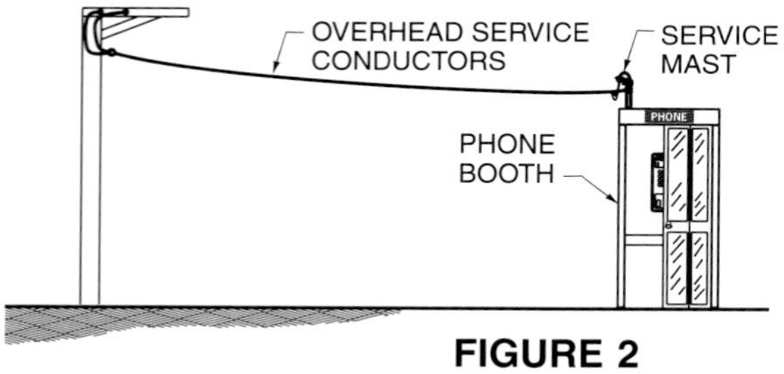

FIGURE 2

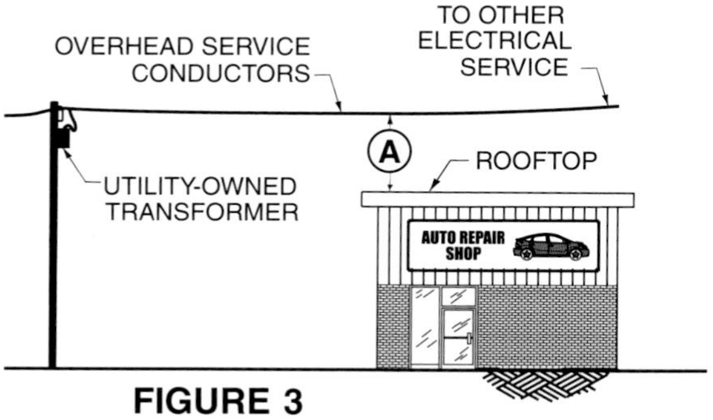

FIGURE 3

230.9 (A) 3 feet **14.** *See Figure 4.* Determine the minimum clearance from the openable windows required for the overhead service conductors at A.

15. *See Figure 5.* Is this installation a violation of Article 230 of the NEC®?

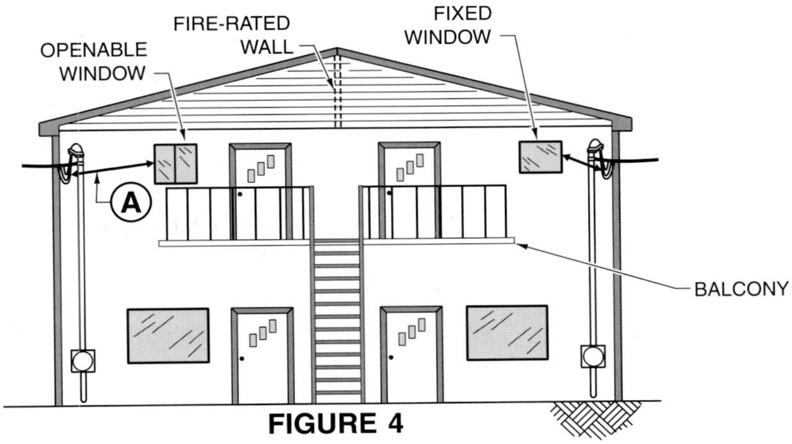

FIGURE 4

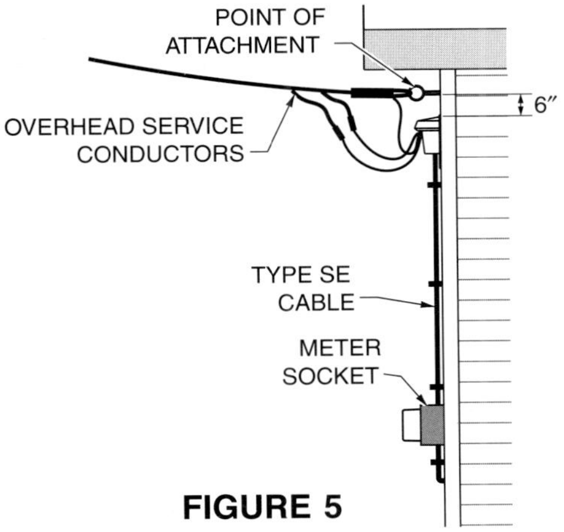

FIGURE 5

Learning the Code

- Conductors are the basic components of the electrical system. The physical construction of the conductors and the manner in which they are installed are critical factors in any electrical system. Apprentices and students of the Code must remember that conductors carry current and must be adequately protected. Protection usually is provided at the point at which the conductor receives its supply and the protection is in accordance with its allowable ampacity.
- Ampacity is often a difficult term and concept for apprentices and students of the Code to grasp. The word is actually unique to the NEC® and is formed from a combination of the words "ampere" and "capacity." Keeping this in mind may make some rules and requirements easier to understand. For example, the ampacity of a conductor is the allowable current (amperes) that the conductor can carry continuously under the conditions of use without exceeding its temperature rating (capacity). Two tables in Article 310 are especially helpful in understanding the concept of ampacity. Table 310.104(A) provides conductor applications for insulated conductors, and Table 310.15(B)(16) lists the allowable ampacities of insulated conductors for 0 V through 2000 V.

- When trying to determine a conductor's allowable ampacity, it is often helpful to consider the entire installation from the point of supply to the utilization equipment or outlet where the conductor terminates. Students often incorrectly think of the conductor as it sits on the shelf.

  For example, according to Table 310.15(B)(16), a 12 AWG copper THW conductor has a listed allowable ampacity of 25 A. But conductors don't sit on the shelf—they are installed in an electrical system, and that 12 AWG conductor is subject to all of the factors of Articles 240 and 310. What is the temperature rating of the equipment that the conductor is terminated in? Don't forget, a conductor has two ends. What are the appropriate conditions of use? Ambient temperature? Number of conductors in the cable or raceway? When you are asked to calculate minimum ampacities for a conductor, remember that all these factors must be considered to determine the actual allowable ampacity.

chapter 5

Conductors and Overcurrent Protection

Conductors carry current throughout the electrical system. All conductors are considered to be copper unless otherwise specified. Conductors may be bare, covered, or insulated and may be solid or stranded. Insulated conductors are the most common conductors installed in electrical systems.

New in the 2011 NEC®

- Clarification for OCPDs rated 800 A or less – 240.4(B)(1)
- Ungrounded overcurrent protection for 3-wire DC circuits – 240.15(B)(4)
- New requirements for CBs without instantaneous trip characteristics – 240.87
- New provisions for protection of conductors in supervised industrial installations – 240.91
- Complete reorganization of Article 310, with renumbering changes to all sections

CONDUCTORS – ARTICLE 100; ARTICLE 310

The heart of all electrical systems is the conductors that carry the current to electrical devices and utilization equipment. There are associated effects whenever current flows through a conductor. For example, a magnetic field that varies in direct proportion to the amount of current flowing in the conductor is formed when current flows.

Heat, another property that is associated with the flow of current in a conductor, is one of the most significant factors to be considered in designing electrical systems. Conductors, in general, are rated on their ability to withstand the effects of heat. The conductor's rating is dependent upon its type of electrical insulation.

Conductors are assigned a specific ampacity that reflects the insulation's ability to handle and dissipate heat under varying conditions. For example, Table 310.15(B)(16) lists the allowable ampacity of a 12 AWG Cu conductor with THHN insulation at 30 A, while a similar 12 AWG Cu conductor with TW insulation has a listed allowable ampacity of only 25 A. In general, all conductors shall be protected against overcurrents in accordance with their listed ampacities, at the point where they receive their supply.

Definitions – Article 100

Conductors used in electrical systems are often defined by the material, if any, that is used to encase the actual electrical conductor. Three classifications exist for electrical conductors: bare, covered, and insulated. **See Figure 5-1.**

Bare. Bare conductors have no outer covering or insulation encasing the actual conductor material. Bare conductors provide no protection for the conductor material, thus the NEC® has very limited applications for bare conductors.

The most common use for bare conductors is for equipment grounding purposes. Copper, aluminum, or copper-clad aluminum may serve as an equipment grounding conductor (EGC) per 250.118(1). The conductor is permitted to be bare and in the form of a busbar or wire.

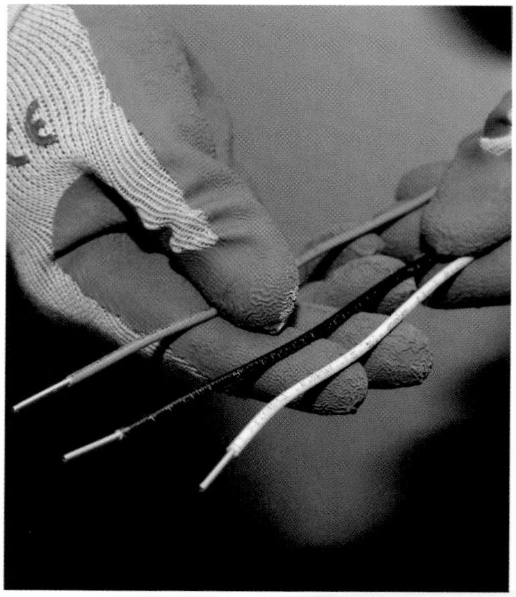

Insulated conductors are typically used for wiring receptacles and switches.

Conductors

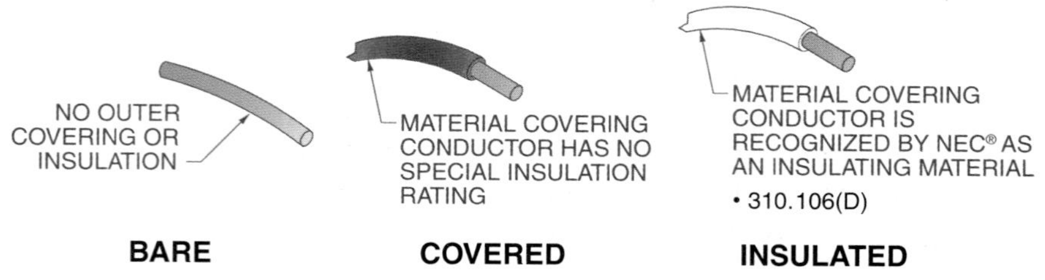
NO OUTER COVERING OR INSULATION
BARE
MATERIAL COVERING CONDUCTOR HAS NO SPECIAL INSULATION RATING
COVERED
MATERIAL COVERING CONDUCTOR IS RECOGNIZED BY NEC® AS AN INSULATING MATERIAL
• 310.106(D)
INSULATED

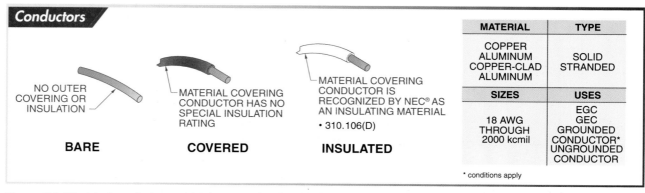

MATERIAL	TYPE
COPPER ALUMINUM COPPER-CLAD ALUMINUM	SOLID STRANDED
SIZES	**USES**
18 AWG THROUGH 2000 kcmil	EGC GEC GROUNDED CONDUCTOR* UNGROUNDED CONDUCTOR

* conditions apply

Figure 5-1. Electrical conductors are bare, covered, or insulated.

Another application for bare conductors is as a grounding electrode conductor (GEC). The GEC is permitted to be constructed of copper, aluminum, or copper-clad aluminum and it may be a bare conductor per 250.62. Section 250.64(A) requires aluminum or copper clad aluminum grounding conductors to be terminated at least 18″ above the earth, where installed outdoors.

For the most part, service-entrance (SE) conductors are required to be insulated. However, the grounded conductor (neutral) may be a bare conductor under specific conditions per 230.41, Ex. A common element in all the provisions for the use of bare conductors is that the hazards associated with the potential electrical flow in these conductors have been lessened by isolation or circuit design.

Covered. Of the three basic types of conductors, covered conductors are utilized the most infrequently in electrical systems. By definition, covered conductors are constructed with a material that encases the electrical conductor, but the material has not been evaluated and is not recognized as having a specific insulation rating. Grounding electrode conductors and equipment grounding conductors are both permitted to be covered conductors, but more frequently these are installed in electrical systems as either bare or insulated conductors. See 250.62.

Perhaps the most frequent use of covered conductors is in conductors installed in free air, in particular overhead service conductors. By their nature, overhead service conductors are installed in a manner that does not make them accessible to unqualified persons. Additionally, because these conductors are installed in free air, their capacity to dissipate heat associated with current flow is greatly enhanced. While covered electrical conductors may offer some degree of electrical insulation, the covering has not been evaluated as an electrical insulation recognized by the NEC®.

Insulated. Insulated conductors are the most common conductors installed in electrical systems. These conductors are constructed with a material that has been identified by the NEC® as a recognized electrical insulation. Unless specifically permitted elsewhere in the NEC®, all conductors shall be insulated per 310.106(D).

There are numerous types of electrical insulations. The type of insulation selected depends upon the conditions of use of the conductor. A conductor for direct burial requires an insulation that is suitable for the conditions it is likely to be exposed to over the life of the conductor. Conductors for direct burial shall be identified for such use per 310.10(F). For example, UF cable is an underground feeder and branch-circuit cable. The outer covering of UF cable shall be suitable for direct burial in the earth per 340.2. *Cable* is a factory assembly with two or more conductors and an overall covering.

Conductors exposed to potentially corrosive conditions shall be provided with an insulation suitable for such exposure. Corrosive conditions include installations where the conductors are exposed to oils, greases, vapors, etc. that break down the integrity of the conductors' insulation. For example, NMC cable is identified in 334.116(B) as being constructed of an overall covering that is corrosion-resistant. See 310.10(G).

Conductors installed in wet locations require insulation that protects conductors from continuous exposure to moisture. A wet location is any location in which a conductor is subject to saturation from any type of liquid or water. Unprotected locations outdoors, installations underground or concrete slabs, and installations in direct contact with the earth are classified as wet locations. See Article 100. Section 310.10(A), (B), and (C) contains the requirements for conductors in dry, damp, and wet locations. Subpart (B) lists the various types of conductors permitted in dry and damp locations. When conductors are installed in wet locations, 310.10(C) requires that the insulated conductors and cables be either moisture-impervious metal sheathed, be of a type listed for use in wet locations, or be of one of the following types: MTW, RHW, RHW-2, TW, THW, THW-2, THHW, THWN, THWN-2, XHHW, XHHW-2, or ZW.

A common element among all electrical insulations is that they provide an opposition or resistance to the flow of electricity. All conductors and cables are marked to identify, among other things, the type of insulation that is used. See 310.120(A)(2). All conductors and cables are required to be marked to identify the maximum rated voltage at which the conductor is listed, the letter or letters necessary to identify the type of wire or cable, the manufacturer's name or trademark or other identifiable markings, and the AWG or circular-mil area size of the conductor. **See Figure 5-2.**

The specific insulation types listed for use in wet locations contain the letter "W" in their prefix. The "W" identifies the conductor as having an insulation that is moisture-resistant and suitable for use in wet locations. Table 310.104(A) can be used to identify the prefix markings found on conductors.

Another important consideration in selecting conductor insulation type is the temperature limitations of the insulation. Table 310.104(A) provides the maximum operating temperature for the particular insulation types. The three basic insulation temperature ratings listed for the conductors used for general wiring are 60°C (140°F), 75°C (167°F), and 90°C (194°F). These ratings designate the maximum temperature that the conductor can be exposed to without the danger of encountering insulation breakdown and damage.

The common factors that can contribute to conductor insulation degradation because of excessive operating temperature are listed in 310.15(A)(3). **See Figure 5-3.** The first factor is the ambient temperature in which the conductor insulation shall operate. The second factor is the internal heat created in the conductor as a result of current flow. The third factor is the dissipation rate of the heat into the surrounding environment. The final factor is the heat generated by adjacent current-carrying conductors. All of these factors shall be considered prior to selecting the type of insulation and determining the conductor's allowable ampacity.

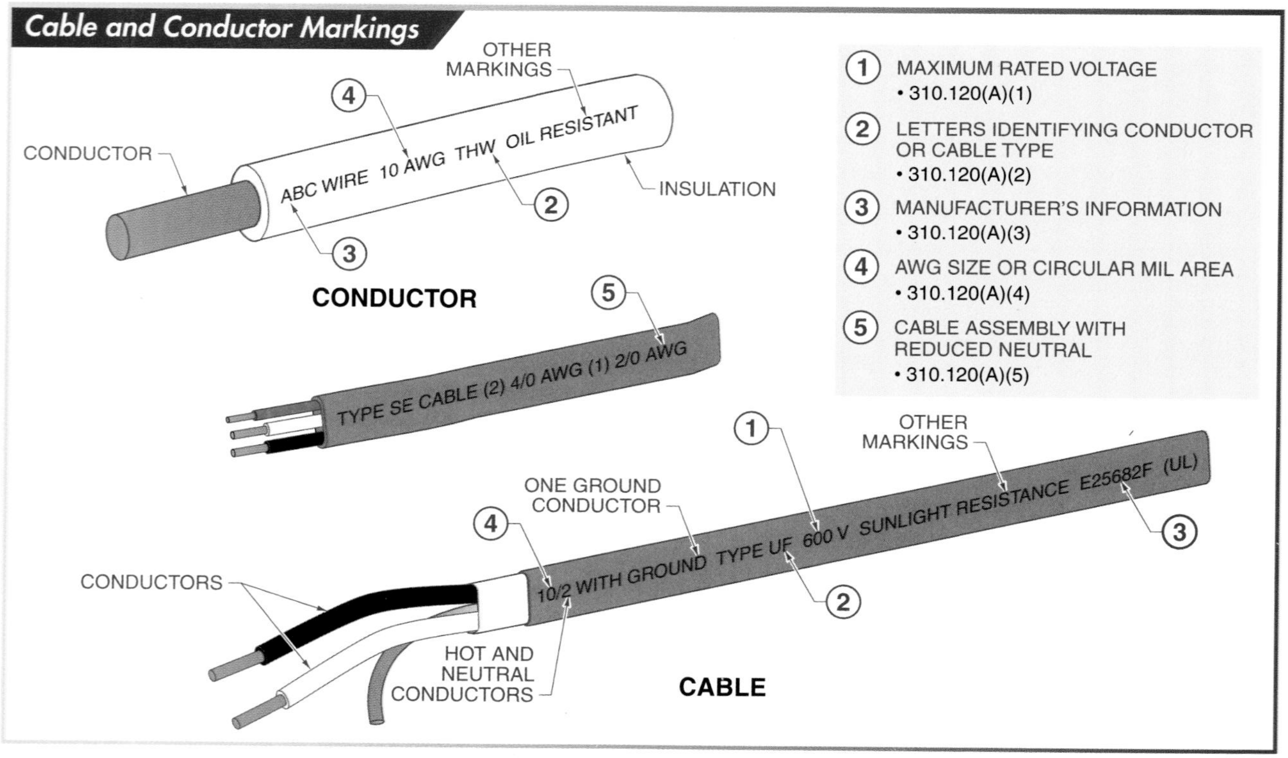

Figure 5-2. All conductors and cables are identified by their markings.

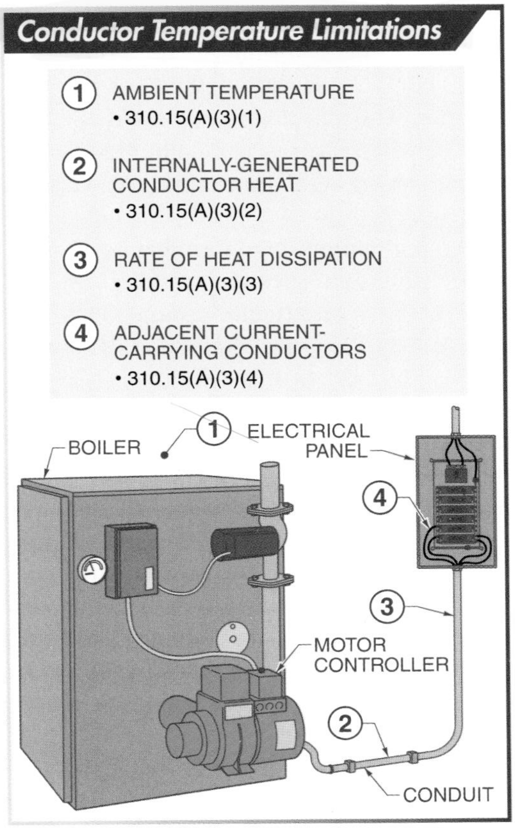

Figure 5-3. Ambient temperature and conductor heat contribute to insulation degradation.

An important Informational Note precedes the general requirements of 310.15(A)(3)(1–4). The Informational Note clarifies the actual temperature rating of a conductor. This rating is the maximum temperature that the conductor is exposed to at any location along its length and the maximum temperature that the conductor can withstand without serious degradation over a prolonged period. This is an important consideration because conductors frequently are exposed to many different temperatures over the length of their installation.

Construction – 310.106; 310.110

The two most common materials used for construction of electrical conductors are copper and aluminum. Copper has excellent conductivity and therefore has a higher allowable ampacity than a comparable aluminum conductor. Aluminum, on the other hand, is not as good a conductor in terms of its allowable ampacity, but is lighter and less expensive to install than copper. The most recent development in conductor material and construction is the advent of copper-clad aluminum.

Copper. Section 110.5 requires all conductors used to carry current to be copper unless otherwise specified by the NEC®. While copper is most extensively used, its cost is a primary disadvantage. The cost of installing electrical distribution systems with copper conductors, particularly in larger sizes, can be very expensive. Another factor to consider when installing electrical conductors is the weight of the conductors. Copper conductors installed in vertical raceways require support at shorter intervals than are required for aluminum conductors. Section 300.19 lists the required intervals and methods for vertical supports to ensure that the weight of the conductors is not passed on to the conductor terminations. **See Figure 5-4.**

Copper conductors are available as hard-drawn, medium-hard-drawn, and soft-drawn. Hard-drawn copper has the greatest strength, but is difficult to work with. Because of the difficulty in bending and shaping hard-drawn copper, its uses are limited. It is used primarily for utility transmission and some service-drop conductor applications.

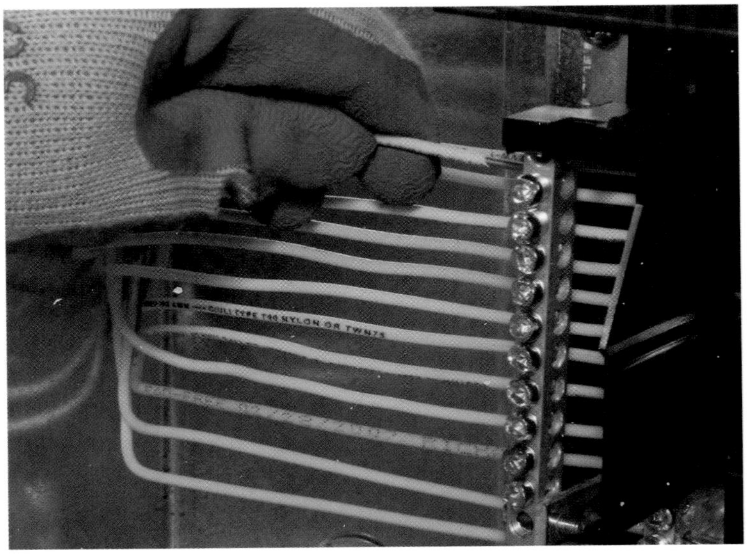

Conductor markings typically include information on the maximum rated voltage, conductor type, and conductor size, as well as manufacturer's information.

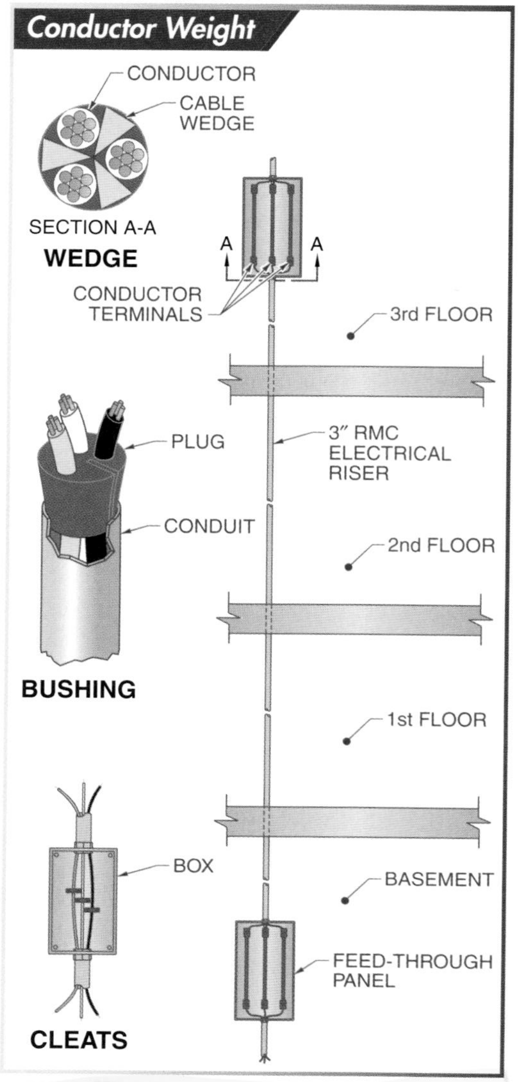

Figure 5-4. The weight of the conductor shall not be passed on to conductor terminations.

Medium-hard-drawn copper is easier to work with than hard-drawn copper but it does not have the tensile strength that hard-drawn copper has. Therefore, it is used primarily for transmission and utility conductors where greater flexibility is required.

Soft-drawn copper is easy to work with and can be installed in many different types of raceways and cable assemblies. Most general building wires used in electrical distribution systems are made of soft-drawn copper.

Aluminum. Aluminum conductors have been used extensively for a number of years in the utility distribution and transmission field. More recently, aluminum conductors have found increasing applications in building electrical distribution systems. Section 310.106(B) requires solid aluminum conductors in sizes 8, 10, and 12 AWG to be constructed of an AA-8000 series electrical grade aluminum alloy. Stranded conductors are permitted in sizes from 8 AWG through 1000 kcmil if constructed from an AA-8000 series electrical grade aluminum alloy.

Despite the fact that aluminum is not as good a conductor as copper, its reduced conductivity is compensated by the considerable cost savings in using aluminum conductors. For the most part, aluminum conductors can be installed with the same installation procedures as used for copper. The primary difference occurs when the conductors are terminated. Because the surface of aluminum conductors oxidizes readily, terminations are generally made with the aid of joint compounds designed to prevent the oxide from re-forming in the installation process. An *oxide* is a thin but highly resistive coating that forms on metal when exposed to the air. If the oxide is not prevented from re-forming, a high-resistance connection could occur which could lead to insulation failure at the termination.

Another consideration when installing aluminum conductors is terminations with dissimilar metals. Conductors of different materials, like copper and aluminum, should not be terminated in a manner that causes the dissimilar metals to come in direct contact with each other unless the termination or splicing device is identified for such use. See 110.14(B). **See Figure 5-5.**

Torque is a turning or twisting force, typically measured in foot-pounds (ft-lb). Manufacturers provide torquing specifications for terminations. The termination should be tightened to those specifications to ensure that the connection is electrically sound. The UL Marking Guide for Molded-Case Circuit Breakers requires that all CBs be marked with their rated tightening torque where field terminations are made. If the tightening torque is dependent on the wire size, the appropriate range of tightening torques shall be provided for each wire size.

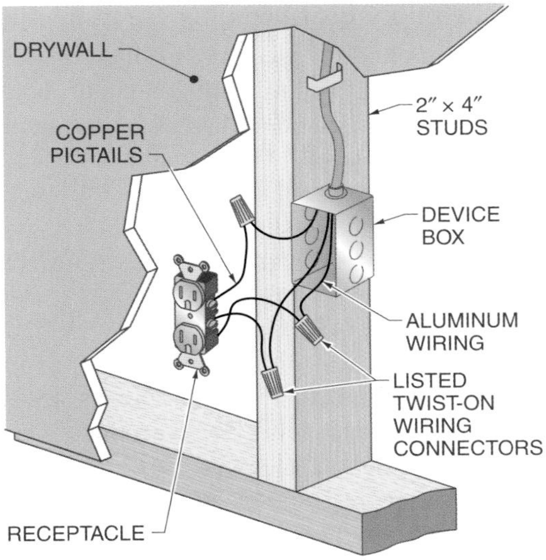

Figure 5-5. Listed twist-on wire connectors are available for directly splicing aluminum to copper.

Copper-Clad Aluminum. Copper-clad aluminum offers a compromise between the increased conductivity and termination qualities of copper conductors with the lighter weight and cost efficiency of aluminum conductors. Copper-clad conductors are constructed using a minimum of 10% copper, which is bonded metallurgically to the aluminum. The ampacity of copper-clad aluminum conductors is selected from the same column used for aluminum conductors. See Table 310.15(B)(16).

Solid or Stranded. Conductors for electrical systems are either solid or stranded. Conductors designed for direct physical contact between aluminum and copper are labeled "AL-CU" and may only be used in dry locations. **See Figure 5-6.** Solid conductors are constructed of a single piece of wire (strand). Stranded conductors are constructed of multiple wires (strands).

Chapter 9, Table 8 lists conductor properties for common building wire. Notice that there are two listings for 8 AWG. Under the heading "conductors" there is a column for stranding. An 8 AWG conductor can be constructed from a single strand (solid) or from seven individual strands, each 0.049″ in diameter (stranded). The choice of solid or stranded conductor lies with the designer and is based largely on the needs of flexibility in the conductor. There are, however, a few provisions in the NEC® that specify either solid or stranded conductors. For example, 310.106(C) requires conductors of size 8 AWG and larger to be stranded where installed in raceways.

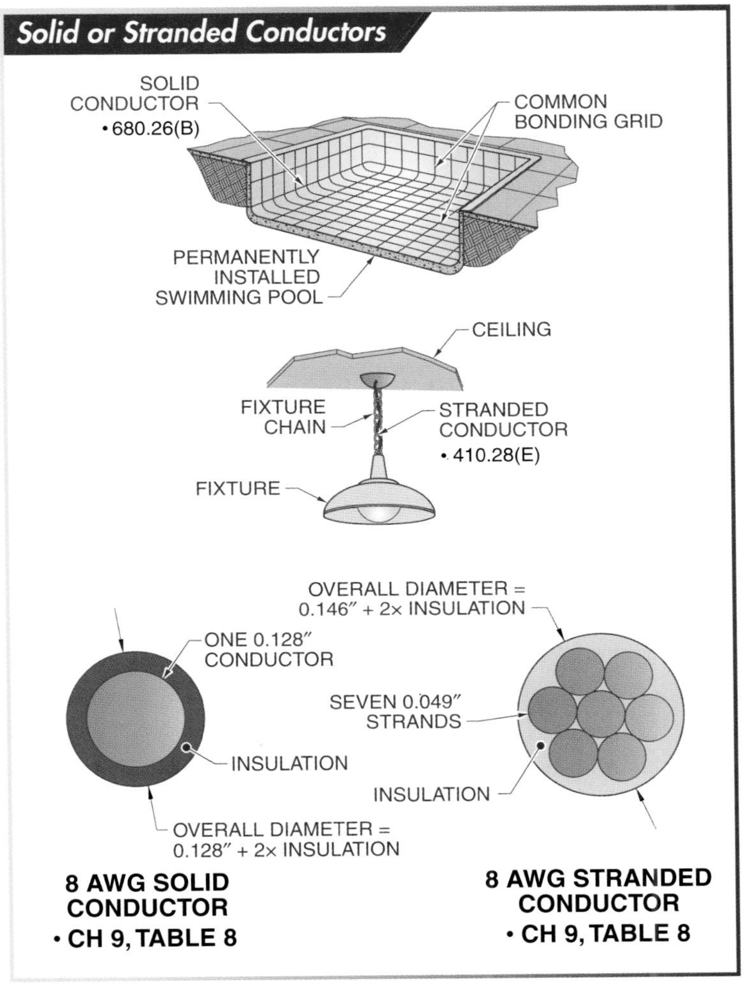

Figure 5-6. Some electrical installations specifically require that solid or stranded conductors be used.

Color Code Identification. The NEC® has adopted various color code requirements for many of the conductors used in electrical distribution systems. **See Figure 5-7.** These requirements ensure proper identification of circuits and conductors, which is critical for the safety of those who are required to maintain the electrical system. Section 310.110 references the applicable color code identification requirements for grounded, equipment grounding, and ungrounded conductors. Subpart (A) references 200.6, which contains the identification requirements for grounded conductors. Subpart (B) covers the identification requirements for the EGC that are located in 250.119. The identification requirements for ungrounded conductors are listed in 310.110(C). Essentially, any color other than white, gray, green, or any combination of these colors with distinguishing markings, is permissible for the identification of ungrounded conductors.

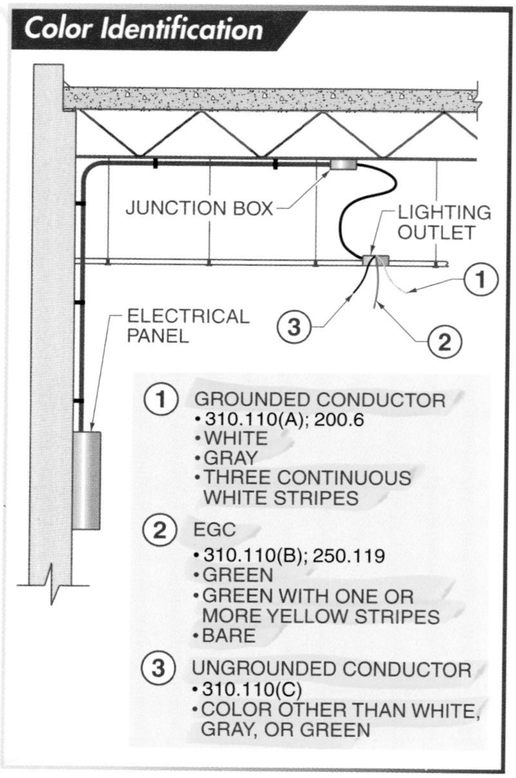

Figure 5-7. Proper conductor identification helps ensure worker safety and maintain the integrity of the electrical system.

In general, the grounded conductor (neutral) is colored white or gray. Depending on the size of the conductor, there are provisions for reidentifying a conductor that is not white or gray, at the conductor termination. A 4 AWG or larger conductor is permitted to be reidentified at the point where it terminates with distinctive white markings to identify it as a grounded conductor. Additionally, where only qualified persons maintain an electrical installation, the grounded conductor in a multiconductor cable is permitted to be reidentified at the termination. See 200.6(A)(B)(E).

When the grounded conductors for two different electrical systems are installed in the same raceway, cable, box, etc., the two system neutrals shall not be mixed. Therefore, the NEC® requires that when grounded conductors are mixed, one system neutral shall be in accordance with 200.6(A) or (B) and the other system neutral shall have a different outer covering conforming to 200.6(A) or (B), or have an outer covering of white or gray with a readily distinguishable colored stripe, other than green, running along the insulation.

There are no specifically mandated color-code requirements for branch-circuit and feeder ungrounded conductors. In the 2005 NEC®, however, additional provisions were added as a means to identify ungrounded branch-circuit and feeder conductors. The provisions require identification of branch-circuit and feeder conductors where more than one nominal voltage system supplies a premises. In the 2011 Code, Section 210.5 was revised to clarify the provisions for identifying ungrounded conductors in branch circuits. All of the provisions were reorganized into three subsections to address the following:

(1) The application of the means of identification for ungrounded conductors.

(2) The means of identification.

(3) The posting or notification requirements for the method utilized to identify the ungrounded conductors.

Typically, the colors brown, orange, and yellow are used as a color identification means for 480Y/277 ungrounded conductors and the colors black, red, and blue are used for 120/240 and 120/208Y ungrounded conductors. While these provisions do not require this color code, they permit the use of the color code as one of the means for identification. The color code should never be used as a means for voltage identification. However, by utilizing color coding as a means of identification, the overall level of safety and reliability of the installation can be enhanced for all those who may service or maintain the electrical distribution system. **See Figure 5-8.**

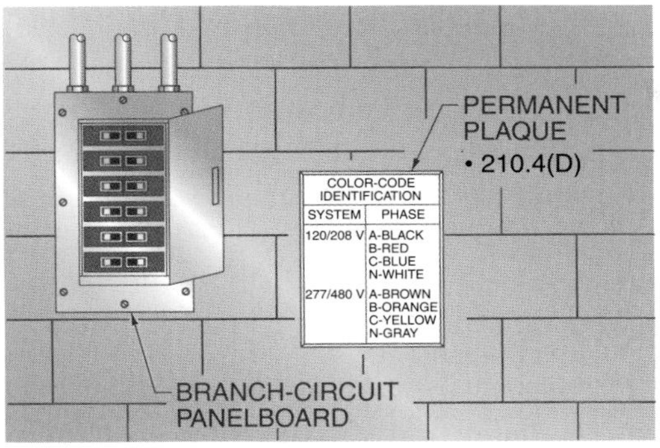

Figure 5-8. Color-code identification directories for ungrounded conductors shall be located at the branch-circuit panelboard.

Another color provision that installers and designers may encounter involves the use of delta 4-wire systems. Section 110.15 requires that the high leg of the delta 4-wire system be identified by an outer finish that is orange in color or other equally effective means.

Equipment Grounding Conductors

Equipment grounding conductors are permitted to be bare conductors. When these conductors are required to be covered or insulated, or if the designer or installer chooses to provide insulation, the outer covering shall be green or green with one or more yellow stripes, per 250.119. As with grounded conductors, there are provisions for reidentification of the equipment grounding conductor under specific conditions. See 250.119(A).

Section 250.119(A) permits reidentification of EGCs, 4 AWG and larger, at the time of installation by permanent means such as stripping the insulation or covering from the exposed conductor, coloring the exposed insulation or covering green, or marking the exposed covering or insulation with green tape or green labels. Section 250.119(B) permits similar methods of reidentification for multiconductor cables provided the conditions of maintenance ensure that only qualified personnel service the installation. **See Figure 5-9.**

Size – 310.106

Table 8, Chapter 9 lists copper, aluminum, and copper-clad aluminum conductors in sizes from 18 AWG through 2000 kcmil. In general, the NEC® does not permit conductors smaller than 14 AWG Cu, 12 AWG Al, or copper-clad aluminum to be used for conductors for general wiring in applications up to 2000 V. Table 310.106(A) lists the minimum conductor size for both copper and aluminum conductors. This size changes depending upon the voltage rating of the conductor. These are minimum sizes and there are many instances in which a particular NEC® provision requires a specific size conductor.

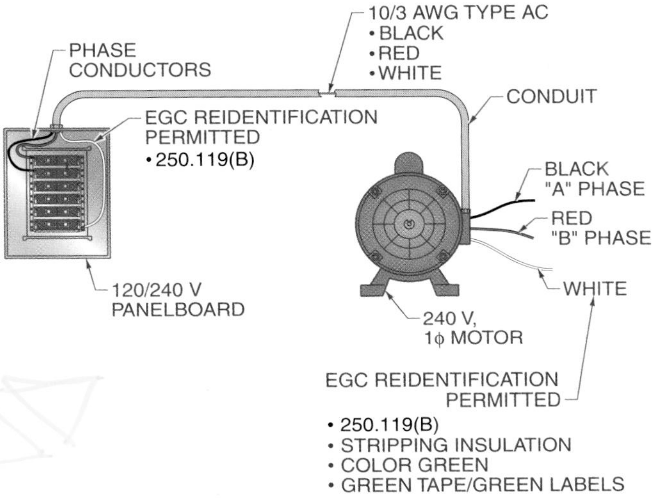

Figure 5-9. Reidentification of multiconductor cables is permitted during installation if only qualified personnel perform service.

Parallel Conductors – 310.10(H)

Parallel conductors are two or more conductors that are electrically connected at both ends to form a single conductor. Despite the wide range of conductor sizes available, there may be installations in which it is desirable to use parallel conductors to meet a specific design consideration or installation requirements. See 310.10(H)(1–6) for specific rules concerning the installation of parallel conductors. Copper, aluminum, and copper-clad aluminum conductors are all permitted to be installed in parallel. **See Figure 5-10.**

Length – 310.10(H)(2)(1). All conductors (each phase, polarity, neutral or grounded conductor) shall be the same length. Paralleling conductors of different lengths may result in an uneven distribution of current in the conductors resulting in excessive heat and possible insulation degradation. Many cable manufacturers recommend cutting the parallel conductors to the same length prior to installation in a raceway to ensure that the conductors remain the same length.

Material – 310.10(H)(2)(2). All conductors (each phase, polarity, neutral or grounded conductor) shall be the same material. Aluminum and copper have different properties that make them incompatible for paralleling.

Size – 310.10(H)(2)(3). All conductors (each phase, polarity, neutral or grounded conductor) shall be the same size. Same-size conductors ensure that the current split and the potential voltage drop between conductors remains the same.

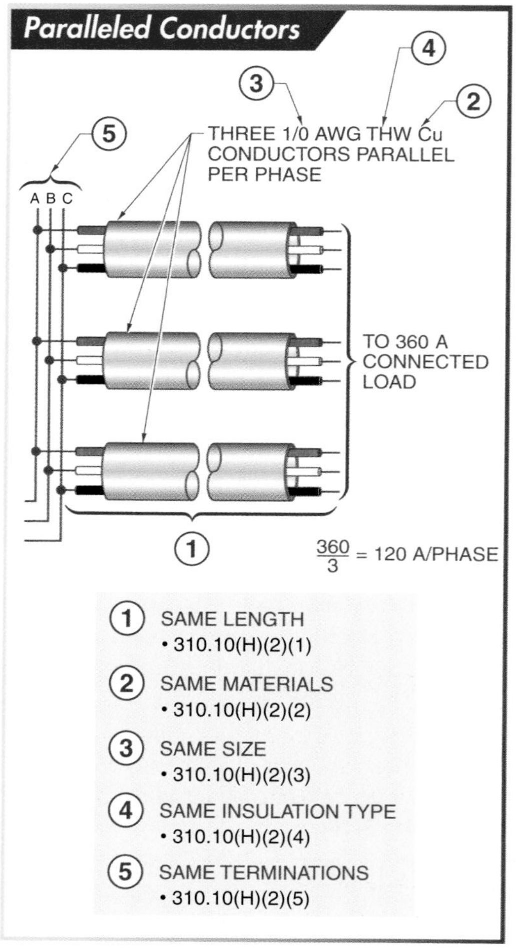

Figure 5-10. Each conductor, of each set of paralleled conductors, shall meet five conditions.

Insulation – 310.10(H)(2)(4). All conductors (each phase, polarity, neutral or grounded conductor) shall be the same insulation type. There are three basic temperature ratings for conductor insulations. Parallel conductors shall have the same temperature rating to ensure that the termination does not reach a temperature beyond the operating temperature of the insulation.

Terminations – 310.10(H)(2)(5). All conductors shall be terminated using the same methods or materials. If termination methods are different, such as when single-barrel lugs and multiple-barrel lugs are used on the same phase, the impedance of the paralleled conductors may be different, resulting in uneven current splits between the paralleled conductors.

These requirements apply to the individual conductors that comprise a single-phase, neutral or grounded conductor and not to all of the conductors of the circuit. Paralleled conductors are derated per Table 310.15(B)(3)(a). **See Figure 5-11.** Equipment grounding conductors must comply with 310.10(H)(5) where installed in parallel; however, they shall be sized per 250.122.

Section 310.10(H)(3) addresses the general provisions for paralleling conductors where the conductors are run in separate cables or raceways. Where this is done, the cables or raceways shall have the same number of conductors and shall have the same characteristics. It is important to note that conductors of one phase, polarity, neutral, grounded circuit conductor, or EGC are not required to have the same physical characteristics as those of another phase, polarity, neutral, grounded circuit conductor, or EGC.

AMPACITY – ARTICLE 100; ARTICLE 310

Ampacity is a term derived from combining the words ampere and capacity. Ampacity is the current that a conductor can carry continuously, under the conditions of use without exceeding the temperature rating. See Article 100. Conditions which may affect a conductor's allowable ampacity, and require derating, include the type of insulation, ambient temperature surrounding the conductor, number of current-carrying conductors in a raceway or cable, and temperature rating.

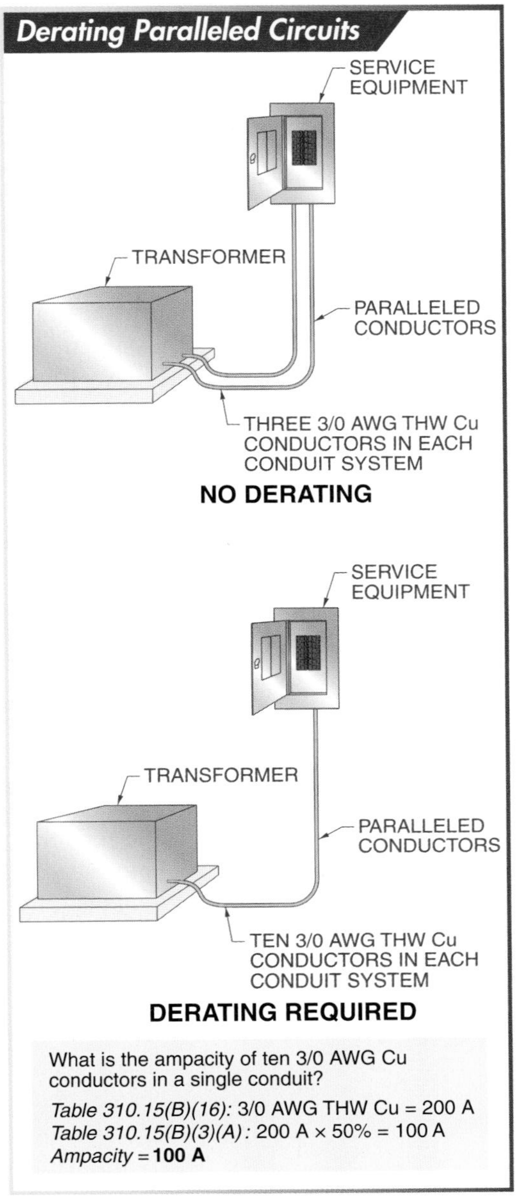

Figure 5-11. Derating of parallel conductors is determined by the design of the electrical distribution system.

The conductor insulation type has a direct effect on the allowable ampacity of the conductor. Generally, the higher the temperature rating of the conductor insulation, the higher the allowable ampacity.

Heat is generated when current flows through a conductor. As the ambient temperature is increased, the ability of the conductor to dissipate this heat is greatly reduced. All ambient temperatures above 30°C (86°F) require a correction of the conductor's allowable ampacity.

Table 310.15(B)(16) lists allowable ampacities based upon not more than three current-carrying conductors in the same raceway or cable. When additional current-carrying conductors are installed, the additional heat generated by these conductors shall be accounted for. Section 310.15(B)(3)(a) provides adjustment factors for installations with more than three current-carrying conductors.

All electrical equipment has temperature ratings associated with the equipment terminations. The temperature rating of the conductor shall be selected in accordance with the temperature rating of any connected termination in the equipment per 110.14(C)(1–2).

Insulation Ratings – Table 310.104(A); Table 310.15(B)(16)

All insulated conductors have a maximum operating temperature at which the insulation of the conductor is not adversely affected. Conductor ampacities are directly related to these operating temperature limitations. Table 310.15(B)(16) lists conductor ampacities for insulated conductors under 2000 V, installed in a raceway, cable, or earth. These values assume an ambient temperature of 30°C (86°F), and not more than three current-carrying conductors in a raceway or cable. While there are many different types of insulations listed, all of them are assigned to one of three basic classifications: 60°C (140°F), 75°C (167°F), and 90°C (194°F). **See Figure 5-12.**

Multiconductor Cables

When using multiconductor cables, such as AC (BX), NM, MC, etc., the derating provisions of 310.15(B)(3) apply. Many electrical installations use these types of cables for branch-circuit wiring for both lighting and receptacle circuits. The most common violation of this rule occurs where all of the cable assemblies come together, such as at a panelboard.

Cables entering an electrical room shall have adequate spacing to allow for proper heat dissipation. The conductor's allowable ampacity shall be calculated per provisions of 310.15(B)(3)(a).

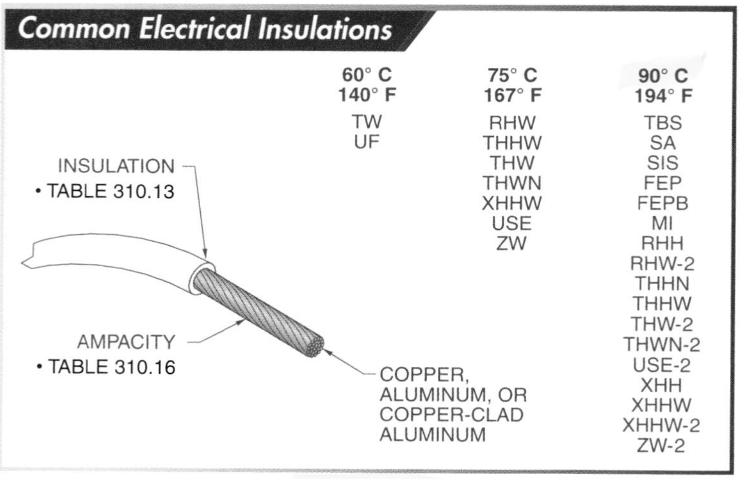

Figure 5-12. Electrical insulations are classified according to their temperature rating.

For example, three different ampacities are listed for an 8 AWG Cu conductor depending upon the insulation used. Any of the 60°C insulations result in a listed table ampacity of 40 A. The 75°C insulations permit a 50 A rating, and the 90°C insulations permit a 55 A rating. These ratings are the listed table values only. The actual allowable ampacity depends on the installation conditions.

Ambient Temperature – Table 310.15(B)(16)

Table 310.15(B)(16) ampacities are based on an assumed ambient temperature of 30°C (86°F). Ambient temperature is the temperature of air around a piece of equipment. The ambient temperature is the maximum temperature that can be found anywhere along the conductor length. It directly affects the conductor's ability to dissipate heat. If the ambient temperature is too high, the heat dissipation rate decreases, resulting in an increased conductor and insulation temperature.

Copper Wire vs. Aluminum Wire

Aluminum wire is more cost efficient than copper wire. Aluminum wire is installed in a manner similar to copper wire. However, the procedure for wire termination is different for aluminum. In the past, the wire connections on aluminum wire had a tendency to loosen as the aluminum expanded and contracted due to temperature changes. Improvements in aluminum wire manufacturing have since been made to alleviate this. However, aluminum is also known to immediately form a high-resistance oxide film when exposed to air due to its inherent nature. Consult with the manufacturer regarding specific oxide inhibitor requirements.

Generally, the higher the ambient temperature, the more difficult it is for a conductor to dissipate heat. When the ambient temperature is above 30°C (86°F), the ampacity of the conductor shall be adjusted to compensate for the decreased heat dissipation. The ambient temperature correction factors are included in the ampacity tables. The correction factors depend upon the type of insulation on the conductor, the type of conductor material (copper, aluminum, or copper-clad aluminum), and the ambient temperature. Correction factors are cross-listed for both Celsius and Fahrenheit to make the application easier. There may be instances in which the ambient temperature remains below the table value of 30°C (86°F). In these cases, an increased ampacity is permitted by the correction factors.

Number of Conductors. The values in Table 310.15(B)(16) are based upon the assumption that no more than three current-carrying conductors are installed in the same raceway, cable, or trench. For installations that require more than three current-carrying conductors, the table values shall be adjusted. See 310.15(B)(3)(a).

Section 310.15(B)(5) covers the conditions under which the neutral conductor shall be counted as a current-carrying conductor. In general, if the neutral carries only unbalanced current from other conductors of the same circuit, it need not be considered a current-carrying conductor. If a major portion of a 3ϕ, 4-wire, wye circuit consists of nonlinear loads, such as those associated with electric-discharge lighting and data processing equipment, then the neutral shall be considered a current-carrying conductor. In addition, 3-wire circuits derived from a 3ϕ wye system have neutral currents equivalent to those in the phase conductors and are counted as current-carrying conductors.

Table 310.15(B)(3)(a) lists the required derating percentages for installations which have more than three current-carrying conductors. These percentages reflect the percent of the Table 310.15(B)(16) values that are permitted to be used in determining the allowable conductor ampacity. Section 310.15(B)(3)(a) also requires that where single conductors or multiconductor cables are bundled together for more than 24″, the conductors shall be derated.

Overcurrent Protective Devices – Photovoltaic (PV) Systems

A photovoltaic (PV) system is used to convert light energy from the sun to electrical energy for powering loads. These systems can power DC loads, or the output can be fed through an inverter to power AC loads. There are three basic types of PV systems: stand-alone systems, interactive (grid-connected) systems, and hybrid systems. Stand-alone systems supply power independently of any other electrical power source. Interactive systems operate in parallel with other electrical power sources, such electric utility systems. An interactive system may also supply electric power to a production or distribution network. Hybrid systems include other power sources, such as wind and hydroelectric generation, in addition to PV sources.

Solar World Industries America

PV systems range from simple to complex. There can be many components, such as PV panels, collector or combiner boxes, battery systems, charge controllers, and inverters. Also, there are various overcurrent protection needs and requirements for different parts of the system.

The basic power-generating component of a PV system is the solar cell. In order to generate useful levels of power, groups of cells are combined to form modules. Modules are then grouped into panels, and several panels form a solar array. A PV power source can consist of one or more arrays. The short-circuit current that can be delivered from a PV panel is only 110% to 115% of the operating current. This is quite different than the conventional AC system supplied by utility or on-site generators. However, parts of PV systems may have to withstand higher short-circuit currents. Many PV systems have battery banks, which can deliver substantial fault current. Also, if the system is connected to the conventional electrical distribution system fed by a utility, fault current can be substantial.

PV system installations are covered by NEC® Article 690. Section 690.2 defines the PV source circuit as "Circuits between modules and from modules to the common connection point(s) of the DC system." Per NEC® 690.9(A), the PV source circuit, PV output circuit, inverter output circuit, storage battery circuit conductors, and equipment shall be protected per Article 240. This requires branch-circuit fuses and circuit breakers (except for PV source circuits).

Any fuse or circuit breaker used in the DC portion of the system must be listed for DC use. 690.9(D) requires all overcurrent devices to be listed for the DC voltage and DC interrupting rating equal to or greater than the required values. Per 690.7(A), the required rated voltage for overcurrent protective devices and other components in the DC PV source and output circuits shall be based on rated open-circuit voltage corrected for expected ambient temperature. The open-circuit voltage is the sum the open-circuit voltage of the series connected PV modules and can be significantly greater than the closed-circuit voltage.

Additional information is available at www.cooperbussmann.com.

Termination Rating. The temperature rating of the equipment terminations also affects a conductor's allowable ampacity. The ampacity of the conductors shall be coordinated so as not to exceed the lowest temperature rating of any component in the electrical circuit, including the electrical equipment per 110.14(C). Although conductors may be selected that are suitable for 90°C conditions, the termination will probably have a lower temperature rating. In these cases, the ampacity of the conductor shall be selected so that it does not exceed the lowest rating of the equipment in the circuit.

Most electrical equipment is rated at 60°C, 75°C, or 90°C. Some equipment contains dual ratings such as 60°C/75°C. Such equipment is permitted to use the 75°C rating provided all other components of the circuit are suitable for 75°C. When the conductor insulation rating exceeds the equipment termination temperature rating, the ampacity of the conductor shall be based on the equipment rating. There is very little equipment suitable for use with 90°C conductors. Conductors rated at 90°C are permitted to be installed with the lower-rated equipment but their ampacities shall be based on the temperature rating of the equipment.

Location. The final factor in determining a conductor's allowable ampacity is the type of location in which the conductor is to be installed. Locations can be classified as either dry, damp, or wet. Conductors installed in dry and damp locations may be subjected to only a moderate degree of moisture and thus no adjustment of the conductor's ampacity is necessary. Conductors installed in wet locations, however, may be unprotected from the weather and subject to continuous wet conditions.

Depending on the insulation type, an adjustment to the ampacity may be required. For example, if THHW is installed in a wet location, the ampacity shall be selected from the 75°C column even though it has 90°C insulation. This is because the wet location reduces the conductor's ability to dissipate heat. The NEC® has recently recognized new insulations that are capable of maintaining their higher ampacities in both wet or dry locations. Table 310.104(A) identifies these insulations with the suffix "-2." For example, RHW-2 and XHHW-2 are capable of maintaining their higher ampacities in both wet or dry locations.

OVERCURRENT PROTECTION – ARTICLE 240

The purpose of overcurrent protection is to protect the conductors from dangerous current levels that could damage the conductors or the conductor insulation per 240.1, Informational Note. Electrical equipment is also provided with overcurrent protection to protect it from high current levels that may occur as a result of short circuits or ground faults.

Allowable Ampacity

The equipment temperature rating shall be considered when determining a conductor's allowable ampacity. Electrical designers and installers may question the benefit of using 90°C conductors such as THHN if there are no equipment terminations that can take advantage of the higher rating of the conductor.

The advantage of using 90°C conductors is that although the ampacity listed in the 90°C column cannot be used, a 90°C conductor that requires derating for either high ambient temperature, more than three conductors in a raceway or cable, or both, can be derated starting with the 90°C ampacity. The selected allowable ampacity shall not exceed the temperature ratings of the other components of the electrical circuit.

Protection by Fuses and Circuit Breakers (CBs) – Article 240

In general, all conductors are protected against overcurrents at the point at which they receive their supply. Overcurrent protection is directly related to the ampacity of the conductors. The two most common devices used to provide overcurrent protection are fuses and circuit breakers (CBs). Ratings for standard fuses and circuit breakers are given in 240.6(A).

Fuses and Inverse-Time Circuit Breakers (ITCBs)

Increase	Standard Ampere Ratings
5	15, 20, 25, 30, 35, 40, 45, 50
10	50, 60, 70, 80, 90, 100, 110
25	125, 150, 175, 200, 225, 250
50	250, 300, 350, 400, 450, 500
100	500, 600, 700, 800
200	1000, 1200
400	1600, 2000
500	2500, 3000
1000	3000, 4000, 5000, 6000

1 A, 3 A, 6 A, 10 A, and 601 A are additional standard ratings for fuses.

Overcurrents. An overcurrent is any current in excess of that for which the conductor or equipment is rated. See Article 100. In terms of application, some equipment is designed to accommodate overcurrents for a period of time depending upon the load. Temporary surge or start-up overcurrents are common for some electrical equipment like motors and generally have no long-term, harmful effects. When overcurrents occur more frequently or persist for longer periods of time, damage to the electrical system may occur. Overcurrents can result from overloads, short circuits, or ground faults. **See Figure 5-13.**

Overloads. An overload is the operation of equipment in excess of normal, full-load rating, or a conductor having an excess of the rated capacity. It is the lowest magnitude of overcurrent. The least potentially damaging overcurrents occur when there is an overload.

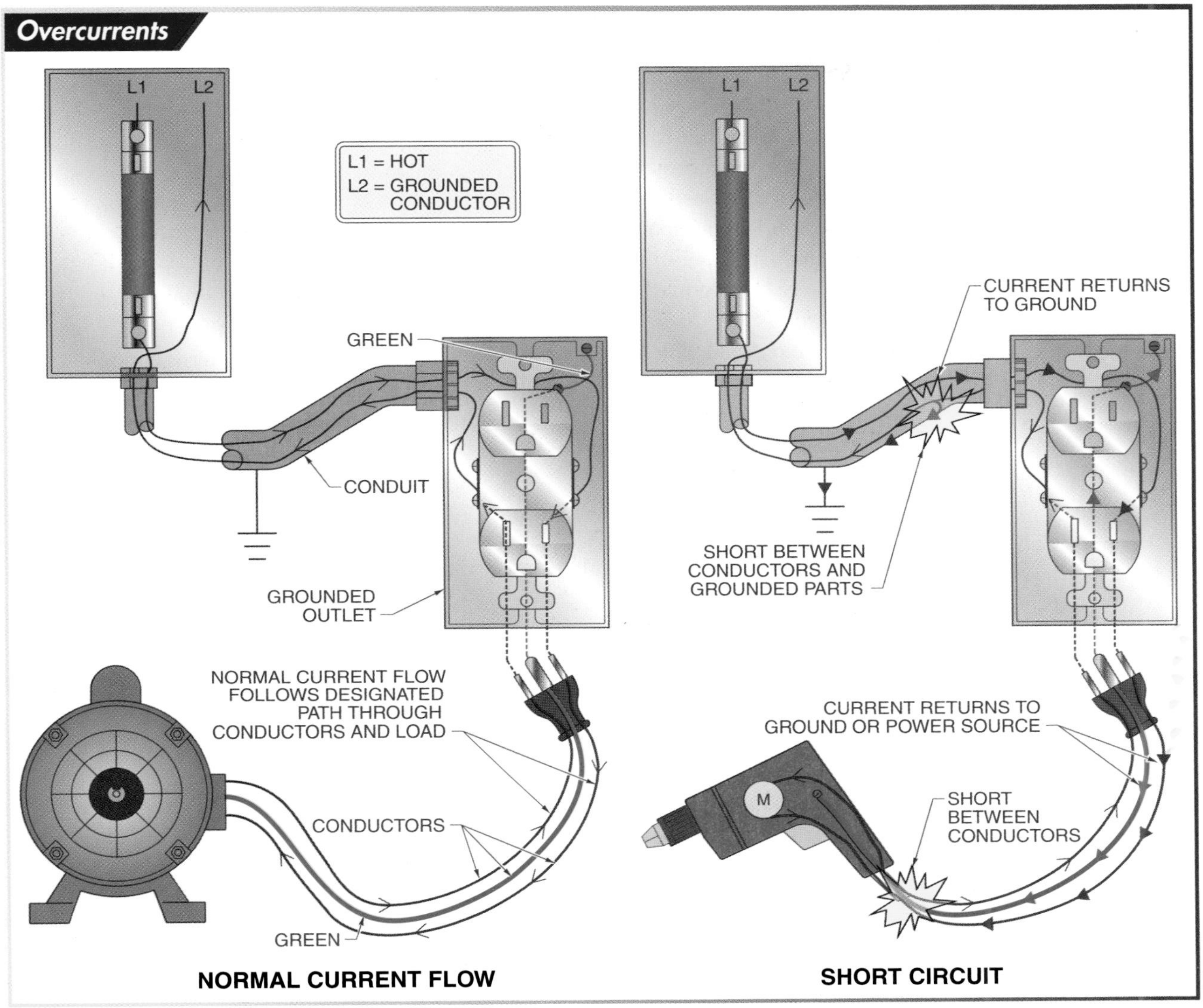

Figure 5-13. Overcurrent is any current in excess of that for which the conductor or equipment is designed.

Overcurrents from overloads most commonly occur as a result of connecting too much load to an electrical system component such as a branch circuit. Assuming the branch circuit is installed properly, the result of the overload is an opening in the branch-circuit Overcurrent Protective Device (OCPD). As a rule of thumb, overcurrents caused by overloads generally run in the range of one to six times the normal circuit current, and the current flow is confined to the normal current path. For the purposes of Article 240, the term overload does not include short circuits or ground faults. See Article 100.

Short Circuits. A *short circuit* is the condition that occurs when two ungrounded conductors (hot wires), or an ungrounded and a grounded conductor of a 1ϕ circuit, come in contact with each other. The most damaging overcurrents occur as the result of short circuits. For example, if the conductors feeding a 240 V water heater were to come in contact with each other, a short circuit, as result of this abnormal current path, would result. Short circuits are the largest-magnitude overcurrents. If the components of the electrical system are not properly protected, they may be severely damaged by a short circuit.

The amount of short-circuit current that flows depends upon the following factors:

- Amperes available at the source of power. Typically, the source of power is a transformer and the most significant factor is the impedance of the transformer. The higher the impedance of the transformer, the lower the available short-circuit current. Section 450.11 requires all transformers which are 25 kVA and larger to be marked with their impedance ratings.
- Length of the circuit. In general, the farther away from the source, the less available short-circuit current. As the circuit length increases, the distance from the power source also increases. Available short-circuit current declines as the distance from the power supply increases.
- Size of conductors and voltage at which they operate. Usually, the larger the conductor size and the higher the voltage, the higher the available short-circuit current. Larger conductor sizes offer a lower impedance, and higher circuit voltages overcome circuit impedances, allowing larger short-circuit currents to pass.

Another factor that is important in determining the amount of short-circuit current that will flow is the design and operation of the overcurrent device that is upstream from the location of the short circuit. Typically, the longer the response or operating time of the overcurrent device, the larger the amount of short-circuit current that will be generated. Different OCPDs have different characteristics or settings that impact the time it takes to respond to the short circuit and operate the device. This is especially important for the protection of personnel that may be exposed to possible arc flash or arc blast associated with the short circuit.

Other factors may also be considered in calculating available short-circuit current. These include: impedance of the fault circuit, impedance of the arc, and the type of equipment supplied by the circuit. For example, motors can contribute to available short-circuit current levels and may need to be considered if they are a significant component of the electrical system. Installers of electrical services shall be aware of the available short-circuit current and ensure that the service equipment is suitable for the available short-circuit current per 110.9 and 110.10.

Ground Faults. A ground fault is an unintentional connection between an ungrounded conductor and any grounded raceway, box, enclosure, fitting, etc. Like short circuits, ground-fault currents follow an abnormal path. In this case, however, the path of current is from an ungrounded conductor to ground. Ground faults can occur due to insulation failures or, more commonly, at conductor terminations.

Ground-fault currents are generally of a lower magnitude than short circuits. Care shall be taken when installing electrical conductors to ensure that insulation is not damaged, as this could contribute to a ground fault.

When 4 AWG and larger ungrounded conductors are installed in raceways that enter boxes, enclosures, cabinets, etc., the conductors shall be protected by the use of an identified insulating fitting per 300.4(G). As with short-circuit currents, the available ground-fault current depends upon the available source current, the impedance of the circuit, the distance from the source that the fault occurs, and the conductor type, size, and voltage rating.

The first requirements for GFCI protection were included in the 1975 NEC®.

Protection of Equipment – 240.3

High-level fault currents and persistent overloads can cause problems for electrical equipment as well as for electrical conductors. All electrical equipment shall be protected from these potentially damaging currents. The NEC® requires that the type and form of the overcurrent protection provided meet the requirements of the particular article covering the equipment. For example, the provisions for overcurrent protection for appliances are discussed in Article 422. Overcurrent protection for motors is discussed in Article 430. See 240.3 and Table 240.3.

Protection of Conductors – 240.4

Overcurrent protection of conductors ensures that conductors are protected in accordance with their allowable ampacity. All conductors shall be protected against overcurrent in accordance with their allowable ampacities unless otherwise required or permitted by 240.4(A–G). Overcurrent protection for flexible cords, cables, and fixture wires shall be per 240.5.

Section 310.15 requires that conductor ampacities be obtained from the Tables 310.15(B)(16) through 310.15(B)(19) or under engineering supervision by calculation. The process of selecting overcurrent protection requires that first the conductor's allowable ampacity be calculated and then the overcurrent protection be selected. See 240.4(A–G) for alternate methods to the general rule of protecting conductors in accordance to their allowable ampacity.

Hazardous Shutdown – 240.4(A). There may be instances in which operation of the overcurrent protection device results in a hazard. Section 240.4(A) permits installations of this nature to omit overload protection, but still requires short-circuit protection. For example, while overload protection is important for most motors, fire pumps by their design continue to operate for as long as possible and are not required to have overload protection. Another example is of a material-handling magnet circuit in which a loss of power could create a hazard by dropping the load held by the magnet.

Overcurrent Devices of 800 A or Less – 240.4(B). Overcurrent protection can be selected once the allowable ampacity is determined. Often, the calculated ampacity does not correspond to a standard-size fuse or circuit breaker. The next higher standard device rating may be used in these cases per 240.4(B), provided the conductors are not part of a branch circuit supplying more than one receptacle for cord-and-plug connected loads and the next higher standard device rating does not exceed 800 A. In cases where the cord-and-plug-connected portable loads are supplied by multioutlet receptacle circuits, and the ampacity of the conductor does not correspond to a standard rating for a fuse or circuit breaker, the next lower standard device rating shall be selected. **See Figure 5-14.**

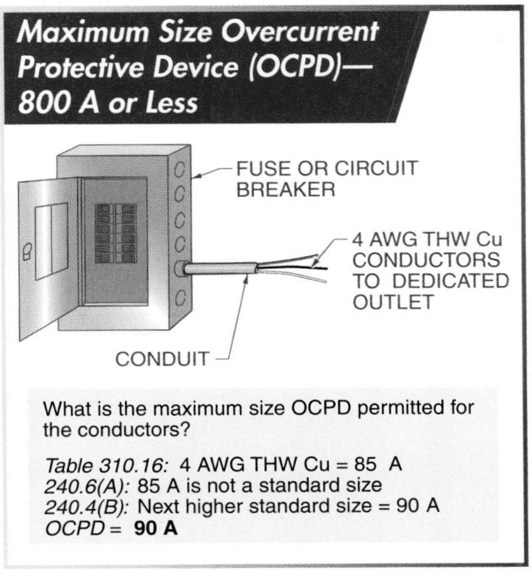

Figure 5-14. OCPDs of 800 A or less which do not correspond to a standard-size device are rounded up.

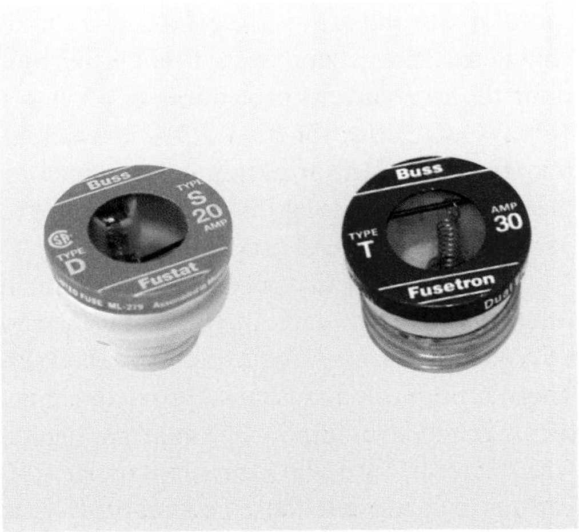

Cooper Bussmann

Plug fuses rated over 15 A are identified with a round configuration of the window.

Overcurrent Devices Over 800 A – 240.4(C). In installations where the calculated conductor ampacity does not correspond to a standard size rating for fuses or circuit breakers, and the rating of the overcurrent device is over 800 A, the next lower standard-size fuses or circuit breakers shall be selected per 240.4(C). **See Figure 5-15.**

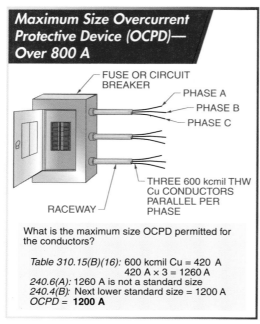

Maximum Size Overcurrent Protective Device (OCPD)— Over 800 A

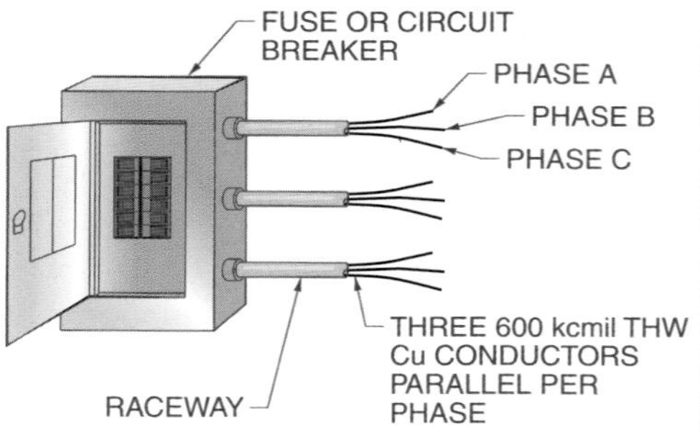

What is the maximum size OCPD permitted for the conductors?

Table 310.15(B)(16): 600 kcmil Cu = 420 A
420 A × 3 = 1260 A
240.6(A): 1260 A is not a standard size
240.4(B): Next lower standard size = 1200 A
OCPD = **1200 A**

Figure 5-15. OCPDs over 800 A which do not correspond to a standard-size device are rounded down.

Small Conductors – 240.4(D). For small conductors, the general overcurrent provisions limit the overcurrent protection to 15 A for 14 AWG Cu, 20 A for 12 AWG Cu, and 30 A for 10 AWG Cu. For aluminum conductors, the overcurrent provisions limit protection to 15 A for 12 AWG Al and 25 A for 10 AWG Al. In the 2008 NEC®, provisions were added for overcurrent protection for 18 AWG and 16 AWG conductors. See 240.4(D)(1) and (2). These limitations and general provisions may be amended if modified by other applicable NEC® provisions. For example, Article 430 permits exceptions to these general provisions because of the need to provide for the high starting currents of motors. See 430.52(C)(1), Ex. 2.

Tap Conductors – 240.4(E). Smaller conductors are tapped from larger feeder or branch-circuit conductors in many installations. In these cases, the conductors are not protected in accordance with their ampacities because the overcurrent protection device is based on the larger conductor. Section 240.4(E) permits taps provided they are in accordance with 210.19(A)(3) and (A)(4), 240.5(B)(2), 240.21, 368.17(B), 368.17(C), and 430.53(D).

For the purposes of Article 240, a tap conductor is defined as a conductor which has overcurrent protection ahead of its point of supply that exceeds the value normally permitted for similar conductors. Service conductors are not considered tap conductors.

Protection of Cords, Cables, and Fixture Wires – 240.5

Like general building conductors, flexible cords and fixture wires shall be protected against overcurrent in accordance with their ampacities. In the case of these conductors, however, ampacities are selected from Tables 400.5(A)(1) and 400.5(A)(2) for flexible cords and Table 402.5 for fixture wires. Table 400.4 lists all of the flexible cords and cables which are permitted to be used without being subject to special investigation.

Flexible cords and cables have very specific uses and are not intended to be used as a general wiring method or a replacement for permanent wiring. Per 240.5(B), flexible cord shall be protected by one of three methods when it is supplied by a branch circuit. Section 240.5(B)(1) lists the permissible overcurrent protection for flexible cords or tinsel cords which is approved for use with a specific listed appliance or portable or permanently installed luminaire. Section 240.5(B)(2) lists the permissible overcurrent protection for fixture wires. Section 240.5(B)(3) permits flexible cord used in listed extension cord sets, or other sets constructed of listed components, to be protected when applied within the provisions of the listing.

Section 240.5(B)(4) contains the provisions for extension cord sets that are assembled in the field. On some construction projects, the electrical contractor or owner may choose to assemble their own extension cord sets. The Occupational Safety and Health Administration (OSHA) has specific requirements for how this can be safely accomplished in the field utilizing listed components, and, where assembled and tested by qualified persons. The NEC® dictates in 240.5(B)(4) that for 20 A circuits the minimum-size AWG conductor shall be 16 and larger.

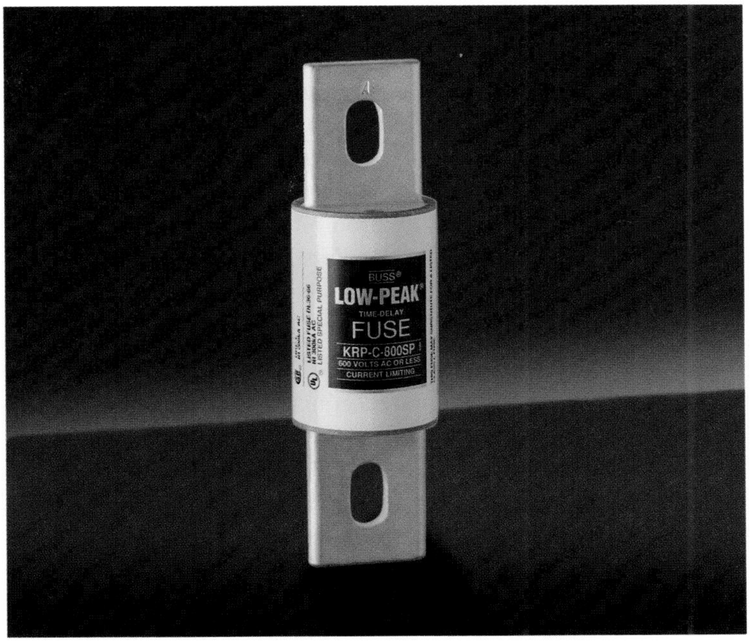

Cooper Bussmann

Current-limiting devices reduce the current flowing in a faulted circuit more quickly than standard OCPDs.

Ampere Ratings – 240.6

The standard ratings for overcurrent protective devices are given in 240.6. Part (A) covers the standard ratings for fuses and inverse-time circuit breakers (ITCBs). Part (B) covers adjustable-trip circuit breakers (ATCBs). Section 240.6(C) contains three provisions which, if met, permit an adjustable-trip circuit breaker to have an ampere rating equal to the adjusted current setting rather than to the maximum setting of the circuit breaker.

Fuses. The standard ratings for 15 A to 50 A fuses increase in increments of 5 (15, 20, 25, 30, 35, 40, 45, and 50). For 50 A to 110 A fuses, the ratings increase in increments of 10 (50, 60, 70, 80, 90, 100, and 110). From 100 A to 250 A fuses, the ratings increase in increments of 25 (100, 125, 150, 175, 200, 225, and 250). For 300 A to 500 A fuses, the ratings increase in increments of 50 (300, 350, 400, 450, and 500). Beyond 500 A, the ratings are not as uniform. The maximum standard rating for a fuse is 6000 A. Additional standard ratings for fuses are 1 A, 3 A, 6 A, 10 A, and 601 A.

Inverse-Time Circuit Breakers. The standard ratings for ITCBs are the same as those for fuses with the exception of 1 A, 3 A, 6 A, 10 A, and 601 A, which apply only to fuses. ITCBs are the most commonly used circuit breakers in the electrical industry.

Adjustable-Trip Circuit Breakers. ATCBs incorporate an adjustable setting for the long-time pickup or trip setting. In general, the rating of these devices shall be determined from the maximum setting available in the ATCB per 240.6(B). For example, if the setting is adjustable over a range of 600 A to 800 A, the rating is based on 800 A. The conductors are then selected on that basis.

However, 240.6(C) permits the actual setting, not the maximum setting, to be utilized if the circuit breaker has restricted access to the adjusting means. Restricted access is defined as being behind removable and sealable covers, or being behind bolted equipment enclosure doors, or being located in locked rooms accessible only to qualified personnel. If these conditions are met, a circuit breaker with an 800 A frame size (maximum rating) that has a long-time pickup setting of 600 A can be considered to have a 600 A rating, and the wire ampacity could be selected based on the 600 A setting instead of the 800 A rating.

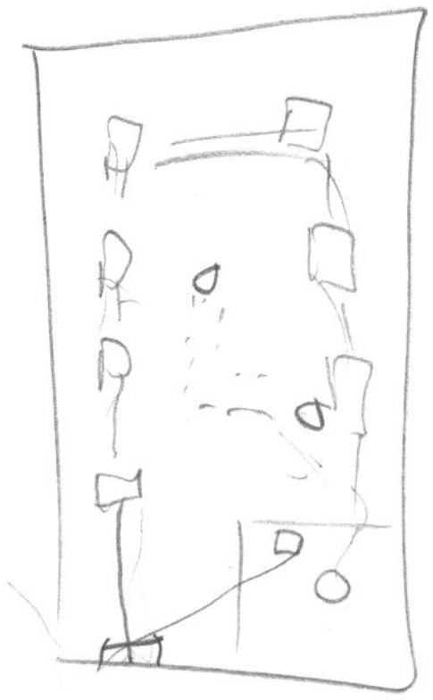

LOCATION OF OVERCURRENT PROTECTIVE DEVICES – ARTICLE 240, PART II

The general rule for the location of overcurrent protection devices is to place the OCPD at the point at which the conductor receives its supply. If the OCPD is properly sized in accordance with the conductor's allowable ampacity, and the device is placed at the point where the conductor receives its supply, then the installation should protect both the conductor and the conductor insulation from potentially dangerous overcurrents. In general, overcurrent protection devices are placed in series with the ungrounded conductor and are not used with the grounded conductor. Section 240.8 prohibits the installation of fuses or circuit breakers in parallel unless they are part of a listed factory assembly.

Ungrounded Conductor – 240.15(A)(B)

Ungrounded conductors shall be installed with an overcurrent protection device installed in series. The overcurrent device may be a fuse, circuit breaker, or a combination of an overcurrent relay and current transformer. When circuit breakers are used as the means of overcurrent protection, they shall simultaneously open all ungrounded conductors of the circuit. Individual single-pole circuit breakers (SPCBs) are not permitted for multiwire branch circuits in dwellings except where permitted by 240.15(B)(1). Individual circuit breakers with identified handle ties are permitted in grounded systems for line-to-line loads for 1ϕ, 3-wire DC circuits and in 4-wire, 3ϕ systems or 5-wire, 2ϕ systems with a grounded circuit conductor not exceeding the voltage limitations in 210.6. See 240.15(B)(2), (3), and (4).

Grounded Conductor – 240.22

In the majority of electrical systems, the grounded conductor is the neutral for the system and carries the return circuit current. For this reason, grounded conductors, in general, are not provided with overcurrent protection. There are, however, two exceptions to this general rule. The first is for installations in which all of the circuit conductors are opened or disconnected from the supply simultaneously, and the design is such that no pole can operate independently of another. The second is for motor overload protection as provided in 430.36. Some electrical systems, like a 3ϕ, 3-wire, corner-grounded delta, operate with a grounded conductor that is a normal current-carrying conductor. **See Figure 5-16.**

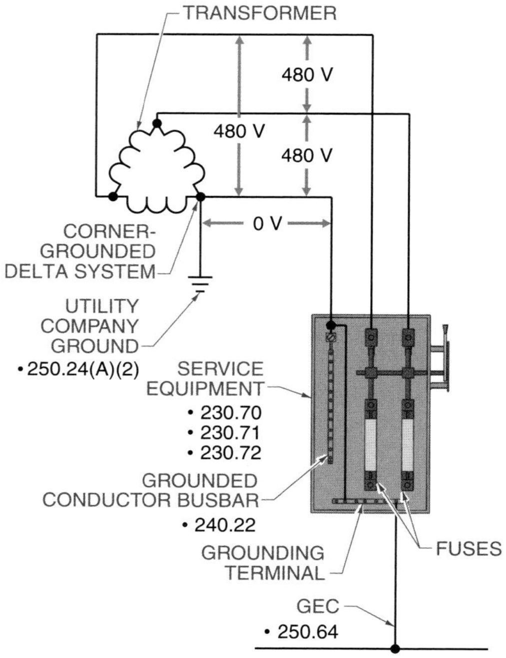

Figure 5-16. All grounded conductors are not neutrals.

Circuit Location – 240.21

Overcurrent protection in general, shall be provided for feeder and branch-circuit conductors at the point at which they receive their supply per 240.21. Subsections 240.21(A–H) do, however, contain provisions which allow conductors to be protected downstream from where they actually receive their supply. These subsections are collectively referred to as "tap rules." These rules, which are essentially exceptions to the general rule for overcurrent protection location, list specific sets of conditions that shall be met in order to make taps and protect the tap conductors at a point other than where it receives its supply. The tap provisions of 240.21 are for feeder taps and for transformer secondary taps. Section 240.21(A), however, does permit branch-circuit taps provided they meet the requirements of 210.19 and have overcurrent protection per 210.20.

10′ Tap Rule – 240.21(B)(1). The most commonly used feeder tap rule in electrical distribution systems is the 10′ tap rule. Section 240.21(B)(1) sets five conditions which shall be met to permit conductors to be run without overcurrent protection. **See Figure 5-17.**

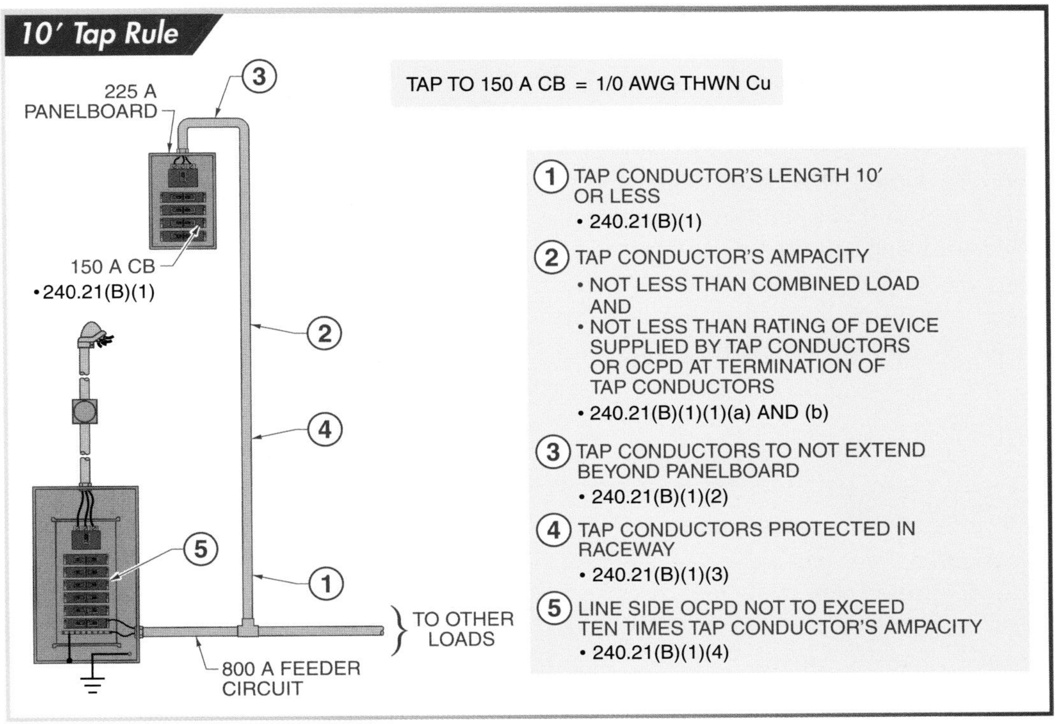

Figure 5-17. The 10′ tap rule permits tap conductors to be supplied from a feeder, without overcurrent protection at the point where they receive their supply, provided five conditions are met. *Note:* The CB lugs must be listed for the termination of more than one conductor.

Section 240.21(B)(1) states that if the total length of the tap does not exceed 10′ the provision of 240.21(B)(1) can be applied. Many of the tap rules contain length requirements that determine how much flexibility the designer or installer has in deviating from the general rule. Section 240.21(B)(1)(1)(a)(b) requires that the ampacity of the tap conductors be at least equal to the calculated load(s) on the circuits supplied, and not less than the rating of the overcurrent device or the device supplied by the tap conductors. Section 240.21(B)(1)(2) states that the tap conductors cannot continue beyond the device, such as panelboard, switchboard, disconnect, etc., they supply. In other words, it is not permissible to tap a tapped conductor.

Section 240.21(B)(1)(3) requires that the tap conductors be physically protected, except at the point at which they terminate, by being installed in a raceway. Section 240.21(B)(1)(4) requires that the line-side OCPD shall not exceed ten times the tap conductor's ampacity. Another way to say this is that the ampacity of the tap conductors is not less than one-tenth of the rating of the OCPD protecting the feeder conductors. This provision does not apply to conductors, no part of which leave the enclosure or vault where the tap is made.

Conductor Material

Copper, aluminum, and copper-clad aluminum are the most common material conductors used for electrical installations. Because of characteristics such as low cost and good conductivity, copper is the most widely used type of conductor material. Copper has better conductive properties than aluminum, and connections made with copper are not as restrictive as those made using aluminum or copper-clad aluminum. Wiring devices that are compatible with copper are readily available, while wiring devices for other types of conductors are more expensive and can require the use of anticorrosive grease as part of the installation process.

25′ Tap Rule – 240.21(B)(2). The 25′ feeder tap rule is also commonly used in electrical distribution systems. The major advantage of the 25′ tap rule over the 10′ tap rule is the additional length that the conductors are permitted to run without overcurrent protection. To compensate for the additional length, the tap conductor's ampacity is more closely matched to the rating of the tapped conductor circuit. There are four conditions that shall be met to use the 25′ tap rule. **See Figure 5-18.**

LP-A

The standard ratings for overcurrent protection devices are given in 240.6.

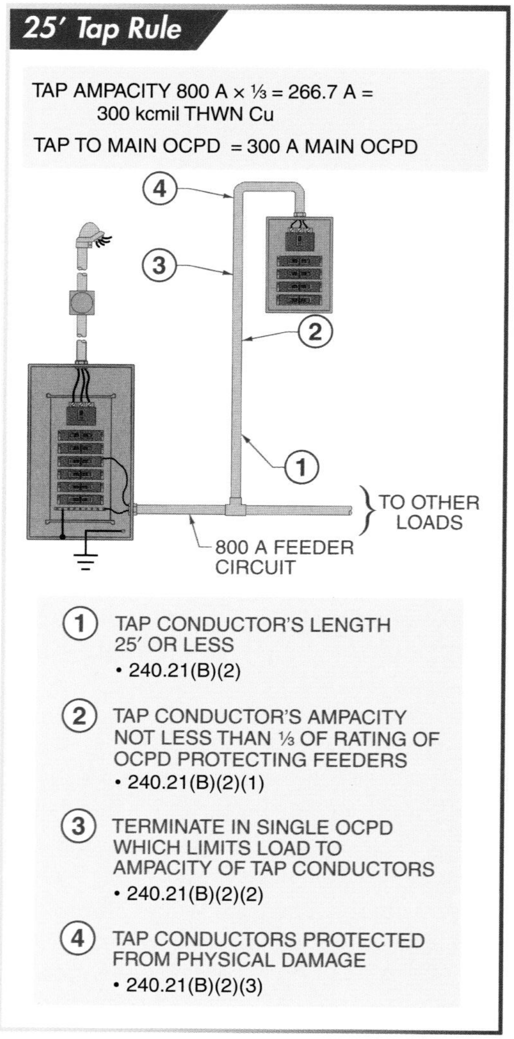
25′ Tap Rule
TAP AMPACITY 800 A × ⅓ = 266.7 A =
300 kcmil THWN Cu
TAP TO MAIN OCPD = 300 A MAIN OCPD
TO OTHER LOADS
800 A FEEDER CIRCUIT
1 TAP CONDUCTOR'S LENGTH 25′ OR LESS
• 240.21(B)(2)
2 TAP CONDUCTOR'S AMPACITY NOT LESS THAN ⅓ OF RATING OF OCPD PROTECTING FEEDERS
• 240.21(B)(2)(1)
3 TERMINATE IN SINGLE OCPD WHICH LIMITS LOAD TO AMPACITY OF TAP CONDUCTORS
• 240.21(B)(2)(2)
4 TAP CONDUCTORS PROTECTED FROM PHYSICAL DAMAGE
• 240.21(B)(2)(3)

Figure 5-18. The 25′ tap rule permits tap conductors to be supplied from a feeder, without overcurrent protection at the point where they receive their supply, provided four conditions are met.

The first condition, found in 240.21(B)(2), limits the total length to 25′. The second condition, found in 240.21(B)(2)(1), requires that the ampacity of the tap conductor be at least ⅓ the rating of the overcurrent device protecting the tapped conductors. The third condition, located in 240.21(B)(2)(2), requires the tap conductors to terminate in a single circuit breaker or set of fuses that limits the connected load to the ampacity of the tap conductors. The fourth condition, located in 240.21(B)(2)(3), requires that the tap conductors be installed in a raceway or otherwise protected from physical damage.

25′ Transformer Feeder Tap Rule – 240.21(B)(3). The 25′ transformer feeder tap rule is for feeder taps supplying a transformer in cases where the total length of the primary plus secondary does not exceed 25′. This tap rule permits the conductors supplying a transformer to be installed without overcurrent protection at the point at which they receive their supply provided five conditions are met. **See Figure 5-19.**

25′ Transformer Feeder Tap Rule

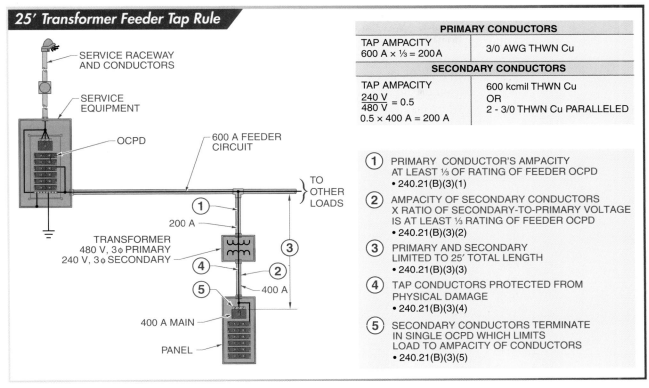

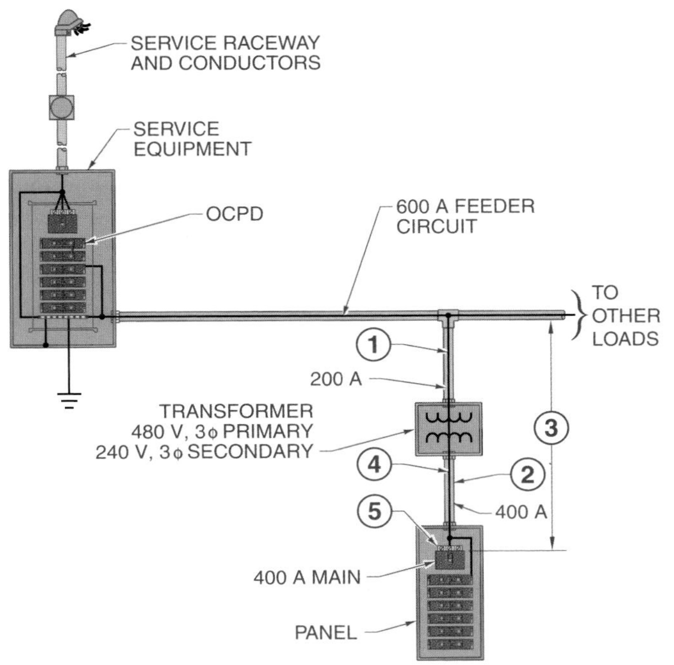

PRIMARY CONDUCTORS	
TAP AMPACITY 600 A × ⅓ = 200 A	3/0 AWG THWN Cu
SECONDARY CONDUCTORS	
TAP AMPACITY $\frac{240\text{ V}}{480\text{ V}} = 0.5$ 0.5 × 400 A = 200 A	600 kcmil THWN Cu OR 2 - 3/0 THWN Cu PARALLELED

① PRIMARY CONDUCTOR'S AMPACITY AT LEAST ⅓ OF RATING OF FEEDER OCPD
• 240.21(B)(3)(1)

② AMPACITY OF SECONDARY CONDUCTORS X RATIO OF SECONDARY-TO-PRIMARY VOLTAGE IS AT LEAST ⅓ RATING OF FEEDER OCPD
• 240.21(B)(3)(2)

③ PRIMARY AND SECONDARY LIMITED TO 25′ TOTAL LENGTH
• 240.21(B)(3)(3)

④ TAP CONDUCTORS PROTECTED FROM PHYSICAL DAMAGE
• 240.21(B)(3)(4)

⑤ SECONDARY CONDUCTORS TERMINATE IN SINGLE OCPD WHICH LIMITS LOAD TO AMPACITY OF CONDUCTORS
• 240.21(B)(3)(5)

Figure 5-19. The 25′ transformer tap rule permits tap conductors to be installed without overcurrent protection at the point where they receive their supply, provided five conditions are met.

The first condition, found in 240.21(B)(3)(1), requires that the ampacity of the primary supply conductors shall be at least ⅓ of the rating of the overcurrent device protecting the feeder conductors. The second condition, found in 240.21(B)(3)(2), requires that the ampacity of the secondary conductors supplied by the transformer shall be at least ⅓ of the rating of the overcurrent device protecting the feeders when multiplied by the ratio of the secondary to primary voltage. The third condition, located in 240.21(B)(3)(3), limits the total length of a primary and a secondary conductor to 25′.

The fourth condition, located in 240.21(B)(3)(4), requires that primary and secondary conductors shall be protected from physical damage. The fifth condition, located in 240.21(B)(3)(5), requires that the conductors shall terminate in a single circuit breaker or set of fuses which limits the load to the conductor's allowable ampacity.

Over 25′ Feeder Tap Rule – 240.21(B)(4). Subpart (4) contains tap rules for a specific set of conditions in a manufacturing building. **See Figure 5-20.** Overcurrent protection, at the point the conductor receives its supply, can be omitted in installations for high-bay manufacturing buildings with walls over 35′ in height. There are nine conditions that shall be met to apply this rule.

Section 240.21(B)(4) requires the conditions of maintenance and supervision shall be such that only qualified persons will service the system. The length of the tap shall not exceed 25′ horizontally and 100′ in total length. The ampacity of the tap conductors shall not be less than ⅓ the rating of the OCPD protecting the feeder conductors, and the tap conductors shall terminate in a single circuit breaker or set of fuses that will limit the load to the ampacity of the tap conductors. In addition, the tap conductors shall be protected from physical damage, they may not contain any splices, and they shall be a minimum size of 6 AWG Cu or 4 AWG Al. Finally, the tap conductors are not permitted to penetrate walls, floors, or ceilings, and the tap itself shall be made at a location which is at least 30′ from the floor.

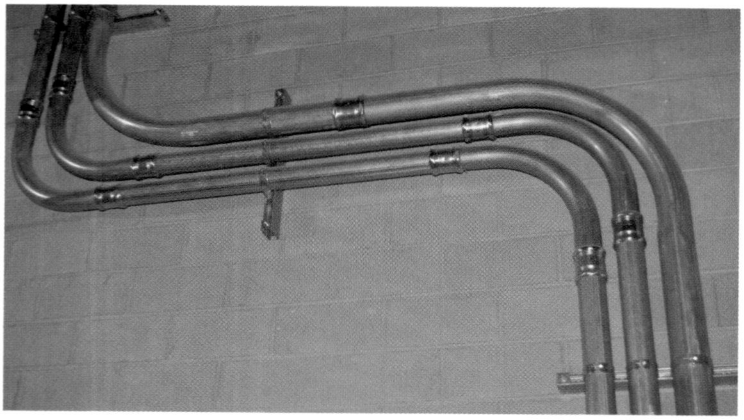

Feeder taps may be permitted without overcurrent protection if they meet certain rules.

Outside Feeder Tap Rule – 240.21(B)(5). Outdoor feeder taps are permitted without overcurrent protection at the tap where the conductors are located outside, except at the point of termination for the conductors. **See Figure 5-21.**

Section 240.21(B)(5) lists four conditions which shall be met to apply the outdoor feeder tap provisions of 240.21. First, the conductors shall be protected from physical damage. Second, the conductors shall terminate in a single overcurrent device that limits the load to the ampacity of the conductors. Third, the OCPD shall be an integral part of or be located adjacent to the disconnecting means. Fourth, the disconnecting means for the conductors shall be installed in a readily accessible location. The equipment may be located either outside the building or structure served, inside nearest the point of entry of the conductors, or where installed per 230.6 nearest the point of entrance of the conductors.

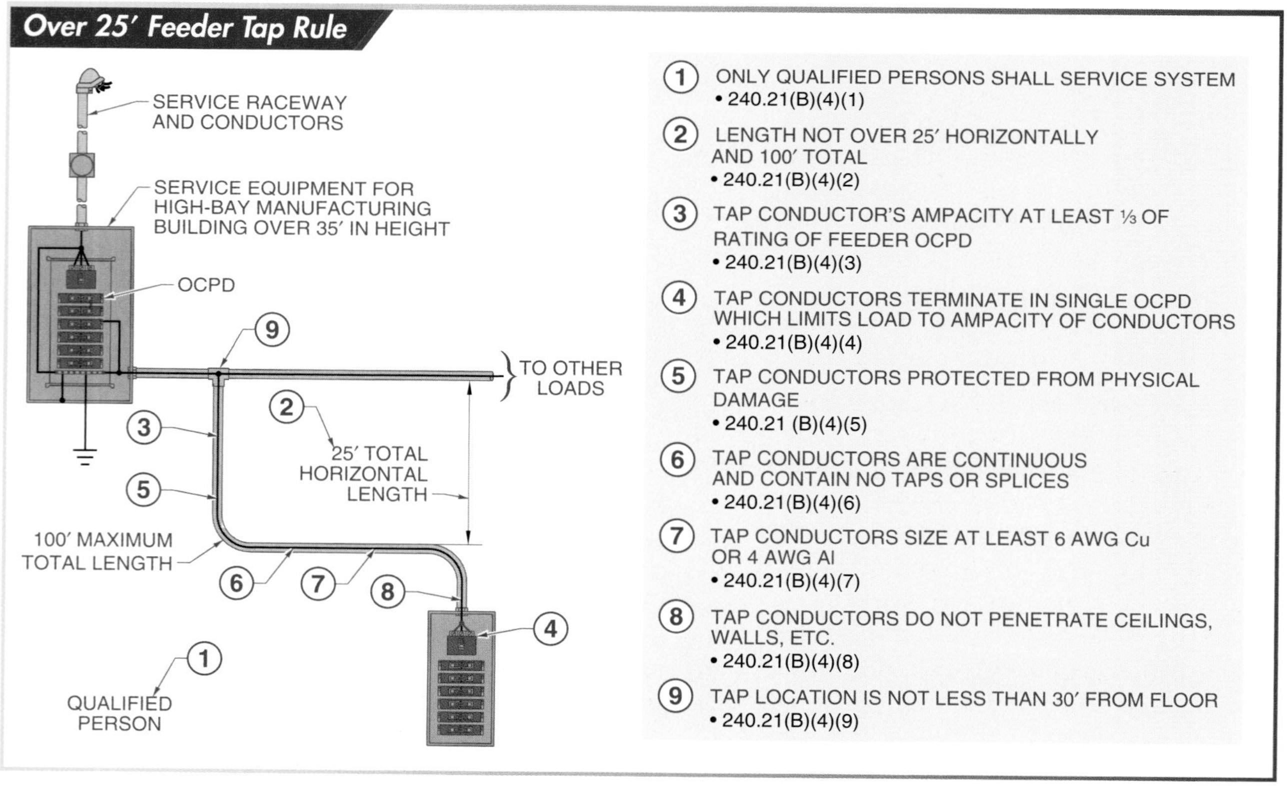

Figure 5-20. The over 25′ tap rule permits tap conductors to be supplied from a feeder, without overcurrent protection at the point where they receive their supply, provided nine conditions are met.

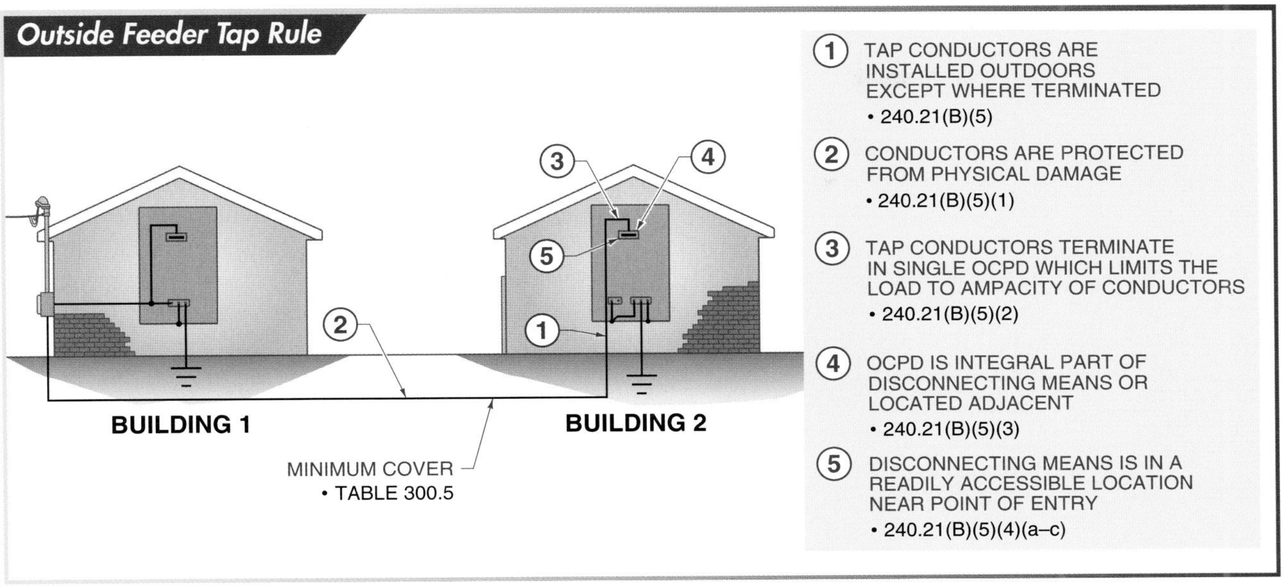
Outside Feeder Tap Rule
1 TAP CONDUCTORS ARE INSTALLED OUTDOORS EXCEPT WHERE TERMINATED
• 240.21(B)(5)
2 CONDUCTORS ARE PROTECTED FROM PHYSICAL DAMAGE
• 240.21(B)(5)(1)
3 TAP CONDUCTORS TERMINATE IN SINGLE OCPD WHICH LIMITS THE LOAD TO AMPACITY OF CONDUCTORS
• 240.21(B)(5)(2)
4 OCPD IS INTEGRAL PART OF DISCONNECTING MEANS OR LOCATED ADJACENT
• 240.21(B)(5)(3)
5 DISCONNECTING MEANS IS IN A READILY ACCESSIBLE LOCATION NEAR POINT OF ENTRY
• 240.21(B)(5)(4)(a–c)
BUILDING 1
BUILDING 2
MINIMUM COVER
• TABLE 300.5

Figure 5-21. The outside feeder tap rule permits tap conductors to be supplied from a feeder, without overcurrent protection at the point where they receive their supply, provided five conditions are met.

Since the conductors are located outside for their entire length, except where they are terminated, they can be treated like service conductors and no length limitation is imposed. Many of the conditions for outside feeder taps are similar to those imposed upon service conductors in Article 230. However, like service conductors, these conductors are not suitably protected against short circuits and overloads. The conditions ensure isolation and protection of the unprotected tap conductors.

Transformer Secondary Conductor Protection – 240.21(C). As with paragraph (B) for feeder types, 240.21(C) contains several overcurrent protection options for taps from a transformer secondary. Section 240.21(C)(1) covers applications where the primary OCPD can be utilized to provide secondary conductor protection. Section 240.21(C)(2) covers provisions for a maximum 10′ secondary conductor length. Section 240.21(C)(3) covers provisions for a maximum 25′ secondary conductor length for industrial installations only. Section 240.21(C)(4) covers provisions for outside secondary conductors. Section 240.21(5) covers secondary conductors from a tapped transformer and 240.21(6) contains provisions for a maximum 25′ secondary conductor.

Service Conductors – 240.21(D). Subpart (D) permits service conductors to be protected against overcurrents. Section 230.91 requires the service overcurrent protection to be an integral part of the service disconnect or be located adjacent to the service disconnecting means.

Premises Location – 240.24

Overcurrent devices shall be installed in a readily accessible location, free from physical damage, and in a location that ensures that all occupants of a building or structure have access to the overcurrent devices protecting the supply conductors for their premises. They shall not be installed near easily ignitible materials or in bathrooms. **See Figure 5-22.**

Accessibility – 240.24(A)(1 – 4). In general, OCPDs shall be installed in a readily accessible location. Readily accessible indicates that the OCPDs shall be capable of being reached quickly and that there are no obstacles preventing ready access. To be considered readily accessible, the grip of the operating handle for any OCPD shall be installed so that the center of the grip or handle, when in its highest position, is not more than 6′-7″ above the floor. There are, however, four conditions that permit exceptions to this main rule. The first permits the OCPDs to be installed in a location that is not readily accessible if they are for use with busduct per 368.17(C). The second permits supplementary OCPDs to be installed in locations that are not readily accessible. Section 240.10 permits supplementary devices provided they are not used as substitutes for branch-circuit devices. The third permits service OCPDs to be installed in locations which are not readily accessible as described in 225.40 and 230.92. The fourth permits OCPDs which are located adjacent to motors and appliances to be installed in locations which are not readily accessible but which are accessible by portable means.

Overcurrent Protective Device (OCPD) Interrupting Rating

Per Article 100, an interrupting rating is the highest current at rated voltage an overcurrent protective device is intended to open to interrupt the flow of current under standard test conditions.

Occupant Access – 240.24(B)(1)(2). In general, the overcurrent devices that protect conductors supplying an occupancy shall be readily accessible to each occupant. In multi-occupancy buildings, the OCPDs shall be installed so that each occupant has ready access to the OCPDs which protect conductors supplying the occupants' space. This helps to ensure the safety of occupants by providing them with individual control over the circuits in their space. There are two conditional provisions for service and feeder overcurrent devices. The first applies to multi-occupancy buildings managed by full-time building supervision. In these types of occupancies, service and feeder OCPDs which supply more than one occupant can be installed where accessible to authorized personnel only. The second permits the OCPDs to be installed and accessible to only authorized personnel in occupancies such as motels and hotels and other guest room occupancies.

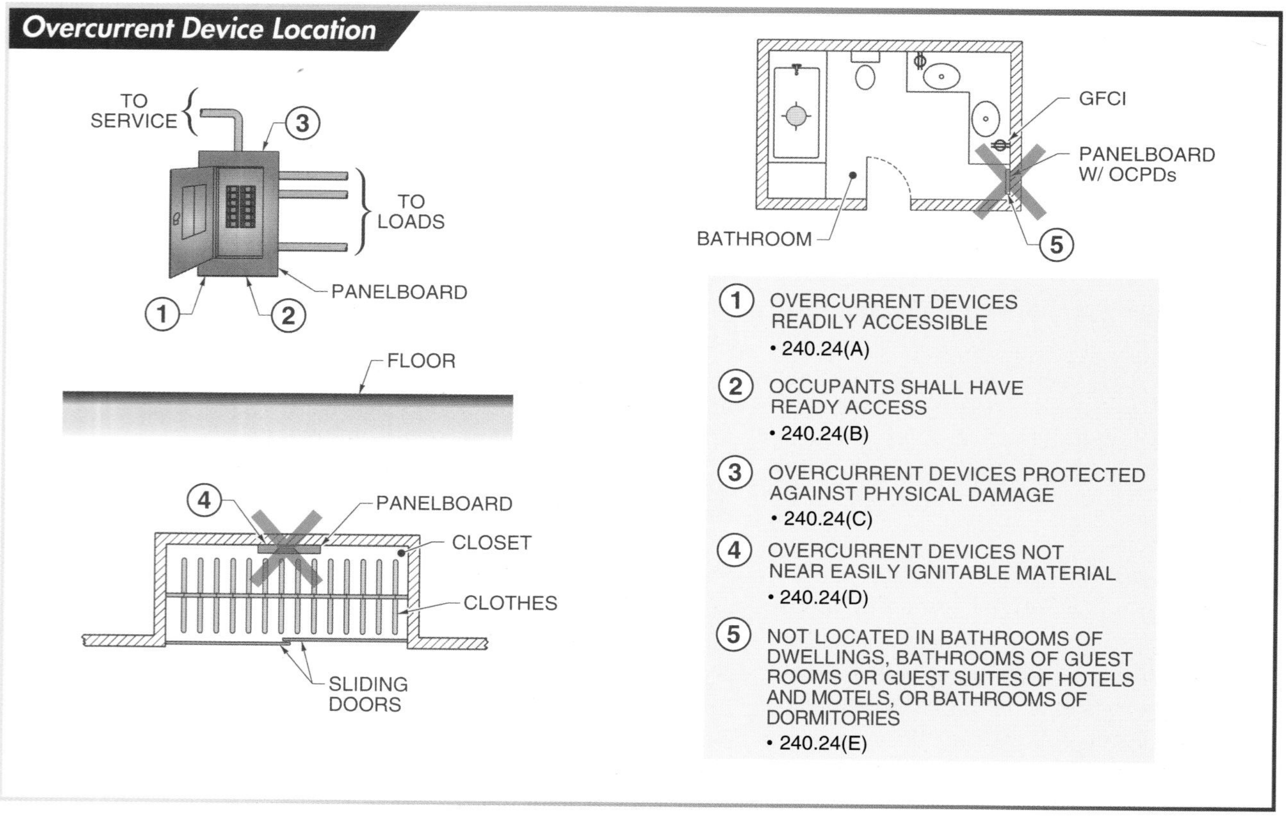
Overcurrent Device Location
TO SERVICE
TO LOADS
PANELBOARD
FLOOR
PANELBOARD
CLOSET
CLOTHES
SLIDING DOORS
GFCI
PANELBOARD W/ OCPDs
BATHROOM
1 OVERCURRENT DEVICES READILY ACCESSIBLE • 240.24(A)
2 OCCUPANTS SHALL HAVE READY ACCESS • 240.24(B)
3 OVERCURRENT DEVICES PROTECTED AGAINST PHYSICAL DAMAGE • 240.24(C)
4 OVERCURRENT DEVICES NOT NEAR EASILY IGNITABLE MATERIAL • 240.24(D)
5 NOT LOCATED IN BATHROOMS OF DWELLINGS, BATHROOMS OF GUEST ROOMS OR GUEST SUITES OF HOTELS AND MOTELS, OR BATHROOMS OF DORMITORIES • 240.24(E)

Figure 5-22. Overcurrent devices are not permitted in clothes closets or bathrooms.

Section 240.24(B)(2) contains the provisions for locating the branch-circuit overcurrent devices in an area that is not accessible to the occupants of guest rooms or guest suites. This is only permissible where electrical service and maintenance are provided by the building management and where there are no permanent provisions for cooking provided in the guest rooms or guest suites.

Physical Damage – 240.24(C). Because OCPDs are a vital link in providing safety to personnel and property, they shall be installed in locations which are not subject to physical damage. This includes locations in which corrosive or other deteriorating agents may be present.

Ignitible Materials – 240.24(D). When OCPDs interrupt the flow of current, an arc can occur depending upon the amount of current flowing in the circuit. Therefore, OCPDs are not permitted to be installed in locations where easily ignitible materials are stored. Examples of such locations are clothes or linen closets.

Each occupant shall have ready access to all OCPDs protecting conductors that supply that occupancy.

Bathrooms – 240.24(E). By definition, bathrooms contain a basin and at least a toilet, tub, or shower. Varying degrees of moisture are always present in bathrooms. Therefore, OCPDs are not permitted to be installed in bathrooms in any of the following: dwelling units, guest rooms or guest suites of motels or hotels, or dormitories. GFCI receptacles are not OCPDs and are permitted in bathrooms, as are supplementary overcurrent devices.

Stairways – 240.24(F). This section was added to the 2008 NEC® and prohibits overcurrent devices from being located over the steps of a stairway. This is primarily a safety issue and is intended to protect personnel who may be required to operate, inspect, or maintain an overcurrent device.

Conductor Identification

When there is more than one voltage system in a building, each ungrounded conductor of a multiwire branch circuit must be identified by phase and system. While different methods can be used for identification, including separate color coding, marking tape, tagging, or other approved methods, the most common method is the use of a color-coding system.

For example, when both a 120/208 V and a 480/277 V system are located in the same building, several color-coding schemes are used with ungrounded conductors in branch circuits. For 120/208 V, black, red, and blue are used for phase conductors, and white or striped for the grounded conductor. For 480/277 V, brown, orange, and yellow are used for phase conductors, and gray is used for the grounded conductor. Marking tape, tagging, or other effective methods such as labels can also be used to identify ungrounded branch-circuit conductors.

OVERCURRENT DEVICES

Overcurrent devices are designed to open the circuit before damage can occur to either conductors or equipment as a result of abnormally high current levels. Overcurrent devices shall be capable of protecting the electrical components against overloads, short circuits, and ground faults. Overcurrent devices are placed in series with ungrounded conductors and, by design, they have very little effect on the circuit during normal operation. Overcurrent devices shall have an interrupting rating suitable for the available short-circuit current, and they shall have a voltage and ampere rating suitable for the electrical system in which they are employed. The two most common overcurrent devices used in electrical distribution systems are fuses and circuit breakers.

An overcurrent protective device is used to provide protection from short circuits and overloads. **See Figure 5-23.** Fuses and circuit breakers (CBs) are overcurrent protective devices designed to automatically stop the flow of current in a circuit that has a short circuit or is overloaded.

Overcurrent Protective Devices

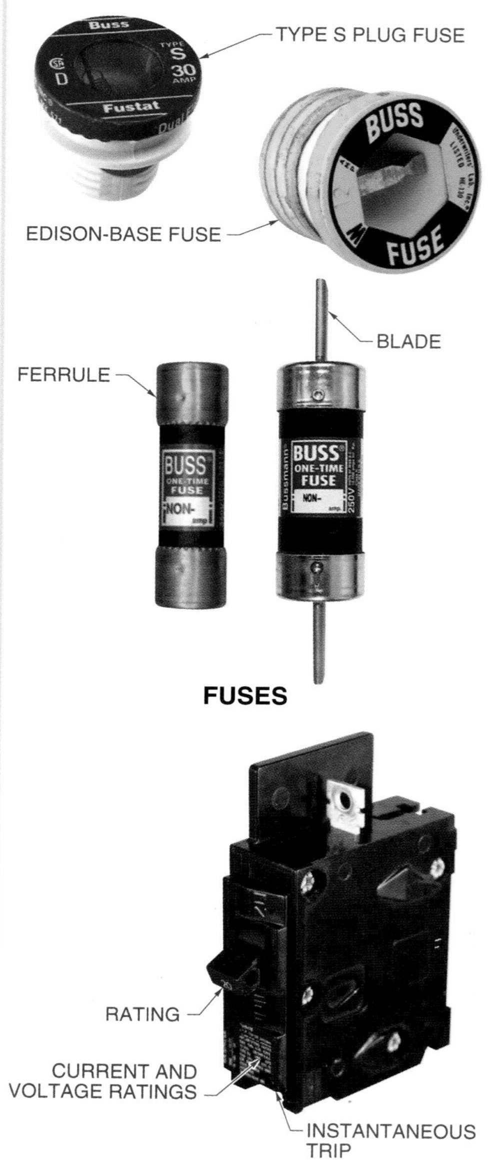

Cooper Bussmann

Figure 5-23. Overcurrent protection for conductors and equipment is provided to open the circuit if the current reaches a specified value.

A *fuse* is an overcurrent protective device with a fusible link that melts and opens the circuit when an overload condition or short circuit occurs. A *circuit breaker (CB)* is an overcurrent protective device with a mechanical mechanism that may manually or automatically open the circuit when an overload condition or short circuit occurs.

Both fuses and circuit breakers are widely used in power distribution systems and individual pieces of equipment. Fuses and circuit breakers have their advantages and disadvantages. The overcurrent protective device selected is normally determined by the application, economics, and individual preference of the person making the choice.

Fuses and circuit breakers include both current and voltage ratings. The listed current rating is the maximum amount of current the overcurrent protective device (OCPD) carries without blowing or tripping the circuit. The current rating of the OCPD is determined by the size and type of conductors, control devices used, and loads connected to the circuit.

The voltage rating is the maximum amount of voltage that may be applied to the OCPD. It safely suppresses the internal arc produced when the OCPD stops the current flow. The voltage rating of the OCPD is greater than (or equal to) the voltage in the circuit.

Every ungrounded (hot) conductor must be protected against short circuits and overloads. A fuse or circuit breaker is installed in every ungrounded conductor. One OCPD is required for low-voltage, 1ϕ circuits (120 V or less) and all DC circuits. The neutral conductor (in AC circuits) or the negative polarity (in DC circuits) does not include an OCPD. Two OCPDs are required for 1ϕ, high-voltage circuits (208 V, 230 V, or 240 V). Both ungrounded conductors are protected by an OCPD. Three OCPDs are required for all 3ϕ circuits (any voltage). All three ungrounded conductors must be protected by an OCPD. **See Figure 5-24.**

Fuse Response Time

The time required for a fusible element to open varies inversely with the magnitude of the current that flows through the fuse or circuit breaker. As current increases, the time required for a fuse to completely open decreases. The time-current characteristic of a fuse determines its rating and type. Time-current characteristic curves are available from manufacturers of fuses and circuit breakers.

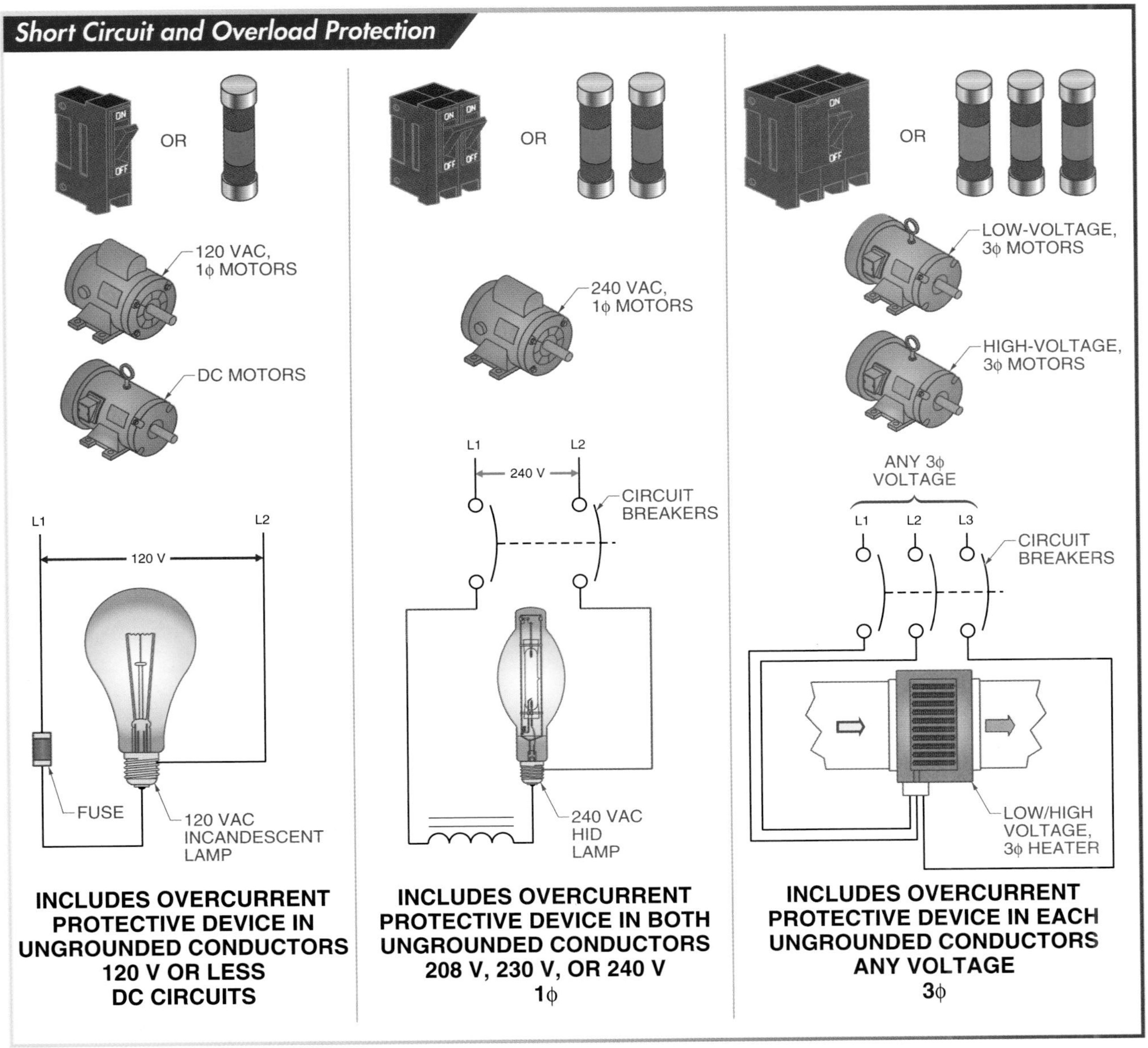

Figure 5-24. All ungrounded (hot) conductors shall be protected against short circuits and overloads.

In general, overcurrent devices must always be located so that they are protected from physical damage. This can be accomplished by locating them in enclosures or cabinets or in panelboards and similar equipment or in rooms that are accessible to qualified personnel only, per 240.30(A).

Plug Fuses – 240.50

A *plug fuse* is a fuse that uses a metallic strip which melts when a predetermined amount of current flows through it. Plug fuses are inserted in series with ungrounded conductors to protect the conductors against overcurrents which could damage the conductor or the connected equipment. At one time, plug fuses were the most common overcurrent protective device used in residential and light commercial electrical installations.

Plug fuses have a voltage rating of 125 V between conductors in most applications. The maximum voltage rating, however, for circuits supplied from a system utilizing a grounded neutral is 150 V to ground. Thus, for a standard 120/240 V, 1ϕ, 3-wire system, plug fuses can be used for the protection of both 120 V and 240 V circuits.

All plug fuses shall be marked to indicate their ampere rating. By design, the plug fuse ampere rating is identified by the shape of the window on the fuse. Plug fuses with ampere ratings of 15 A or less have a hexagonal window, while those with ratings greater than 15 A have a round window. The two basic types of plug fuses are the Edison-base fuse and the Type S fuse. **See Figure 5-25.**

Edison-Base Fuses – 240.51. An *Edison-base fuse* is a plug fuse that incorporates a screw configuration which is interchangeable with fuses of other ampere ratings. Edison-base fuses are the older of the plug fuse designs. Edison-base fuses are classified at not over 125 V and are designed to be used in circuits with ampere ratings of 30 A and below. Because the Edison-base fuse is not designed to be noninterchangeable with fuses of other ampere ratings, they are permitted to be installed only as replacements for existing installations where there is no evidence of overfusing.

Type S and T Fuses

Types S and T fuses are heavy-duty time delay fuses used for circuits that have motor loads or circuits with motors that cycle on and off frequently. Types T and S fuses have a longer time delay than TL or SL fuses. As with TL and SL fuses, the main difference between types T and S fuses is the base.

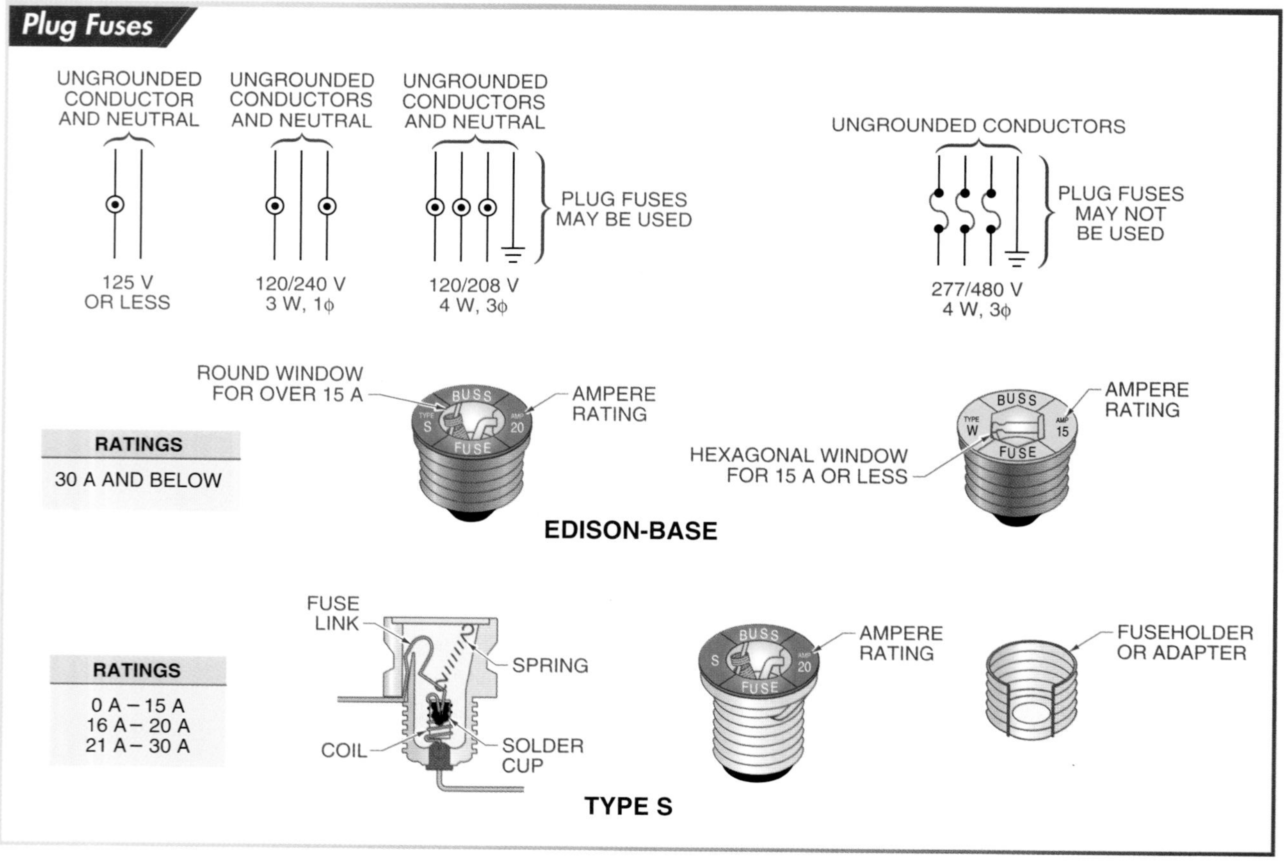

Figure 5-25. Plug fuses shall not be installed in circuits operating at over 150 V to ground.

Edison-Base Plug Fuses

One of the greatest dangers associated with the use of Edison-base plug fuses is that they are interchangeable from one ampere rating to another. The homeowner, often unaware of the potential danger, may replace a fuse that persistently "blows" with another fuse with a greater ampacity. For example, a 15 A circuit installed with 14 AWG Cu conductors is overloaded, resulting in frequent blown fuses. The homeowner, in an effort to solve the problem, replaces the properly sized 15 A fuse with a 30 A fuse. Although the rating is twice that of the 15 A fuse, the 30 A device fits, and the homeowner may mistakenly believe the problem is solved.

Electrical installers, when performing work in occupancies protected by Edison-base plug fuses, should ensure that the homeowner is aware of any overfusing and should encourage the homeowner to consider Type S plug fuses, which do not permit overfusing.

Type S Fuses – 240.53. A *Type S fuse* is a plug fuse that incorporates a screw and adapter configuration which is not interchangeable with fuses of another ampere rating. Like Edison-base plug fuses, Type S fuses are designed for circuits not exceeding 125 V. Type S fuses have three ampere rating classes: 0 A–15 A, 16 A–20 A, and 21 A–30 A. The fuses are noninterchangeable with fuses of a higher or lower ampere rating which protects the circuit from the possibility of overfusing. Additionally, Type S fuses are designed with fuseholders and adapters so that the adapters fit Edison-base fuseholders and are nonremovable. Once a Type S adapter is installed, a Type S fuse must be used.

Electrical equipment is provided with overcurrent protection to protect it from high current levels caused by short circuits or ground faults.

Cartridge Fuses – 240.60

A *cartridge fuse* is a fuse constructed of a metallic link or links which is designed to open at predetermined current levels to protect circuit conductors and equipment. Cartridge fuses are classified according to their performance, operation, and construction characteristics. Some of the more common classifications are RK1, RK5, G, L, T, J, H, and CC.

Cartridge fuses are available in many more types and in more ampere ratings than are plug fuses. Ampere ratings for cartridge fuses range from 0 A to 6000 A. Cartridge fuses are available in either the one-time type or the renewable type. One-time cartridge fuses must be replaced after they operate, whereas renewable-type cartridge fuses contain a fusible link which can be replaced after it operates.

Cartridge fuses can be constructed with a ferrule or knife-blade configuration. Both of the types are inserted into fuseholders which make the connection from the line to the load. **See Figure 5-26.**

Cartridge fuses can also be classified as a current-limiting type. Current-limiting fuses are designed to operate very quickly to protect electrical system components. By definition, a current-limiting device shall operate within ½ cycle to prevent high-magnitude short-circuit currents from damaging electrical components. Cartridge fuses can also be classified as to their operational characteristics. Cartridge fuses can be classified as either non-time delay (NTDF) or time delay (TDF).

Non-time delay fuses are fast acting and provide excellent overcurrent protection. Time delay fuses are designed to be used with utilization equipment, like motors, that are subject to temporary start-up or surge currents. Time delay fuses allow overcurrents to exist for short periods of time to avoid nuisance tripping due to equipment start-up characteristics. **See Figure 5-27.**

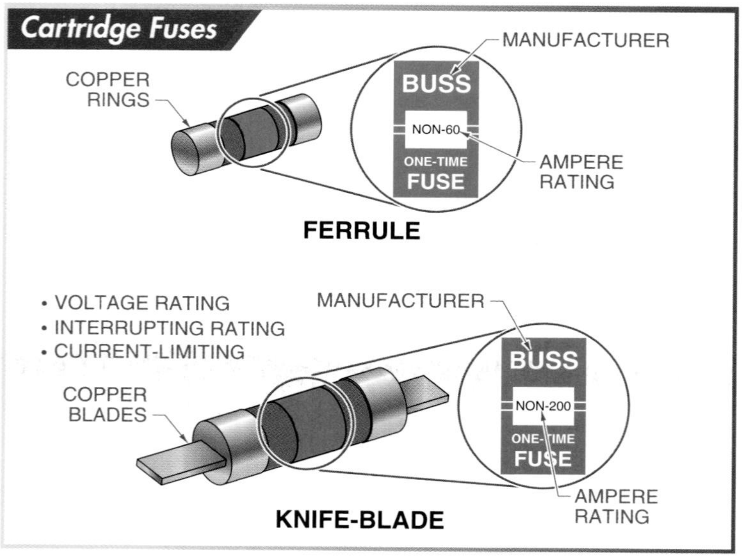

Figure 5-26. Cartridge fuses are designed with either a ferrule or knife-blade configuration.

A *non-time delay fuse (NTDF)* is a fuse that may detect an overcurrent and open the circuit almost instantly. They contain a fusible link that melts and opens the circuit at a set temperature. The fusible link is heated by the current. The higher the current, the higher the temperature. Under normal operation and when the circuit is operating at or below its ampere rating, the fusible link simply functions as a conductor. However, if an overcurrent occurs in the circuit, the temperature of the fusible link reaches a melting point at the notched section. The notched section melts and burns back, causing an open in the fuse. If the overcurrent is caused from an overload, only one notched link opens. If the overcurrent is caused from a short circuit, several notched links open. NTDFs are also known as single-element fuses.

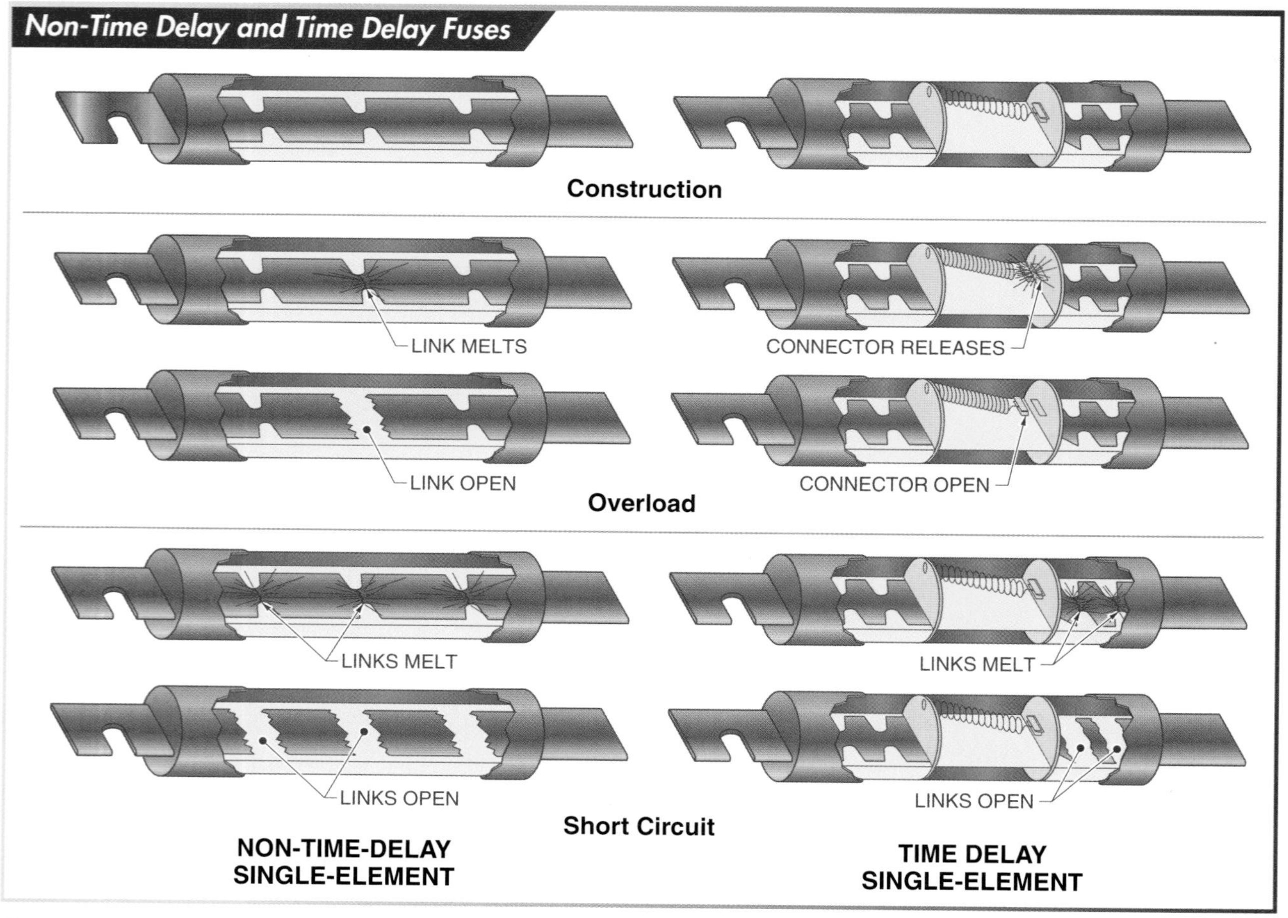

Figure 5-27. Non-time delay fuses (NTDFs) contain a fusible link that melts and opens a circuit at a set overcurrent. Time delay fuses (TDFs) can detect and remove a short circuit almost instantly, but allow small overloads to exist for a short period.

Low-current rated fuses have only one element, while high-current rated fuses have many elements. Very high-current rated fuses (several hundred amperes) are developed by using many elements connected in parallel inside the same fuse. The heat and pressure created by the arc(s) are safely contained within the fuse.

After removing an overcurrent, the fuse is discarded and replaced. Non-time delay fuses provide a very effective way of removing overloads and short circuits, but may also open on harmless low-level overloads. To prevent this, a time delay may be built into the fuse for use in circuits that have temporary and safe low-level overload currents.

A *time delay fuse (TDF)* is a fuse that may detect and remove a short circuit almost instantly, but allows small overloads to exist for a short period of time. Time delay fuses include two elements. One element removes overloads, and the other element removes short circuits. The short-circuit element is the same type of fuse link element used in the non-time delay fuse. The overload element is a spring-loaded device that opens the circuit when solder holding the spring in position melts. The solder melts after an overload exists for a short period of time (normally several seconds). The overload element protects against temporary overloads up to approximately 800%. TDFs are also known as dual-element fuses.

TDFs are used in circuits that have temporary, low-level overloads. Motor circuits are the most common type of circuit that includes low-level overloads. Motors draw an overload current when starting. As the motor accelerates, the current is reduced to a normal operating level. Therefore, TDFs are always used in motor applications.

Voltage Ratings – 240.60(A). All fuses shall have voltage ratings suitable for the electrical circuits in which they are used. Cartridge fuses, in general, shall not be used in circuits exceeding 300 V between conductors. However, 240.60(A)2 permits the maximum voltage to be 300 V line-to-neutral, where the 1ϕ circuit is supplied from a 3ϕ, 4-wire, solidly grounded neutral system. Additionally, cartridge fuses shall not be used in circuits greater than that for which they are rated. Typical voltage ratings of cartridge fuses are 250 V, 300 V, and 600 V.

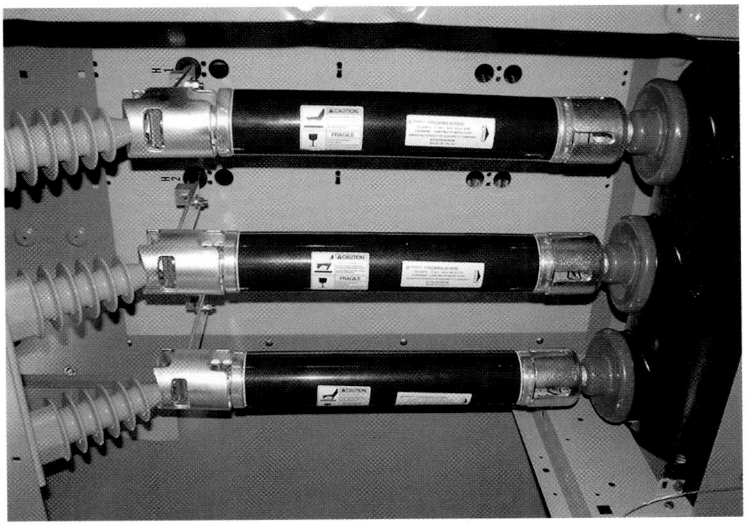

High-voltage fuses are used by utility companies for protection of distribution systems.

Markings – 240.60(C). All cartridge fuses shall be marked, by means of a label or printing on the fuse, to indicate the ampere rating, voltage rating, and the manufacturer's name or trademark. Cartridge fuses shall also be marked to indicate if they are current limiting.

A *current-limiting fuse* is a fuse that opens a circuit in less than one-half of a cycle to protect the circuit components from damaging short-circuit currents. Each cartridge fuse shall also be marked to indicate the interrupting or RMS rating, where other than 10,000 A. The overcurrent device shall have an interrupting rating suitable for the nominal voltage and fault current available at the line side of the terminals per 110.9 and 110.10. Cartridge fuses with no markings are rated for 10,000 A. Other interrupting ratings for cartridge fuses are 100,000 A, 200,000 A, and 300,000 A.

The use of current-limiting fuses grows each year. An awareness of and concern for the hazards of electrical arc blasts and flashes has focused increased emphasis on electrical equipment design considerations as a means to mitigate these potential hazards. For example, 110.16, *Arc-Flash Hazard Warning,* requires field markings of certain types of electrical equipment in other than dwelling units to warn of potential electric arc flash hazards. The use of overcurrent protection devices that employ means for current limitation is one such design consideration that can dramatically impact worker safety by significantly reducing the arc flash and blast potential.

Circuit Breakers – 240.80

Circuit breakers (CBs) are OCPDs which use a mechanical mechanism to protect a circuit from short circuits and overloads. Like fuses, CBs are connected in series with the circuit's conductors. When the circuit current exceeds the rating of the CB, it opens and prevents current from flowing in that part of the circuit.

Circuit breakers are designed to perform the same basic functions as fuses. Circuit breakers offer the added benefit of not damaging themselves and they are resettable after an overload is cleared. CBs that clear short circuits and ground faults can be subjected to extremely high current levels. Many manufacturers include testing requirements in their instructions for CBs that have cleared short circuits or ground faults. Circuit breakers are designed to be trip free. Essentially, this design permits the internal operation of the CB to occur regardless of whether the handle is locked in the open position or not. CBs are available in four basic types: inverse time, adjustable-trip, nonadjustable-trip, and instantaneous trip.

Operation. Circuit breakers can be designed to be operated either manually, automatically, or both manually and automatically. Although the design characteristics may be different, all CBs contain elements that sense a fault or overload condition and open or interrupt the flow of current. The design elements determine the operational characteristics of the CB.

Thermal CBs contain a spring-loaded electrical contact which opens the circuit. The spring is used to open and close the contacts with a fast snap action. A handle is added to the contact assembly so the contacts may be manually opened and closed. The contacts are automatically opened on an overcurrent by a bimetal strip and/or an electromagnetic tripping device. The contacts have one stationary contact and one movable contact. The movable contact is attached to a spring that provides a fast snap action when tripped. **See Figure 5-28.**

The bimetal strip is made of two dissimilar metals that expand at different rates when heated. The strip bends when heated and opens the contacts. The bimetal strip is connected in series with the circuit and is heated by the current flowing through it. The higher the circuit current, the hotter the bimetal strip becomes. Likewise, the higher the current, the shorter the time required to trip the CB.

Like the bimetal strip, the electromagnetic device is connected in series with the circuit. As current passes through the coil, a magnetic field is produced. The higher the circuit current, the stronger the magnetic field. When the magnetic field becomes strong enough, the magnetic field opens the contacts.

Whenever the CB contacts are opened, an arc is drawn across the movable and stationary contacts. A short circuit (higher current) produces a much hotter arc than an overload. Therefore, a CB that has tripped numerous times from short circuits may be damaged to the point of requiring replacement.

Magnetic CBs utilize the magnetic effect created when current flows in a conductor. One of the most common CB designs combines a magnetic element with a thermal element. These thermal-magnetic CBs utilize both the thermal and magnetic effects associated with current flow to sense and operate the CB. Generally, the thermal element is used for overload protection while the magnetic element provides faster protection for ground faults and short circuits.

Fuses and circuit breakers are available in a variety of styles, depending on the application.

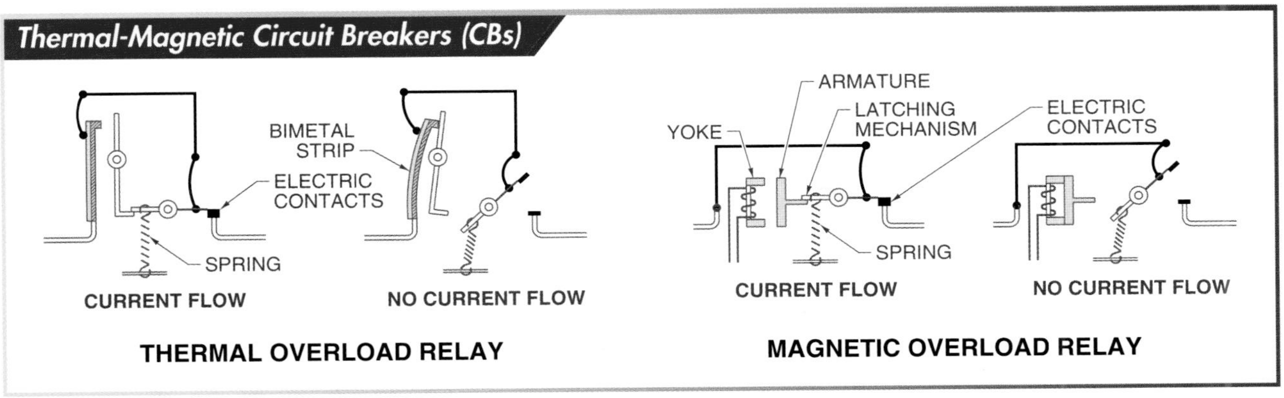

Figure 5-28. Thermal-magnetic CBs contain both a bimetal strip and a magnetic element to sense overloads and faults.

Recent developments in CB design have led to electronically operated "smart" CBs. Smart CBs incorporate electronic components to accomplish the monitoring, sensing, and tripping mechanisms of the CB. Smart CBs are true-rms (root-mean-square) symmetrical current-sensing devices that utilize a microprocessor to gain closer control over the performance of their operations. Typical settings and adjustments for these CBs include continuous amperes, long time delay setpoint, short time pickup, short time delay, instantaneous pickup, and ground-fault pickup.

Inverse-Time CBs. An *inverse-time circuit breaker (ITCB)* is a circuit breaker with an intentional delay between the time when the fault or overload is sensed and the time when the CB operates. The relationship between the trip action and the delay is such that the smaller the overload or fault, the longer the delay. Conversely, the larger the overload or fault, the faster the device operates.

Inverse-time CBs are used extensively in electrical distribution systems and for many different types of applications. They are available as single-, two-, and three-pole configurations. ITCBs provide protection from normal overloads and against damaging currents from short circuits and ground faults. This is accomplished in a thermal-magnetic CB by using the thermal element for overload protection and the magnetic element for short-circuit and ground-fault protection. **See Figure 5-29.**

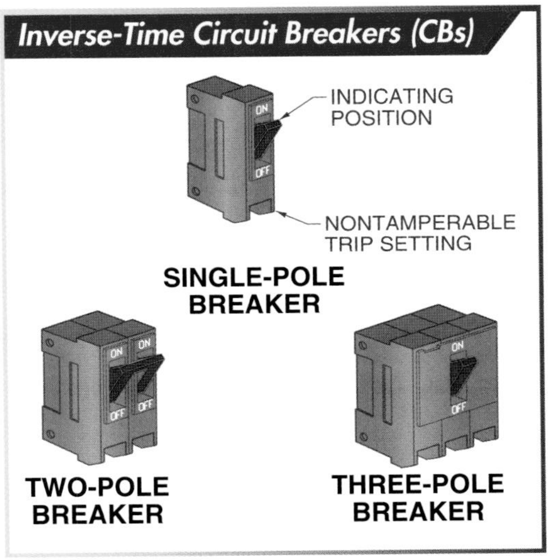

Figure 5-29. Inverse-time CBs are available in single-, two-, and three-pole configurations.

Adjustable-Trip CBs. An *adjustable-trip circuit breaker (ATCB)* is a circuit breaker with a trip setting that can be changed by adjusting the ampere setpoint, trip time characteristics, or both, within a particular range. These CBs are used in industry where utilization equipment may be frequently changed and different overcurrent protection characteristics are required. For example, an ATCB with an 800 A frame size may be installed to protect a feeder for a downstream subpanel. The initial setpoint of the CB is 600 A. If the load on the subpanel increases and the feeder conductors are sized for the additional load, the setpoint of the ATCB can be increased to the new value. This provides great flexibility to the electrical installation at a considerable cost savings.

Nonadjustable-Trip CBs. A *nonadjustable-trip circuit breaker (NATCB)* is a fixed circuit breaker designed without provisions for adjusting either the ampere trip setpoint or the time-trip setpoint. These CBs are used when utilization equipment with known and constant operational characteristics require protection. For example, if a piece of utilization equipment with a full-load ampere rating of 60 A is installed, an NATCB can be installed because the operational characteristics of the equipment are constant, and there is no need for increasing the size or rating of the circuit supplying the equipment.

Instantaneous-Trip CBs. An *instantaneous-trip circuit breaker (ITB)* is a circuit breaker with no delay between the fault or overload sensing element and the tripping action of the device. Because a great deal of utilization equipment, such as motors, is subject to surge or start-up currents much higher than normal operating levels, ITBs are sized well above the ampere rating for normal operation. For example, motors with ITBs for the branch-circuit, short-circuit, and ground-fault protection required by 430.52 may have the CB sized at up to 1300% of the full-load current when starting current is a problem. ITBs may only be used if adjustable and in combination with a listed combination motor controller per 430.52(C)(3).

Circuit Breaker Markings – 240.83

All CBs shall clearly indicate whether they are in the open or closed position per 240.81. The open position is OFF and the closed position is ON. For vertically mounted switches, the ON position of the CB shall be up. CBs shall be marked to indicate the ampere rating, the interrupting rating, and the voltage rating of the device. Additionally, CBs which are permitted to be used as switches shall be marked to indicate such use. **See Figure 5-30.**

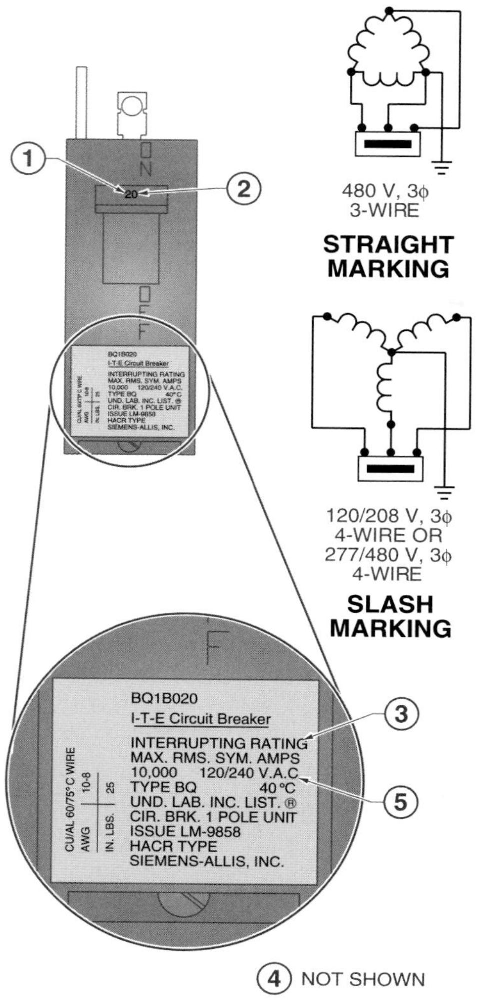

(1) AMPERE RATING SHALL BE DURABLE AND VISIBLE
• 240.83(A)

(2) 100 A OR LESS AND 600V OR LESS CBs SHALL BE MARKED WITH AMPERE RATING IN THE HANDLE OR ESCUTCHEON AREA
• 240.83(B)

(3) INTERRUPTING RATING SHALL BE MARKED (EXCEPT FOR 5000 A)
• 240.83(C)

(4) CBs USED AS SWITCHES SHALL BE LISTED AND MARKED "SWD" OR "HID"
• 240.83(D)

(5) VOLTAGE RATINGS SHALL BE MARKED
• 240.83(E)

Figure 5-30. Circuit breakers are required to be clearly marked with information necessary to ensure their proper application.

Ampere Rating – 240.83(A). CBs shall be durably marked to indicate their ampere rating and shall be installed so that the ampere rating is visible. For CBs rated at 600 V and 100 A or less, this can be accomplished by having the ampere rating marked or printed directly into the handle. For other CBs, it is permissible to make the marking visible by removing a panelboard trim or cover.

Interrupting Rating – 240.83(C). The *interrupting rating* is the maximum amount of current that an overcurrent protective device can clear safely. All CBs shall be marked to indicate interrupting rating. All CBs without markings have an interrupting rating of 5000 A. The overcurrent device shall have an interrupting rating suitable for the nominal voltage and fault current available at the line side of the terminals per 110.9 and 110.10.

CBs Used as Switches – 240.83(D). In some electrical installations, it is practical to control the lighting circuits directly from the panelboard or load center. CBs used for switches in 120 V or 227 V fluorescent lighting circuits shall be listed and shall contain the marking "SWD, or HID." CBs used as switches in high-intensity discharge lighting circuits shall be listed and shall contain the marking "HID."

Voltage Markings – 240.83(E). A critical requirement for the proper installation of CBs in electrical distribution systems is that the CBs have a voltage rating suitable for the voltages in the system. Section 240.83(E) requires that CBs be marked with a voltage rating which is not less than the nominal system voltage. This voltage rating should be an indicator of the CB's capacity to safely interrupt the available phase-to-ground or phase-to-phase fault current.

Circuit Breaker Application – 240.85. Voltage markings are also provided on CBs to ensure that the application of the CB is correct for the type of electrical distribution system. Section 240.85 lists the permissible applications for CBs based upon their voltage markings. CBs are generally marked with either a straight or slash voltage rating. CBs with straight voltage ratings are suitable for use only in circuits in which the nominal voltage between conductors is not greater than the voltage rating of the CB.

Slash voltage markings indicate that the CB is suitable for use in electrical systems in which the nominal voltage between two conductors is not greater than the highest value of the voltage slash rating, and the voltage to ground is not greater than the lowest voltage slash rating. For example, a CB with a 277/480 V slash rating is permitted to be installed in any circuit where the nominal voltage of any conductor to ground does not exceed 277 V and the nominal voltage between any two conductors does not exceed 480 V.

Circuit Breaker Series Ratings – 240.86. In the 2005 NEC®, provisions were added to permit the use of circuit breakers on circuits that have a higher available fault current than the marked rating of the circuit breaker provided that several key provisions were met. This was a controversial Code change that had been discussed and debated for several Code cycles. Section 240.86(A) limits the application to existing installations only and only where a licensed professional engineer has been employed to make the appropriate selection. The series ratings are also not permitted to be used under certain conditions, such as where motors are connected to the load side of the higher-rated overcurrent device and on the line side of the lower-rated overcurrent device. Series ratings are also prohibited where the sum of the full-load currents exceeds 1% of the interrupting rating of the lower-rated CB. Where the conditions warrant the use of series ratings, it is hoped that an additional level of safety can be provided by their use. One example is where older electrical equipment may be installed that is not adequately rated for the present available fault-current levels.

QUICK QUIZ®

Refer to Chapter 5 Quick Quiz® on CD-ROM.

Additional information is available at ATPeResources.com

Conductors and Overcurrent Protection

REVIEW QUESTIONS

Name ______________________ **Date** ______________________

T F **1.** Flexible cords and cables shall be protected against overcurrent in accordance with their ampacities.

______________ **2.** Plug fuses and fuseholders shall not be installed in circuits exceeding ___ V between conductors.

T F **3.** All conductors installed in electrical distribution systems shall be insulated.

T F **4.** THW conductors are permitted to be installed in wet locations.

T F **5.** Conductors used for direct burial shall be identified for such use.

______________ **6.** Insulated conductors installed in wet locations shall be ___ for the use.

______________ **7.** Edison-base fuses are for ___ in existing locations only.

T F **8.** When used for equipment grounding purposes, only covered or insulated conductors are permitted.

______________ **9.** No conductor shall be installed in such a manner that the operating temperature of its ___ is exceeded.

______________ **10.** When conductors are run in parallel in separate raceways, the raceways shall have the same ___ characteristics.

______________ **11.** Plug fuses of 15 A and lower are identified by a(n) ___ configuration of the fuse window.

T F **12.** Circuit breakers shall clearly indicate whether they are in the ON or OFF position.

T F **13.** Overcurrent protective devices are permitted to be installed in clothes closets if they are located at least 12″ from combustible materials.

T F **14.** Circuit breakers used as switches for a 120 V fluorescent lighting circuit shall be listed and marked SWD or HID.

______________ **15.** Cartridge fuses rated ___ V or less shall be permitted to be installed in circuits at or below their voltage rating.

______________ **16.** Circuit breakers that are not marked have an interrupting rating of ___ A.

______________ **17.** Conductors exposed to corrosive conditions shall have an insulation that is ___ for the application.

______________ **18.** Ampacity of feeder taps not over 25′ shall not be less than ⅓ of the rating of the overcurrent protective device protecting the ___ conductors.

______ **19.** In general, fuses and circuit breakers shall not be connected in ___.

green **20.** Equipment grounding conductors shall be provided with an outer finish that is ___ in color.

______ **21.** A(n) ___ is a circuit breaker with no delay between the fault and the tripping action.

gray **22.** In general, the grounded conductor (neutral) is colored white or ___.

90°C **23.** The three basic classifications of insulation ratings are 60°C (140°F), 75°C (167°F), and ___°C (194°F).

higher **24.** The next ___ standard size OCPD of 800 A or less is permitted to be used where the OCPD does not correspond to a standard size.

T F **25.** Overcurrent devices are permitted in clothes closets in bathrooms.

(T) F **26.** Copper and aluminum conductors shall not be terminated in a manner that causes the dissimilar metals to come in direct contact with each other.

T F **27.** Paralleled conductors shall all be the same length.

heat **28.** ___ is generated when current flows through a conductor.

______ **29.** A(n) ___ is any current in excess of that for which the conductor or equipment is rated.

______ **30.** ___ AWG and larger ungrounded conductors installed in a raceway that enters a box shall be protected by an identified insulating fitting.

T F **31.** The least potentially damaging overcurrent occurs when there is an overload.

T F **32.** The maximum standard rating for a fuse is 5000 A.

T F **33.** Type S fuses are designed for circuits not exceeding 100 V.

T F **34.** Insulated conductors are the most common conductors installed in electrical systems.

______ **35.** Locations can be classified as either dry, ___, or wet.

Paralleled Conductors Shall Be:

_______________ 1. Same length

_______________ 2. Same material

_______________ 3. Same size

_______________ 4. Same insulation type

_______________ 5. Same terminations

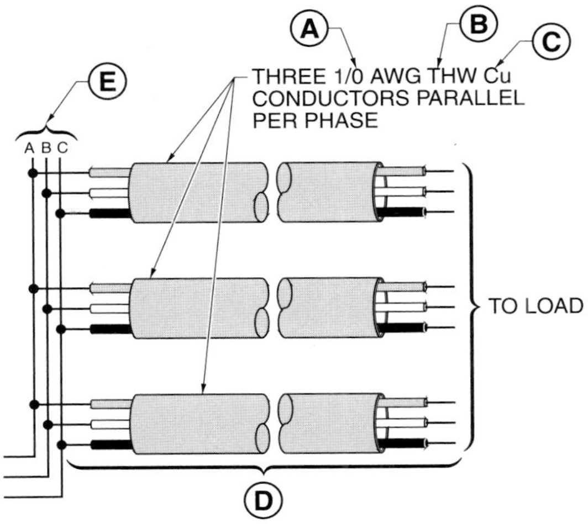

Matching

_______________ 1. Temperature of the air around a piece of equipment.

_______________ 2. Any current in excess of that for which the conductor or equipment is rated.

_______________ 3. A small-magnitude overcurrent which may operate the OCPD.

_______________ 4. Condition that occurs when two ungrounded conductors, or an ungrounded and a grounded conductor of a 1ϕ circuit, come into contact with each other.

_______________ 5. An unintentional connection between an ungrounded conductor and any grounded raceway, box, fitting, etc.

_______________ 6. Any location in which a conductor is subjected to excessive moisture.

_______________ 7. A turning or twisting force, typically measured in foot-pounds.

_______________ 8. Two or more conductors that are electrically connected at both ends to form a single conductor.

_______________ 9. The current that a conductor can carry continuously under the conditions of use.

_______________ 10. An OCPD with a link that melts and opens the circuit when an overload or short-circuit current occurs.

A. overload
B. ground fault
C. torque
D. wet location
E. ampacity
F. overcurrent
G. short circuit
H. ambient temperature
I. parallel conductors
J. fuse

Cable Markings

________________ **1.** Maximum rated voltage

________________ **2.** Letters identifying insulation

________________ **3.** Manufacturer's information

________________ **4.** AWG size or CM area

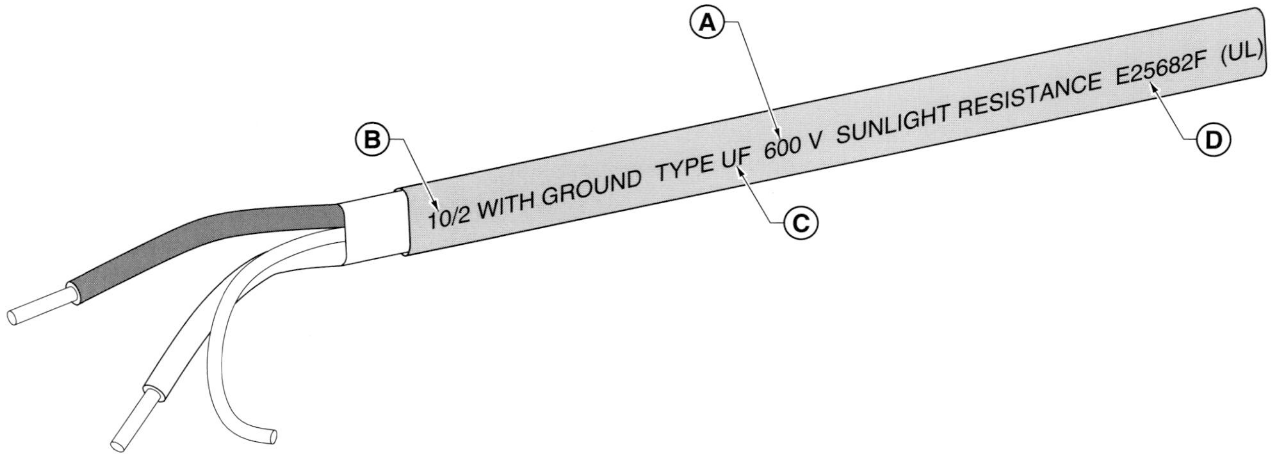

Overcurrent Protective Devices

________________ **1.** Blade fuse

________________ **2.** Ferrule fuse

________________ **3.** Type S plug fuse

________________ **4.** Edison-base fuse

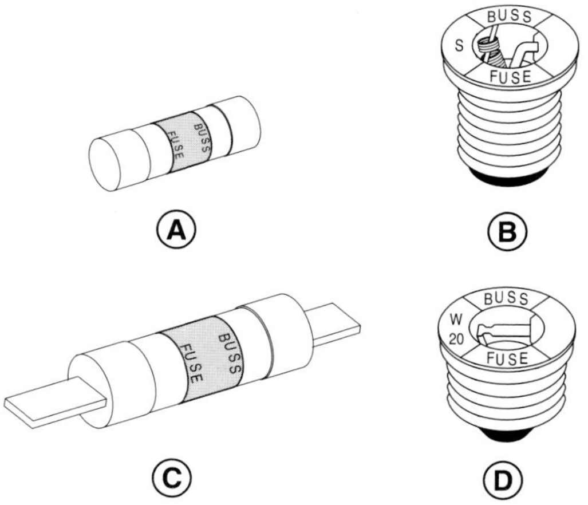

Conductors and Overcurrent Protection

TRADE COMPETENCY TEST

Name ______________________ Date ______________________

NEC®	Answer	
______	______	**1.** Determine the ampacity of three 8 AWG THW Cu conductors installed in the same conduit in an ambient temperature of 30°C. All of the conductors are current-carrying conductors.
______	______	**2.** Determine the ampacity of twelve 10 AWG THHN Cu conductors installed in the same conduit in an ambient temperature of 30°C. Nine of the conductors are current-carrying conductors.
______	______	**3.** Determine the ampacity of six 6 AWG THWN Al conductors installed in the same conduit in an ambient temperature of 30°C. All of the conductors are current-carrying conductors.
______	______	**4.** Determine the ampacity of three 12 AWG THHN Cu conductors installed in the same conduit in an ambient temperature of 104°F. All of the conductors are current-carrying conductors.
______	______	**5.** Determine the ampacity of twelve 1/0 AWG THW Cu conductors installed in the same conduit in an ambient temperature of 96°F. All of the conductors are current-carrying conductors.
______	______	**6.** Determine the ampacity of three 4 AWG THHN Cu conductors installed in the same conduit in an ambient temperature of 86°F. All of the conductors are current-carrying conductors. The conductors terminate on equipment rated for 75°C terminations.
______	______	**7.** Four 6 AWG THW Cu conductors are installed in the same conduit to supply fastened-in-place equipment. Three of the conductors are current-carrying conductors. The ambient temperature around the conduit is 30°C. What is the largest-size OCPD permitted to protect these conductors?
______	______	**8.** *See Figure 1.* Determine the maximum-size OCPD permitted for the installation.
______	______	**9.** *See Figure 2.* Determine the maximum-size OCPD permitted for the installation.

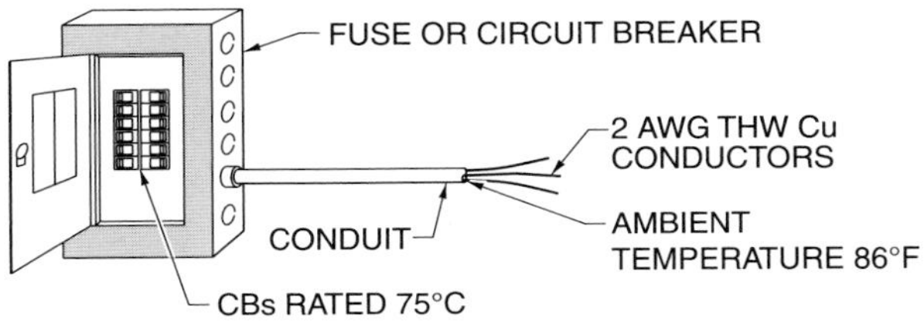

FIGURE 1

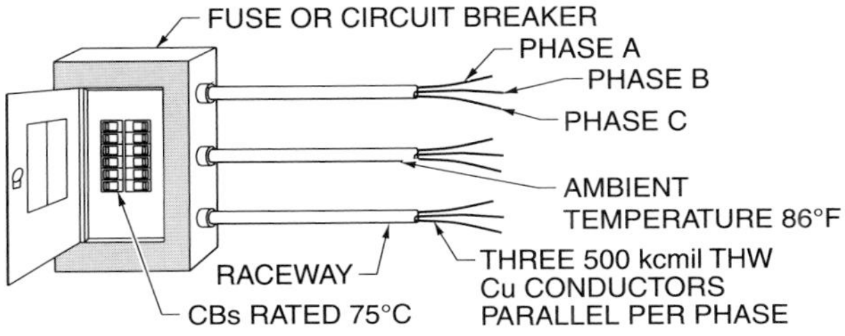

FIGURE 2

_______________ _______________ **10.** *See Figure 3.* Determine the minimum size of the tap conductors for the installation.

_______________ _______________ **11.** *See Figure 4.* Determine the minimum size of the tap conductors for the installation.

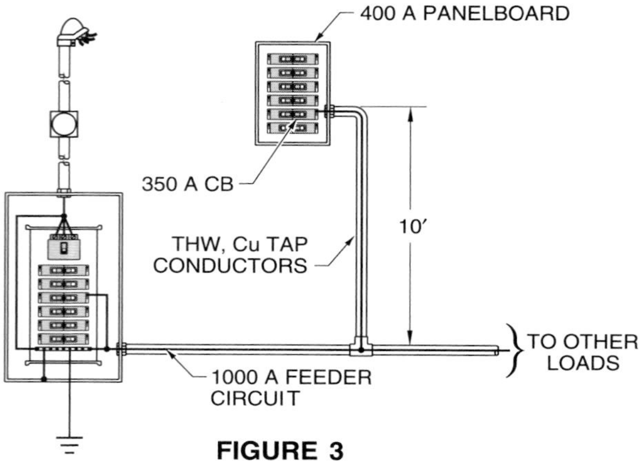

FIGURE 3

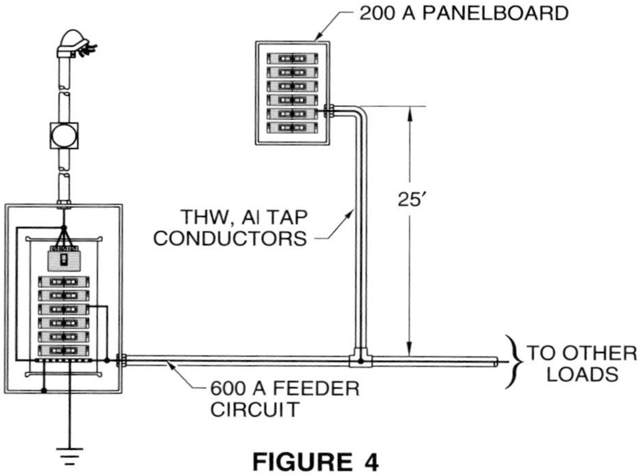

FIGURE 4

_______________ _______________ **12.** *See Figure 5.* Determine the minimum size of the tap conductors for the installation.

_______________ _______________ **13.** *See Figure 6.* Determine the minimum size of the tap conductors for the installation.

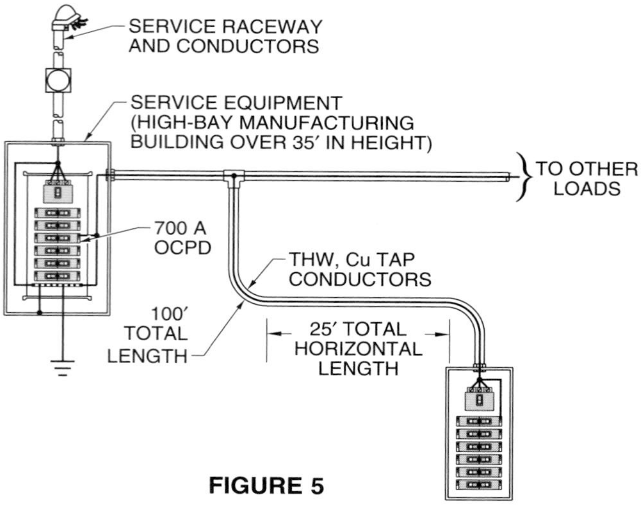

FIGURE 5

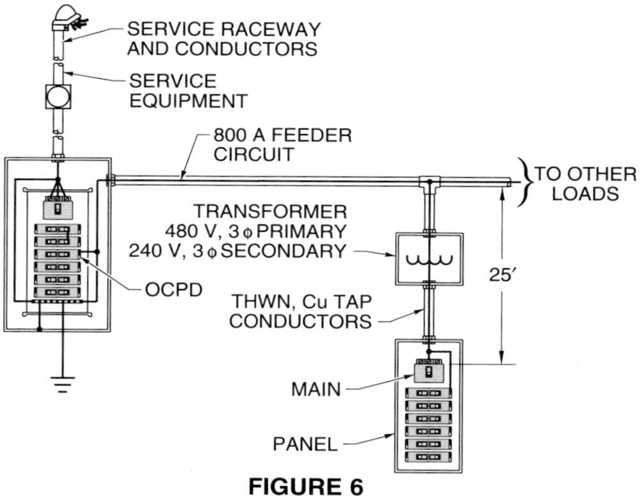

FIGURE 6

______________ ______________ **14.** A 30 A ITCB is installed as a branch-circuit OCPD in a 200 A branch-circuit panelboard. The interrupting rating of the 30 A CB is not marked on the CB. The available short-circuit current at the line side of the CB is 7500 A. Does this installation violate the overcurrent protection provisions of Article 240?

______________ ______________ **15.** An 800 A OCPD protects the installation of three parallel 500 kcmil Cu conductors per phase. At a suitable wireway, a 10′ field tap is made from the 500 kcmil conductors. Determine the minimum-size THW Cu conductors required for the tap conductors.

Learning the Code

- One of the most misunderstood articles in the NEC® is Article 250, Grounding & Bonding. Unfortunately, when an apprentice or student misunderstands a Code rule or requirement, it can result in an incorrect, and sometimes unsafe, electrical installation. The key to understanding Article 250 and correctly applying its provisions lies in obtaining a fundamental understanding of the terminology used in Article 250. The best first step any apprentice or student of the Code can take is to review the Article 100 and Article 250 definitions of grounding and bonding.
- For years little effort was made to distinguish between grounding and bonding. Over the past few Code cycles clearer lines were drawn, culminating in the 2008 NEC®, which goes to great lengths to differentiate between the two terms. Pay close attention to how the terms are used in Article 250. They are often used in a manner that appears to suggest they are interchangeable, but they are not.

- Two tables are critical to understanding Article 250. The first is Table 250.66, which is primarily used in sizing the grounding electrode conductor for AC systems. The second table is Table 250.122, which is primarily used in sizing the equipment grounding conductor for grounding raceways and equipment. Both tables, however, have additional applications. For example, the equipment bonding jumpers on the supply side of a service are sized per Table 250.66, while equipment bonding jumpers on the load side of the service are sized per Table 250.122.

  One other important note regarding these tables: make certain you understand which conductors are sized only from the table and which conductors go beyond the sizing of the table by the use of the 12½% rule. For example, a grounding electrode conductor sized from Table 250.66 would never have to be larger than 3/0 copper or 250 kcmil aluminum. If, however, Table 250.66 was used to size an equipment bonding jumper on the line side of a service, 250.102(C) would require that where the service-entrance phase conductors exceeded 1100 kcmil copper or 1750 kcmil aluminum, the bonding jumper size shall be at least 12½% of the area of the largest phase conductor. In this case, the final conductor size can exceed the maximum sizes listed in the table.

chapter 6

Grounding

A ground is a conducting connection between electrical circuits or equipment and the earth. Grounding electrical systems protects equipment by ensuring that the maximum phase-to-ground voltage is not exceeded. Article 250 covers grounding.

New in the 2011 NEC®

- New definition of supply-side bonding jumper – 250.2
- Revised requirements for ground detectors – 250.21(B), (C)
- Clarification of requirements for grounded conductor brought to service equipment – 250.24(C)
- New Informational Notes added for grounding separately derived AC systems – 250.30(C)
- Revised description of metal underground water pipe – 250.52(A)(1)
- Reformatted and restructured requirements for concrete-encased electrodes – 250.52(A)(3)

GROUNDING – ARTICLE 250

Proper grounding of electrical systems is one of the most critical concerns facing installers and designers of electrical systems. Despite this, no other conductor is so often installed improperly. Sections 250.4(A)(1) and (2) detail some of the many reasons why electrical systems and equipment are grounded.

Electrical systems are grounded to limit and stabilize voltages to ground. Unintentional line surges, lightning strikes, or contact with higher-voltage lines may result in voltages being placed on the electrical system that could damage or destroy electrical components and equipment. A *line surge* is a temporary increase in the circuit or system voltage or current that may occur as a result of fluctuations in the electrical distribution system.

Grounding electrical systems protects equipment by ensuring that the maximum phase-to-ground voltage is not exceeded. *Phase-to-ground voltage* is the difference of potential between a phase conductor and ground. For example, the phase-to-ground voltage in a 120/240 V, 1ϕ, 3-wire system is 120 V. *Phase-to-phase voltage* is the maximum voltage between any two phases of an electrical distribution system. In the 120/240 V, 1ϕ, 3-wire system, the phase-to-phase voltage is 240 V. In the 2008 NEC®, an important Informational Note was added after 250.4(A)(1). The Informational Note emphasizes the need to keep all bonding and grounding conductors as short as possible to limit the imposed voltage on the conductors.

Electrical systems are grounded to ensure that a low-impedance ground path for fault current is present. A low-impedance ground path is a ground path that contains very little resistance or opposition to the flow of fault current. The impedance of the ground path is kept low by ensuring that all electrical connections and any other connections or terminations in the ground path are electrically continuous and done in a manner that ensures very little opposition to the flow of current. The low-impedance ground path helps to ensure that overcurrent protective devices open under ground-fault conditions.

> **Lightning Strikes**
>
> *Protection from surges caused by lightning strikes is especially important in the southeastern states, which have the highest lightning-strike density in the United States at 104 to 130 flashes per square mile per year.*

Ohm's law states that the amount of current (I) that flows in any circuit is determined by the voltage (E) and resistance (R) of the circuit ($I = E \div R$). Ground-fault circuits are no different. If there is a high resistance or impedance in the ground-fault path, the amount of current that flows is smaller and it may not permit the overcurrent protective device to operate. **See Figure 6-1.**

Ground faults can, and oftentimes do, occur. Proper equipment grounding is required to limit the duration of the fault and protect electrical equipment and personnel. Two provisions are necessary to ensure that an effective grounding path is present per 250.4(A)(5). Splices are permitted in grounding conductors but they must be done in a manner that maintains a low circuit impedance. The grounding path shall be capable of handling any fault currents that are likely to be placed upon it. The size of the grounding conductor shall be adequate for the available fault current. The ground path shall be of a sufficiently low impedance to permit the ready operation of the OCPD.

This is the primary purpose of EGCs and the most important factor in ensuring personnel safety. Section 250.4(A) covers the general requirements for grounding and bonding for grounded electrical systems. Section 250.4(B) covers the requirements for ungrounded electrical systems.

Grounding Terminology – Article 100

Terms used in grounding are similar to one another and may be inadvertently misused or misinterpreted if care is not taken. For example, ground, grounded, effectively grounded, grounded conductor, grounding conductor, equipment grounding conductor (EGC), and grounding electrode conductor (GEC) are grounding terms that, while similar, are distinct. Each of these terms is defined in Article 100.

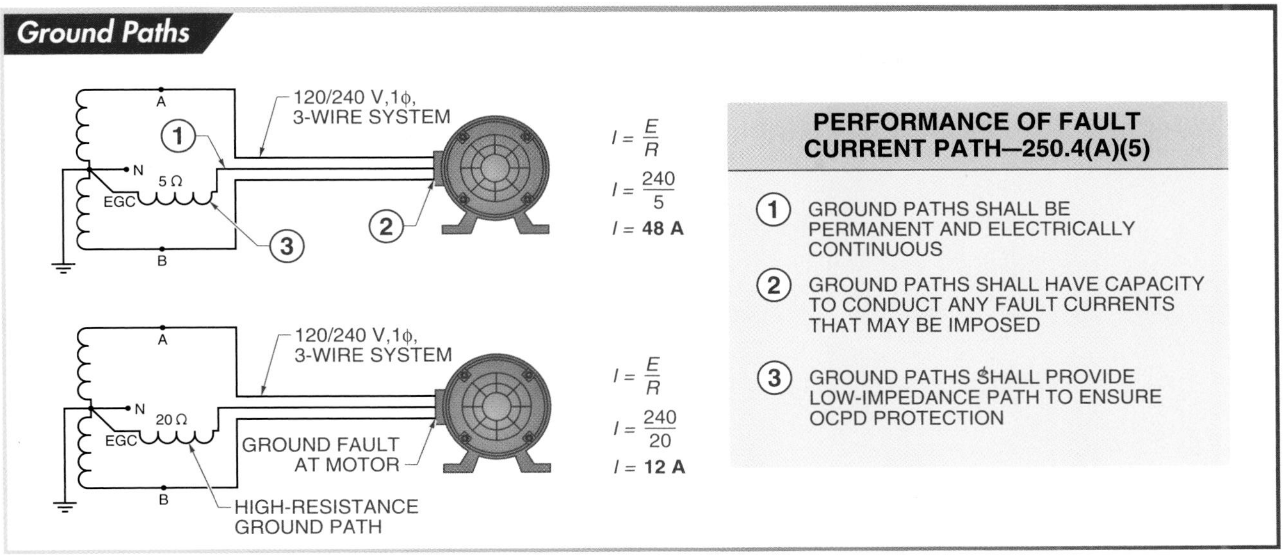

Figure 6-1. Increasing the impedance of the ground-fault path reduces the amount of current that flows in the circuit.

Ground. *Ground* is the earth. For many years ground was defined as a conducting connection between electrical circuits or equipment and the earth. **See Figure 6-2.** This connection can be an intentional connection or it can result accidentally. Intentional connections are made to achieve the benefits of grounding as noted in 250.4(A).

Accidental grounds result in an unintentional connection to the earth or something connected to the earth. For example, an ungrounded conductor's insulation may get nicked while being installed in a metal raceway. An unintentional ground occurs, and a fault current flows, depending upon the impedance of the ground path. *Impedance* is the total opposition to the flow of current in a circuit. In AC circuits, impedance consists of the total resistance, inductance, and capacitance of the circuit.

Often the connection for a ground is not made directly to the earth but by means of an electrode that is, in turn, connected to the earth. In some parts of the world, the ground connection is known as earthing.

Grounded. *Grounded* is connected to the earth or to a conductive body connected to the earth. Equipment, circuits, or systems that have been grounded have been connected to the earth directly or to some type of electrode which is, in turn, connected to the earth. For example, 250.80 requires that service raceways and enclosures shall be grounded. These raceways and enclosures shall be connected to the earth or a conducting body connected to the earth.

Grounded Conductor. A *grounded conductor* is a conductor that has been intentionally grounded. The grounded conductor can be a system conductor or a circuit conductor. **See Figure 6-3.** The grounded conductor is commonly a neutral conductor. Not all electrical distribution systems, however, use the grounded conductor as a neutral. For example, corner-grounded delta systems contain a grounded conductor that is not a neutral conductor. Therefore, it is incorrect to refer to all grounded conductors as neutral conductors, although that is the case in the majority of electrical distribution systems.

Electrodes Permitted for Grounding

Per 250.52, electrodes permitted for grounding include the metal frame of a building or structure, concrete-encased electrodes, underground ground rings that encircle a building, rod or pipe electrodes, and plate electrodes.

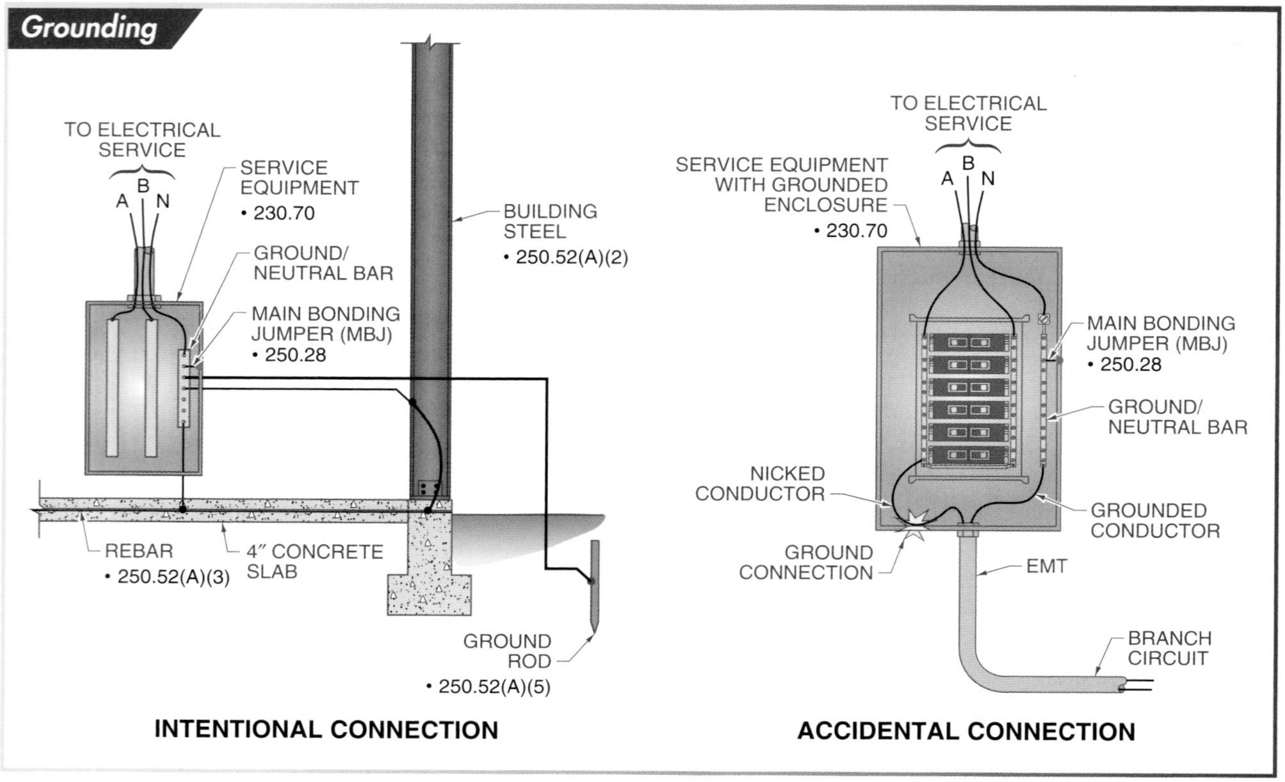

Figure 6-2. A ground is the earth, represented as a conducting connection between electrical circuits or equipment and the earth.

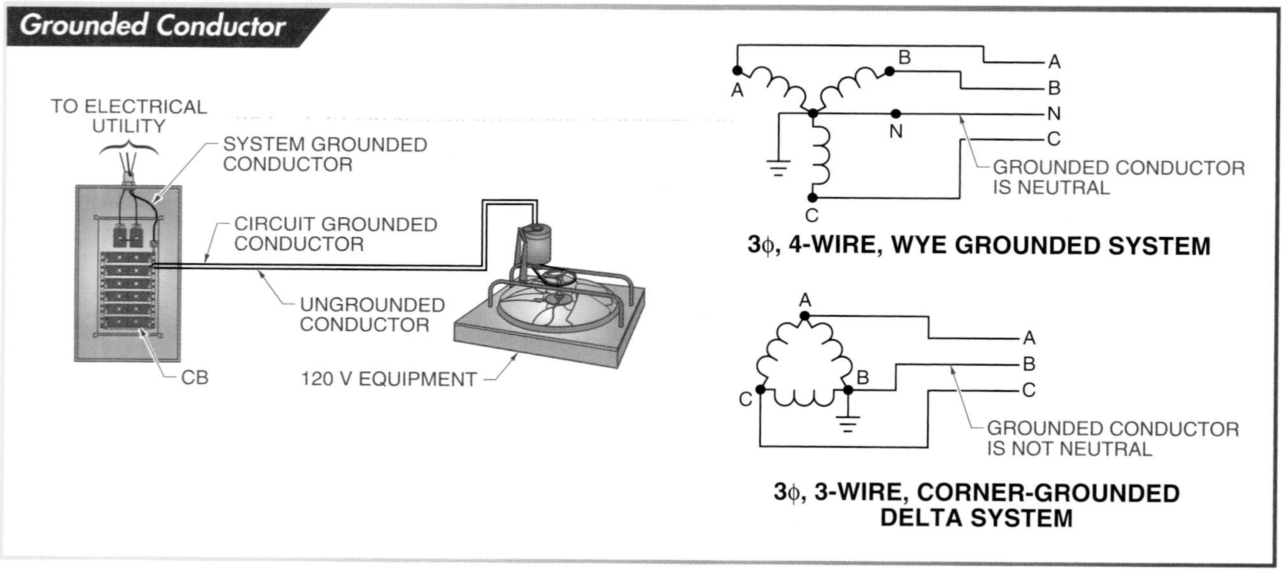

Figure 6-3. A grounded conductor is a conductor that has been intentionally connected or grounded.

Grounding Conductor. A *grounding conductor* is the conductor that connects electrical equipment or the grounded conductor to the grounding electrode. The grounding conductor can be a GEC or an EGC. In either case, the connection is made between electrical equipment or the grounded conductor and the grounding electrode(s). **See Figure 6-4.**

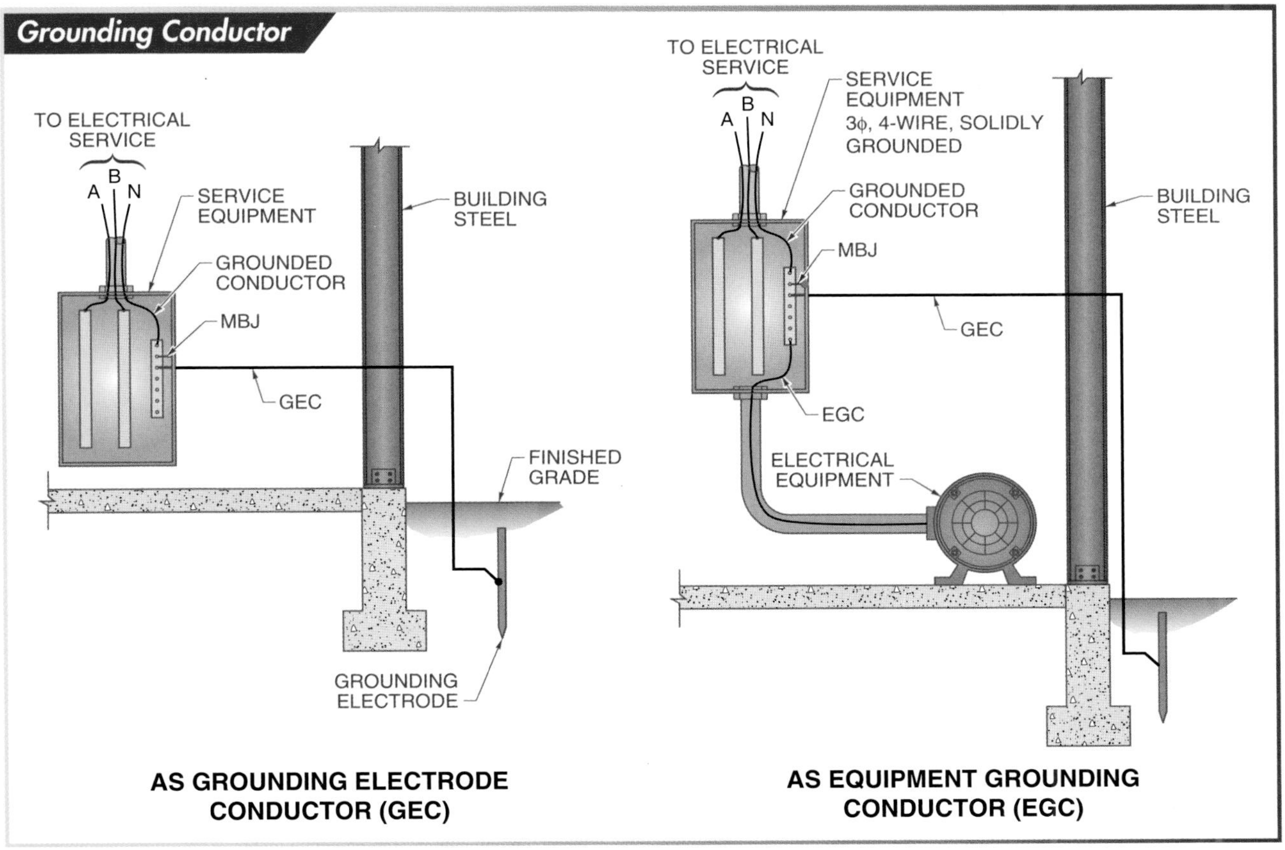

Figure 6-4. A grounding conductor is the conductor that connects electrical equipment or the grounded conductor to the grounding electrodes.

Equipment Grounding Conductor (EGC). An *equipment grounding conductor (EGC)* is an electrical conductor that provides a low-impedance path between electrical equipment and enclosures and the system grounded conductor and grounding electrode conductor. This connection usually occurs at the service equipment or at the source of power and helps to ensure the operation of the OCPD when fault conditions occur. EGCs may be in the form of a separate wire, metal conduit, metal tubing, metal cable sheath, etc. EGCs may also take on the additional function of bonding as discussed in Article 250, Part V, Bonding.

The most important factor in reducing the risk of shock and electrocution in electrical distribution systems is the proper installation of the EGC. The EGC connects metal frames and enclosures of equipment and other non-current-carrying metal parts of equipment to the grounded conductor and/or the EGC. This connection can occur at the service equipment or at the source of a separately derived system, such as a transformer.

Lightning strikes are usually the most severe form of voltage surge to occur on a system.

The EGC is connected to the grounded system conductor to facilitate the operation of the OCPD by providing a low-impedance path for fault current to flow. This conductor is essential for life safety, yet most equipment can function quite normally with a poor EGC or even without an EGC. **See Figure 6-5.**

During the 2002 NEC® cycle, a new definition was added to Article 100 for a special type of bonding jumper. A *system bonding jumper* is the bonding jumper that is used to connect the grounded circuit conductor and the supply-side bonding jumper and/or the equipment grounding conductor at a separately derived system.

Although it functions like a main bonding jumper, the new term was added to avoid any confusion in the field as to the requirements for bonding and grounding of new or separately derived systems. In the 2011 NEC®, a new definition was added to recognize supply-side bonding jumpers. This term is used for conductors that are installed on the supply side of a service, or that are located within a service equipment enclosure for separately derived systems, that are used to provide electrical continuity between all metal parts required to be electrically connected.

Grounding Electrode Conductor (GEC). A *grounding electrode conductor* is the conductor that connects the grounding electrode(s) to the system grounded conductor and/or the equipment grounding conductor. Like the EGC, this connection can occur at the service equipment or at the source of a separately derived system. The connection is also permitted to be made to a point on the grounding electrode system. The GEC may be constructed of aluminum, copper, or copper-clad aluminum; may be solid or stranded; and either insulated, covered, or bare per 250.62. **See Figure 6-6.**

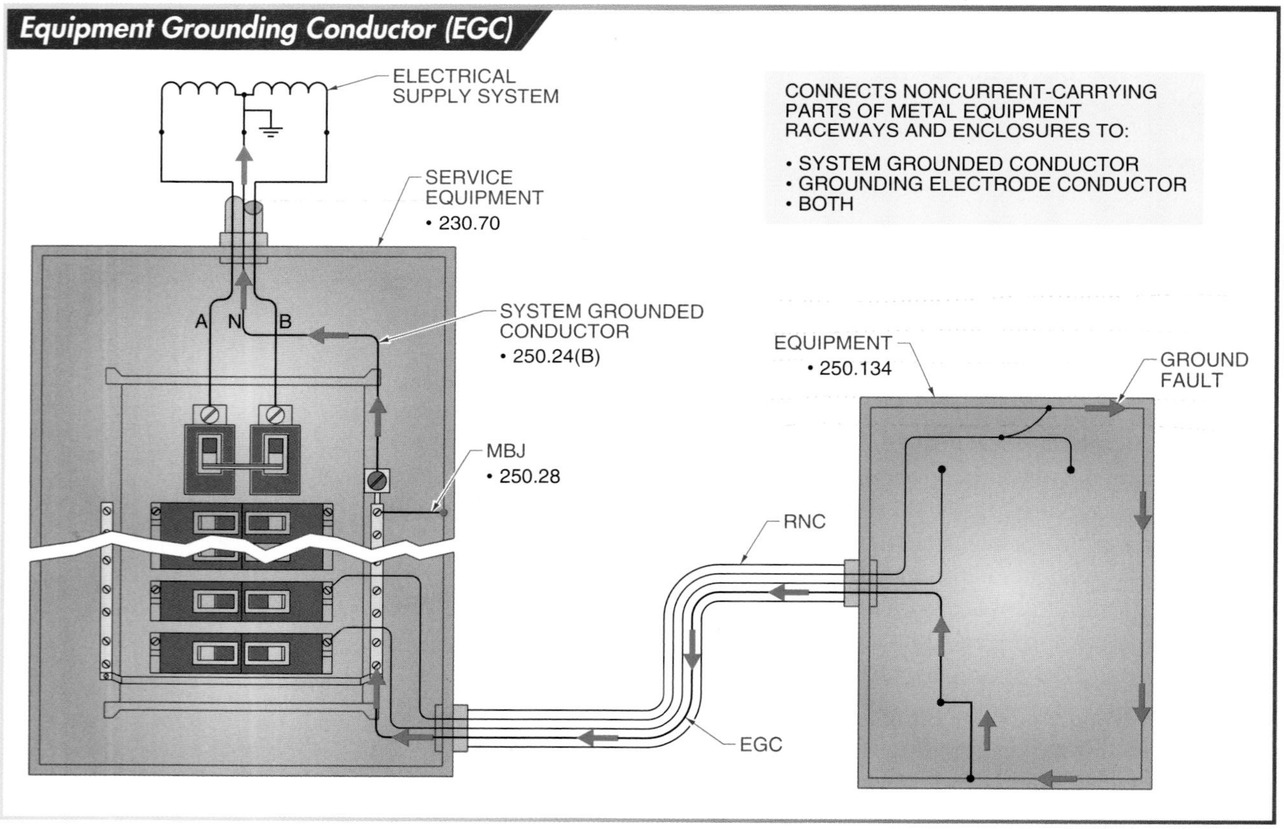

Figure 6-5. An equipment grounding conductor (EGC) is an electrical conductor that provides a low-impedance path between electrical equipment and enclosures and the system grounded conductor and grounding electrode conductor (GEC).

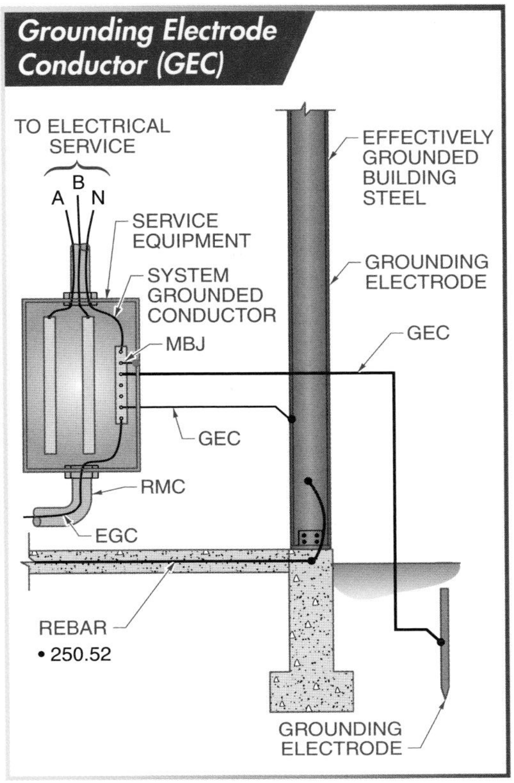

Figure 6-6. A GEC is the conductor that connects the grounding electrode(s) to the grounded conductor and/or the EGC.

GROUNDING SYSTEMS

Despite the benefits of grounding, there are electrical systems and circuits that are not grounded. Electrical systems can be classified as those that are required to be grounded, those that are permitted to be grounded, and those that are prohibited from being grounded. The first step in designing electrical systems is to determine if the system is to be grounded. Sections 250.162 for direct-current (DC) systems and 250.20 for alternating-current (AC) systems contain the requirements for system grounding.

It is important to note the difference between system grounding and circuit grounding. Section 250.22, for example, lists the electrical circuits that are not permitted to be grounded. Once a determination has been made that the system is to be grounded, the NEC® is used to determine the manner in which the grounding occurs. Systems are grounded as follows:

- AC Systems – 250.20; 250.24
- DC Systems – 250.162
- Separately Derived AC Systems – 250.30
- Generators – 250.34

AC Systems – 250.20; 250.24

AC systems can be installed as either grounded or ungrounded systems. The voltage at which the system operates plays an important part in determining if the system is required to be grounded. Section 250.20(A) contains three conditions under which AC systems operating at less than 50 V are required to be grounded. The first condition, per 250.20(A)(1), involves systems that are supplied by transformers of circuits that operate at over 150 V to ground. These systems are required to be grounded to prevent the higher voltages of the primary from being imposed on the secondary side of the transformer in the case of a winding fault.

The second condition under which an AC system of less than 50 V is required to be grounded occurs when the system is supplied from a transformer that is ungrounded. Per 250.20(A)(2), the low-voltage system is required to be grounded because the supply system is ungrounded. Section 250.20(A)(3) requires AC systems of less than 50 V to be grounded when they are installed outside as overhead conductors. **See Figure 6-7.**

Exothermic welding is an acceptable means for connecting the grounding conductor to a grounding electrode.

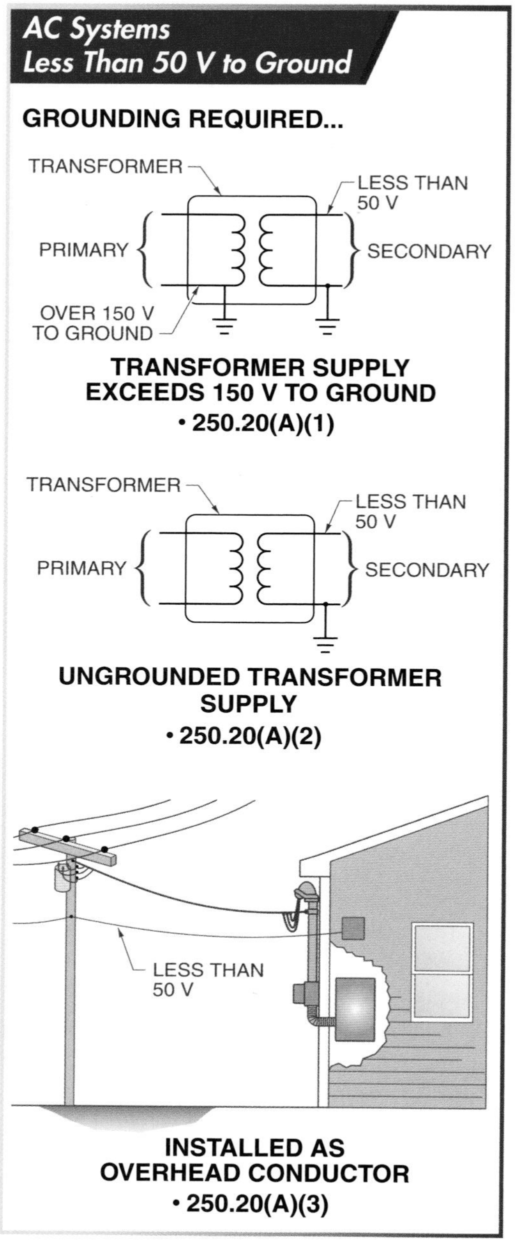

Figure 6-7. Three AC systems of less than 50 V are required to be grounded.

The NEC® does not require other AC systems operating at less than 50 V to be grounded. The decision as to whether to ground the system is left to the designer of the electrical system.

Section 250.20(B) provides the requirements for grounding AC electrical systems operating between 50 V and 1000 V. AC systems shall be grounded when the system can be grounded in a manner that the maximum voltage to ground does not exceed 150 V per 250.20(B)(1). Examples of these types of systems are 120 V, 1ϕ, 2-wire systems and 120/240 V, 1ϕ, 3-wire systems.

Section 250.20(B)(2) requires 3ϕ, 4-wire, wye-connected AC systems in which the neutral is used as a circuit conductor to be grounded. The most common electrical systems to fall under this condition are the 277/480 V, 3ϕ, 4-wire system and the 120/208 V, 3ϕ, 4-wire system. This system is frequently used to distribute power in commercial and industrial locations such as high-rise office buildings and manufacturing plants.

Grounding AC Systems—50 V and Less

AC systems of less than 50 V must be grounded when supplied by transformers, if the transformer supply system exceeds 150 V to ground or if the transformer supply system is ungrounded. AC systems of less than 50 V must also be grounded if the system is installed as overhead conductors outside of buildings.

Section 250.20(B)(3) requires 3ϕ, 4-wire, delta-connected systems with the midpoint of one phase used as a circuit conductor to be grounded. This condition requires 120/240 V, 3ϕ, 4-wire delta systems to be grounded. These systems are commonly used in commercial and light industrial installations where the 3ϕ power requirements are large and the 120 V lighting and receptacle requirements are small.

Some highly specialized electrical installations permit AC systems which operate between 50 V and 1000 V to be installed without grounding. Section 250.21(A)(1) permits electric systems used exclusively to supply electric furnaces used in industry for the purposes of melting, refining, or tempering to be ungrounded. Another exemption from the grounding requirements is for industrial, adjustable-speed drive systems.

Section 250.21(A)(2) permits systems supplying rectifiers which supply these adjustable-drive systems to be ungrounded if they are a separately derived system. Separately derived systems supplied from transformers with a primary voltage of less than 1000 V may also be installed ungrounded. Section 250.21(A)(3) permits this type of installation if the system is used solely for control circuits, only qualified persons service the installation, loss of the control power cannot be tolerated.

Sections 250.21(A)(3) and 250.21(A)(4) contain additional special applications for AC systems of 50 V to 1000 V that are not required to be grounded. Section 250.21(A)(3) covers highly specialized control circuit applications that are supplied from transformers that have a primary voltage rating of less than 1000 V. There are three conditions that must be met to permit these systems to operate ungrounded. See 250.21(A)(3)(a–c). Per 250.21(A)(4), other systems are also permitted to be installed ungrounded in accordance with 250.20(B).

The majority of the electrical systems in the United States are required to be grounded per 250.20(B). Section 250.24 specifies the required grounding connections for these AC-supplied systems. Section 250.24(A) requires that all AC-supplied systems which are required to be grounded shall have a grounding electrode conductor connected to the grounded service conductor at the service meeting the requirements of 250.24(A)(1–5). Therefore, each AC system which is required to be grounded shall have a grounding electrode system (GES).

Section 250.24(A)(1) also requires that the system grounded conductor shall be connected to the GEC at any accessible point on the load end of the service drop or lateral up to and including the grounded conductor terminal in the service disconnecting means. **See Figure 6-8.**

Section 250.24(A)(2) also requires that an additional connection be made between the system grounded conductor and the GEC where the service is supplied by a transformer located outside the building. Connection of the grounded conductor and ground or the EGC on the load side of the service disconnecting means is prohibited per 250.24(A)(5). If a connection were to be made on the load side of the service disconnecting means, a parallel connection between the grounded conductor and the grounding conductor would occur. Since some current flow always occurs in a parallel circuit, this connection might result in an undesirable current flow on the grounding conductor.

Under some conditions, however, this connection is permitted to be made. For example, 250.30 permits this connection for separately derived systems. This exception is necessary because when a separately derived system is installed, a new ground reference must be established. The load-side connection is permitted when a common service is used to supply two or more separate buildings per 250.32. Under the requirements of 250.32(B), this connection may need to be re-established at the remote building for grounded systems.

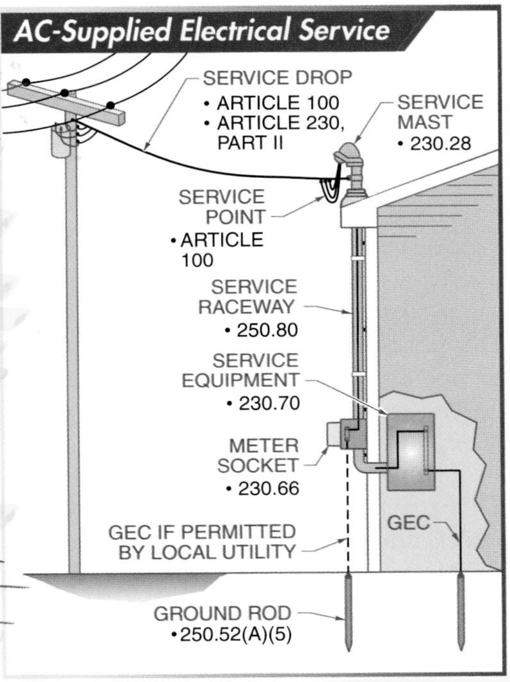

Figure 6-8. GECs may be connected to the grounded conductor within the meter enclosure if the local utility permits connections within the metering equipment.

The most common installation, however, in which a connection is made between the grounded conductor and the grounding conductor on the load side of the service disconnecting means is for electric ranges, clothes dryers, counter-mounted cooking units, and wall-mounted ovens, and metering enclosures per 250.140 and 250.142. For ranges and clothes dryers, the exception for the use of the grounded circuit conductor for grounding equipment applies only to existing branch circuits. The second exception is for meter enclosures and is restricted to enclosures located immediately adjacent to the service disconnecting means.

Grounding Connections

Section 250.24(A)(1) clearly permits the connection between the grounded conductor and the GEC at any point on the load side of the service drop or lateral up to and including the grounded terminal block in the service disconnecting means. Many installers and designers choose to make this connection within the meter socket. With the ground rod installed on the outside of the house, an easy connection can be made without entering the house.

While this is permissible by the NEC®, the local utility company may not permit this connection. Many utility companies do not permit any connections to be made within the metering equipment, regardless of who owns the equipment. Check local utility service requirements before deciding where to make this connection.

Section 250.24(C) contains installation provisions for the grounded conductor of AC systems operating at less than 1000 V. The grounded conductor shall be run to each service disconnecting means and it shall be bonded to the service disconnecting means enclosure. The initial determination of the size of the grounded conductor is done per 220.61 based on the neutral load. After that calculation is performed, the conductor size is determined using the rules found in 310.15. Section 250.24(C)(1) states that the grounded conductor cannot be smaller than the GEC required by table 250.66.

On the line side of the service disconnecting means, the grounded conductor performs the additional task of the EGC. Therefore, the EGC shall be of a sufficient size that it can adequately handle any line-side fault current that may occur.

For electrical installations which utilize parallel service-entrance phase conductors, the minimum size of the grounded conductor is based on the total circular-mil area of the conductors. For example, if three 500 kcmil conductors are installed in parallel, the total equivalent area is 1500 kcmil (500 kcmil × 3 = 1500 kcmil). **See Figure 6-9.**

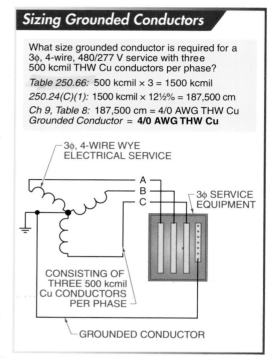

Sizing Grounded Conductors

What size grounded conductor is required for a 3ϕ, 4-wire, 480/277 V service with three 500 kcmil THW Cu conductors per phase?

Table 250.66: 500 kcmil × 3 = 1500 kcmil

250.24(C)(1): 1500 kcmil × 12½% = 187,500 cm

Ch 9, Table 8: 187,500 cm = 4/0 AWG THW Cu

Grounded Conductor = **4/0 AWG THW Cu**

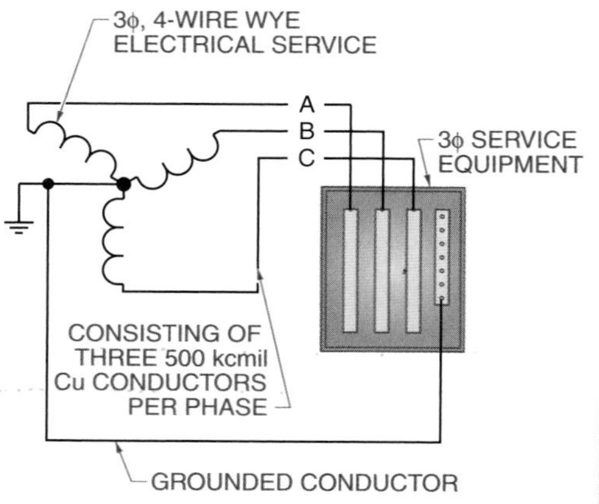

Figure 6-9. The grounded conductor shall never be smaller than 12½% of the largest service-entrance conductor.

Any premises wiring that is supplied from an AC service that is grounded must have a GEC installed. The GEC must be connected to the grounded service conductor at any accessible point from the load end of the service drop or lateral to the service disconnecting means. Section 250.24(D) requires that the GEC also connect all equipment grounding conductors, service-equipment enclosures, and to any grounding electrodes that may be installed. Part III of Article 250 contains the requirements for the grounding electrode system and the grounding electrode conductor.

DC Systems – 250.160

DC systems are treated separately by the NEC®. Sections 250.162(A) and (B) contain the provisions for DC systems that are required to be grounded. In general, two-wire

DC systems and circuits shall be grounded where such systems supply premises wiring and the systems or circuits operate at greater than 50 V but not more than 300 V. *Premises wiring* is basically all interior and exterior wiring installed on the load side of the service point or the source of a separately derived system. See Article 100. There are, however, three exceptions to 250.162(A) for grounding two-wire DC systems.

DC systems installed with ground detection equipment that serve only industrial equipment in limited areas need not be grounded per 250.162(A), Ex. 1. There are several types of ground detectors available, but they all work on the principle that a ground detection system lamp arrangement identifies when an unintentional ground occurs. Ground detection system lamp arrangements vary, but they all employ a light bank, consisting of individual light bulbs representing each system phase, that changes in lamp intensity or turns the lamps ON and OFF, depending upon the status of the system.

Per 250.162(A), Ex. 2, specialized installations are permitted in which a rectifier-derived system is supplied by an AC system installed per 250.20. For example, a 2-wire DC rectifier-derived circuit could operate ungrounded provided that the AC supply system is installed per 250.20. Per 250.162(A), Ex. 3, a 2-wire DC system is permitted to operate ungrounded if it supplies fire alarm circuits installed per Article 760, Part III. These circuits are limited to a maximum current of 30 mA.

The provision for grounding 3-wire DC systems is contained in 250.162(B). A neutral of a 3-wire DC system that supplies premises wiring shall be grounded. Once a determination has been made to ground a DC system, 250.164 is reviewed to determine how and where the grounding connection shall be made.

Section 250.164 contains the requirements for the point of the grounding connection for DC systems. The location of the point depends upon whether the DC system source is located on the premises or off the premises. Section 250.164(A) requires the point of connection for off-premises DC sources to be at one or more supply stations. The grounding connection for off-premises DC sources shall not be made at individual services or at any point on the premises wiring system.

For on-premises DC sources, the grounding connection can be made at any one of three locations per 250.164(B)(1–3). The first permits the connection to be made directly at the DC source. The second option is to make the grounding connection at the first system disconnecting means of the overcurrent device. The last option permits the grounding connection to be made by other means if the equipment is listed and identified for the use and there is an assurance that equivalent system protection has been provided.

DC supply systems, while rare, are still used for specialized applications. The AC system is almost universally accepted as the best means for distributing electrical power. There are, however, more and more applications for DC circuits, but the vast majority of these circuits are derived or converted from AC sources.

Equipment Grounding—PV Systems

Equipment grounding requirements for PV systems are covered in Sections 690.43 and 690.45. All exposed non-current-carrying metal parts of module frames, support structures, enclosures, or other equipment in PV systems must be grounded. Equipment grounding is required regardless of system voltage, even for small 12 V or 24 V systems not otherwise required to have a grounded current-carrying conductor. Damaged PV arrays can energize their metal frames or support structures, but effective equipment grounding diverts this fault current safely to the ground.

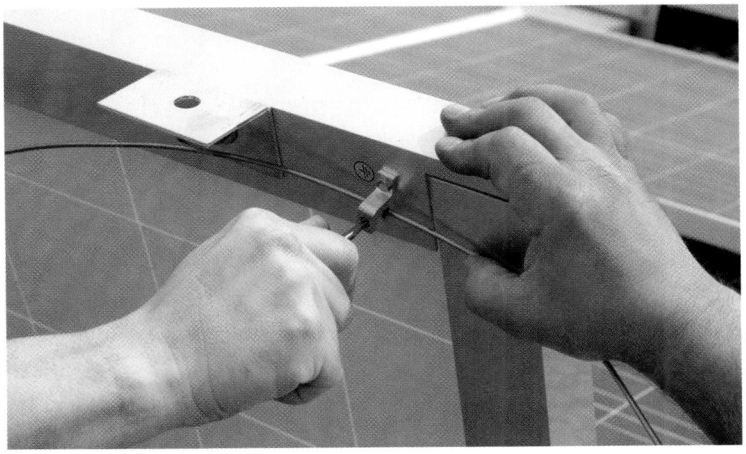

In a PV system, equipment grounding may be accomplished with continuous runs of bare conductor that are secured to each module with a special connector.

Separately Derived Systems – 250.30

Separately derived systems are a very common component of today's electrical distribution systems. A *separately derived system* is a system that supplies premises with electrical power derived or taken from storage batteries, solar photovoltaic systems, generators, transformers, or converter windings. See Article 100. By definition, a separately derived system can have no direct electrical connection to the conductors of the supplying system. This also applies to the grounded conductor. If the grounded conductor is solidly connected, then there is not a separately derived system. **See Figure 6-10.**

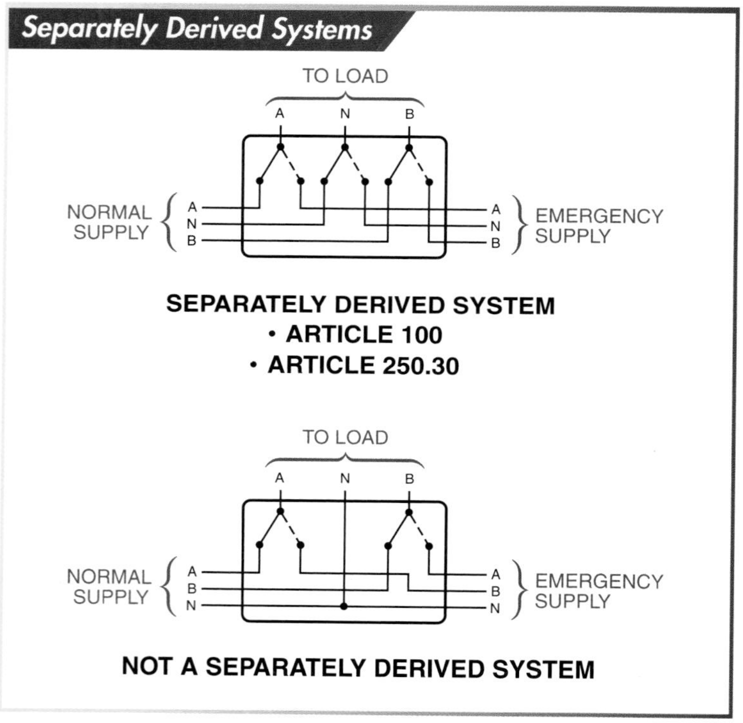

Figure 6-10. Separately derived systems do not have any direct electrical connection between the supply system and the separately derived system.

The vast majority of separately derived systems which electricians work on are derived from transformers. Transformers are available in a wide variety of types for various applications. A *transformer* is a device that converts electrical power at one voltage or current to another voltage or current. Transformers accomplish this task without any rotating parts and essentially without the use of significant amounts of energy.

No matter what the source of the separately derived system is, the grounding requirements are very similar. Section 250.30 contains the requirements for grounding separately derived systems. Subsection 250.30(A) states that a separately derived system that is required to be grounded shall comply with eight parts (1–8).

Ground references cannot be transferred through windings of any type. The grounding connections that were in place for the service must be re-established. Per 250.30(A)(1), an unspliced system bonding jumper that complies with 250.28(A) through (D) is required to be installed between the EGC of the derived system and the grounded conductor. This connection shall be made at any point from the source of the separately derived system up to and including the first system disconnecting means or OCPD. **See Figure 6-11.**

The system bonding jumper is not permitted to leave the enclosure where it originates. In cases where the source is located outside the structure or building that is being supplied, the system bonding jumper must be installed at the location of the grounding electrode connection and the installation must meet the provisions of 250.30(C).

The system bonding jumper shall be sized per 250.28(D). It shall not be smaller than the sizes shown in Table 250.66 for GECs. Unlike the GEC, the size of the bonding jumper does not stop increasing at a specified point. For conductors larger than 1100 kcmil Cu or 1750 kcmil Al, the bonding jumper is required to be not less than 12½% of the area of the largest derived phase conductor. **See Figure 6-12.**

Section 250.30(A)(2) was added to the 2011 NEC® and addresses installation where the source of a separately derived system and the first disconnecting means are located in separate enclosures. In this application, a supply-side bonding jumper must be installed with the circuit conductors from the source to the first disconnecting means. The supply-side bonding jumper must be of a type that complies with the requirements of 20.102(C) and is not required to be larger than the derived ungrounded conductors.

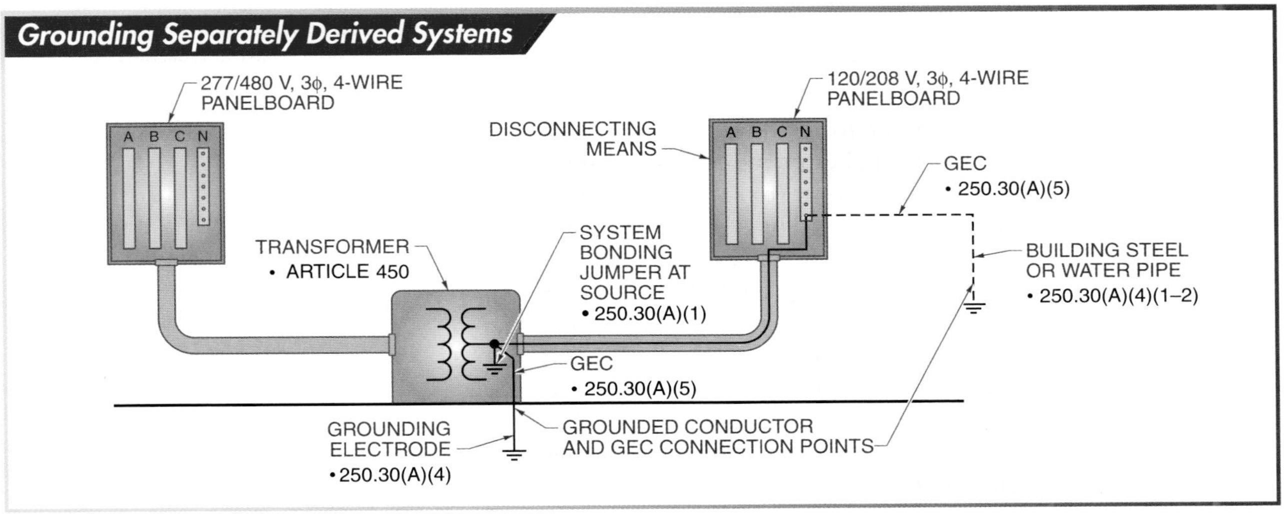

Figure 6-11. The GEC and grounded conductor can be connected at the source of the separately derived system or at the first disconnecting means. A bonding jumper at both locations shall be permitted if doing so does not establish a parallel path for the grounded conductor per 250.30(A)(1), Ex. 2.

Single Separately Derived Systems – 250.30(A)(5). The 2005 NEC® was revised to separate the requirements for grounding electrode conductor installation between single and multiple separately derived systems.

Per 250.30(A)(5), for single separately derived systems a GEC is required to connect the newly established grounded conductor to the grounding electrode. This connection must be made at the same point on the separately derived system where the system bonding jumper is connected. In some installations, such as a panelboard with main lugs only, the connection is made at the source of the separately derived system. The GEC is sized per 250.66. Instead of basing the calculation on the size of the largest service-entrance conductors, the calculation is based on the size of the largest derived phase conductors. **See Figure 6-13.**

Per 250.30(A)(4), the selection of the grounding electrode for separately derived systems is somewhat different than that of AC-supplied services. Instead of four components which must be connected together, if available, the GEC for separately derived systems can consist of one of two components. The component is selected based upon its proximity to the GEC connection to the system. Section 250.30(A)(4) requires that the grounding electrode be as near as practicable and preferably in the same area as the grounding electrode connection to the system.

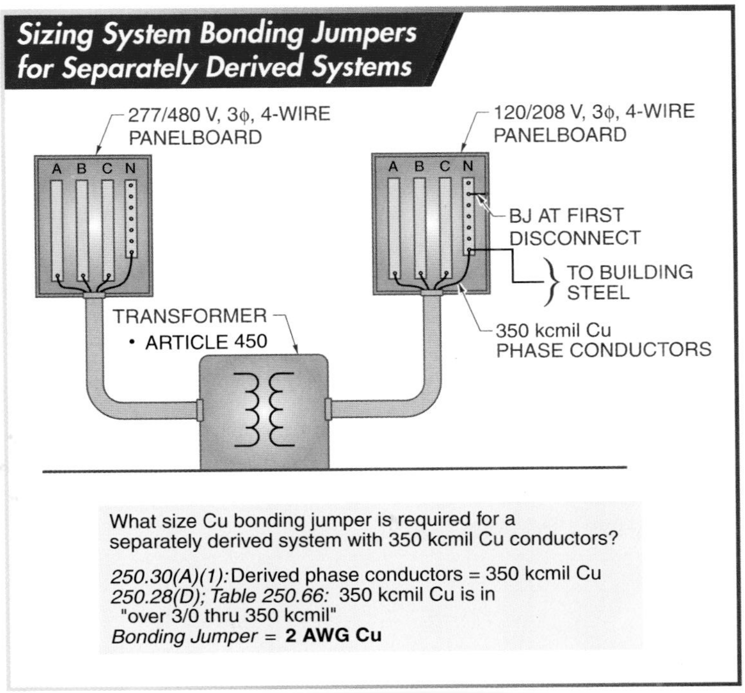

Figure 6-12. Bonding jumpers for separately derived systems are sized per 250.28(D)(1–3).

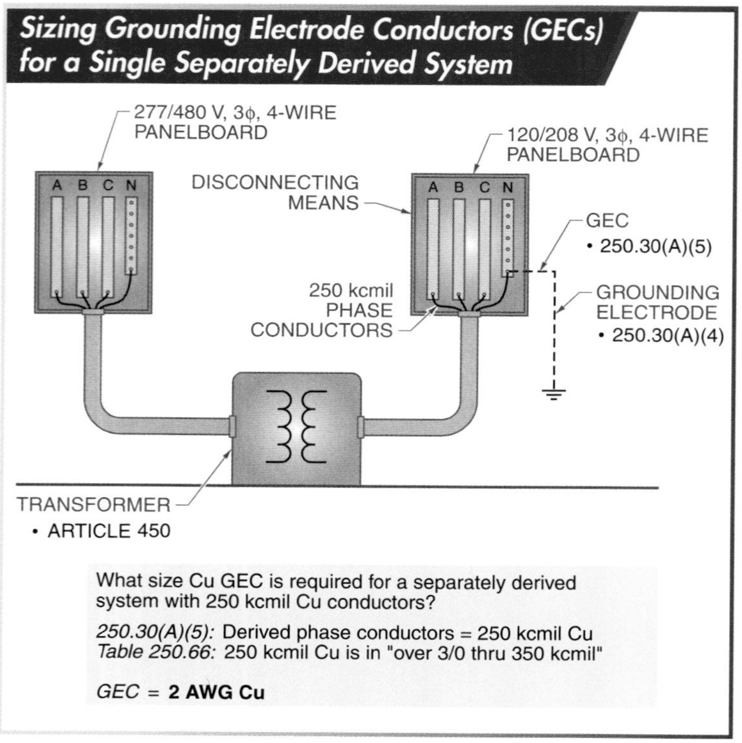

Figure 6-13. GECs for separately derived systems are sized based on the largest derived phase conductors per 250.30(A)(5).

For example, if the separately derived system is created by a transformer on the 17th floor of a building, the grounding electrode selected should be as close as possible to that system on the 17th floor. The actual grounding electrode must be the nearest of either 1) the metal water pipe or 2) the structural metal grounding electrode. The metal water pipe must be as specified in 250.52(A)(1) and the structural steel must conform with 250.52(A)(2).

> **Grounding of Permanent Generators**
>
> *A permanent generator is considered a separately derived system and should be grounded to prevent injury to personnel and damage to equipment being powered. Grounding of permanent generators is specified in Article 250.35.*

If neither the steel nor the water piping proves to be suitable, other electrodes per 250.30(A)(4), Ex. 1, such as concrete-encased electrodes, ground rings, or rod and pipe electrodes shall be used. No matter what type of grounding electrode is selected, it shall be as near as possible to, and preferably in the same area as, the point where the connection is made to the system. **See Figure 6-14.**

Multiple Separately Derived Systems – 250.30(A)(6). It is very common that a building or structure may contain multiple separately derived systems. Section 250.30(A)(6) permits taps to be made to a common grounding electrode. The individual tap conductors connect the grounding electrode to the grounded conductor of the separately derived system. The common grounding electrode conductor is not permitted to be smaller than a 3/0 AWG copper or 250 kcmil aluminum conductor. The individual tap conductors are sized from Table 250.66. The basis for the sizing is the size of the derived ungrounded conductors for the system it serves.

Section 250.30(A)(6)(c) contains the provisions for making the physical connection to the common grounding electrode conductor. Three options are permitted including the use of listed connectors and exothermic welding connection methods. See 250.30 (A)(4) for the requirements for the installation of the grounding electrode(s).

Generators – 250.34

A *generator* is used to convert mechanical power to electrical power. Generators are another source of power for separately derived systems. While not as common as transformers, they are commonly used for supplying emergency power to buildings where required. For example, buildings in which the interruption of power could pose a life safety concern frequently are required to use emergency sources of power. Examples of these types of buildings are schools, office buildings, supermarkets, restaurants, hospitals, and medical facilities. See Article 445.

Emergency power provided by a generator may also be required in buildings with critical computer system operations. In order to be considered a separately derived system, all load conductors including the grounded conductor must be switched from the normal service conductors to the generator supply conductors.

Portable Generators – 250.34(A). The frame of a portable generator is not required to be connected to a GES. The frame is permitted to serve as the grounding electrode if two conditions are met.

Per 250.34(A)(1), the portable generator shall supply equipment or loads that are either directly connected to the frame of the generator or are cord-and-plug connected through receptacles which are directly connected to the frame of the generator. In either case, the frame is permitted to serve as the GEC. If necessary, both conditions are permitted to occur at the same time. In this case, equipment mounted directly on the generator frame and equipment which is cord-and-plug connected to receptacles mounted on the generator can be supplied from the portable generator.

Per 250.34(A)(2), the frame of the generator shall be bonded to the equipment grounding terminals of the receptacles and to any noncurrent-carrying parts of equipment that is served from the generator. Bonding the receptacle terminals to the frame of the generator ensures that any equipment which is cord-and-plug connected to the generator is also connected to the generator frame which serves as the GEC. **See Figure 6-15.**

Vehicle-Mounted Generators – 250.34(B). Vehicle-mounted generators are treated in a similar fashion to portable generators. The main difference is that the frame of the generator is not in direct contact with the earth. The frame of the vehicle is not required to be connected to the GES if three conditions are met.

Per 250.34(B)(1), the vehicle frame shall be bonded to the frame of the generator. Per 250.34(B)(2), as with portable generators, the generator shall supply equipment or loads that are either directly connected to the frame of the vehicle or are cord-and-plug connected through receptacles which are directly connected to the frame of the vehicle. Equipment can be supplied by cord-and plug-connected means and direct connection to the vehicle frame at the same time.

Per 250.34(B)(3), the frame of the generator shall be bonded to the equipment grounding terminals of the receptacles and to any noncurrent-carrying parts of equipment that is served from the generator.

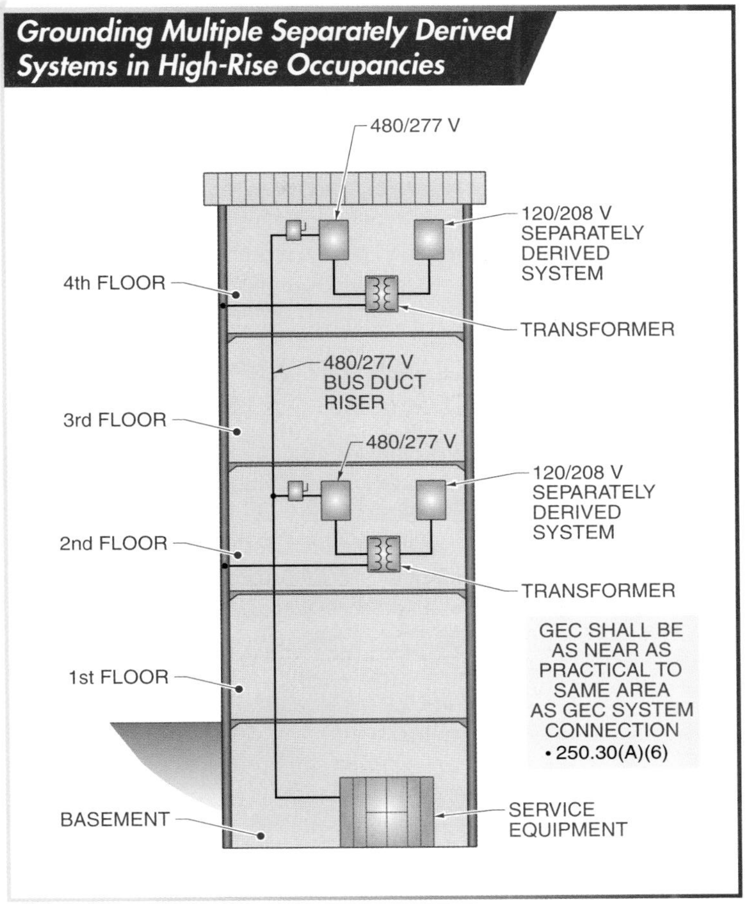

Figure 6-14. A grounding electrode system shall be created for each separately derived system, or a common electrode shall be used per 250.30(A)(6).

Figure 6-15. The requirements for grounding portable generators are included in Section 250.34.

When a portable or vehicle-mounted generator supplies a system which contains a conductor that is required to be grounded per 250.26, the conductor shall be grounded by bonding it to the generator frame when the generator is part of a separately derived system. If a portable generator system supplies premises wiring, it shall be grounded per the requirements for separately derived systems. See 250.34(C).

GROUNDING EQUIPMENT AND ENCLOSURES – 250.80; 250.86

The primary reason enclosures and noncurrent-carrying metal parts of equipment are grounded is to limit the voltage to ground and facilitate the operation of the OCPD under ground-fault conditions. Most equipment operates regardless of the state of the grounding connection. Grounding of enclosures and equipment is done for safety considerations and is the primary factor in reducing the risk of electrical shock.

Enclosures

An *enclosure* is the case or housing of equipment or other apparatus which provides protection from live or energized parts. See Article 100. Outdoor equipment is sometimes provided with fences or walls to accomplish the same goal. In these cases, the fences and walls are also considered to be enclosures. Enclosures also protect equipment and apparatus from physical damage.

Service Enclosures – 250.80. Service enclosures provide protection from severe physical damage and protection from accidental contact with energized parts that comprise part of the electrical service. All service enclosures and raceways for the protection of service conductors and service equipment shall be grounded.

A service conductor is a conductor that extends from the service point to the service disconnecting means. Service equipment is all of the necessary equipment to control the supply of electrical power to a building or structure. See Article 100. These definitions, taken together, require that essentially all of the components of the electrical service are required to be grounded.

Grounding on the line (supply) side of the service equipment is accomplished differently from grounding of equipment on the load side. Grounding on the load side is primarily through the use of an EGC. There is no separate EGC on the line side. Per 250.92(B)(1), the grounded conductor is permitted to be used to ground any noncurrent-carrying part of equipment and enclosures on the supply side of the service. The grounded conductor performs the dual function of serving as the neutral conductor and the EGC.

There is, however, one exception to the requirement that all service enclosures and raceways be grounded. Per 250.80, Ex., the use of a metal elbow to be installed for underground nonmetallic raceway installations without being grounded is permitted. This is commonly done where nonmetallic raceways are installed and a metal elbow is used to ensure that the raceway is not damaged during installation of the conductors. The elbow shall be installed so that no part of it is within 18″ of finished grade. **See Figure 6-16.**

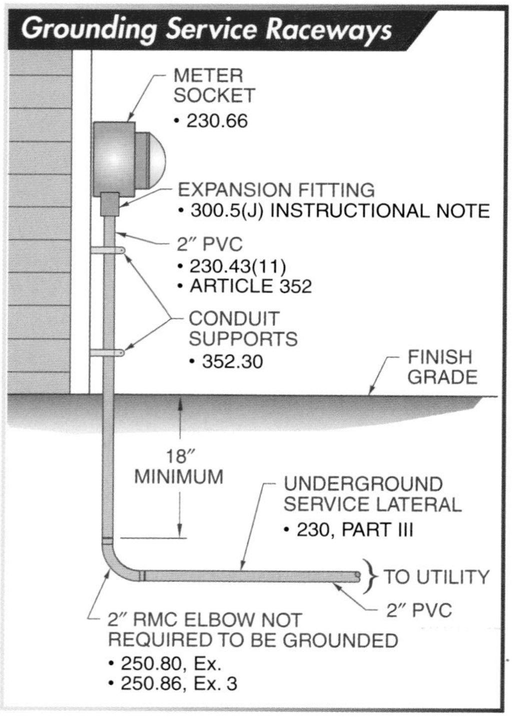

Figure 6-16. Metal elbows installed in nonmetallic raceways are not required to be connected to the grounded system or electrode conductor when they are at least 18″ below grade.

Other than Service Enclosures – 250.86. This section requires all metal enclosures and raceways, other than those covered under 250.112(I), to be connected to the EGC. There is no limitation to service conductors only. Any enclosure or raceway that houses or protects a branch-circuit or a feeder shall be connected to the EGC. This protects personnel from an electrical shock hazard. For example, if a raceway protecting a branch circuit is not grounded and, because of an insulation failure, it becomes energized, there is a good probability that the impedance of the ground-fault path would be too great to facilitate the operation of the OCPD. The earth shall never be used as the sole EGC.

Per 250.86, Ex. 1, metal enclosures and raceways that are installed as part of an addition to an existing installation of open wiring, knob-and-tube, or older NM cables that do not provide an equipment ground are not required to be connected to the EGC. The total length of the circuit shall not exceed 25′, the wiring method shall be free from contact with the ground or other grounded materials, and the installation shall be guarded from contact.

Per 250.86, Ex. 2, short sections of metal enclosures or raceways are permitted to be installed without being grounded. This gives relief from grounding to those installations in which the enclosure or raceway is merely used to provide support or mechanical protection. However, the term "short section" is not defined, and this could lead to confusion in the field and, more importantly, installations that pose a threat to life safety. Always check with the AHJ if there are any questions concerning grounding. **See Figures 6-17** and **6-18.**

Per 250.86, Ex. 3, the same provisions for permitting a metal elbow to be installed in a nonmetallic raceway that are found in 250.80, Ex. for service raceways and enclosures is also permitted for other conductor enclosures and raceways. The elbow must be located such that a minimum of 18″ of cover is provided to any point on the elbow or the elbow must be encased in not less than 2″ of concrete.

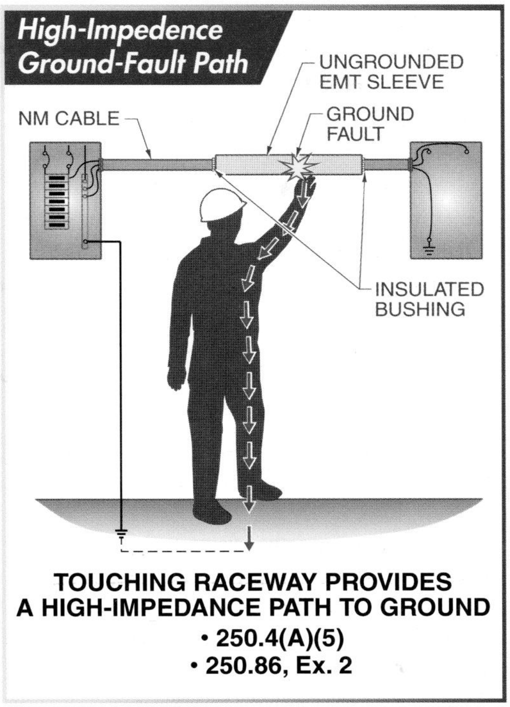

Figure 6-17. Short sections of raceway could pose a threat to personnel.

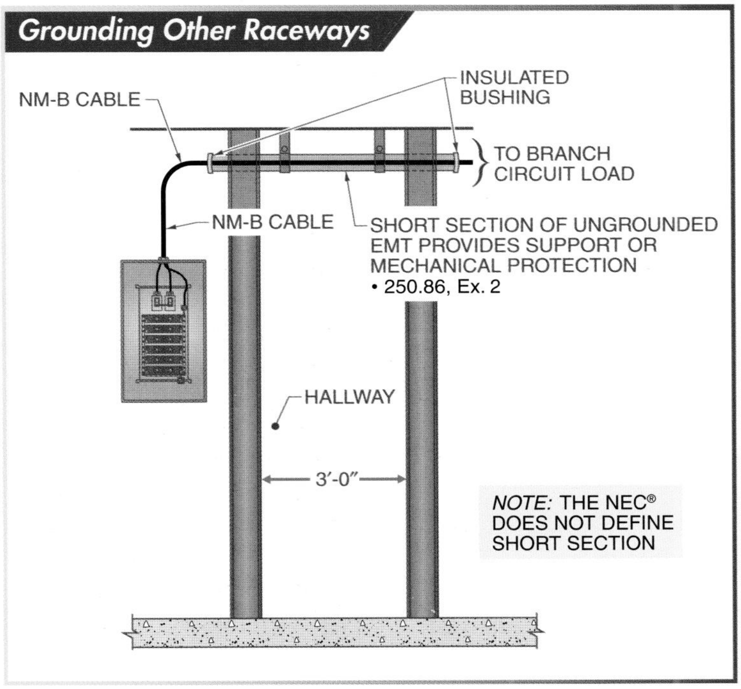

Figure 6-18. Short sections of metal enclosures or raceways are permitted to be installed without being connected to the EGC.

Fixed Equipment – 250.110

Equipment is any material, device, fixture, apparatus, appliance, etc., used in conjunction with electrical installations. See Article 100. Grounding of equipment is divided into two basic sections depending on whether the equipment is fixed in place or cord-and-plug connected.

Section 250.110 lists six general conditions where equipment which contains exposed noncurrent-carrying metal parts and which is fastened in place or connected by a permanent wiring method shall be connected to the EGC. No specific equipment is mentioned. The requirements apply generally to all electrical installations. **See Figure 6-19.**

The first condition, found in 250.110(1), establishes a zone measuring 5′ horizontally and 8′ vertically from ground or grounded objects in which exposed equipment shall be grounded. These dimensions establish a "reach zone" under which someone could contact an exposed energized metal part of a piece of equipment and ground or any other grounded metal objects. Objects beyond this zone do not pose a significant risk to personnel safety because they cannot be reached while maintaining contact with an energized metal part.

Atlas Technologies, Inc.

Section 250.110 lists general conditions where equipment containing exposed noncurrent-carrying metal parts shall be connected to the EGC.

The second condition, located in 250.110(2), requires that all equipment located in a wet or damp location shall be connected to the EGC unless the equipment is isolated. Equipment is considered to be isolated when it is not readily accessible to persons without the use of special means. A wet location is any location in which a conductor is subject to saturation from any type of liquid or water. A damp location is a partially protected area subject to some moisture. See Article 100. Equipment installed in a damp or wet location poses an increased hazard for persons who may use the equipment or otherwise come in contact with the equipment.

The third condition, stated in 250.110(3), requires that all fixed electrical equipment in contact with metal be connected to the EGC. If the metal were to become energized, it would sustain a voltage to ground and pose a threat of electrical shock for persons in contact with the equipment.

The fourth condition, given in 250.110(4), requires that all fixed equipment located in hazardous (classified) locations shall be connected to the EGC per Article 500 through Article 517. Because of the presence of, or the potential for, hazardous atmospheres, special grounding requirements may exist when installing electrical equipment in hazardous (classified) locations. Always check the requirements of the specific occupancy to determine if grounding is required.

The fifth condition, found in 250.110(5), requires that all fixed equipment supplied from metallic wiring methods be connected to the EGC. For example, fixed equipment supplied by metal-clad cable or metal raceways shall be connected to the EGC. This requirement does not apply to short sections of metal enclosures where their function is support or mechanical protection. See 250.86, Ex. 2.

The sixth condition, given in 250.110(6), requires grounding of fixed equipment when it operates with any terminal at over 150 V to ground. This is a broadly worded provision. There are three exceptions for specialized electrical installations to this general rule. See 250.110, Ex. 1–3.

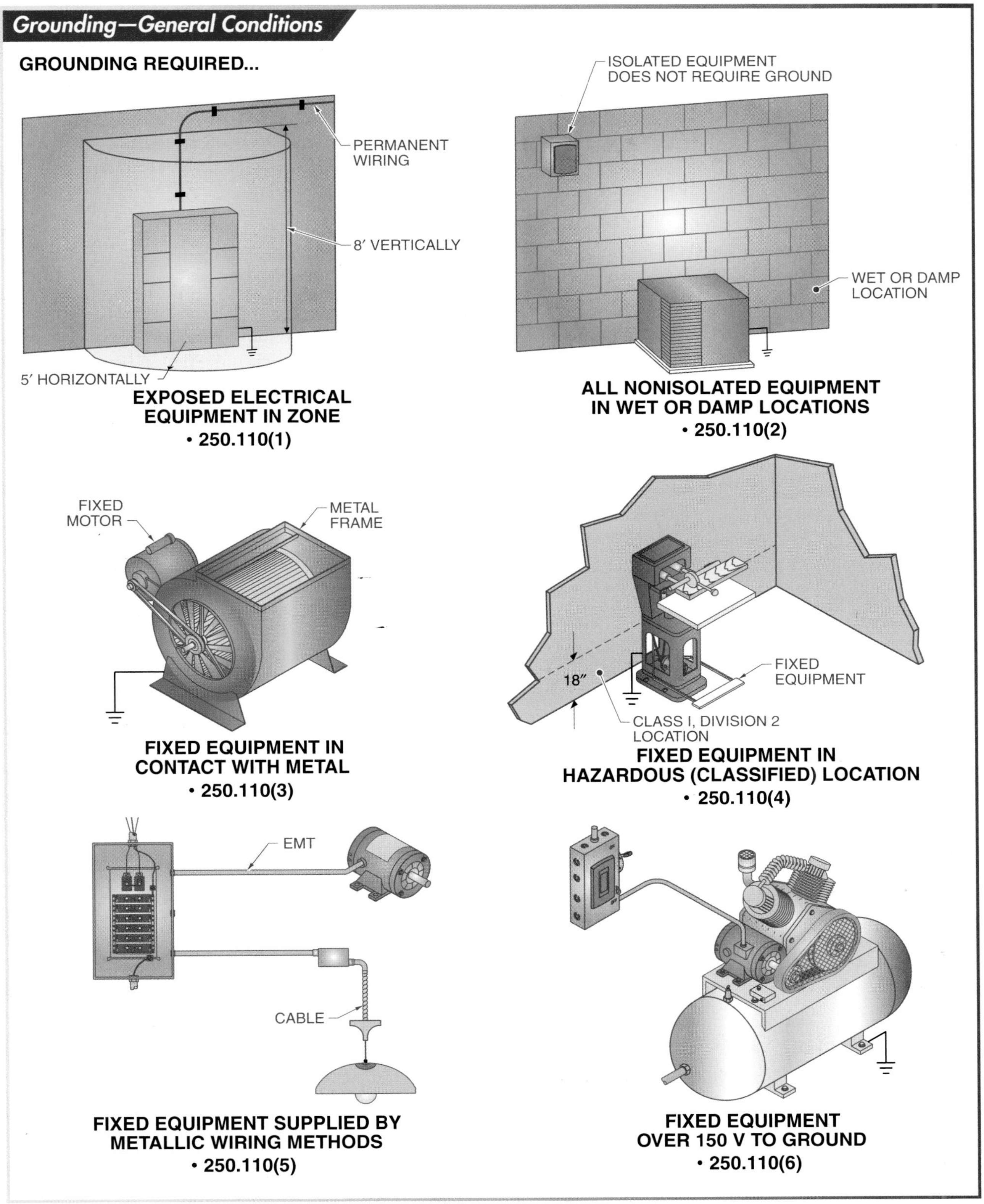

Figure 6-19. Permanently wired, fastened-in-place equipment with exposed noncurrent-carrying metal parts shall be connected to the EGC.

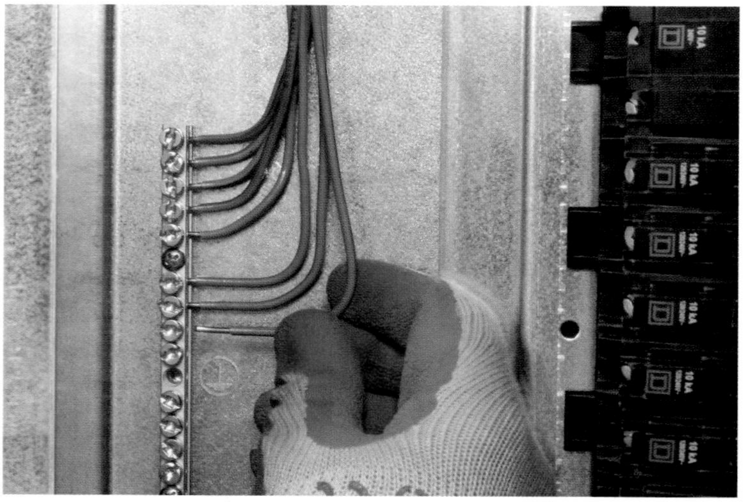

A grounding busbar in a panelboard is used to ground branch circuits.

Specific Conditions – 250.112. This section lists specific equipment that shall be connected to the EGC if it contains exposed, noncurrent-carrying metal parts that are likely to become energized. There are no voltage limitations to these requirements. **See Figure 6-20.** The following shall be connected to the EGC, regardless of the voltage at which they operate:

- (A) Switchboard frames and structures
- (B) Pipe organs
- (C) Motor frames
- (D) Motor controller enclosures
- (E) Elevators and cranes
- (F) Garages, theaters, and motion picture studios
- (G) Electric signs
- (H) Motion picture equipment
- (I) Remote control, signaling, and fire alarm circuits
- (J) Luminaires (lighting fixtures)
- (K) Skid-mounted metal equipment
- (L) Motor-operated water pumps
- (M) Metal well casings

Cord-and-Plug-Connected Equipment – 250.114. Grounding cord-and-plug-connected equipment poses some design considerations beyond those required for fixed or fastened-in-place equipment. Cord-and-plug-connected equipment is more susceptible to physical damage, which could affect the integrity of the equipment grounding conductor. One potential solution to this problem has been the advent of double-insulated equipment, which in some cases can be used in lieu of a separate EGC. Subsection 250.114(1–4) contains four conditions under which exposed, normally noncurrent-carrying metal parts of cord-and-plug-connected equipment shall be connected to the EGC. **See Figure 6-21.**

The first two conditions are the same as those for fixed equipment. Cord-and-plug-connected equipment located in hazardous classified locations shall be grounded per 250.114(1). Cord-and plug-connected equipment operating at over 150 V to ground shall be grounded per 250.114(2). The second and third provisions apply to cord-and-plug-connected equipment installed in residential occupancies (3) and other than residential occupancies (4).

Subsection 250.114(3) lists five groups of specific equipment that shall be connected to the EGC when installed in residential occupancies. Per 250.114, Ex., listed appliances and listed equipment which employ a system of double insulation do not require connection to the EGC. The double insulation or equivalent system shall be listed and shall be distinctively marked. While the grounding requirement for any tool or appliance meeting this exception is removed, double-insulated tools are not permitted to be used in lieu of GFCI protection where so required. For example, a 120 V, listed, double-insulated, hand-held, motor-operated tool used for construction purposes on a residential job site is still required to be protected with GFCI protection per 590.6.

Subsection 250.114(4) covers requirements for grounding cord-and-plug-connected equipment in other than residential occupancies. These provisions apply to commercial and industrial classifications as well as any other nonresidential classification the AHJ may adopt. The same exception for listed, double-insulated tools and appliances applies, but the scope of 250.114(4) is greater than that for the residential occupancies. It applies to listed stationary and fixed motor-operated tools and to listed light industrial motor-operated tools as well.

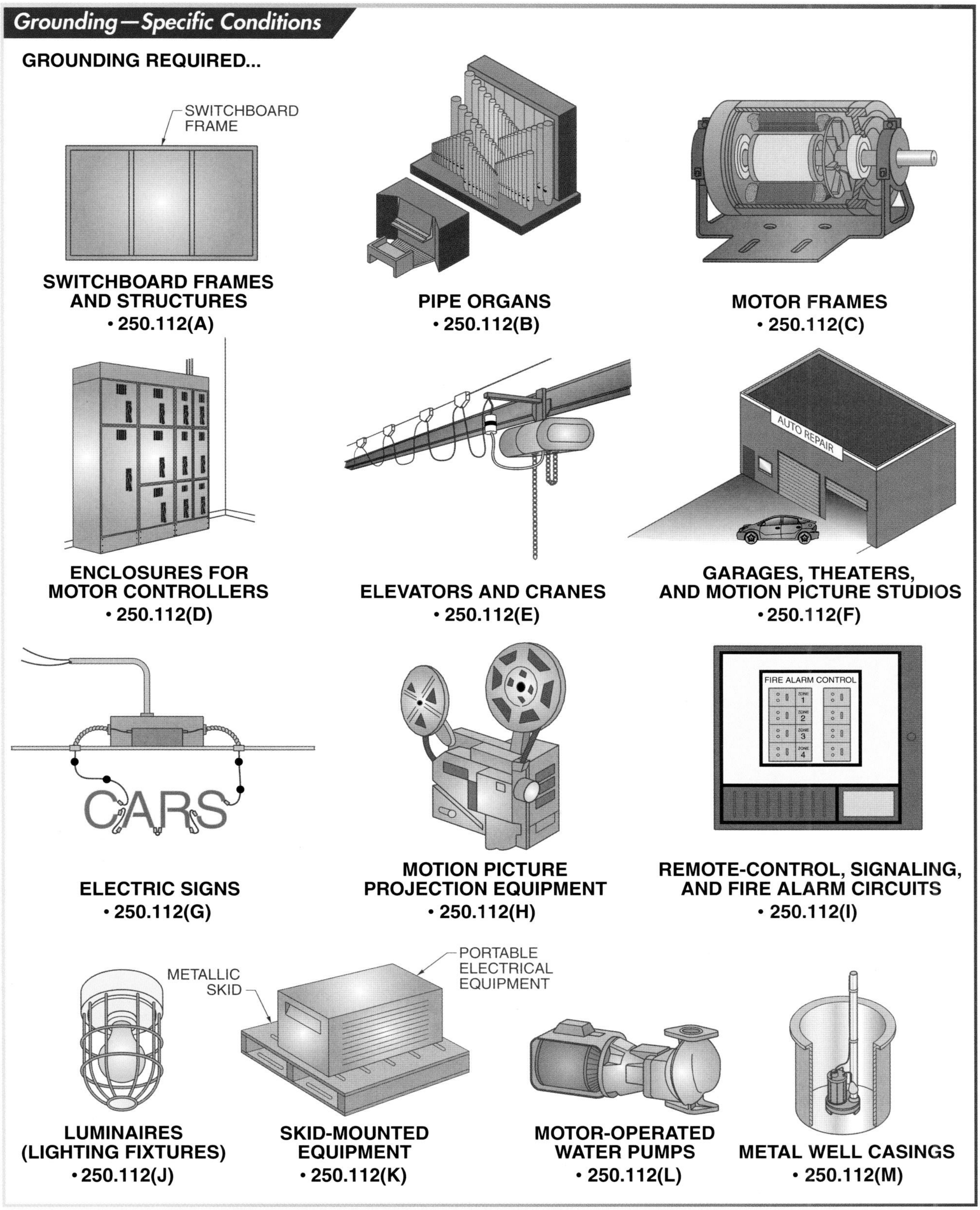

Figure 6-20. Permanently wired, specific fastened-in-place equipment with exposed noncurrent-carrying metal parts shall be connected to the EGC regardless of voltage.

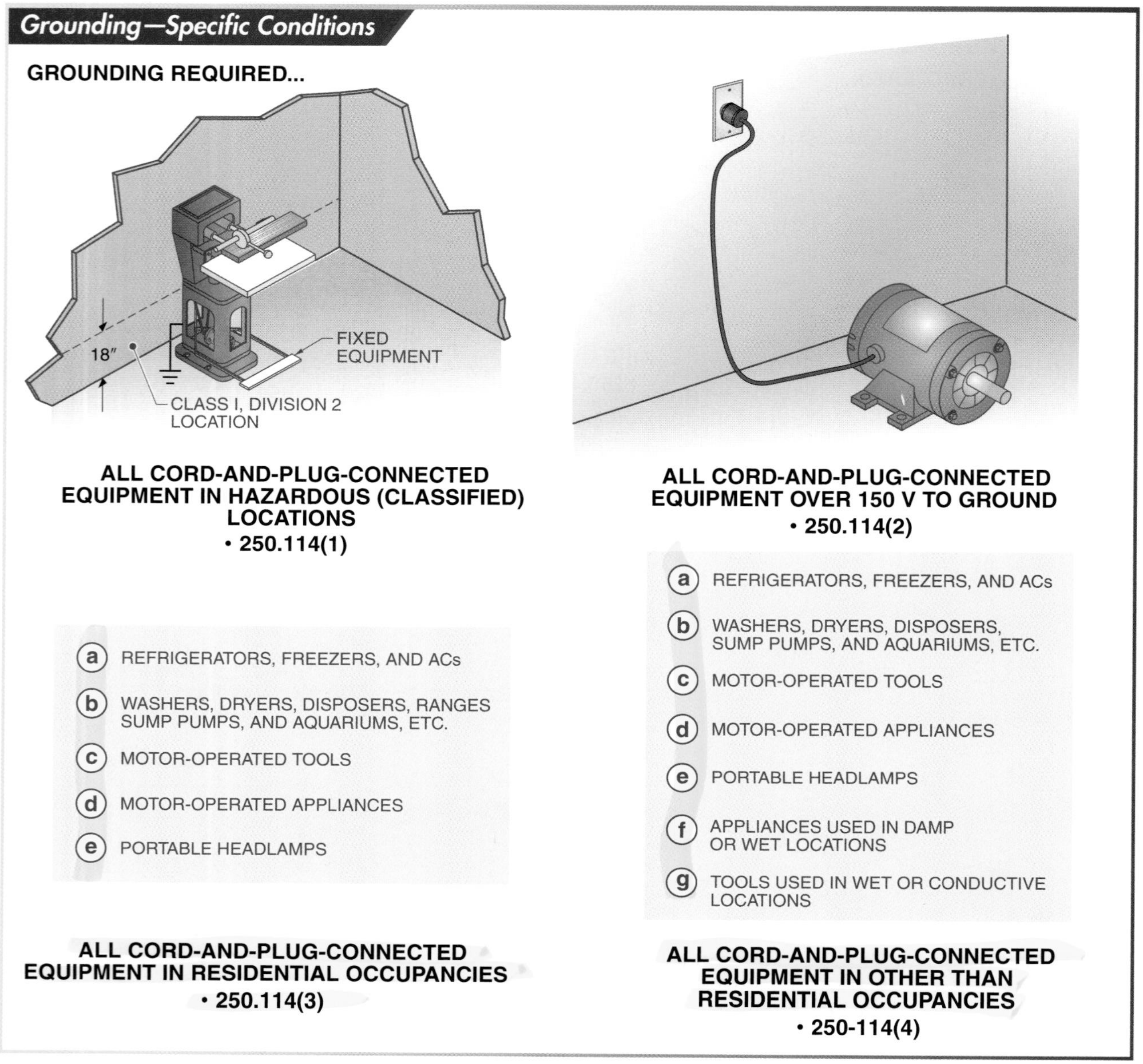

Figure 6-21. Cord-and-plug-connected equipment in hazardous locations, over 150 V to ground applications, residential occupancies, and other-than-residential occupancies shall be connected to the EGC.

It is permissible to use ungrounded portable handlamps and tools in other-than-residential occupancies per 250.114(4)(G), Ex. These ungrounded handlamps and tools may be used in wet or corrosive locations and if they are supplied from an isolating transformer. The ungrounded secondary of the transformer is not permitted to exceed 50 V.

Grounding Exceptions

Listed tools, appliances, and equipment covered in 250.114(2–4) are not required to be grounded if protected by a system of double-insulation or equivalent. All double-insulated equipment must be distinctively marked as such.

GROUNDING METHODS–ARTICLE 250, PART VII

Once a determination has been made as to whether a system is to be grounded, Article 250, Part VII should be reviewed to determine the method by which equipment should be grounded. For grounded systems, an effective grounding path, per 250.4(A)(5), shall have sufficient capacity for the available fault current, and it shall have a low enough impedance to permit the OCPD to operate. These guidelines apply for systems and equipment.

Services

Electrical services are grounded by several connections that usually are made within or at the service equipment. The EGC shall be bonded to the grounded service conductor and the GEC per 250.130(A). **See Figure 6-22.**

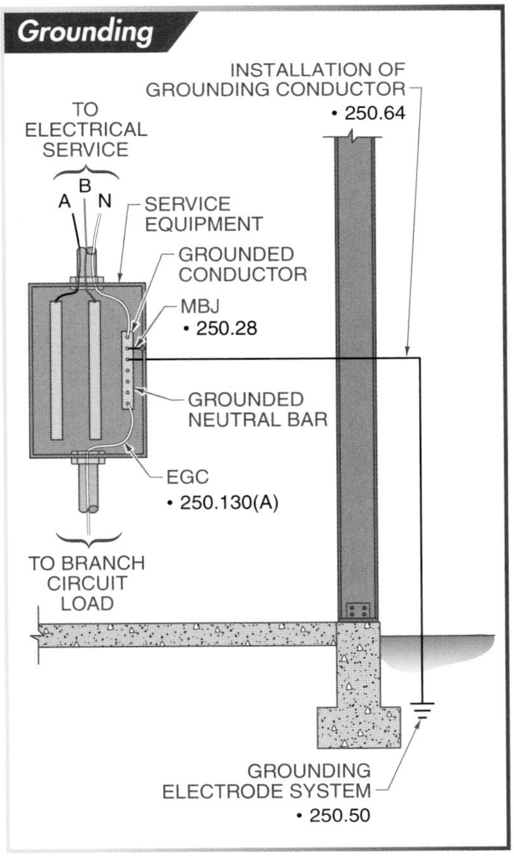

Figure 6-22. Most electrical system grounding connections are made at or within the service equipment.

Section 250.130(A) reiterates the connection of the GEC, EGC, and the grounded service conductor at the service equipment. This connection is made with an unspliced main bonding jumper (MBJ) per 250.28. The MBJ connects these conductors to the service disconnecting means at the service equipment. The MBJ is the heart of the grounding system at the service equipment.

Equipment

Sections 250.134, 250.138, and 250.140 cover requirements for grounding different types of equipment. Equipment, much like systems, can be installed in conditions that require it to be grounded, permit it to be grounded, or even prohibit it from being grounded.

Fixed Equipment –250.134. This section requires that all noncurrent-carrying metal parts of equipment, enclosures, etc., which are required to be grounded, shall be grounded by one of two methods. The first method, given in 250.134(A), permits the use of a recognized EGC found in 250.118. **See Figure 6-23.** Fourteen different listings under this section could be used as the EGC for fixed equipment. For example, a fixed piece of equipment which operates at 240 V is fed by two ungrounded conductors installed in a ¾″ run of electrical metallic tubing. EMT is item (4) on the list and is permitted to serve as the EGC for this circuit.

Information technology (IT) equipment, such as a data center, must be protected with a grounding conductor.

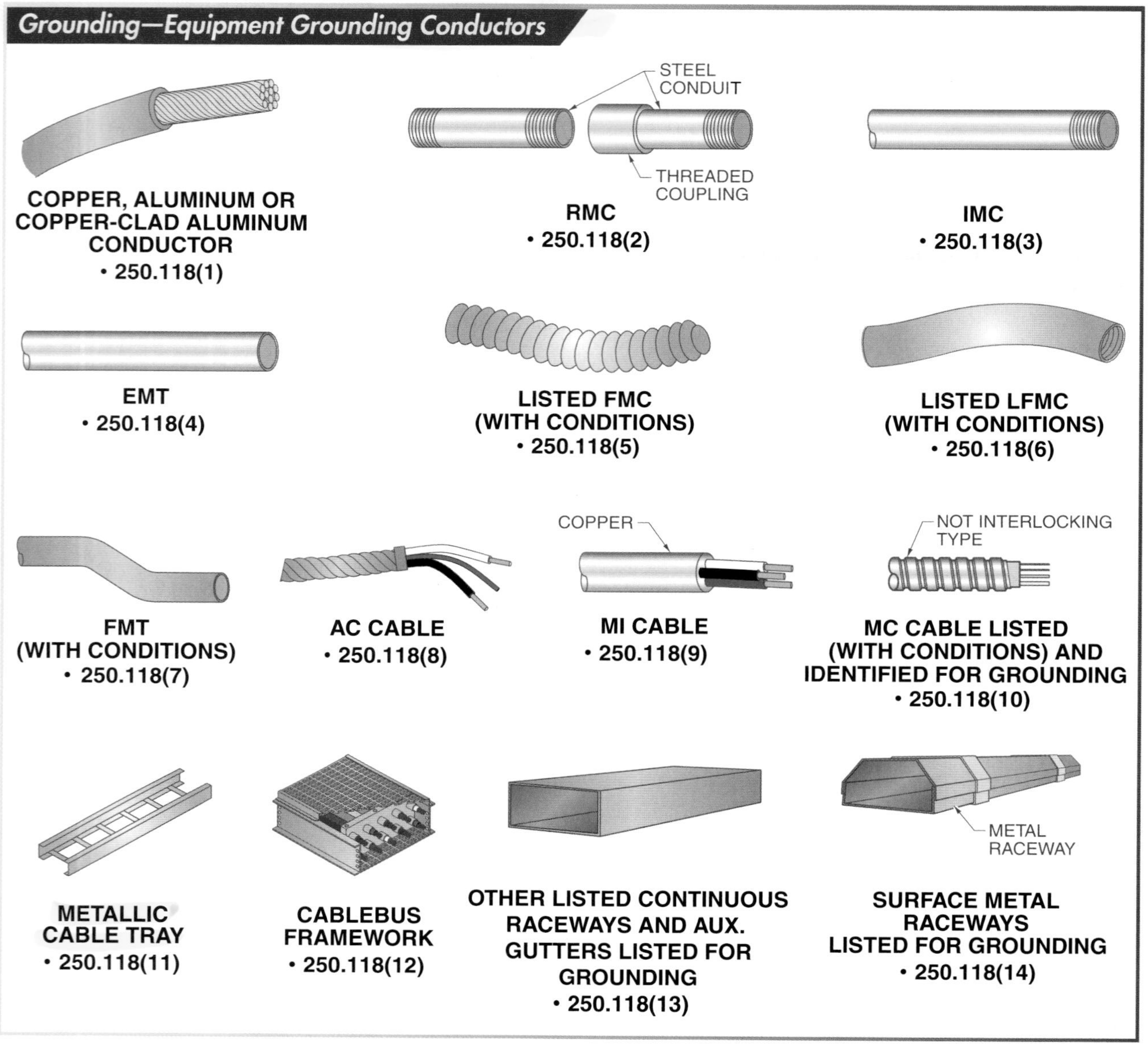

Figure 6-23. Fourteen different types of EGCs are provided for in 250.118.

The second method, stated in 250.134(B), permits the use of an EGC which is run with or contained within the same circuit. All circuit conductors, including the EGC where so required, shall be run in the same raceway, cable, trench, cord, etc., per 300.3(B). This helps to ensure that the impedance of the ground path remains low enough to permit operation of the OCPD.

Per 250.119, the EGC may be a bare, covered, or insulated conductor. In general, if the conductor is insulated, the color shall be green or green with one or more yellow stripes. Re-identification of the EGC is permitted for conductors larger than 6 AWG (see 250.119(A)) or for conductors of any size within multiconductor cable where the conditions of maintenance ensure that only qualified persons will service the installation. See 250.119(B). Re-identification shall be in the form of stripping the insulation or covering from the exposed conductor, coloring or painting the exposed insulation or covering green, or by otherwise effectively marking the exposed insulation or covering with green tape or labels. See 250.119(A).

Cord-and-Plug-Connected Equipment – 250.138. Two general methods may be used to ground metallic noncurrent-carrying parts of cord-and-plug-connected equipment. The most common method employed to ground cord-and-plug-connected equipment is by use of a separate EGC run with the supply conductors per 250.138(A). The EGC is part of the cable assembly or cord and terminated in a grounding-type attachment plug. For fixed equipment, the EGC is permitted to be bare, covered, or insulated, and the same color code requirements shall be followed.

The second method listed for grounding cord-and-plug-connected equipment is by use of a separate strap or flexible wire per 250.138(B). The wire or strap may be insulated or bare. If the wire or strap is part of the equipment, it shall be protected against physical damage.

Identifying EGCs

Electrical designers and installers should be careful when applying the re-identification provisions of 250.119. Subpart (A) only applies to conductors larger than 6 AWG. The three options for re-identification cannot be applied to smaller conductors. The re-identification shall occur at all points where the conductor is accessible. This includes all splice points as well as the point at which the conductor terminates.

The re-identification provision of Subpart (B) only applies to multiconductor cables. While there is no minimum size of the conductor, as there is in Subpart (A), all of the conditions of the provision shall be met. This provision cannot be applied to conductors installed in raceways. If a contractor needs to install an EGC in PVC, and it is not larger than 6 AWG, it shall be bare, covered, or insulated. If it is covered or insulated, it shall have an outer finish which is green or green with one or more yellow stripes.

Ranges and Clothes Dryers – 250.140. The 1996 NEC® contained a major change in 250.60, *Frames of Ranges and Clothes Dryers.* Since the 1940s, the NEC® had permitted the frames of ranges, including counter-mounted cooking units and wall-mounted ovens, and clothes dryers to be grounded by the use of the grounded conductor provided all of the following conditions were met:

- The supply circuit was 120/240 V, 1ϕ, 3-wire or 120/208 V, 3ϕ, 4-wire wye.
- The grounded conductor was at least 10 AWG Cu or 8 AWG Al.
- The grounded conductor was insulated, or if uninsulated, was part of an SE cable branch-circuit originating at the service equipment.
- The equipment was bonded to the grounding contacts of any receptacle which is part of the equipment.

The change no longer permits the use of the grounded conductor for this application. The change does, however, only apply to new branch-circuit installations. New installations require the frames to be grounded per 250.134 and 250.138. **See Figure 6-24.**

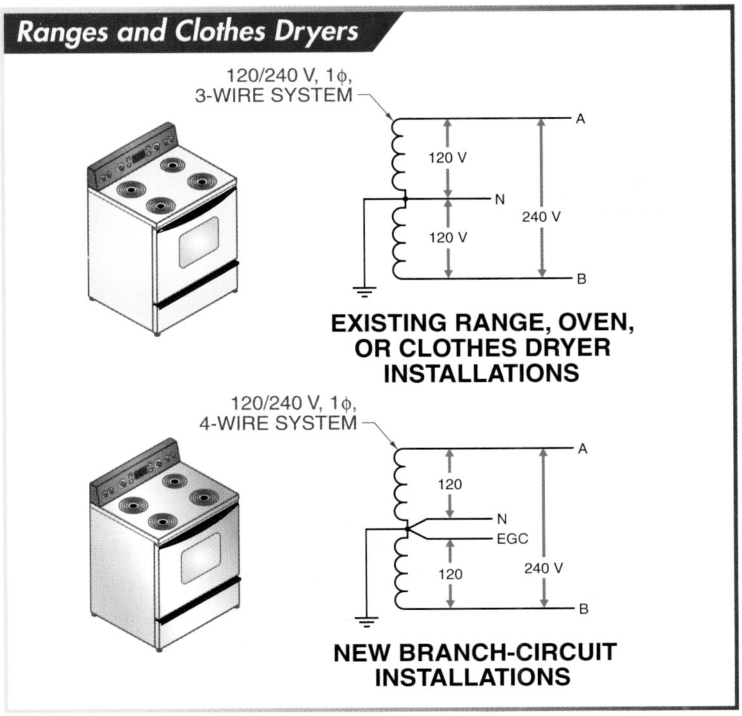

Figure 6-24. New branch-circuit installations for ranges, clothes dryers, and ovens require an EGC.

grounding conductors shall be green!

Grounded conductor
White - 120/220
gray - 480/277

BONDING – ARTICLE 250, PART V

Part V of Article 250 contains the bonding requirements for service equipment, circuits over 250 V to ground, main and equipment bonding jumpers, interior metal piping, and structural steel. Bonding is sometimes used interchangeably with grounding. The two terms should not be used interchangeably because they both seek to accomplish different goals.

Five terms in Article 100 use "bonding" as a root word. *Bonding* is establishing continuity and conductivity through a connection. Bonding is typically a connection between various metal parts. This connection shall be electrically continuous and shall have sufficient capacity to handle any fault currents that may occur.

Although these requirements are similar to those of an effective grounding path, the difference lies in the purpose of the connection. Bonding is done to form an equipotential plane among all metal parts. An *equipotential plane* is an area in which all conductive elements are bonded or otherwise connected together in a manner which prevents a difference of potential from developing within the plane. Bonding ensures that there cannot be a difference of potential between metal parts.

Installing Old Appliances

Because the old rule was in place for over 50 years, some time may be required before designers and installers begin to recognize the branch-circuit requirements for new installations. Installers should be particularly concerned when asked to install branch circuits or connect any of these appliances. For example, if someone moves to a new house and brings their old appliances with them, the new installation is to be wired for a 4-wire connection while the appliances are wired for a 3-wire connection. Depending upon how the branch circuit is run and whether the appliance is hard-wired or cord-and-plug connected, the installer may need to reconfigure the appliances, devices, and attachment plugs for the new connection.

Services

Electrical services have special bonding requirements which are contained in 250.92 and 250.94. The primary reason for these special requirements for services is the absence of adequate overcurrent protection on the line side of service equipment. Electrical utility power systems frequently do not provide secondary overcurrent protection. The only protection provided by the electrical utility is for the primary of the transformer. Care should be exercised when bonding electrical service equipment because the available fault current is typically greater there than anywhere else in the electrical distribution system.

Bonding Service Equipment – 250.92(A). Part (A) lists the components of the service that shall be bonded together if they enclose, contain, or support service conductors. These include:

- All service raceways or cable assemblies employing metallic sheath, cable trays, cablebus, etc.
- All enclosures for service conductors, including meter sockets, boxes, etc.

Bonding Service Equipment – 250.92(B). This section lists four methods that can be used to accomplish the required bonding for service equipment. **See Figure 6-25.** The four methods that can be used for bonding for service equipment are:

- Grounded service conductor. See 250.8.
- Threaded connections. Fittings, such as threaded hubs or bosses, are permitted for bonding purposes when the fittings are made wrenchtight. Threaded connections are commonly used with RMC and IMC.
- Threadless connections. Where made up tight for metal raceways and metal-clad cables.
- Other devices. Bonding is permitted to be accomplished by use of bonding or grounding-type locknuts.

Bonding jumpers meeting the requirements of Article 250 shall be used to bond around concentric or eccentric knockouts. Bonding jumpers must also be used around any impaired connections, including reducing washers or knockouts. This is the most common bonding method for service equipment. Standard locknuts shall not be the sole means for bonding at the service.

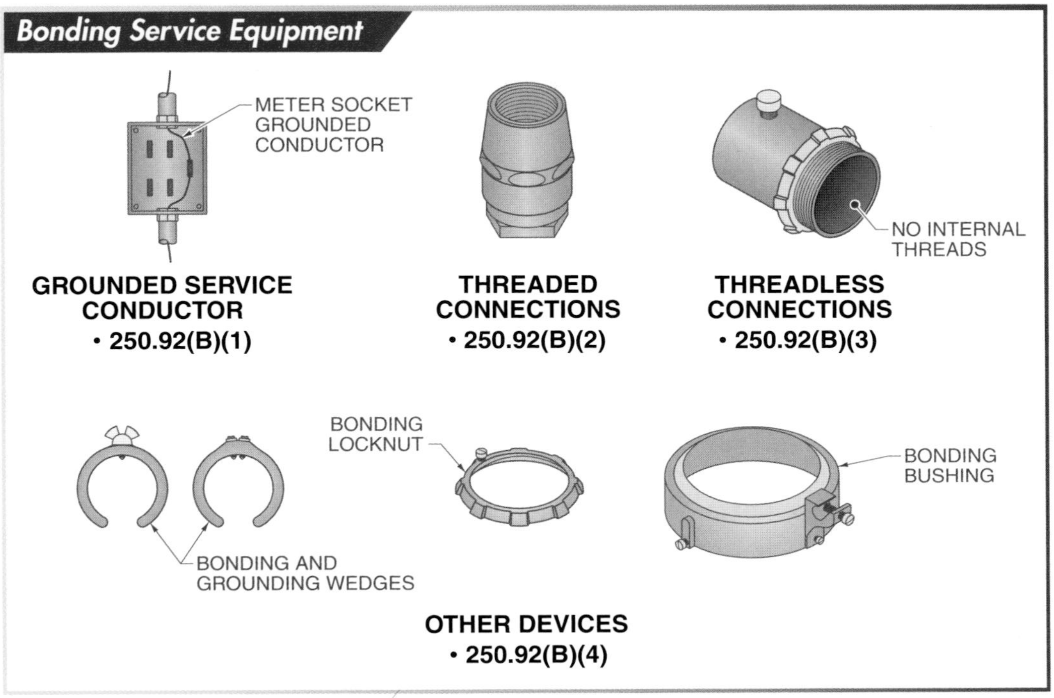

Figure 6-25. Four methods may be used to bond service equipment.

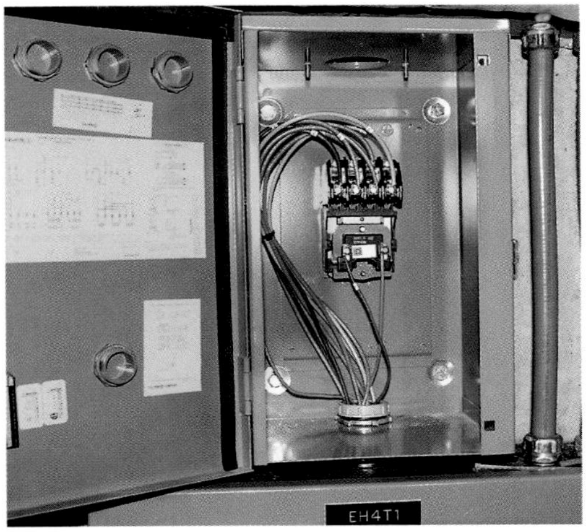

Grounding of enclosures reduces the risk of electrical shock.

Section 250.94 contains the requirements for intersystem bonding. This section requires an external bonding termination for connecting intersystem bonding and grounding conductors. For example, in a building or structure that has intersystem bonding provisions, communication circuits are required to have the grounding conductor connected to the intersystem bonding termination per 800.100(B)(1)(2). Section 250.94 requires that the intersystem bonding termination be accessible for inspection and termination, and have the capacity to connect at a minimum of at least three intersystem bonding conductors.

Three methods are provided for the intersystem bonding termination. The first method permits the use of a set of listed grounding and bonding terminals that are electrically connected to the meter enclosure. The second method is by use of a bonding bar located near the service equipment enclosure, meter enclosure, or service raceway. The bonding bar must be connected to the internal EGC of this equipment with a minimum of a 6 AWG copper conductor. The third method for the intersystem bonding termination is by use of a bonding bar located near the GEC. Again, the connection of the bonding bar to the GEC must be by a minimum of a 6 AWG copper conductor.

Note that 250.94, Ex., provides an exception for existing buildings or structures where other intersystem bonding and grounding conductors already exist. The Informational Note that follows this exception states that a 6 AWG copper conductor with one end bonded to the grounded metallic nonflexible raceway or equipment can be used to meet the requirements of this provision. The 6 AWG conductor must provide at least 6″ or more of free conductor outside of the equipment and it must be accessible for intersystem bonding. If installed in this manner, the copper conductor can be an approved means for complying with the requirements of the exception to this section.

Bonding Over 250 Volts – 250.97

An often overlooked requirement is the bonding of metal raceways and cable for circuits of over 250 V to ground. This section requires that electrical continuity be ensured by either threaded connections, threadless connections, bonding jumpers, or other devices. These are the same methods used for bonding the service equipment with the exception of the grounded conductor. See 250.92(B). **See Figure 6-26.**

There is, however, an exception for bonding circuits of over 250 V to ground. There are four alternative means to accomplish bonding for circuits which operate at over 250 V to ground per 250.97, Ex. These means are permitted provided that oversized, concentric, or eccentric knockouts are not encountered or when the knockouts are encountered but have been tested and the box or the enclosure is listed for the use.

The first means is to use threadless couplings and cable connectors for cables with metallic sheaths. Cable assemblies such as MC, AC, and MI are covered by this section. The second means is to use double locknuts, one on the inside and one on the outside of the box or enclosure. This provision applies to RMC and IMC only. The third means is to use fittings that are designed with a shoulder that seats firmly against the box or enclosure. Fittings for EMT, FMC, and MC are covered by this section. The fourth means is to use listed fittings that are identified for the purpose. This category covers some of the newer fittings that incorporate a "snap-tite" or "speed-lock" design which does not use a separate locknut.

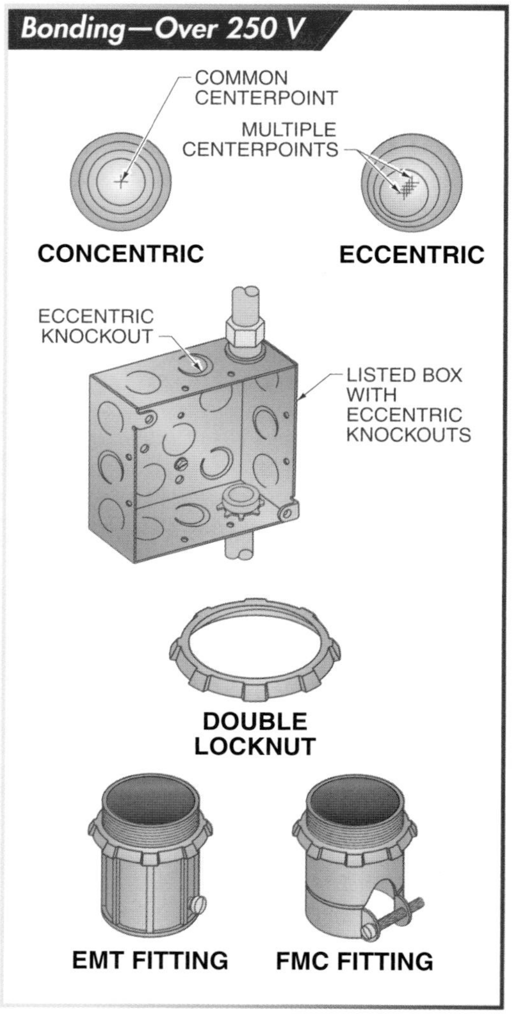

Figure 6-26. Listed boxes, with concentric or eccentric knockouts, may be used for bonding circuits of over 250 V to ground.

Main and Equipment Bonding Jumpers – 250.28; 250.102

The *main bonding jumper (MBJ)* is the connection at the service equipment that ties together the equipment grounding conductor, the grounded conductor, and the grounding electrode conductor. The MBJ can be a simple machine screw, as in the case of a 100 A panelboard, or it can be a 4/0 AWG conductor for a 1000 A service.

The MBJ may be constructed of copper or other corrosion-resistant materials per 250.28(A). It should not be installed downstream or in subpanels located in the same building, as this could result in multiple or parallel grounding paths, posing a threat of electrical shock.

When the MBJ is a screw only, it shall be colored green. This distinguishes it from the other screws on the grounded conductor terminal block. The MBJ is attached by a green screw or exothermic welds, listed pressure connectors, listed clamps, or other listed means. See 250.8 for applicable provisions. **See Figure 6-27.**

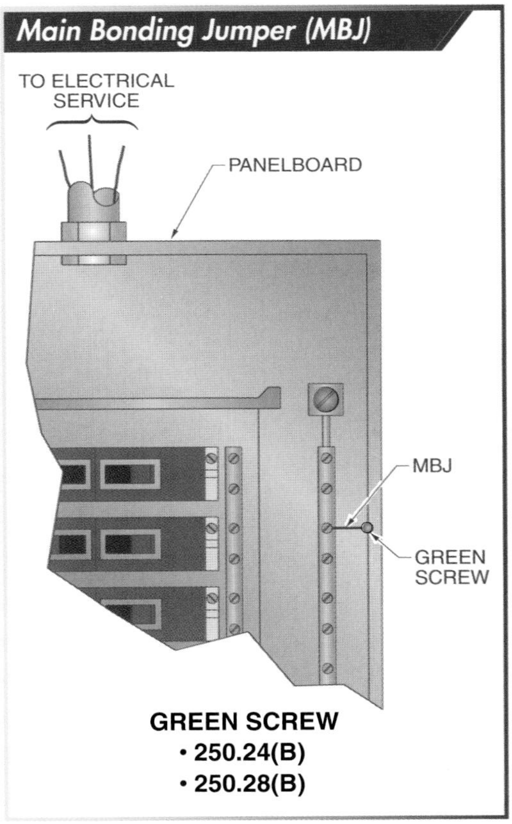

Figure 6-27. The main bonding jumper (MBJ) is the connection at the service equipment that ties together the EGC, the grounded conductor, and the GEC.

The *equipment bonding jumper (EBJ)* is a conductor that connects two or more parts of the equipment grounding conductor. See Article 100. **See Figure 6-28.** Like MBJs, EBJs are required to be constructed of copper or other corrosion-resistant materials. Section 250.102 contains the provisions for equipment bonding jumpers.

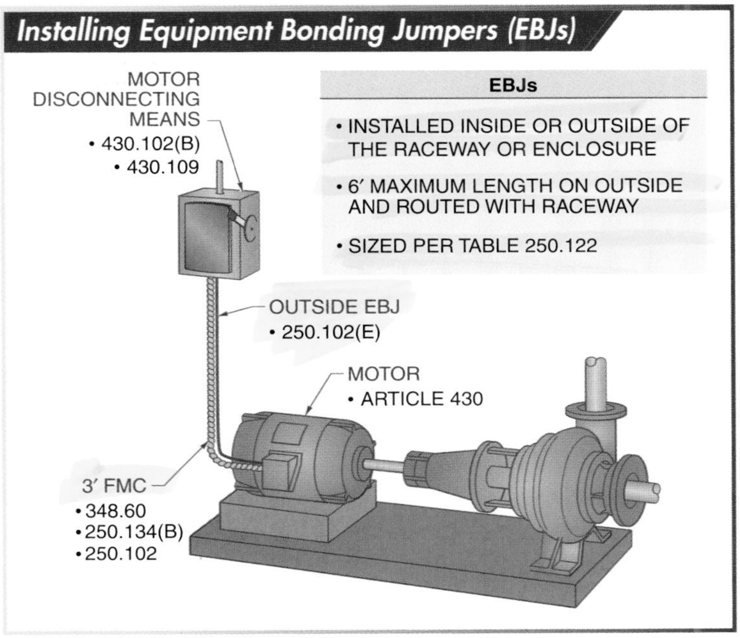

Figure 6-28. The equipment bonding jumper (EBJ) is a conductor that connects two or more parts of the EGC.

Equipment bonding jumpers are permitted by 250.102(E) to be installed inside or outside of a raceway or enclosure.

EBJs are permitted to be installed either on the inside or the outside of the raceway or enclosure per 250.102(E). The total length, when installed on the outside, shall not exceed 6′, and they shall be installed so that they closely follow the raceway. EBJs should not be installed by wrapping the jumper around a raceway. Such installations would add to the overall inductance of the circuit and act like a choke on any potential current flow.

Size – Supply-Side Equipment Bonding Jumper and Main Bonding Jumpers – 250.102(C); 250.28(D). EBJs on the line side or the supply side of the service and the MBJ are sized per Table 250.66. The size of the EBJ and MBJ are based upon the size of the largest phase conductor for the service.

Where the size of the service-entrance conductors exceeds 1100 kcmil Cu and 1750 kcmil Al, the size of the line-side EBJ or MBJ shall not be smaller than 12½% of the phase conductors. If the service-entrance conductors are run in parallel, then the EBJs shall also be run in parallel. Unlike grounded conductors, EBJs, when run in parallel, shall be full-size conductors based on the largest conductor in each raceway or cable. **See Figure 6-29.**

Size – Load-Side Bonding Jumpers – 250.102(D). Load-side EBJs are sized per 250.102(D) and selected based upon the values listed in Table 250.122. The bonding jumper size is determined by the size or rating of the OCPD ahead of the equipment. If the EBJ is to be used for more than one circuit in the raceway or cable, the size is selected from Table 250.122 based upon the largest OCPD protecting a circuit within the raceways or cables. **See Figure 6-30.**

Interior Metal Piping and Structural Steel – 250.104. Metal underground water piping with at least 10′ of piping in direct contact with the earth shall be part of the grounding electrode system (GES). If the water piping is not effectively grounded, such as when plastic piping is used outside the building, 250.104(A)(1) still requires that the interior metal water piping be bonded to the grounded conductor and GEC at the service equipment. The point at which the bonding jumper is attached shall be accessible. As with line-side bonding jumpers, the size of the bonding jumper shall be based on Table 250.66.

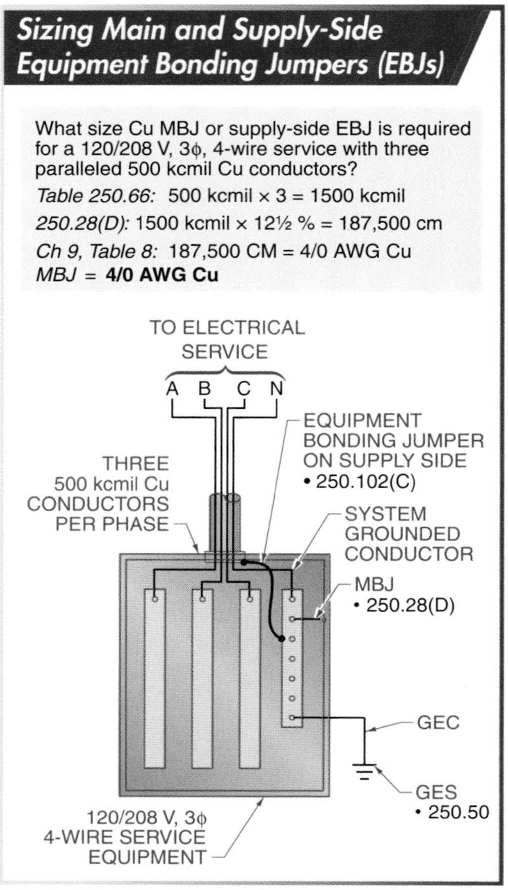

Figure 6-29. Line-side EBJs and MBJs are sized to handle the available fault current that the system can deliver.

A 1996 NEC® change required that when separately derived systems are installed and the metal water piping is not part of the GES, the nearest point on the interior metal water piping system shall be bonded to the grounded conductor of the separately derived system. See 250.104(D)(1–3). Section 250.104(B) contains the additional requirement that other interior metal piping systems shall also be bonded to the service equipment enclosure, service grounded conductor, or the GEC. The size of the bonding jumper for these applications is based on Table 250.122.

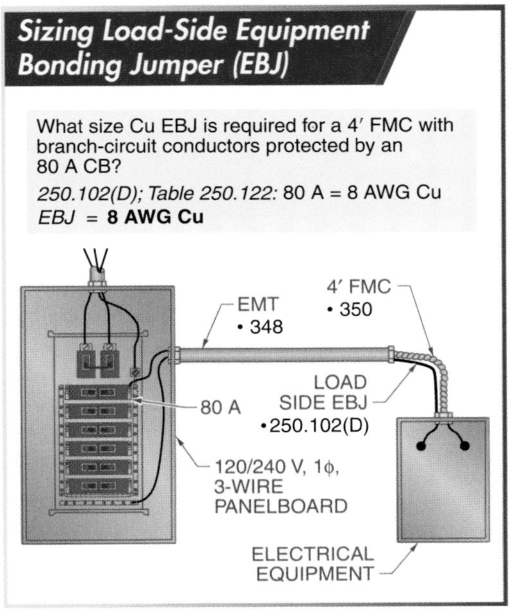

Figure 6-30. Load-side EBJs are sized by the rating or setting of the OCPD in the circuit ahead of equipment, conduit, etc.

The EBJ size is selected based upon the rating of the OCPD protecting the circuit that may energize the piping. For example, if a piping system contains motor-operated valves, then that piping system could become energized and it should be bonded. Bonding all piping and metal ductwork provides an additional level of safety per 250.104(B), Informational Note.

Bonding requirements have been added in 250.104(C) for exposed interior structural steel. This steel, if it forms a metal frame of a building, shall be bonded to the service equipment enclosure, the service grounded conductor, or the GEC. This section applies to structural steel that is interconnected to form a metal frame. The provisions of this section do not apply to a steel beam supported by concrete columns. **See Figure 6-31.**

Bonding Separately Derived Systems – 250.104(D). Bonding and grounding requirements are related but are treated differently in the Code. The treatment of separately derived systems is an example of how the requirements are different but related. As covered, 250.30 contains very comprehensive requirements for grounding a separately derived system. Section 250.30(A)(4) specifies that the nearest of either the metal water piping or structural metal steel in the same area of the separately derived system shall be the grounding electrode for the newly derived system. If the structural steel was the nearest it would be used to comply with this requirement. If metal water piping were present, it still would be required to be bonded per 250.104(D)(1).

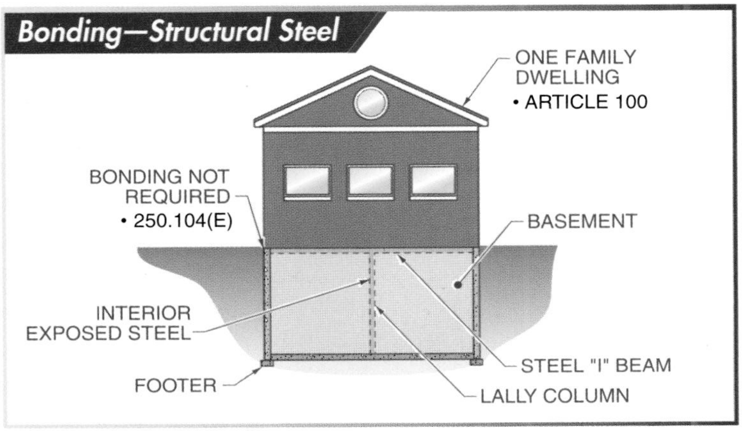

Figure 6-31. Single beams in dwelling units are not required to be bonded.

Although the piping will not serve as a grounding electrode, it still must be bonded to the newly derived system in the event that there is a ground fault to the water piping from a circuit supplied from the newly derived system. Per 250.104(D)(2) the same bonding requirement applies for exposed structural steel that is interconnected to form the building. Sections 250.52(A)(1–2) provide restrictions to the use of metal water piping and building as electrodes.

GROUNDING ELECTRODE SYSTEM – ARTICLE 250, PART III

Section 250.24(A)(D), which details the grounding connections required for an AC service that is to be grounded, requires that the GEC be connected to a grounding electrode that complies with Article 250. Part III contains the requirements for the grounding electrode system. **See Figure 6-32.**

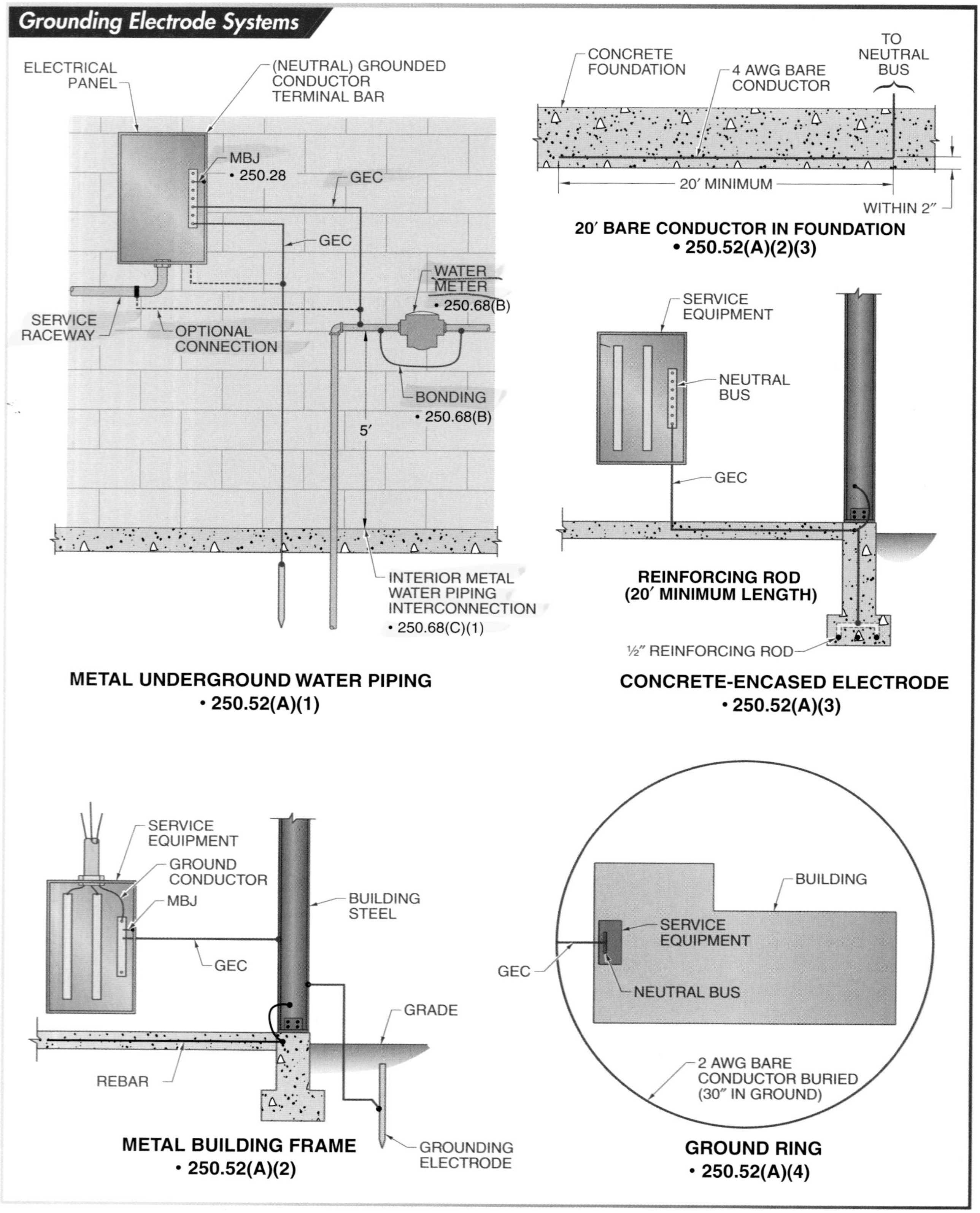

Figure 6-32. Each AC electrical system that is required to be grounded shall be connected to a GES per Article 250, Part III.

Construction and Installation – 250.50, 250.52, 250.53

A GES must be formed at each building or structure to comply with the requirements of 250.50. Prior to the 2005 NEC®, the components that comprised the GES were determined based on their "availability." This led to many application and interpretation problems in the field. The most common problem was the requirement for concrete-encased electrodes.

The 2005 NEC® revised Part III of Article 250 to clarify how the GES is to be constructed and added an exception that provided that the concrete-encased electrode need not be part of the GES if it would require disturbing the concrete already poured. Section 250.50 now clearly states that if any of the grounding electrodes in 250.52(A)(1–7) are present at a building or structure, they shall be bonded together to form the GES. If none of them exist, then at least one grounding electrode from 250.52(A)(4–8) must be installed and used as the GES.

Metal Water Piping – 250.52(A)(1). The first component that is connected to GES is metal underground water piping. The piping shall be in direct contact with the earth for a minimum distance of 10′. With the increasing popularity of nonmetallic water piping, always verify that there is at least 10′ in contact with the earth.

Interior metal wiring piping shall not be used as a part of the GES unless it is the first 5′ of piping that enters the building. This ensures that downstream electrical systems are not isolated by the insertion of nonmetallic piping. See 250.68(C)(1).

All insulating fittings or joints in the piping systems require bonding to ensure electrical continuity. Water meters, filtering devices, and similar equipment can interrupt the continuity of the piping system and require proper bonding.

Per 250.53(D)(2), a metal underground water pipe must be supplemented by an additional grounding electrode. The additional electrode is permitted to be any one of the types listed in 250.52(A)(2)(1–5). A supplemental electrode is not required where a single rod, pipe, or plate is utilized and where the electrode has a resistance to ground of 25 Ω or less. If a supplemental electrode is used, 250.53(D)(2) requires that it be bonded to the GEC, the system grounded conductor, the grounded service raceway, or any grounded service enclosure.

Grounding Electrode Burial Depth

Per 250.53(F), underground ground rings must be buried at a minimum depth of 30″ (750 mm). Per 250.53(G), rod and pipe electrodes must be installed to a minimum depth of 8′ (2.44 m), unless a rock bottom is encountered. Rod and pipe electrodes must be at an angle not to exceed 45°.

Structural Steel – 250.52(A)(2). This section contains the requirements for the second component of the GES, the metal building frame. The metal structural steel that forms the building frame is required to be part of the GES if it is connected to the earth by any one of four means. The first means of connection to the earth is where 10′ or more of a single structural metal member is in direct contact with the earth. This means also applies where the structural member is encased in concrete that is in direct contact with the earth.

minimum 5/8 for ground rod

Water meters and similar equipment must be bonded to ensure electrical continuity.

The second means is where the structural metal frame of the building or structure is connected to the reinforcing bars of a concrete-encased electrode or ground ring. Sections 250.52(A)(3) and (A)(4) cover the requirements for concrete-encased electrodes and ground rings. The third means of connection is by bonding the structural frame member to one or more of the electrodes covered by 250.52(A)(5) or (A)(7). These sections provide the requirements for rod and pipe electrodes and plate electrodes.

The fourth means is through other approved means of connection to earth. By definition, an approved means would be any method that was authorized by the authority having jurisdiction. The building steel should be cleaned with a wire brush to remove any nonconductive coatings and to ensure a low-impedance connection per 250.12.

Concrete-Encased Electrode – 250.52(A)(3). A concrete-encased electrode or reinforcing bar (rebar) is one of the best grounding electrodes. The rebar shall be encased in at least 2″ of concrete, be located at the bottom of the pour so as to maintain good contact with the earth, and be at least 20′ long (in one or more pieces) and ½″ in diameter. This section also permits 20′ of bare copper at least 4 AWG in size as a substitute for the rebar.

Some rebars may be coated with nonconductive coverings which make them unsuitable as a grounding electrode. The connection to the concrete-encased electrode is not required to be accessible, but the connection to the electrode shall be made by means of exothermic weld, listed lugs, listed pressure connectors, listed clamps, or other listed means. Connectors for connection to a buried electrode shall be listed for direct burial uses. See 250.70.

Made Electrodes – 250.52(A)(4–8). There may be buildings in which none of the electrodes specified by 250.52(A)(1–3) are available, or in which the only electrode available is the metal underground water pipe. In these cases, a supplemental electrode shall be installed. Section 250.52(A)(4–8) contains the requirements for made or other electrodes. An electrode is simply a means to permit current to enter the earth. Made electrodes can be designed to function much like the electrodes listed in 250.52(A)(1–3). As with these electrodes, made electrode connections shall be free of nonconductive coverings that may affect the impedance of the grounding path. **See Figure 6-33.**

Ground Ring – 250.52(A)(4). This section lists the requirements for ground rings. Ground rings shall consist of at least 20′ of 2 AWG or larger bare copper. The ring shall encircle the building and be buried in a trench at least 2½′ in depth. See 250.53(F). The most common application for this type of electrode is in supplemental grounding systems such as those frequently used for grounding computer systems.

Rod and Pipe Electrodes – 250.52(A)(5). The most commonly encountered made electrodes are those constructed of rod and pipe. The minimum length of the ground rod shall not be less than 8′. The minimum diameter is ½″ for stainless steel or other nonferrous rods and ⅝″ for iron or steel rods.

Pipe electrodes can be constructed of iron or steel provided they are galvanized or otherwise provided with corrosion protection. Pipe electrodes shall be a minimum ¾″ trade size. At least 8′ of the electrode shall be in contact with the earth. If rock is encountered during the installation, precluding driving the rod straight down, the rod or pipe can be installed on an angle, not to exceed 45° from vertical. If this is not possible, the rod or pipe can be buried in a trench at least 2½′ in depth. See 250.53(G).

Rod and Pipe Electrodes

The minimum length required for rod and pipe electrodes is 8′ per 250.52(A)(5). At least 8′ of the electrode shall be in contact with the earth per 250.53(G). Therefore, the electrode is driven to a point at least flush with the finished grade. The depth to which the rod or pipe electrode is driven affects the impedance of the ground path. The deeper the electrode is driven, the lower the impedance. Do not cut off a ground rod when rock or other obstructions are encountered. This can have a serious effect on the integrity of the ground path.

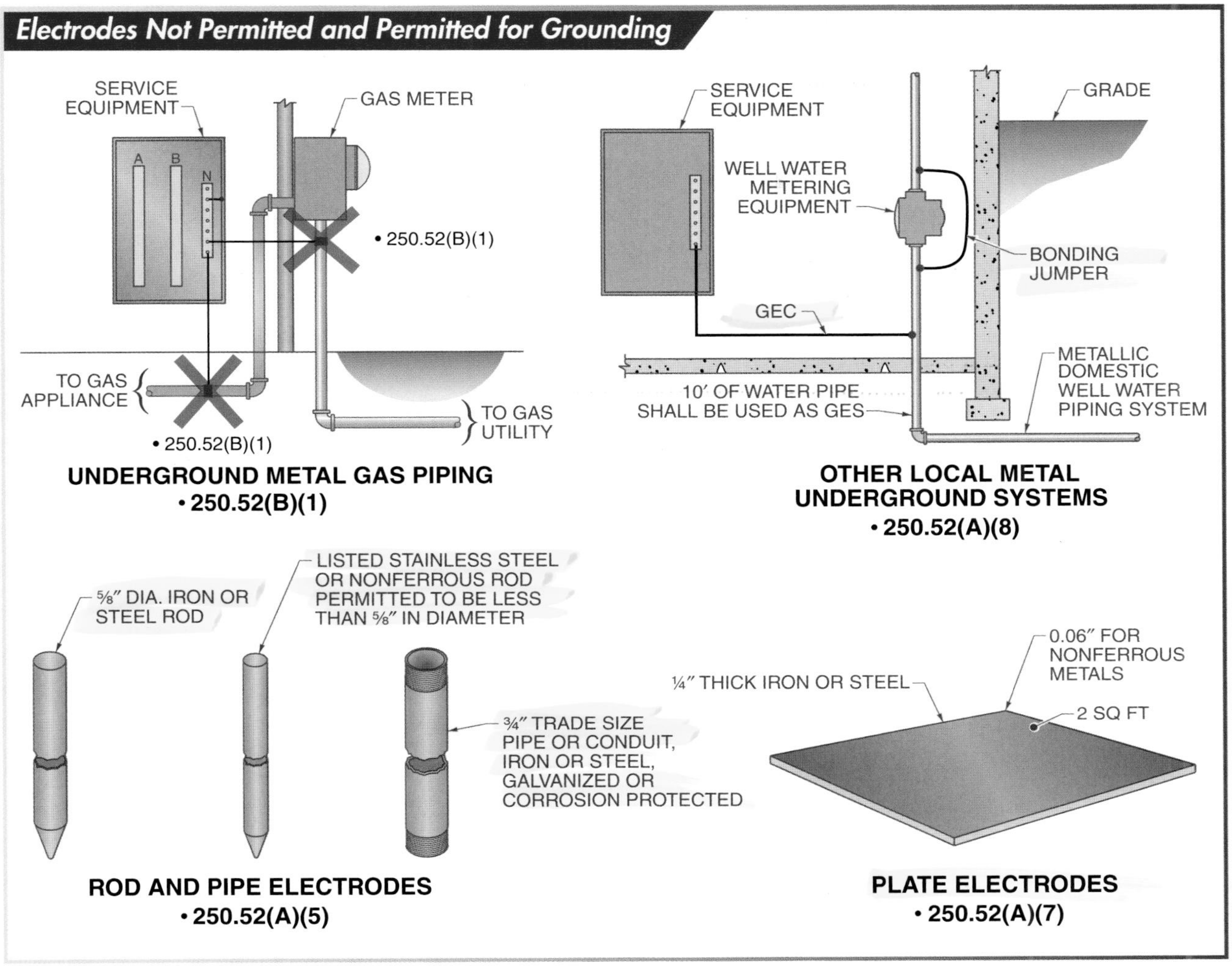

Figure 6-33. All grounding electrodes must be bonded together to form the GES.

Other Electrodes – 250.52(A)(6). Grounding electrodes historically have been referred to as those that exist as part of the building, such as the metal frame of the building and concrete-encased electrodes, and those that are "man made," such as rod and pipe and plate electrodes. Section 250.52(A)(6) permits other types of electrodes to also be employed provided they are listed. Examples of some of the "other" listed electrodes are chemical- or electrolyte-based ground rod enhanced systems.

Plate Electrodes – 250.52(A)(7). Plate electrodes shall have at least 2 sq ft in area in contact with the earth and at least 1/4″ thick for iron or steel electrodes and 0.06″ thick for nonferrous metals. The minimum burial depth required is 2½′ below the surface of the earth. See 250.53(H).

Local Metal Underground Systems – 250.52(A)(8). The last items that must be connected to the GES, if present, are local metal underground piping systems or structures. These can be common metal water piping systems such as those used on domestic well-water systems, or more elaborate structural metal underground systems used in tank fields or with well casings.

Prohibited Electrodes

Section 250.52(B)(1) specifically prohibits metal underground gas piping systems from being used as a grounding electrode. Under fault conditions, the GEC and the electrode itself will carry current. The gas piping is required to be bonded per 250.104(B). It is not, however, permitted to be used as a grounding electrode. Per 250.52(B)(2), aluminum electrodes are not permitted in the grounding electrode system.

GROUNDING ELECTRODE CONDUCTORS – ARTICLE 250, PART III

The GEC is used to connect the EGC and the grounded conductor at the service or separately derived system to the grounding electrode. Four major items should be considered when installing or designing the GEC. The first is the construction of the GEC. The second is the installation requirements for the GEC. The third is the sizing calculations for the GEC, and the fourth is the methods of connection which are permitted.

Construction – 250.62

The GEC is constructed of copper, aluminum, or copper-clad aluminum per 250.62. The solid or stranded conductors are permitted to be insulated, covered, or bare. The conductor shall have good corrosion-resistant properties and should be suitable for the conditions under which it may be installed.

Bonding of Other Piping Systems

Per 250.104(B), other metal piping systems, including gas piping, installed in or attached to a building that are likely to become energized must be bonded to the service, a grounding electrode conductor of the proper size, or to one or more of the grounding electrodes permitted by 250.52. All points of attachment of any bonding jumpers must be accessible. Bonding of all metal piping and air ducts within the premises provides additional safety to building occupants and equipment.

The GEC shall be one continuous length per 250.64(C). Per 250.64(C)(1), splices are permitted only where listed irreversible compression-type connectors or exothermic welding processes are used. Per 250.64(C)(2), sections of busbar are permitted to be connected together to form the GEC.

There are two additional permitted means for splicing the GEC. Per 250.64(C)(3) bolted, welded, or riveted metal frames of structures and buildings are permitted, as threaded, welded, brazed, or bolted-flange connections of metal water piping are permitted per 250.64(C)(4). Per 250.64(D), the GEC may be spliced or tapped when multiple service enclosures are used in accordance with 230.40, Ex. 2. GECs may be spliced if the splicing method consists of either exothermic welding or irreversible compression-type connectors per 250.64(C)(1).

Section 250.64(C)(2) permits sections of busbars to be connected together to form a GEC.

Installation – 250.64

Per 250.64(B), the GEC is not required to be installed in a raceway if it is of adequate size and not subject to severe physical damage. GECs are permitted to be installed on or run through framing members. GECs which are subject to severe physical damage shall be a 4 AWG Cu or 4 AWG Al or larger conductor or it shall be protected. If not subject to severe physical damage, 6 AWG and larger conductors shall be permitted to be installed along the surface of the building if adequately supported. RMC, IMC, PVC, EMT, or cable armor is permitted to be used for protection against severe physical damage.

Where used outdoors, an aluminum or copper-clad aluminum GEC shall not be used within 18″ of the earth. Additionally, aluminum or copper-clad aluminum cannot be used in direct contact with masonry or other corrosive conditions. See 250.64(A).

Per 250.64(E), the metal enclosures or raceways in which a GEC is installed shall be electrically continuous. This requires that both ends of the GEC enclosure be bonded to either the equipment or electrode or to the GEC itself. **See Figure 6-34.**

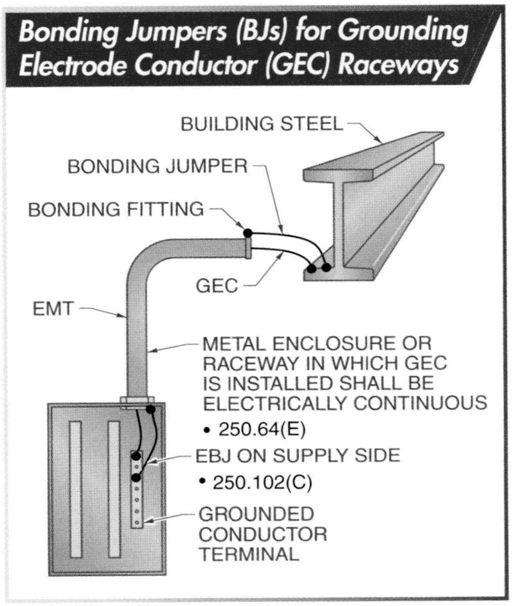

Figure 6-34. Per 250.64(E), the bonding jumper for a grounding electrode conductor raceway shall be the same size or larger than the required GEC.

Size – 250.66

The size of the GEC is selected from Table 250.66 and is based upon the largest service-entrance conductor for one phase. If the service-entrance conductors are paralleled, the equivalent area for a set of phase conductors shall be used as a basis for sizing the GEC. **See Figure 6-35.**

While the purpose of the grounding electrode is to provide a path into the earth for current flow, many factors affect its ability to dissipate current flow into the earth. Some electrodes dissipate current flow better than others. For example, rebar is typically a better electrode than a ground rod. The rebar provides a greater surface area from which the electrons can be dissipated.

Other conditions may also affect the rate of dissipation. With ground rods, the deeper the ground rod enters the earth, the greater the rate of dissipation. In addition, the higher the moisture content of the earth, the higher the rate of dissipation from the ground rod.

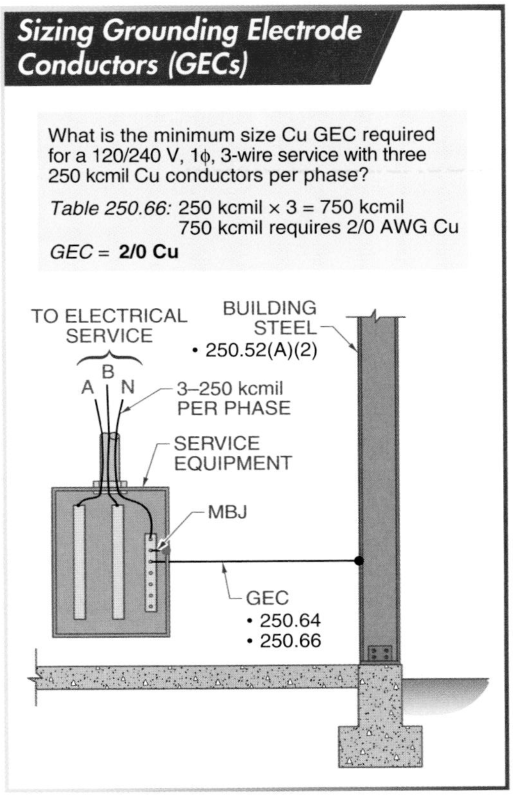

Figure 6-35. GECs are never required to be larger than 3/0 AWG Cu or 250 kcmil Al.

A continuous grounding conductor may be fastened to the frame of an electrical device.

The size of the GEC reflects, to some degree, the ability of the grounding electrode to dissipate current flow into the earth. For example, 250.66(A)(B)(C) permits smaller-size GECs for specific types of grounding electrodes. For sole connections to made electrodes, 250.66(A) permits the GEC to be 6 AWG Cu or 4 AWG Al. The GEC is not required to be larger than 4 AWG Cu per 250.66(B) for sole connections to concrete-encased electrodes. For ground rings, 250.66(C) permits the GEC, which is a sole connection, to be the same size as the conductor used for the ground ring. **See Figure 6-36.**

Grounding Electrode Connections – Article 250, Part III

The provisions for making GEC connections are given in Article 250, Part III. Per 250.68(A), any mechanical components used in terminating the GEC or bonding jumper to a grounding electrode must be accessible.

Per 250.68(A), Ex. 1, the connection for concrete-encased, driven, or buried electrodes is not required to be accessible. Bonding jumpers around insulated joints to ensure electrical continuity where equipment is installed shall be of sufficient length to permit the removal of the equipment without removal of the bonding jumper. This safety requirement protects service personnel from possible current flow on the water piping system. Connections to electrodes shall be made with listed fittings per 250.70. The connection shall be protected against physical damage per 250.10. Per 250.68(A), Ex. 2, the connections need not be accessible if exothermic or compression-type connections are used for fire proofed structural metal terminations.

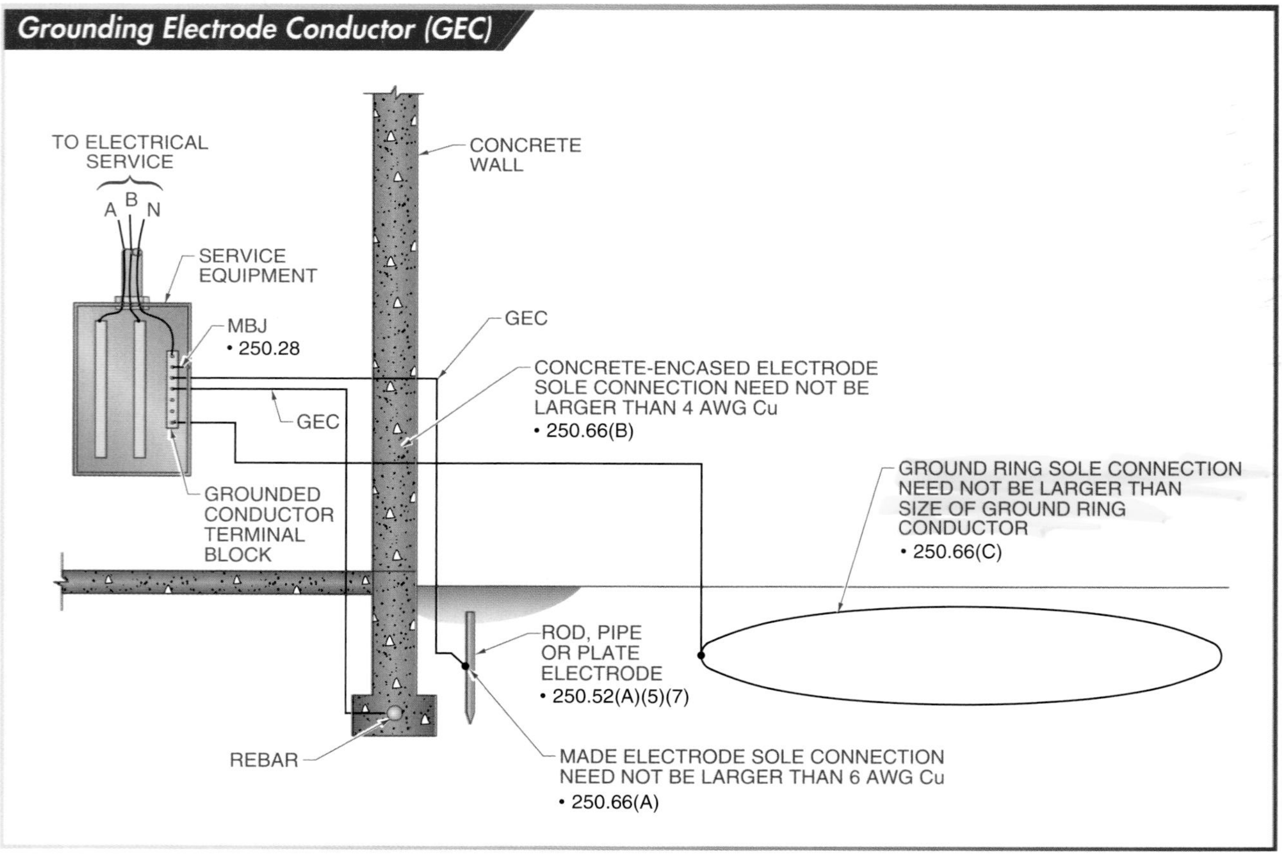

Figure 6-36. The GEC to a ground rod or other made electrode need not be larger than 6 AWG Cu when it is the sole connection to the grounding electrode.

Clamp-On Ground Resistance Measurements

Clamp-on ground resistance meters directly measure the resistance of a grounding electrode or small grounding grid. They can be used at any time, regardless of ground condition (frozen, wet, or dry). Clamp-on ground resistance meter measurements do not require any part of the grounding system to be disconnected (opened) during the test.

Although a clamp-on ground resistance meter appears to be a standard clamp-on ammeter and has the same clamp-on feature, the two meters are quite different. When placed around a conductor, a standard current clamp-on meter measures the strength of the magnetic field produced by current flow. The clamp-on ground resistance meter, however, includes a transmitter that sends out a signal through the conductor (typically the earth grounding system) and a receiver that receives the signal back and uses the information to determine (calculate) the resistance of the grounding system.

Ideal Industries, Inc.

A high resistance reading indicates a problem. If the measured ground resistance is high, the grounding system will operate poorly. High resistance may indicate any of the following:

- Open, corroded, or damaged grounding system or connections – Open or loose connections and other damage can occur anyplace in the system (above or below the ground). Look for the obvious problems first, such as loose connections, corrosion, and damaged conductors.
- Insufficient grounding electrode, grid, system size, or incorrect type of system for soil conditions – Look for proper grounding practices, size, and number of electrodes typically used in the area.
- Changing soil resistance – Grounding systems are required to last a long time, but soil conditions can change over the years. Look for any construction changes since the system was installed. Another cause of high ground resistance is a falling water table that leaves the grounding system in drier soil compared to when it was installed.

Additional information is available at www.idealindustries.com.

EQUIPMENT GROUNDING CONDUCTORS – ARTICLE 250, PART VI

The same items are considered before designing or installing EGCs in electrical distribution systems as are considered when designing or installing GECs. The EGC is used to connect the noncurrent-carrying metal parts of equipment, raceways, enclosures, etc., to the grounded conductor and the GEC at the service equipment or source of a separately derived system. This connection provides a low-impedance path back to the source to permit the operation of the OCPD.

EGC Types – 250.118

Section 250.118 lists all of the permissible types of EGCs. EGCs are constructed of copper or other corrosion-resistant materials. The conductors, which can be in the form of a wire or busbar, may be solid or stranded and either insulated, covered, or bare. However, there are several instances where the EGC is not permitted to be a bare conductor. For example, the EGC to a remote building that houses livestock shall be insulated or covered per 547.5(F).

Section 250.118(5–7) contains two provisions which permit FMC, FMT, and LFMC under specific conditions to be used as EGCs. Listed FMC can be utilized as an EGC when four conditions are met. First, the FMC shall be terminated in fittings that are listed for grounding. Second, the circuit conductors contained within the FMC shall be protected by an OCPD rated at 20 A or less. Third, the FMC's total combined length shall not exceed 6′. Fourth, the FMC shall not have been installed for flexibility. If the FMC was selected because the equipment moves or is subjected to vibration, a separate EGC would have to be installed within the FMC. See 250.118(5).

Similar conditions exist for listed LFMC. One of the differences from FMC, however, is the size restrictions. For trade sizes of 3/8″

through ½″, the circuit conductors shall be protected at not over 20 A. For trade sizes ¾″ through 1¼″, the circuit conductors shall be protected at no more than 60 A. See 250.118(6)(b)(c).

For FMT, the fittings shall be listed for grounding, the circuit conductors shall be protected at not over 20 A, and the total length of the FMC shall not exceed 6′. See 250.118(7). The 6′ length limitation for all FMC, LFMC, and FMT relates to the total length, which includes any combination of these raceways in the same installation. For example, a 3′ length of FMC and a 4′ length of LFMC in the same installation is a violation. **See Figure 6-37.**

Grounding—Flexibility Required

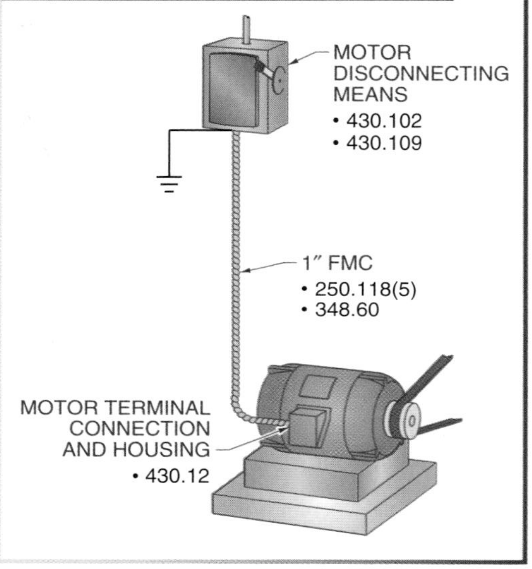

Figure 6-37. A separate EGC shall be provided when FMC is used to connect equipment where flexibility is required.

Installation – 250.120

When the EGC consists of a raceway, metallic cable tray, cable armor, cable sheath, or a wire which is part of a cable or installed in a raceway, the EGC shall be installed in a manner which meets all of the applicable provisions for the type of raceway or cable that is used. Care shall be taken to ensure that the EGC terminations are made with fittings suitable for the raceway or cable system used.

Bonding Grounding Electrode Conductor (GEC) Raceways

Failing to bond both ends of the GEC raceway is a common NEC® violation which may dramatically affect the integrity of the grounding path. Studies have shown that when a raceway is not properly bonded, and a fault current is applied to the GEC, the current does not split evenly between the GEC and the raceway. The net effect is the fault current is choked by the inductive effect of a single conductor in a metal raceway.

GECs, however, may be sized as large as 3/0 AWG Cu and 250 kcmil Al. With these larger sizes, bonding the raceway can be especially difficult. One solution to the problem is to use PVC as the raceway for enclosing the GEC. The nonmetallic raceway alleviates any induced current problem resulting from single conductors in a raceway. Make sure that a nonmetallic raceway is not subject to severe physical damage.

Section 250.120(C) requires that where an EGC that is smaller than 6 AWG and is not run with the circuit conductors as permitted by 250.130(C) and 250.134(B), the EGC shall be protected from physical damage by an identified raceway or cable armor. Such protection is not required where the EGC is installed within hollow spaces of the framing members of structures or buildings or subject to physical damage.

Per 250.120(B), where the separate EGC is aluminum or copper-clad aluminum, it shall not be installed in any of the following conditions:

- Direct contact with masonry or the earth
- Where subject to corrosive conditions
- Outdoors within 18″ of the earth

Size – 250.122

The size of the EGC shall be based upon the size or rating of the OCPD ahead of the equipment which is supplied. Table 250.122 lists the OCPD ratings and the appropriate size EGC for copper, aluminum, or copper-clad aluminum. For example, for a 30 A branch-circuit run in a nonmetallic sheathed cable to serve equipment, the EGC is based upon the 30 A OCPD rating and requires either a 10 AWG Cu or 8 AWG Al EGC.

The EGCs shall be run in parallel when conductors are run in parallel in multiple raceways or cables per 310.10(H). Each of the EGCs shall be a full-size conductor based on the ampere rating of the OCPD protecting the circuit conductors. See 250.122(F). If multiple circuits are run in the same raceway, the size of the EGC shall be based on the largest OCPD protecting the circuit conductors per 250.122(C). **See Figure 6-38.** If the circuit conductors are increased in size, similar adjustment shall be made in the size of the EGC. See 250.122(B).

Sizing Equipment Grounding Conductors (EGCs) for Multiple Circuits in a Common Raceway

What size Cu EGC is required for three branch circuits installed in the same PVC raceway? The ratings of the three branch circuits are 20 A, 40 A, and 60 A.

250.122(C): Largest OCPD = 60 A
Table 250.122: 60 A OCPD requires 10 AWG Cu
EGC = **10 AWG Cu**

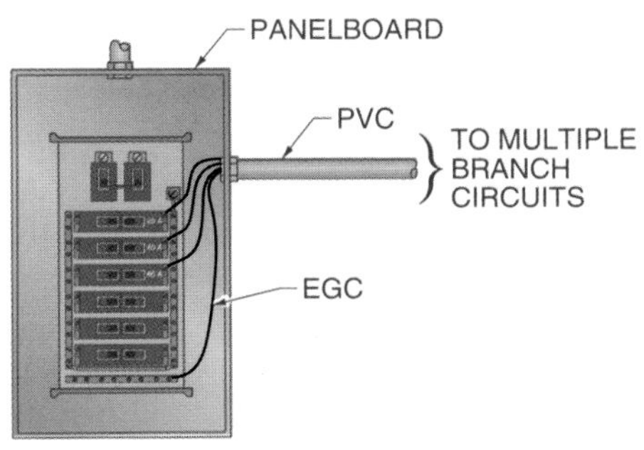

Figure 6-38. If multiple circuits are run in a common raceway, only one EGC, based on the largest size OCPD protecting the conductors, is required.

Connections – 250.148

Whenever several EGCs are present in a box or an enclosure, the conductors shall be spliced together or connected within the box with devices suitable for the use. If devices or fixtures are fed from the box, care shall be taken to ensure that the continuity of the EGCs is not interrupted when the device or fixture is removed. Additionally, if a metal box is used, provisions shall be made to connect the one or more EGCs to the box by means of a dedicated grounding screw, other equipment listed for grounding, or a listed grounding device. See 250.148(C).

Refer to Chapter 6 Quick Quiz® on CD-ROM.

Using MC Cable as an Equipment Grounding Conductor (EGC)

Installers should be very careful when using the outer sheath of MC cable as an EGC. Per 250.118(10), the metallic sheath or the combined metallic sheath and grounding conductors for MC cable may be used as an EGC.

MC cable is constructed in three basic types: interlocking metal tape, smooth, and corrugated metallic sheath. Only the smooth and corrugated types are suitable for equipment grounding purposes. The interlocking tape type requires the use of a separate EGC. Do not use the interlocking tape type in applications which require two EGCs such as for isolated ground receptacles or patient care receptacles in health care facilities. This section was revised during the 2002 NEC® cycle to clarify that only the combined metallic sheath and a separate grounding conductor are suitable for the interlocked metal tape-type MC cable.

Installers of Type MC Cable must ensure that they have selected the correct type of MC Cable for the specific grounding application. The combined metallic sheath and uninsulated EGC of interlocked metal tape-type MC cable that is listed and identified as an EGC is an acceptable type of EGC. Likewise, the metallic sheath or the combined metallic sheath and EGCs of the smooth or corrugated tube-type MC that is listed and identified as an EGC is also an acceptable type of EGC.

Electrical equipment located in wet or damp locations must be grounded.

Additional information is available at ATPeResources.com

If nonmetallic boxes are used, the EGCs shall be arranged so that any device or fitting in the box can be grounded if so required. See 250.148(D). Terminals for the EGC in wiring devices shall be identified by either a green-colored screw, green-colored terminal screw, or a green-colored pressure wire nut. If the terminal point is not visible, it shall be marked with either the word "green," the letters "G" or "GR," or the grounding symbol. See 250.126.

chapter 6

Grounding

REVIEW QUESTIONS

Name ______________________ **Date** ______________________

(T) F 1. Electrical systems are often grounded to limit and stabilize voltages to ground.

(T) F 2. The amount of current that flows in any circuit is determined by the voltage and resistance of the circuit.

grounding 3. A(n) ___ conductor is the conductor that connects electrical equipment or the grounded conductor to the grounding electrode.

50 4. AC systems of less than ___ V to ground shall be grounded when installed as overhead conductors.

premesis 5. ___ wiring is basically all interior and exterior wiring installed on the load side of the service point or the source of a separately derived system.

service 6. The ___ conductors extend from the service point to the service disconnecting means.

D 7. A wet location is a(n) ___.
- A. installation in concrete slabs contacting earth
- B. underground installation
- C. location saturated with water or other liquids
- D. all of the above

C 8. A damp location is a(n) ___.
- A. partially protected area
- B. area subject to some moisture
- C. all of the above
- D. none of the above

D 9. Electrical service equipment which shall be bonded together includes ___.
- A. service raceways with GEC
- B. service conductor enclosures
- C. metallic raceways protecting the metallic sheath
- D. all of the above

main bonding jumper 10. The ___ is the connection at the service equipment that ties together the EGC, the grounded conductor, and the GEC.

grounding electrical conductor 11. The ___ connects the EGC and the grounded conductor at the service or separately derived system to the grounding electrode.

8 12. The minimum length of a nonlisted ground rod is ___′.

5/8 13. The minimum diameter of a nonferrous nonlisted ground rod is ___″.

30 **14.** Plate electrodes shall be installed not less than ___″ below the surface of the earth.

T (F) **15.** The grounded conductor shall never be smaller than 10½% of the largest service-entrance conductor.

T (F) **16.** The only choice for the grounding electrode of a separately derived system is the grounded building steel.

(T) F **17.** The EGC may be a bare, covered, or insulated conductor.

10 **18.** Metal underground water piping with at least ___′ of piping in direct contact with the earth shall be part of the GES.

OCPD **19.** The size of the EGC is based upon the rating of the ___ ahead of the supplied equipment.

T (F) **20.** Aluminum electrodes may be used as part of the GES.

Matching

E **1.** A temporary increase in the circuit or system voltage or current.

A **2.** The maximum voltage between any two phases of an electrical system.

G **3.** The connection between the grounded circuit conductor and the EGC at the service.

B **4.** The earth.

F **5.** A device that converts electrical power at one voltage or current to another voltage or current.

C **6.** A device that converts mechanical power to electrical power.

H **7.** The case or housing of equipment or other apparatus which provides protection from live or energized parts.

I **8.** Any device, fixture, apparatus, appliance, etc., used in conjunction with electrical installations.

D **9.** Joining metal parts to enhance electrical continuity and conductivity.

J **10.** A conductor that has been intentionally connected or grounded.

A. phase-to-phase voltage
B. ground
C. generator
D. bonding
E. line surge
F. transformer
G. main bonding jumper
H. enclosure
I. equipment
J. grounded conductor

Grounding Electrode Systems

5 1. The GEC at A shall terminate within ___′ of the metal water pipe entering the building.

(T) F 2. An optional connection point for the GEC at A could be the service raceway.

20 3. The minimum length of the bare conductor at B is ___′.

20 4. The minimum length of the reinforcing rod at C is ___′.

1/2 5. The minimum diameter of the reinforcing rod at C is ___″.

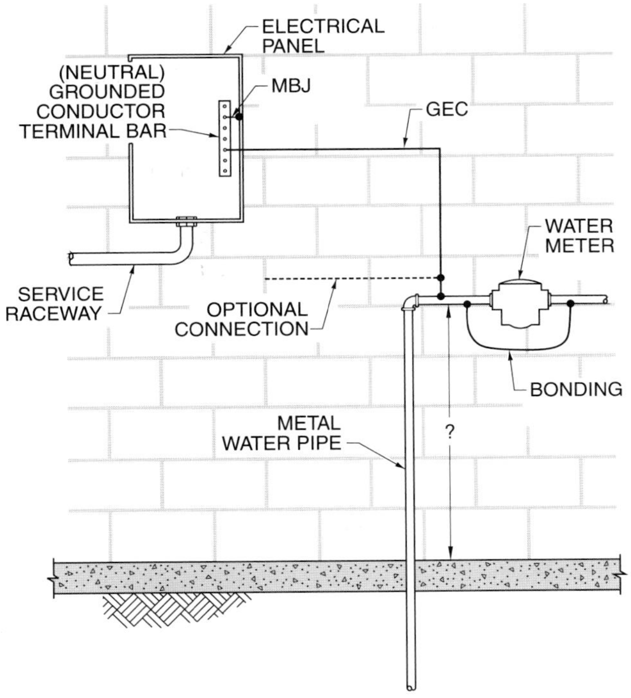

Ⓐ

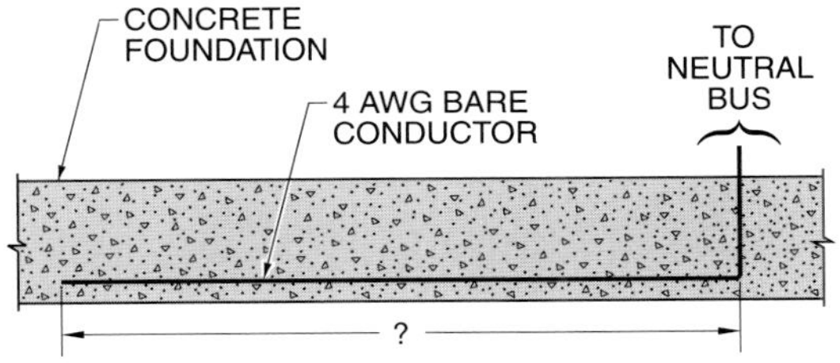

Ⓑ

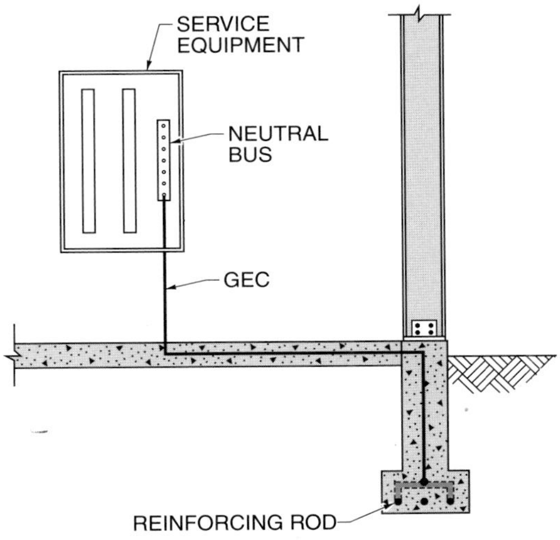

Ⓒ

Made Electrodes

T (F) **1.** The grounding electrode at A is permitted.

(T) F **2.** The grounding electrode at B is permitted.

5/8 **3.** The minimum diameter at C is ___″.

5/4 **4.** The minimum diameter at D is ___″.

3/4 **5.** The minimum diameter at E is ___″.

2 **6.** The made electrode at F shall be at least ___ sq ft in area.

.06 **7.** The made electrode at F shall be a minimum of ___″ thick when made of iron or steel.

______ **8.** The made electrode at F shall be a minimum of ___″ when made of nonferrous metals.

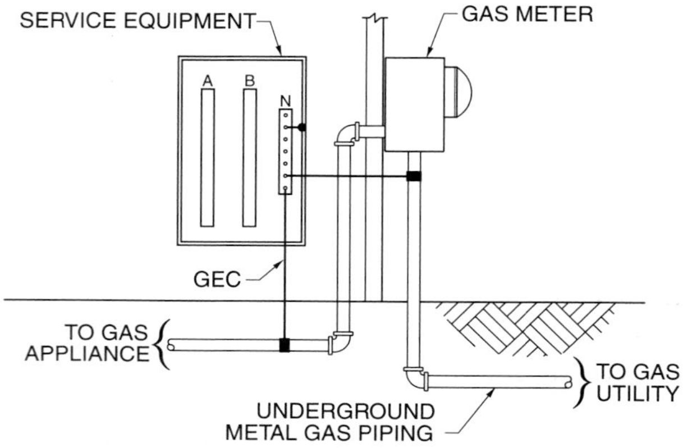

Ⓐ

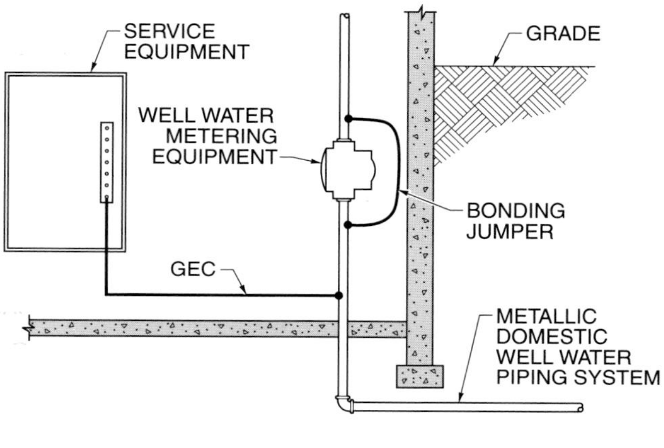

Ⓑ

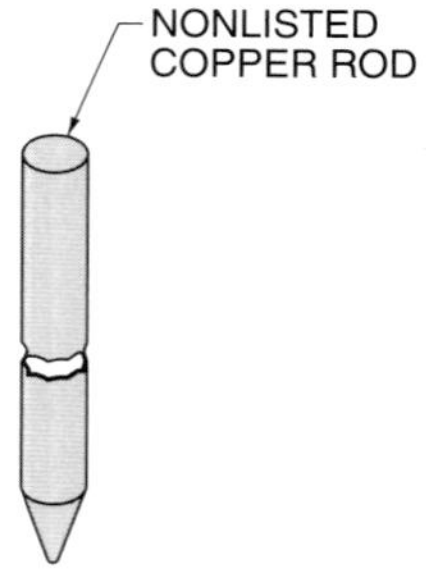

Ⓒ

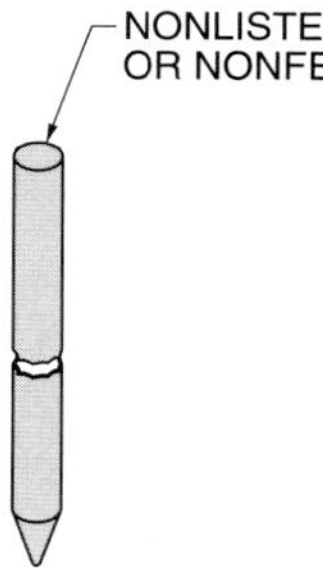

Ⓓ

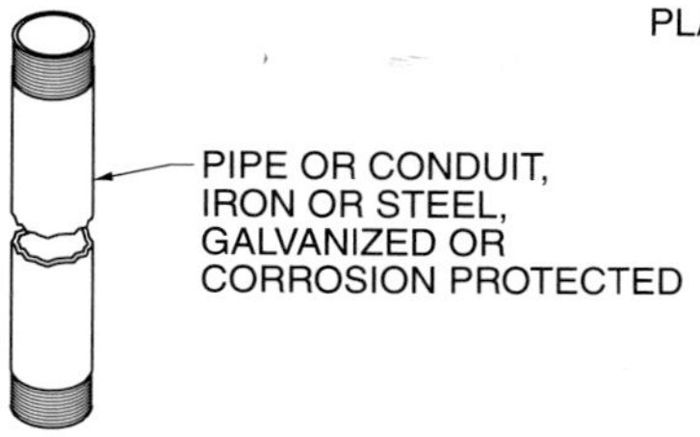

Ⓔ

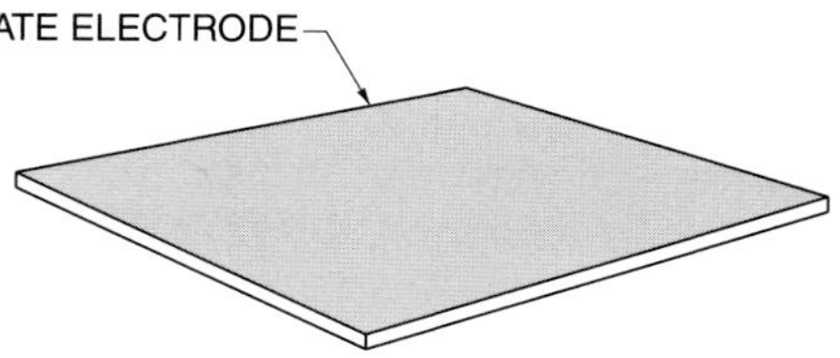

Ⓕ

chapter 6

Grounding

TRADE COMPETENCY TEST

Name ____________________ Date ____________________

NEC®	Answer	
______	______	1. Determine the minimum-size Cu grounded conductor required for a 120/208 V, 3ϕ, 4-wire service which consists of two parallel 1/0 AWG THW Cu conductors per phase.
______	______	2. What size grounded conductor is required for a 277/480 V, 3ϕ, 4-wire service with three 400 kcmil THW Cu conductors per phase?
______	______	3. *See Figure 1.* Determine the minimum-size bonding jumper required for the installation.
______	______	4. Determine the minimum-size Al main bonding jumper required for a 277/480 V, 3ϕ, 4-wire service which consists of two parallel 4/0 AWG THW Cu conductors per phase.
______	______	5. *See Figure 2.* Determine the minimum-size Cu main bonding jumper required for the installation.
______	______	6. Determine the minimum-size Cu GEC required for a 120/208 V, 3ϕ, 4-wire service which consists of four 1/0 AWG THW Cu conductors per phase. The grounding electrode is effectively grounded building steel.
______	______	7. Determine the minimum-size Cu GEC required for a 120/240 V, 1ϕ service which consists of three 4/0 AWG Al service-entrance conductors. The grounding electrode is a ⅝″ Cu ground rod. The GEC is the sole connection to the ground rod.

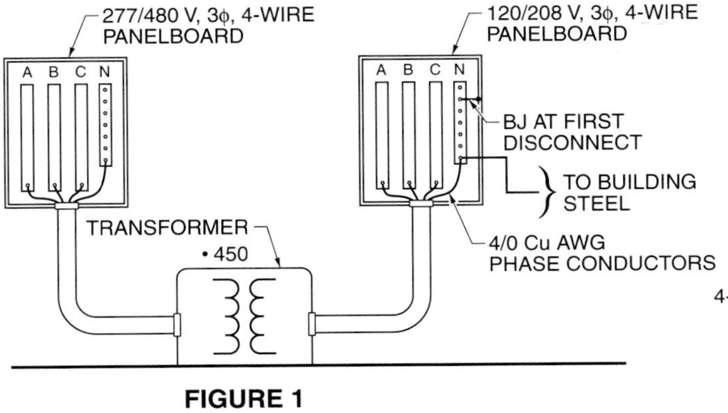

FIGURE 1

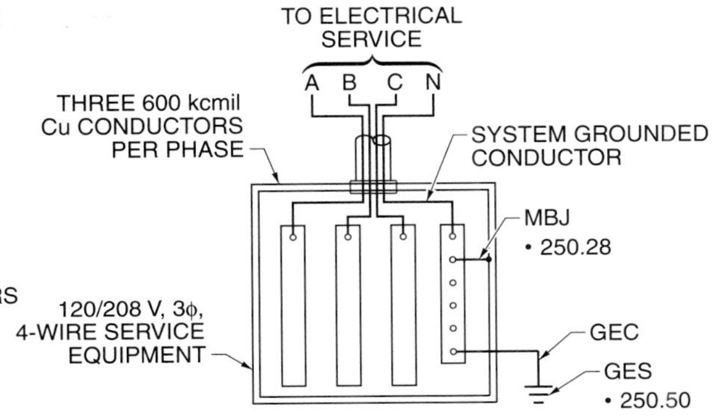

FIGURE 2

______________ ______________ **8.** What is the minimum-size Cu GEC required for a 120/208 V, 3ϕ, 4-wire service with three 300 kcmil Cu conductors per phase?

______________ ______________ **9.** *See Figure 3.* Determine the minimum-size Cu GEC for the separately derived system.

______________ ______________ **10.** *See Figure 4.* Determine the size of the Al EBJ required for the installation.

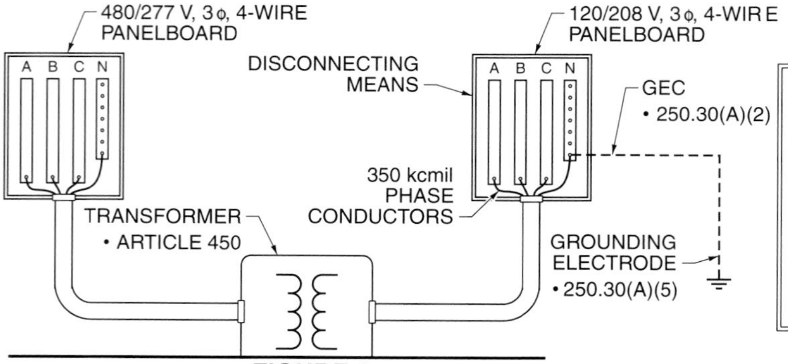

FIGURE 3

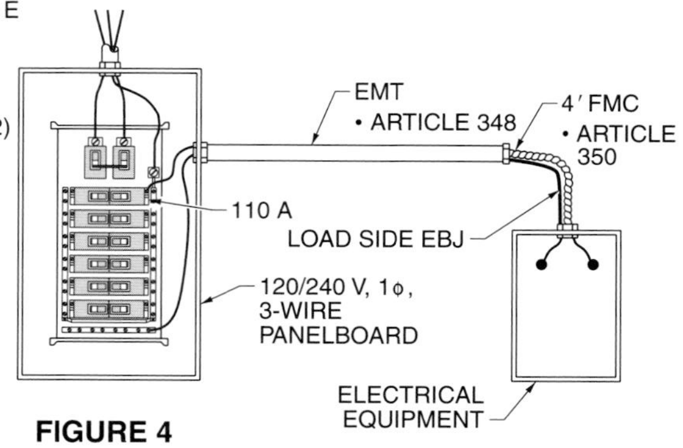

FIGURE 4

______________ ______________ **11.** Determine the minimum size Cu EGC required for a 50 A branch circuit installed in RNMC.

______________ ______________ **12.** Determine the minimum number and minimum size of Al EGCs required for an installation which consists of two parallel 2″ RNMCs, each of which has the conductor backed up by a 225 A OCPD.

______________ ______________ **13.** What size Cu EGC is required for three branch circuits installed in the same PVC raceway? The ratings of the three branch circuits are 20 A, 60 A, and 100 A.

______________ ______________ **14.** *See Figure 5.* Determine the minimum-size Cu bonding jumper required for the GEC raceway.

______________ ______________ **15.** A 4′ piece of 1″ FMC is installed for flexibility to supply a motor that has 15° of movement and is protected by a 20 A OCPD. All fittings used are listed for grounding and the total length of FMC in the circuit is less than 6′. A separate EGC is not provided for the FMC. Does this installation violate the requirements of the NEC®?

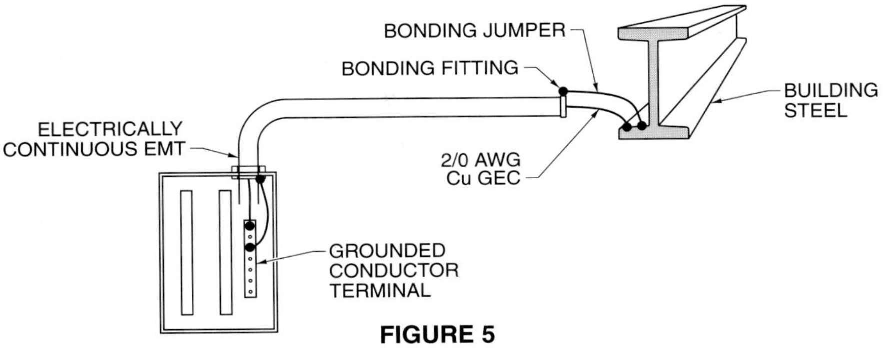
BONDING JUMPER
BONDING FITTING
ELECTRICALLY CONTINUOUS EMT
2/0 AWG Cu GEC
BUILDING STEEL
GROUNDED CONDUCTOR TERMINAL

FIGURE 5

Learning the Code

- Article 300 is another important article for installers of electrical systems. Like Article 100, it is helpful to think of this article as a "bread and butter" article. Article 300 contains important provisions that are referenced throughout the Code.
- Apprentices and students of the Code should become familiar with Table 300.5, which lists the minimum cover requirements for underground conductors, cables, and raceways. As with any Code table, be certain to review the applicable notes included with the table. In this case, Note 1 defines what is meant by the term "cover."

- Section 300.11 is an especially important section of Article 300. This section covers the provisions for securing and supporting raceways and cable assemblies in both fire-rated and non-fire-rated ceiling or roof assemblies. These requirements have received a lot of attention over the past three Code cycles. The general trend that has emerged is that an independent means of support, other than those used to support a ceiling assembly, shall be used to support raceways and cable assemblies. Additionally, there are identification requirements to ensure that the electrical inspector or Code official can clearly distinguish the support wires.
- Another important section in Article 300 is 300.22. This section covers the requirements for wiring in fabricated ducts for environmental air, plenums, and other air-handling spaces. The challenge in correctly applying these requirements is to understand the various classifications of these locations. Section 300.22(B) is for wiring directly within a duct or plenum specifically fabricated for environmental air.

  A plenum is a compartment or chamber to which one or more air ducts are connected in forming a part of the air distribution system. Environmental air is the conditioned or fresh air taken into and distributed through a building. Section 300.22(C) covers the requirements for other spaces. For example, a ceiling space above a drop ceiling that is used for return air in an air distribution system is an "other space." The key here is to understand that the specific wiring method that is selected must match the defined space. The wiring methods are different and are not interchangeable.

chapter 7

Wiring Methods

Article 300 covers wiring methods for all installations of 600 V, nominal, or less, AC and DC. All wiring methods shall be used within their voltage and temperature limitations and shall be suitable for their location. Wiring methods shall provide protection from physical damage and support per Article 300.

New in the 2011 NEC®

- *Clarification of means of measuring "lowest point" for under roof decking – 300.4(E)*
- *New Exceptions added to requirements for underground cables under buildings – 300.5(C)*
- *Revised Exception No. 2 for conductors of the same circuit – 300.5(I) Ex. 1*
- *Clarification of the provisions for spread of fire or products of combustion – 300.21*
- *Revisions and reorganization of rules for wiring in ducts not used for air handling – 300.22*

WIRING METHODS – ARTICLE 300

Article 300 covers requirements for all wiring installations unless modified by other articles. Article 300 is concerned with the following aspects of the electrical system:

- Protecting the electrical conductors from physical damage, water damage, and corrosion, and when they are installed in underground installations
- Securing and supporting raceways, cable assemblies, boxes, and cabinets
- Supporting conductors in vertical raceways
- Preventing the heating effect of inductive currents in metallic parts of an electrical system
- Securing the integrity of fire-resistant-rated walls
- Preventing the spread of toxic fumes in an air handling system in the event of a fire to the electrical system

Article 300 does not apply to wiring that is part of equipment, such as motors, motor control centers, or other factory-assembled control equipment.

Conductors of Same Circuit – 300.3(B)

Unless specifically permitted by another code provision, all conductors of the same circuit, including the grounded conductor and the EGCs, shall be run together in the same raceway, cable tray, auxiliary gutter, cablebus assembly, trench, cable, or cord. However, 300.3(B)(2) permits EGCs and EBJs to be run as a single conductor on the exterior of the raceway under certain conditions. **See Figure 7-1.**

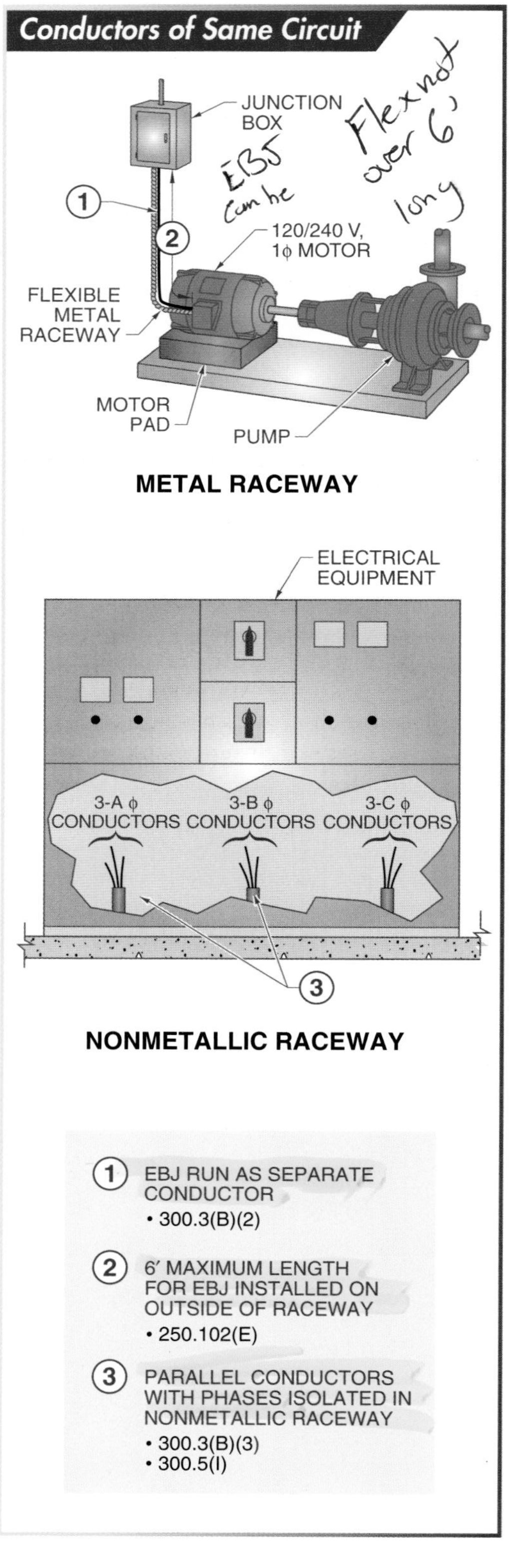

Figure 7-1. The equipment bonding jumper (EBJ) may be run as a separate conductor on the exterior of the raceway.

> **Article 300**
>
> *Article 300 deals with wiring methods and materials. Part I covers general requirements for 600 V, nominal, or less. Part II covers requirements for over 600 V, nominal. It is important to remember that requirements in the under 600 V part cannot be applied to code provisions in the over 600 V part. The two parts must be applied only to their respective applications.*

Another exception to the general rule, found in 300.3(B)(3), is when the conductors are run in parallel and the phases are isolated in a nonmetallic raceway per 300.5(I) and 300.20(B). There is no inductive heating within these raceways because they are nonmetallic. The raceways shall not enter a metallic electrical enclosure individually. This causes inductive heating of the metal where the individual phase conductors penetrate the enclosures.

Conductors of Different Systems – 300.3(C)

Conductors of different systems (600 V or less, AC and DC) may occupy the same raceway, wiring enclosure, or equipment. All conductors shall have an insulation rating equal to the highest circuit voltage applied to any conductor within that raceway or wiring enclosure. For example, 300.3(C)(I) permits the use of power and control wires from two different systems to be installed in a single raceway to supply a motor load. **See Figure 7-2.**

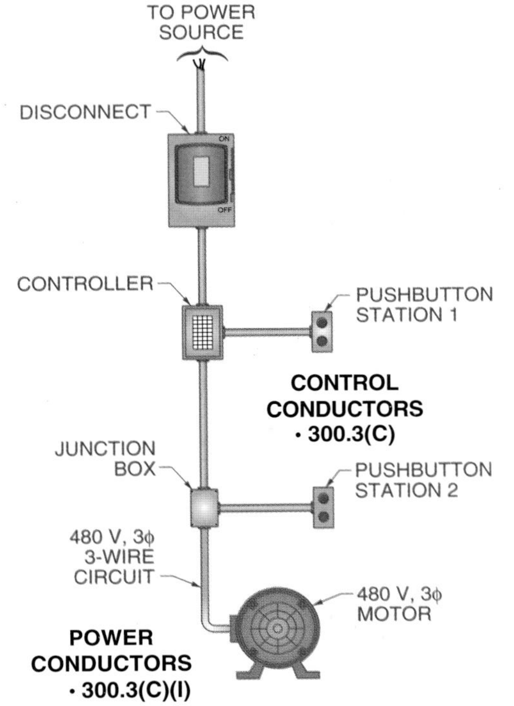

Figure 7-2. Conductors of different systems (600 V or less, AC and DC) may occupy the same raceway.

Protection from Physical Damage – 300.4

Where conductor raceways and cables are subject to physical damage, they shall be adequately protected. The conductors inside raceways and cable assemblies require protection against the penetration of nails and screws. Holes that are drilled into framing members to permit the installation of raceways or cables shall be at least 1¼″ from the nearest edge of the wood member. This ensures that nails or screws do not penetrate installed raceways or cables. **See Figure 7-3.**

Cables and raceways shall not be laid on a suspended ceiling.

Bored Holes and Notches in Wood – 300.4(A). When a cable or raceway-type wiring method is installed through a bored hole of a wooden framing member, it shall be at least 1¼″ from the nearest edge of the joist or rafter per 300.4(A)(1). If the cable or raceway is less than 1¼″ from the edge, a steel plate or bushing at least 1/16″ thick shall be used to protect the cable or raceway from being damaged by nails or screws per 300.4(A)(1).

The steel plate shall be wide enough and long enough to cover the wiring area. Plates less than 1/16″ thickness are permitted if they are listed and marked and provide equivalent protection against nail or screw penetrations. Steel plates or bushings are not required for installations containing IMC, RMC, PVC, or EMT. See Ex. 1 and 2.

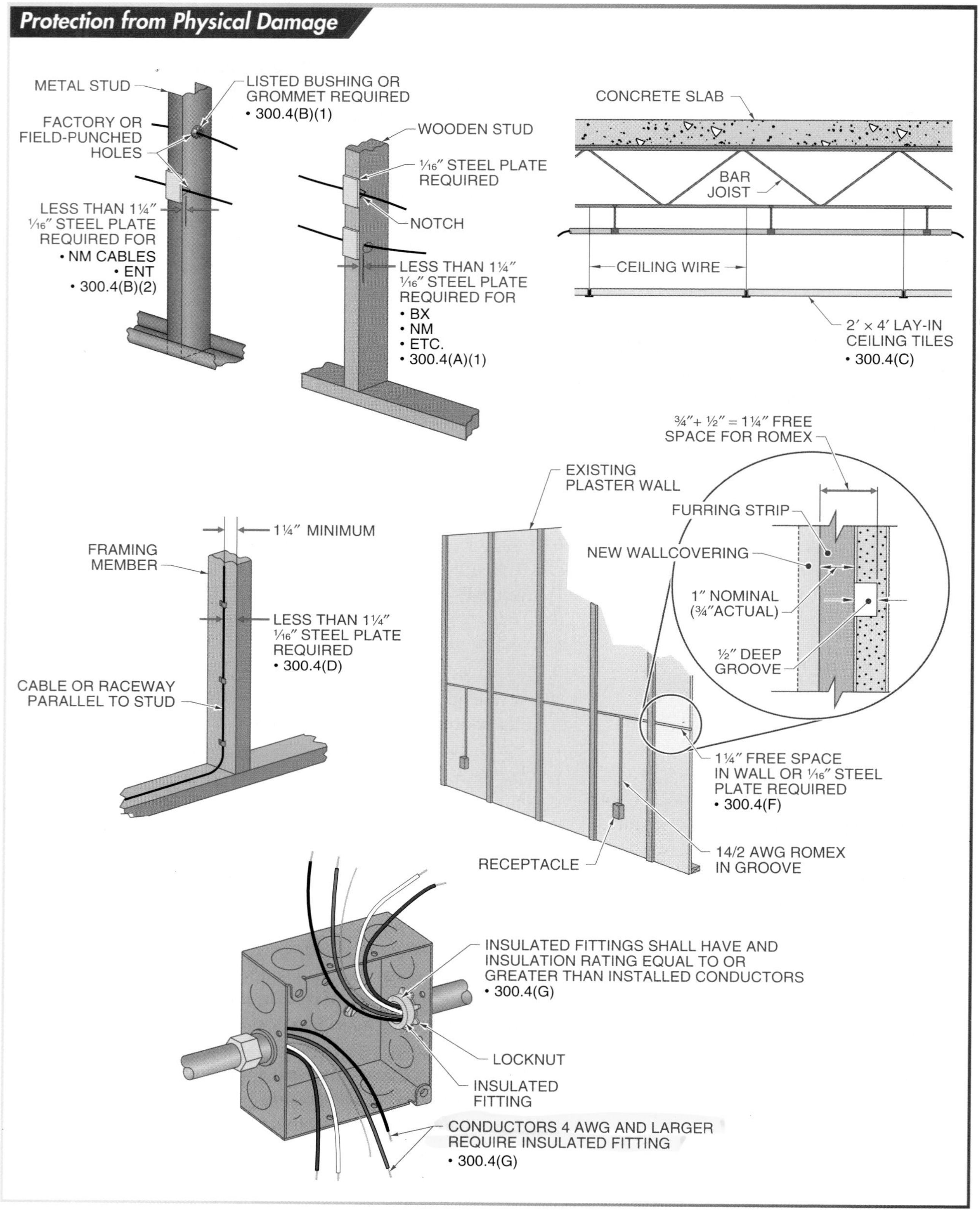
Protection from Physical Damage
METAL STUD
FACTORY OR FIELD-PUNCHED HOLES
LISTED BUSHING OR GROMMET REQUIRED
• 300.4(B)(1)
LESS THAN 1¼″ 1/16″ STEEL PLATE REQUIRED FOR
• NM CABLES
• ENT
• 300.4(B)(2)
WOODEN STUD
1/16″ STEEL PLATE REQUIRED
NOTCH
LESS THAN 1¼″ 1/16″ STEEL PLATE REQUIRED FOR
• BX
• NM
• ETC.
• 300.4(A)(1)
CONCRETE SLAB
BAR JOIST
CEILING WIRE
2′ × 4′ LAY-IN CEILING TILES
• 300.4(C)
¾″+ ½″ = 1¼″ FREE SPACE FOR ROMEX
EXISTING PLASTER WALL
FURRING STRIP
NEW WALLCOVERING
1″ NOMINAL (¾″ACTUAL)
½″ DEEP GROOVE
1¼″ MINIMUM
FRAMING MEMBER
LESS THAN 1¼″ 1/16″ STEEL PLATE REQUIRED
• 300.4(D)
CABLE OR RACEWAY PARALLEL TO STUD
1¼″ FREE SPACE IN WALL OR 1/16″ STEEL PLATE REQUIRED
• 300.4(F)
RECEPTACLE
14/2 AWG ROMEX IN GROOVE
INSULATED FITTINGS SHALL HAVE AND INSULATION RATING EQUAL TO OR GREATER THAN INSTALLED CONDUCTORS
• 300.4(G)
LOCKNUT
INSULATED FITTING
CONDUCTORS 4 AWG AND LARGER REQUIRE INSULATED FITTING
• 300.4(G)

Figure 7-3. Conductors subject to physical damage shall be protected.

Section 300.4(A)(2) permits cables or raceways to be laid in notches in wooden studs, joists, rafters, or other wooden members provided the conductors are protected against nails or screws by a steel plate at least 1⁄16″ thick. Plates less than 1⁄16″ thickness are permitted if they are listed and marked and provide equivalent protection against nail or screw penetrations. See exception. The notch cannot weaken the building structure. Steel plates or bushings are not required for installations containing IMC, RMC, PVC, or EMT.

Cables and Nonmetallic Tubing Through Metal Framing Members – 300.4(B). Where nonmetallic-sheathed cables pass through cut or drilled slots or holes in metal members, the cable shall be protected by listed bushings or listed grommets covering all metal edges and securely fastened in the opening prior to installation of the cable per 300.4(B)(1). This applies to factory and field slots and holes.

A steel sleeve, steel plate, or steel clip shall be used to protect nonmetallic-sheathed cables or ENT cable or tubing where nails or screws are likely to penetrate per 300.4(B)(2). The steel plate shall be at least 1⁄16″ thick. Plates less than 1⁄16″ thickness are permitted if they are listed and marked and provide equivalent protection against nail or screw penetrations. See exception.

Cables Through Spaces Behind Panels Designed to Allow Access – 300.4(C). Cable or raceway-type wiring methods, installed above suspended ceiling panels, shall not be laid on the suspended ceiling. The wiring methods shall be supported per their applicable articles.

Cables and Raceways Parallel to Framing Members – 300.4(D). Where cables and raceways are installed parallel to framing members such as joists, rafters, furring strips, or studs, there shall be 1¼″ minimum clearance between the outside surface of the raceway and the nearest edge of the framing member. If this distance is less than 1¼″, a 1⁄16″ steel plate shall be used to protect the raceway from nails, screws, etc. Plates less than 1⁄16″ thickness are permitted if they are listed and marked and provide equivalent protection against nail or screw penetrations. See Exception. This applies to exposed and concealed locations.

Roof Decking – 300.4(E). Where certain raceways and cables are installed under metal-corrugated sheet roof decking there is potential for physical damage to the cables and raceways. Section 300.4(E) requires in either exposed or concealed locations that adequate protection be provided, with a minimum spacing of 1½″ from the nearest surface of the roof decking to any cable or raceway. For the purposes of this provision, the 1½″ is to be measured from the lowest surface of the roof decking to the top of the cable, raceway, or box. IMC and RMC are not required to comply with this provision because their construction provides adequate protection against nail or screw penetrations.

Cables and Raceways in Shallow Grooves – 300.4(F). Cables and raceways installed in shallow grooves and covered by drywall, paneling, etc. shall be protected by 1¼″ minimum free space for the full length of the groove in which the cable or raceway is installed. If this distance falls below 1¼″, a 1⁄16″ steel plate shall be used to protect the raceway from nails, screws, etc.

Insulated Fittings – 300.4(G). It is important to adequately protect conductors from physical damage when they are installed in raceways. Section 300.4(G) requires that where conductors 4 AWG and larger are installed in raceways and these conductors enter junction boxes, enclosures, cabinets, etc., the conductors must be protected from physical damage. Such protection is typically provided by the installation of identified fiber bushings or a fitting that provides a smooth surface for the conductors to rest upon. Such fittings are not necessary where the installation of the conductors is such that conductors are separated from the fitting or raceway by insulating material that is securely fastened in place.

Where used, the conduit bushing or fittings must be constructed of an insulating material that has a temperature rating equal to that of the installed conductors. Note that the insulating fitting is not permitted to be the only means that is employed to secure the fitting or the raceway.

Structural Joints – 300.4(H). Many structural building designs call for the use of expansion joints because of contraction or deflection of the structural components of the building. To protect raceways that are installed in these locations where the raceway crosses a structural joint, a listed expansion/deflection fitting must be used. In the event that such a fitting is not available, other approved means may be used.

Underground Installations – 300.5

The general requirements for installing cables, conduits, and other raceway systems underground are covered by 300.5. The main concerns are for the protection of the conductors, splices, and taps and the prevention of moisture from coming in contact with energized parts of the electrical system. The burial depths depend on the location of the installation and the wiring method used.

Minimum Burial Depths of Cables and Raceways. Table 300.5 establishes the minimum cover requirements that cables and raceways shall have when installed in the ground. *Cover* is the shortest distance measured between a point on the top surface of any direct-buried conductor, cable, conduit, or other raceway and the top surface of a finished grade, concrete, or similar cover. For example, the minimum cover for a direct-buried cable in a location not specified in Table 300.5 is 24″ from the top surface of the wiring method to the top surface of the finished grade, concrete, or similar cover. The minimum cover for RMC under a one-family dwelling driveway is 18″ as listed in Table 300.5. **See Figure 7-4.**

EMT is not shown in Table 300.5. However, this does not mean that EMT cannot be buried in earth or concrete. Section 358.10(B) permits EMT to be installed in concrete and in direct contact with the earth where protected by corrosion protection and approved as suitable for the condition. It shall not be installed in cinder concrete or cinder fill, unless protected on all sides by a layer of noncinder concrete at least 2″ thick or unless the tubing is at least 18″ under the fill per 358.12(3).

Cables or raceways buried in a trench below a 2″ concrete slab or equivalent shall maintain the required burial depths shown in Table 300.5. Cables or conductors that are installed under a building have no minimum burial depth provided the cables and conductors are installed in an approved raceway.

Cables or raceways installed under a 4″ minimum thick concrete exterior slab without vehicular traffic shall meet the requirements shown in Table 300.5. The concrete slab shall extend at least 6″ beyond the cables or raceways. All cables and raceways installed under streets, highways, roads, driveways, and parking lots shall be buried a minimum depth of 24″.

Cables and raceways installed under one- and two-family dwelling unit driveways and outdoor parking areas shall be buried a minimum of 18″. These driveways and parking areas are to be used for dwelling-related purposes only. Underground, residential branch circuits that are rated 20 A, 1ϕ, 120 V or less and are GFCI-protected have a minimum burial depth of 12″.

Cables and raceways installed in or under airport runways, including adjacent areas where trespassing is prohibited, shall be buried to a minimum depth of 18″. Running conduit on the surface of the ground is not prohibited as long as the conduit is securely fastened in place and not exposed to physical damage such as vehicular traffic. Depth requirements may be reduced when cables or conductors rise for termination or splices or when other access is required per Table 300.5, Note 3.

Raceways shall maintain the required burial depth per Table 300.5.

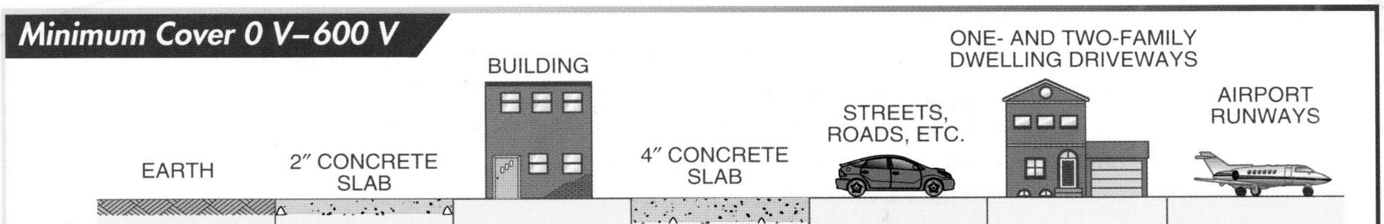

	EARTH	2″ CONCRETE SLAB	BUILDING	4″ CONCRETE SLAB	STREETS, ROADS, ETC.	ONE- AND TWO-FAMILY DWELLING DRIVEWAYS	AIRPORT RUNWAYS
1	24″	18″	0″	18″	24″	18″	18″
2	6″	6″	0″	4″	24″	18″	18″
3	18″	12″	0″	4″	24″	18″	18″
4	12″	6″	0″	6″†‡	24″	12″	18″
5	6″	6″	0″*	6″†‡	24″	18″	18″

* in raceway only
† direct burial
‡ in raceway

MINIMUM COVER REQUIREMENTS—TABLE 300.5

1. DIRECT-BURIAL CABLES OR CONDUCTORS
2. RMC OR IMC
3. NONMETALLIC RACEWAYS LISTED FOR BURIAL WITHOUT CONCRETE ENCASEMENT AND EMT OR OTHER APPROVED RACEWAYS
4. RESIDENTIAL BRANCH CIRCUITS 120 V OR LESS WITH GFCI PROTECTION AND MAXIMUM OVERCURRENT OF 20 A
5. IRRIGATION CONTROL LANDSCAPE LIGHTING 30 V OR LESS USING UF OR OTHER IDENTIFIED CABLE IN RACEWAY

Figure 7-4. All cables and raceways approved for direct burial shall meet minimum cover requirements.

In general, any underground cable installed beneath a building is required to be installed in a raceway per 300.5(C). There are two exceptions to this general rule. The first permits Type MI cable to be installed under a building without being installed in a raceway provided the cable is embedded in concrete, fill, or other masonry and meets the requirements of 332.10(6) and 332.10(10). The second exception permits Type MC cable that is listed for direct burial or that is encased in concrete to be installed beneath a building without a raceway. This installation must also comply with the provisions of 330.10(A)(5) and 330.10(11). See 300.5(C) Ex. 1 and 2.

Protection from Damage – 300.5(D). All cables and conductors emerging from the ground shall be protected from physical damage to a maximum depth of 18″ below finished grade. Cables and conductors installed on the side of a pole or building shall be provided with physical protection up to a height of 8′ above finished grade. This protection may be provided by using RMC, IMC, Schedule 80 PVC, or equivalent as determined by the AHJ. Per 300.5(D)(3), service laterals buried 18″ or more, which are not encased in concrete, shall be identified with a warning ribbon placed in the trench at least 12″ above the installation. **See Figure 7-5.**

Buried Splices and Taps – 300.5(E). Splices and taps are permitted to be buried in a trench without the use of a splice box, provided splicing devices identified for the use are used, with listed materials per 110.14(B).

Backfill – 300.5(F). Care must be taken so as not to damage cables and raceways when backfilling trenches. Heavy rocks or any sharp or corrosive materials shall not be used in the backfill material. Where necessary, running boards, sleeves, or some form of granular material shall be used to prevent damage to the cable or conduit.

Raceway Seals – 300.5(G). Seals shall be used on either end of conduits where moisture may come in contact with energized live parts. An Informatinal Note after the section states that seals may also be required where hazardous gases or vapors are present.

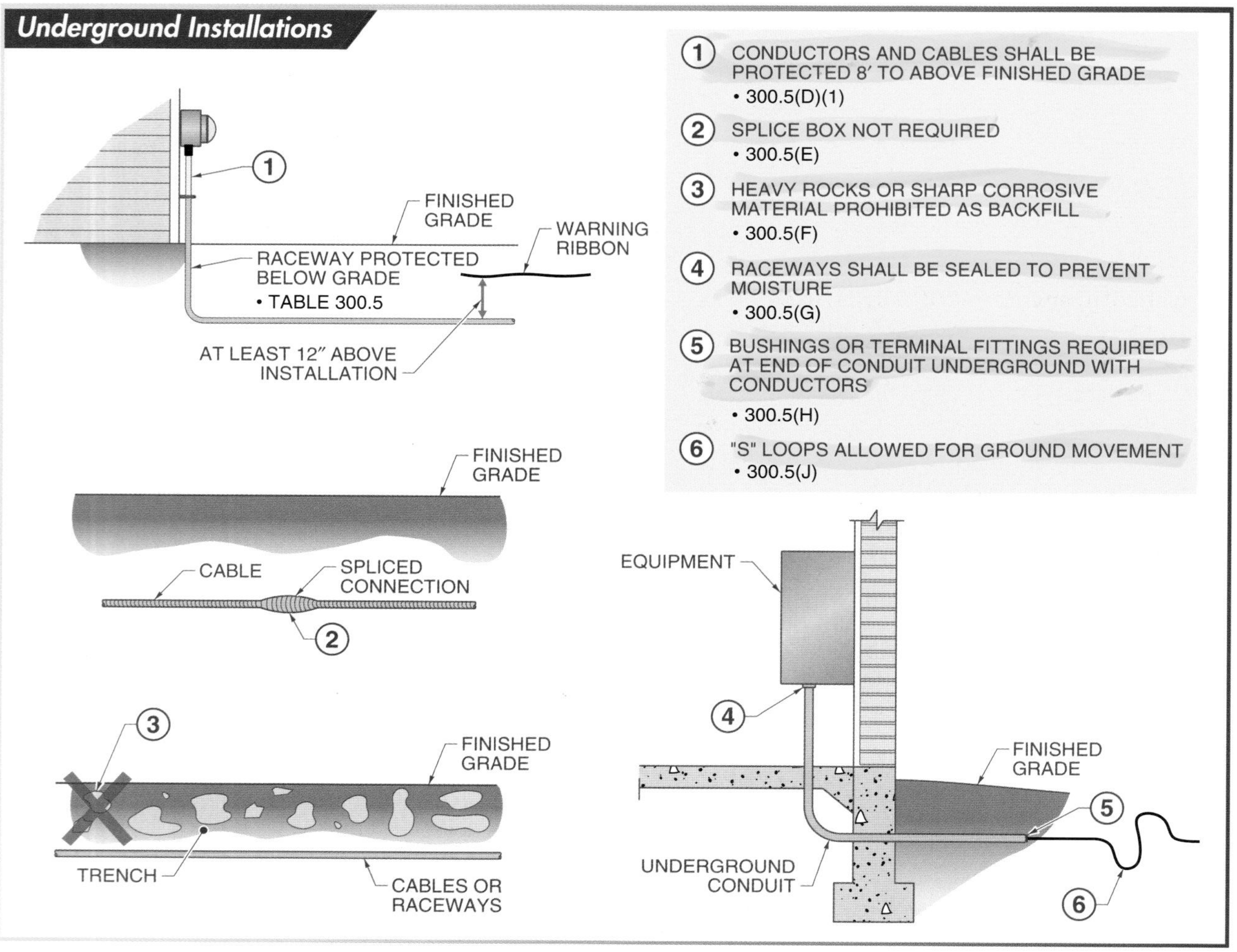

Figure 7-5. Underground installation of direct-buried cables or raceways shall be protected.

Underground Bushings – 300.5(H). A bushing or terminal fitting shall be used on the end of a conduit where direct-burial cables leave the conduit. A seal may be used provided it offers the same protection as a bushing.

Conductors – 300.5(I). As a general rule, all circuit conductors of the same circuit, including the grounded circuit conductor and all equipment grounding conductors when used, must be installed in the same raceway or cable. In the event that direct-buried cables are installed in a trench, all of the circuit conductors must be run together in the trench as well. This provision applies to both multiconductor underground cables and to single underground conductor cables.

Metal conduit installed underground must have protection from corrosion.

There are two exceptions to this general rule. The first applies to conductors installed in parallel raceways, multiconductor cables, and direct-buried single conductor cables. The second applies to isolated phase, polarity, grounded conductor, and equipment grounding and bonding conductor installations in nonmetallic raceways or cables. See 300.5(I) Ex. 1 and 2.

Ground Movement – 300.5(J). Direct-buried cables, raceways, and conductors that are subject to ground movement shall be arranged to prevent damage to the equipment to which these cables, raceways, and conductors are connected. The ground movement may be caused by settlement or frost heaves. Section 300.5(J), FPN recognizes "S" type loops in underground direct-burial cable to raceway transitions and expansion joints in raceway risers to fixed equipment installations.

Directional Boring – 300.5(K). In the 2002 Code a section was added to address provisions for the installation of cables and raceways with directional boring equipment. Due to the speed at which these machines operate, it is impossible to use standard installation practices, such as glues and adhesives for PVC-type raceways. The Code requires any cable or raceway installed with directional boring equipment to be approved for the purpose. Listed or identified cables and raceways was one means the electrical inspector could use to approve the installation. See Article 100 definition of "approved."

Protection Against Corrosion – 300.6

All of the metallic components that make up an electrical system shall be of materials suitable for the environment in which they are to be installed. This includes the metal raceways, cable armor, boxes, cable sheathing, cabinets, elbows, couplings, fittings, supports, and support hardware. Boxes or cabinets marked "raintight," "rainproof," or "outdoor type" can be installed outdoors.

Nonmetallic Equipment – 300.6(C). Although most of the electrical products and equipment manufactured today are designed for installation in many types of corrosive environments, 300.6(C) contains two special provisions for nonmetallic equipment protection. The first is for exposure to sunlight. Because of the deleterious effect that sunlight may have on the products, any nonmetallic equipment, where exposed to sunlight, must be listed as being sunlight resistant or shall be identified as being sunlight resistant. This ensures that the product has been tested to withstand, to a specified degree, the effects of the sunlight exposure.

Fittings, connectors, conduit, and boxes must only be used in applications for which they are designed.

The second concern is for chemical exposure. If nonmetallic equipment or products may be exposed to chemical vapors, solvents, or chemicals, 300.6(C)(2) requires the product to be listed or identified as resistant to the particular chemical exposure.

Indoor Wet Locations – 300.6(D). In dairies, breweries, canneries, and other locations where the walls are frequently washed down, the electrical system shall not entrap water or moisture between system parts and the surface to which they are mounted. This is true for the entire electrical system including boxes, fittings, and conduits. There shall be at least ¼″ airspace between the electrical system and the wall or the supporting surface. **See Figure 7-6.**

Raceways Exposed to Different Temperatures – 300.7. Installed raceways that are exposed to different temperatures present two problems. **See Figure 7-7.** The first is that condensation may build up inside the raceway system. The second is the expansion and contraction of the raceway system.

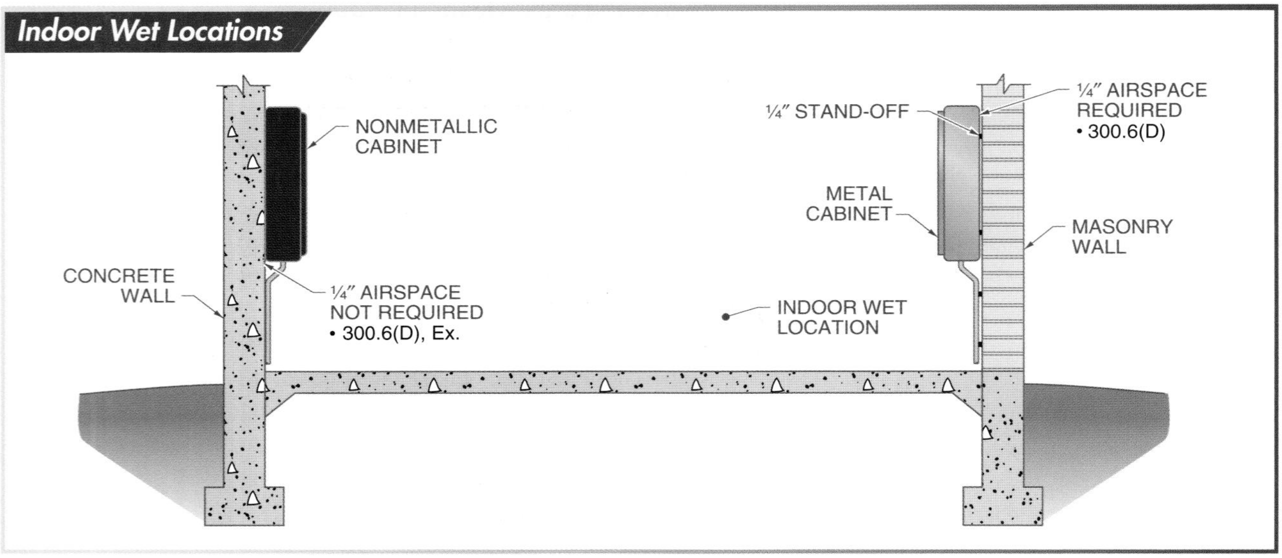

Figure 7-6. A ¼″ airspace is required in wet locations so that there is no entrapped water or moisture between the electrical system and the surface.

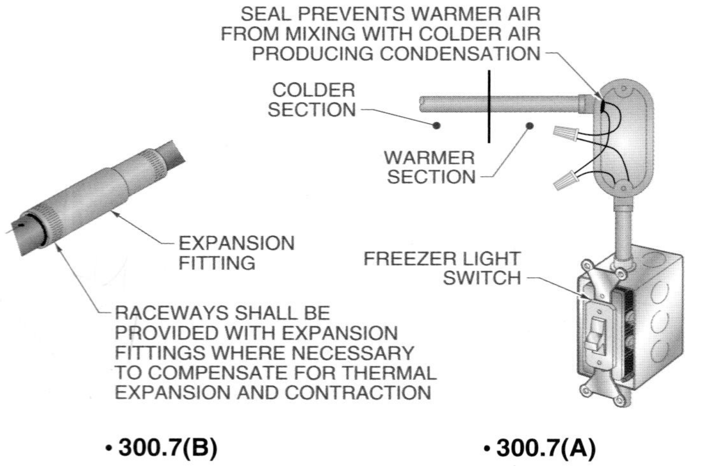

Figure 7-7. Installed raceways, portions of raceways, or sleeves that are exposed to different temperatures shall avoid condensation and allow for expansion and contraction.

Section 300.7(A) requires seals to be installed in the raceway system whenever the raceway changes from one environmental temperature to another such as where passing from the interior to the exterior of the building. Air traveling through a raceway and going from a warm environment into a cooler environment condenses in the raceway as it travels through the cooler environment. Accumulation of this condensed moisture could travel into equipment that contains live parts. The required seal shall be installed where the raceway changes from one environment to the other.

Section 300.7(B) requires expansion joints in the raceway system to compensate for thermal expansion and contraction. Expansion joints allow for the movement of the conduit within the expansion fitting. This relieves any strain on other connections within the conduit system.

Electrical Continuity of Metal Raceways and Enclosures – 300.10

All noncurrent-carrying metallic parts of an electrical system shall be bonded together with a permanent and continuous bonding method. The metal raceways, cable armor, and other metallic enclosures shall be electrically connected to provide a continuous low-impedance path. This low-impedance path is critical in allowing enough fault current to flow to facilitate the opening of the overcurrent protective device.

If the raceway system were nonmetallic, all noncurrent-carrying metal parts of that system would have to be bonded to the EGC. If a noncurrent-carrying metal part were to come in contact with a live conductor, this low-impedance path for fault current would have to be present to allow the overcurrent device to trip or open. **See Figure 7-8.**

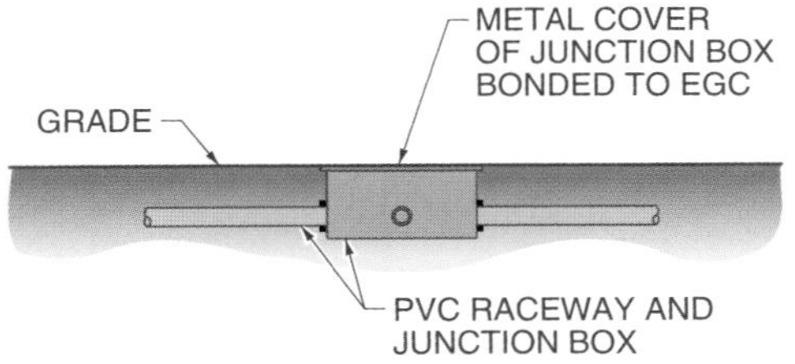

Figure 7-8. All noncurrent-carrying metallic parts of an electrical system shall be bonded together with a permanent and continuous bonding method.

Securing and Supporting Raceways, Cable Assemblies, Boxes, and Cabinets – 300.11(A)

Wiring systems installed within the cavity of a fire-rated floor-ceiling or roof-ceiling assembly shall have an identifiable independent means of being secured and supported. **See Figure 7-9.** These wiring systems shall not be secured or supported by the ceiling assembly or the ceiling support wires. The ceiling assembly may support the wiring system if the ceiling assembly and support of the wiring system were both tested together as part of the fire-rated assembly.

Wiring located within non-fire-rated assemblies also is not permitted to be secured to the ceiling or ceiling support wires. An independent support means must be provided. Where independent support wires are used in both fire-rated and non-fire-rated assemblies, they shall be distinguishable by color tagging or other effective means.

Raceways Used as Means of Support – 300.11(B). Raceways shall not support other raceways, cables, or nonelectric equipment unless used in one of the following conditions:

(1) Where the raceway or means of support is identified for the purpose.

(2) Where the raceway contains power conductors for electrically controlled equipment it may support Class 2 cables used solely for the controlling of the circuit.

(3) Where the raceway is used to support properly installed boxes or fixtures. See 314.23 and 410.36(E).

Cable trays are used to support insulated conductors for power distribution and communication.

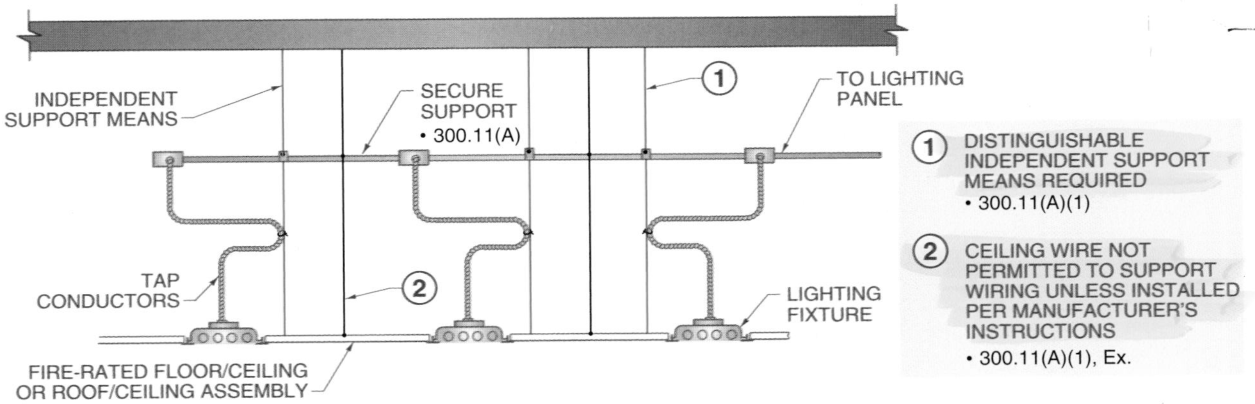

Figure 7-9. Wiring systems installed above a roof-ceiling assembly shall have an independent means of support.

Mechanical Continuity of Raceways and Cables – 300.12

Metal or nonmetallic raceways, cable armors, and cable sheaths shall be continuous between all cabinets, boxes, fittings, or other enclosures or outlets. However, per 300.12, Ex. 1, a short section of raceway used to provide support or protection of cable assemblies from physical damage does not have to be continuous. Section 300.12, Ex. 2, permits raceways and cables installed in the open bottom of equipment, such as panelboards, MCCs, and switchboards, to be noncontinuous.

Mechanical and Electrical Continuity of Conductors – 300.13(A)

Conductors in raceways shall be continuous between all outlets, boxes, devices, etc. There shall be no splice or tap within a raceway unless permitted elsewhere in the Code. Raceways that provide removable or hinged covers may have splices or taps contained within them. See the following Code references where splices or taps may be permitted within a raceway: 300.15, 368.56(A), 376.56, 378.56, 384.56, 386.56, 388.56, and 390.7.

Removing Devices in Multiwire Circuits – 300.13(B). The grounded conductor of a multiwire branch circuit shall never be opened. If the grounded conductor were opened and the loads were not equal, then the voltage across these loads would be changed to something other than 120 V. The more the unbalance, the more dramatic the voltage change. **See Figure 7-10.**

When a multiwire branch circuit is run from outlet box to outlet box, the terminal screws of the device, such as a duplex receptacle, shall not be used to make the grounded conductor continuous. All of the grounded conductors of a multiwire branch circuit within an outlet box shall be spliced together with a tail provided for the terminal screw of the device. If this device were replaced while the circuit was energized, the grounded conductor would not become open. **See Figure 7-11.**

Length of Conductors at Outlet Box – 300.14

At least 6″ of free conductor, measured from the point in the box where it emerges from its raceway or cable sheath, shall be left at each outlet, junction, or switch point. If the opening to an outlet, junction, or switch point is less than 8″ in any dimension, each conductor shall be long enough to extend at least 3″ outside the opening. This rule ensures that if extension rings are used, a minimum of 3″ of free conductor will extend past the last extension ring that may be installed. The free conductor is to be used for terminating receptacles, fixtures, or other devices. Section 300.14, Ex. does not require the 6″ of free conductor if the conductors are not spliced or terminated at this point.

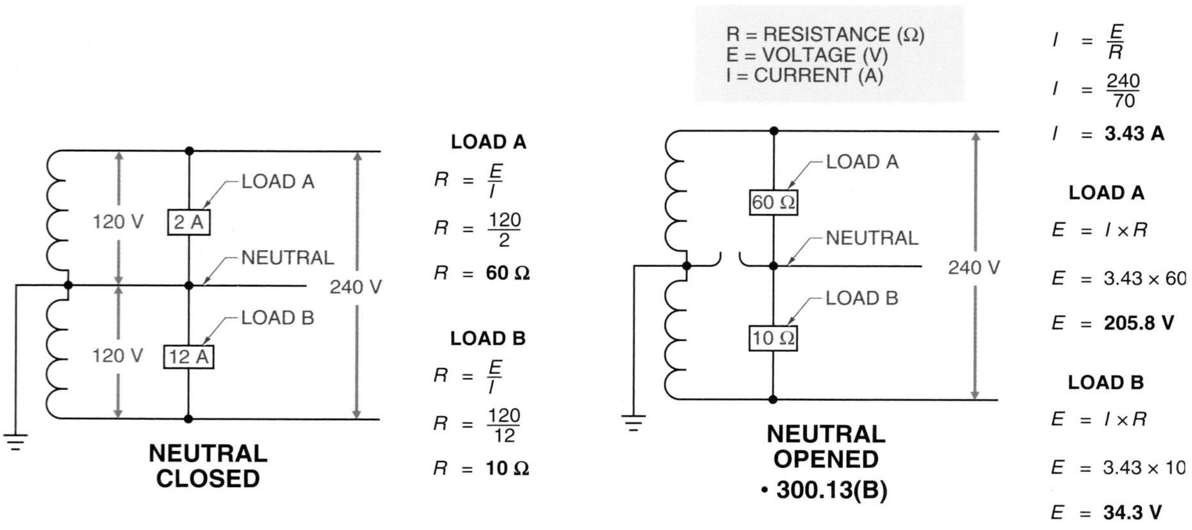

Figure 7-10. The grounded conductor of a multiwire branch circuit shall not be dependent on device connection.

Raceways exposed to different temperatures shall be provided with a means to prevent the circulation of warm air to a colder section of the raceway.

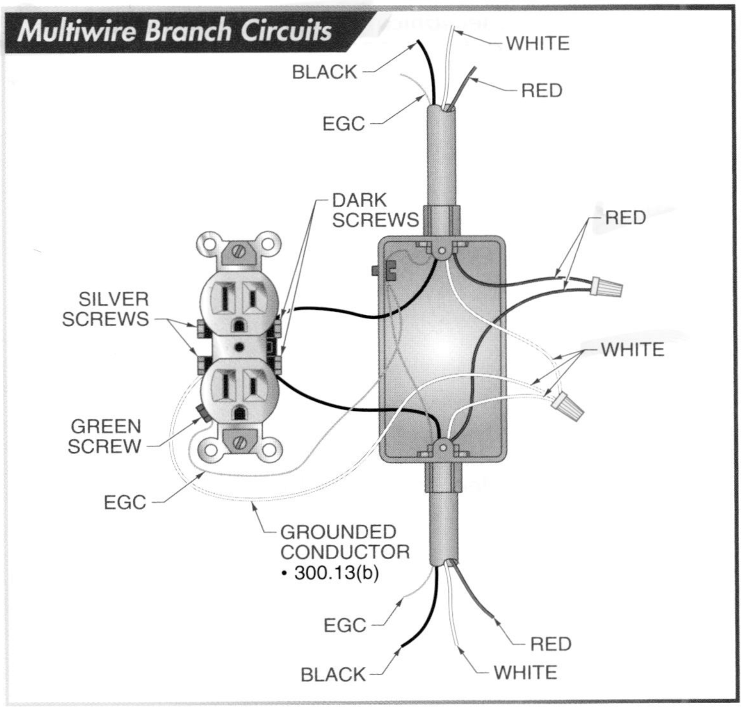

Figure 7-11. The grounded conductors shall be spliced together to prevent an open neutral and shall not rely on the device for continuity.

Conductor Installation Considerations

The small details can make the difference between successful installations and having to remove damaged conductors. In preparing for a conductor pull, it is just as important to cover the small details as it is to ensure that the conductor does not exceed maximum sidewall pressure, minimum bending radiuses, or maximum pulling tensions. General field practices are provided to aid in preparing for large and small conductor installations.

Conductors installed in conduits have installation parameters, such as maximum pulling tensions, sidewall pressure, clearance, and jamming, that must be considered. These installations also involve some general considerations, such as field handling, storage, training of ends, and junction box sizes. These and other considerations can make the difference between a good installation and one resulting in damaged conductors.

Mechanical stresses during installation are generally more severe than those encountered while in service. In order to keep stresses to a minimum, calculations should be made to assess the difficulty of the pull. When a marginal situation is encountered, the entire pull should be reviewed. This review may include more rigorous calculations or trial pulls. A final decision should be made based on installation factors known to the end user and installer. The sizes of the conduit are determined based on the calculations of the clearances, jamming, and fill. Pulling tensions may then be evaluated by determining the maximum tension based on the pulling device used and the maximum tension that can be applied to the conductors.

Additional information is available at www.southwire.com.

Boxes, Conduit Bodies, or Fittings Required – 300.15

A box, conduit body, or fitting shall be installed at every splice point, outlet, switch point, junction point, or pull point unless permitted in 300.15(A) through (L). **See Figure 7-12.** Section 300.15(A) requires a box or conduit body for the connection of conduit, EMT, AC cable, MC cable, MI cable, NM cable, and other cables.

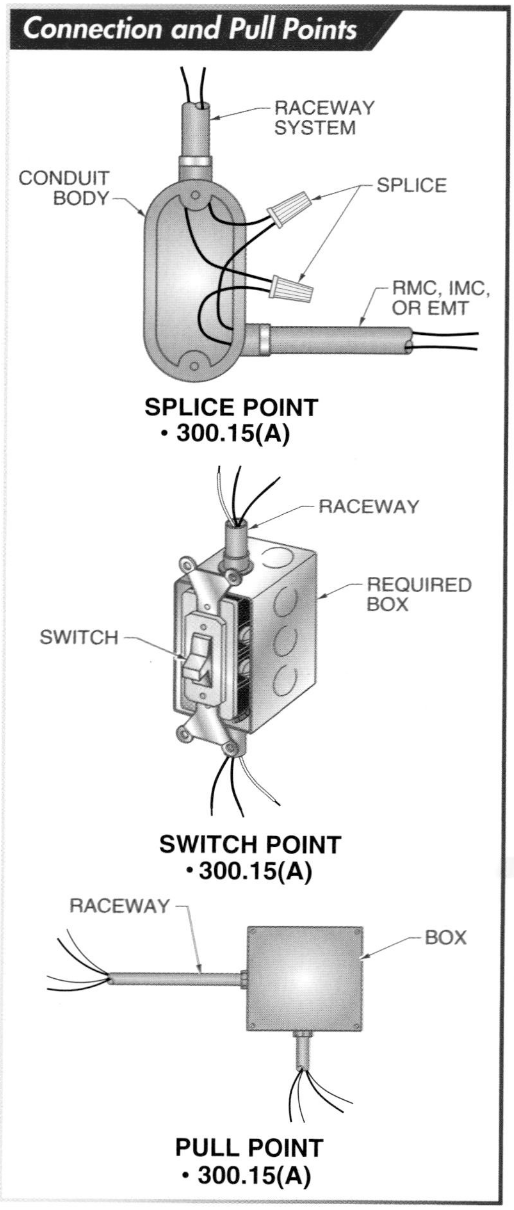

Figure 7-12. A box, conduit body, or fitting shall be installed at each connection or pull point.

Where cables enter or exit conduit or tubing used for support or protection of the cable, it is not necessary to use a box or conduit body. Section 300.15(C) allows a fitting to be used by itself in this situation. Section 300.15(F) allows identified fittings to be used in the connection of conduit and cables. The fitting shall be identified for this use and the conductors shall not be spliced at this point. **See Figure 7-13.**

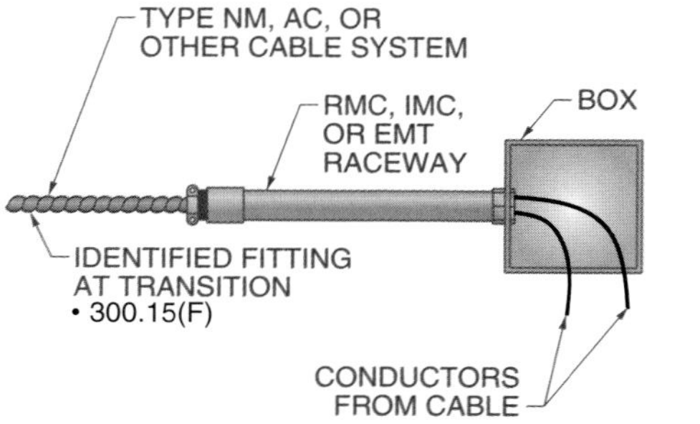

Figure 7-13. A box is not required at the transition from a cable to a raceway system.

Section 300.15(G) permits splices and taps that are listed for direct burial to be installed without the use of a box or conduit body. There is no requirement for a box based on the length of a raceway. Section 300.15(J) allows fixtures that are used as raceways in lieu of boxes and conduit bodies.

Fittings and connectors shall only be used for applications for which they are designed. For example, it is a violation of the NEC® to terminate MC cable in a connector listed for a BX cable only.

Number and Size of Conductors in a Raceway – 300.17. The number and size of conductors in any raceway shall not be more than permits the dissipation of heat given off by the conductors when they carry current. The installation and withdrawal of conductors from a raceway system should be possible without causing damage to the conductors or to their insulation.

The general rule for conduit fill as found in Chapter 9, Table 1 still applies. When a conduit contains three or more conductors and is longer than 24″, the cross-sectional area of the conductors shall not exceed 40% of the cross-sectional area of the conduit or tubing. Chapter 9, Appendix C lists the maximum number of conductors permitted in the various types of raceways based on the internal diameter of the raceway. For example, IMC has a thinner wall thickness when compared to RMC. Therefore the ID is larger for IMC as opposed to RMC.

> **Wire-Pulling Lubricants**
>
> *Wire-pulling lubricants are required for most installations. Wire-pulling lubricants are typically a gel composed of 95% or more water and are available in 1 qt, 1 gal., 5 gal., and 55 gal. containers. Wire-pulling lubricant manufacturers usually provide charts, based on the length of the pull and diameter of the raceway, to help determine the amount of lubricant required. Typically, more wire-pulling lubricant is required for installations that are performed in temperatures of 80°F or higher, have high conduit fill rates, have stiff conductor insulation, or consist of several bends. Wire-pulling lubricant can be applied as it is fed into a raceway.*

Raceway Installations – 300.18. Raceway systems, including all required boxes, shall be installed before the conductors can be pulled. The primary function of the raceway system is to provide protection for the conductors. There have been many reports of damage to conductors that were pulled into incomplete raceway and enclosure systems. Metal raceways shall not be supported, connected, or terminated through welding unless otherwise allowed in the NEC®. This section does not, however, prohibit the use of prewired raceways where otherwise permitted by the code.

SUPPORTING CONDUCTORS IN A VERTICAL RACEWAY – 300.19

In long vertical runs, conductors shall not rely on the termination points for their sole weight support. The conductors shall be supported at the top of the raceway or as close as possible to the top.

Spacing Conductor Supports – 300.19(A)

The intervals at which intermediate cable supports are required are determined by weight of the conductor. The larger the conductor, the more it weighs, thereby requiring more frequent supports. The conductor material is also a factor to consider. Copper is heavier than aluminum, therefore it requires supports at more frequent intervals. Table 300.19(A) shows the required spacing for conductor supports in vertical runs. For example, 250 kcmil Al conductors require supports every 135′, while 250 kcmil Cu conductors require supports every 60′.

Article 300 covers requirements for all wiring installation unless modified by other articles.

Support Methods – 300.19(C)

Vertical conductors may be supported by insulating wedges installed in the end of the raceway. This helps prevent damage to the insulation due to the weight of the conductors.

Another recognized method of providing support for vertical conductors is to install boxes at required intervals per Table 300.19(A). Insulating supports or cleats are installed to provide the necessary support for each conductor.

The third method of providing support for vertical conductors is to deflect the conductors by at least 90°. The conductors would have to be run horizontally a distance not less than twice the diameter of the conductors. The conductors would have to be carried on two or more insulating supports. The conductors should be secured to these supports by tie wires. 300.19(C)(4) permits alternative methods that are as effective as those included in 300.19 (C)(1–3).

PREVENTING HEATING EFFECT OF INDUCTIVE CURRENTS IN METALLIC PARTS – 300.20

AC conductors produce a fluctuating magnetic field around the conductor. If these conductors are in close proximity to ferrous metals, the fluxing magnetic field causes the molecules in the ferrous metal to move back and forth, producing a heating effect in the ferrous metal. If conductors of opposite polarities are brought together, their fluxing magnetic fields cancel one another out. Therefore conductors of opposite polarities which are run in close proximity to one another do not produce a heating effect on ferrous materials.

Conductors Grouped Together – 300.20(A)

AC conductors installed in metal raceways or enclosures shall be arranged so as not to cause induction heating of the metal raceways or enclosures. To accomplish this, all phase conductors, grounded conductors (if used), and EGCs shall be grouped together. **See Figure 7-14.** Since all of the circuit conductors are grouped together, the magnetic fields around each individual conductor cancel out. This eliminates the inductive heating effect on the ferrous metal.

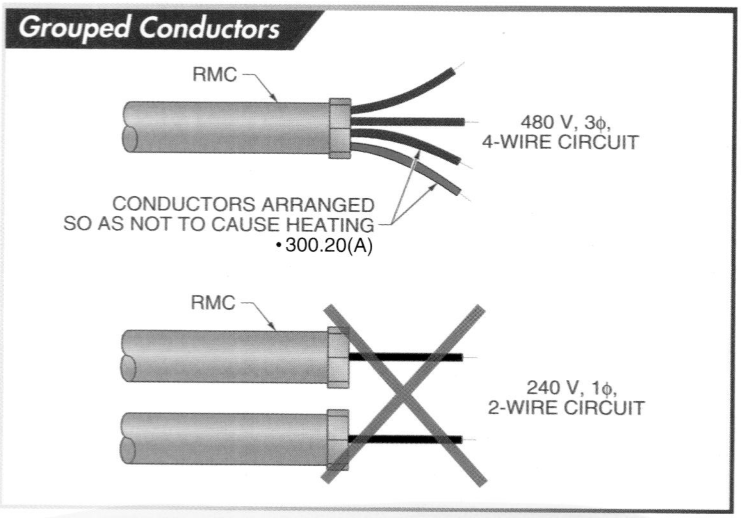

Figure 7-14. All AC phase conductors, grounded conductors, and EGCs shall be grouped together.

Single Conductors Passing Through Metal – 300.20(B)

Single conductors carrying AC current must not pass through metal with magnetic properties unless the inductive heating effect is minimized by cutting slots in the metal between the individual holes through which the individual conductors pass. **See Figure 7-15.**

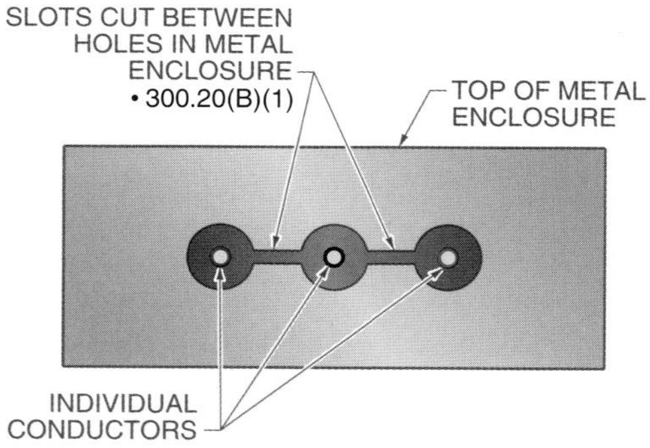

Figure 7-15. The inductive heating effect of metal is minimized by cutting slots between the holes.

If conductors carrying AC currents are run into a metal enclosure, they should all enter the enclosure through the same opening. **See Figure 7-16.**

SECURING INTEGRITY OF FIRE-RESISTANT-RATED WALLS – 300.21

The installation of electrical systems in hollow spaces, vertical shafts, and air-handling ducts shall be made without decreasing the integrity of the fire resistance rating of the installation. The concern here is the spread of fire or products of combustion. Whenever an electrical system penetrates a fire-resistant-rated wall, floor, or ceiling, the opening around the penetration shall be firestopped using approved methods to restore the fire resistance rating. **See Figure 7-17.**

PREVENTING SPREAD OF TOXIC FUMES IN AN AIR-HANDLING SYSTEM – 300.22

Electrical wiring and equipment may give off toxic fumes in the event of a fire. The requirements for the installation of electrical equipment and its associated wiring systems within ducts, plenums, and other air-handling spaces are covered in 300.22. The concern here is to limit the use of materials that could contribute smoke and products of combustion during a fire in an area that handles environmental air. The four categories of air-handling installations are ducts for dust, loose stock, or vapor removal; ducts specifically fabricated for environmental air; other space used for environmental air; and data processing or information technology systems.

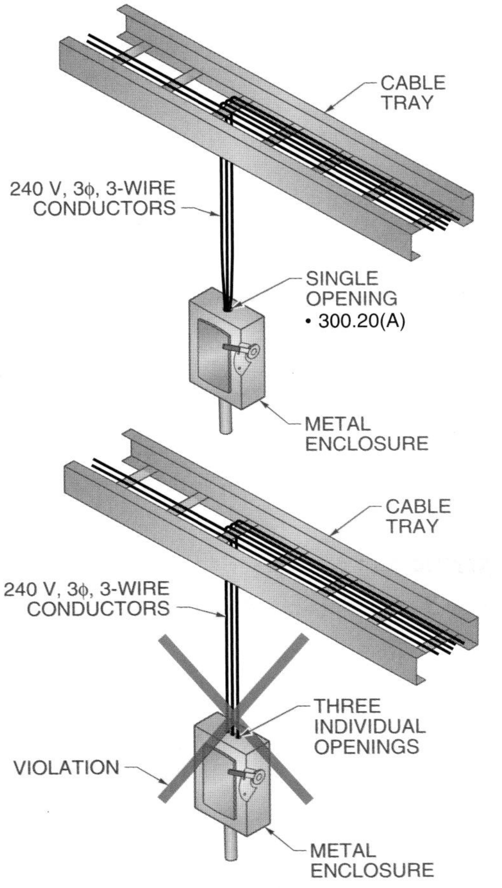

Figure 7-16. Conductors carrying AC current should enter the enclosure through the same hole.

Wiring in Ducts for Dust, Loose Stock, or Vapor Removal – 300.22(A)

Wiring systems are not permitted to be installed in ducts used to handle dust, loose stock, or flammable vapors. No wiring system shall be installed in any shaft containing ducts used for vapor removal or for ventilation of commercial-type cooking equipment.

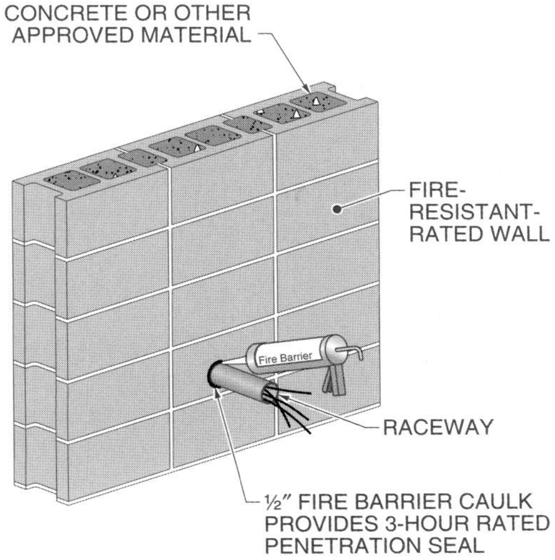

Figure 7-17. Openings in fire-resistant walls shall be firestopped using approved methods.

Wiring in Ducts Specifically Fabricated for Environmental Air – 300.22(B)

The wiring systems permitted in ducts specifically fabricated to handle the supply and return of conditioned air are:

- MI cable
- MC cable, the smooth or corrugated sheath type, not the MC cable with an interlocking outer jacket and without nonmetallic covering
- EMT
- FMT
- IMC
- RMC without overall nonmetallic covering

Flexible metal conduit is permitted in these ducts and plenums provided the length does not exceed 4′ and it is used to connect physically adjustable equipment and devices. The connectors used with this flexible conduit shall effectively close any openings in the connection. **See Figure 7-18.**

Equipment and devices shall only be permitted in these fabricated ducts if the equipment and devices are associated with the movement or sensing of the contained air. Luminaires (lighting fixtures) installed to facilitate the maintenance and repair of this equipment shall be the enclosed, gasketed type.

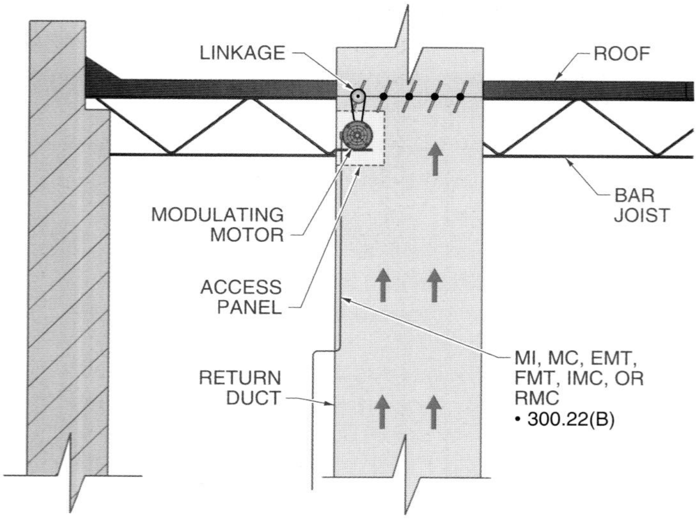

Figure 7-18. MI, MC, EMT, FMT, IMC, or RMC may be run in ducts or plenums used for environmental air.

Wiring in Other Spaces Used for Environmental Air – 300.22(C)

Plenum

"Other spaces" refers to space not specifically fabricated for environmental air-handling purposes. An example of "other spaces" is the space over a hung ceiling used for environmental air-handling purposes. **See Figure 7-19.**

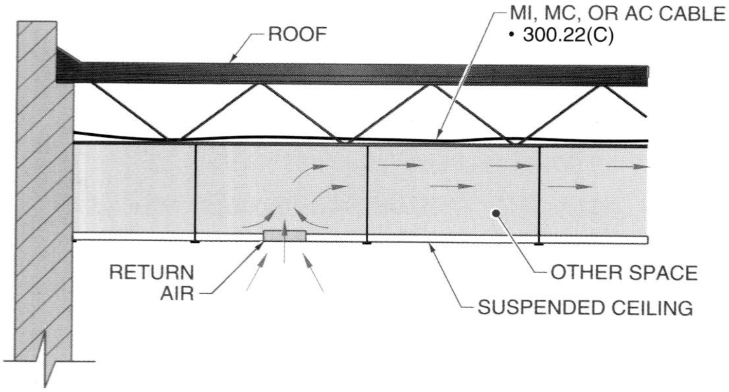

Figure 7-19. An "other space" is the space over a hung ceiling used for environmental air.

The wiring methods permitted in other spaces are:

- Totally enclosed, nonventilated, insulated busway without provisions for plug-in connections
- MI cable
- MC cable without an overall nonmetallic covering
- AC cable
- Other factory-assembled multiconductor control or power cable that is listed for the use
- Listed prefabricated cable assemblies of metallic manufactured wiring systems without nonmetallic sheath

Section 300.22(C)(1) permits the installation of factory-assembled cords and cables such as SJ, SO, and other types listed in Article 400, provided they are installed in EMT, FMT, IMC, and RMC without an overall nonmetallic covering, FMC, or where there are accessible surface metal raceway or metal wireway or solid bottom metal tray with metal covers.

The installation of other type cables and conductors shall be installed in one of the following: EMT, flexible metallic tubing, IMC, RMC, flexible metal conduit, and where accessible, surface metal raceway or metal wireway with metal covers or solid bottom metal cable tray with solid metal covers.

Section 300.22(C), Ex. permits NM cable, for example, to pass through a closed-in stud or joist space used as a return in a forced-air heating/air conditioning system. **See Figure 7-20.** This exception applies only to dwelling units.

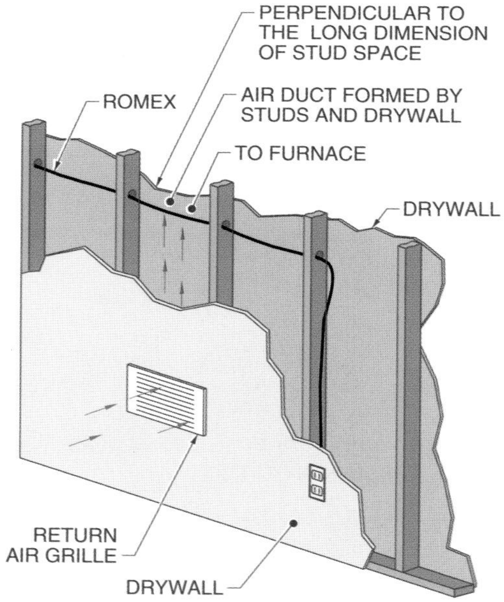

Figure 7-20. NM cable may pass through a closed-in stud or joist space used as a return in a forced-air heating/air conditioning system.

Wiring in Spaces Used for Information Technology Equipment – 300.22(D)

Wiring in air-handling areas beneath raised floors of computer/data processing equipment is permitted per Article 645, formerly entitled "Electronic Computer/Data Processing Equipment," and now named "Information Technology Equipment." For a room to fall into this category, it shall comply with all six conditions required by 645.4. **See Figure 7-21.** Only when these six conditions are met is a room considered a dedicated Information Technology Equipment room. Section 300.22(D) allows wiring methods installed in air-handling spaces of information technology equipment rooms to be installed per Article 645. Section 645.5(D) imposes six conditions in allowing information technology equipment under a raised floor.

Supply circuits and interconnecting cables under raised floors in information technology equipment rooms are permitted provided six conditions are met. Per 645.5(E)(1), the raised floor must be of approved construction to support the computer equipment which will be placed on top of it. These floor assemblies are available with UL listings. They come with vertical floor stands that are mounted on a heavy-duty base plate which sits on top of the subfloor. On the top of these floor stands is a platform which accepts the 2′ × 2′ fire-resistant floor panels. These floor stands are adjustable in height, enabling the installer to guarantee a level floor. For more information see NFPA 75-2009 (ANSI).

Wiring for information technology equipment beneath raised floors is permitted per Article 645.

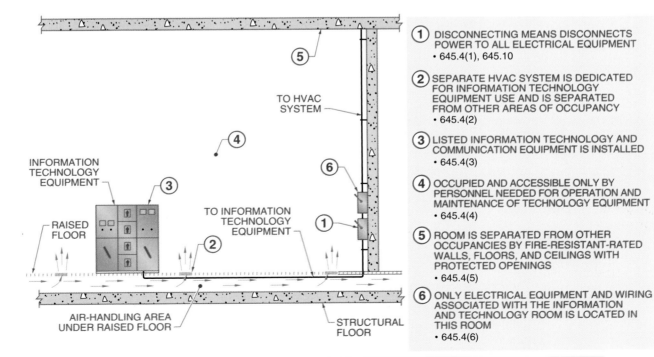

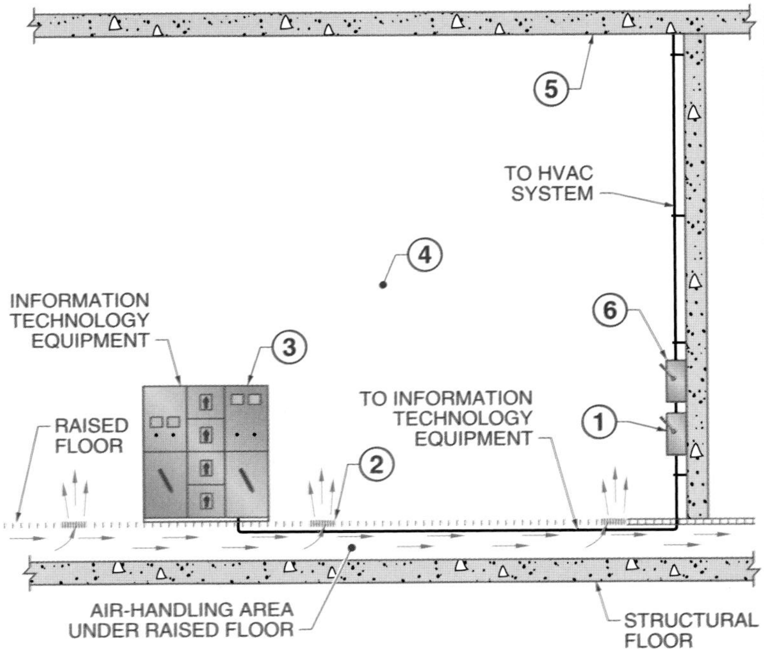

① DISCONNECTING MEANS DISCONNECTS POWER TO ALL ELECTRICAL EQUIPMENT
• 645.4(1), 645.10

② SEPARATE HVAC SYSTEM IS DEDICATED FOR INFORMATION TECHNOLOGY EQUIPMENT USE AND IS SEPARATED FROM OTHER AREAS OF OCCUPANCY
• 645.4(2)

③ LISTED INFORMATION TECHNOLOGY AND COMMUNICATION EQUIPMENT IS INSTALLED
• 645.4(3)

④ OCCUPIED AND ACCESSIBLE ONLY BY PERSONNEL NEEDED FOR OPERATION AND MAINTENANCE OF TECHNOLOGY EQUIPMENT
• 645.4(4)

⑤ ROOM IS SEPARATED FROM OTHER OCCUPANCIES BY FIRE-RESISTANT-RATED WALLS, FLOORS, AND CEILINGS WITH PROTECTED OPENINGS
• 645.4(5)

⑥ ONLY ELECTRICAL EQUIPMENT AND WIRING ASSOCIATED WITH THE INFORMATION AND TECHNOLOGY ROOM IS LOCATED IN THIS ROOM
• 645.4(6)

Figure 7-21. Wiring is permitted in air-handling areas beneath raised floors of information technology equipment.

In addition to the raised floor meeting structural standards, the area under the floor shall be accessible. Most of these floors are constructed by using 2′ × 2′ panel sections which lie in a grid structure. These panels may be removed with a plunger suction device for access to the wiring system. Per 645.5(E)(2), the following wiring methods may be used to supply the branch-circuit supply conductors: RMC, RNMC, IMC, EMT, ENT, metal wireway, nonmetallic wireway, surface metal raceway with metal cover, nonmetallic surface raceway, FMC, LFMC, LFNC, MI cable, MC cable, or AC cable.

Per 645.5(E)(4), ventilation in the underfloor area must be used for information technology equipment only. The HVAC system can be either dedicated to the information technology equipment use and separated from other areas of occupancy, or at the point of penetration to the information technology equipment room, be provided with dampers that operate in unison with the smoke detectors or by the disconnecting means in 645.10.

Per 645.5(E)(5), the openings in the raised floor that permit the cables to enter and leave shall offer protection to the cables from abrasions. This protection could be offered by providing a bushed fitting at the opening. Per 645.5(E)(6), cables other than those covered in (E)(2) and (E)(3) shall be listed as Type DP (data processing).

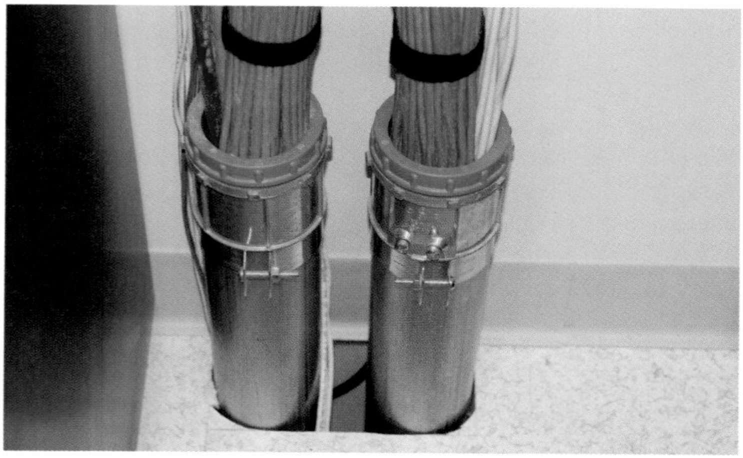

Cables passing through openings in a raised floor must be protected from abrasions.

Wiring Methods Over 600 V

As with many of the NEC® articles, Article 300 has a separate part that contains the provisions for over 600 V installations. Part II of Article 300 contains special provisions for conductors, wiring methods, and underground installations for over 600 V applications. Remember that the provisions of the Part II are special and cannot be applied to under 600 V installations in Article 300. Likewise, the under 600 V provisions of Part I cannot be applied to over 600 V installations.

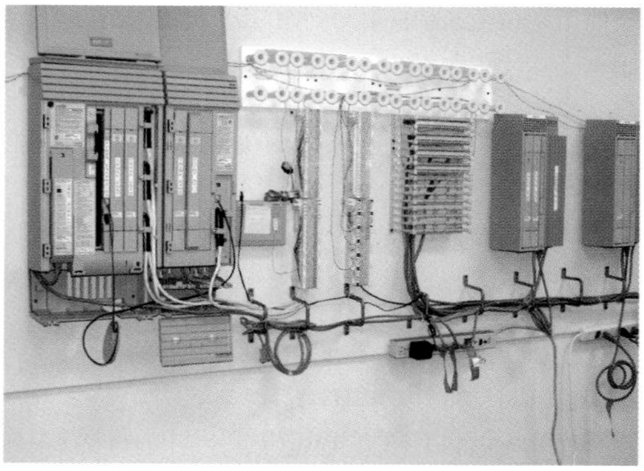

Article 645 only applies to information technology equipment that is installed within an information technology equipment room.

Over 600 V Conductor Requirements – 300.31, 300.34. A special concern for over 600 V installations is accessibility. Section 300.31 requires suitable covers on all boxes, fittings, and enclosures to ensure they are protected against accidental contact by any persons.

Section 300.34 contains provisions to ensure that a safe and adequate bending radius is provided for all conductors that operate at over 600 V. The higher voltage levels can have a damaging effect on any improper cable installations or conductor installations that might damage the insulation of the conductor.

The conductors shall not be installed where they are bent to a radius less than eight times the overall diameter for nonshielded conductors and twelve times for shielded or lead-covered conductors. If the over 600 V installation requires the use of multiconductor or multiplexed conductors, the minimum bending radius is the greater of either twelve times the diameter of the individually shielded conductors or seven times the overall diameter of the cable.

Over 600 V Wiring Methods – 300.37. Only specific wiring methods are permitted for over 600 V above- ground installations. Rigid metal conduit, intermediate metal conduit, electrical metallic tubing, and rigid nonmetallic conduit RTRC and PVC are acceptable raceways.

Conductors may also be installed in cable trays, busways, or cablebus or as metal-clad cable provided it is suitable for the purpose. Because of the increased voltage levels, particular concern must be given to ensuring that the conductors are adequately protected against physical damage. Section 300.37 does, however, permit exposed runs of Type MV cable, bare conductors, and bare busbars provided the location is accessible to qualified persons.

It is critical in these installations that the access be strictly limited to qualified persons because of the immediate damage of the exposed conductors operating over 600 V. Article 100 defines "qualified person" as one who has all of the necessary skills and knowledge of the electrical equipment and installation and who has also had safety training to recognize and avoid the hazards involved.

Underground Installations – 300.50

Part II of Article 300 contains specific requirements for the installation of underground conductors that operate at over 600 V. There are other related articles that should be reviewed as well. Part III of Article 110 contains general requirements for electrical installations over 600 V. In addition, Part IV contains provisions for tunnel installations over 600 V. These special provisions apply to high-voltage electrical equipment and distribution wiring that is portable or mobile and installed or utilized in tunnel applications.

Section 310.10(F) also contains provisions for direct-buried cables. This section requires that cables for such use be identified. Article 100 defines "identified" as being suitable for the specific purpose, function, environment, or use. Typically, the AHJ will review product listing and marking information to determine if a product is identified for the use. Section 310.10(E) also requires that, in general, cables rated over 2000 V be shielded.

The specific provisions for shielding are contained in 310.10(E) and 300.50(A)(1). Whenever shielded cables are installed, it is important to ensure that the shields are properly grounded in accordance with the requirements of Article 250 and that the shield provides the necessary ground-fault current path. See 250.4(A)(5) and 250.4(B)(4).

Section 300.50(B) contains several important provisions for installations in wet locations. It is important to remember that any connection or splice that is made in an underground location is considered to be in a wet location and must be approved for wet locations.

In a similar way, the interior of any raceway or enclosure that is installed underground is also considered to be a wet location. Any cable or conductor that is installed in an underground raceway or enclosure must therefore be listed for use in wet locations and must meet all of the provisions of 310.10(C). This section specifies the insulation types permitted for insulated conductors and cables.

Section 300.50(C) requires that underground conductors operating at over 600 V be protected against physical damage. Section 300.50(E) requires that when backfilling is performed, large rocks and similar materials must not be placed in the trench or excavation where they could damage the raceway.

Protection must also be provided where the conductors emerge from the ground. Protection for conductors emerging from the ground must be by enclosure in a listed raceway. If the conductors emerge from the ground and run up a pole, the raceway must be of rigid metal conduit, intermediate metal conduit RTRC-XW, Schedule 80 PVC, or equivalent. The protection must be provided from the minimum cover depth required by Table 300.50 up to a point at least 8′ above finished ground.

QUICK QUIZ®

Refer to Chapter 7 Quick Quiz® on CD-ROM.

The primary function of the raceway system is to provide protection for the conductors.

Additional information is available at ATPeResources.com

Table 300.50 provides the minimum cover requirements for raceways and cables that contain conductors operating at over 600 V. This table is very similar to Table 300.5 for conductors operating at 0 V–600 V, but the general minimum cover depths are significantly increased. As with Table 300.5, cover is defined as the shortest distance measured from the top surface of any direct-buried cable or raceway and the top surface of finished grade. See Table 300.50, Notes. Table 300.50 contains minimum cover depths for both general and specific conditions. The conditions however, must be related to the circuit voltage and the specific raceway type or direct-buried cable that is employed.

Section 300.50(D) does permit direct-buried cables to be tapped or spliced. This is useful where the cables may have been damaged, such as where operating equipment may have come in contact with the direct-buried cables. Splice boxes must be provided at the point where the direct-buried cables are tapped or spliced, and the materials used must be suitable for the application. Many splicing kits are available that provide the necessary watertight connection and mechanical protection.

chapter 7

Wiring Methods

REVIEW QUESTIONS

Name ______________________ Date ______________________

T F **1.** Article 300 applies to wiring that is part of equipment.

T F **2.** Conductors of different systems (600 V or less, AC and DC) may occupy the same raceway.

___C___ **3.** Holes that are drilled into framing members to permit the installation of raceways or cables shall be at least ___″ from the nearest edge of the wood member.

A. 1/4 C. 1 1/4
B. 1 D. none of the above

___B___ **4.** Cables or raceways in notches of framing members shall be protected from nails or screws by a steel plate at least ___ ″ thick.

A. 1/32 C. 1/8
B. 1/16 D. none of the above

___24″___ **5.** The minimum cover for a direct-buried cable in a location not specified in Table 300.5 is ___″.

T F **6.** Per Table 300.5, RMC shall be buried under 12″ of earth beneath a highway.

T F **7.** Seals shall be installed in the raceway system whenever condensation is known to be a problem.

___B___ **8.** All cables and conductors emerging from the ground shall be protected from physical damage to a maximum depth of ___″.

A. 12 C. 24
B. 18 D. 36

___B___ **9.** In an indoor wet location, the electrical panelboard metal enclosure shall be at least ___″ from the masonry wall.

A. 1/8 C. 3/8
B. 1/4 D. 1/2

T F **10.** Splices or taps are never permitted within a raceway.

T F **11.** Raceway systems are always permitted to be used to support other raceways, cables, or nonelectric equipment.

T F **12.** The continuity of the conductor of a multiwire branch circuit shall never depend upon device connections.

______A **13.** At least ___″ of free conductor shall be left at each outlet, junction, or switch point.

A. 6
B. 12
C. 18
D. 24

______C **14.** An example of a(n) ___ space is the space over a hung ceiling used for environmental air.

A. "roof"
B. "duct"
C. "other"
D. none of the above

T (F) **15.** Conductors of opposite polarities that are run in close proximity to one another produce a heating effect on ferrous metal.

T (F) **16.** AC conductors installed in metal raceways or enclosures shall be arranged to provide induction heating of the metal raceways or enclosures.

(T) F **17.** Copper conductors in a vertical raceway require more supports than an equivalent length of aluminum conductors.

______40% **18.** When a conduit contains three or more conductors, the cross-sectional area of the conductors shall not exceed ___% of the cross-sectional area of the conduit.

______slots **19.** The inductive heating effect of metal is minimized by cutting ___ between the holes for conductors.

______12 **20.** Underground, residential branch circuits that are rated 20 A, 120 V or less and are GFCI protected shall have a minimum burial depth of ___″.

______18 **21.** Cables and raceways in or under airport runways shall be buried to a minimum depth of ___″.

______C **22.** Cables and raceways installed under streets, highways, roads, driveways, and parking lots shall be buried to a minimum depth of ___″.

A. 12
B. 18
C. 24
D. none of the above

______Cover **23.** ___ is the shortest distance measured between a point on the top surface of any direct-buried conductor, cable, conduit, or other raceway and the top surface of finished grade, concrete, or similar cover.

______D **24.** Cables and conductors installed on the side of a pole or building shall be protected up to ___′ above finished grade.

A 5
B. 7½
C. 10
D. none of the above

______D **25.** Boxes or cabinets marked "___" can be installed out-of-doors.

A. raintight
B. rainproof
C. outdoor type
D. all of the above

Burial Depths to Wiring Method/Circuit

______ 1. A = 2 and 5

______ 2. B = 2, 4, and 5

______ 3. C = 4

______ 4. D = 3

______ 5. E = 4

______ 6. F = 3

______ 7. G = 1

______ 8. H = 1, 2, 3, and 5

______ 9. I = 1

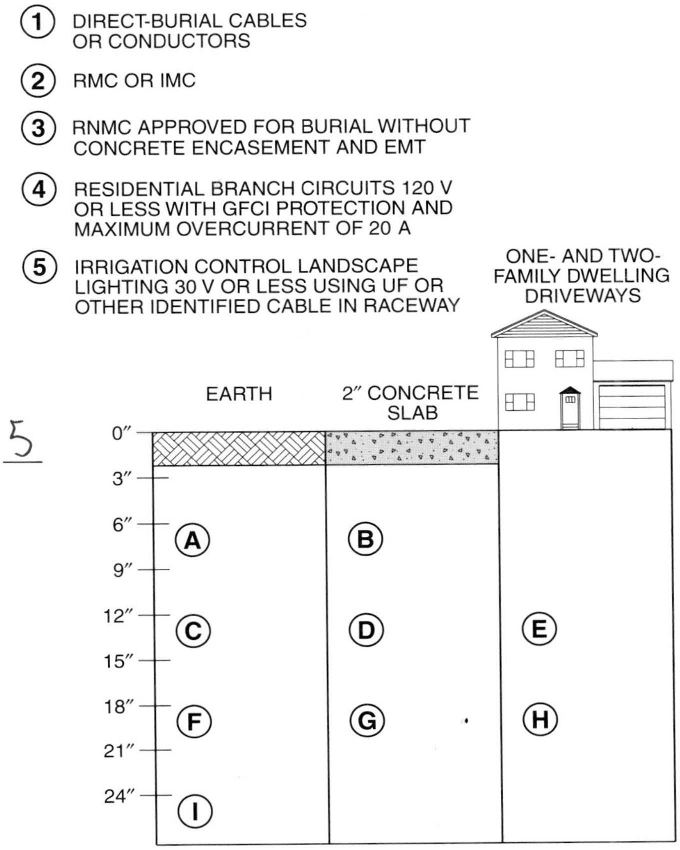

Neutral Closed

______ 1. The load at A is 40 Ω.

______ 2. The load at B is 10 Ω.

______ 3. The load at C is 30 Ω.

______ 4. The load at D is 12 Ω.

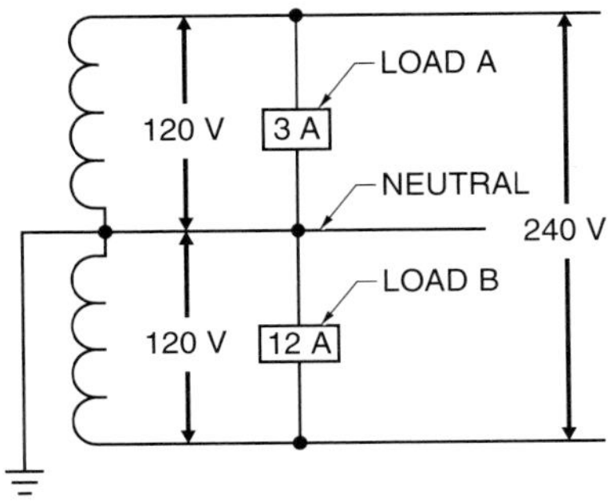

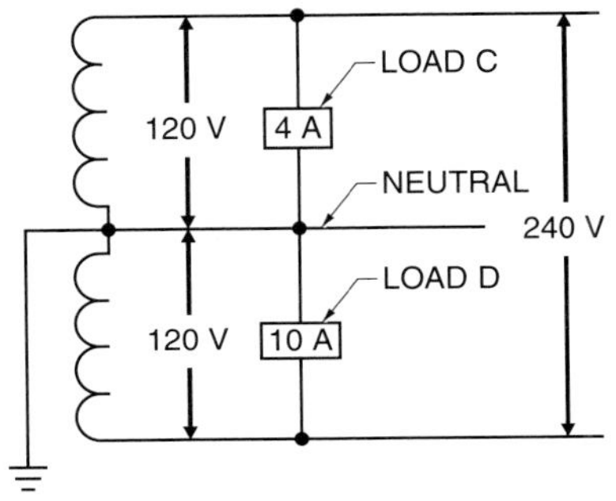

Neutral Opened

192 **1.** The load at A is ___ V.

48 **2.** The load at B is ___ V.

180 **3.** The load at C is ___ V.

60 **4.** The load at D is ___ V.

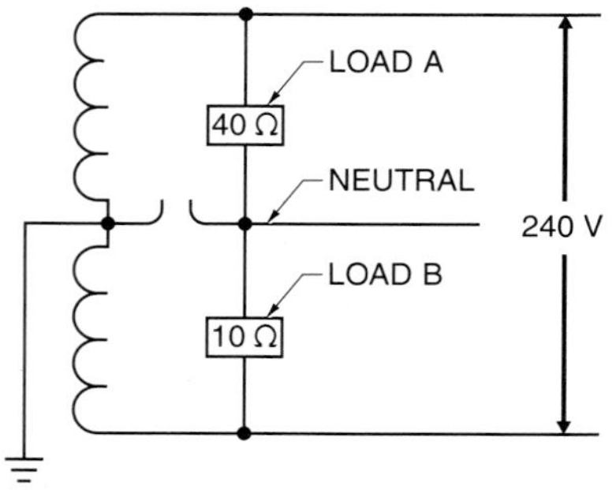

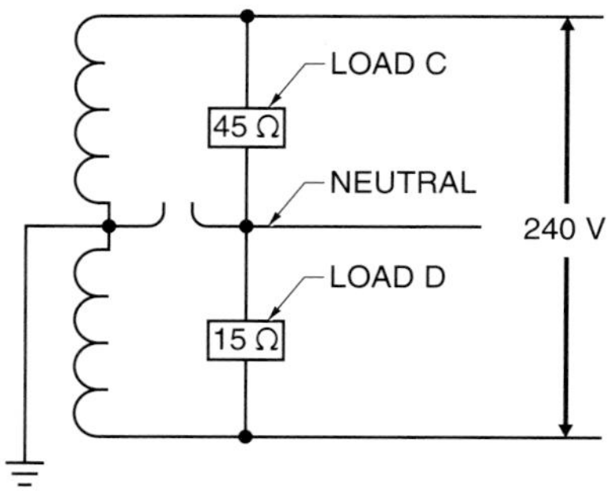

Underground Installations

A **1.** Conductors and cables shall be protected to 8′ above finished grade.

C **2.** Splice box not required.

B **3.** Heavy rocks or sharp corrosive material prohibited as backfill.

D **4.** Raceways shall be sealed to prevent moisture.

E **5.** Bushings required at end of conduit underground with conductors.

F **6.** "S" loops allowed for ground movement.

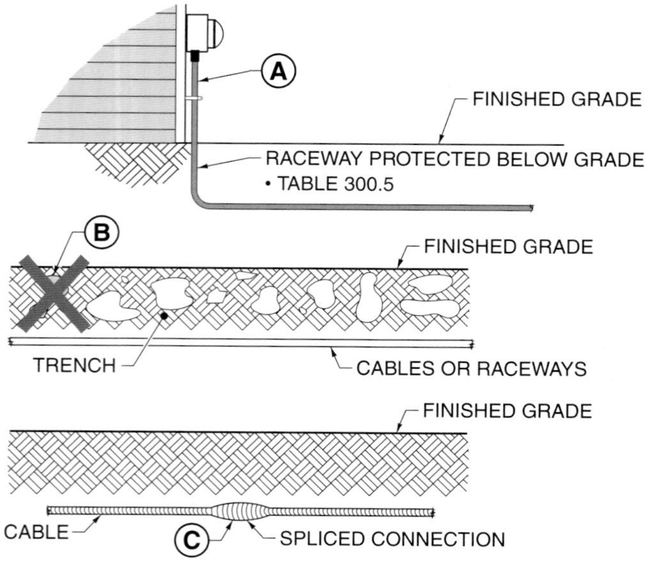

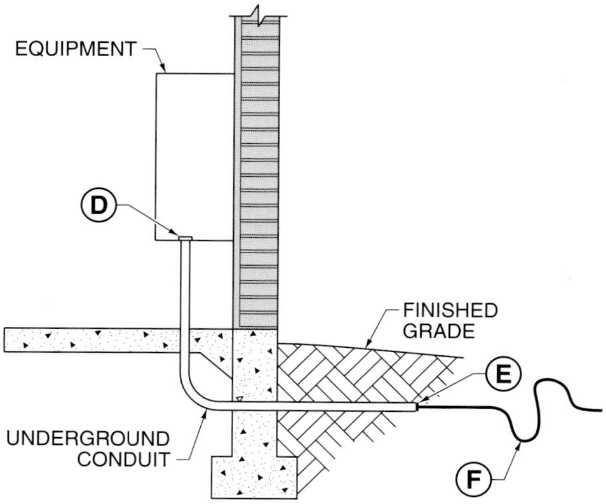

chapter 7

Wiring Methods

TRADE COMPETENCY TEST

Name ______________________ Date ______________________

NEC®	Answer	
________	________	**1.** Determine the minimum burial depth for a Type UF cable installed to supply an outdoor receptacle in a one-family dwelling. The branch circuit is rated at 20 A, 120 V and provided with GFCI protection.
________	________	**2.** Determine the minimum burial depth for listed Schedule 80 PVC conduit installed under the driveway of a two-family dwelling unit used only for dwelling unit purposes.
________	________	**3.** A direct-burial cable emerges from the ground and is installed on the service of a wooden utility pole. Determine the minimum burial depth and maximum height to which the cable shall be protected from physical damage.
________	________	**4.** *See Figure 1.* An electrical service is installed in an area with frost heaves and a good degree of ground movement. The electrical contractor installs the service using Schedule 80 PVC conduit per 230.43(11) and includes expansion joints where the conduit emerges from the ground. Does this installation violate Article 300 of the NEC®?

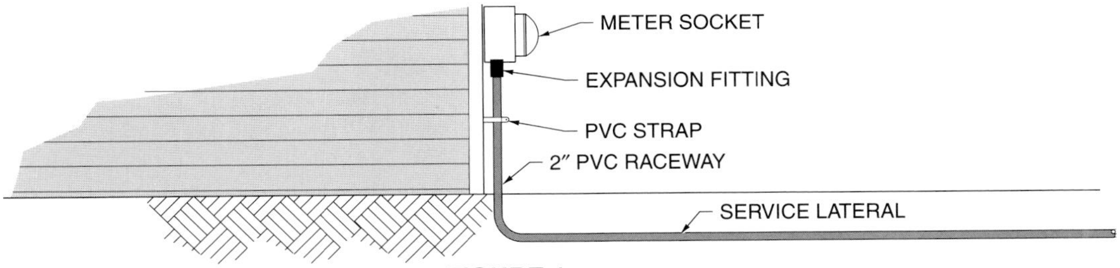

FIGURE 1

________	________	**5.** An electrical contractor installs branch-circuit conductors in RNMC in a one-family dwelling. The contractor supports the horizontal run of RNMC in notches cut in the 2″ × 4″ framing members which are spaced on 24″ centers. The conduit is supported within 3′ of the termination point. Steel plates are not used to cover the notches in the framing members. Does this installation violate Article 300 of the NEC®?

_______________ _______________ **6.** When installing a 5′ length of Type MC cable, the electrician runs out of cable connectors. Instead of a connector for Type MC cable, the electrician uses a fitting listed for use with type NM cable only. Does this installation violate Article 300 of the NEC®?

_______________ _______________ **7.** A 75′ run of EMT is installed in a commercial location. The raceway system is completed up to the last junction box. The electrician installs the conductors in the raceway system and leaves enough length on the conductors so that they can be installed into the rest of the raceway when it is completed. The conductors are wrapped and protected from physical damage. Does this installation violate Article 300 of the NEC®?

_______________ _______________ **8.** Determine the minimum number of conductor supports required for a 180′ vertical raceway installed in a commercial high-rise building. The conductors are 500 kcmil Al.

_______________ _______________ **9.** Determine the minimum number for conductor supports required for a 140′ vertical raceway which contains four 1/0 AWG Cu conductors.

_______________ _______________ **10.** An electrical contractor installs a 4′ length of Type AC cable in an environmental air plenum. The cable connects a modulating motor which is used to control the flow of the contained air. A panel is provided in the plenum to access the motor and the connections. Does this installation violate Article 300 of the NEC®?

_______________ _______________ **11.** A single 5′ length of LFMC is used to connect a lay-in type luminaire (lighting fixture) in a suspended ceiling. The ceiling space is used for environmental air-handling purposes. Does this installation violate Article 300 of the NEC®?

_______________ _______________ **12.** In order not to support Type AC cable with the ceiling support wires per 300.11(A)(1), an electrician lays the AC cables directly on the ceiling grid system. The electrician uses clamps to hold the cables in place and to permit removal of the ceiling tiles per 300.23. Does this installation violate Article 300 of the NEC®?

_______________ _______________ **13.** A cable is installed in the notch of a wooden joist. What is the minimum thickness of the steel plate used to cover the notch and protect the cable?

_______________ _______________ **14.** *See Figure 2.* Does the illustration violate Article 300 of the NEC®?

_______________ _______________ **15.** *See Figure 3.* Does the illustration violate Article 300 of the NEC®?

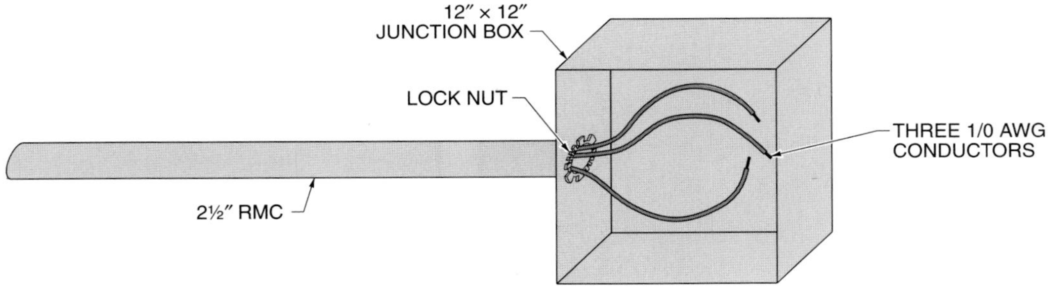

FIGURE 2

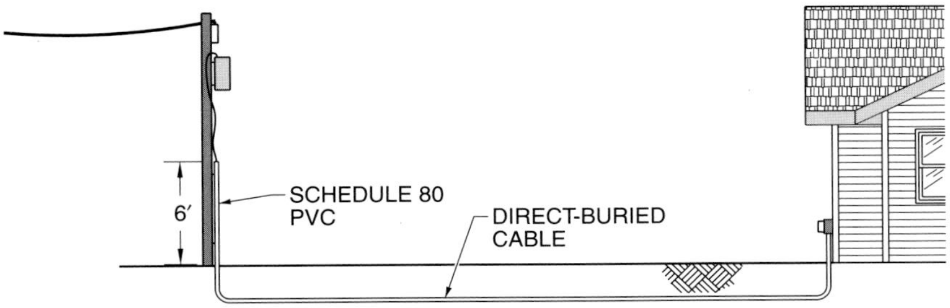

FIGURE 3

Learning the Code

- The "nuts and bolts" of the electrical distribution system are the actual wiring materials, raceways, cables, and boxes. It is critical that apprentices and students of the Code understand the importance of selecting and applying the right wiring method and wiring materials for the specific application.

  An important tool for accomplishing this goal is the structure and the format of many of these NEC® Chapter 3 wiring methods. For example, look at Article 344, Rigid Metal Conduit (RMC), and Article 352, Rigid Polyvinyl Chloride Conduit (PVC). Notice that the section format is fairly consistent. The ".1" section covers the scope of the article, and the ".2" section contains definitions. Listing requirements are covered in the ".6" section, while the ".20" section covers minimum sizes. Once apprentices and students of the Code become familiar with this format, it is much easier to locate Code provisions across many of the articles.

- The first step in correctly applying any NEC® article is to review the article scope, which is typically the first section of the article. For wiring methods, the next step should be to review the ".10" Uses Permitted section, and the ".12" Uses Not Permitted section, where one is included. The most common mistake made in selecting and installing electrical systems is the use of wiring methods that are not suitable for the specific application. The Uses Permitted and Uses Not Permitted sections are key to ensuring that wiring methods are correctly applied and correctly installed.
- Frequently, Code and licensing examinations focus on outlet and junction-box sizing requirements. Apprentices and students of the Code should pay close attention to 314.28 for pull and junction-box sizing requirements. In addition, two tables are used to ensure that the box selected can safely handle the number of conductors that may be installed within it. Table 314.16(A) lists the maximum number of conductors permitted for various sizes and types of metal boxes. Table 314.16(B) contains the volume allowances per conductor size for use in calculating box fill where different-size conductors are present.

Wiring Materials—Raceways and Boxes

Raceways are metallic or nonmetallic enclosed channels for conductors. Rigid metal conduit is the universal raceway. Boxes are metallic or nonmetallic electrical enclosures. They are used for equipment, devices, and pulling or terminating conductors.

New in the 2011 NEC®

- *Restructured requirements for outlet boxes, pull boxes, junction boxes, and conduit bodies – 314.27(A)*
- *A new subdivision for power distribution boxes added to Article 314 – 314.28(E)*
- *New requirements for installing Type MV cable by qualified persons – 328.14*
- *Clarification of the use of Type NM cable in detached garages and storage buildings – 334.10*
- *Clarification of the use of uninsulated conductors in Type SE cable – 338.10(B)(2)*
- *Revised provisions for determining the allowable ampacity of Type SE cable in interior locations – 338.10(B)(4)*
- *Revised requirements for the use of FMC and LFMC where flexibility is required – 348.30(A) and 350.30(A)*

RACEWAY SYSTEMS

A *raceway* is a metal or nonmetallic enclosed channel for conductors. A raceway includes all of the enclosures used for running conductors between different components of an electrical system per Article 100. This includes switches, panels, controllers, and cabinets. Examples of raceways include RMC, RNC, IMC, LFMC, FMT, FMC, ENT, EMT, wireways, and busways. *Note:* When the NEC® uses the word conduit, it means only those raceways that contain the word "conduit" in their titles.

A *raceway system* is an enclosed channel of metal or nonmetallic materials used to contain the wires or cables of an electrical system. Raceway systems are composed of raceways. Raceways are commercially available in a variety of lengths. They provide protection for conductors and give the system flexibility in that existing conductors can be removed and new conductors installed. New conductors may be added to an existing system provided there is enough room. Some examples of raceways are rigid metal conduit, rigid nonmetallic conduit, intermediate metal conduit, liquidtight flexible metallic conduit, flexible metal conduit, electrical nonmetallic tubing, electrical metallic tubing, underfloor raceways, cellular concrete floor raceways, cellular metal floor raceways, surface raceways, wireways, and busways.

Each of these raceways has its own article in the NEC®. The sections entitled "Uses Permitted" and "Uses Not Permitted" should be reviewed for the particular raceway being considered. Some of these articles require the raceway to be listed. If that is the case, the raceway shall be installed in accordance with any instructions included in the listing or labeling per 110.3(B).

Intermediate Metal Conduit (IMC) – Article 342

Intermediate metal conduit (IMC) is a raceway of circular cross section designed for protection and routing of conductors. The raceway is made of steel, is threadable, and is frequently used because of its excellent protective qualities, large internal diameter, and ease of installation. Section 342.6 requires that IMC and all associated fittings and associated factory elbows and couplings be listed. The raceway wall thickness provides excellent protection against physical damage in most atmospheric conditions.

If IMC is going to be installed in corrosive environments, 342.10(B) requires corrosion protection judged suitable for the condition. If IMC is going to be installed in direct contact with cinder fill, 342.10(C) requires that a layer of concrete at least 2″ thick be placed around the IMC or the conduit placed at least 18″ below the cinder fill. In some cases, the IMC can be designed and manufactured to provide adequate protection. For example, IMC is available with a PVC coating for use in corrosive environments.

IMC is made in trade sizes ½″ through 4″. The number of conductors permitted to be installed in IMC shall not exceed the permitted fill specified in Table 1 of Chapter 9. Where the Code allows a cable type to be installed in a raceway, it may be installed within IMC provided the allowable percentage fill does not exceed that permitted by Table 1, Chapter 9.

Section 342.30 contains the securing and supporting requirements for IMC. The NEC® makes a distinction between securing and supporting requirements. For example, a raceway run through framing trusts may be supported but not necessarily secured. IMC is required to be secured within 3′ of each box, cabinet, conduit body, or termination. If the building design does not permit ready support, such as in a case where beam spacing is at 4′ intervals, IMC support requirements can be extended, but in no case greater than 5′ from the termination. The general support requirement for IMC requires IMC to be supported at intervals that do not exceed 10′. See 342.30(B) for special provisions for vertical and horizontal runs of IMC.

Rigid Metal Conduit (RMC) – Article 344

Rigid metal conduit (RMC) is a threadable conduit made of metal. It is the universal raceway. **See Figure 8-1.** It is permitted under all atmospheric conditions and in all types of occupancies per 344.10(A). When installing RMC, dissimilar metals that could cause galvanic action shall be avoided. *Note:* Aluminum fittings and enclosures are permitted to be used with RMC and vice versa where they are not subject to severe corrosive influences.

Rigid Metal Conduit (RMC)

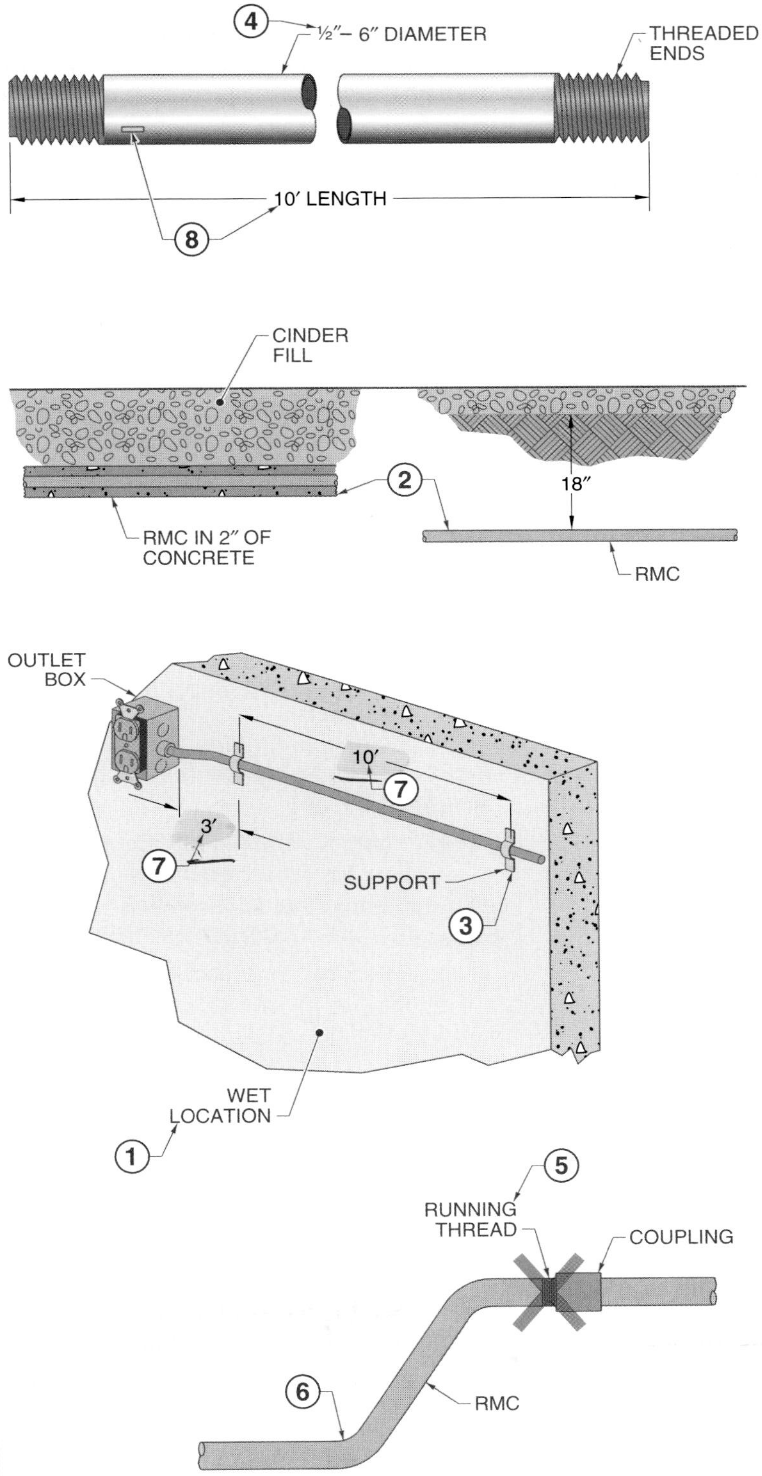

SIZES • 344.20, INFORMATIONAL NOTE	
• 10′ LENGTHS	
INCHES	METRIC DESIGNATOR
½	16
¾	21
1	27
1¼	35
1½	41
2	53
2½	63
3	78
3½	91
4	103
5	129
6	155

(1) RMC PERMITTED IN ALL ATMOSPHERIC CONDITIONS
- 344.10(A)

(2) RMC IN OR UNDER CINDER FILL SHALL BE ENCASED IN 2″ CONCRETE OR BE 18″ BELOW CINDER FILL
- 344.10(C)

(3) SUPPORTS, STRAPS, SCREWS, ETC. SHALL BE CORROSION RESISTANT
- 344.10(B) AND (D)

(4) RMC IS AVAILABLE IN 10′ LENGTHS FROM ½″ TO 6″ DIAMETERS
- 344.20(A)(B)

(5) RUNNING THREADS PROHIBITED
- 344.42(B)

(6) BENDS SHALL NOT TOTAL MORE THAN 360° IN ONE RUN
- 344.26

(7) ~~RMC SHALL BE SUPPORTED WITHIN 3′ OF BOXES AND AT INTERVALS OF NOT MORE THAN 10′~~
- ~~344.30(A)(B)~~

(8) RMC IS SHIPPED IN 10′ LENGTHS AND SHALL BE IDENTIFIED AND MARKED
- 344.120, 344.130

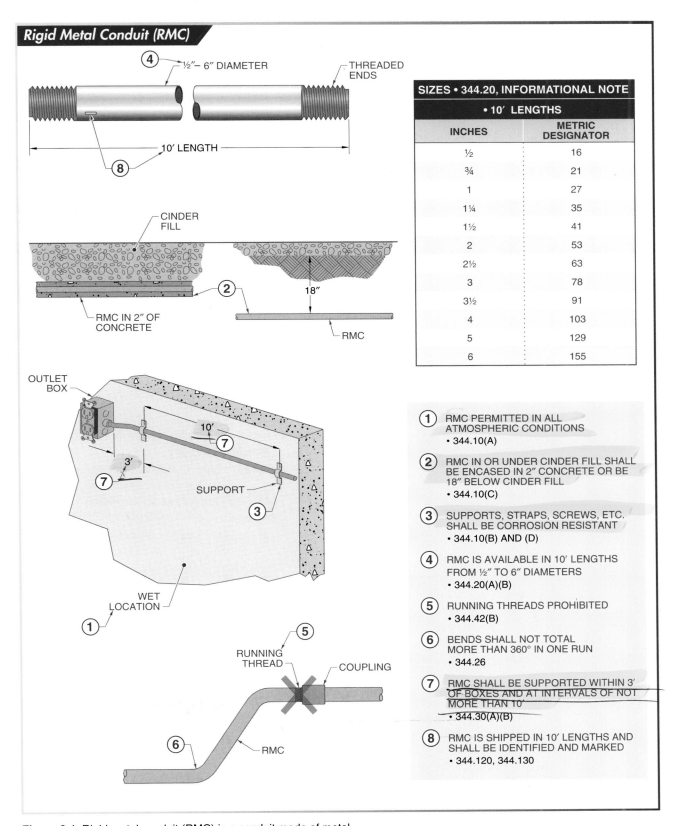

Figure 8-1. Rigid metal conduit (RMC) is a conduit made of metal.

RMC shall not be run in or under cinder fill unless it is contained in an envelope of at least 2″ of noncinder concrete or is buried at least 18″ below the cinder fill per 344.10(C). All supports, bolts, straps, etc. shall be corrosion-resistant per 344.10(B)(1) and (2) and 344.10(D).

RMC is available in 10′ lengths with a coupling on one end. Per 344.130, lengths shorter or longer than 10′ may be shipped. This is considered a special order by the manufacturer and a large quantity should be ordered or the price is too expensive. It is rare that the manufacturer provides lengths other than the standard 10′ length. RMC is available in sizes from ½″ in diameter up to 6″ in diameter per 344.20. Sizes may be given in inches or millimeters.

Because of its wall thickness, RMC is typically bent with a mechanical conduit bender.

Running Threads Prohibited

An attempt is made to screw a length of RMC with an offset onto a straight length of RMC installed in a narrow ditch. The offset prevents the installer from turning the RMC onto the straight length of conduit. An approved split coupling or Erickson coupling is needed to join these two lengths of conduit as the NEC® does not permit the use of running threads.

A running thread is twice as long as a standard thread. The coupling is run all the way up this extended thread. The two lengths of conduit are then aligned and the coupling on the running thread is backed off onto the length with the standard thread. This exposes half of the running threads, which would eventually lead to a corrosion problem.

The general support requirements for RMC, per 344.30, require RMC to be supported every 10′ and secured within 3′ of every outlet, junction box, or conduit termination. Section 430.245(B) permits ⅜″ RMC to be used between a motor and its junction box when the junction box is not part of the motor housing. Section 344.42(B) prohibits the use of running threads when installing RMC.

Nipples. A *nipple* is a short piece of conduit or tubing that does not exceed 24″ in length. **See Figure 8-2.** Nipples are used to join enclosures that are mounted close together. The wiring passes through the nipple from one enclosure to the other. The requirements for conductor fill for nipples are found in Chapter 9, Notes to Table 4.

Nipples are permitted to be installed between boxes, cabinets, and similar enclosures. The conductor fill for a nipple may be as high as 60% of its total cross-sectional area. The requirement for derating conductor ampacity in Section 310.15(B)(3)(a) does not apply to nipples.

Rigid Polyvinyl Chloride (PVC) Conduit – Article 352

Rigid polyvinyl chloride (PVC) conduit is a conduit made of materials other than metal designed for the installation of electrical conductors and cables. **See Figure 8-3.** There are several products that come under the regulation of Article 352. These materials are recognized as having suitable physical characteristics for direct burial in the earth. Some of the materials used in the manufacture of these products include fiber, asbestos cement, soapstone, rigid polyvinyl chloride, fiberglass epoxy, high-density polyethylene (for underground use), and rigid polyvinyl chloride (for aboveground use).

PVC conduit is available in thin wall (Schedule 40) and thick wall (Schedule 80). PVC water pipes were given a number designation depending on the wall thickness. The thinnest wall thickness was given the designation number 10. The next thicker size was 20, then 30, 40, and up to 100. When PVC was presented to the electrical industry, the two sizes adopted were 40 and 80.

Nipples

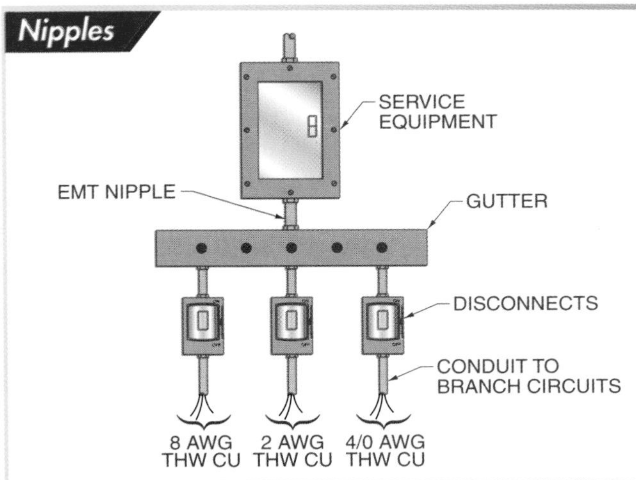

What size EMT nipple is required for three 8 AWG THW, three 2 AWG THW, and three 4/0 AWG THWN Cu conductors?

Ch 9, Table 5: 8 AWG THW = 0.0437″ × 3 = 0.1311 sq in.
2 AWG THW = 0.1333″ × 3 = 0.3999 sq in.
4/0 AWG THW = 0.3718″ × 3 = 1.1154 sq in.
1.6464 sq in.

Ch 9, Table 4: 1½″ EMT = 2.036 sq in. at 100%
2″ EMT = 3.356 sq in. at 100%
2.036 sq in. × 60% = 1.221 sq in. (Too small)
3.356 sq in. × 60% = 2.013 sq in.
(Exceeds 1.682 sq in.)

Nipple = **2″ EMT**

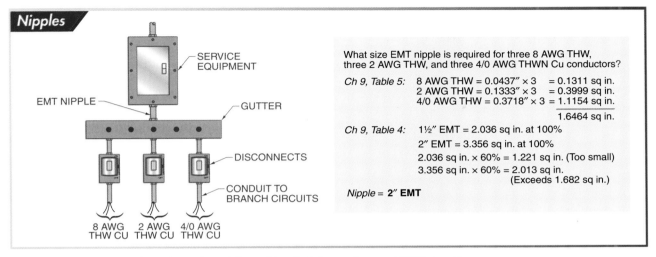

Figure 8-2. A nipple is a short piece of conduit or tubing that does not exceed 24″ in length.

Schedule 40 is manufactured from fibers of polyvinyl chloride. It is waterproof, rustproof, and does not corrode in most locations. Because Schedule 40 does not have the strength of metal conduits, it shall not be used where subject to physical damage. Schedule 80 has a heavier wall thickness and is suitable for most locations where PVC is used. PVC must be supported as required in Table 352.30. Generally, the maximum spacing between conduit supports increases with the size of the conduit. PVC is required to be secured within 3′ of every termination point.

If the length change in a run of PVC exceeds ¼″ due to thermal expansion, an expansion fitting shall be provided as required by 352.44. Table 352.44 gives the expansion characteristics of PVC conduit.

Per 352.26, a total run of PVC shall not exceed 360° of bends (including kicks and offsets). A *bend* is any change in direction of a raceway. **See Figure 8-4.** There shall not be more than the equivalent of four 90° bends between pull points. A *kick* is a single bend in a raceway. Kicks are often used for a raceway entrance and exit in an enclosure. An *offset* is a double bend in a raceway, each containing the same number of degrees. Offsets are commonly used when raceways are routed around obstacles, other raceways, etc. Bends can be any angle necessary to meet job requirements.

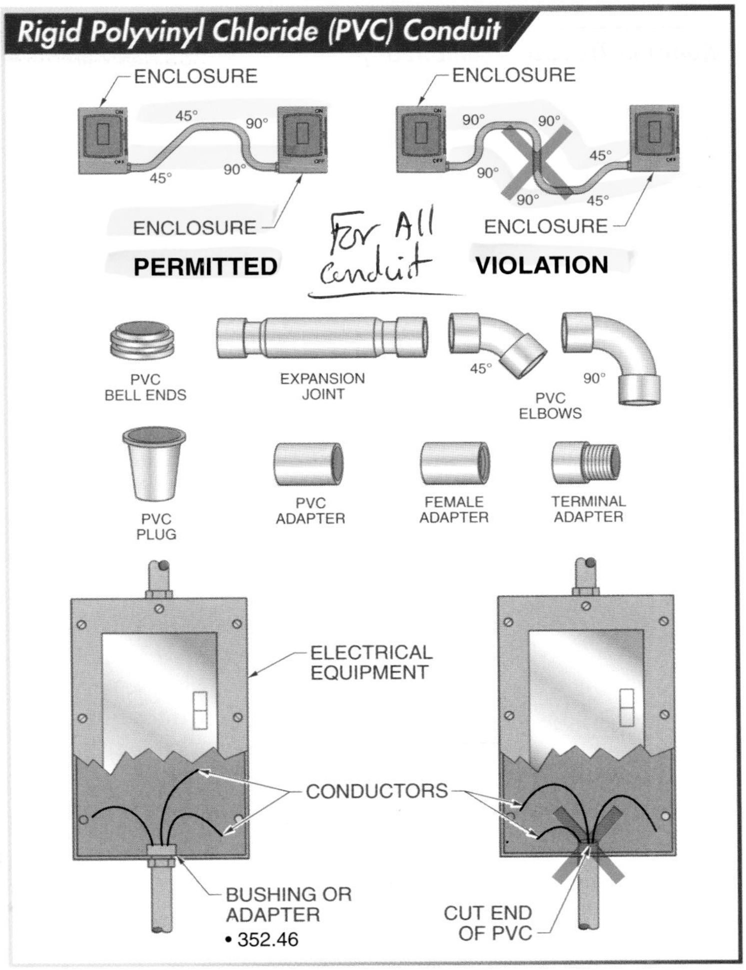

Figure 8-3. Rigid polyvinyl chloride (PVC) conduit is a conduit made of materials other than metal.

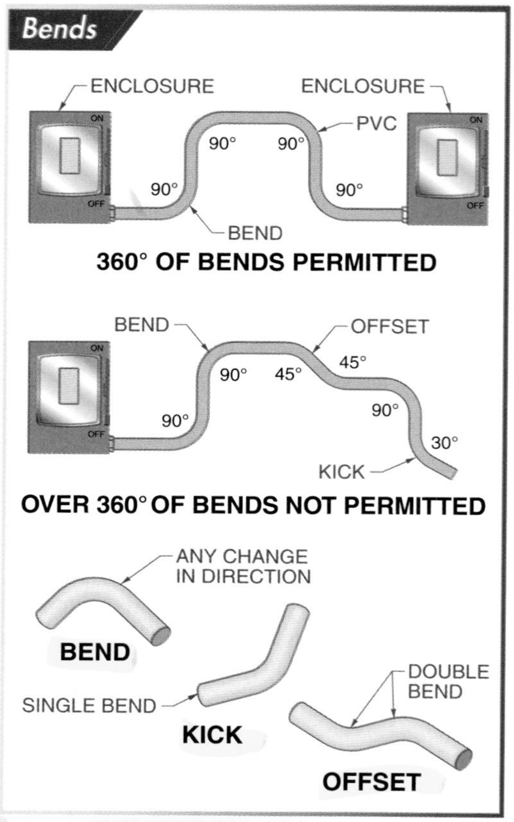

Figure 8-4. A total run of PVC shall not exceed 360° of bends.

Joining PVC Conduit

All schedules of PVC conduit are fabricated and installed by the same methods, despite the differences in wall thickness. Sections are joined together with a solvent-welding technique. The solvent is a powerful chemical and must be used in a well-ventilated area. Mating surfaces should be dry and clean before applying the solvent. The solvent works by dissolving the surfaces of the conduit and fittings in contact with each other. After the parts are joined, the parts need to be rotated to ensure complete contact of the solvent applied and the mating surfaces of the joined parts.

Permitted Uses – 352.10. PVC was first recognized for use in the NEC® in the late 1960s. Today PVC is widely used for many applications in the electrical industry. Section 352.10 lists the permitted uses for the types of PVC covered by Article 352. As a general note, the use of PVC in extremely cold environments should be carefully considered. The extreme cold can make PVC extremely brittle and therefore susceptible to physical damage. Section 352.6 requires that PVC and all associated fittings and factory elbows be listed.

By definition, a concealed location is one that is rendered inaccessible by the structure or finish of the building. Section 352.10(A) permits the use of PVC in concealed locations such as walls, floors, and ceilings. PVC can be used in dry, damp, or wet locations. PVC is one of the few raceways that can be installed in cinder fill without the use of additional protective measures. It is also permitted for exposed work. If PVC is installed exposed in areas subject to physical damage, it must be identified for such use. Schedule 80 PVC is identified for such use.

PVC conduits have a listed temperature rating associated with their particular construction. Where conductors or cables are installed in a PVC conduit, it is important that the temperature rating associated with the conductors or cables does not exceed the listed temperature rating of the PVC conduit. In other words, conductors and cables with temperature ratings higher than that of the PVC conduit are permitted to be installed in the conduit, but the conductors or cables must not be operated at a temperature higher than the listed temperature rating of the conduit.

Uses Not Permitted – 352.12. Section 352.12 lists the prohibited uses for rigid nonmetallic conduit. As mentioned above, temperature considerations should always be considered when selecting PVC for a specific application. In general, PVC should not be installed where there are significant temperature extremes. PVC is prohibited from use where it is subject to ambient temperatures in excess of 122°F unless it is otherwise listed for such use. PVC is only permitted in hazardous locations under very specific conditions. PVC is not permitted to be installed in theaters or similar locations except as permitted in 518.4 and 520.5.

Bushings – 352.46. A *bushing* is a fitting placed on the end of a conduit to protect the conductor's insulation from abrasion.

A bushing or adapter shall be provided on the end of PVC when it enters a box, fitting, or other enclosure.

Grounding – 352.60. The ".60" section is reserved for the grounding provisions for most conduits and tubing. Section 352.60 requires that for PVC, where grounding is required, a separate EGC must be installed. There are two special exceptions to this general rule, but note that PVC is not a permitted EGC per 250.118. These exceptions cover special provisions within Article 250 only. See 352.60, Ex. 1 and 2.

High Density Polyethylene (HDPE) Conduit–Article 353

High density polyethylene conduit (Type HDPE conduit) is conduit constructed of high density polyethylene that is resistant to moisture and chemical atmospheres. Type HDPE conduit is a newer type of nonmetallic raceway that was recognized in the 2005 NEC®. The conduit is a listed circular raceway consisting of couplings, connectors, and fittings for the installation of electrical conductors. The high density material is resistant to moisture and corrosive agents and is designed to offer excellent protection against physical damage.

Section 353.10 lists six generally permitted uses for Type HDPE conduit. Section 353.20 permits HDPE conduit in sizes from ½″ trade size to 4″ trade size. It is designed for use in either discrete lengths or, as is commonly found in the field, in continuous lengths from a reel. Type HDPE is permitted for use in severe corrosive environments and in cinder fill. Type HDPE is commonly installed in the earth or concrete in directburial applications. HDPE is permitted to be used above ground where encased in not less than 2″ of concrete, and where not otherwise prohibited by 353.12.

HDPE conduits have a listed temperature rating associated with their particular construction. Where conductors or cables are installed in an HDPE conduit, it is important that the temperature rating associated with the conductors or cables does not exceed the listed temperature rating of the HDPE conduit. In other words, conductors and cables with temperature ratings higher than that of the HDPE conduit are permitted to be installed in the conduit, but the conductors or cables must not be operated at a temperature higher than the listed temperature rating of the conduit.

Section 353.12 lists four prohibited uses for Type HDPE conduit. The conduit is not permitted to be installed where it is exposed. It is also prohibited within a building or in any hazardous location. Type HDPE conduit cannot be installed where the ambient temperature is in excess of 50°C unless it is listed for temperatures beyond that.

Type HDPE conduit is permitted for use in applications where it is encased in concrete.

Nonmetallic Underground Conduit with Conductors (NUCC) – Article 354

Several Code cycles ago, an article was added to the Code to cover a special type of raceway that was being installed with the conductors already contained within the raceway. Section 300.18 requires all raceways to be installed complete, prior to the installation of the conductors. The section was modified to permit these types of "prewired" raceway assemblies. Nonmetallic underground conduit with conductors (Type NUCC) is one such prewired raceway. *Nonmetallic underground conduit with conductors (Type NUCC)* is a factory assembly of conductors or cables that are contained within a nonmetallic, smooth wall circular raceway. NUCC is a listed raceway assembly that is constructed of a material that is resistant to moisture and corrosive agents.

Typically, NUCC is installed directly from large reels that permit it to be wound without damage or distortion to the raceway assembly. Section 354.100(C) requires that the individual conductors and cable installed within the NUCC be listed and meet the provisions of 310.10(C). Specifically, the conductors must be of a type identified in 310.10(C) for use in wet locations. The outer jacket of the NUCC is required by 354.120 to be marked every 10′ with the manufacturer's name, trademark, or other distinctive marking.

Section 354.10 limits the use of NUCC to five specific applications. NUCC can be installed for direct-burial applications, where encased or embedded in concrete, in cinder fill, and in underground locations that are subject to severe corrosive influences and in aboveground installations, not otherwise prohibited by 354.12 and where encased in not less than 2″ of concrete. See 354.10(1–5). Section 354.12 prohibits the use of NUCC in exposed locations and inside buildings. Section 354.12, Ex., does permit the conductor or cable portion of NUCC to extend within the building for termination purposes. NUCC is also prohibited from general use in hazardous locations.

Section 354.20 limits the use of NUCC to trade sizes ½″ through 4″. Because of the construction of NUCC, special handling and bending provisions must be strictly followed during the installation of the raceway assembly. Section 354.24 states that all bends shall be manually made so that the assembly is not damaged. The radius for bends in the NUCC must be in accordance with those provided in Table 354.24. For example, Table 354.24 requires the minimum bending radius for a 2″ NUCC to be 26″. As with all raceways, the maximum number of bends between termination points of NUCC is 360°, or the equivalent of four quarter bends.

Electrical Metallic Tubing (EMT) – Article 358

Electrical metallic tubing (EMT) is a lightweight tubular steel raceway without threads on the ends. It is the lightest raceway manufactured. The lengths are joined together by using set screws or compression fittings. **See Figure 8-5.**

The wall thickness of EMT is much thinner than RMC. A piece of ½″ EMT has an ID of 0.620″ and an OD of 0.706″ while a piece of ½″ RMC has an ID of 0.632″ and an OD of 0.840″. The EMT is about 40% thinner in wall thickness. There are six restrictions to the use of EMT:

(1) Where subject to severe physical damage.
(2) In corrosive atmospheres.
(3) Where buried in cinder fill.
(4) In hazardous locations, except as permitted elsewhere in the Code.
(5) For the support of fixtures or other equipment except conduit bodies no larger than the largest trade size of the EMT.
(6) Where practicable, contact with dissimilar metals must be avoided.

Per 358.26, the total number of bends in a run of EMT shall not exceed 360°.

Electrical Metallic Tubing (EMT)

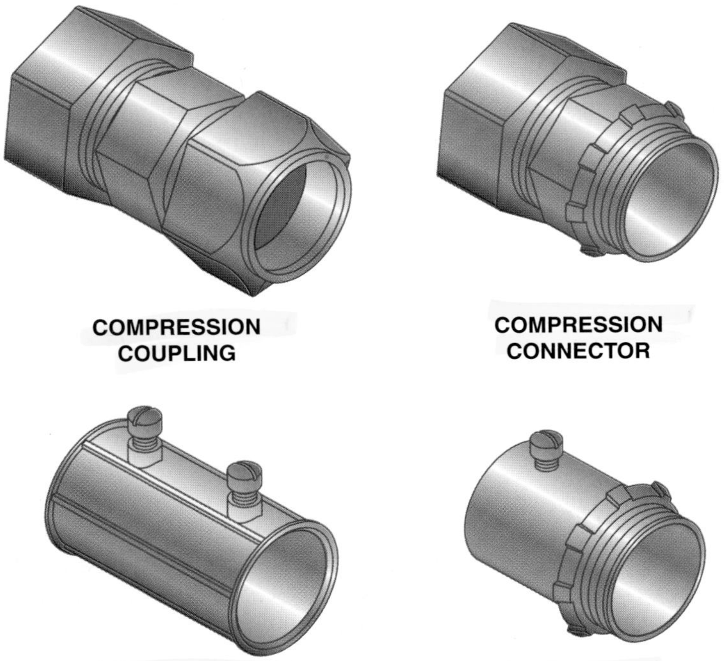

SIZES • 358.20, INFORMATIONAL NOTE • 10′ LENGTHS	
INCHES	METRIC DESIGNATOR
½	16
¾	21
1	27
1¼	35
1½	41
2	53
2½	63
3	78
3½	91
4	103

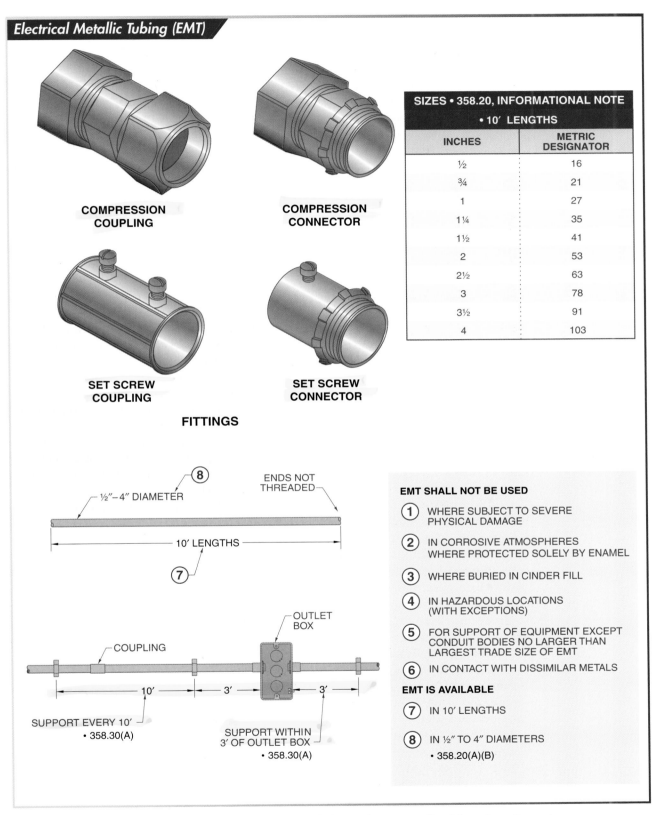

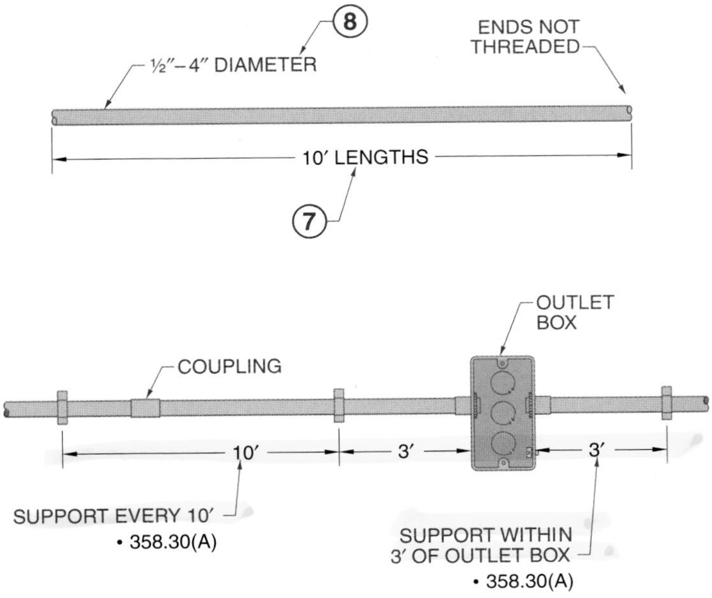

EMT SHALL NOT BE USED

1. WHERE SUBJECT TO SEVERE PHYSICAL DAMAGE
2. IN CORROSIVE ATMOSPHERES WHERE PROTECTED SOLELY BY ENAMEL
3. WHERE BURIED IN CINDER FILL
4. IN HAZARDOUS LOCATIONS (WITH EXCEPTIONS)
5. FOR SUPPORT OF EQUIPMENT EXCEPT CONDUIT BODIES NO LARGER THAN LARGEST TRADE SIZE OF EMT
6. IN CONTACT WITH DISSIMILAR METALS

EMT IS AVAILABLE

7. IN 10′ LENGTHS
8. IN ½″ TO 4″ DIAMETERS
   • 358.20(A)(B)

Figure 8-5. Electrical metallic tubing (EMT) is a lightweight tubular steel raceway without threads on the ends.

The total number of bends in a run of EMT shall not exceed 360° per 358.26. EMT shall be supported every 10′ and within 3′ of every outlet box per 358.30(A). There are two exceptions to this support requirement. The first is 358.30(A), Ex. 1, which states that where the building structure does not permit fastening within 3′ of the outlet box, it can be extended to 5′ of the outlet box. The second is 358.30(A), Ex. 2, which states that when concealed in finished buildings or prefinished wall panels and supporting is impracticable, unbroken lengths of EMT may be fished. The fittings used with EMT are either of the set screw or compression type. EMT is shipped in 10′ lengths. Couplings are not provided on one end as is the case with RMC.

Electrical Nonmetallic Tubing (ENT) – Article 362

Over the past 10 to 15 years, electrical nonmetallic tubing has made a tremendous impact on the choice of wiring methods for electrical contractors and electricians who install residential and commercial electrical distribution systems. *Electrical nonmetallic tubing (ENT)* is a nonmetallic corrugated raceway. ENT is easy to install, pliable, and economical. It is constructed of a material that is flame retardant, resistant to moisture, and resistant to corrosive agents. ENT is available in trade sizes from ½″ to 2″.

One of the primary disadvantages of ENT is that the material that goes into its construction can be toxic or result in excessive smoke generation when it is subjected to products of combustion. Section 362.100 requires that ENT be constructed of material that does not exceed the ignitibility, flammability, smoke generation, and toxicity characteristics of rigid (nonplasticized) polyvinyl chloride.

Section 362.10, for example, lists the uses permitted for ENT. The first listed permitted use is in exposed or concealed locations within any building not exceeding three floors above grade. The first floor is defined as a floor that has 50% or more of its exterior wall surface area level with or above finished grade. The primary concern is life safety. As the building size increases in floors above finished grade, there is an increased concern that the use of ENT could impact the life safety of those habitants that might have to exit the building in the event of a fire.

For buildings exceeding three floors above grade, 362.10(2) limits the use of ENT to concealed locations within walls, floors, and ceilings that provide a thermal barrier of a material that has at least a 15-minute finish rating. An exception to this provision permits the use of ENT within walls, floors, and ceilings, exposed or concealed in buildings exceeding three floors above grade provided that the building has a fire sprinkler system that meets the requirements of NFPA 13-2007, *Standard for the Installation of Sprinkler Systems*.

ENT is also permitted in locations that have severe corrosive agent exposure, in accordance with 300.6, and in concealed dry and damp locations that are not otherwise prohibited by 362.12. ENT can be installed above suspended ceilings where the suspended ceiling offers a 15-minute finish rating and the building does not exceed three stories above grade, or above suspended ceilings in buildings that do exceed three stories above grade provided the building has a fire sprinkler system that meets the requirements of NFPA 13-2007, *Standard for the Installation of Sprinkler Systems*. ENT is commonly installed encased in poured concrete or embedded in concrete. ENT is also permitted to be installed in wet locations or in a concrete slab on or below grade where the fittings are listed for the purpose. ENT is permitted as prewired assemblies in trade sizes ½″ through 1″.

ENT has a listed temperature rating associated with its construction. Where conductors or cables are installed in an ENT, the temperature rating associated with the conductor or the cable shall not exceed the listed temperature rating of the ENT. In other words, conductors and cable with temperature ratings higher than that of the ENT are permitted to be installed in the ENT, but the conductors or cables must not be operated at a temperature higher than the listed temperature rating of the ENT.

Electrical Nonmetallic Tubing (ENT) Color-Coding

Electrical nonmetallic tubing (ENT) is available in different colors to facilitate identification of the system in which it is being installed.

Section 362.12 establishes nine prohibited uses for ENT. Except as may otherwise be permitted, ENT cannot be used in any hazardous (classified) location. ENT cannot be used to support luminaires (lighting fixtures) or other equipment and cannot be used where the ambient temperature exceeds 122°F unless it is listed for higher temperatures. ENT shall not be used for direct burial in the earth or where the voltage of the circuit exceeds 600 V. In general, ENT shall not be used in exposed locations or where subject to physical damage. ENT is also prohibited from use in theaters and similar locations or where exposed to the direct rays of the sun unless it is identified as being sunlight resistant.

Because ENT is a pliable, corrugated raceway, strict support and securing provisions are included in 362.30. In general, ENT shall be securely fastened in place at intervals not exceeding 3′ and within 3′ of each outlet box, device box, junction box, or fitting where it terminates. There are, however, three exceptions to this general rule. These are intended to give latitude for connections to luminaires and where the ENT is fished.

ENT is required by 362.100 to be clearly and durably marked every 10′ with the manufacturer's name, trademark, or other distinctive marking. ENT can be supplied as a prewired raceway assembly in continuous lengths on a coil, reel, or in a carton. The type, size, and quantity of the conductors used in a prewired ENT assembly will be identified by means of a printed tag or label attached to each end of the prewired assembly or by attachment directly to the coil, reel, or carton.

Flexible Metal Conduit (FMC) – Article 348

Flexible metal conduit (FMC) is a raceway of metal strips which are formed into a circular cross-sectional raceway. **See Figure 8-6.** The metal strips are helically wound, formed, and interlocked. FMC and any associated fittings shall be listed as required by 348.6. It is permitted to be used in exposed and concealed locations. FMC is commonly known in the field as Greenfield. FMC is not permitted to be installed under the following seven conditions:

(1) In wet locations.

(2) In hoistways other than per 620.21(A)(1).

(3) In storage battery rooms.

(4) In hazardous locations except as permitted elsewhere in the code.

(5) Where exposed to materials that could deteriorate the conductors such as gasoline and oil products.

(6) Underground or in poured concrete.

(7) Where subject to physical damage.

Flexible Metal Conduit (FMC)

SIZES • 348.20	
DIAMETER	LENGTH OF COIL
⅜″	100′
½″	100′
¾″	100′
1″	50′
1¼″	50′
1½″	25′
2″	25′

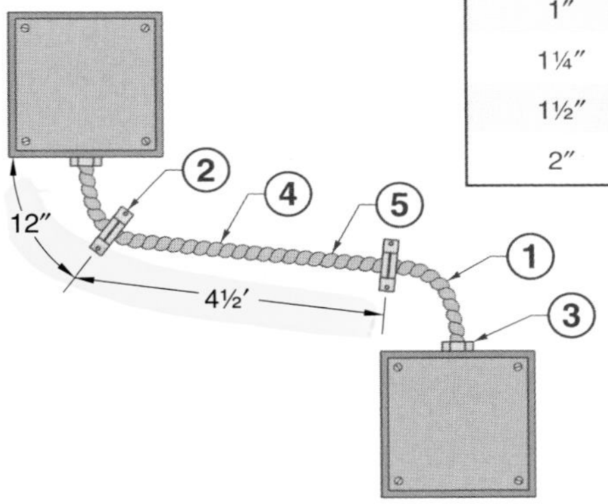

(1) NO MORE THAN 360° OF BENDS BETWEEN PULL POINTS
• 348.26

(2) FMC SHALL BE SUPPORTED WITHIN 12″ OF EACH BOX, CABINET, ETC. AND NO MORE THAN 4½′ APART
• 348.30(A)

(3) FITTINGS SHALL BE LISTED
• 348.6

(4) FMC PERMITTED IN EXPOSED AND CONCEALED LOCATIONS
• 348.10

(5) FMC NOT PERMITTED:
• 348.12
(1) WET LOCATIONS
(2) HOISTWAYS EXCEPT PER 620.21(A)(1)
(3) STORAGE BATTERY ROOMS
(4) HAZARDOUS LOCATIONS (WITH EXCEPTIONS)
(5) WHERE EXPOSED TO OIL OR GASOLINE PRODUCTS
(6) UNDERGROUND OR EMBEDDED IN CONCRETE
(7) SUBJECT TO PHYSICAL DAMAGE

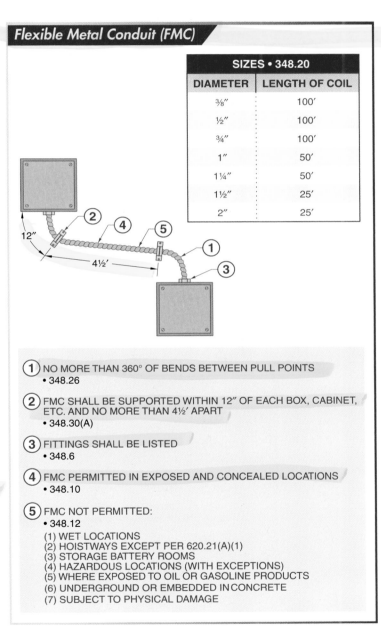

Figure 8-6. Flexible metal conduit (FMC) is a raceway of metal strips which are formed into a circular cross-sectional raceway.

FMC is manufactured in diameters from 3⁄8″ through 4″. It is shipped in three lengths based on its diameter. FMC in diameters of 3⁄8″, 1⁄2″, and 3⁄4″ is shipped in 100′ coils. FMC in diameters of 1″ and 1¼″ is shipped in 50′ coils. FMC in diameters of 1½″ and larger is shipped in 25′ coils. Nationally recognized testing laboratories do not list FMC as an EGC. They do, however, recognize FMC per 250.118(5)(a–d). If the total length in any ground path is 6′ or less, the conduit is terminated in fittings listed for grounding, and the overcurrent protective devices are rated at 20 A or less.

FMC shall be supported every 4½′ and be securely fastened in place within 12″ of each box, cabinet, or conduit body per 348.30(A). There are four exceptions to this rule:

- Ex. 1 – Where FMC is fished and supporting is impractical.
- Ex. 2 – Lengths of FMC not exceeding 3′–5′ may be used at terminals where flexibility is required.
- Ex. 3 – Lengths of FMC not exceeding 6′ may be used from a fixture terminal connection for tap connections to luminaires per 410.117(C).
- Ex. 4 – Lengths not exceeding 6′ for connections to luminaires within accessible ceilings.

Type AC cable is a fabricated assembly of insulated conductors in a flexible metallic enclosure.

Section 348.42 prohibits the use of angle connectors where they are concealed. An angle connector for FMC makes a 90° bend in a very short radius. Angle connectors have a backplate which is removed when installing the conductors. Once the conductors are installed through the angle connector, the backplate is reattached. It is very difficult, if not impossible, to fish conductors through an angle connector with the backplate attached. FMC angle connectors shall be accessible after installation.

Liquidtight Flexible Metal Conduit (LFMC) and Liquidtight Flexible Nonmetallic Conduit (LFNC) – Article 350, Article 356

Liquidtight flexible metal conduit (LFMC), Article 350, and liquidtight flexible nonmetallic conduit (LFNC), Article 356, are raceways of circular cross section. **See Figure 8-7.** *Liquidtight flexible metal conduit (LFMC)* is a raceway of circular cross section with an outer liquidtight, nonmetallic, sunlight-resistant jacket over an inner helically wound metal strip. The most common name for this product is "Sealtite," which is a registered name for the product manufactured by the Anaconda Co., Metal Hose Division. There are associated couplings, connectors, and fittings that shall be used when running this raceway. Sizes smaller than 1⁄2″ are prohibited, except for 3⁄8″ size where used for flexible connections to motors and 3⁄8″ size as part of an approved assembly or for fixture whips not over 6′. See 350.20(A).

LFMC shall be supported every 4½′ and within 12″ of every outlet per 350.30(A). There are four exceptions to this rule:

- Ex. 1 – Where LFMC is fished and supporting is impractical.
- Ex. 2 – Lengths of LFMC not exceeding 3′–5′ at terminals where flexibility is necessary.
- Ex. 3 – Lengths of LFMC not exceeding 6′ for luminaire whips per 410.117(C).
- Ex. 4 – Lengths not exceeding 6′ for connections to luminaires within accessible ceilings.

The outer jacket of LFMC permits use in wet locations or where exposed to mineral oils, grease, and other chemicals. LFMC is not, however, intended to be installed in locations where it is exposed to petroleum products unless identified for such use.

LFMC shall not be used in areas where subject to physical damage per 350.12(1). Per 350.12(2), it shall not be used where any combination of ambient and conductor temperatures produces an operating temperature in excess of that for which the material is approved. LFMC is permitted for direct burial in the earth per 350.10(3) if it is listed and marked for such use.

Per 250.118(6), LFMC is permitted to be used as an EGC in sizes of 1¼″ or less, provided the following conditions are met:

- The LFMC is terminated in listed fittings.
- For trade sizes ⅜″ and ½″, the conductors within the LFMC are protected by an overcurrent device rated 20 A or less.
- For trade sizes ¾″ through 1¼″, the conductors within the LFMC are protected by an overcurrent device rated 60 A or less.
- There is no more than 6′ of LFMC, FMC, or FMT in the ground-fault current path.
- The LFMC is not installed for flexibility or to minimize the transmission of vibration from equipment.

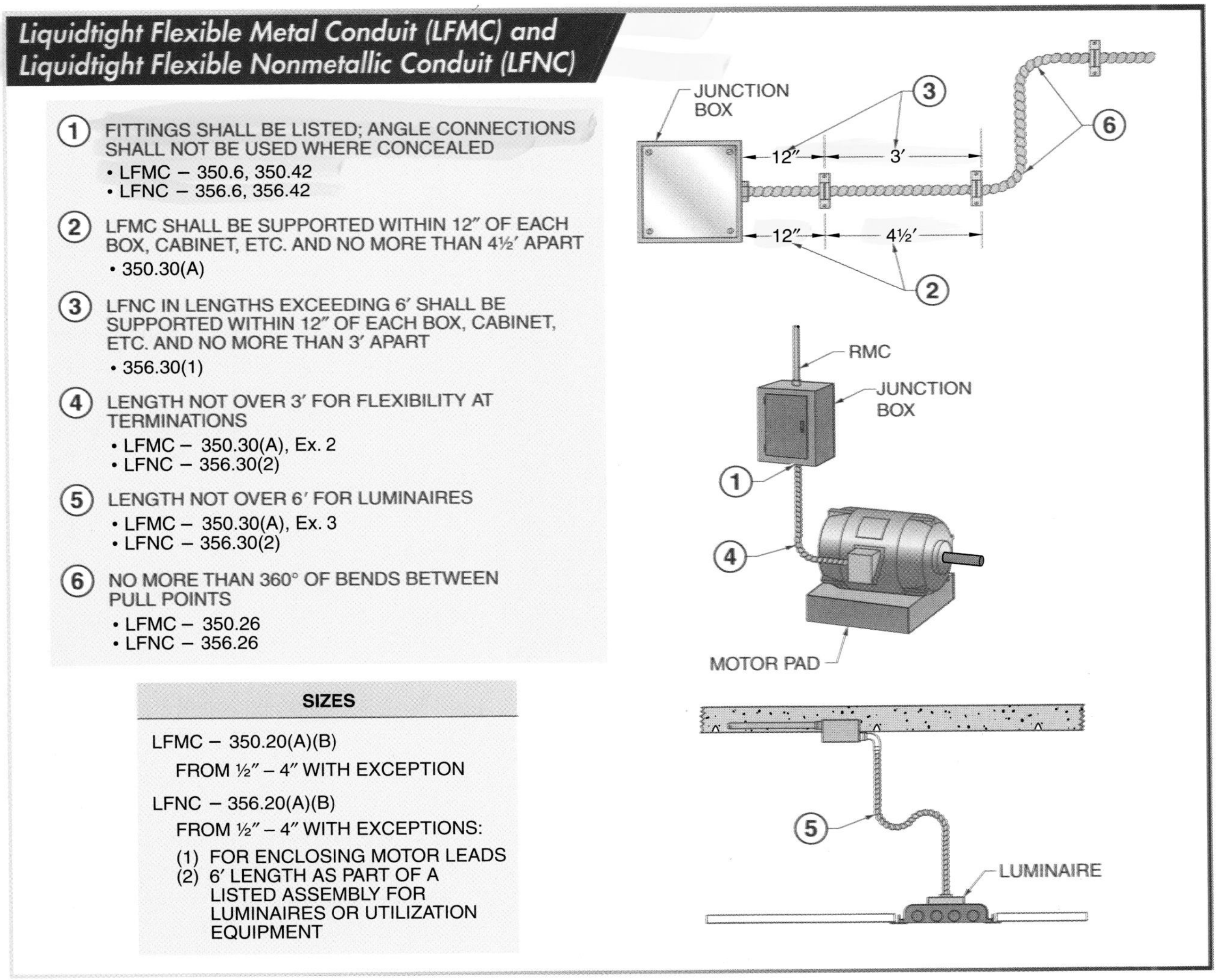

Figure 8-7. Liquidtight flexible metal conduit (LFMC) and liquidtight flexible nonmetallic conduit (LFNC) are raceways of circular cross section.

Underfloor raceways are permitted to be installed beneath the surface of concrete or other flooring material.

Liquidtight flexible nonmetallic conduit (LFNC) is a raceway of circular cross section with an outer jacket and an inner core and reinforcement that varies based on intended use. The provisions for LFNC are included in Article 356. The first type has a smooth, seamless inner core and cover bonded together, making it look somewhat like a heavy-duty garden hose. The second type has a smooth inner surface. On the outside, it looks like LFMC. The third type has a corrugated internal and external surface and is similar in appearance to ENT. None of these three types has a helically wound metal strip on the inside. One of the main applications for LFNC is on industrial machinery. For example, the wiring associated with a lathe, milling machine, or punch press may be run in LFNC.

Listed

During the 1996 NEC® cycle, CMP 8 included the word "listed" for several raceway systems. This listing requirement now ensures that these raceway systems are manufactured to a published standard which states the required physical and electrical characteristics upon which the requirements are based. It also ensures that the follow-up service of the listing agency helps maintain this consistency.

It is interesting that CMP 7 has not followed the same path with cable wiring methods. With the exception of type NM and type UF, most of the cable wiring methods do not have a listing requirement.

Strut-Type Raceways – Article 384

A *strut-type channel raceway* is a surface raceway formed of moisture-resistant and corrosion-resistant metal. These channel raceways may be galvanized, stainless, enameled, PVC coated, or aluminum. These raceways are intended to be mounted to the surface or suspended from a structure. Covers shall be either metallic or nonmetallic. **See Figure 8-8.** Section 384.10 permits these raceways to be used:

(1) Where exposed.

(2) In dry locations.

(3) In locations subject to corrosive vapors where protected by finishes judged suitable for the conditions.

(4) Where voltage does not exceed 600 V.

(5) As power poles.

(6) In Class I, Division 2 locations where wiring is per 501.10(B)(3).

(7) As unbroken extensions through walls, floors and partitions.

(8) Indoor locations only where protected from corrosion solely by enamel.

Section 384.12(1) prohibits the use of strut-type channel raceways where the raceway is concealed. Section 384.22 specifies that the number of conductors permitted in strut-type raceways shall not exceed the percentage fill shown in Table 348.22. The derating factors of 310.15(B)(3)(a) do not apply to conductors installed in strut-type raceways when all of the following conditions are met:

(1) The cross-sectional area of the raceway exceeds 4 sq in.

(2) There are no more than 30 current-carrying conductors.

(3) The sum of the cross-sectional areas of all the conductors does not exceed 20% of the interior cross-sectional area of the strut-type channel raceway.

Section 384.30(A) requires that strut-type channel raceways be secured at least every 10′ and within 3′ of each outlet box, junction box, or other channel raceway termination.

Section 384.60 contains the special provisions for grounding strut-type channel raceways. In general, a means to connect an EGC must be provided for strut-type channel raceways where a transition is provided to or from the strut-type raceway to other wiring methods. Note that 250.118(13) does generally permit strut-type channel raceways to serve as the EGC. Additionally, if the strut-type raceway includes a metal snap-fit cover that is used to ensure continuity of the raceway and is part of the raceway listing, the raceway cover is not permitted for providing the necessary continuity for a receptacle mounted in the raceway cover. In this case, a separate EGC would need to be installed.

Surface Metal and Nonmetallic Raceways – Article 386, Article 388

A *surface raceway* is an enclosed channel for conductors which is attached to a surface. **See Figure 8-9.** Surface raceways are available in several sizes. Each size has a complete line of boxes and accessories needed to install that size surface raceway as a system. There are a number of products that fall under the requirements of Article 386. These systems are very convenient to use when adding to or extending an existing system in an installation where fishing cables is difficult. These raceways are for installation in dry locations.

Section 386.10(4) permits surface metal raceways to pass through dry walls, dry partitions, and dry floors provided the length passing through is unbroken. Access to the conductors shall be maintained on both sides of the wall, partition, or floor.

There are five specific prohibited applications for surface metal raceways. Unless the AHJ has specifically otherwise approved the use, surface metal raceways are not permitted to be installed where they are subject to physical damage. Surface metal raceways cannot be installed where the voltage is 300 V or more between conductors unless the metal thickness is 0.040″ or greater. Surface metal raceways are not permitted to be installed where corrosive vapors are present. Surface metal raceways cannot be used in hoistways or where concealed.

Section 386.56 permits splices and taps in surface metal raceways that have removable covers that provide access to the conductors after installation. Conductors, including splices and taps, shall not fill the raceway to more than 75% of its cross-sectional area.

Section 386.60 requires a means to connect an EGC to the surface metal raceway enclosures when a transition is made to another wiring method. For example, a Romex cable terminated in the back of a metallic wiremold outlet enclosure requires a means to connect the EGC of the cable to the metal enclosure.

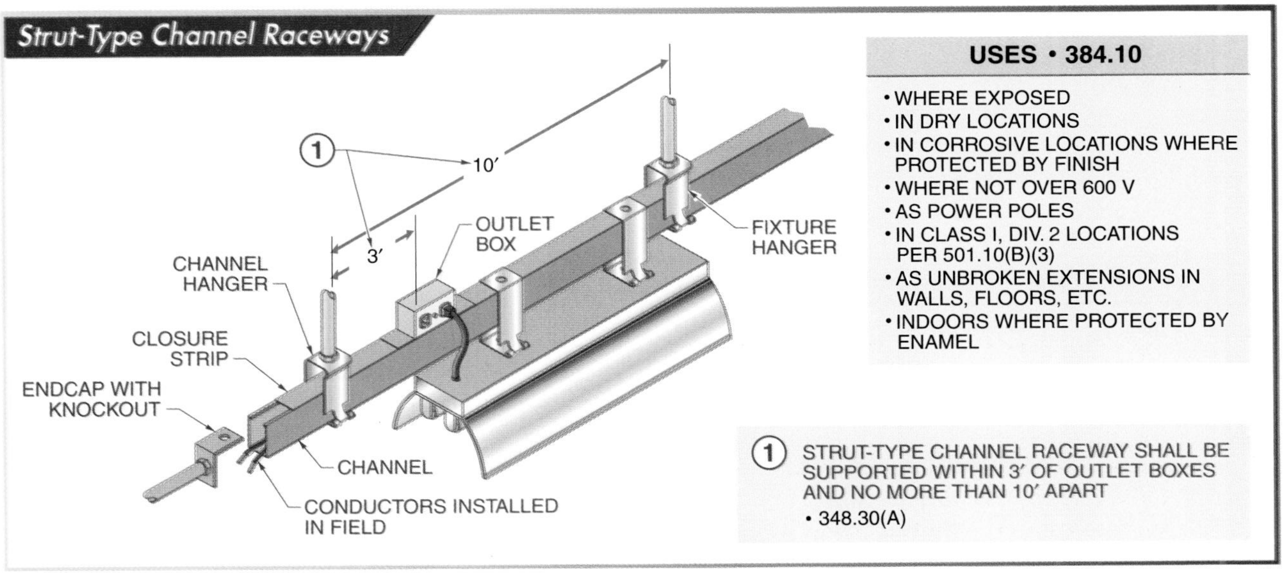

Figure 8-8. A strut-type channel raceway is a surface raceway formed of moisture-resistant and corrosion-resistant metal.

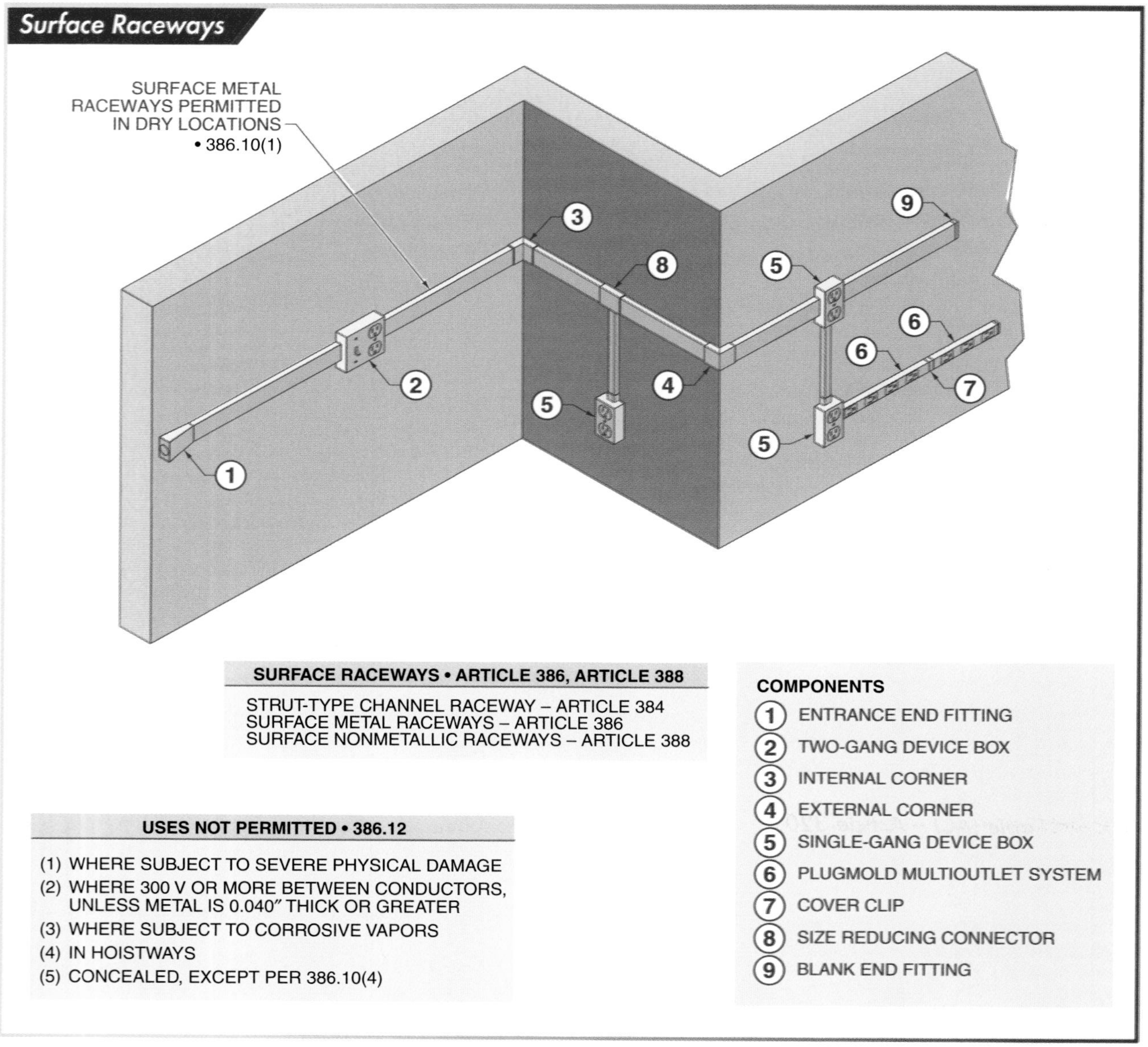

Figure 8-9. A surface raceway is an enclosed channel for conductors which is attached to a surface.

Installing Surface Metal Raceway Systems

When installing a surface metal raceway system, verify that the metal is mechanically continuous and all connections are clean and tight. This aids in ensuring a low-impedance path in the event of a ground fault.

Multioutlet Assembly – Article 380. A multioutlet assembly is a surface, flush, or freestanding raceway which contains conductors and receptacles. **See Figure 8-10.** There is no provision for field installation of additional conductors except where the product is marked to indicate the number, type, and size of additional conductors which may be field installed. Multioutlet assemblies are similar to surface raceways. The difference is that they contain conductors and receptacles which are assembled in the field or at the factory. Multioutlet assemblies are popular for work benches, laboratories, countertop areas, and for upgrading older systems to meet today's requirements.

Section 380.2(B) lists six applications not permitted for multioutlet assemblies. Generally, they shall not be installed:

(1) Where concealed.

(2) Where subject to severe physical damage.

(3) Where voltage exceeds 300 V.

(4) Where subject to corrosive vapors.

(5) In hoistways.

(6) In hazardous locations.

Section 380.76 permits the extension of metal multioutlet assemblies through (not within) dry partitions. Provisions shall be made so that the cap or cover is removable on all exposed portions and no outlet is located within the partitions.

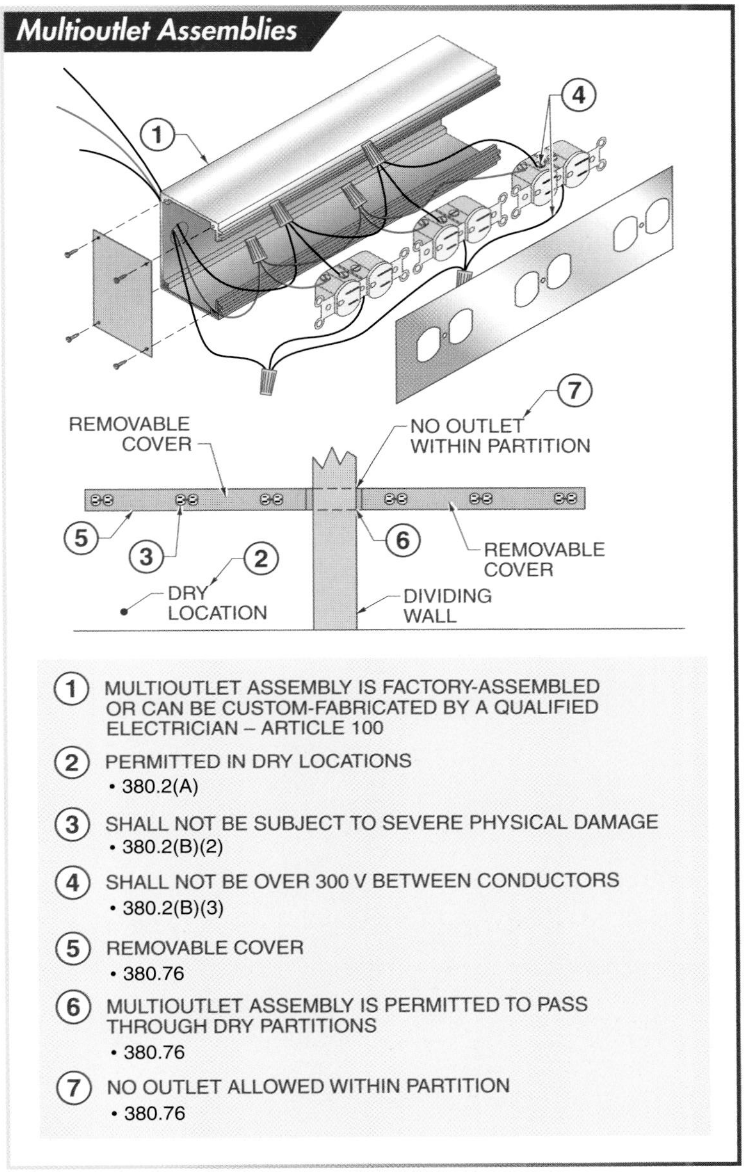

Multioutlet Assemblies
REMOVABLE COVER
NO OUTLET WITHIN PARTITION
REMOVABLE COVER
DRY LOCATION
DIVIDING WALL
1 MULTIOUTLET ASSEMBLY IS FACTORY-ASSEMBLED OR CAN BE CUSTOM-FABRICATED BY A QUALIFIED ELECTRICIAN – ARTICLE 100
2 PERMITTED IN DRY LOCATIONS • 380.2(A)
3 SHALL NOT BE SUBJECT TO SEVERE PHYSICAL DAMAGE • 380.2(B)(2)
4 SHALL NOT BE OVER 300 V BETWEEN CONDUCTORS • 380.2(B)(3)
5 REMOVABLE COVER • 380.76
6 MULTIOUTLET ASSEMBLY IS PERMITTED TO PASS THROUGH DRY PARTITIONS • 380.76
7 NO OUTLET ALLOWED WITHIN PARTITION • 380.76

Figure 8-10. A multioutlet assembly is a metal raceway with factory-installed conductors and attachment plug receptacles.

CABLE ASSEMBLIES

A *cable assembly* is a flexible assembly containing multiconductors with a protective outer sheath. The purpose of this outer sheath is to protect the conductors contained therein. Four of the more popular types of cable assemblies are armored cable (AC), metal-clad cable (MC), nonmetallic-sheathed cable (NMC), and service-entrance cable (SE).

Armored Cable (AC) – Article 320

Armored cable (AC) is a factory assembly that contains the conductors within a jacket made of a spiral wrap of steel. **See Figure 8-11.** A bare bonding strip is placed inside the armor with the conductors. This bare bonding strip is in intimate contact with the outer metal jacket for its entire length to ensure a low-impedance path on the outer armor in the event of a ground fault. AC cable may only be used in dry locations.

Section 320.30(B) requires that AC cable be supported every 4½′ and within 12″ of outlets and fittings unless otherwise permitted in the Code. Section 320.40 requires a connector to be used when terminating AC cable in a box, cabinet, or piece of equipment. The purpose of this connector is to protect the conductors within the cable from abrasion. An insulating bushing (anti-short) shall be used with the connector. This bushing slips over the conductors and under the armor of the cable for additional protection.

Section 320.30(D) allows for unsupported sections of AC cable under three different situations:

(1) Where fished and concealed in finished buildings and support is impracticable.

(2) No more than 24″ at terminals where flexibility is necessary.

(3) No more than 6′ from the last point of support for connecting to a luminaire (lighting fixture) within an accessible ceiling.

Armored Cable (AC)

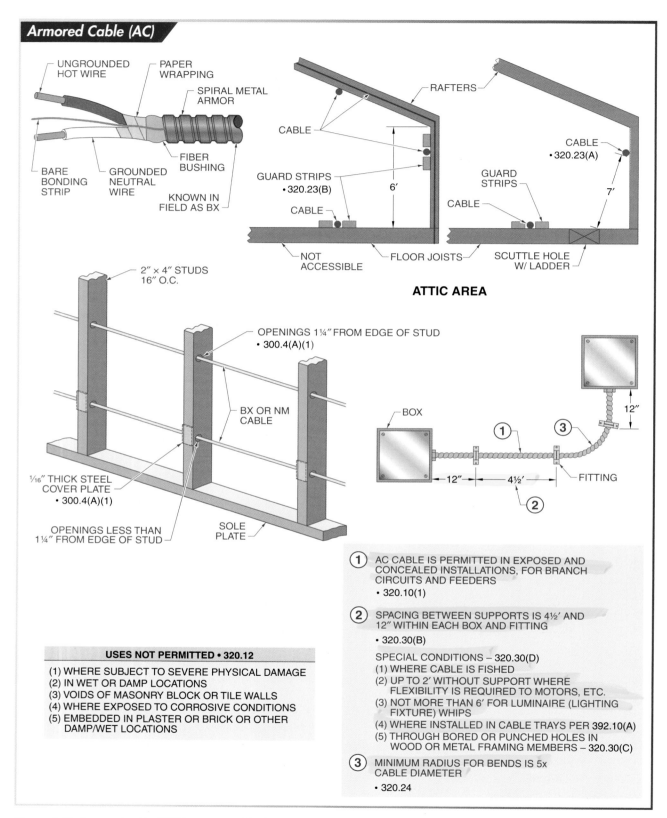

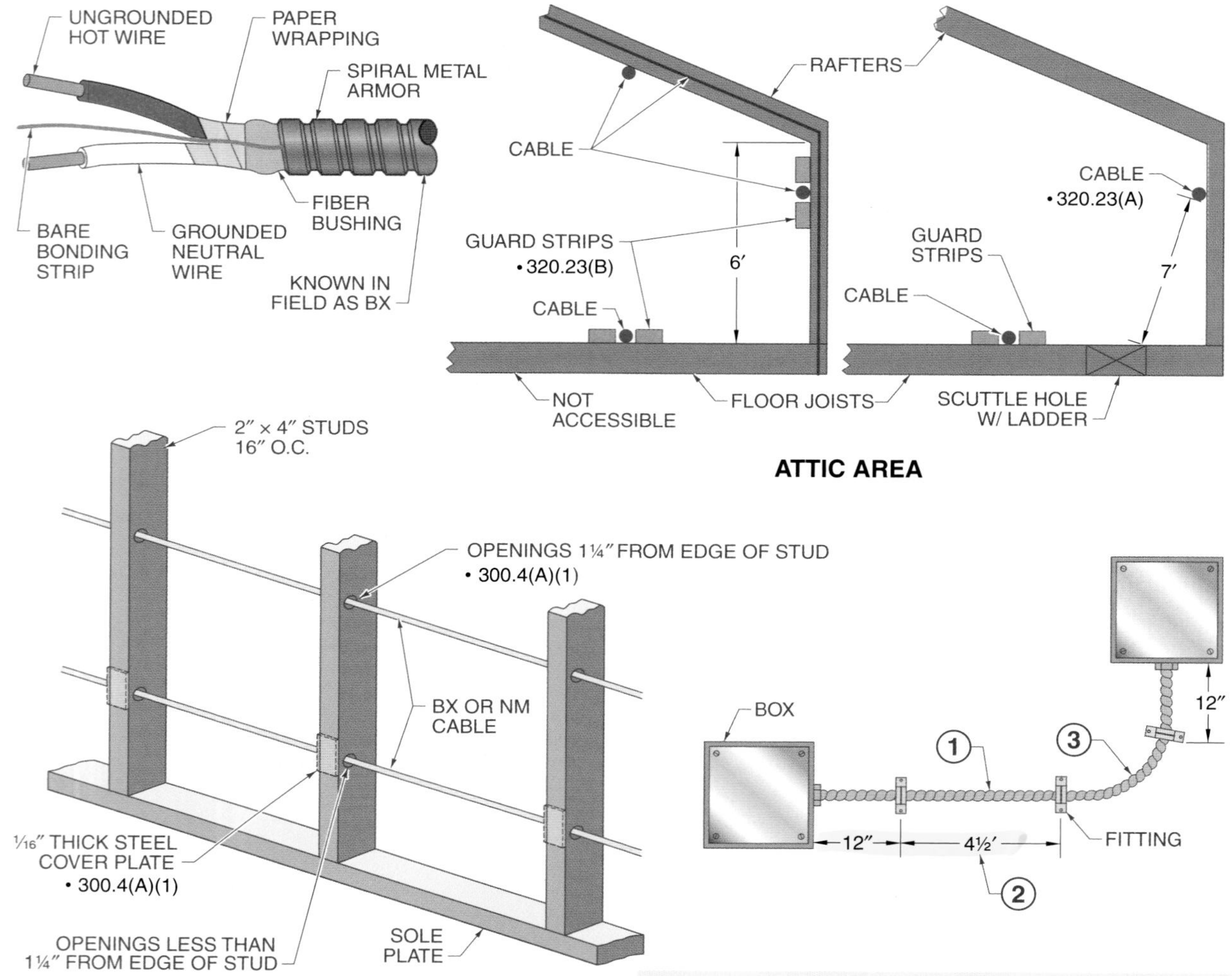

USES NOT PERMITTED • 320.12
(1) WHERE SUBJECT TO SEVERE PHYSICAL DAMAGE
(2) IN WET OR DAMP LOCATIONS
(3) VOIDS OF MASONRY BLOCK OR TILE WALLS
(4) WHERE EXPOSED TO CORROSIVE CONDITIONS
(5) EMBEDDED IN PLASTER OR BRICK OR OTHER DAMP/WET LOCATIONS

(1) AC CABLE IS PERMITTED IN EXPOSED AND CONCEALED INSTALLATIONS, FOR BRANCH CIRCUITS AND FEEDERS
• 320.10(1)

(2) SPACING BETWEEN SUPPORTS IS 4½′ AND 12″ WITHIN EACH BOX AND FITTING
• 320.30(B)

SPECIAL CONDITIONS – 320.30(D)
(1) WHERE CABLE IS FISHED
(2) UP TO 2′ WITHOUT SUPPORT WHERE FLEXIBILITY IS REQUIRED TO MOTORS, ETC.
(3) NOT MORE THAN 6′ FOR LUMINAIRE (LIGHTING FIXTURE) WHIPS
(4) WHERE INSTALLED IN CABLE TRAYS PER 392.10(A)
(5) THROUGH BORED OR PUNCHED HOLES IN WOOD OR METAL FRAMING MEMBERS – 320.30(C)

(3) MINIMUM RADIUS FOR BENDS IS 5x CABLE DIAMETER
• 320.24

Figure 8-11. Armored cable (AC) is an assembly that contains the conductors within a jacket made of a spiral wrap of steel.

Section 320.23(B) requires the AC cable be installed at least 1¼″ back from the finished edge of framing members. This protects the cable from nails and drywall screws per 300.4(D).

Section 320.15 requires exposed runs of AC cable, except as provided for in section 300.11(A), to be installed closely to the finish of the building or along running boards. Exposed runs may also be run on the underside of joists, where supported at each joist and not exposed to physical damage.

Section 320.23 permits the use of AC cables in accessible attics. AC cables installed across the top of floor joists require protection by using guard strips that are at least as high as the cable. Guard strips shall also be used when AC cable is installed across the face of rafters or studs that are less than 7′ above the attic floor.

AC cable installed in attics that are inaccessible (without stairways or pull-down ladders) only require guard strips within 6′ of the scuttle-hole opening. Within inaccessible attics, AC cables may be run along the sides of joists, rafters, and studs or fished through drilled holes without guard strips. Guard strips are usually 1″ × 2″ wooden strips nailed directly to the structure.

Medium Voltage Cable – Article 328

Medium voltage cable (Type MV cable) is a single or multiple copper, aluminum, or copper-clad aluminum conductor cable that is constructed of a solid dielectric insulation rated 2001 V or higher. Type MV cable is used for voltages up to 35,000 V. Section 328.10 permits the use of Type MV cable:

(1) In both wet and dry locations.

(2) In raceways.

(3) In cable trays where identified for the intended use and installed per Article 392.

(4) Through direct burial in accordance with 300.50.

(5) In messenger-supported wiring. See Article 396.

(6) For exposed runs in accordance with 300.37.

Section 328.12 prohibits the use of Type MV cable where it is exposed to direct sunlight. Cable jackets and insulations can be damaged by exposure to direct sunlight. For this reason, they are not permitted to be used in that application unless they have been identified for the use. Sunlight-resistant marking ensures that the cables or insulated conductors have been tested and identified for that use. See 310.10(D). The use of Type MV cable is also limited in cable trays and where directly buried. Article 392 provides the rule for cable tray use, and any direct burial of Type MV cable must be in accordance with 300.50.

Section 328.14 was added to the 2011 NEC® and contains an important provision regarding the installation of Type MV cable. Because of the nature of the work associated with the installation of Type MV cable, only qualified persons may install, terminate, and test it. As defined in Article 100, a qualified person is one who is familiar with the construction and operation of electrical equipment and installations and who has received appropriate safety training to ensure that they can recognize and avoid the hazards involved.

Section 328.80 specifies that the ampacity of Type MV cable must be determined by 310.60. Section 310.60(B)(1) requires that where more than one calculated or tabulated ampacity can be applied for a given circuit of conductors rated 2001 V to 35,000 V, the lowest value shall apply. There is an exception to 310.60 that permits the higher ampacity to be applied where there is a transition between different points in the circuit and where the distance equals 10′ or 10% of the circuit length calculated at the higher ampacity. The ampacity shall be the lesser of the two values.

Redundant Grounding

Redundant grounding is required in several places in the NEC®. Redundant grounding is grounding with two separate grounding paths. For example, in patient care areas per 517.13, the metal raceway or recognized outer metallic sheath of a smooth tube or corrugated tube MC cable counts as one grounding path, and an insulated EGC contained within the metal raceway or recognized metallic sheath counts as the second path. Interlocking type MC cable does not meet NEC® requirements for redundant grounding.

Metal-Clad (MC) Cable – Article 330

Metal-clad (MC) cable is a factory assembly of one or more conductors with or without fiber-optic members. **See Figure 8-12.** MC cable does not have an internal bonding strip. The conductors are enclosed in a metal sheath with one of the following configurations per 330.2: smooth sheath, corrugated sheath, or interlocked sheath.

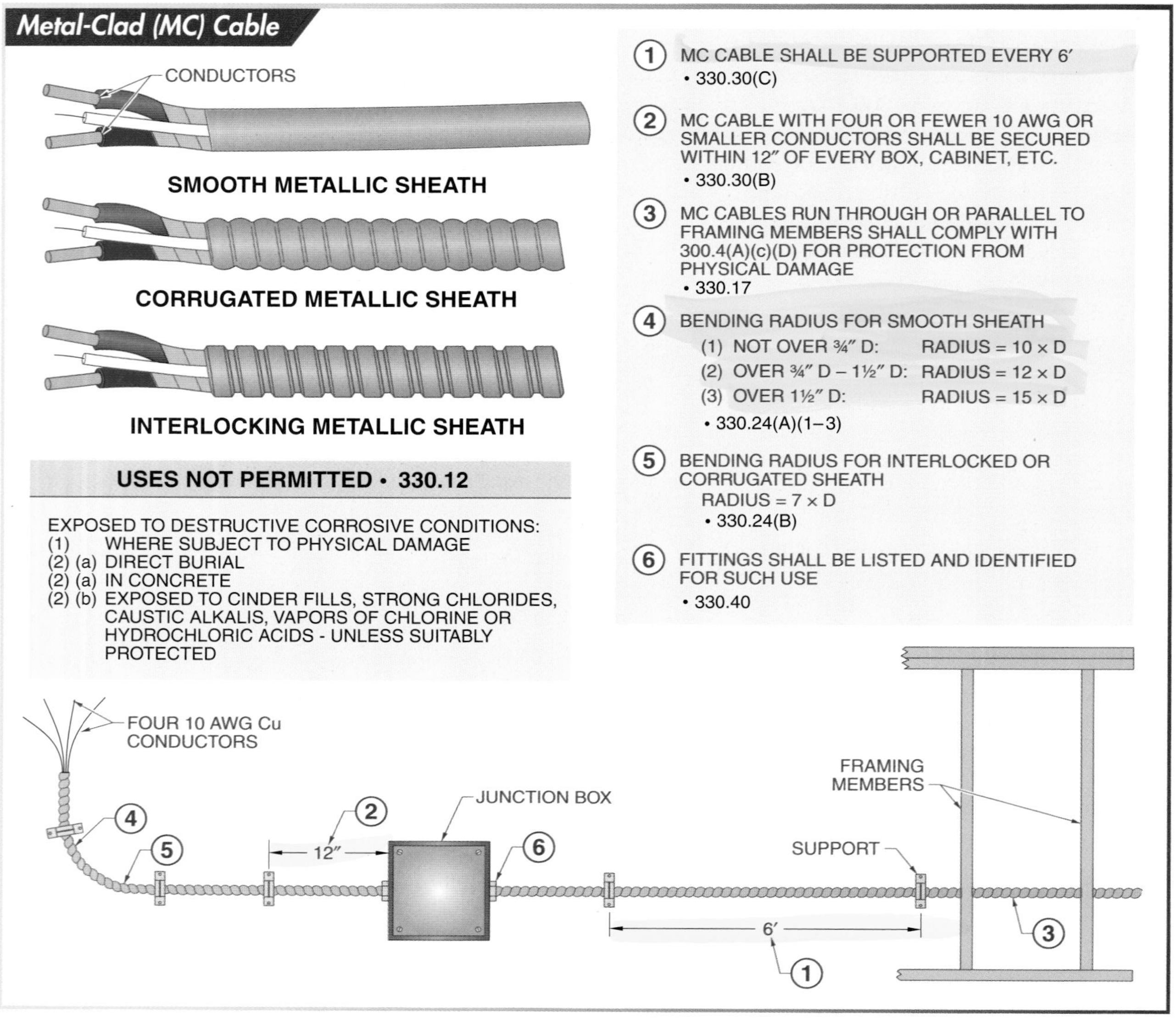

Figure 8-12. Metal-clad (MC) cable is a factory assembly of one or more conductors with or without fiber optic members.

Smooth Sheath—The outer metallic sheath is smooth and continuous. This type of sheath makes an excellent EGC provided it is terminated with connectors listed for this type of cable and where it meets the requirements of 250.118(10). Since the outer sheath is metallic, smooth, and continuous, it is a little difficult to bend.

Corrugated Sheath—This type has some of the metallic sheath grooved out to make it more flexible. It looks like a piece of flexible gas tubing used to hook up gas appliances. The outer sheath is a continuous piece of metal and makes an excellent EGC.

Interlocking Sheath—This type has an outer jacket which looks like AC cable. The outer jacket is spiraled and has the electrical characteristics of a coil. When used in an AC circuit, it offers inductive reactance which causes a high impedance, thus opposing the flow of ground-fault current. With this type of MC cable, the manufacturer always provides an insulated green EGC for the grounding.

Type MC cable is manufactured with conductor sizes of 18 AWG up to 1000 kcmil. MC cable is listed up to 5000 V by UL. MC cable is available with insulations rated as high as 15,000 V.

Section 330.10(A)(1) permits MC cable to be used for services, feeders, and branch circuits. It may be used for power, lighting, control and signal circuits, indoors or outdoors, exposed or concealed, in any approved raceway, as open runs, as aerial cable on a messenger, in hazardous locations where specifically permitted in the Code, and in any wet location (provided the cable is approved for that location) with a metallic covering that is impervious to moisture or with insulated conductors under the metallic covering approved for wet locations.

Section 330.12 prohibits the use of MC cable where exposed to destructive corrosive conditions such as direct burial in the earth, in concrete, or where exposed to cinder fills, strong chloride, caustic alkalis, or vapors of chlorine or hydrochloric acids. Section 330.30(C) requires MC cable to be supported and secured at intervals not exceeding 6′. Section 330.30(B) requires cables containing four or fewer conductors sized no larger than 10 AWG to be secured within 12″ of every box, cabinet, or fitting.

The radius of the curve of the inside bend of MC cable varies with the type of outer covering and the diameter of the cable per 330.24.

(A) Smooth sheath—(1) Not more than ¾″ in diameter shall have a minimum bending radius of a minimum of 10 times the diameter of the cable. (2) More than ¾″ in diameter but not more than 1½″ in diameter shall have a minimum bending radius of a minimum of 12 times the diameter of the cable. (3) More than 1½″ in diameter shall have a minimum bending radius of a minimum of 15 times the diameter of the cable.

(B) Interlocked or corrugated sheath—Shall have a bending radius of not less than seven times the external diameter of the metallic sheath.

(C) Shielded conductors—Shall have a bending radius of twelve times the overall diameter of one of the individual conductors or seven times the overall diameter of the multiconductor cable, whichever is greater.

Per 330.104, MC cable conductors shall be of copper, aluminum, copper-clad aluminum, nickel, or nickel-coated copper, solid or stranded. The minimum conductor size shall be 18 AWG copper, nickel, or nickel-coated copper, and 12 AWG aluminum or copper-clad aluminum.

Mineral-Insulated, Metal-Sheathed Cable – Article 332

Mineral-insulated, metal-sheathed cable (Type MI cable) is a factory assembled cable construction that consists of one or more conductors that are insulated with a highly compressed refractory mineral insulation. The conductors and insulation are enclosed within a liquidtight and gastight continuous copper or alloy outer sheath. The conductors can be constructed of copper, nickel, or nickel-coated copper. The outer sheath not only provides mechanical protection and protection against the entrance of moisture, but it also is permitted by 332.108 to be used as the equipment grounding conductor if the outer sheath is constructed of copper. If the outer sheath is made of steel, a separate equipment grounding conductor must be used.

Because of the construction specifications of the cable assembly, Type MI cable is being used in more diversified applications. Section 332.10 list eleven permitted uses for Type MI cable:

(1) Services, feeders, and branch circuits.

(2) Power, lighting, and signal circuits.

(3) Dry, wet, or continuously moist locations.

(4) Indoor or outdoor locations.

(5) In exposed or concealed locations.

(6) Embedded in plaster, concrete, or masonry.

(7) In hazardous locations where specifically permitted in the Code.

(8) Where exposed to oil and gasoline.

(9) Where exposed to corrosive vapors.

(10) In underground runs, if suitably protected.

(11) In or attached to cable tray.

Section 332.12 prohibits the use of Type MI cable in two cases. The first case is where it is installed in underground runs and not suitably protected. The protection must be from physical damage and corrosive conditions. The second case is where the cable is exposed to destructive or corrosive environments. Type MI cable installed in these locations must have supplemental protection for the outer jacket, such as a PVC coating.

Special precautions must be provided to support Type MI cable. It is generally not permitted to be installed unsupported unless it is fished between access points in concealed spaces in finished buildings where supporting is not practicable. See 332.30(B). The general support requirements for Type MI cable are every 6′. See 332.30.

Section 332.80 states that the ampacity of Type MI cable must be in accordance with 310.15. Type MI cable uses special end fittings or termination fittings to ensure that the mineral insulation remains intact and moisture free. The conductor temperature at the end fitting must not exceed the temperature rating of the listed end seal fitting.

Nonmetallic-Sheathed Cable – Article 334

Nonmetallic-sheathed (NM) cable is a factory assembly of two or more insulated conductors having an outer sheath of moisture-resistant, flame-retardant, nonmetallic material. **See Figure 8-13.** NM is an economical wiring method used in residential units and small commercial locations. It is also known in the field as "Romex."

NM comes with a bare EGC. The circuit conductors are manufactured in sizes 14 AWG through 2 AWG. It is available in copper, aluminum, or copper-clad aluminum. It is commonly shipped in 250′ rolls.

Until 1984, NM was manufactured with TW insulation which has a temperature rating of 60°C for Type A. Since 1984, the conductors have been made with THHN insulation which has a temperature rating of 90°C for Type B. This change was initiated due to higher R values of insulation being installed in the attics of dwelling units. The Type A NM cable did not dissipate the heat under the thicker type insulation and, as a result, many of these cables failed. Caution should be taken not to use the ampacity from the 90°C column in Table 310.15(B)(16) when installing Type B NM cable. The value from the 60°C column shall be the maximum ampacity permitted. See 334.80.

Uses Permitted – 334.10. Nonmetallic-sheathed cable is permitted for use in one-family, two-family, and multifamily dwelling units and in other structures except as prohibited by 334.12. The three types of nonmetallic-sheathed cable are NM, NMC, and NMS.

Type NM cable has a flame-retardant and moisture-resistant outer jacket that is restricted to indoor wiring. Type NMC cable has a flame-retardant, moisture-, fungus-, and corrosion-resistant outer jacket. NMC cable is permitted to be installed in outdoor locations. Type NMS cable was new to the 1996 NEC®. It is used in closed-loop and programmed power distribution systems.

The use of NM cable in any building or structure designated in NFPA 220 as Type I or II is prohibited by 334.12(A)(1) unless the NM, NMC, or NMS cables are installed in raceways permitted in Type I or II construction. The first level is not counted if it is used for vehicle parking, storage, or similar use.

NM, NMC, and NMS shall not be used in commercial garages, theaters, motion picture studios, hoistways, or hazardous locations. NM and NMS shall not be embedded in plaster or used in corrosive locations.

Type NM cable, such as Romex, is available in 25′, 50′, 100′, and 250′ rolls for ease of installation.

Installation – Article 334, Part II. Section 334.15 defines the provisions for installing NM when exposed. The cable shall closely follow the surface of the building finish or running boards. It shall be protected whenever it passes through a floor in an exposed work area. This protection may be provided by a piece of conduit, EMT, or Schedule 80 PVC. The protective sleeve shall extend a minimum of 6″ above the floor.

Cables in unfinished basements shall be installed through bored holes in joists or on running boards. Three-conductor 8 AWG or two-conductor 6 AWG or larger cables may be installed on the lower edges of the joist. Per 334.30, NM cable shall be secured every 4½′ and within 12″ of every outlet or fitting.

Nonmetallic-Sheathed (NM) Cable

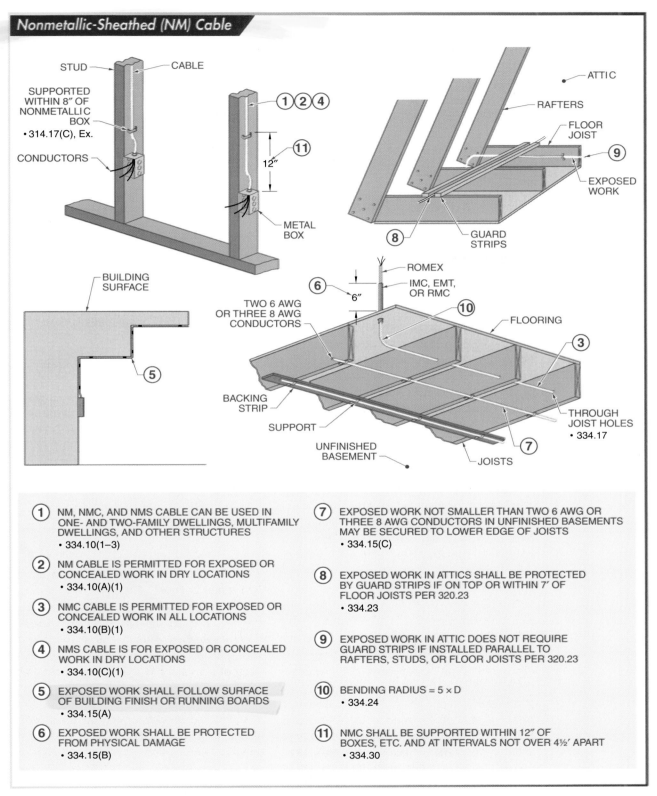

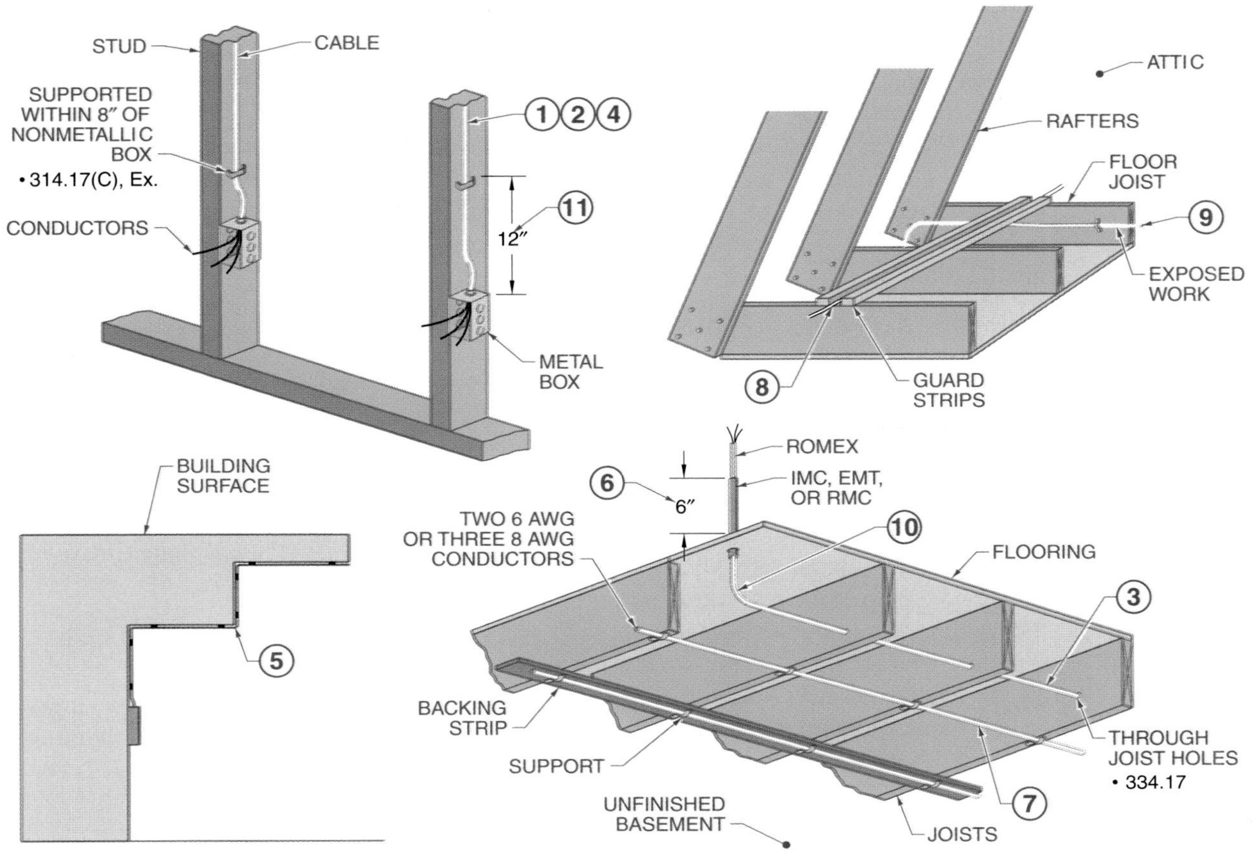

① NM, NMC, AND NMS CABLE CAN BE USED IN ONE- AND TWO-FAMILY DWELLINGS, MULTIFAMILY DWELLINGS, AND OTHER STRUCTURES
• 334.10(1–3)

② NM CABLE IS PERMITTED FOR EXPOSED OR CONCEALED WORK IN DRY LOCATIONS
• 334.10(A)(1)

③ NMC CABLE IS PERMITTED FOR EXPOSED OR CONCEALED WORK IN ALL LOCATIONS
• 334.10(B)(1)

④ NMS CABLE IS FOR EXPOSED OR CONCEALED WORK IN DRY LOCATIONS
• 334.10(C)(1)

⑤ EXPOSED WORK SHALL FOLLOW SURFACE OF BUILDING FINISH OR RUNNING BOARDS
• 334.15(A)

⑥ EXPOSED WORK SHALL BE PROTECTED FROM PHYSICAL DAMAGE
• 334.15(B)

⑦ EXPOSED WORK NOT SMALLER THAN TWO 6 AWG OR THREE 8 AWG CONDUCTORS IN UNFINISHED BASEMENTS MAY BE SECURED TO LOWER EDGE OF JOISTS
• 334.15(C)

⑧ EXPOSED WORK IN ATTICS SHALL BE PROTECTED BY GUARD STRIPS IF ON TOP OR WITHIN 7′ OF FLOOR JOISTS PER 320.23
• 334.23

⑨ EXPOSED WORK IN ATTIC DOES NOT REQUIRE GUARD STRIPS IF INSTALLED PARALLEL TO RAFTERS, STUDS, OR FLOOR JOISTS PER 320.23

⑩ BENDING RADIUS = 5 × D
• 334.24

⑪ NMC SHALL BE SUPPORTED WITHIN 12″ OF BOXES, ETC. AND AT INTERVALS NOT OVER 4½′ APART
• 334.30

Figure 8-13. Nonmetallic-sheathed (NM) cable is a factory assembly of two or more insulated conductors having an outer sheath of moisture-resistant, flame-retardant, nonmetallic material.

Two Romex cables shall not be stapled on top of one another. Cables run through holes in wood or metal joists, rafters, or studs shall be considered to be supported and secured.

Metal or nonmetallic boxes may be used with NM cable. The cable shall be supported within 12″ of a metal box. NM cable shall be supported within 8″ of a nonmetallic box with at least ¼″ of the cable's outer jacket inside the box.

Power and Control Tray Cable – Article 336

Power and control tray cable (Type TC cable) is a factory assembled cable consisting of one or more insulated copper, aluminum, or copper-clad aluminum conductors. The conductors are contained within a flame-retardant, nonmetallic jacket that may or may not contain bare or covered grounding conductors. The insulated conductors are generally in sizes 18 AWG through 1000 kcmil when made of copper, and sizes 12 AWG through 1000 kcmil when made of aluminum or copper-clad aluminum.

Section 336.10 lists eight permitted uses for Type TC. The cable is permitted generally for lighting, control, and signal circuits. It is permitted to be installed in cable trays, raceways, or outdoor locations where supported by a messenger wire. Type TC is permitted in Article 725 for Class 1 circuits and for non-power-limited fire alarm circuits in accordance with 760.49.

Section 336.10(7) is a special-use application for Type TC that permits its use outside of a cable tray system or raceway in industrial establishments under specified conditions. Type TC is permitted, under these conditions, to be installed where the cable is continuously supported and protected against physical damage. Essentially, the application is as open wiring where properly protected and secured. The cable must be secured at intervals not exceeding 6′. Section 336.10(8) requires that where Type TC is installed in wet locations, the cable must be resistant to moisture and corrosive agents.

Section 336.12 lists four prohibited uses for Type TC. First, Type TC may not be installed where it could be subjected to physical damage. Second, Type TC must not be installed outside a raceway or cable tray system except as permitted in 336.10(4) and 336.10(7). Third, the cable cannot be installed in the direct rays of the sun unless identified as sunlight resistant. Fourth, Type TC cannot be used in direct burial applications unless it has been identified for such use.

Section 336.80 requires that the ampacity for Type TC cable, where installed in cable tray, be determined in accordance with 392.80 for conductor sizes 14 AWG and larger and with 402.5 for conductor sizes 18 AWG through 16 AWG. Where Type TC is installed in raceways or where used as messenger supported wires, the ampacity shall be in accordance with 310.15.

Service-Entrance (SE and USE) Cable – Article 338

Service-entrance (SE) cable is a single or multiconductor assembly with or without an overall covering. **See Figure 8-14.** There are two types of service-entrance cable recognized by the NEC®, Type SE and Type USE. Both are rated up to 600 V.

Service entrance cable runs from the weatherhead to the electric meter and then to the panelboard.

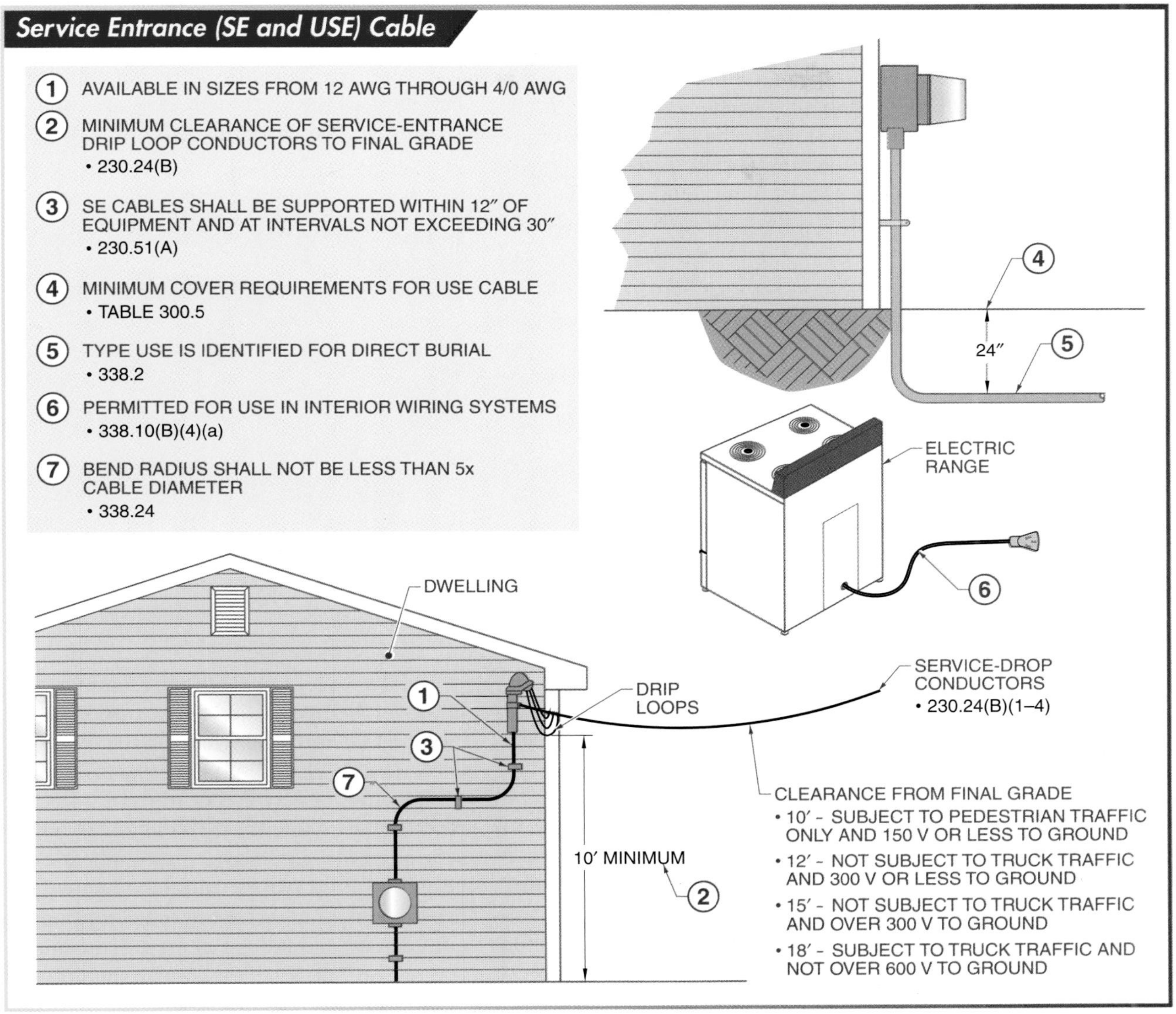

Figure 8-14. Service-entrance (SE) cable is a single or multiconductor assembly with or without an overall covering.

SE cable is available in two styles, U and R. Type U is a flat cable with an uninsulated concentric grounded conductor. Type R is a round cable with an insulated grounded conductor. Both styles are listed for above-ground installations.

The outer jacket or finish of SE cable is suitable for use where exposed to the sun. SE cable is available in sizes from 12 AWG to 4/0 AWG. The cable comes with two or three insulated conductors and one bare conductor.

Section 338.10(A) permits SE cable to be used for service-entrance wiring as service-entrance conductors. With this use, the grounded conductor may be insulated or bare. Section 338.10(B)(4) permits SE cable to be used for general interior wiring in buildings provided it complies with Part II of Article 334 excluding 334.80, ampacity. SE cable used indoors for feeders or branch circuits shall have an insulated grounded conductor. It is permissible to use the uninsulated conductor for equipment grounding purposes. Where Type SE cable is used for exterior wiring as a feeder or branch circuit, it must be in accordance with Part I of Article 225.

Where Type SE cable is installed in interior locations in thermal insulation, 338.10(A)(4)(a) requires that the ampacity of the conductors in the Type SE cable be in accordance with the 60°C temperature rating. For derating purposes, the maximum conductor temperature rating may be used as a starting point for any adjustment or correction calculations, but the final derated ampacity may not exceed that for a 60° C-rated conductor.

Section 338.24 requires that the protective coverings of the cable not be damaged when the cable is bent during installation. The radius of the curve of the inner edge of any bend shall not be less than five times the diameter of the cable.

USE cable has a moisture-resistant outer jacket. It is listed for burial directly in the earth. USE cable is moisture-resistant. It is not fire retardant and is not suitable for use in premises or aboveground except to terminate at the service equipment or metering equipment.

Underground Feeder and Branch-Circuit Cable – Article 340

Underground feeder and branch-circuit cable (Type UF cable) is a factory assembly of one or more insulated conductors contained within an overall covering of nonmetallic material that is suitable for direct-burial applications. The conductors in Type UF cable are permitted to be in sizes 14 AWG copper, or 12 AWG aluminum or copper-clad aluminum, through 4/0 AWG. The overall covering is flame retardant, resistant to moisture and corrosion, and resistant to fungus. The conductors of Type UF cable shall be of the moisture-resistant types that are listed in Table 310.104(A) for branch-circuit wiring. If the Type UF cable is going to be installed as a substitute for Type NM, the conductor insulation for the Type UF cable conductors must be rated at 90°C. See 340.112.

Section 340.10 lists seven permitted uses for Type UF cable. The most common application for Type UF cable is for underground or direct-buried applications. Type UF cable is permitted to be installed as single-conductor cables provided the provisions of 300.3 that require all conductors of the same circuit to be run together are met. Type UF cable is a versatile wiring method that is permitted for use in wet, dry, or corrosive conditions. It can be installed as Type NM cable provided the applicable installation provisions of Article 334 are met.

Article 555.13(B)(3) covers wiring over and under navigable water.

Increasingly, per 690.31, many solar photovoltaic installations are using Type UF cable as permitted in 340.10(5). Type UF cable is also permitted as nonheating leads for heating cables, as permitted in 424.43. Another less common application is where supported by cable trays. This application is permitted only for the multiconductor type cable and is used under limited conditions in commercial and industrial establishments.

Section 340.12 lists eleven specific applications that are not permitted for Type UF cable. Generally, Type UF cable is not permitted to be installed in commercial garages, theaters, motion picture studios, or similar locations. Type UF cable is also prohibited for use as service-entrance cable, in storage battery rooms, in hoistways, or in hazardous locations unless otherwise permitted in the Code.

Type UF cable cannot be embedded in concrete, masonry, or poured cement unless it is for use as nonheating leads for fixed electric space heating equipment. Type UF cable is also prohibited where subject to physical damage, where exposed to direct sunlight unless specifically identified as being sunlight resistant, and where used as overhead cable, except as provided in Article 396 as messenger supported wiring. Section 334.80 requires that the ampacity for the conductors of Type UF cable be based on that of the 60°C conductors in accordance with 310.15.

OTHER WIRING SYSTEMS

Cable trays, wireways, and busways are found primarily in commercial and industrial installations. The advantage of these systems, as opposed to a conduit system, is the ease of installation and the ability to alter the loads that are served from these systems.

Cable Tray Systems (CTS) – Article 392

A *cable tray system (CTS)* is an assembly of sections and associated fittings which form a rigid structural system used to support cables and raceways. **See Figure 8-15.** These systems are available in different configurations including ladder, trough, channel, solid bottom, and ventilated trough.

Cable tray systems are assembled in the field. They are available in aluminum, steel, or nonmetallic. Nonmetallic cable trays shall be of a flame-retardant material. Cable trays are classified as a support system for cables and raceway systems. The cable tray itself is not considered a raceway system.

Cable trays are typically used to route structured cabling systems within a facility.

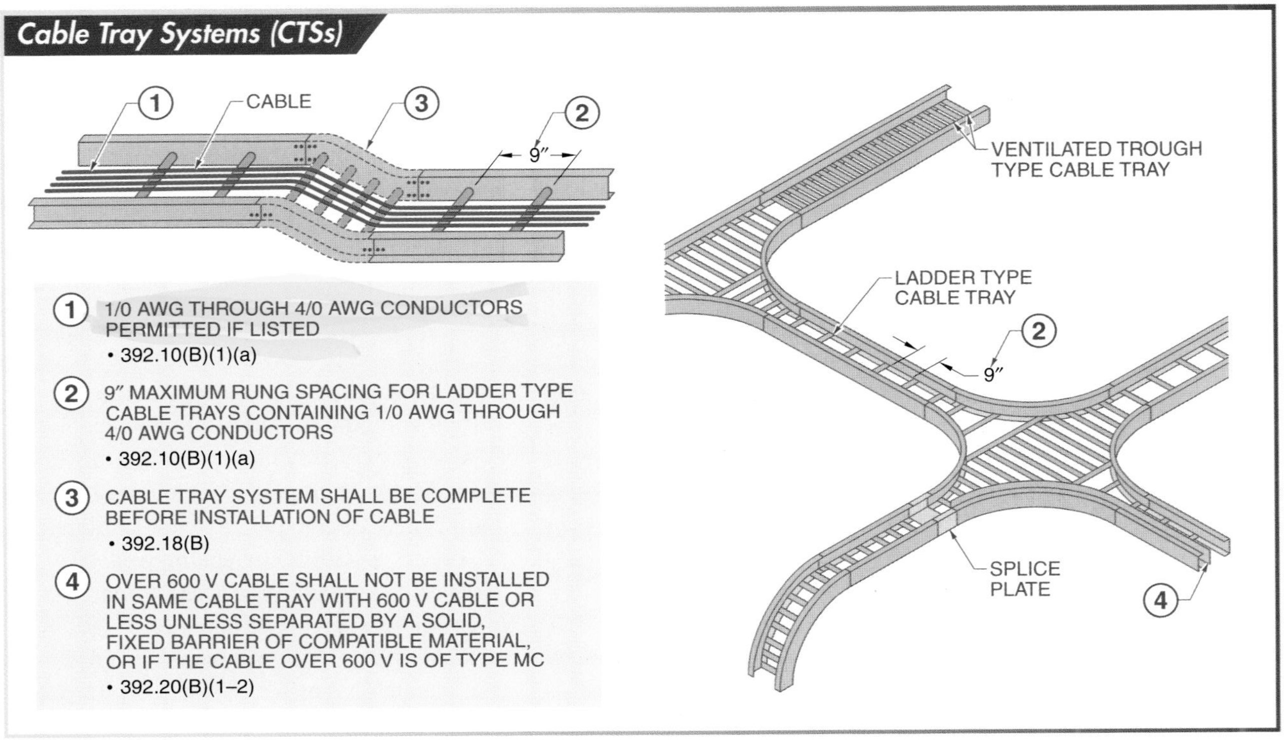

Figure 8-15. A cable tray system (CTS) is an assembly of sections and associated fittings which form a rigid structural system used to support cables and raceways.

Wiring Methods Permitted – 392.10(A). There are over 25 wiring methods permitted in cable tray systems. The wiring methods must be installed per the conditions described in their respective articles and sections. Section 392.10(B) permits ladder type, ventilated trough type, and ventilated channel type CTS to be installed in industrial establishments when running single and multiconductor cable. Section 392.10(B)(1) permits single conductors in smaller sizes, 1/0 AWG through 4/0 AWG, to be installed in cable trays provided they are marked and listed for such use. Installing conductor sizes 1/0 AWG through 4/0 AWG in ladder type trays requires that the rungs of the ladder tray be spaced no more than 9″ apart.

Wiring Methods Not Permitted – 392.12. There are a few specific locations where the use of cable tray systems is prohibited. Section 392.12 prohibits the use of cable tray systems in a few specific locations such as hoistways or any location where the cable tray system would be subjected to severe physical damage. Note that there is no definition of what constitutes "severe" physical damage. The best practice is to discuss the installation and use of the cable tray system with the AHJ before the system is installed. Additionally, cable tray systems are not permitted to be installed in ducts, plenums, or other air-handling spaces unless they are used to support approved wiring methods for these spaces in accordance with 300.22(C)(2).

Installation as a Complete System – 392.18. CTS shall be installed as complete systems. The electrical continuity of the tray system shall be maintained. Section 392.60(B) permits the use of steel and aluminum cable tray systems as the EGC when the cable tray and fittings are identified for the purpose and the total cross-sectional area of both side rails meets the requirements of Table 392.60(A). The concern here is that there is enough mass of metal to carry the available fault current. All cable tray sections must be legibly and durably marked to show the cross-sectional area. Bonding shall be in accordance with sections 250.96 and 250.102.

Sizing Cable Trays for Multiconductor Cables – 392.22. Multiconductor cables rated 2000 V or less are permitted to be installed in ladder, ventilated trough, solid bottom, or ventilated channel cable trays. **See Figure 8-16.**

Section 392.22(A)(1)(a) requires that ladder or ventilated trough cable trays have a width equal to or greater than the sum of all the diameters of the cables where all of the cables are 4/0 AWG or larger. These cables shall be installed side by side in a single layer. Section 392.22(A)(1)(b) requires that cables smaller than 4/0 AWG in size meet the maximum allowable fill requirements of Column 1 in Table 392.22(A). These smaller cables do not have to be placed side by side in a single layer. Section 392.22(A)(1)(c) requires installations that have a combination of cable sizes smaller and larger than 4/0 AWG to meet the maximum allowable fill requirements of Column 2 in Table 392.22(A). Section 392.22(A)(2) requires that signal and control cables not exceed 50% of the cross-sectional area of ladder or ventilated trough cable trays that have an inside depth of 6″ or less.

For solid-bottom cable trays where all of the cables are 4/0 AWG or larger, 392.22(A)(3) requires that the diameter of all cables 4/0 AWG and larger shall not exceed 90% of the cable tray width and that the cables shall be installed in a single layer. Section 392.22(A)(3)(b) requires that cables smaller than 4/0 AWG in size shall meet the maximum allowable fill requirements of Column 3 in Table 392.22(A) where all of the cables are smaller than 4/0 AWG. These cables do not have to be installed in a single layer. Section 392.22(A)(3)(c) requires installations containing cables smaller and larger than 4/0 AWG to comply with the maximum allowable fill requirements of Column 4 in Table 392.22(A). Section 392.22(A)(4) requires that signal and control cables shall not exceed 40% of the cross-sectional area of solid-bottom cable trays that have an inside depth of 6″ or less.

Section 392.22(A)(5) requires that for ventilated channel cable trays, the total cross-sectional areas of all cables shall not exceed 1.3 sq in. for 3″-wide trays, 2.5 sq in. for 4″-wide trays, and 3.8 sq in. for 6″-wide trays.

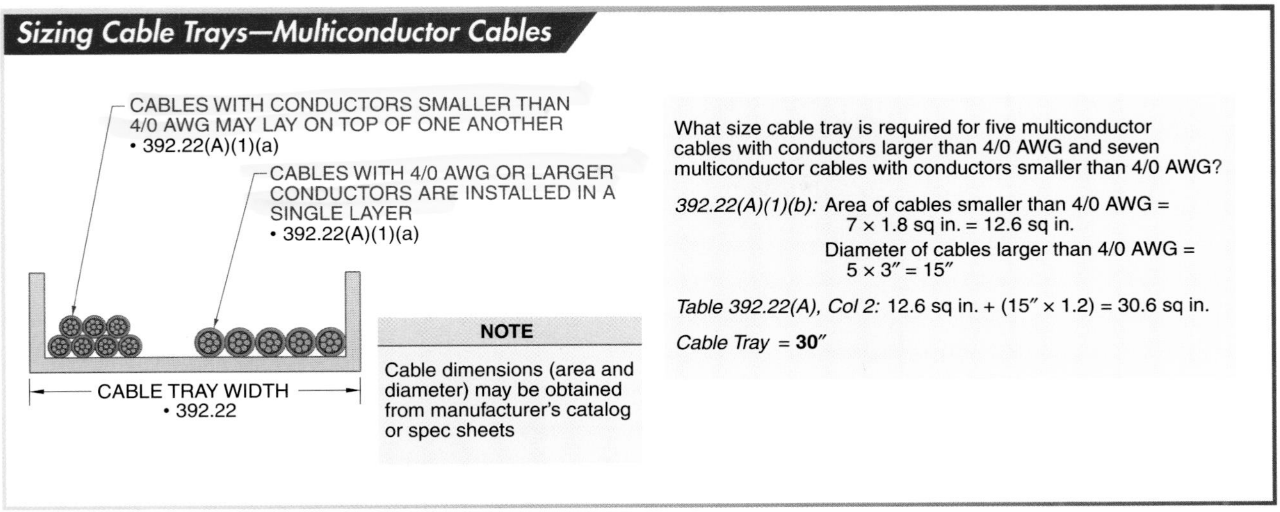

Sizing Cable Trays—Multiconductor Cables
CABLES WITH CONDUCTORS SMALLER THAN 4/0 AWG MAY LAY ON TOP OF ONE ANOTHER • 392.22(A)(1)(a)
CABLES WITH 4/0 AWG OR LARGER CONDUCTORS ARE INSTALLED IN A SINGLE LAYER • 392.22(A)(1)(a)
CABLE TRAY WIDTH • 392.22
NOTE
Cable dimensions (area and diameter) may be obtained from manufacturer's catalog or spec sheets
What size cable tray is required for five multiconductor cables with conductors larger than 4/0 AWG and seven multiconductor cables with conductors smaller than 4/0 AWG?
392.22(A)(1)(b): Area of cables smaller than 4/0 AWG = 7 × 1.8 sq in. = 12.6 sq in.
Diameter of cables larger than 4/0 AWG = 5 × 3″ = 15″
Table 392.22(A), Col 2: 12.6 sq in. + (15″ × 1.2) = 30.6 sq in.
Cable Tray = 30″

Figure 8-16. Cable trays for multiconductor cables are sized by Table 392.22(A).

Sizing Cable Trays for Single-Conductor Cables – 392.22(B). Single conductors rated 2000 V or less are permitted in ladder, ventilated trough, and ventilated channel cable trays. The single conductors shall be evenly distributed across the cable tray. Cable trays are sized from Columns 1 and 2 of Table 392.22(B)(1) based on the square inch area and diameter of conductors per Chapter 9, Table 5.

Section 392.22(B)(1)(a) requires that ladder or ventilated trough cable trays containing 1000 kcmil or larger cables shall have a minimum width equal to the sum of the diameters of all the cables and be installed in a single layer. Section 392.22(B)(1)(b) requires that ladder or ventilated trough cable trays containing 250 kcmil up to 900 kcmil cables shall have a minimum width equal to the sum of the cross-sectional areas of all single conductor cables as permitted in Column 1 of Table 392.22(B)(1).

Section 392.22(B)(1)(b) contains the provisions for determining the number of conductors permitted in a single cable tray section where all of the conductors are sizes 250 kcmil through 900 kcmil. In this case, the sum of the cross-sectional areas of all single-conductor cables is not permitted to exceed the maximum permitted cable fill in Column 1 of Table 392.22(B)(1). If the single-conductor cable sizes are 1000 kcmil or greater, the sum of the cross-sectional area of all cables, per 392.22(B)(1)(c), is sized in accordance with the values given in Column 2 of Table 392.22(B)(1). Additionally, if any of the single-conductor cables are 1/0 through 4/0 AWG, the sum of the diameters of all single-conductor cables shall not be permitted to exceed the actual width of the cable tray. See 392.22(B)(1)(d).

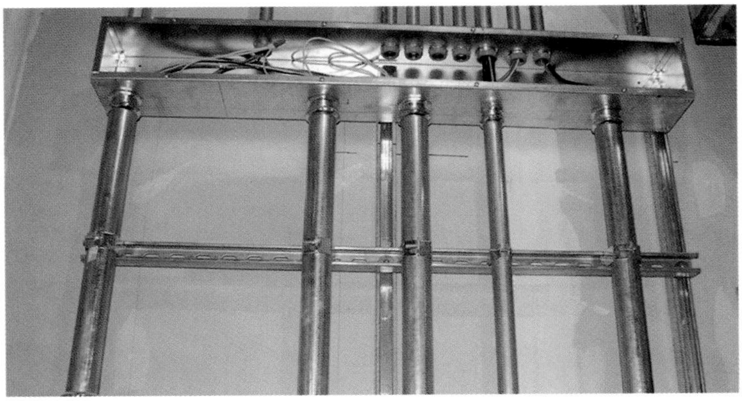

Wireways are installed as a complete system before conductors are installed.

Wireways – Article 376, Article 378

A *wireway* is a metallic or nonmetallic trough with a hinged or removable cover designed to house and protect conductors and cables. **See Figure 8-17.** Wireways shall be installed as a complete system before the conductors are installed. They are sized by Chapter 9, Table 5. **See Figure 8-18.**

Wireways

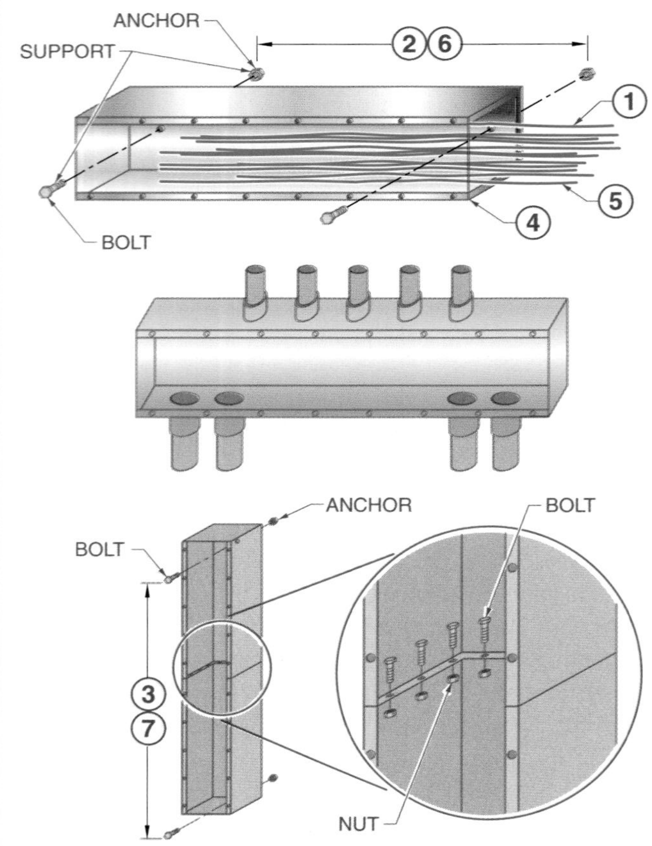

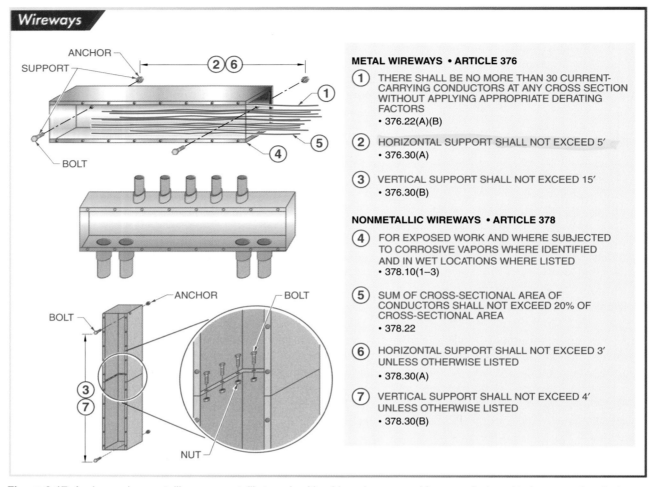

METAL WIREWAYS • ARTICLE 376

1. THERE SHALL BE NO MORE THAN 30 CURRENT-CARRYING CONDUCTORS AT ANY CROSS SECTION WITHOUT APPLYING APPROPRIATE DERATING FACTORS
   • 376.22(A)(B)
2. HORIZONTAL SUPPORT SHALL NOT EXCEED 5′
   • 376.30(A)
3. VERTICAL SUPPORT SHALL NOT EXCEED 15′
   • 376.30(B)

NONMETALLIC WIREWAYS • ARTICLE 378

4. FOR EXPOSED WORK AND WHERE SUBJECTED TO CORROSIVE VAPORS WHERE IDENTIFIED AND IN WET LOCATIONS WHERE LISTED
   • 378.10(1–3)
5. SUM OF CROSS-SECTIONAL AREA OF CONDUCTORS SHALL NOT EXCEED 20% OF CROSS-SECTIONAL AREA
   • 378.22
6. HORIZONTAL SUPPORT SHALL NOT EXCEED 3′ UNLESS OTHERWISE LISTED
   • 378.30(A)
7. VERTICAL SUPPORT SHALL NOT EXCEED 4′ UNLESS OTHERWISE LISTED
   • 378.30(B)

Figure 8-17. A wireway is a metallic or nonmetallic trough with a hinged or removable cover designed to house and protect conductors and cables.

Sizing Wireways

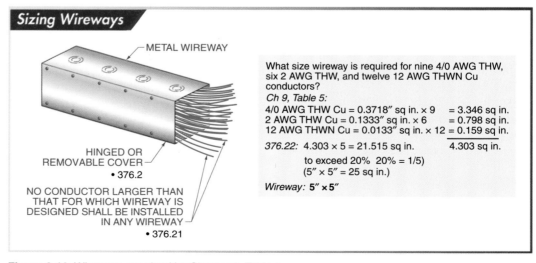

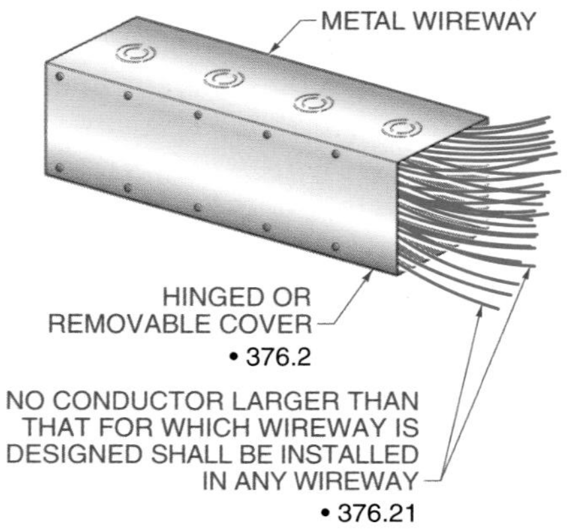

What size wireway is required for nine 4/0 AWG THW, six 2 AWG THW, and twelve 12 AWG THWN Cu conductors?

Ch 9, Table 5:

4/0 AWG THW Cu = 0.3718″ sq in. × 9 = 3.346 sq in.
2 AWG THW Cu = 0.1333″ sq in. × 6 = 0.798 sq in.
12 AWG THWN Cu = 0.0133″ sq in. × 12 = 0.159 sq in.
4.303 sq in.

376.22: 4.303 × 5 = 21.515 sq in.
to exceed 20% 20% = 1/5)
(5″ × 5″ = 25 sq in.)

Wireway: **5″ × 5″**

Figure 8-18. Wireways are sized by Chapter 9, Table 5.

Metallic Wireways – Article 376. Metallic wireways shall be permitted for exposed work, in concealed locations in compliance with 376.10, and in certain hazardous locations. Wireways installed in wet locations shall be listed for the purpose.

The sum of the cross-sectional area of all of the contained conductors shall not exceed 20% of the interior cross-sectional area of the wireway. The derating factors contained in Table 310.15(B)(3)(a) do not apply to the signaling and control conductor. See 376.22.

Splices and taps in wireways are permitted per 376.56(A) provided they are accessible. The conductors, including splices and taps, shall not occupy more than 75% of the cross-sectional area at that point.

Section 376.30(A) requires supports of horizontal wireways not to exceed 5′. Wireways longer than 5′ shall have a support at each end or joint. The distance between supports shall not exceed 10′. Section 376.30(B) requires that vertical runs of metallic wireways shall be supported at intervals not exceeding 15′. Vertical runs shall not have more than one joint between supports.

Nonmetallic wireways must be supported per Article 378.

Nonmetallic Wireways – Article 378. Nonmetallic wireways shall be used for exposed work only per 378.10(1) except as permitted in 378.10(4). Nonmetallic wireways may be used in areas subject to corrosive vapors where identified for the use and in wet locations where listed for the purpose. Per 378.10(3), Informational Note, the installer is cautioned that extreme cold may cause nonmetallic wireways to become brittle and more susceptible to damage from physical contact. Section 378.12 prohibits the use of nonmetallic wireways in the following five areas:

(1) Where subject to physical damage.

(2) In any hazardous location, except as otherwise permitted in the Code.

(3) Where exposed to sunlight unless listed and marked as suitable for the purpose.

(4) Where subject to ambient temperatures other than those for which it is listed.

(5) For conductors whose insulation temperature limitations exceed those for which it is listed.

The sum of the cross-sectional area of all the contained conductors in a nonmetallic wireway shall not exceed 20% of the cross-sectional area of the wireway per 378.22. Conductors for signaling circuits or conductors of motor control circuits used only for starting duty shall not be considered current-carrying conductors. The derating factors of Table 310.15(B)(3)(a) shall apply to the current-carrying conductors up to and including the 20% fill.

Section 378.56 permits splices and taps in nonmetallic wireways. The conductors, including splices and taps, shall not fill the nonmetallic wireway to more than 75% of its area at that point.

Horizontal runs of nonmetallic wireways shall be supported at least every 3′, unless listed for other support intervals. The maximum distance between supports shall never exceed 10′. Section 378.30(B) requires vertical runs of nonmetallic wireways to be supported at intervals not exceeding 4′, unless listed for other support intervals. There shall not be more than one joint between supports.

Section 378.44 requires expansion fittings in straight runs of nonmetallic wireways where the length change is expected to be 0.25″ or greater. Table 352.44(A) gives the expansion characteristics of PVC rigid nonmetallic conduit. PVC nonmetallic wireway has identical characteristics.

Busways – Article 368

A *busway* is a sheet metal enclosure that contains factory-assembled aluminum or copper busbars which are supported on insulators. **See Figure 8-19.** Busways are available for either indoor or outdoor use. If used in an outdoor location, the busway shall be listed for outdoor use. UL lists busways up to 600 V. Their ampacity can be as high as 6500 A for copper bus and 5000 A for aluminum bus. Most systems are in the 200 A to 1000 A range. There are small busway systems available with ranges between 20 A to 250 A. Some systems are designed to accept disconnect switches with overcurrent devices that can be plugged into the busway in order to feed loads as required. The three basic types of busways are busways that provide an opening for plugging in a disconnect with an overcurrent device, busways that will not accept a plug-in disconnect, and trolley busways.

A trolley busway has an open bottom. The trolley runs on wheels and electrical contact is made by brushes or rollers on the busbars. The trolley system feeds portable equipment in industrial installations.

Section 368.10(A) permits busways to be installed only in open locations where they are visible. The only exception is for installations through walls and floors per 368.10(C). They have to be installed so that the joints between sections and fittings are accessible for maintenance. Section 368.10(B) permits these busways to be installed behind panels where means of access are provided, and one of two provisions is followed:

(1) The space behind the access panels is not used for air-handling purposes.

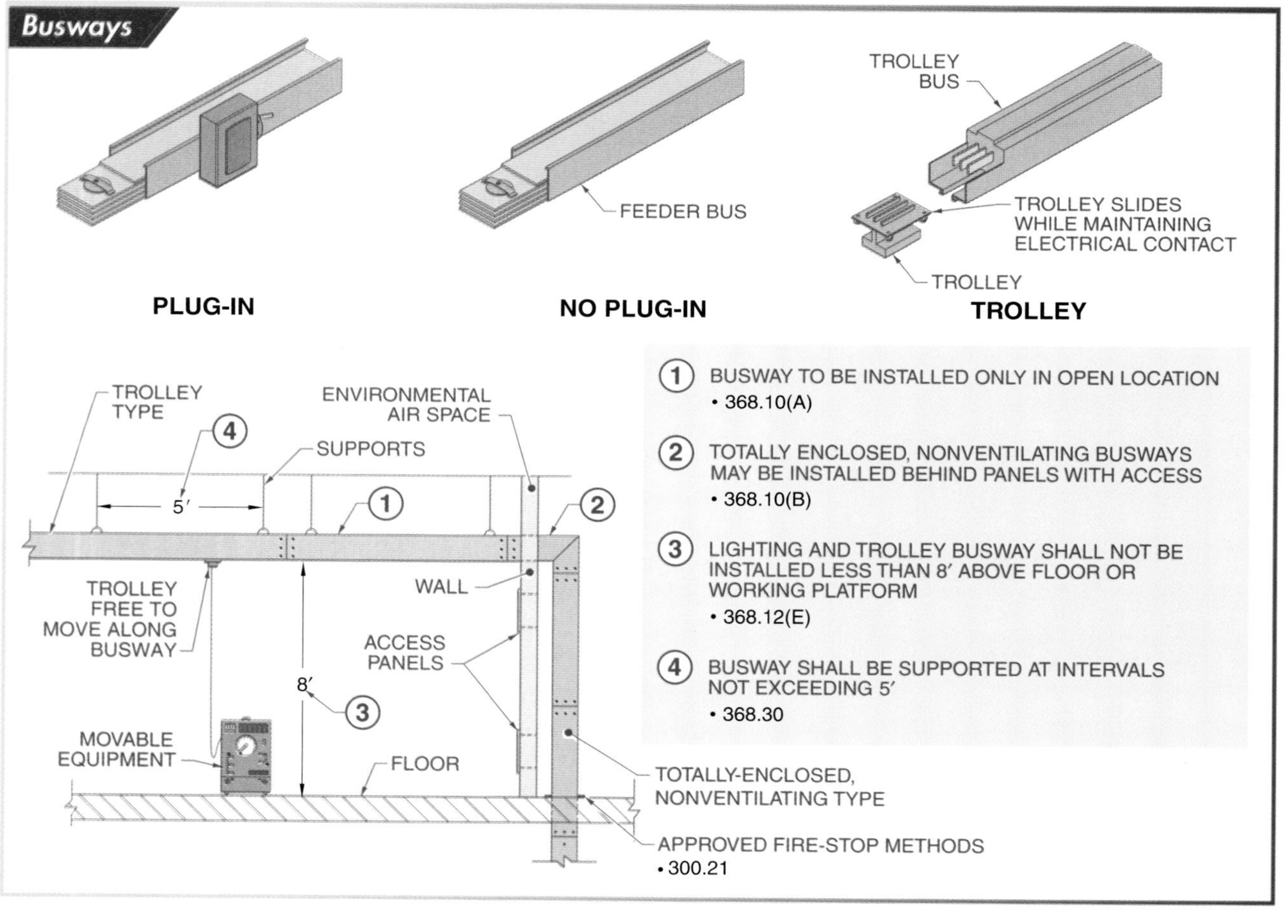

Figure 8-19. A busway is a sheet-metal enclosure that contains factory-assembled aluminum or copper busbars which are supported on insulators.

(2) The space behind the access panels is used for environmental air, in which case there shall be no provisions for plug-in connections and the conductors shall be insulated.

Section 368.12(A–E) prohibits the use of busways in the following five conditions:

(A) Where subject to severe physical damage or corrosive vapors.

(B) In hoistways.

(C) In any hazardous location, unless approved for such use.

(D) Outdoors or in damp or wet locations unless identified for the use.

(E) Lighting busways and trolley busways shall not be installed less than 8′ above the floor or working platform unless provided with a cover identified for the purpose.

Section 368.30 requires that busways shall be supported at intervals not exceeding 5′ unless otherwise designed and marked.

Busways are permitted to be used for service-entrance conductors where the voltage is 600 V nominal or less per 230.43(9). Unbroken lengths of busways are permitted to extend through dry walls per 368.10(C)(1). It is permissible to extend busways vertically through dry floors provided the busway is of the totally enclosed type under certain conditions. See 368.10(C)(2)(a–b).

Section 368.17 requires that busways have over-current protection in accordance with the allowable current rating of the busway. Where the allowable current rating of the busway does not correspond to a standard ampere rating of the overcurrent device, the next higher standard size of the overcurrent device may be used. However, if this rating exceeds 800 A, then the lower size shall be used.

Branches from Busways – 368.56. Branches from busways are permitted, provided one of the following methods is used:

- Type AC cable
- Type MC cable
- Type MI cable
- RMC
- IMC
- EMT
- FMC
- PVC
- ENT
- LFMC
- LFNC
- Surface metal and nonmetallic raceways and busways
- Strut-type channel raceway

Per Article 368.58, a dead end of a busway shall be closed.

Where a nonmetallic raceway is used, the EGC in the nonmetallic raceway shall be connected to the busway in accordance with 250.8 and 250.12. The concern here is that grounding and bonding conductors are terminated properly by cadwelding or using listed materials. *Cadwelding* is a welding process used to make electrical connections of copper to copper or copper to steel in which no outside source of heat or power is required. Powdered metals (copper oxide and aluminum) are dumped from a container into a graphite crucible and ignited by means of a flint igniter. The reduction of the copper oxide by the aluminum (exothermic reaction) produces molten copper and aluminum oxide slag. The molten copper flows over the conductors in the graphite mold, melting them and welding them together.

Section 368.56(B) permits suitable cord and cable assemblies approved for extra-hard or hard usage and listed bus drop cable to be used for the connection of portable or stationary equipment, provided the following four conditions are met:

(1) The cord or cable shall be attached to the building by an approved means.

(2) There shall be no more than 6′ of cord or cable between the busway plug-in device and the tension take-up device. Section 368.56(B)(2), Ex., permits industrial establishments that have maintenance and supervision performed by qualified persons to exceed this 6′ length. Although there is not a length limitation between the busway plug-in device and the tension take-up device, the cord or cable shall be supported every 8′.

(3) The cord or cable shall be installed as a vertical riser between the tension take-up device and the equipment served.

(4) Strain-relief cable grips shall be provided for the cord or cable at the busway plug-in device and at the equipment terminations.

Reduction in Size of Busway – 368.17. Section 368.17(B) requires overcurrent protection at the point where busways are reduced in ampacity. **See Figure 8-20.** Section 368.17(B), Ex., permits only industrial establishments to reduce the ampacity of busways without providing overcurrent protection provided the following three conditions are met:

(1) The length of the reduced busway does not exceed 50′.

(2) The reduced busway has an ampacity of at least ⅓ of the rating or setting of the overcurrent device ahead of where the reduction takes place.

(3) The reduced busway is free from contact with combustible material.

Busways are permitted to be used as a feeder to supply both feeders and branch circuits per 368.17(C). The overcurrent protective device for the busway used as a branch circuit is the same as any other wiring method used to supply a branch circuit. The rating or setting of the overcurrent protective device protecting the busway shall determine the ampere rating of the branch circuit.

Auxiliary Gutters – Article 366

An auxiliary gutter and associated fittings are identical to wireways and associated fittings. UL registers one listing for both of these products, *Listed Wireway or Auxiliary Gutter.* The only difference between them is their intended use. An *auxiliary gutter* is a sheet-metal or nonmetallic enclosure equipped with hinged or removable covers that is used to supplement wiring space. These gutters supplement wiring space at meters, panelboards, switchboards, and similar types of equipment. **See Figure 8-21.** They are manufactured in 1′ to 10′ lengths in various widths and depths. Different fittings are available such as couplings, elbows, end plates, and tees.

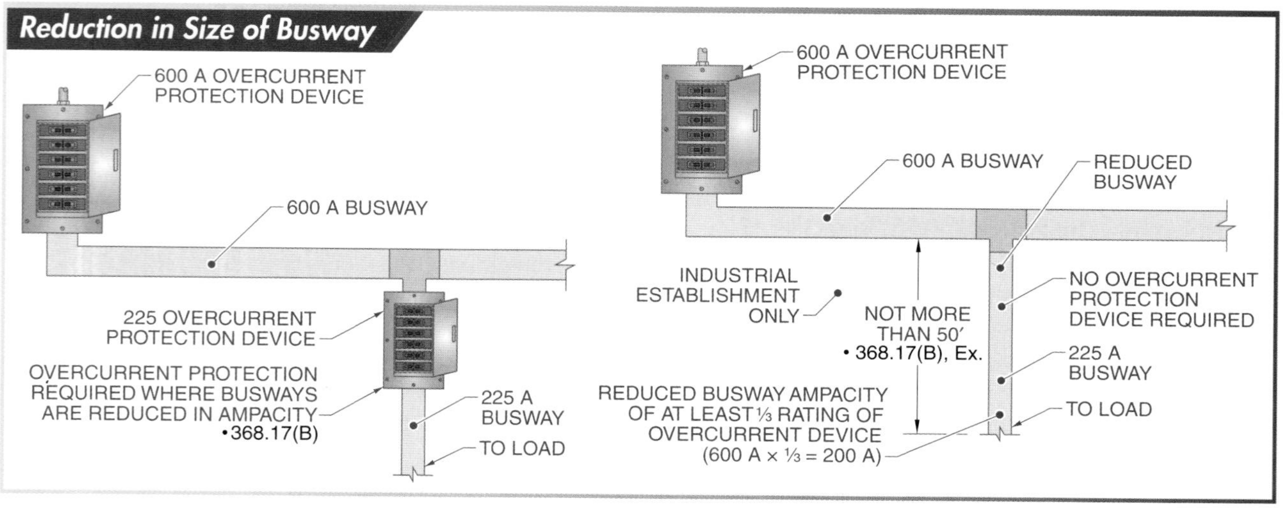

Figure 8-20. Overcurrent protection shall be provided at the point where busways are reduced in ampacity.

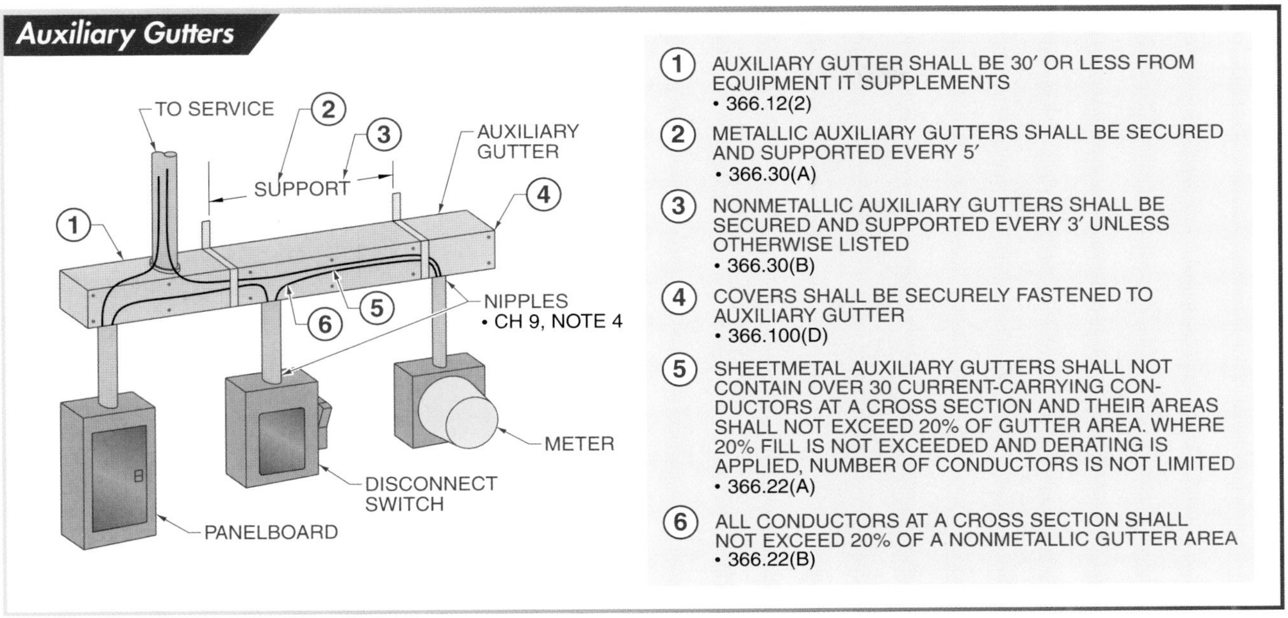
Auxiliary Gutters
TO SERVICE
SUPPORT
AUXILIARY GUTTER
NIPPLES
• CH 9, NOTE 4
METER
DISCONNECT SWITCH
PANELBOARD
1 AUXILIARY GUTTER SHALL BE 30′ OR LESS FROM EQUIPMENT IT SUPPLEMENTS • 366.12(2)
2 METALLIC AUXILIARY GUTTERS SHALL BE SECURED AND SUPPORTED EVERY 5′ • 366.30(A)
3 NONMETALLIC AUXILIARY GUTTERS SHALL BE SECURED AND SUPPORTED EVERY 3′ UNLESS OTHERWISE LISTED • 366.30(B)
4 COVERS SHALL BE SECURELY FASTENED TO AUXILIARY GUTTER • 366.100(D)
5 SHEETMETAL AUXILIARY GUTTERS SHALL NOT CONTAIN OVER 30 CURRENT-CARRYING CONDUCTORS AT A CROSS SECTION AND THEIR AREAS SHALL NOT EXCEED 20% OF GUTTER AREA. WHERE 20% FILL IS NOT EXCEEDED AND DERATING IS APPLIED, NUMBER OF CONDUCTORS IS NOT LIMITED • 366.22(A)
6 ALL CONDUCTORS AT A CROSS SECTION SHALL NOT EXCEED 20% OF A NONMETALLIC GUTTER AREA • 366.22(B)

Figure 8-21. An auxiliary gutter is any enclosure equipped with hinged or removable covers that is used to supplement wiring space.

An auxiliary gutter is prohibited by 366.12 from extending more than 30′ beyond the equipment that it supplements. Auxiliary gutters are not permitted to enclose switches, OCPDs, appliances, or other similar equipment. Section 366.12, Ex., eliminates the 30′ extension restriction for elevator installations as permitted in 620.35.

BOXES, CONDUIT BODIES, AND FITTINGS – ARTICLE 314

This article covers the installation and use of all boxes and conduit bodies. Article 314 also includes the installation requirements for fittings used to join raceways and to connect raceways and cables to boxes and conduit bodies. Section 300.15 requires that, in general, a box or conduit body be installed at each conductor splice connection point, outlet, switch point, junction point, or pull point.

Types of Boxes

A *box* is a metallic or nonmetallic electrical enclosure used for equipment, devices, and pulling or terminating conductors. **See Figure 8-22.** An *outlet box* is a point on the wiring system where current is taken to supply utilization equipment. For example, an outlet box may contain a luminaire (lighting fixture). A *device box* is a box which houses an electrical device. For example, a device box may contain a switch or receptacle.

Nonmetallic boxes shall be permitted only when using one of the following wiring methods: open wiring on insulators, concealed knob-and-tube wiring, nonmetallic-sheathed cable, and nonmetallic raceways per 314.3. There are two exceptions to this rule. Section 314.3 Ex. 1 allows metal raceways or metal-jacket cables to be used with nonmetallic boxes so long as an internal bonding means is provided between all entries.

Section 314.3, Ex. 2, permits metal raceways or metal-jacket cables to be used with nonmetallic boxes with threaded entries. An integral bonding means with a provision for attaching an equipment grounding jumper inside the box shall be provided between all threaded entries. All metal boxes shall be grounded per Article 250. See 314.4.

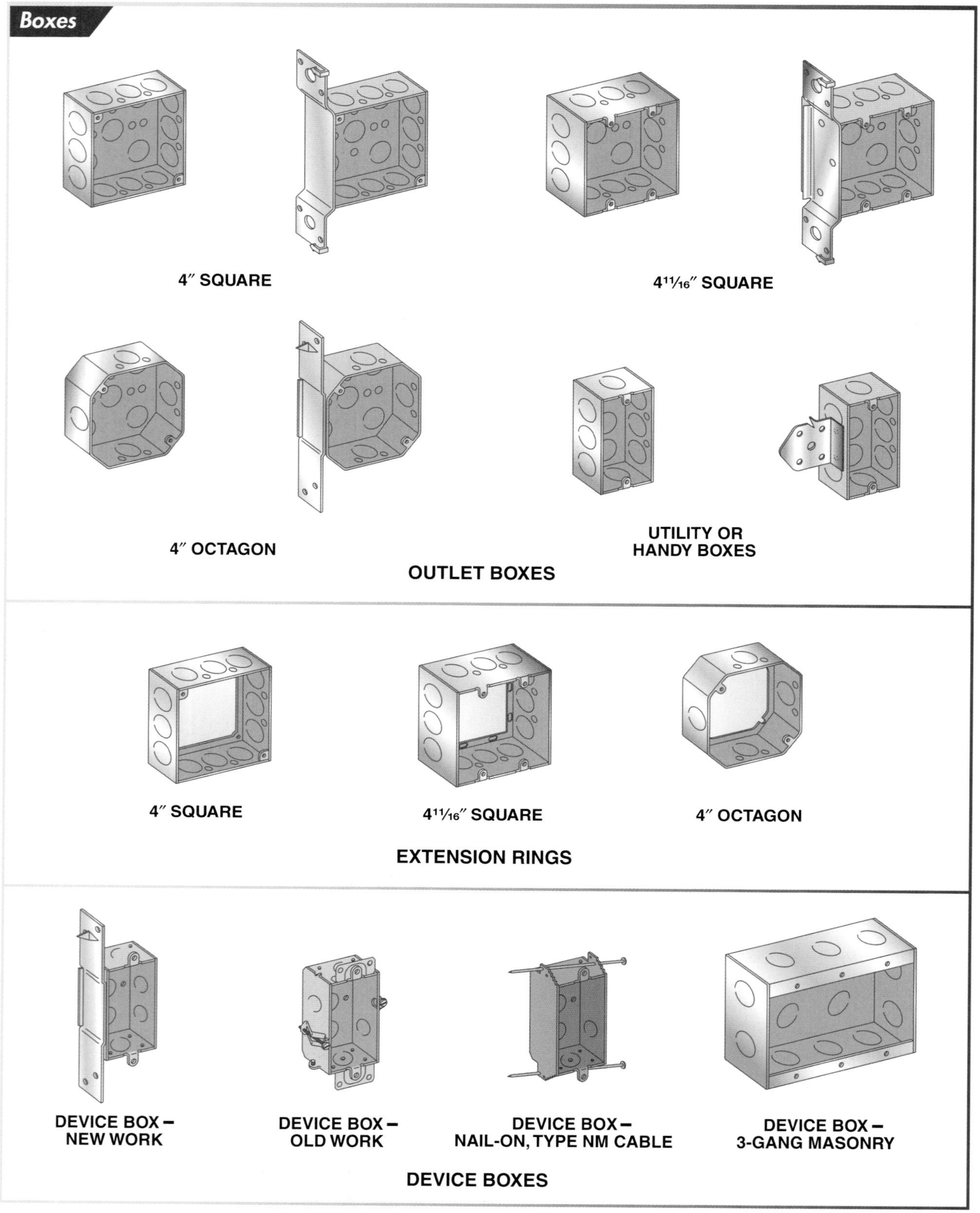

Figure 8-22. A box is a metallic or nonmetallic electrical enclosure used for equipment, devices, and pulling or termination conductors.

Steel Conduit: Recycled Content

Unlike other materials, steel contains recycled material and is also fully recyclable. In fact, according to the Steel Recycling Institute (SRI), steel is the world's most recycled material.

The amount of recycled material in a steel product depends upon the process used by the individual steel supplier. If the conduit is made from a steel coil that comes from a steel mini-mill, then the scrap or recycled content of the steel is about 80% because mini-mills use a technology called "electric arc furnace" (EAF). If the conduit is made from a steel coil that comes from an integrated mill, then the scrap or recycled content is about 30% because integrated mills use a technology called "basic oxygen furnace" (BOF).

Both the EAF and the BOF methods provide an enhanced environmental benefit. One is not environmentally superior to the other, since they are both complementary parts of the total interlocking infrastructure of steelmaking, product manufacture, scrap generation, and recycling.

The additional "green" aspect of steel products is what the SRI terms the "reclamation rate." Currently Leadership in Energy and Environmental Design (LEED®) only considers the recycled content of a product, not how much can be reclaimed at the end of the service life or from job-site scrap. Steel conduit is exceptional relative to reclamation rate. Four major factors account for this:

- The service life is very long. (There is steel rigid conduit still in use after more than sixty years.)
- Conductors can be removed and new conductors installed.
- Additional circuits can frequently be added in the same conduit.
- When the conduit is finally discarded, it is almost totally recyclable.

Additional information is available at www.steelconduit.org and www.recycle-steel.org.

An *outlet* is any point in the electrical system where current supplies utilization equipment. An outlet box provides that access point in the electrical system. Wall cases and handy boxes are "field" names given to two types of outlet boxes that are used to terminate branch circuits of current-consuming devices such as fixtures, appliances, and other pieces of utilization equipment. Boxes are also used for terminations to switches, receptacles, and other non-current consuming devices.

Outlet and device boxes are available in different sizes for mounting switches, receptacles, and other devices. The mounting holes on these boxes are drilled and tapped with No. 6 × 32 threads which are standard for each box. Handy boxes are used for exposed work. Boxes come in a variety of types and sizes and in various depths ranging from 1¼″ to 3½″. See Table 314.16(A).

Section 314.27(A) requires that boxes used with luminaire (lighting fixture) outlets be designed for the purpose and permits the luminaire (lighting fixture) to be attached to the box. Section 314.27(B) requires that floor boxes be listed for the application where they are used for receptacles located in the floor. Section 314.27(B), Ex., allows for the installation of other type boxes in elevated floors of show windows and similar locations. The AHJ shall determine if these installations are free from physical damage, moisture, and dirt.

Section 314.27(C) prohibits outlet boxes from being the sole support for ceiling fans unless the box is listed, marked as suitable for the purpose, and supports a fan that weighs 70 lb or less. Outlet boxes and systems that are intended for fans that weigh more than 35 lb shall be marked to indicate the maximum permitted weight. Square and octagonal boxes may also be used as outlet or device boxes. They come in a variety of depths and are available with ½″ and ¾″ knockouts.

Pull boxes and junction boxes are points in the electrical system which provide access to the raceways entering and leaving the boxes. **See Figure 8-23.** A *pull box* is a box used as a point to pull or feed electrical conductors into the raceway system. A *junction box* is a box in which splices, taps, or terminations are made. Conduits and tubings generally shall not have bends totaling more than 360° between pull points. See, for example, 344.26 for rigid metal conduit and 358.26 for electrical metallic tubing. If a conduit run has a total of more than 360° of bends and offsets, a pull box must be installed. Vertical runs require support for the conductors per Table 300.19(A). A pull or junction box provides a place to install these supports.

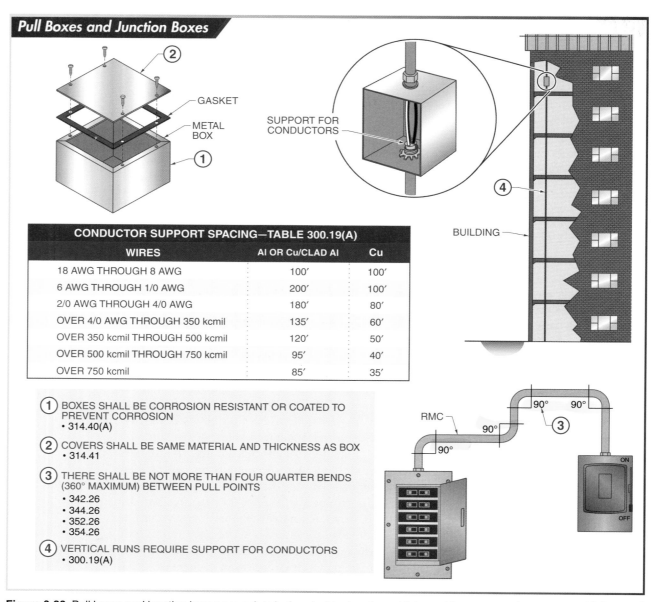

Pull Boxes and Junction Boxes

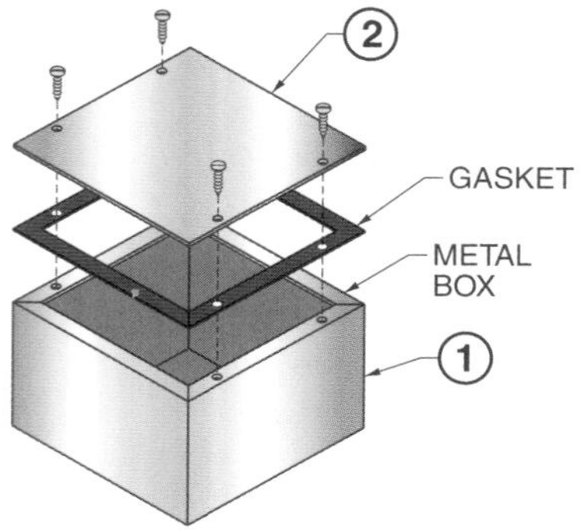

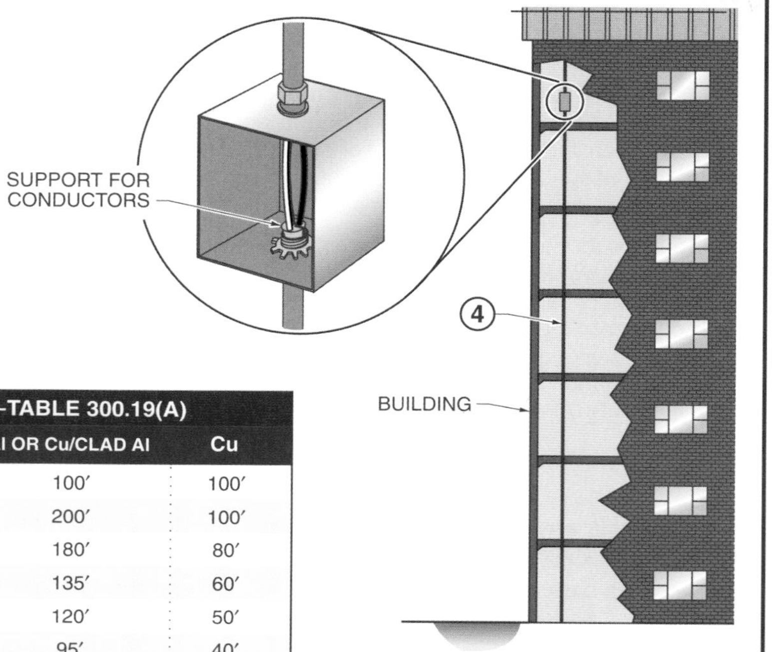

CONDUCTOR SUPPORT SPACING—TABLE 300.19(A)		
WIRES	Al OR Cu/CLAD Al	Cu
18 AWG THROUGH 8 AWG	100′	100′
6 AWG THROUGH 1/0 AWG	200′	100′
2/0 AWG THROUGH 4/0 AWG	180′	80′
OVER 4/0 AWG THROUGH 350 kcmil	135′	60′
OVER 350 kcmil THROUGH 500 kcmil	120′	50′
OVER 500 kcmil THROUGH 750 kcmil	95′	40′
OVER 750 kcmil	85′	35′

(1) BOXES SHALL BE CORROSION RESISTANT OR COATED TO PREVENT CORROSION
- 314.40(A)

(2) COVERS SHALL BE SAME MATERIAL AND THICKNESS AS BOX
- 314.41

(3) THERE SHALL BE NOT MORE THAN FOUR QUARTER BENDS (360° MAXIMUM) BETWEEN PULL POINTS
- 342.26
- 344.26
- 352.26
- 354.26

(4) VERTICAL RUNS REQUIRE SUPPORT FOR CONDUCTORS
- 300.19(A)

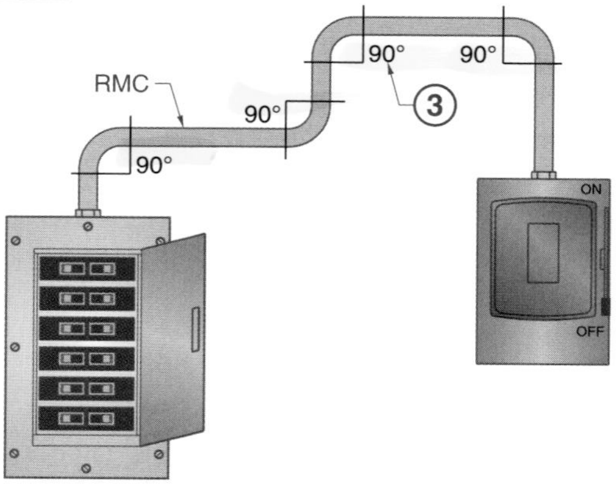

Figure 8-23. Pull boxes and junction boxes are points in the electrical system which provide access to the raceways entering and leaving the boxes.

Pull and junction boxes are sized for straight or angle pulls. **See Figure 8-24.** The minimum-size pull and junction boxes to be used with raceways or cables containing conductors 4 AWG or larger shall be determined based on the type of pull. The length of boxes containing straight pulls shall not be less than eight times the trade diameter of the largest raceway per 314.28(A)(1).

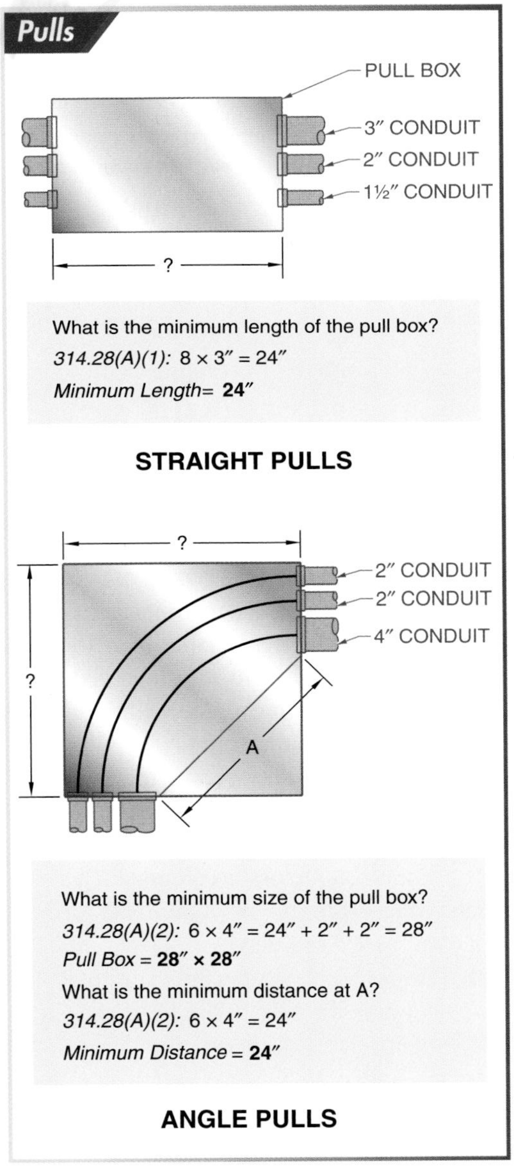

Pulls
PULL BOX
3″ CONDUIT
2″ CONDUIT
1½″ CONDUIT
?
What is the minimum length of the pull box?
314.28(A)(1): 8 × 3″ = 24″
Minimum Length= 24″
STRAIGHT PULLS
?
2″ CONDUIT
2″ CONDUIT
4″ CONDUIT
?
A
What is the minimum size of the pull box?
314.28(A)(2): 6 × 4″ = 24″ + 2″ + 2″ = 28″
Pull Box = 28″ × 28″
What is the minimum distance at A?
314.28(A)(2): 6 × 4″ = 24″
Minimum Distance = 24″
ANGLE PULLS

Figure 8-24. Pull and junction boxes are sized for straight or angle pulls.

Section 314.28(A)(2) considers the requirements for sizing boxes and conduit bodies where the conductors are pulled at an angle or in a U configuration. The distance between the side of the box in which the raceways enter and the opposite side of the box shall be at least six times the trade diameter of the largest raceway in that row, plus the diameters of all the other raceways in that row. The distance between raceway entries containing the same conductors shall not be less than six times the trade size diameter of the raceway containing these conductors. This rule ensures the conductor is not damaged during installation.

Conduit Bodies – 314.29. A *conduit body* is a conduit fitting that provides access to the raceway system through a removable cover at a junction or termination point. Conduit bodies are named based upon their similarity to letters of the alphabet. **See Figure 8-25.**

Conduit bodies get their letter designation from their shape. All conduit bodies have an opening to provide access to the conductors. When a conduit body is placed on a flat surface with the opening in the up position, the shape of the letter designation is apparent. For example, the T and the X conduit bodies resemble the letters *T* and *X*.

The conduit bodies that are used as 90° Ls get their designation in another way. If the hub of the short end of the L is held in the hand like a pistol, the location of the access opening determines the letter designation. If the opening is to the *left*, it is an LL conduit body. If the opening is to the *right*, it is an LR conduit body. If the opening is on top, it is an LB conduit body, meaning the access cover goes on the *back* of the conduit body. The C conduit body has a hub at both ends; therefore, a person can *see* right through it. The E designation comes from one entrance hub on the *end*.

Section 314.29 requires that conduit bodies be installed so that the wiring contained in them is accessible without removing any part of the building. Section 314.40(A) requires that conduit bodies be corrosion resistant or galvanized, enameled or otherwise coated inside and out to prevent corrosion. The wall of a conduit body shall not be less than 3⁄32″ thick.

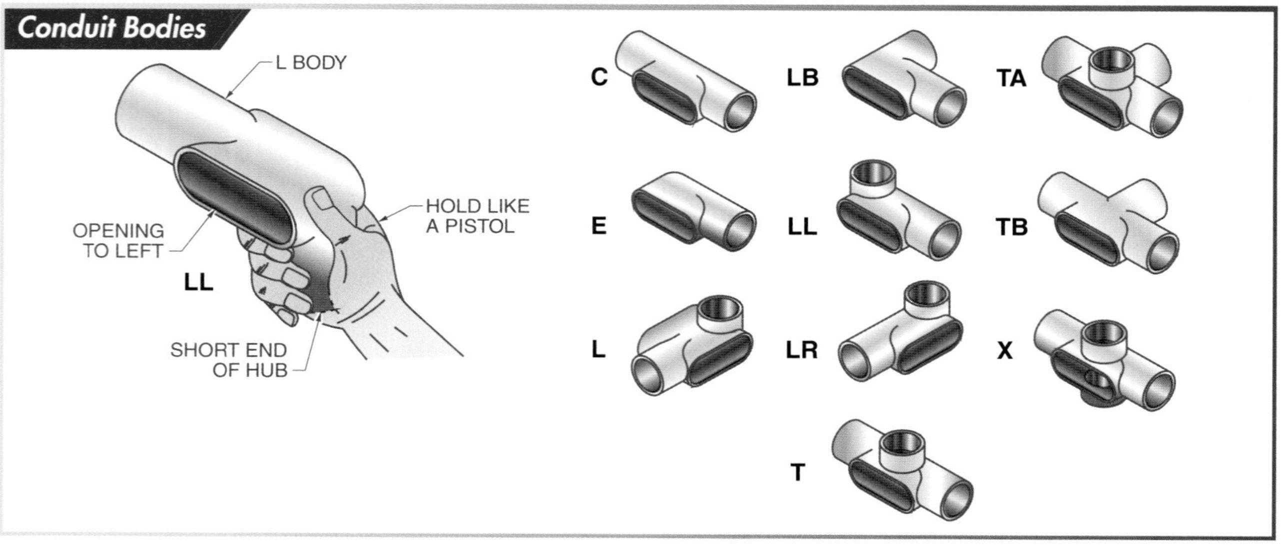

Figure 8-25. A conduit body is a conduit fitting that provides access to the raceway system through a removable cover at a junction or termination point.

Number of Conductors Permitted in a Box – 314.16

Boxes shall be large enough to provide enough free space for all of the enclosed conductors. This free space is needed so that any heat given off by the conductors in the form of I^2R losses can be dissipated. Box size is calculated by the number and size of conductors contained within the box. **See Figure 8-26.**

Table 314.16(A) lists the number of conductors of the same size that may be installed in a certain volume box. For example, a 4″ square box 1½″ deep has a minimum capacity of 21 cu in. This is sufficient to contain ten 14 AWG conductors or eight 10 AWG conductors, provided there are no cable clamps, support fittings, devices, or EGC contained within the box per 314.16(B)(2), (3), (4), and (5).

To determine the minimum-size box needed when there are different-size conductors contained within the box, see Table 314.16(B) and calculate the total cubic inches required for all of the conductors in the box. For example, if a box contained four 12 AWG conductors and six 10 AWG conductors, the box would require a minimum capacity of at least 24 cu in. (4 × 2.25 cu in. = 9 cu in.; 6 × 2.5 cu in. = 15 cu in.; 9 cu in. + 15 cu in. = 24 cu in.). Refer to Table 314.16(A) to find the proper size box. A 4″ square × 2⅛″ deep or any of the 4$^{11}/_{16}$″ square boxes would satisfy this requirement.

Conductor Fill – 314.16(B)(1). Each conductor originating outside the box and terminated or spliced inside the box counts as one. Each conductor passing through the box without a splice or termination counts as one. A conductor that does not leave the box is not counted. Section 314.16(B)(1), Ex., permits an EGC and up to four fixture wires smaller than 14 AWG which enter a box from a domed fixture or similar canopy and terminate within that box to be omitted from the count.

When different-size conductors are used, the conductor fill in cubic inches shall be computed per Table 314.16(B). After the volume for the conductors is calculated, a box is selected according to its cubic inch capacity listed in Table 314.16(A).

Section 314.16(A) permits the total volume in cubic inches to include the cubic inch volume marked on any assembled sections added to the box, such as plaster rings, domed covers, extension rings, etc. Section 314.16 prohibits using a box with less volume than calculated according to the conductor volume.

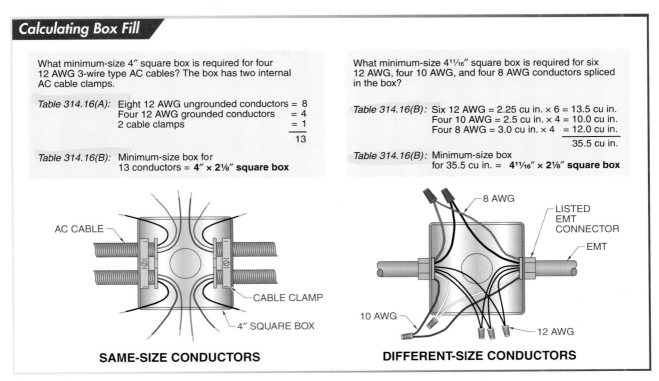

Calculating Box Fill

What minimum-size 4″ square box is required for four 12 AWG 3-wire type AC cables? The box has two internal AC cable clamps.

Table 314.16(A):	Eight 12 AWG ungrounded conductors	= 8
	Four 12 AWG grounded conductors	= 4
	2 cable clamps	= 1
		13
Table 314.16(B):	Minimum-size box for 13 conductors = **4″ × $2\frac{1}{8}$″ square box**	

What minimum-size $4\frac{11}{16}$″ square box is required for six 12 AWG, four 10 AWG, and four 8 AWG conductors spliced in the box?

Table 314.16(B):	Six 12 AWG = 2.25 cu in. × 6 = 13.5 cu in.
	Four 10 AWG = 2.5 cu in. × 4 = 10.0 cu in.
	Four 8 AWG = 3.0 cu in. × 4 = 12.0 cu in.
	35.5 cu in.
Table 314.16(B):	Minimum-size box for 35.5 cu in. = **$4\frac{11}{16}$″ × $2\frac{1}{8}$″ square box**

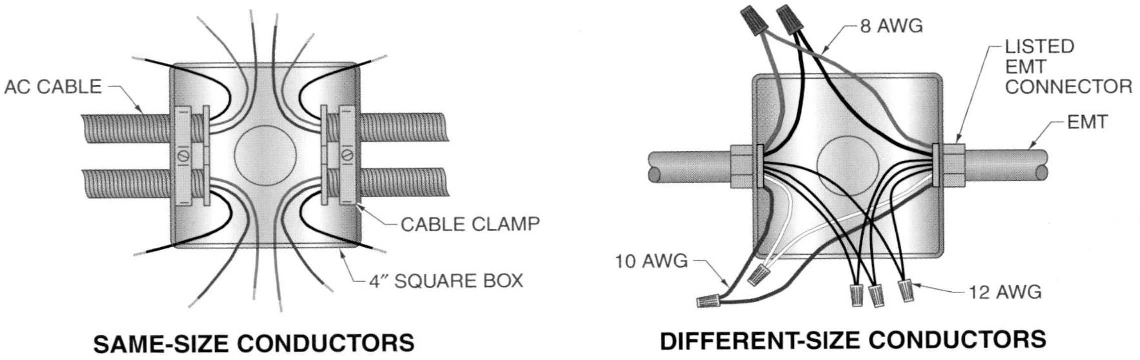

Figure 8-26. Box size is calculated by the number and size of conductors contained within the box.

Section 300.15 requires a box conduit body to be installed at each conductor splice connection point, outlet, switch point, junction point, or pull point.

Clamp Fill – 314.16(B)(2). A deduction of one conductor from Table 314.16(A) shall be made when one or more internal cable clamps are inside the box. A single volume allowance per Table 314.16(B) shall be made based on the largest conductor in the box. Some box designs build the clamping mechanism directly into the box. An allowance is not required for a cable connector with its clamping mechanism outside the box.

Support Fitting Fill – 314.16(B)(3). A deduction of one conductor from Table 314.16(A) is made when one or more fixture studs or hickeys are inside the box. A single volume allowance per Table 314.16(B) shall be made for each fitting based on the largest conductor in the box.

Device or Equipment Fill – 314.16(B)(4). A deduction of two conductors from Table 314.16(A) is made for each yoke or strap containing one or more devices or pieces of equipment. The double volume allowance per Table 314.16(B) is made for each yoke or strap based on the largest conductor connected to the device or equipment on that yoke or strap. Some of the newer electronic devices and equipment require a device box wider than that required for a single device. In these cases, a double volume allowance shall be made for each gang that is required for mounting.

EGC Fill – 314.16(B)(5). A deduction of one conductor from Table 314.16(B) is made when one or more EGCs or EBCs enter a box. Per 250.146(D), a second deduction of one conductor is made if an additional set of

EGCs is present. This refers to an isolated EGC permitted to reduce electrical noise. A single volume allowance per Table 314.16(B) is made based on the largest EGC in the box. An additional volume allowance is made where an additional set of EGCs is present as permitted by 250.146(D). This second allowance is based on the largest EGC of the second set.

Box Supports – 314.23

Boxes and enclosures shall be rigidly and securely fastened in place. Boxes and enclosures may be installed under one of eight different types of installations. The box is considered one of the anchors of the electrical system since raceways and cables terminate at this point. Since many types of raceways and outer sheaths of cable systems act as the EGC, it is imperative that the box is securely anchored in place.

Surface and Structural Mounting – 314.23 (A)(B). Boxes shall be rigidly fastened to the surface upon which they are mounted. Section 314.23(B) requires that enclosures be supported from a structural member of the building or by using a metal, polymeric, or wood brace. Support wires shall provide a rigid support. Metal braces shall be protected against corrosion and not be less than 0.020″ thick and uncoated. Wood braces shall not be less than 1″ × 2″. Wood braces in wet locations shall be treated for the conditions. Polymeric braces shall be identified suitable for the use.

Nonstructural Mounting – 314.23(D). Framing members of suspended ceiling systems are permitted as support for boxes if the framing member is adequately supported and securely fastened to the building structure and the enclosure is limited in size to not more than 100 in^3. Boxes shall be fastened to the framing member by a mechanical means such as bolts, screws, or clips identified for the use. **See Figure 8-27.**

Note that the provisions of 314.23 only apply to enclosures within the scope of Article 314. Section 300.11 must be followed for support of wiring within fire-rated and non-fire-rated assemblies.

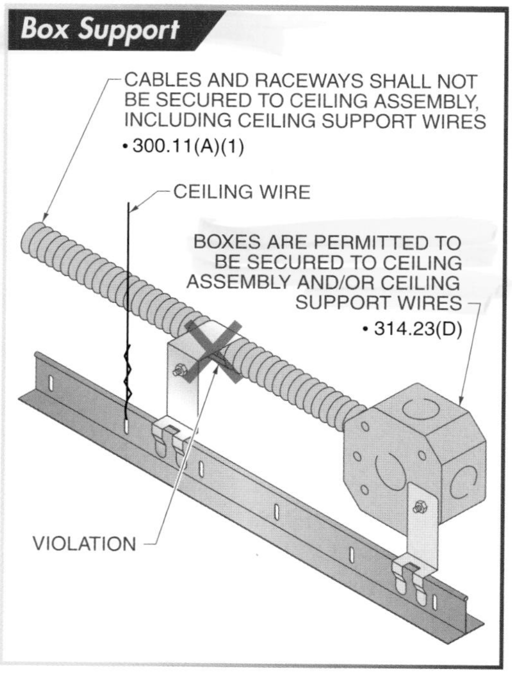

Figure 8-27. Framing members of suspended ceiling systems are permitted as support for boxes if the framing member is adequately supported and securely fastened to the building structure.

Raceway Supported Enclosures without Devices or Fixtures – 314.23(E). Boxes that are not over 100 cu in. and have threaded entries or hubs may be supported where two or more conduits are threaded wrench tight into the enclosure or hubs. These conduits shall be supported within 3′ of the box on two or more sides so as to provide a rigid and secure installation. **See Figure 8-28.** Section 314.23(E), Ex., permits RMC, IMC, EMT, PVC, and Type RTRC conduit to support conduit bodies, provided the conduit bodies are not larger than the largest trade size of the conduit or tubing.

Raceway Supported Enclosures with Devices or Fixtures – 314.23(F). Boxes that are not over 100 cu in. and have threaded entries or hubs are permitted to support fixtures, contain devices, or both, provided they are supported by two or more conduits and are threaded wrench tight into the enclosure or hubs and each conduit is supported within 18″ of the enclosure.

Pendant Boxes – 314.23(H). Boxes shall be permitted to be supported on a pendant by one of two methods. Section 314.23(H)(1) permits boxes to be supported from a multiconductor cord or cable in an approved manner. The installation shall ensure that the conductors are protected against strain by use of a strain-relief connector threaded into a box with a hub. Section 314.23(H)(2) permits rigid or intermediate conduit stems to also support pendants.

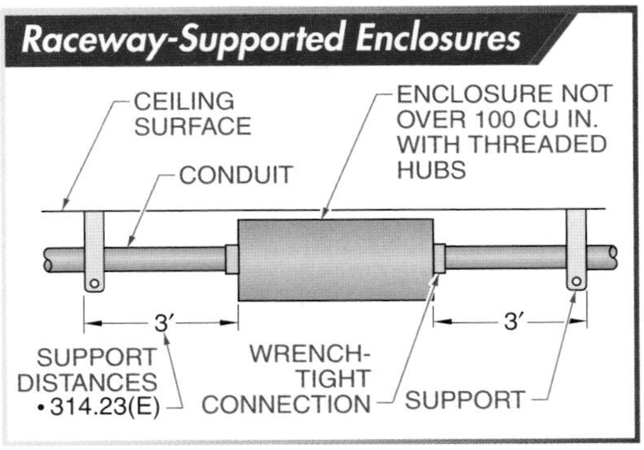

Figure 8-28. Conduits for boxes not over 100 cu in. with threaded hubs shall be supported within 3′ where two or more conduits are threaded wrench tight into the box.

Outlet Boxes – 314.27. Section 314.27(A–D) contains important provisions for the installation and use of outlet boxes. In general, 314.27(A) permits outlet boxes to support lighting fixtures and lampholders. For wall outlet installation, 314.27(A)(1) requires that the boxes be marked on the interior to indicate the maximum weight of the lighting fixture that is permitted to be supported by the box if it is other than 50 lb. Section 314.27(A)(2) requires that every outlet that is used exclusively for lighting be equipped with an outlet box that is designed for the support of a lighting fixture weighing a maximum of 50 lb. If a lighting fixture weighs in excess of 50 lb, an independent means of support beyond the box itself must be installed.

Floor Boxes – 314.27(B). Section 314.27(B) permits receptacles to be installed in floor boxes provided the box is specifically listed for this application. The exception to this rule permits other boxes located in elevated floors of show windows and similar locations to be used provided the AHJ judges them to be free from physical damage, moisture, and dirt.

Boxes at Fan Outlets – 314.27(C). Section 314.27(C) prohibits outlet boxes from being used as the sole support for ceiling fans unless the box or box system is listed, marked as being suitable for the purpose, and the fan it supports weighs 70 lb or less.

A ceiling fan is a dynamic load. A *dynamic load* is a load that produces a small but constant vibration. On a standard ceiling outlet box, the No. 8 × 32 mounting screws may have a tendency to back out or loosen completely due to the dynamic load. Therefore, the NEC® does not permit an outlet box to be the sole support for ceiling fans that weigh in excess of 70 lb.

It is common in many residential installations to install a spare, separately switched, ungrounded conductor to ceiling outlet boxes where ceiling fans may be installed at a later date. Section 314.27(C) requires that where this is done in single or multifamily dwellings, the outlet box or outlet box system must be listed for the sole support of a ceiling fan.

QUICK QUIZ®

Refer to Chapter 8 Quick Quiz® on CD-ROM.

Additional information is available at ATPeResources.com

Wiring Materials—Raceways and Boxes

REVIEW QUESTIONS

Name Bradly Lowry ______________ Date ______________

Raceway ______ **1.** A(n) ___ is an enclosed channel for conductors.

Nipple ______ **2.** A(n) ___ is a short piece of conduit or tubing that does not exceed 24″ in length.

360 ______ **3.** A total run of PVC shall not exceed ___° of bends.

multi-outlet ______ **4.** A(n) ___ assembly is a metal raceway with factory-installed conductors and attachment plug receptacles.

Wire way ______ **5.** A(n) ___ is a metallic or nonmetallic trough with a hinged or removable cover designed to house and protect conductors and cables.

T (F) **6.** Aluminum fittings and enclosures shall not be used with RMC.

T (F) **7.** AC cable shall be installed at least ¾″ back from the finished edge of framing members. 1¼″

(T) F **8.** MC cable with a smooth sheath shall have a bending radius of a minimum of 10 times the diameter of the cable.

(T) F **9.** The use of NM cable in any building or structure designated in NFPA 220 as Type I or II is prohibited unless the NM cables are installed in raceways that are permitted in Type I and II construction.

(T) F **10.** A trolley busway has an open bottom.

C fitting ______ **11.** A(n) ___ conduit body has a hub at both ends.

(T) F **12.** Type USE cable is listed for direct burial.

30 ______ **13.** An auxiliary gutter shall not extend more than ___′ beyond the equipment it supplements.

offset ______ **14.** A(n) ___ is a double bend in a raceway.

one ______ **15.** MC cable is a factory assembly of ___ or more conductors.

(T) F **16.** Two Romex cables shall not be stapled on top of one another.

(T) F **17.** Cable tray systems shall be installed as complete systems.

T (F) **18.** Busways shall be used for indoor work only.

outlet ______ **19.** A(n) ___ is any point in the electrical system where current supplies utilization equipment.

junction ______ **20.** A(n) ___ box is a box in which splices, taps, or terminations are made.

Support

12 1. MC cable shall be supported within ___″ at A.

6 2. MC cable shall be supported every ___′ at B.

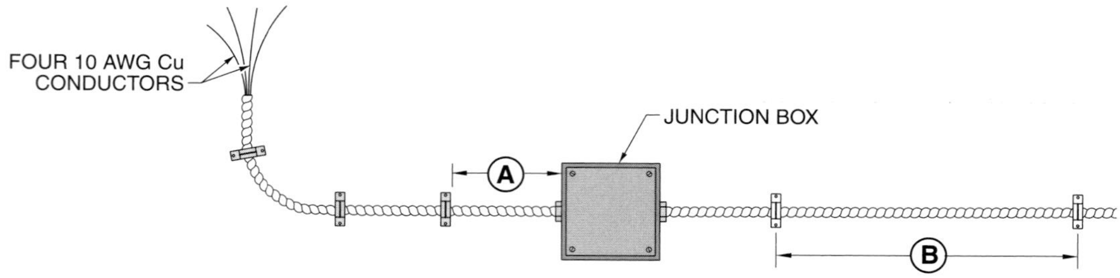

12 3. AC cable shall be supported within ___″ at C.

4½ 4. AC cable shall be supported every ___′ at D.

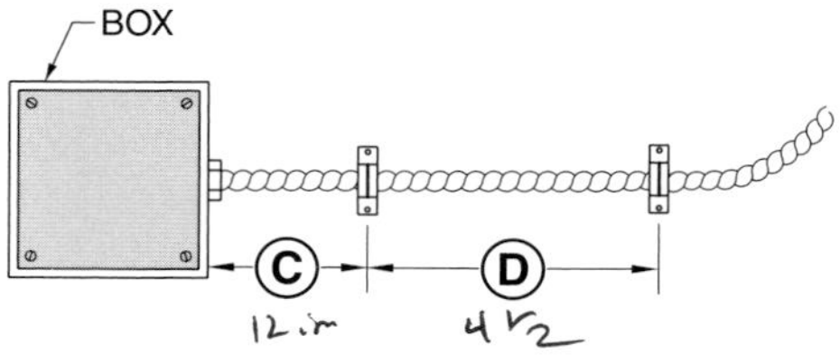

12 5. FMC cable shall be supported within ___″ at E.

4½ 6. FMC cable shall be supported every ___′ at F.

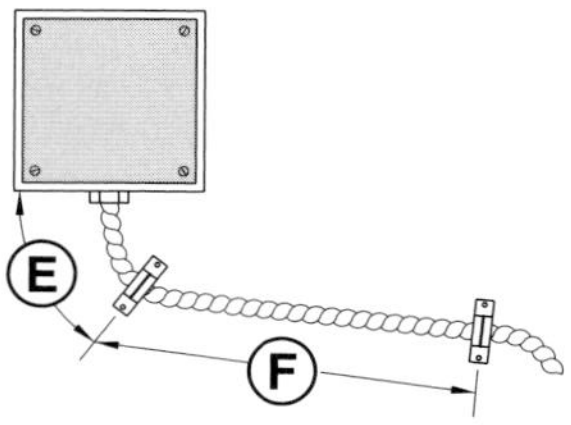

10 7. EMT shall be supported every ___′ at G.

3 8. EMT shall be supported within ___′ at H.

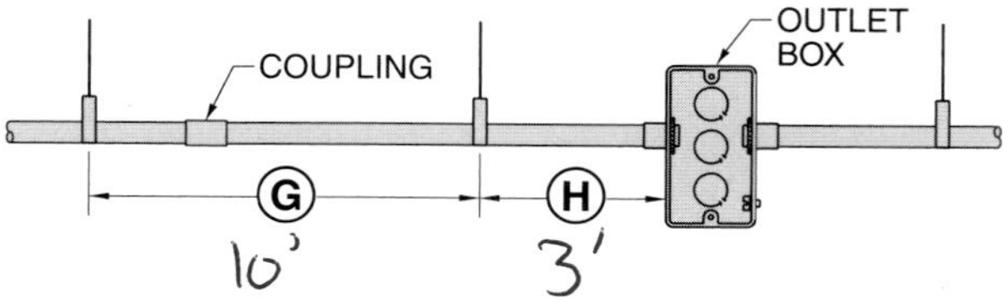

chapter 8

Wiring Materials—Raceways and Boxes

TRADE COMPETENCY TEST

Name Bradly Lawry **Date** ____________

NEC®	Answer	
	2 1/8	**1.** *See Figure 1.* Determine the minimum-size EMT nipple required for the installation.
		2. Determine the minimum cubic inch capacity required for a 4″ square outlet box which contains three 12 AWG and six 8 AWG THW Cu conductors. No internal clamps are provided in the outlet box.
		3. Determine the minimum-size 4″ square box required for four 14/3 AWG AC cables. The box has two internal AC clamps.
		4. *See Figure 2.* Determine the minimum-size $4^{11}/_{16}''$ box required for the installation.

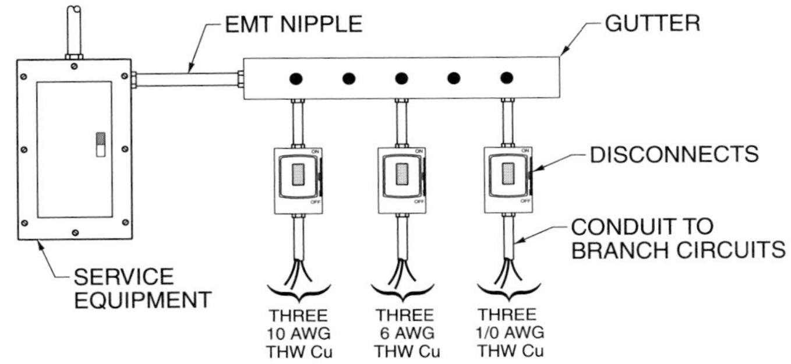

FIGURE 1

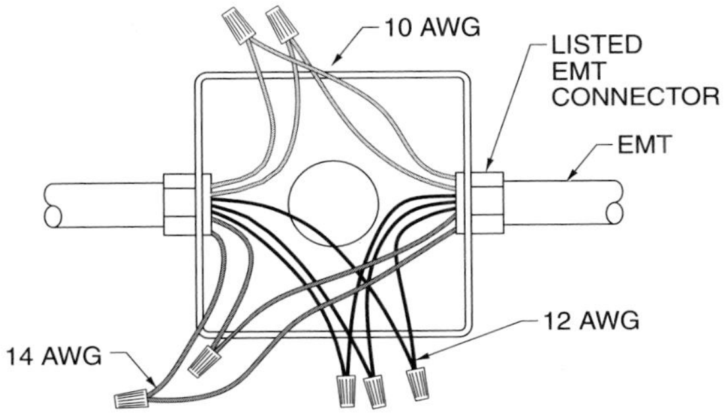

FIGURE 2

	8″ × 3″ × 24″ l	**5.** Determine the minimum length of a pull box which contains a 3″ RMC entering and leaving the box through opposite sides.
	4″ × 4″ × 32″ l	**6.** *See Figure 3.* What is the minimum length of the pull box?

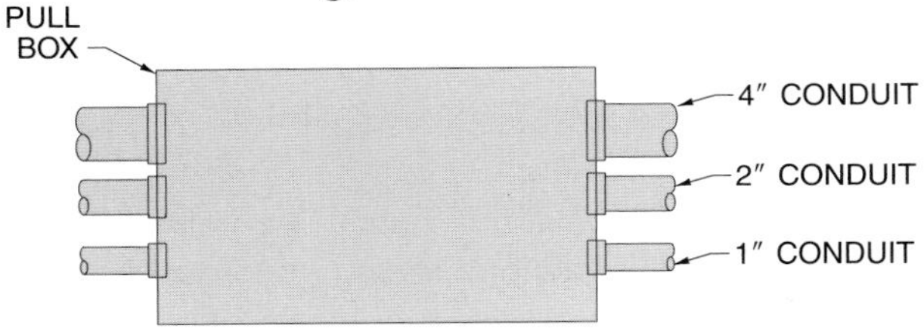

FIGURE 3

______ 18" square 7. *See Figure 4.* What is the minimum size of the pull box?

______ 15" 8. *See Figure 4.* What is the minimum distance of A?

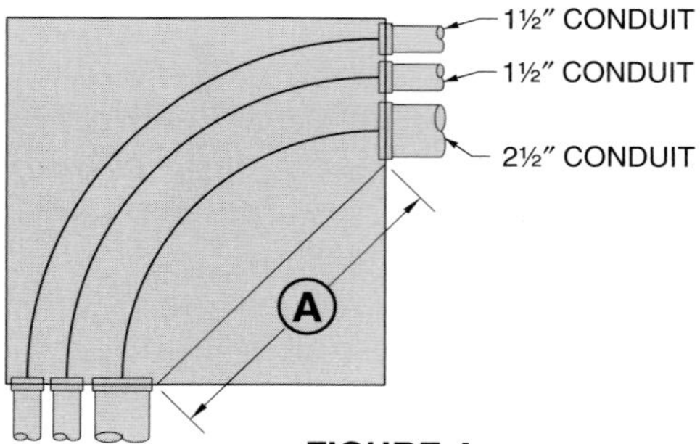

FIGURE 4

______ 9. Determine the minimum bending radius for a smooth-sheath Type MC cable with an external diameter of 1.25″.

______ 10. Determine the minimum bending radius for Type AC cable with an external diameter of 0.80″.

______ 5 11. *See Figure 5.* Determine the minimum number of conduit supports required for the installation.

______ 6' 12. Determine the maximum spacing between conduit supports for 2½″ PVC installed horizontally on a wall.

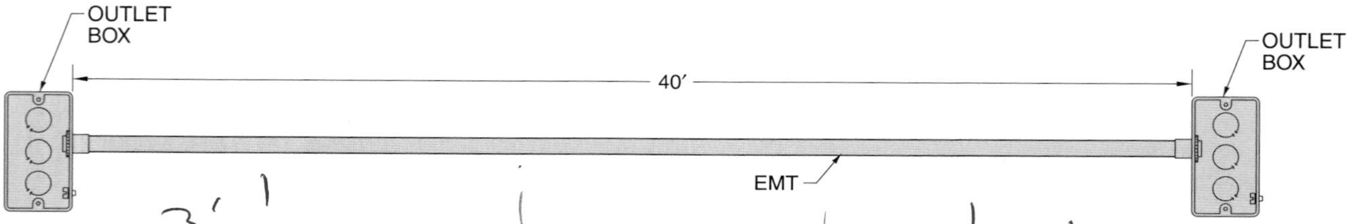

FIGURE 5

Learning the Code

- Before beginning to study Article 404, apprentices and students of the Code should carefully review the Article 100 definitions related to switches. There are approximately six different defined terms listed in Article 100 that are related to the term "switch." It is important that the type of switch in use be clearly defined so that the provisions of Article 404 can be properly applied.
- A common problem associated with the installation of switches is the proper grounding of the switch and the switch plate. Section 404.9(B) clearly states that both the switch itself and the faceplate must be grounded. Two methods are listed to ensure that the snap switch is part of an effective ground-fault current path.

- Another common installation problem with snap switches that is frequently addressed in Code tests and licensing examinations is the voltage permitted between adjacent devices. For example, when installing a 277 V switch in an outlet box with a 120 V receptacle, Section 404.8(B) requires that the devices must be arranged such that the voltage between adjacent devices does not exceed 300 V. One solution in cases where the voltage exceeds 300 V is to install an identified barrier between the devices that is securely fastened in place.
- Switchboards and panelboards are covered in Article 408. Part II of the article covers switchboards and Part III covers panelboards. The overcurrent protection requirements for panelboards were greatly simplified in the 2008 NEC®. The distinction between power panelboards and lighting and appliance panelboards was removed. Section 408.36 now simply states that each panelboard shall be protected by an overcurrent protective device that has a rating not greater than that of the panelboard.

Wiring Materials—Switches, Switchboards, and Panelboards

Switches are devices which open or close electrical circuits. Only ungrounded conductors shall be opened by switches, no matter what the configuration of the switch may be. Switchboards and panelboards are installed to control light and power circuits. They shall be grounded.

New in the 2011 NEC®

- New requirements for all switches installed to control lighting loads – 404.2(C)
- New Exceptions added for grounding of snap switches and mounting straps – 404.9(B)
- New definition of child care facility added to Article 406 – 406.2
- New provisions added for AFCI-protected receptacles when used as replacements – 406.4(D)(4)
- New and revised requirements for tamper-resistant receptacles in dwelling units – 406.12 Ex.
- New and revised requirements for tamper-resistant receptacles in child care facilities – 406.14
- New marking and identification requirements for switchboards and panelboards – 408.3(F) and 408.4(B)

SWITCHES – ARTICLE 404

The provisions of Article 404 apply to all switches, switching devices, and circuit breakers (CBs) used as switches. In Article 100 there are six definitions for a switch. Switches defined are the bypass isolation switch, general-use snap switch, general-use switch, isolating switch, motor-circuit switch, and transfer switch. A *switch* is a device, with a current and voltage rating, used to open or close an electrical circuit. This definition does not include an isolation switch. A circuit breaker (CB) is an overcurrent protective device with a mechanical mechanism that may manually or automatically open the circuit when an overload condition or short circuit occurs.

Switch Connections – 404.2(A)

Only ungrounded conductors shall be opened by switches, no matter what the configuration of the switch might be. A switched conductor run in a metallic raceway or cable assembly shall have the return conductor in the same metallic raceway or cable assembly. The current in each conductor is the same, but the flow of current is in opposite directions, canceling out the inductive heating effect. **See Figure 9-1.**

A switched conductor run in conduit shall have the return conductor run in the same conduit. Travelers or shunts for 3-way and 4-way switches cannot be run alone in conduit or any metal wiring system. The grounded conductor (neutral) could be considered the return. If so, it is run in the same raceway as the travelers or shunts. With both polarities in the same enclosure or raceway, the inductive heating is minimized.

Grounded Conductors – 404.2(B)

Switching of the grounded conductor is prohibited. There are two conditional exceptions to this rule. Per 404.2(B), Ex., the switching of a grounded conductor is permitted provided all of the circuit conductors are disconnected simultaneously. The exception also allows a switch or circuit breaker to disconnect a grounded conductor provided the device is arranged so that the grounded conductor cannot be disconnected until all the ungrounded conductors of the circuit have been disconnected. **See Figure 9-2.**

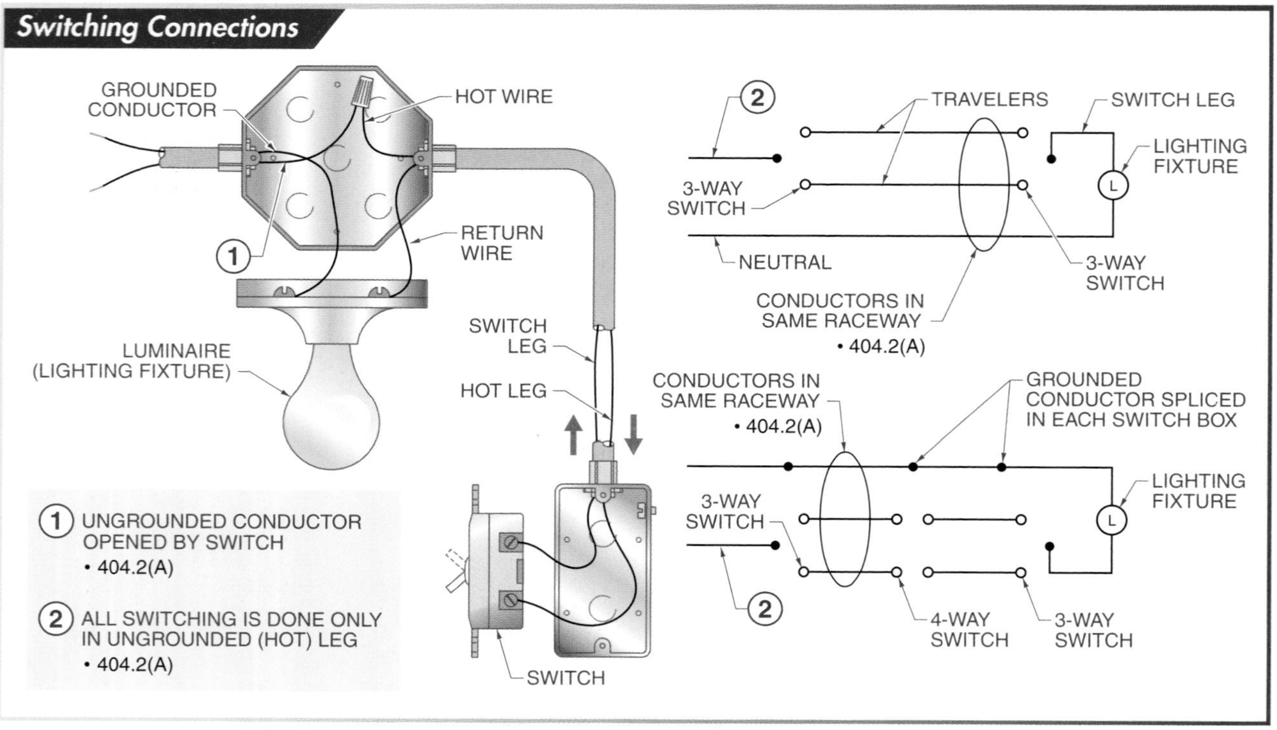

Figure 9-1. Only ungrounded conductors shall be opened by switches.

With more electronic switches and switching control devices being installed in both residential and commercial locations, 404.2(C) was added to the 2011 NEC® to ensure that a grounded conductor is available at switch outlets. Where the switch is intended to control lighting loads, and the source of supply is a grounded general-purpose branch circuit, the grounded circuit conductor must be provided at the switch location. This will facilitate the use of these electronic controls, which require the presence of the grounded conductor to operate.

There is an exception to this rule that addresses two general types of installations. The first is where the conductors for switches controlling the lighting loads are supplied from a raceway. The second is for cable assemblies where the cable assembly enters the box through a framing cavity that is open at the top or bottom of the same floor level, or through a wall, floor, or ceiling that is unfinished on one side. In both of these cases, the grounded conductor can be omitted from the switch enclosure because it can be added, if necessary, at a later date.

Enclosures – 404.3

Enclosed switches minimize fire hazards and protect switch components from mechanical abuse. Enclosed switches also protect people from electrical shock. Where the enclosure contains switches or circuit breakers it shall be listed for that purpose and be of the externally operable type. The enclosure shall provide adequate wire-bending space at the terminals per Tables 312.6(A) and (B). The concern here is to provide adequate space within the enclosure to bend the conductors when making terminations. The two types of termination configurations are L bends and S bends. **See Figure 9-3.**

In an L bend, the conductor does not enter or leave the enclosure through the wall opposite its terminals. In this bend, the conductors take a 90° or L bend. For an L bend, the distance between the terminal lugs and the opposite wall of the enclosure shall conform to Table 312.6(A).

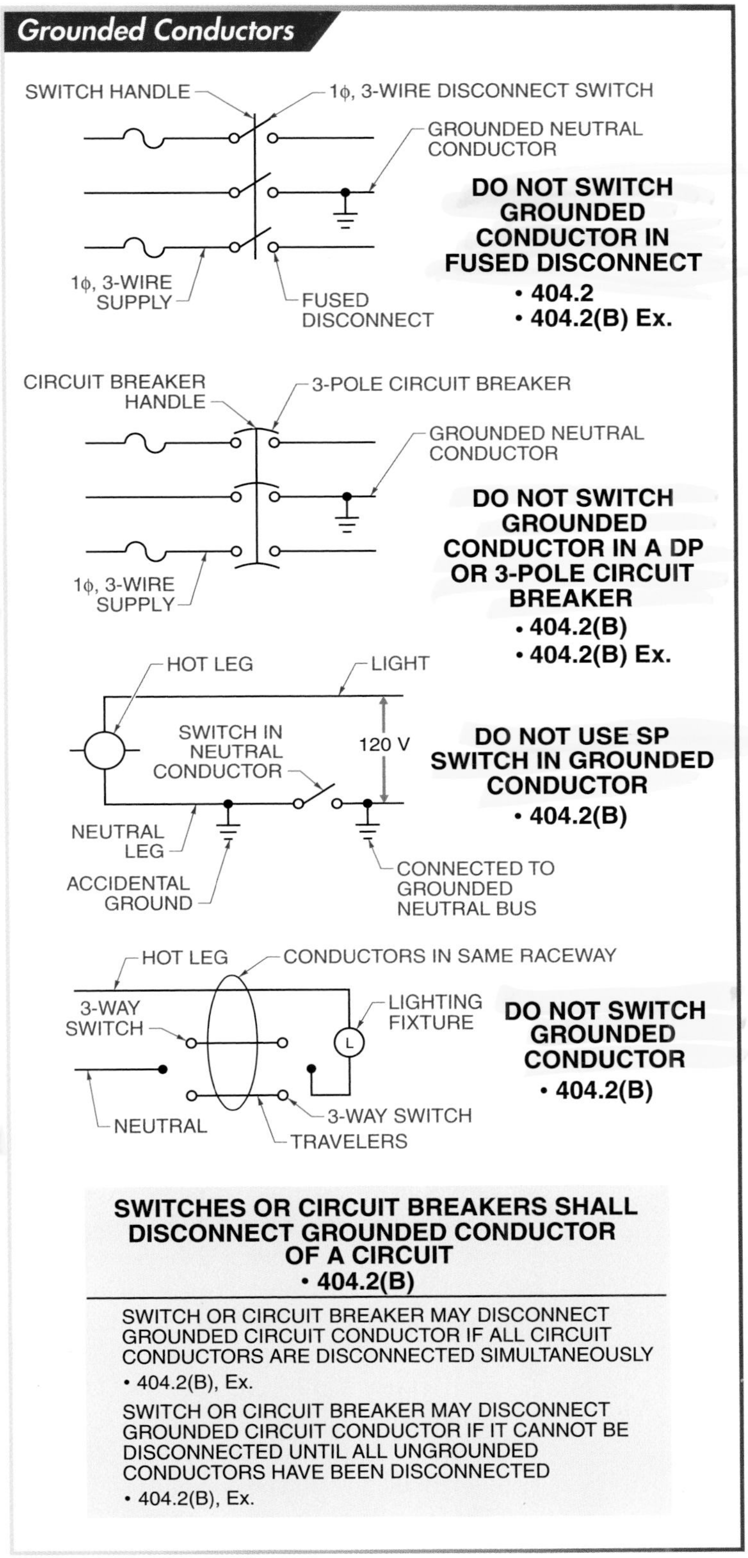

Figure 9-2. A switch or circuit breaker shall not be used to open the grounded conductor.

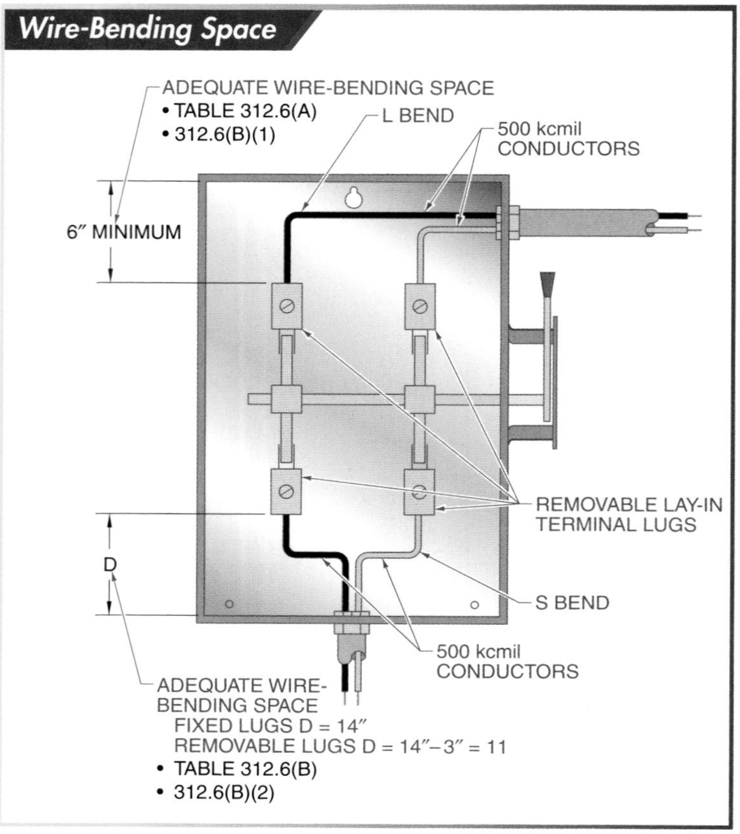

Figure 9-3. Enclosed switches shall provide adequate wire-bending space.

In an S bend, the conductor enters or leaves the enclosure through the wall opposite the terminals. In this bend, more space is required to bend the conductors so that they can terminate properly at the terminal lugs. In an offset or S bend, the distance between the terminal lugs and the opposite wall of the enclosure shall conform to Table 312.6(B). For removable and lay-in wire terminals intended for only one wire, bending space shall be permitted to be reduced by the number of inches shown in parentheses in the table.

As a general rule, 404.3(B) prohibits the use of enclosures as junction boxes, auxiliary gutters, or raceways for conductors that are feeding through or tapping off to other switches or OCPDs. The only case where this is permissible is where the installation complies with 312.8. To comply with the provisions of 312.8, three conditions must be met. First, the total of all conductors installed at any cross section of the enclosure must not exceed 40% of the area of that space. Second, the total area of all conductors, splices, and taps installed at any one point must not exceed 75% of the cross-sectional area of the space. Third, a label must be installed on the enclosure to warn of the potential hazard and to identify the closest disconnecting means for any feed-through conductors.

Wet Locations – 404.4. Switch enclosures installed in wet locations shall be of the weatherproof type. A 1/4″ space shall be provided between the enclosure and the wall to which it is mounted. This 1/4″ airspace prevents the water from accumulating at the back of the enclosure, helping to prevent rust and corrosion. Nonmetallic cabinets and cutout boxes may be installed without the 1/4″ airspace between the enclosures and the wall or other surface used for support per 312.2(A), Ex. **See Figure 9-4.** Switches are not permitted to be installed within a shower or tub space unless they are installed as part of a listed tub or shower assembly.

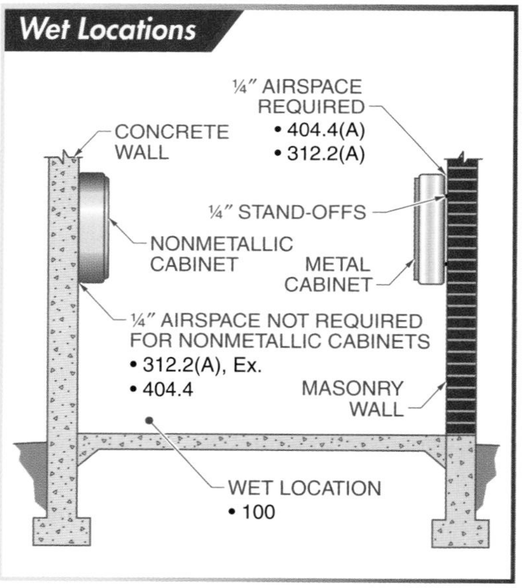

Figure 9-4. Switch enclosures installed in wet locations shall be of the weatherproof type.

Position of Knife Switch – 404.6

An improperly installed knife switch could be inadvertently closed due to gravity or an improperly wired switch could expose live parts in the OFF position, both of which could cause an accident. Per 404.6(A), single-throw knife switches shall be installed with the hinged end down. In this position, gravity will not tend to close the switch blades when left in the open position. Per 404.6(B), double-throw knife switches shall be mounted vertically or horizontally. Double-throw knife switches mounted vertically shall be equipped with a locking device to prevent gravity from closing the blades. **See Figure 9-5.**

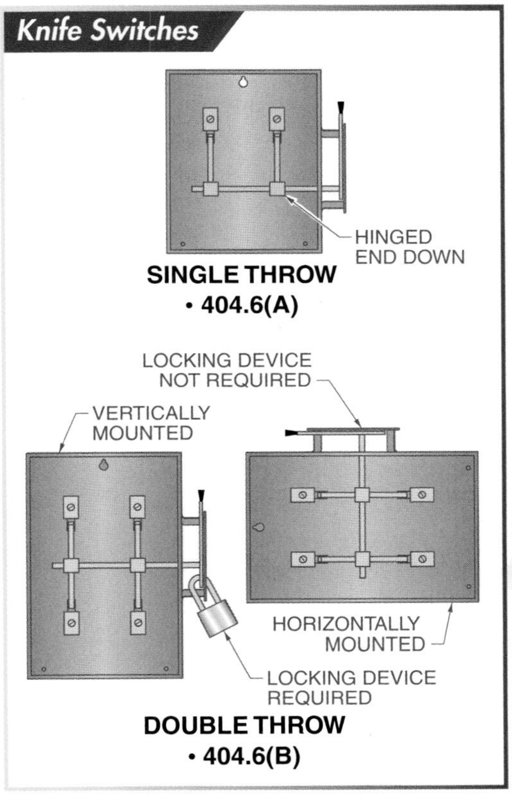

Figure 9-5. Knife switches shall be installed so that gravity does not close the blade.

Per 404.6(C), single-throw knife switches shall be connected in such a way that the blades are de-energized when the switches are in the open position. The load side of switches connected to circuits or equipment which may provide a backfeed source of power shall have a warning sign installed. The sign shall be permanently installed on the switch enclosure or immediately adjacent to open switches. The sign shall read, "WARNING – LOAD SIDE OF TERMINALS MAY BE ENERGIZED BY BACKFEED."

ON-OFF Position – 404.7. Switches or circuit breakers mounted in an enclosure shall clearly indicate whether they are in the open (OFF) or closed (ON) position. Where these switches or circuit-breaker handles are operated vertically, the up position of the handle shall be the ON position. See Ex. 1 and 2 for two special switching applications where the up position of the handle may not be the ON position.

Knife Switch Contacts

In a knife switch, metal contacts are used to open or close a circuit. The metal contact material is selected based on its conductivity, hardness, mechanical strength, cost, toxicity, and resistance to corrosion. Corrosion resistance is important because metals often form insulating oxides that can prevent the switch from operating properly. Also, the hardness of the metal is important to prevent abrasive wear.

Accessibility and Grouping – 404.8

All switches and circuit breakers used as switches shall be installed so that they are operated from a readily accessible location. The center of the grip of the operating handle, when in the highest position, shall not be more than 6′-7″ above the floor or working platform. **See Figure 9-6.** There are three exceptions to this general rule.

Per 404.7, vertically mounted switches shall be ON in the up position and OFF in the down position.

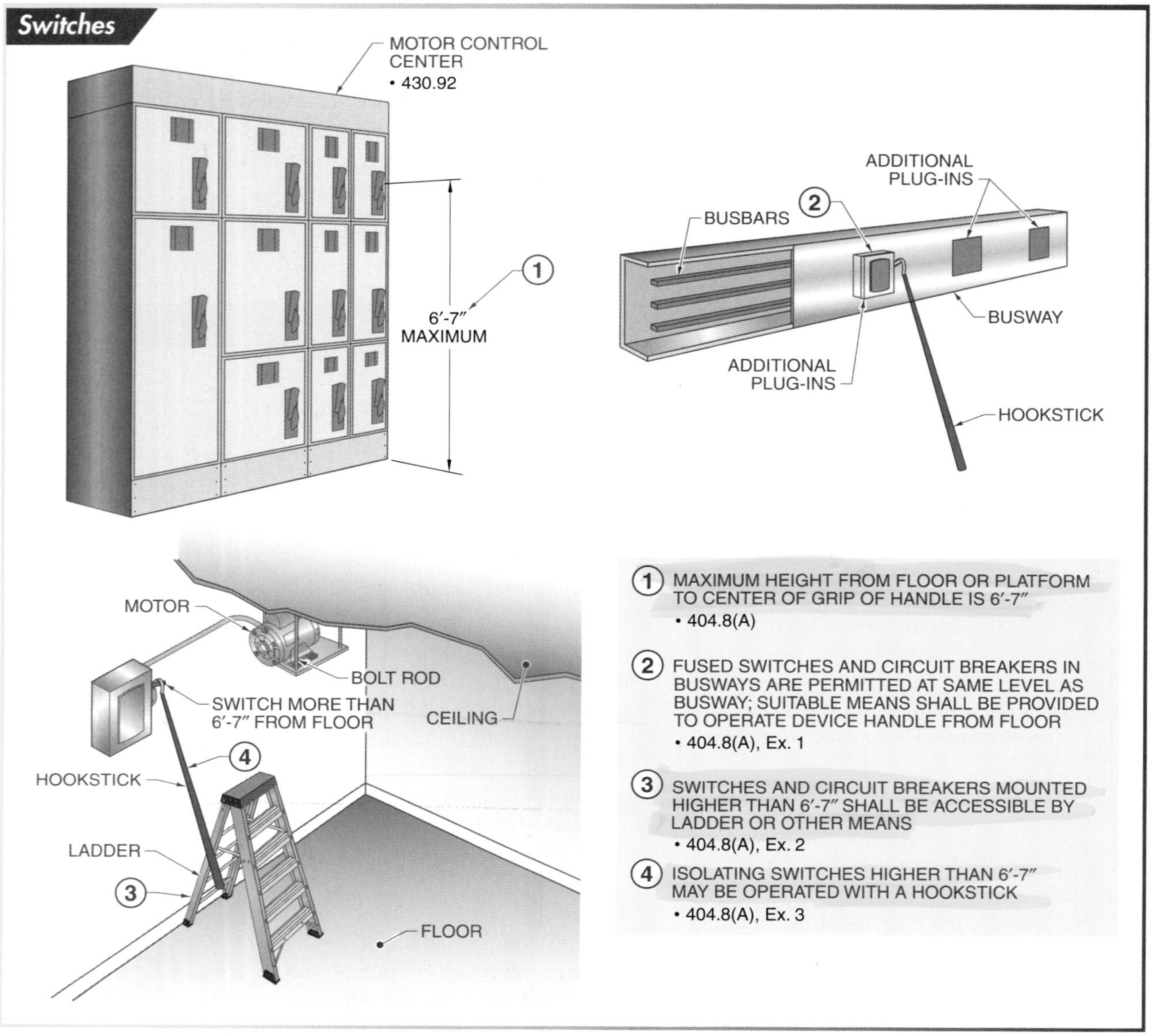

Figure 9-6. All switches and circuit breakers used as switches shall be installed so that they are operated from a readily accessible location.

Per 404.8(A), Ex. 1, fused disconnect switches and circuit breakers shall be mounted at the same level as busways if a suitable means is provided to disconnect the switch or circuit breaker from the floor. Per 404.8(A), Ex. 2, motors, appliances, and other equipment mounted higher than 6′-7″ above the floor are permitted to have a switch or circuit breaker mounted adjacent to the equipment. The switch or circuit breaker shall be accessible by a ladder or other portable means. Per 404.8(A), Ex. 3, hookstick-operated isolating switches are permitted at heights greater than 6′-7″ above the floor.

Voltage Between Adjacent Switches – 404.8(B). Snap switches are permitted to be grouped or ganged in an enclosure provided the voltage between adjacent snap switches does not exceed 300 V. If the voltage between adjacent snap switches exceeds 300 V, an identified and securely installed barrier shall be installed in the enclosure between the adjacent snap switches. **See Figure 9-7.**

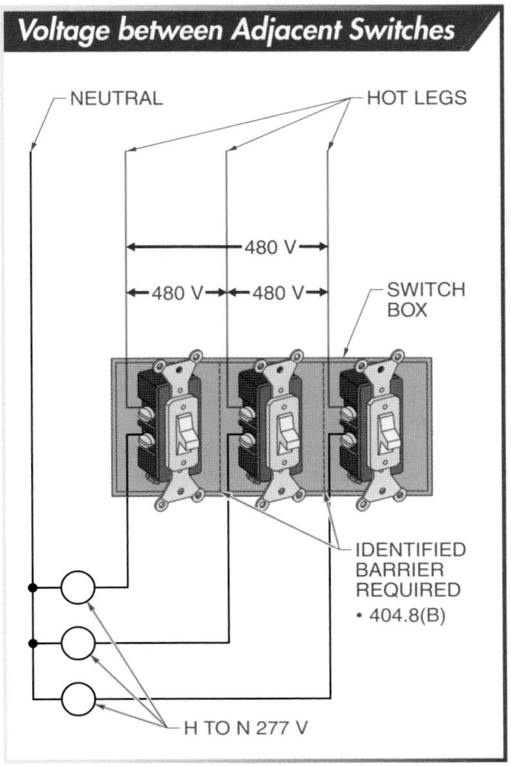

Figure 9-7. Secure barriers are required if the voltage between adjacent snap switches exceeds 300 V.

Provisions for Snap Switch Faceplates – 404.9. Faceplates for flush-mounted snap switches shall be installed so that they are flat against the wall and completely cover the wall opening. Ungrounded metal boxes that can be reached by a person standing on a grounded, conductive floor or near other grounded, conductive items shall be provided with a nonmetallic faceplate. **See Figure 9-8.**

Section 404.9(B) contains the general provisions for grounding general-use snap switches. In general, snap switches shall be grounded by the use of an EGC. Means must be provided to ensure that metal faceplates are grounded, regardless of whether a metal faceplate is actually installed. Note that this requirement includes dimmer-type and other similar control-type switches. Two conditions are presented to ensure that snap switches are part of an effective ground-fault current path. The first condition is by the use of metal screws to attach the switch to a metal box or metal cover that is connected to an EGC. This can also be accomplished in nonmetallic boxes where an integral means is provided for connecting to an EGC. The second condition is where an EGC or an equipment bonding jumper is used to connect directly to the EGC terminal of the snap switch.

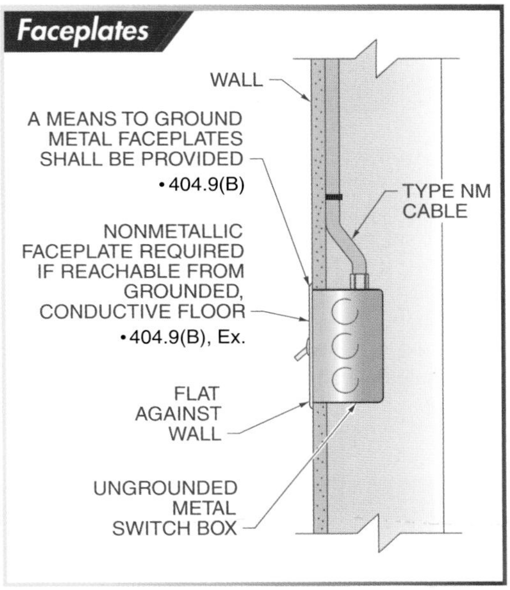

Figure 9-8. Faceplates shall be flat against the wall and completely cover the wall opening.

Grounding Enclosures – 404.12

Metal enclosures for switches and circuit breakers shall be grounded per Article 250. Exposed non-current-carrying metal parts of fixed equipment likely to become energized shall be grounded. Where a grounding means exists, nonmetallic boxes for switches are required to be installed with a wiring method that contains or provides for an EGC. **See Figure 9-9.**

Rating and Use of Snap Switches – 404.14

Snap switches shall be used within their ratings. This includes both tumbler and toggle switches. There are four types of snap switches: AC general-use, AC-DC general-use, CO/ALR, and AC specific-use rated for 347 V. **See Figure 9-10.**

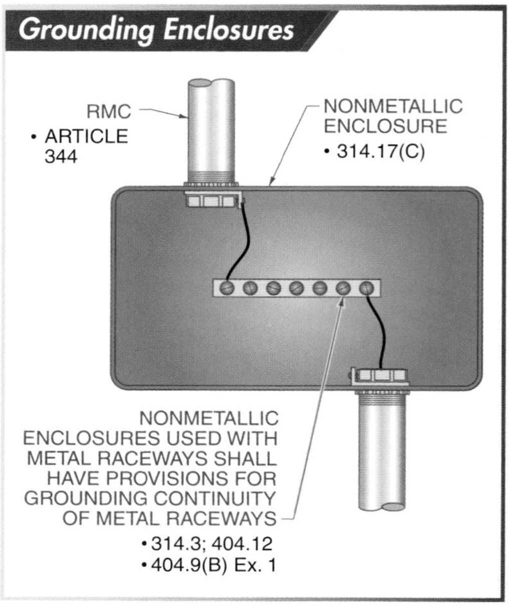
Grounding Enclosures
RMC
• ARTICLE 344
NONMETALLIC ENCLOSURE
• 314.17(C)
NONMETALLIC ENCLOSURES USED WITH METAL RACEWAYS SHALL HAVE PROVISIONS FOR GROUNDING CONTINUITY OF METAL RACEWAYS
• 314.3; 404.12
• 404.9(B) Ex. 1

Figure 9-9. Exposed non-current-carrying parts of fixed equipment shall be grounded.

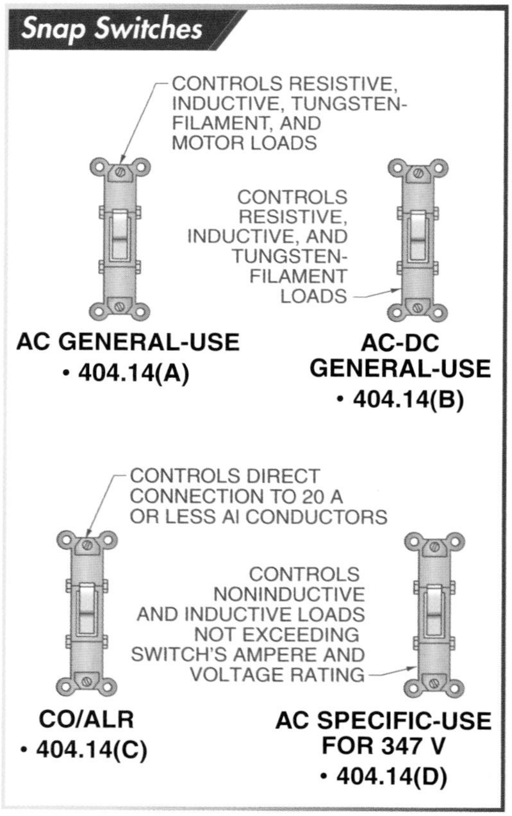
Snap Switches
CONTROLS RESISTIVE, INDUCTIVE, TUNGSTEN-FILAMENT, AND MOTOR LOADS
CONTROLS RESISTIVE, INDUCTIVE, AND TUNGSTEN-FILAMENT LOADS
AC GENERAL-USE
• 404.14(A)
AC-DC GENERAL-USE
• 404.14(B)
CONTROLS DIRECT CONNECTION TO 20 A OR LESS Al CONDUCTORS
CONTROLS NONINDUCTIVE AND INDUCTIVE LOADS NOT EXCEEDING SWITCH'S AMPERE AND VOLTAGE RATING
CO/ALR
• 404.14(C)
AC SPECIFIC-USE FOR 347 V
• 404.14(D)

Figure 9-10. Snap switches shall be used within their ratings.

AC General-Use Snap Switch – 404.14(A). AC general-use snap switches are recognized as being suitable for use on AC circuits only and for controlling the following three types of loads:

1. Resistive and inductive loads, including electric-discharge lamps. The load shall not exceed the ampere and voltage rating of the switch.
2. 120 V tungsten-filament lamp loads. The load shall not exceed the ampere and voltage rating of the switch.
3. Motor loads. The load shall not exceed 80% of the ampere rating of the switch and the voltage of the motor shall not exceed the voltage rating of the switch.

AC-DC General-Use Snap Switch – 404.14(B). AC-DC general-use snap switches are suitable for use on AC or DC circuits to control the following three types of loads:

1. Resistive loads. The load shall not exceed the ampere and voltage rating of the switch.
2. Inductive loads. The load shall not exceed 50% of the ampere rating of the switch at the applied voltage. Switches rated in horsepower are suitable for controlling motor loads within their horsepower and voltage rating.
3. Tungsten-filament lamp loads. The load shall not exceed the ampere rating of the switch at the applied voltage provided the switch has a "T" rating.

Per 404.14(B)(2), snap switches can be used with manual motor starter units.

CO/ALR Snap Switches – 404.14(C). Snap switches which are connected directly to aluminum conductors and rated 20 A or less shall be listed and marked CO/ALR. Per UL's *General Information for Electrical Construction Materials*, terminals of 15 A and 20 A switches not marked CO/ALR are for use with copper and copper-clad aluminum conductors only. Terminals marked CO/ALR are for use with aluminum, copper, and copper-clad aluminum conductors.

Screwless pressure terminal connectors of the conductor push-in type are for use only with copper and copper-clad aluminum conductors.

Terminals of switches rated 30 A and above not marked AL/CU are for use with copper conductors only. Terminals of switches rated 30 A and above marked AL/CU are for use with aluminum, copper, and copper-clad aluminum conductors.

AC Specific-Use Snap Switches Rated for 347 V – 404.14(D). This section has been added to the rating and use of snap switches in an effort to harmonize the NEC® with the CEC. Listed snap switches rated 347 V AC are permitted for controlling noninductive and inductive loads not exceeding the ampere and voltage rating of the switch.

Dimmer Switches – 404.14(E). Dimmer switches are a convenient and safe way to control incandescent lighting loads. Section 404.14(E) was added to the Code to clarify that general-use dimmer switches are only permitted for the control of incandescent lighting loads. Some electricians and electrical contractors had mistakenly thought that the general-use dimmers could be used for other loads, such as motors or other types of utilization equipment. By their product listing this is prohibited unless the dimmer switch is specifically listed for use with other types of loads.

Cord-and-Plug-Connected Loads – 404.14(F). For installations where a snap switch is used to control a cord-and-plug- connected piece of equipment that is supplied from a general-purpose branch circuit, 404.14(F) requires that the rating of the snap switch be not less than the rating of the maximum permitted ampere rating or setting of the OCPD protecting the receptacle or cord connector. There is one exception to this general rule and applies to installations where a snap switch is used to control only one receptacle on a branch circuit. In this case, the switch shall be permitted to be rated at not less than the rating of the receptacles. An Informational Note included in this section refers to 210.50(A) and 400.7(A)(1) for equivalency comparisons for receptacle outlet and cord connectors supplied by a permanently connected cord pendant.

Per 404.14(E) general-use dimmer switches are only permitted for the control of incandescent lighting loads unless listed for other loads.

SWITCHBOARDS AND PANELBOARDS – ARTICLE 408

Switchboards and panelboards are installed to control light and power circuits. They are designed with busbars and circuit breakers or fuse holders which supply power distribution through feeder circuits and branch circuits. Lighting panelboards include a permanently installed main circuit breaker or single set of fuses used as mains. See 408.14.

A *switchboard* is a single panel or group of assembled panels with buses, overcurrent devices, and instruments. Switchboards are not intended to be installed in cabinets. They are commonly accessible from the front and rear.

A *panelboard* is a single panel or group of assembled panels with buses and overcurrent devices, which may have switches to control light, heat, or power circuits. Panelboards are placed in a cabinet or cutout box which is accessible only from the front. Panelboards are mounted in or on partitions or walls.

Support and Arrangement of Busbars and Conductors – 408.3

Busbars and conductors used in a switchboard or panelboard shall be installed so as to be free from physical damage. **See Figure 9-11.** The busbars and conductors shall be held firmly in place. Other than the required interconnections and control wiring, only those conductors that are intended for termination in a vertical section of a switchboard shall be located in that section.

A switchboard may include a single panel or group of assembled panels with buses, overcurrent devices, and instruments.

Conductors and Busbars on a Switchboard or Panelboard – 408.3(A). Barriers are required in all listed switchboards that are used as service-entrance equipment. These barriers isolate the service-entrance conductors or busbars and their termination points from the rest of the switchboard. Load conductors shall exit vertically from the section in which they originate. However, 408.3(A)(3), Ex., permits conductors to travel horizontally through vertical sections of a switchboard where barriers are installed to isolate these conductors from the live busbars. Per 408.3(B), busbars and conductors shall be arranged to avoid overheating from inductive effects.

Overheating and Inductive Effects – 408.3(B). Conductors or busbars carrying AC current produce a fluxing magnetic field around the conductor or busbar. When ferrous metal (iron or steel) surrounds a single current-carrying conductor, there is an interaction between the molecules of the ferrous material and the fluxing magnetic field. The molecules of the ferrous material vacillate back and forth, producing a heating effect on the ferrous material. In some cases, the heat produced may be hot enough to ignite combustible materials. For this reason, 408.3(B) requires that the arrangement of busbars and conductors be such that overheating due to inductive effects is avoided.

Overheating of Conductors

A thermal imager can be used to identify overheating due to inductive effects around conductors or caused by loose connections on terminals in switchboards and panelboards.

Used as Service Equipment – 408.3(C). Switchboards or panelboards used for service equipment shall have a main bonding jumper bonded to the casing of the panelboard or switchboard frame. The main bonding jumper shall be placed in the panelboard or one of the sections of the switchboard and shall connect the grounded conductor of the service-entrance conductors to the switchboard or panelboard enclosure frame. See 250.28(D) for sizing the main bonding jumper.

Terminals – 408.3(D). Terminals in switchboards and panelboards shall be arranged so that it is not necessary to reach over or behind an ungrounded line bus in order to make connections. The concern is the safety of the worker who has to modify or make changes to an existing switchboard or panelboard.

Barriers in Switchboard

Barriers provide some protection to those performing maintenance on this equipment. It is not uncommon to work on switchboards while the main busbars are energized. Proper PPE must be worn when work is performed on energized equipment. For additional safety information on the importance of selecting and utilizing the proper PPE and following the appropriate safe work practices while working in switchboards, see NFPA 70E®—Electrical Safety in the Workplace 2009 Edition.

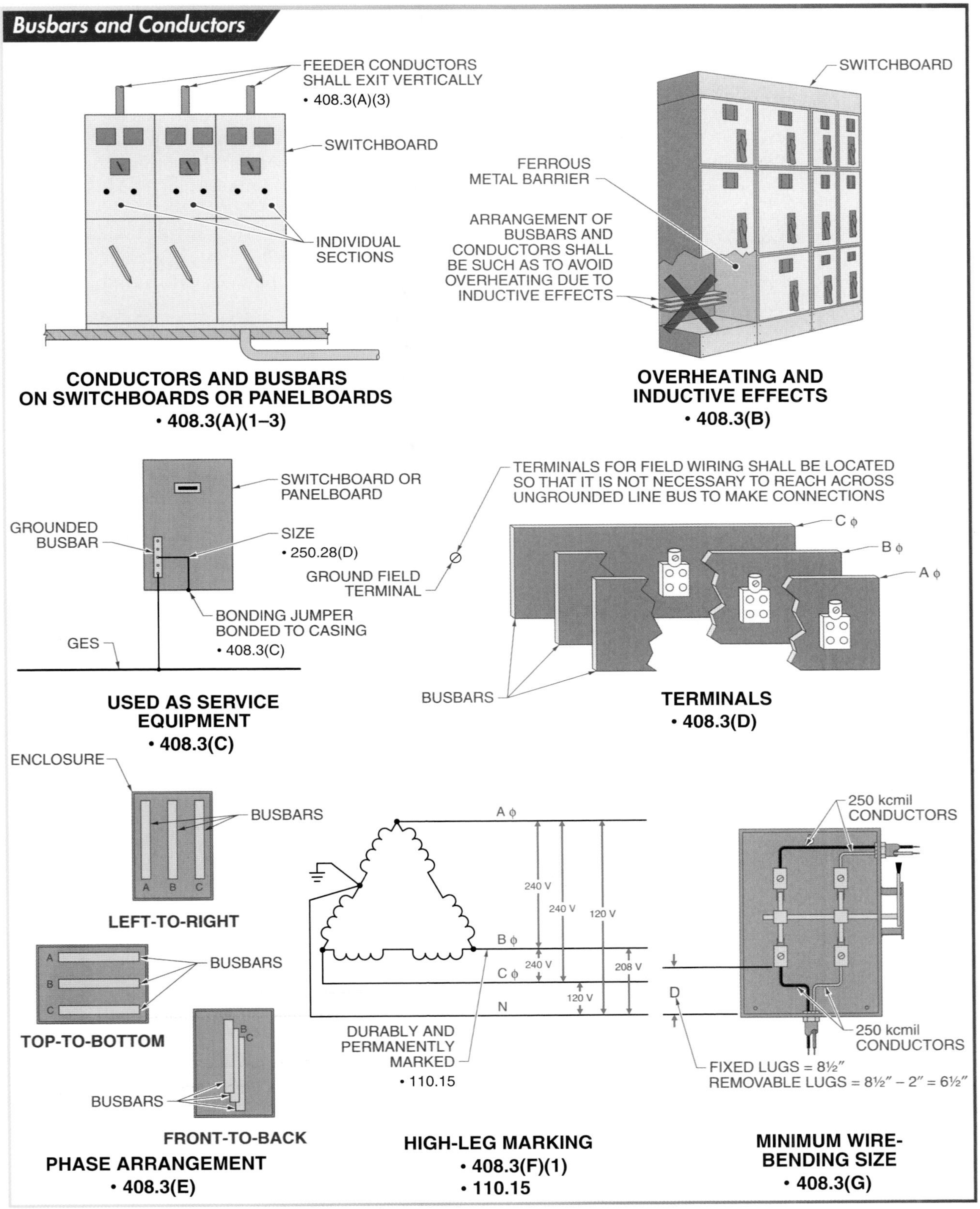

Figure 9-11. Busbars and conductors on a switchboard or panelboard shall be installed so as to be free from physical damage.

Phase Arrangement – 408.3(E). On a 3ϕ, 4-wire, delta-connected system, the B phase shall be the phase having the higher voltage to ground. The phase arrangement on 3ϕ buses shall be A, B, C from front to back, top to bottom, or left to right as viewed from the front of the switchboard or panelboard.

High-Leg Marking 408.3(F)(1) – 110.15. The requirements for marking busbars and phase conductors having a higher voltage to ground are located in 110.15. Where switchboards or panelboards are supplied from a 4-wire delta system with a high-leg, a permanently installed field marking must be made. The marking shall read "Caution: ___ Phase has ___ Volts to Ground."

The B phase shall be durably and permanently marked with an orange outer finish or by other effective means. For example, a 4-wire delta system is used to supply large 3ϕ loads and small 1ϕ loads. These systems are also known as 3ϕ, 4-wire, 120/240 V systems. Per 408.3(E), the B phase is required to be the high-leg (wild leg or stinger leg). In order to accomplish this, the transformer in which two ends go to the A phase and C phase shall be the transformer that is center-tapped and grounded. This causes the A phase to read 120 V to the grounded midpoint and C phase to read 120 V to the grounded midpoint. From B phase to the grounded midpoint is 208 V. This is the vector sum of the 240 V of one full winding and 120 V of the half-winding that is center-tapped and grounded.

To calculate the high-leg voltage, take the voltage between the A phase or C phase and the grounded midpoint (usually 120 V). Multiply this voltage by 1.73 ($\sqrt{3}$) to find the voltage of the high-leg to ground (120 V × 1.73 = 208 V).

Per 110.15, the high-leg or B phase shall be identified by the color orange. This alerts other installers who tap into the 4-wire delta system and take out a 1ϕ, 3-wire, 120/240 V system to avoid the high-leg.

Wire-Bending Space – 408.3(G). The minimum wire-bending space at terminals in panelboards and switchboards shall comply with 312.6. The concern is to provide adequate wiring space between the connecting terminal in the equipment and the opposite wall of the enclosure. If there is not adequate room, the installer may have to make sharp bends in the conductors while shaping them to enter the connection terminals. Minimum wire-bending space prevents sharp bends in conductors. **See Figure 9-12.**

Wire-Bending Space

Figure 9-12. Minimum wire-bending space prevents sharp bends in conductors.

Sharp bends in conductors often cause damage to the insulation. Per 312.6(B)(1), the wire-bending space in Table 312.6(A) shall be used where the conductor does not enter or leave the enclosure through the wall opposite its terminal. Per 312.6(B)(2), the wire-bending space in Table 312.6(B) shall be used where the conductor enters or leaves the enclosure through the wall opposite its terminal.

Per Article 408.3(G), terminals in switchboards and panelboards shall have minimum wire-bending space.

Circuit Identification – 408.4. Section 110.22 requires that each disconnecting means must be marked to clearly identify its purpose. The marking is not required if the disconnecting means location is such that its purpose is "evident." Good engineering practice, however, should ensure that every disconnecting means is legibly and durably marked to avoid any potential misidentification in the field.

Section 408.4 reinforces the general requirement in 110.22 and ensures that any circuit modifications are added to the panelboard or switchboard directory. The requirement is included to ensure that one circuit can be distinguished from another. This is extremely important for persons who may be required to maintain or make modifications to the original circuit installation. Panelboards and switchboards are provided with a circuit directory, typically on the inside of a door, that must be filled out at the time of the circuit installation or modification.

In addition to the field marking requirements for circuit directories for switchboards and panelboards, 408.4(B) requires that an additional type of identification be provided for all switchboards and panelboards installed in other than one- or two-family dwellings. This identification requirement is for switchboard and panelboard feeders and requires that a field marking be made to indicate the device or equipment where the power supply originates. This is an important safety provision and makes it easier for workers who must maintain or troubleshoot the electrical systems to properly de-energize the source of power to comply with applicable safety standards and procedures.

Unused Openings – 408.7. Section 110.12(A) requires that any unused openings for cables or raceways in any box, gutter, cabinet, enclosure, equipment casing, etc., be effectively closed. While there are products specifically designed for this purpose, the Code only mandates that the openings be effectively closed to afford protection substantively equivalent to the enclosure or equipment wall. Section 408.7 applies this general principle to panelboards and switchboards. Specifically, any unused opening for a circuit breaker or switch must be closed. In this case, however, the means to close the opening must be identified or approved and provide a level of protection substantially equivalent to the wall of the equipment or enclosure.

NFPA 70E®

For maximum safety, personal protective equipment and safety requirements for test instrument procedures must be followed as specified in NFPA 70E®, OSHA Standard Part 1910 Subpart 1—Personal Protective Equipment (1910.132 through 1910.138), and other applicable safety mandates. Personal protective equipment includes protective clothing, head protection, eye protection, ear protection, hand and foot protection, back protection, knee protection, and rubber insulated matting.

SWITCHBOARDS – ARTICLE 408, PART II

Switchboards that have any exposed live parts shall be installed in permanently dry locations that are under supervision and accessible only to qualified persons. Switchboards shall be located where the probability of damage from equipment or processes is reduced to a minimum. Per 408.17, switchboards shall not be placed in locations where it is possible for them to spread fire to adjacent combustible materials. Where switchboards are installed in damp or wet locations, they must comply with 312.2. Switchboards should also be located such that the risk of fire from adjacent combustible materials is minimized.

Clearances – 408.18

A space of not less than 3′ shall be provided between the top of a switchboard and a combustible ceiling. This is not required if a noncombustible shield is provided between the switchboard and the ceiling or if the switchboard is of the totally enclosed type. **See Figure 9-13.**

Per 408.18(B), the clearances around switchboards shall comply with 110.26. The minimum space required between the bottom of the switchboard enclosure and the busbars, their supports, or other obstructions is 8″ for insulated busbars and 10″ for uninsulated busbars.

Per 408.5, conduits or other raceways including their fittings shall not rise more than 3″ above the bottom of switchboards or floor-mounted panelboards. Sufficient space shall be provided to permit the installation of conductors when raceways enter the bottom of these enclosures.

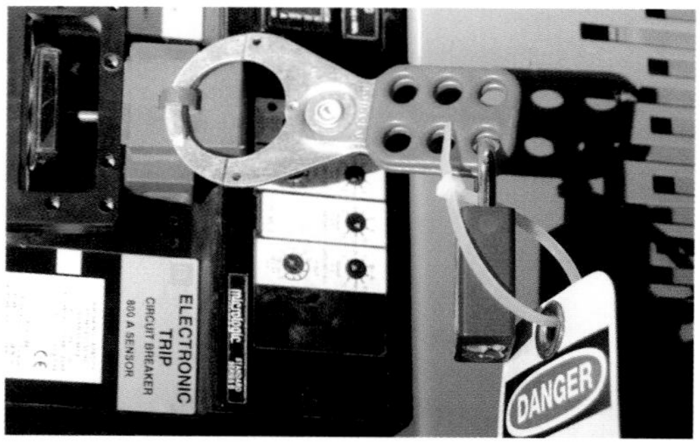

During maintenance and installation operations, switchboards are locked out and tagged out to prevent inadvertent re-energization of the circuit.

PANELBOARDS – ARTICLE 408, PART III

Panelboards, per 408.30, shall have a rating equal to or higher than the minimum feeder capacity required for the load calculated according to Article 220. If the calculated load on a panelboard is 175 A, the panel shall be sized at a rating of at least 175 A (minimum load). The next available standard-size panelboard is 200 A.

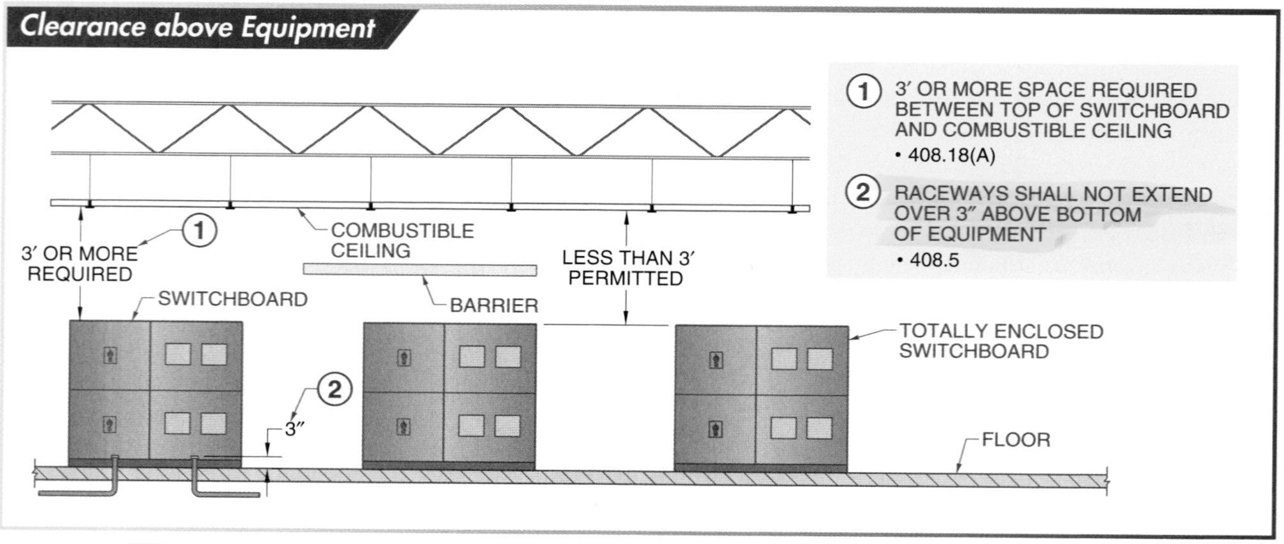

Figure 9-13. Minimum space above the top of a switchboard and combustible ceiling is based on open or totally enclosed switchboards and if barriers are present.

Arc Flash Protection Boundary

The Arc Flash Protection Boundary is the distance from exposed energized conductors or circuit parts where bare skin would begin to receive the on-set of a second-degree burn, equivalent to 1.2 cal/cm2. According to OSHA 1910.269(l)(6)(ii), an employer must train employees about the dangers of being exposed to flames or electric arcs when they work near such hazards. This would include training employees how to calculate the Arc Flash Protection Boundary for specific tasks and circumstances.

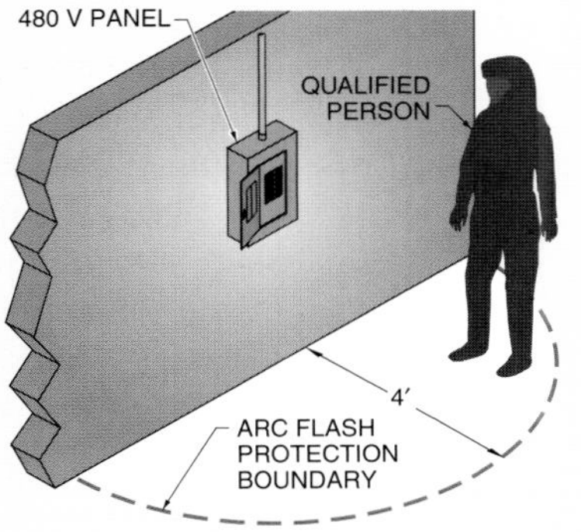

The 1995 edition of NFPA 70E identified the Arc Flash Protection Boundary for the first time. The 2000 edition included Table 130.7(C)(9) and Table 130.7(C)(10) to guide employees in choosing arc flash protective equipment and clothing. These tables have been expanded and improved in the 2004 and 2009 editions to cover tasks based on specific types of equipment, which include the following:

- panelboards and other equipment rated 240 V and below
- panelboards and switchboards rated above 240 V up to 600 V using molded-case circuit breakers or insulated-case circuit breakers
- 600 V class motor-control centers
- 600 V class switchgear with power circuit breakers or fused switches
- other 600 V class equipment from 277 V through 600 V
- National Electrical Manufacturers Association (NEMA) class E2 motor starters, 2300 V through 7200 V
- metal-clad switchgear, 1000 V through 38,000 V
- arc-resistant switchgear Type 1 or Type 2
- other equipment rated 1000 V through 38,000 V

Each type of low-voltage equipment has specific notes attached to it that set the limits for the available short-circuit current and the operating time of the protective device. If the available short-circuit current or operating time of the equipment exceeds either of these limits, the tables cannot be used and an arc flash hazard analysis must be performed.

Additional information is available at www.nfpa.org.

Panelboards shall be durably marked by the manufacturer with the voltage and current ratings and the number of phases for which they are designed. The manufacturer's name or trademark shall also be marked on the panel so that it is visible after the installation, without disturbing the interior parts or wiring. All panelboards shall be provided with a circuit directory located on the face or inside cover of the panel door to identify the loads. All panelboard circuit modifications shall be legibly marked on the panelboard directory. See 408.4 and 110.22.

Overcurrent Protection – 408.30, 408.36. In general, all panelboards must have a rating that is not less than the minimum load calculated for the feeder that supplies the panelboard. Article 220 contains the requirements for determining the minimum load on the panelboard. Additionally, protection must be provided for each panelboard by the use of an overcurrent protective device (OCPD) that does not exceed the rating of the panelboard. Section 408.36 requires the location of this OCPD to be within or at any point on the supply side of the panelboard.

There are three exceptions to this general rule. The first exception removes the need for individual protection for panelboards used as service equipment with multiple service disconnecting means. Provisions for this application are covered in 230.71. The second exception also removes the need for individual protection of the panelboard where the panelboard is provided with protection on the supply side by two main circuit breakers or two sets of fuses that have a combined rating not greater than that of the panelboard. The last exception applies to panelboards in existing residential occupancies. Again, in these cases, individual protection for the panelboard is not required.

Historically, panelboards were classified by the Code as "lighting and appliance" or "power" panelboards. Both designations and categories were removed from the 2008 NEC®. Since then, each panelboard, as covered above, must be provided with an overcurrent protective device, the rating of which must not exceed that of the panelboard. Panelboards are no longer limited to a maximum of 42 overcurrent protective devices (fuses or circuit breakers).

Supplied through a Transformer – 408.36(B). Panelboards supplied through transformers shall have their overcurrent protection located on the secondary side of the transformer. **See Figure 9-14.** Per 408.36(B), Ex., a 1ϕ transformer with a 2-wire primary and a 2-wire secondary is permitted to have the OCPD installed on the primary side only. See 240.21(C)(1).

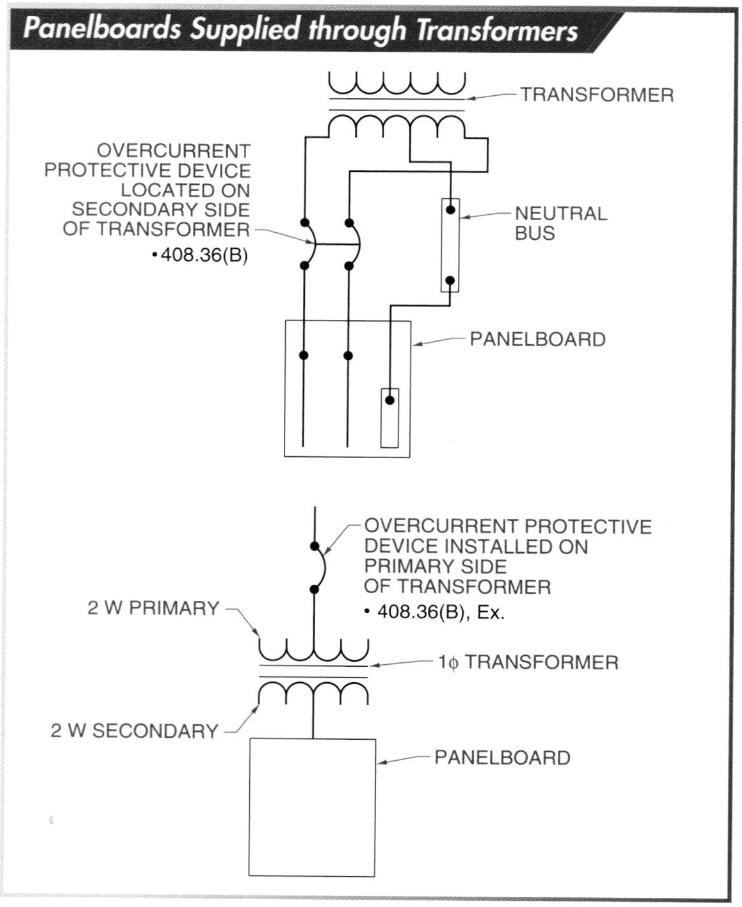

Figure 9-14. Panelboards supplied through transformers shall have OCPDs located on the secondary side.

Back-Fed Devices – 408.36(D). Plug-in OCPDs or plug-in main lug assemblies which are connected for backfeed (the plug-in stabs are the load side of the OCPD) shall be mechanically secured in their installed positions. This requirement covers all plug-in-type protective devices (circuit breakers or fuses). The intent is to eliminate the hazard of exposed energized plug-in stabs of an OCPD that could possibly become dislodged from their plug-in or connected position. **See Figure 9-15.**

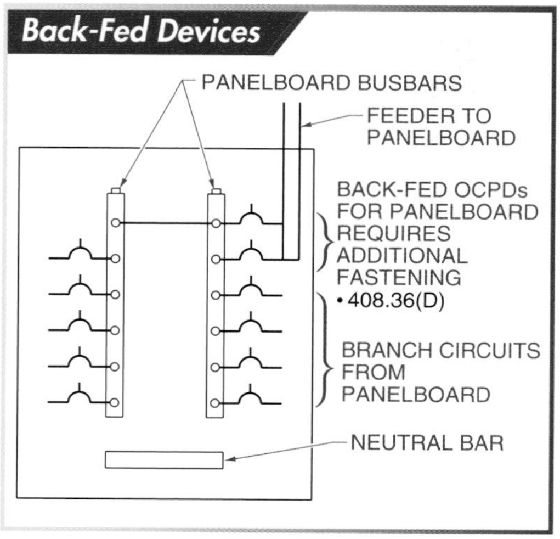

Figure 9-15. Plug-in OCPDs that are connected for backfeed shall be mechanically secured.

Grounding of Panelboards – 408.40

All metal panelboard cabinets and frames shall be connected together and grounded. This requirement may be satisfied if these enclosures are supplied by metal raceways or metal sheath cable assemblies which are recognized as EGCs. If a nonmetallic wiring method is used, such as Romex or PVC, an EGC shall be run with the feeder conductors.

If a panelboard is used as service equipment, the EGC terminal bar and the neutral terminal bar shall be bonded to the panelboard case along with the GEC per 250.130. In other than service equipment, the EGC terminal bar and the neutral terminal bar shall not be connected and bonded together in the panel enclosure. The neutral bar shall be isolated from the panelboard case and the EGC terminal bar shall be bonded to the panelboard case. **See Figure 9-16.**

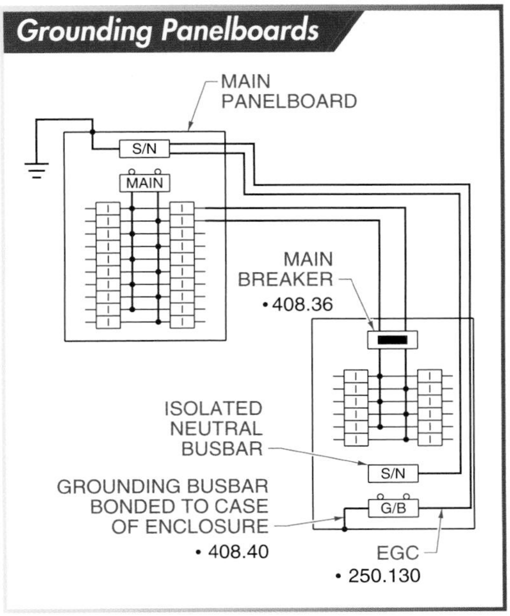

Figure 9-16. All panelboard cabinets and frames shall be grounded.

The neutral (grounded conductor) of an electrical system carries the unbalanced current of that system, while the only time an EGC is expected to carry current is in the event of a ground fault. A ground fault occurs when an ungrounded conductor (hot wire) inadvertently comes in contact with a portion of the metallic electrical system not intended to carry current. Per 250.86, metal enclosures and raceways shall be grounded. Exposed non-current-carrying metal parts of fixed equipment normally supplied by or enclosing conductors or components that are likely to become energized shall be grounded per 250.110.

If the EGC system came in contact with a grounded conductor (neutral) system that was carrying current, there would be a parallel path for this current to return to the service equipment bonding point, such as the metallic raceway or the metal armor of an AC cable. This would create the possibility of a potential difference between all of the exposed dead metal parts of electrical equipment and any other grounded metallic parts within that facility, such as metallic pipe and appliances within the plumbing system.

Panelboard Construction Specifications – Article 408, Part IV

Per 408.50, the panels of switchboards are required to be made of moisture-resistant, noncombustible material. Insulated or bare busbars shall be rigidly and securely mounted in place per 408.51. Section 408.52 requires circuits which provide switchboards with pilot lights, instrument devices, and switchboard devices with potential coils to be protected by standard overcurrent devices rated 15 A or less.

Per 408.52, Ex. 1, OCPDs rated more than 15 A are permitted where interruption to the circuit may create a hazard. Per 408.52, Ex. 2, special types of enclosed fuses rated 2 A or less are permitted.

Wire-Bending Space within an Enclosure Containing a Panelboard – 408.55. The concern here is the same as in 312.6, which is the deflection of conductors in cabinets and cutout boxes. Adequate space is required where the conductors are shaped and terminated at the equipment. The wire-bending space at the top and bottom of a panelboard shall comply with the dimensions in Table 312.6(B), regardless of the position of the conduit entries. The width of the side wiring gutters may be in accordance with the lesser distances of Table 312.6(A), based on the largest conductor to be terminated in that space. There are four exceptions to the general wire-bending space requirements in 408.55. All four exceptions allow modifications to the minimum bending space requirements in Table 312.6 if certain conditions are met. See 408.55, Ex. 1–4.

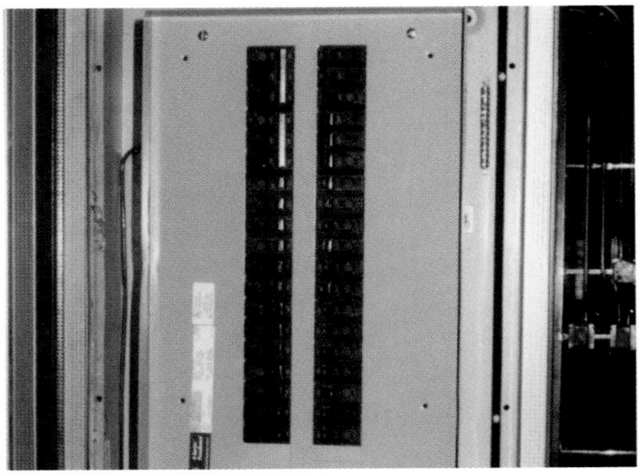

Panelboards shall be durably marked by the manufacturer with the voltage and current ratings.

QUICK QUIZ®

Refer to Chapter 9 Quick Quiz® on CD-ROM.

Additional information is available at ATPeResources.com

Wiring Materials—Switches, Switchboards, and Panelboards

chapter 9

REVIEW QUESTIONS

Name ______________________ Date ______________________

Switch 1. A(n) ___ is a device, with a current and voltage rating, used to open or close an electrical circuit.

ON 2. In general, vertically mounted switches shall be ___ in the up position.

Switch board 3. A(n) ___ is a single panel or group of assembled panels with buses, overcurrent devices, and instruments.

panel board 4. A(n) ___ is a single panel or group of assembled panels with buses and overcurrent devices, which may have switches to control light, heat, or power circuits.

3 5. A space of not less than ___′ shall be provided between the top of a switchboard and a combustible ceiling.

weatherproof 6. A surface-mounted switch installed in a wet location shall be enclosed in a(n) ___ enclosure.

T F 7. All panelboard cabinets and frames of panelboards shall be grounded.

T F 8. The neutral of an electrical system is always carrying the unbalanced current of that system.

T F 9. In general, only ungrounded conductors shall be opened by switches.

continuous 10. A(n) ___ load is any load that operates for three or more hours.

circuit breaker 11. A(n) ___ is an overcurrent protective device with a mechanical mechanism that may manually or automatically open the circuit when an overload condition or short circuit occurs.

______ 12. The two types of wire termination configurations are ___.

A. U and L
B. U and S
C. L and S
D. none of the above

______ 13. Snap switch terminals marked CO/ALR are for use with ___ conductors.

A. aluminum
B. copper
C. copper-clad aluminum
D. all of the above

______ 14. The arrangement of busbars and conductors in a switchboard shall be such as to avoid overheating due to ___ effects.

1/4″ 15. A metal switch enclosure installed in a wet location shall be spaced ___″ off the wall to which it is mounted.

__________ **16.** Double-throw knife switches mounted vertically shall be equipped with a locking device to prevent ___ from closing the blades.

__________ **17.** Conduits or other raceways including their fittings shall not rise more than ___″ above the bottom of switchboards or floor-mounted panelboards.

Secondary **18.** Panelboards supplied through transformers generally shall have their overcurrent protection located on the ___ side of the transformer.

barrier **19.** An identified ___ is required to be securely installed between adjacent snap switches rated over 300 V.

T F **20.** Switchboards and panelboards are installed to control light and power circuits.

Phases

T F **1.** The busbars at A are correctly arranged.

T F **2.** The busbars at B are correctly arranged.

T F **3.** The busbars at C are correctly arranged.

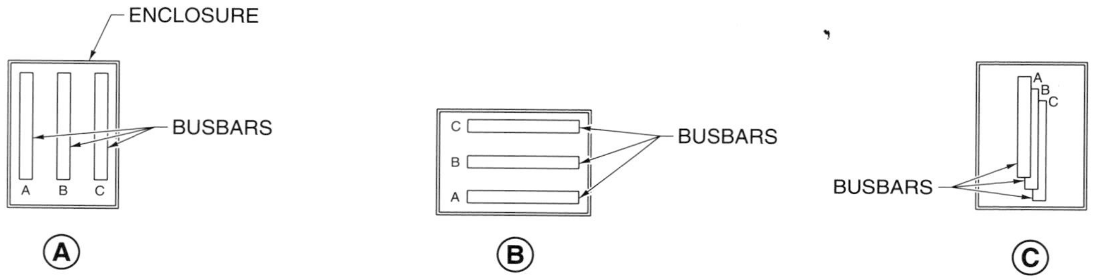

Wiring Materials—Switches, Switchboards, and Panelboards

chapter 9

TRADE COMPETENCY TEST

Name ______________________ **Date** ______________________

NEC®	Answer	
________	6"	**1.** *See Figure 1.* Determine the minimum wire-bending space required at A.
________	12"	**2.** *See Figure 1.* Determine the minimum wire-bending space required at B.
________	________	**3.** *See Figure 1.* Determine the minimum wire-bending space required at B if the installation is done with removable and lay-in wire terminals intended for only one wire.
________	B	**4.** *See Figure 2.* Which phase is required to be identified as the high-leg?
________	6'-7"	**5.** Determine the maximum permissible height above the finished floor to the center of the handle of a motor disconnect switch which is located in the top row of a motor control center.

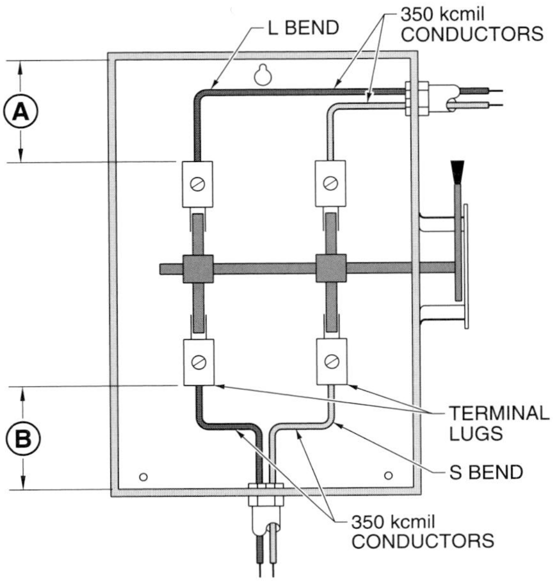

FIGURE 1

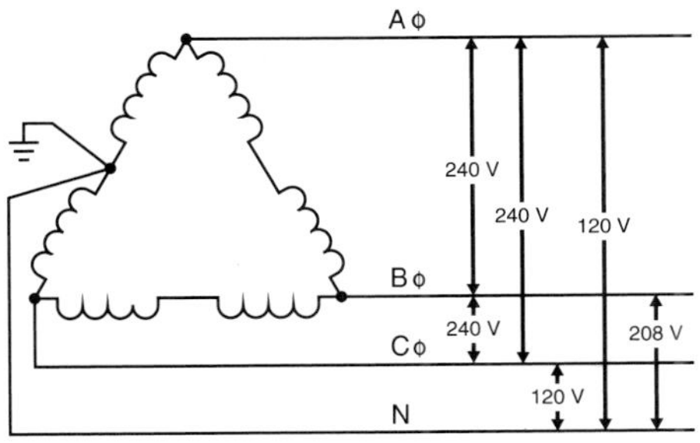

FIGURE 2

______ 300V ✓ **6.** Determine the maximum voltage between adjacent snap switches which are installed in a ganged enclosure that is not provided with identified and secure barriers.

______ 3′ **7.** *See Figure 3.* Determine the minimum clear space required between the switchboard and combustible ceiling.

______ ______ **8.** *See Figure 4.* A 42 circuit panelboard is installed in a commercial office building. The circuit directory is located and maintained by the building maintenance staff in a central location that is readily accessible. Does this meet the requirements of Article 408?

______ ______ **9.** A service switchboard is installed in an electrical equipment room to supply a commercial building. The top of the switchboard is 7′ above the finished floor. Determine the minimum headroom about the switchboard per Article 408 of the NEC®.

______ 3″ **10.** A 4″ PVC conduit is run underground into an electrical service room. The conduit emerges from the concrete floor beneath an open-bottom switchboard. Determine the maximum height, including fittings, which the conduit is permitted to rise above the bottom of the enclosure.

______ ______ **11.** What is the minimum space required between the bottom of the switchboard enclosure and the energized busbars for the installation in Problem 10? The busbars are uninsulated.

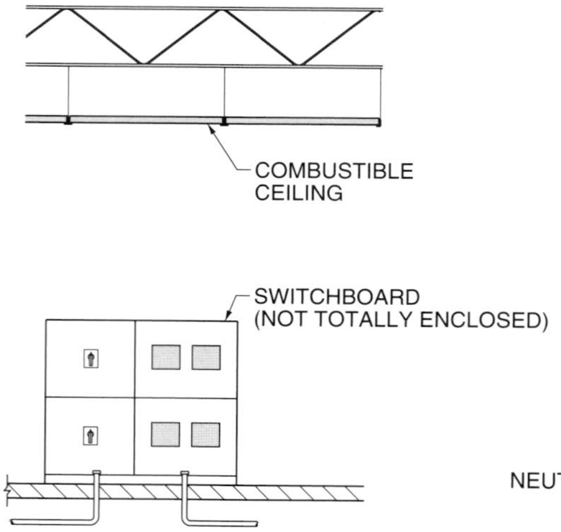

FIGURE 3

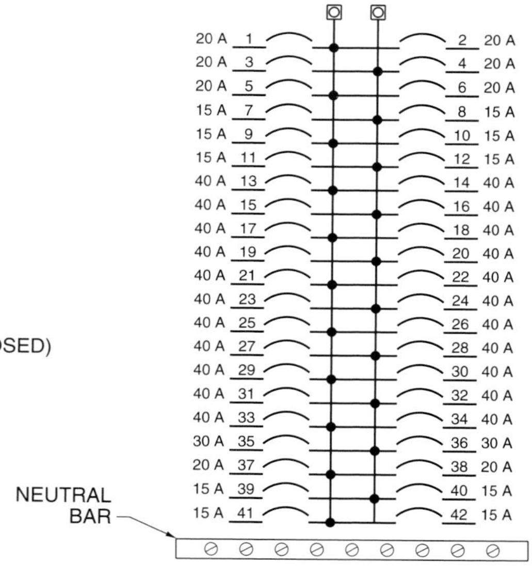

FIGURE 4

Learning the Code

- Part of the arrangement of the NEC® is the chapter designations. Chapter 4 of the NEC® covers articles 400 through 490. For the most part, these articles cover equipment for general use. For example, Article 430 covers motors and Article 450 covers transformers. In some cases, it's a stretch to think of the article as pertaining to equipment for general use. For example, Article 400 covers flexible cords and cables, and Article 402 covers fixture wires. Nevertheless, it is often helpful to think about the entire Code in terms of the individual chapters.

Greenlee Textron, Inc.

- Like Table 310.104(A) for conductor applications for insulated conductors, Table 400.4 is an important table that contains information about flexible cord or cable design, construction, and application. As always, do not forget to review the notes that accompany the table.
- Article 410 of the NEC® covers the rules and requirements for lighting fixtures. In the 2005 NEC® the term "luminaire" was inserted to replace the term "lighting fixture." When you see "luminaire," think "lighting fixture." Luminaire is simply the term used internationally.
- Article 422 covers appliances. Appliances are utilization equipment, typically of the type that is found in dwelling units and commercial installations. One of the most important parts of this article is the disconnecting requirements contained in Part III of the article. The requirements for disconnecting means are important safety provisions that are designed to protect personnel who operate, inspect, or maintain such equipment.

Equipment for General Use

Appliances and utilization equipment are often connected with flexible cords and cables. Flexible cords are commonly used to connect equipment such as lighting fixtures, dishwashers, vacuum cleaners, food waste disposers, etc. Flexible cables are commonly used to connect larger equipment and equipment that has great flexibility needs because of frequent movement.

New in the 2011 NEC®

- *Reorganization of requirements for the use of lighting fixtures (luminaires) as a raceway – 410.64*
- *Clarification of the provisions for lighting fixtures (luminaires) installed in suspended ceilings – 410.110*
- *The addition of installation provisions for LED drivers and power supplies – 410.116(B)*
- *New disconnecting means requirements for fluorescent luminaires – 410.130(G)*
- *New definition of vending machine added to Article 422 – 422.2*
- *Clarification of the requirements for disconnecting means for motor-operated appliances rated over ⅛ HP – 422.31(C)*

FLEXIBLE CORDS AND FLEXIBLE CABLES – ARTICLE 400

An *appliance* is any utilization equipment which performs one or more functions, such as clothes washing, air conditioning, cooking, etc. Appliances are generally constructed in standard sizes and types and are used in residential or commercial occupancies as opposed to industrial occupancies. Utilization equipment is any electrical equipment that uses electrical energy for electronic, electromechanical, chemical, heating, lighting, or similar purposes. Appliances and utilization equipment may be connected with flexible cords and cables.

A *flexible cord* is an assembly of two or more insulated conductors, with or without braids, contained within an overall outer covering and used for the connection of equipment to a power source. A *flexible cable* is an assembly of one or more insulated conductors, with or without braids, contained within an overall outer covering and used for the connection of equipment to a power source.

A flexible cord consists of two or more insulated conductors and shall be suitable for its intended use and location.

The primary difference between flexible cords and flexible cables is their designated use. Flexible cords are used to supply equipment such as luminaires (lighting fixtures), dishwashers, vacuum cleaners, food waste disposers, etc. Flexible cables are used for the connection of larger equipment and equipment that is frequently moved or interchanged. Typical equipment connected with flexible cables includes elevators, stage and lighting equipment, and electric vehicle charging equipment.

Table 400.4 contains a list of flexible cords and cables. Information on the construction of the cord or cable, type of insulation and covering, and permitted uses for each cord or cable is provided. Cords or cables, other than those listed in Table 400.4, shall be subjected to a special investigation to determine their suitability for the intended use.

Markings – 400.3; 400.6

Each flexible cord or cable and its associated fittings shall be suitable for the intended use and location. For example, cords intended for use in a wet location shall be suitable for wet locations. The "Use" column in Table 400.4 indicates which cords are suitable for use in wet locations. Likewise, if a cord must be identified for hard or extra-hard use, as is required in 590.4(C) for temporary branch-circuit wiring, the cord shall be identified in Table 400.4 as suitable for such use.

Marking of flexible cords and cables may be accomplished by means of a printed tag attached to the coil, reel, or carton of the cord or cable per 400.6(A). **See Figure 10-1.** The information required on the printed tag is:

- Maximum rated voltage for which the cord or cable is listed.
- Proper letters for identifying the cord or cable.
- Type of cord or cable.
- Manufacturer's name, trademark, or other distinctive markings.
- AWG size or circular mil area of the conductors. (Some specific cords and cables are required to be surface-marked at intervals not exceeding 24″ with the type, size, and number of conductors.)

Other optional cord or cable markings may be surface marked per 400.6(B). These markings can be used to indicate special cable or cord characteristics such as LS for limited smoke characteristics or sunlight resistant for cords or cables that can be exposed to the direct rays of the sun.

Flexible Cords and Cables—Marking

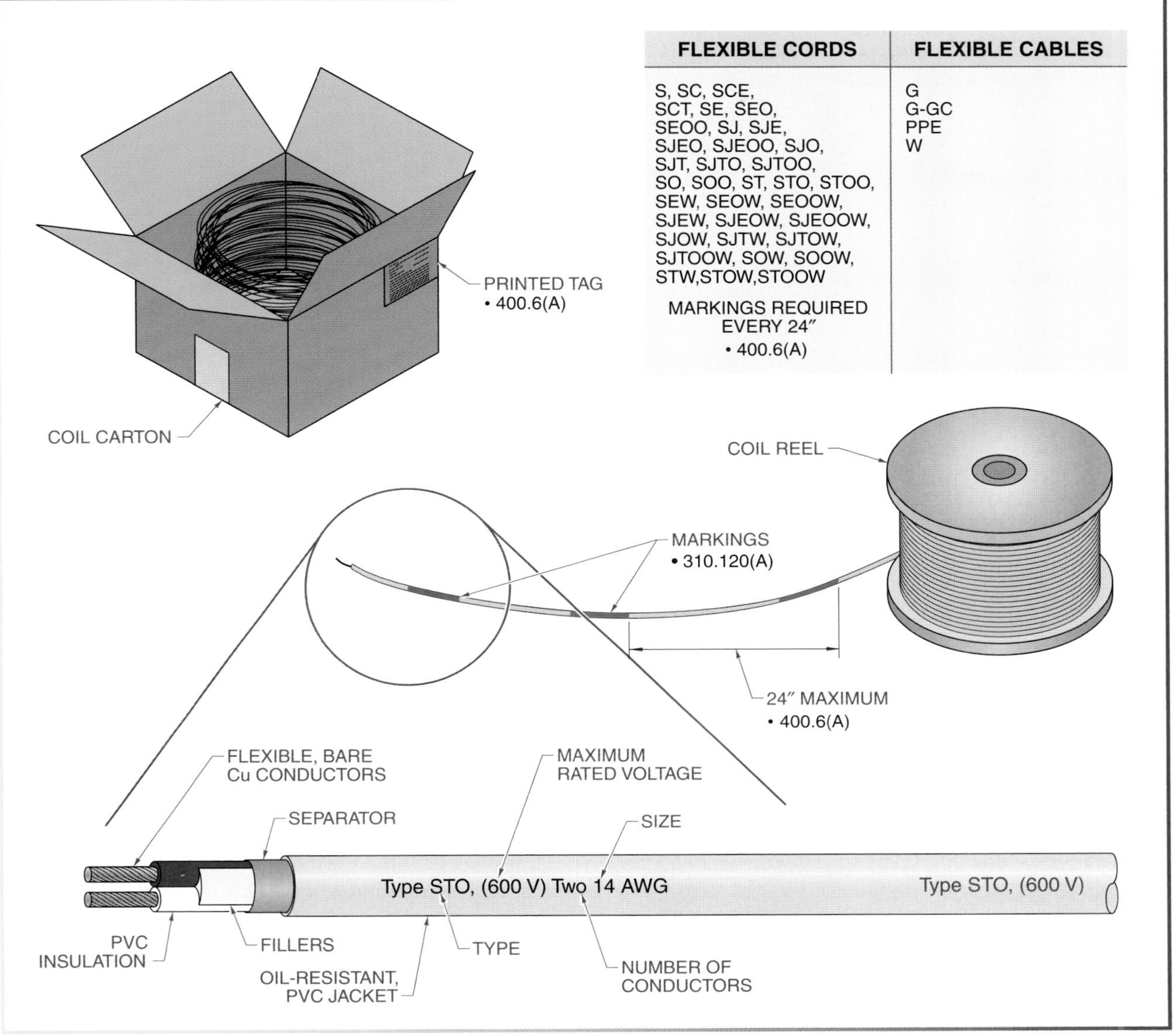

FLEXIBLE CORDS	FLEXIBLE CABLES
S, SC, SCE, SCT, SE, SEO, SEOO, SJ, SJE, SJEO, SJEOO, SJO, SJT, SJTO, SJTOO, SO, SOO, ST, STO, STOO, SEW, SEOW, SEOOW, SJEW, SJEOW, SJEOOW, SJOW, SJTW, SJTOW, SJTOOW, SOW, SOOW, STW,STOW,STOOW	G G-GC PPE W
MARKINGS REQUIRED EVERY 24″ • 400.6(A)	

Figure 10-1. Flexible cords and cables shall be marked.

Ampacities – 400.5

Flexible cords and cables, like any other conductor, have a maximum allowable ampacity which shall not be exceeded. As with other conductors, the allowable ampacity depends upon the insulation used and the number of conductors contained within the flexible cord or cable. Table 400.5(A)(1) provides the allowable ampacities for flexible cords and cables. Two allowable ampacities are listed for many of them.

The allowable ampacities under Column A apply to multiconductor cords connected to utilization equipment in which only three of the conductors are current-carrying. Column B applies to multiconductor cords connected to utilization equipment in which only two of the conductors are current-carrying. Because the number of current-carrying conductors is limited to two, Column B permits higher allowable ampacities.

The same rules that apply to other conductors for determining when they are current-carrying also apply to flexible cords and cables. If a neutral conductor only carries unbalanced current from other conductors of the same circuit, it need not be considered current-carrying. If the circuit is a 3-wire circuit consisting of two phase wires and the neutral of a 4-wire, wye connected, 3ϕ system, the neutral shall be considered a current-carrying conductor. In addition, if a major portion of the load on a 3ϕ, 4-wire wye circuit consists of nonlinear loads, such as electric-discharge lighting and computer equipment, the neutral shall also be considered a current-carrying conductor.

Where more than three current-carrying conductors are in a cord or cable, Table 400.5(A)(3) provides the applicable derating values based upon the number of current-carrying conductors. As with single conductors installed in a raceway, the EGC is not considered a current-carrying conductor for the purposes of determining the number of conductors in the flexible cord or cable for ampacity adjustment.

Table 400.5(A)(2) can be used to determine the ampacity of flexible cable types SC, SCE, SCT, PPE, G, G-GC and W. Ampacities shall be selected from Columns D, E, and F depending upon the number of current-carrying conductors within the flexible cable. The same derating provisions which apply for Table 400.5(A)(1) apply for Table 400.5(A)(2).

Overcurrent Protection – 400.13

Overcurrent protection for flexible cords and cables is treated separately from other conductors. Per 400.13, flexible cords not smaller than 18 AWG shall be considered protected by the overcurrent devices specified by 240.5. Tinsel cords or other flexible cords of sizes smaller than 18 AWG, with similar construction characteristics, and approved for use with specific appliances can also be protected against overcurrents by overcurrent protective devices specified in 240.5.

In general, flexible cords and cables shall be protected against overcurrents per ampacities listed in Tables 400.5(A)(1) and 400.5(A)(2). If a flexible cord or tinsel cord, approved for use with a specific listed appliance or portable lamp, is connected to a branch circuit installed per Article 210, the minimum cord size is determined by the specific branch-circuit ratings. **See Figure 10-2.**

If flexible cords are used in listed extension cords, 240.5(B)(3) states that the cord shall be considered to be protected when applied within the cord's listing requirements. The use of extension cords is covered by Article 590, Temporary Installations.

Installation – 400.7 through 400.14

As with any wiring method, the most important step, prior to the installation of the flexible cords and cables, is to review the permitted uses. For flexible cords and cables, 400.7(A) contains the permitted uses and 400.8 contains the uses not permitted. Electrical installers and design personnel should review these sections to determine if the use of flexible cords and cables is permitted for the specific installation.

Another installation practice which requires review is splices. In general, 400.9 requires that flexible cords shall not contain splices. Hard-service and junior hard-service cord are, however, permitted to be repaired if the conductors are spliced in accordance with 110.14(B) and the splice restores the original insulation and outer jacket properties of the cord. Likewise, 400.10 requires that flexible cords shall be installed in a manner that provides strain-relief for the cord or the cable at joints and terminals.

Uses Permitted – 400.7. Ten uses are permitted for flexible cords and cables per 400.7(A). **See Figure 10-3.** Uses permitted are:

(1) Pendants—A *pendant* is a hanging luminaire that uses flexible cords to support the lampholder.

(2) Fixtures—Flexible cables are used to wire hanging fixtures.

(3) Appliances or portable lamps—Flexible cords are also permitted to supply luminaires and other portable lamps, mobile signs or appliances. An attachment plug is required and shall be supplied from a receptacle outlet.

Flexible Cords—Overcurrent Protection

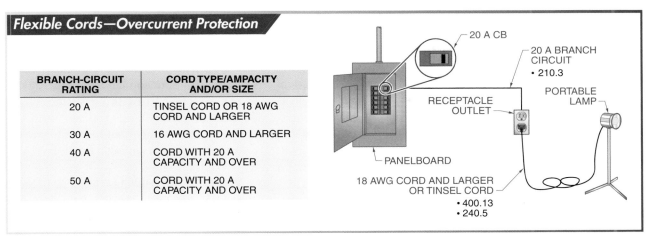

BRANCH-CIRCUIT RATING	CORD TYPE/AMPACITY AND/OR SIZE
20 A	TINSEL CORD OR 18 AWG CORD AND LARGER
30 A	16 AWG CORD AND LARGER
40 A	CORD WITH 20 A CAPACITY AND OVER
50 A	CORD WITH 20 A CAPACITY AND OVER

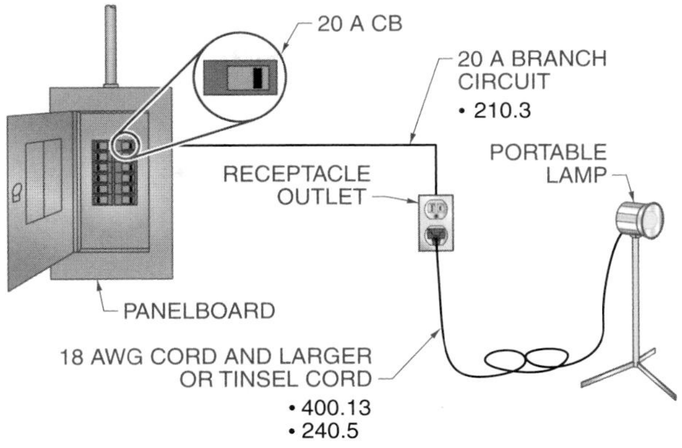

Figure 10-2. Branch-circuit OCPDs are permitted to provide overcurrent protection for flexible cords and cables if of sufficient size or ampacity.

Flexible Cord and Cable Uses

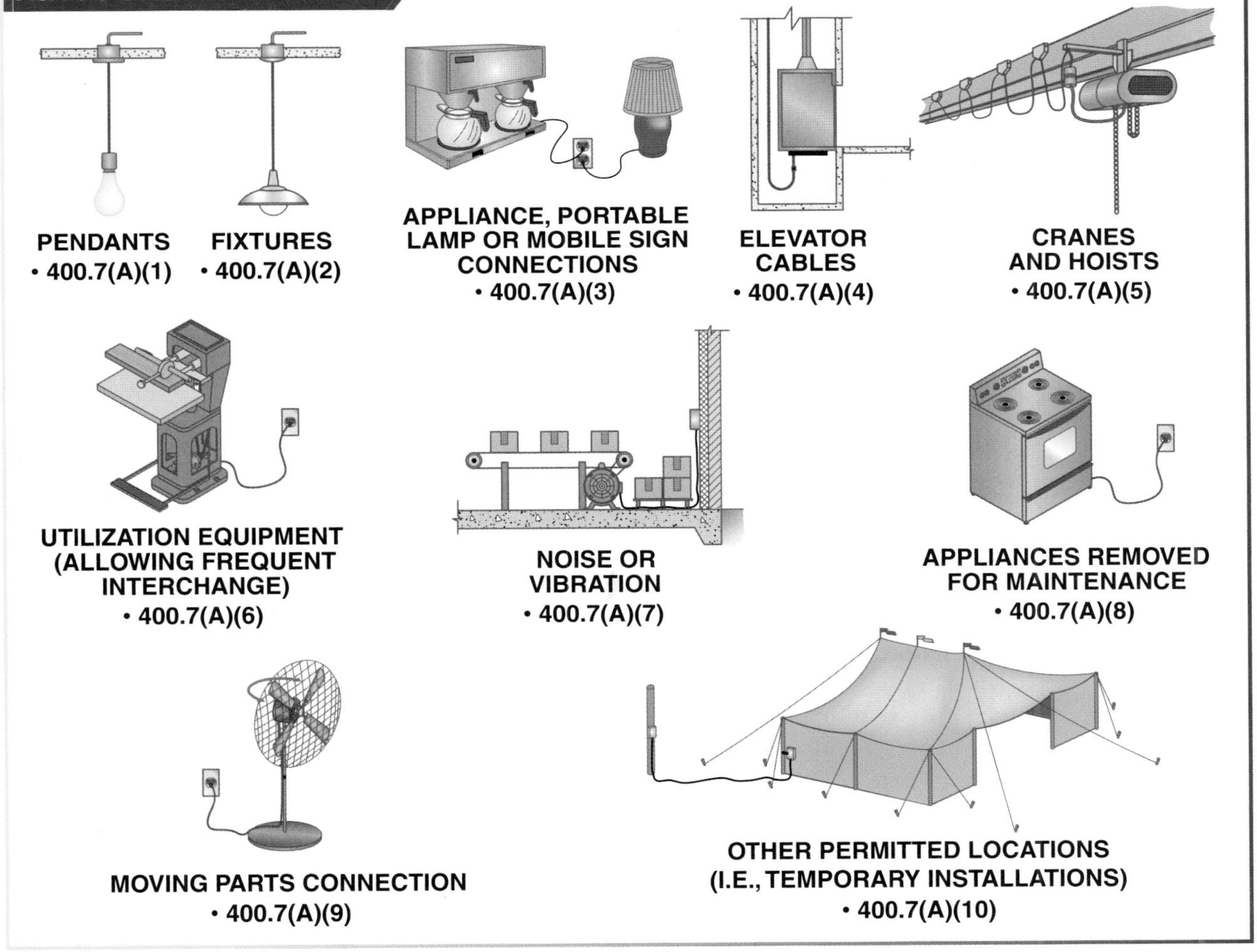

Figure 10-3. Ten uses are permitted for flexible cords and cables.

(4) Elevator cables—Flexible cables are required to follow elevator cars.

(5) Cranes and hoists—Flexible cables follow the moving parts of cranes and hoists.

(6) Utilization equipment with frequent interchange—Flexible cables provide flexibility in equipment requiring frequent interchange. An attachment plug is required and shall be supplied from a receptacle outlet. See Article 100 for definition of utilization equipment.

(7) Noise or vibration transmission prevention—Machines subject to vibration are commonly wired with flexible cables.

(8) Appliances removed for maintenance—One of the most common uses for flexible cords and cables is for the supply of appliances that need to be removed for repair or periodic maintenance. An attachment plug is required and shall be supplied from a receptacle outlet.

(9) Moving parts—The electrical connection of moving parts requires great flexibility in the wiring method and flexible cords and cables meet the requirements.

(10) Other permitted uses—Flexible cords and cables where permitted elsewhere in the Code.

For example, sections 590.4(B) and 590.4(C) permit flexible cords and cables to be used for feeder and branch-circuit wiring where temporary power is required. Such flexible cords and cables shall be identified in Table 400.4 for hard usage or extra-hard usage.

Uses Not Permitted – 400.8. Seven uses for flexible cords and cables are specifically not permitted per 400.8. **See Figure 10-4.** Uses not permitted are:

(1) Substitute for building wiring—If an extension to an existing branch circuit is required to supply a new appliance, the wiring method for the extension of the branch circuit could not be flexible cord or cable as this would be a substitute for fixed wiring.

(2) Passed through walls, ceilings, or floors—To ensure that flexible cords and cables are not subject to physical damage, they are not permitted to be passed through walls, ceilings, or floors.

(3) Passed through doorways, windows, or similar openings—To ensure that flexible cords and cables are not subject to physical damage, they are not permitted to be run through doorways, windows, or similar openings.

(4) Attached to building surfaces—To ensure that flexible cords and cables are not subject to physical damage, they are not permitted to be attached to building floors, walls, ceilings, etc.

(5) Concealed behind building walls, ceilings, or floors—Flexible cords and cables are not permitted to be concealed behind building walls, ceilings, or floors, or to be located above suspended ceilings.

(6) Installed in raceways unless permitted elsewhere in the NEC®—For example, 680.23(B)(2) permits the flexible cord supplying a wet-niche fixture to be installed within a raceway which extends from the forming shell to the deck box or other suitable junction box.

(7) Where subject to physical damage, flexible cords and cables shall not be used because their outer jacket offers very little mechanical protection.

Per 400.8(4) Ex., flexible cords and cables may be attached to the building structure for branches from busways. The installation shall be per 368.56(B). Branches from busways shall be made with suitable cord and cable assemblies of the hard or extra-hard usage type. Suitable means shall be provided to ensure strain relief and protection from physical damage. See 368.56(B)(1–4).

Splices – 400.9. Generally, flexible cords and cables shall be installed in continuous lengths without splices or taps. Requiring that the cord or cable be continuous ensures that the integrity of the conductors is not affected. If the cord types are either hard-service or junior hard-service cords, 400.9 permits splices for repairs. Frequently, in the course of use, these types of cables are damaged. This permits the cable to be repaired instead of replaced, provided the conductors are 14 AWG and larger and the completed splice retains the insulation, outer sheath properties, and usage characteristics of the original cord. All splices shall be per 110.14(B).

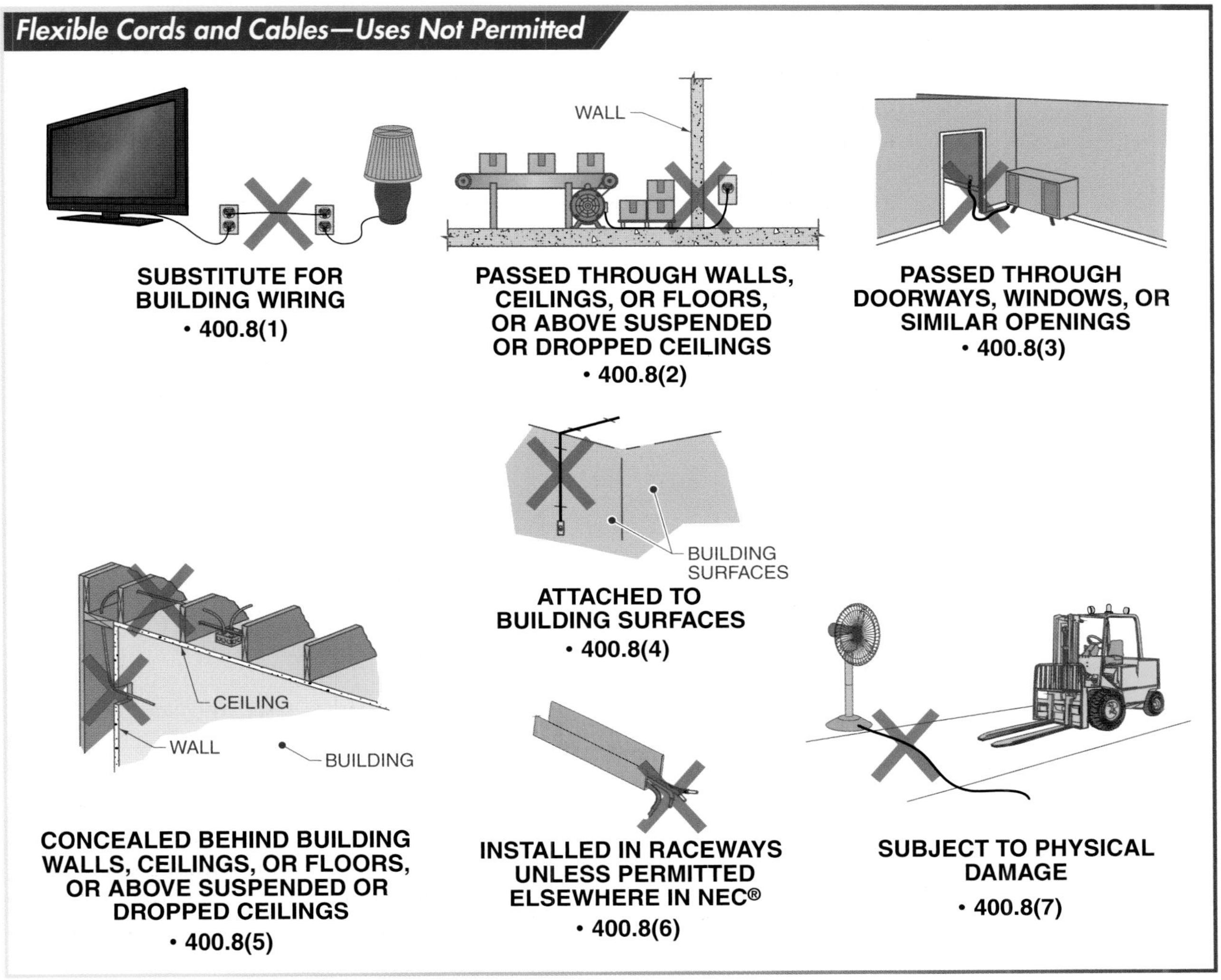

Figure 10-4. Seven uses for flexible cords and cables are specifically not permitted.

Terminations – 400.10. Suitable devices or installation methods shall be used to ensure that tension is not transmitted from flexible cords or cables to the termination point. **See Figure 10-5.** Strain-relief fittings are an example of the type of fitting designed for this purpose. Installation methods that meet the requirements of this section are knotting the cord, winding with tape, or fittings designed for the purpose. Additionally, 400.14 requires that where flexible cords or cables pass through holes in covers of junction boxes or other similar enclosures, a suitable bushing or fitting shall be installed to protect the cord or cable.

Flexible Cords and Cables – Construction Specifications – Article 400, Part II

Part II of Article 400 contains all of the construction specifications for flexible cords and cables. Many of the construction specifications for flexible cords and cables are contained within the applicable product standard. Section 400.20 requires that flexible cords be tested and examined at the factory and properly labeled before they are shipped. Installers should always check the label on the flexible cord or cable and review permitted uses and applications in accordance with Article 400 and Table 400.4

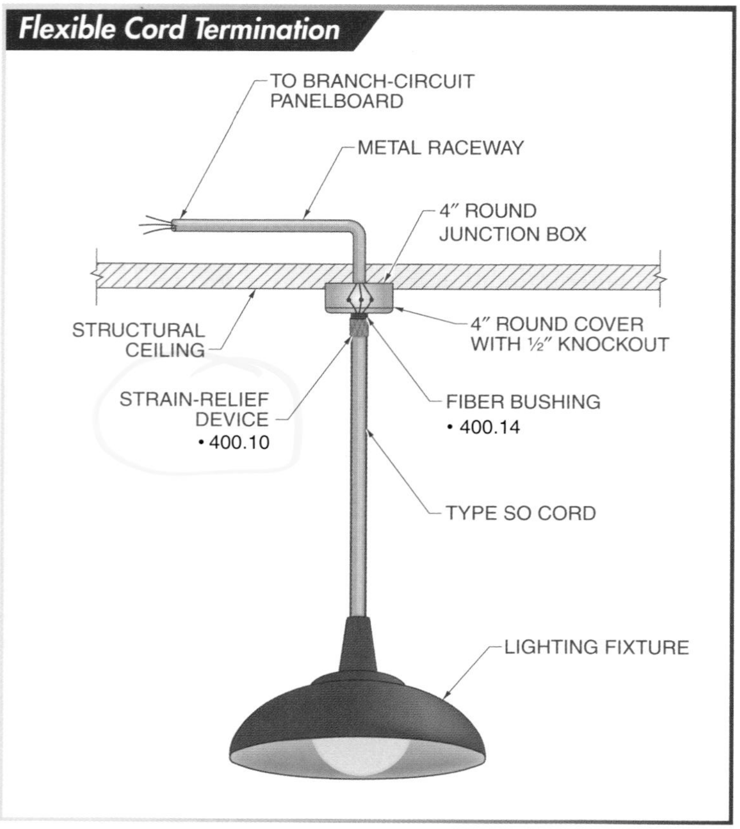

Figure 10-5. Flexible cords shall be provided with strain-relief devices or other suitable means to ensure that the weight is not transmitted to the termination or splice.

> **Flexible Cords and Cables**
>
> *Section 400.7(A) lists 10 specific uses permitted for flexible cords and cables. Despite this, designers and installers of electrical systems should be very cautious when deciding to use flexible cord or cable as a wiring method. Part of the problem stems from the fact that the first item listed under uses not permitted is "as a substitute for the fixed wiring of the structure." Some electrical inspectors give very little latitude for the use of flexible cords and cables. These inspectors maintain that other wiring methods, such as FMC or LFMC, can accomplish everything that flexible cord or cable can and the use of cable constitutes a "substitution" for fixed wiring. Always check with the AHJ before choosing flexible cords or cable as a wiring method.*

In terms of the construction of flexible cords and cables, 400.21 requires that the individual conductors of both flexible cords and flexible cables have flexible stranding. The minimum sizes for the conductors within the flexible cords and cables must be in accordance with the provisions of Table 400.4. For example, Table 400.4 specifies that Type SJ cord, a junior hard service cord, is available with conductors size 10 to 18 AWG and in configurations from two to six conductors.

Another important part of the construction is the proper means for identification of the grounded and equipment grounding conductors. Section 400.22 lists the methods that are used to identify the grounded conductor. In general, the grounded conductor must be distinguishable from the other conductors. Section 400.22(A–F) lists some of the methods. Options include the use of a colored braid, colored insulation, colored separator, tinned conductors, or surface marking.

Section 400.23 covers the requirements for identification of the equipment grounding conductor. The EGC in flexible cords or cables is treated the same in terms of color identification as any other equipment grounding conductor. Generally, conductors having a continuous green color or continuous green with one or more yellow stripes are acceptable. The methods for identifying the EGC are limited to either colored braid or colored insulation or covering. See 400.23(A) and (B).

Flexible Cords and Cables – Over 600 V – Article 400, Part III

Part III of Article 400 applies to portable multiconductor cables that are used to connect mobile equipment and machinery. These are highly specialized applications that require the appropriate cable design and construction. Generally, conductor size must be at least 12 AWG or larger. The conductors must be made with flexible stranding because of the increased demand on the cables during their use in the field. If the cables are used at over 600 V, the cables must be shielded to ensure that the insulation can withstand the higher voltage that may be impressed upon it. Section 400.32 requires that all of the shields be connected to an EGC. Splicing cords that operate at over 600 V is an application that requires specific procedures and identified products. Generally, only permanently molded and vulcanized types of splicing kits that are in accordance with 110.14(B) are permitted. See 400.36.

LUMINAIRES (LIGHTING FIXTURES) – ARTICLE 410

Article 410, Luminaires (lighting fixtures), Lampholders and Lamps lists the requirements for all types of luminaires (lighting fixtures) including incandescent filament lamps, arc lamps, electric-discharge lamps, and pendants. **See Figure 10-6.** A *luminaire* is a complete lighting unit consisting of a light source such as a lamp or lamps together with the parts designed to distribute the light, to position and protect the lamps and ballast (where applicable), and to connect the lamps to the power supply. See Article 100.

The NEC® does have other provisions for special lighting systems. Article 411 covers lighting systems which operate at 30 V or less. Arc lamps, when used in theaters and similar locations, are also covered in Article 520. Article 680 contains the special provisions for luminaires (lighting fixtures) installed in or near swimming pools, fountains, spas, and similar locations.

Luminaires up to 50 lb are permitted to be supported by the outlet box.

Fixture Locations – Article 410, Part II; 410.10

Fixtures, lampholders, and lamps are required to be constructed with no live parts normally exposed to contact. However, as with most electrical equipment, special precautions are required when the electrical equipment is installed in some types of locations. For example, per 410.10(A), precautions should be taken when luminaires are installed in wet or damp locations. These locations present additional life-safety concerns for persons who may be required to operate the fixtures. Likewise, luminaires installed in bathrooms can create additional hazards because of the constant presence of moisture.

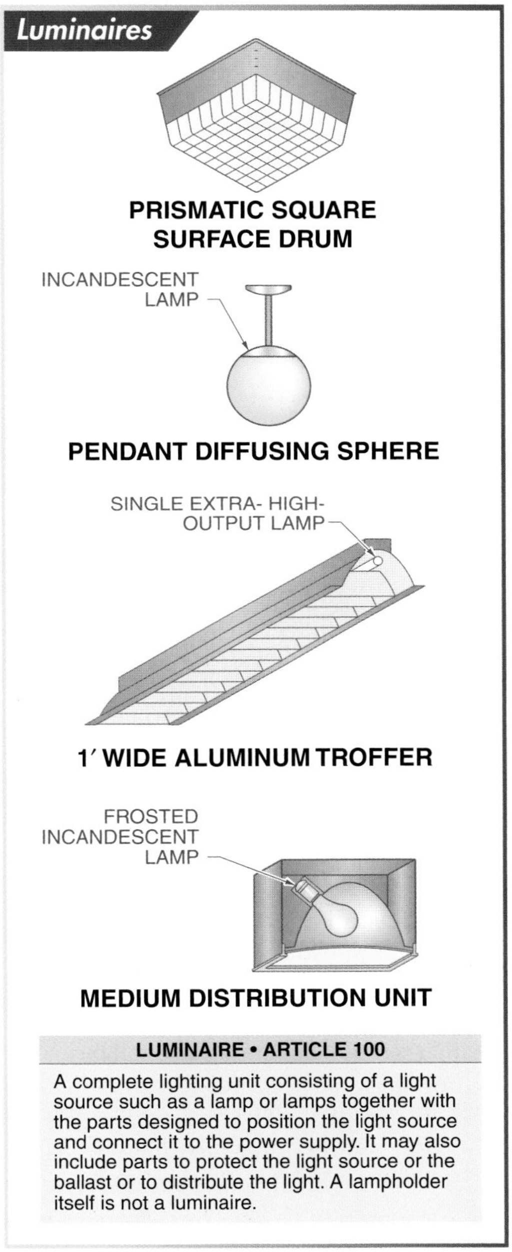
Luminaires
PRISMATIC SQUARE SURFACE DRUM
INCANDESCENT LAMP
PENDANT DIFFUSING SPHERE
SINGLE EXTRA- HIGH-OUTPUT LAMP
1′ WIDE ALUMINUM TROFFER
FROSTED INCANDESCENT LAMP
MEDIUM DISTRIBUTION UNIT
LUMINAIRE • ARTICLE 100
A complete lighting unit consisting of a light source such as a lamp or lamps together with the parts designed to position the light source and connect it to the power supply. It may also include parts to protect the light source or the ballast or to distribute the light. A lampholder itself is not a luminaire.

Figure 10-6. The term for lighting fixture is luminaire.

Energy-Efficient LED Lighting

The use of LEDs as an energy-efficient lighting solution, also known as solid state lighting (SSL), offers great promise as the next generation of lighting for everyday environments. Unlike traditional sources that use vacuum or gas-filled tubes, such as incandescent, fluorescent, and HID lighting, LEDs produce light solely by the rapid movement of electrons through a semiconductor material.

LEDs generate light so efficiently that they are becoming an attractive alternative lighting source. Compared to traditional sources, LEDs draw less power and produce a lower heat output. This results in a significant reduction for two of the largest energy consumers in our working and living environments: HVAC and lighting. LEDs typically offer the additional advantage of having a longer service life than traditional sources.

LEDs are directly controlled by a "driver," which in turn is controlled by a control source. The LED driver regulates the current flowing through the LED during operation to produce the maximum output and prevent premature failure. The driver also acts as an interface between the LED and a standard control source.

In order to select the appropriate control source, the specific requirements of the fixture/driver must be understood. Fixtures may be primarily designed for simple ON/OFF switching, dimming, color changing, or effect. Because of the vast array of fixture/driver combinations, there is no one standard control solution for all combinations. The intent of the output dictates the requirements of the control, especially in the case of color changing or other effects. Leviton offers several control solutions for LED technology.

Philips Solid-State Lighting Solutions

Additional information is available at www.leviton.com.

Another location that has continually resulted in problems with luminaires has been clothes closets. Although some light source is often needed in these spaces, great care is required in the selection of the luminaire and its placement because of the presence of easily ignitable materials. See 410.16.

Precaution should be taken when installing luminaires in hazardous (classified) locations. All electrical equipment installed in hazardous (classified) locations shall conform to the applicable provisions of Article 500 through Article 517.

Wet and Damp Locations – 410.10(A). The presence of moisture in wet and damp locations increases the potential for hazard from use of electrical equipment. Fixtures installed in either wet or damp locations shall be identified for such use. Fixtures identified for use in wet locations shall be marked *Suitable for Wet Locations*. Fixtures identified for use in damp locations shall be marked *Suitable for Wet Locations* or *Suitable for Damp Locations*. Fixtures identified for wet locations can be used in damp locations, but fixtures identified for damp locations cannot be used in wet locations. These markings ensure that the fixture is constructed for the type of location. For example, fixtures marked for wet or damp location use are constructed so that water does not enter or accumulate in wiring compartments or other places where electrical parts are included.

Bathrooms and Showers – 410.10(D). Nowhere is the potential for electric shock from luminaires greater than in the bathroom. The presence of moisture and the fact that persons may be immersed in water present a great risk to personnel. Therefore, 410.10(D) establishes a zone above bathtubs from which cord-connected fixtures, hanging fixtures, lighting track, pendants, and ceiling fans shall not be located. Any luminaire installed in this zone must be listed for damp locations or wet locations if the fixtures are subjected to the shower spray. The zone extends 8′ vertically from the top of the bathtub rim and 3′ horizontally. The zone includes the space directly over the tub or shower stall. **See Figure 10-7.**

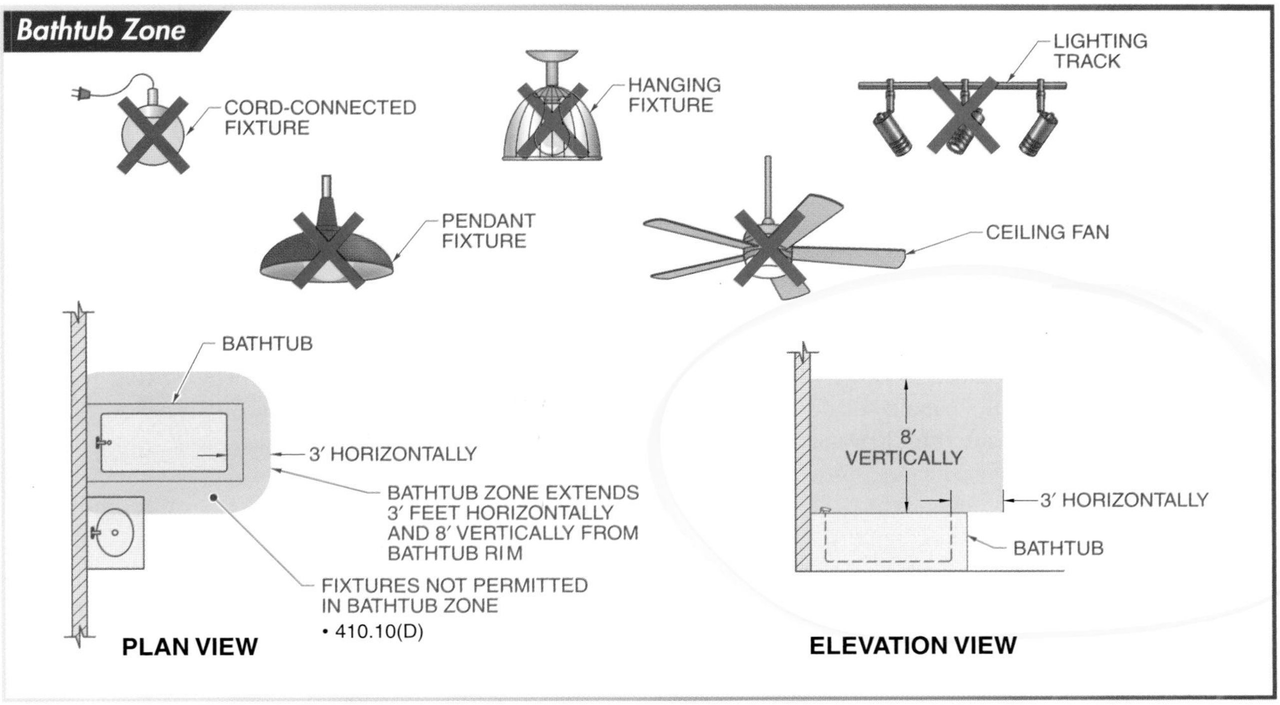

Figure 10-7. Cord-connected fixtures, hanging fixtures, lighting track, pendants, and ceiling fans are not permitted to be installed in the bathtub zone.

Clothes Closets – 410.16. The use of luminaires in clothes closets has been a problem for many years. This is largely because the space used for storage is often overfilled, resulting in easily ignitable materials coming in direct contact with the fixture lamp. Pendants, lampholders and incandescent luminaires with open or partially enclosed lamps are not permitted to be installed in clothes closets. **See Figure 10-8.**

Fixtures installed in showers must be identified for use in wet locations per 410.10(D).

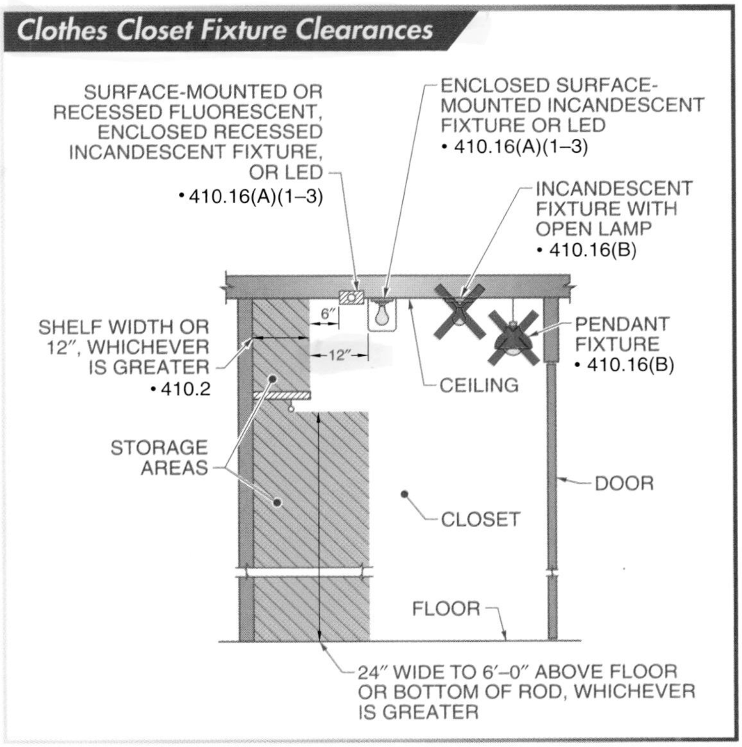

Figure 10-8. Pendants and incandescent luminaires with open or partially enclosed lamps are not permitted to be installed in clothes closets.

Fully enclosed incandescent fixtures or LED luminaires are permitted provided they are surface-mounted on the wall above the door or on the ceiling and have a minimum of 12″ free space from the nearest storage space point or they are recessed in the wall or on the ceiling with a minimum clearance of 6″ free space from the nearest storage space point. Fluorescent luminaires, because of their cooler operating temperatures, are a better choice for installation in clothes closets.

Surface-mounted fluorescent fixtures can be installed above the door or on the ceiling with a minimum clearance of only 6″ free space from the nearest storage space point. Recessed fluorescent fixtures can be installed in the wall or on the ceiling provided there is a minimum clearance of 6″ free space from the nearest storage space point. See 410.16(C)(1–5).

Fixture Outlet Boxes – Article 410, Part III

By definition, an outlet is a point on the wiring system from which current is taken to supply utilization equipment. Part III of Article 410 contains the provisions for luminaire outlets, canopies and pans. These provisions are intended to ensure that adequate wiring space is provided where luminaires are installed at outlets.

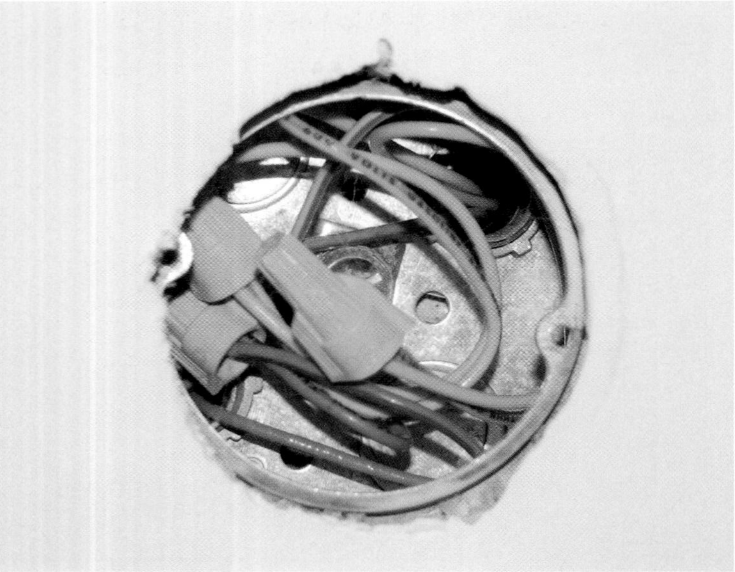

Per 410.20, luminaire outlet boxes must provide adequate space for conductors.

Several concerns are addressed in Part III. The provisions are intended to ensure that the maximum operating temperature of the conductors within the outlet box is not exceeded. Section 410.21 limits the branch-circuit conductors in the outlet that is an integral part of the luminaire to only those supplying power to the luminaire. There are some luminaires, however, that are designed and listed for use as a raceway and for through-wiring. In these cases, the additional branch-circuit wiring is permitted.

Section 410.22 requires that every outlet box be provided with a cover unless the outlet box is covered with a luminaire canopy, lampholder, receptacle, or similar equipment. An important safety concern related to this is the requirement that any combustible material be removed between the wall or ceiling finish and the luminaire canopy or pan. In some cases, the outlet box itself can be covered with a noncombustible material to remove the concern about fire. See 410.23.

Section 410.24 addresses installations where the luminaires, including LEDs, are installed separately from the outlet box. In these applications, the outlet box must be connected to the luminaire by metal or nonmetallic raceways or Type MC, AC, or MI cable. Flexible cord is also a permissible wiring method if it meets the requirements listed in 410.62 (B) or (C).

Finally, a means must be provided to access the conductors within the outlet box. This is especially important where the luminaire is installed such that it covers the outlet or pull box. A suitable means must be provided within the luminaire to access the conductors.

Fixture Supports – Article 410, Part IV

Luminaires shall be adequately supported. Support requirements vary depending upon the type of fixture selected for the installation. In some cases fixtures can be supported directly from the outlet box. If the fixture exceeds a certain weight, it shall be supported independently of the outlet box. Fixtures installed in suspended ceilings also shall be adequately supported. Generally, this is accomplished by the use of listed fixture clips which attach the fixture to the ceiling assembly. See 410.36(B).

Per 410.5, luminaires, portable luminaires, lampholders, and lamps shall have no live parts normally exposed to contact.

Lampholders – 410.30(A). A *lampholder* is a device designed to accommodate a lamp for the purpose of illumination. Lampholders consist of a lamp socket, lamp support mechanism, and a protective covering. **See Figure 10-9.** Section 410.30(A) prohibits fixtures that weigh more than 6 lb, or exceed 16″ in any direction, from being supported by the screw shell of the lampholder.

Metal Poles – 410.30(B). Metal poles are commonly used in outdoor locations to support luminaires and as a raceway to enclose supply conductors. Section 410.30(B) permits metal and nonmetallic poles to be used to support fixtures and as a raceway to enclose supply conductors provided six conditions are met. First, the metal poles shall be provided with an accessible handhole. In addition to providing support for the fixture, the metal pole also serves as an enclosure for the supply conductors. An accessible handhole of at least 2″ × 4″, with a cover suitable for use in wet locations, ensures access to the supply conductors or cable terminations.

Handholes can be omitted if the fixture pole height is 8′ or less, the supply wiring does not have a splice or pull point, and the interior of the pole is accessible by removing the fixture. If a hinged base is provided on the metal pole, handholes can be omitted if the metal pole is 20′ or less in height. **See Figure 10-10.**

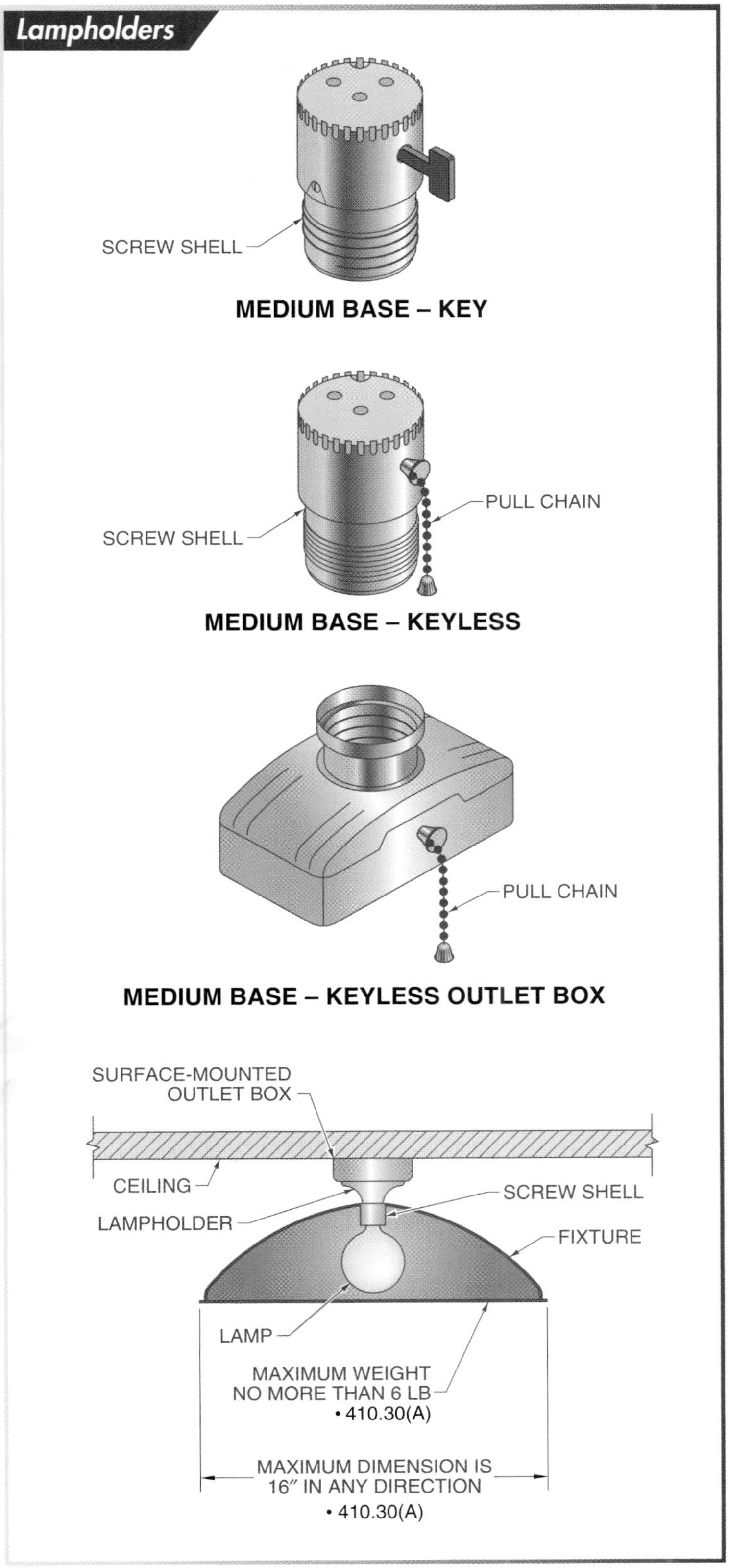

Figure 10-9. Fixtures and lampholders shall be securely supported.

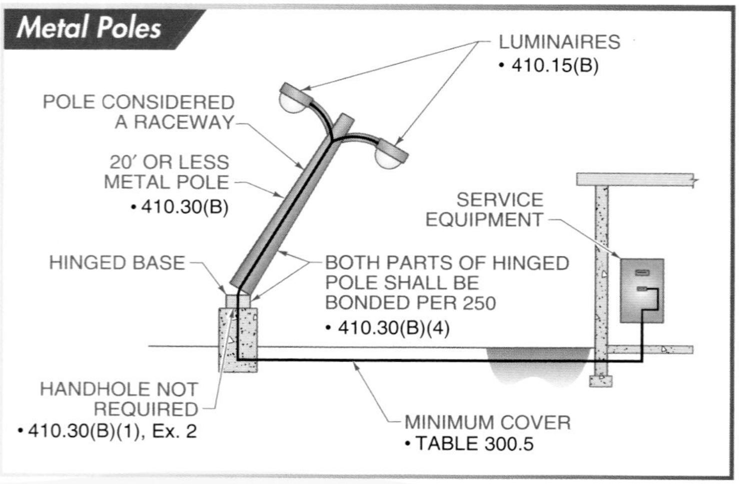

Figure 10-10. Handholes are not required in 20′ or less metal poles that support luminaires (lighting fixtures) if the base is hinged.

Rapid-Start Circuits

A rapid-start circuit is a fluorescent lamp-starting circuit that has separate windings to provide continuous heating voltage to lamp cathodes. Lamp starting-time is reduced because cathodes are continuously heated. A rapid-start circuit brings the lamp to full brightness in about two seconds. It uses lamps that have short, low-voltage electrodes, which are automatically preheated by the lamp ballast. A rapid-start circuit is the most common circuit used in fluorescent lighting.

Second, if a raceway riser or cable is not installed within the pole, a threaded fitting or conduit nipple shall be welded to the pole for the supply connection. Third, metal poles shall be provided with an equipment grounding terminal. If a handhole is provided, the equipment grounding terminal shall be accessible from the handhole. If a hinged base is provided, the equipment grounding terminal shall be accessible within the base section. Fourth, all poles with a hinged base shall have the base section bonded to the pole section. Fifth, metal raceways or other EGCs shall be bonded to the pole with an EGC listed in 250.118 and sized per 250.122. Sixth, poles can be used to support fixtures and enclose supply conductors if the conductors within the pole are properly supported. See 300.19 for requirements for supporting conductors in vertical raceways.

Outlet Boxes – 410.36(A). The vast majority of luminaires installed are supported directly from the outlet box from which they are supplied. Section 410.36(A) permits fixtures up to 50 lb to be supported by the outlet box. See 314.27(A) and (B) Outlet boxes are covered by Article 314. Support requirements for boxes are contained in 314.23, and the general requirements for outlet boxes are contained in 314.27. Outlet boxes used for luminaires shall be designed for the purpose. The outlet box shall be installed in a manner that permits the luminaire to be attached to the box.

An independent means of support shall be provided for luminaires that weigh over 50 lb. This can be accomplished by directly mounting the fixture assembly to the building structure or by use of rods, pipe, or straps which attach to the building structure. Independent support of the outlet box is not required if the outlet box is listed for the weight to be supported.

Suspended Ceilings – 410.36(B). Another common means for supporting luminaires is by the use of the framing members of suspended ceilings. Lay-in type fixtures are available which take the place of the standard 2′ × 2′ or 2′ × 4′ ceiling tiles. Section 410.36(B) requires that fixtures installed in suspended ceilings shall be securely fastened to the ceiling framing members. Fastening shall be by mechanical means such as screws, bolts, or rivets. Listed clips are also permitted. Hurricane clips were, at one time, the most common means used to secure fixtures to the ceiling framing members. New fixtures have been designed with the supporting mechanism built into the fixture. **See Figure 10-11.**

Trees – 410.36(G). In some cases, installing metal poles for the support of outside luminaires is impractical. Section 410.36(G) permits trees to support outdoor luminaires and their associated equipment such as photovoltaic control devices. **See Figure 10-12.** However, several other NEC® provisions shall be considered when using trees for the support of luminaires.

Section 225.26 prohibits trees from being used to support overhead conductor spans. This is because the trees are growing and moving continually. Connecting overhead spans directly to the trees would put a tremendous amount of stress on the conductors. Another option is to supply the luminaires from underground raceways or direct-buried cables.

Section 300.5(D) requires that direct-buried conductors and cables which emerge from the ground shall be protected from the required burial depth to a point at least 8′ above finished grade. See also 590.4(J), which prohibits vegetation to be used for the support of overhead spans of temporary wiring.

Grounding of Luminaires – Article 410, Part V

Article 250 and Part V of Article 410 contain the provisions for grounding of luminaires. As with all utilization equipment, the installation must provide a level of protection against accidental or unintentional contact with live parts. Section 410.42 requires all exposed metal parts of luminaires to be connected to the EGC. In addition to grounding, guarding or isolation and insulation methods can also be utilized. Guarding or isolation methods provide protection by limiting or restricting the approach to the equipment. Insulation is a method that provides protection through the use of acceptable levels of insulation to avoid an electrical shock hazard. Insulation must be accomplished with a material that is of a specific composition and construction that is recognized within the Code as electrical insulation. See Article 100.

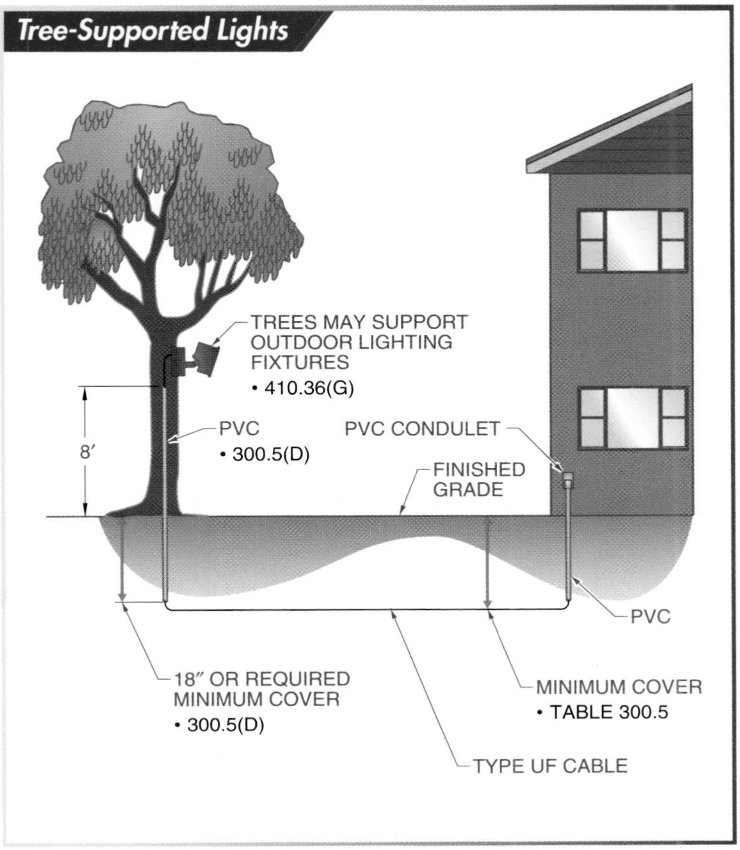

Figure 10-12. Trees are permitted to support outdoor lighting fixtures and their associated equipment.

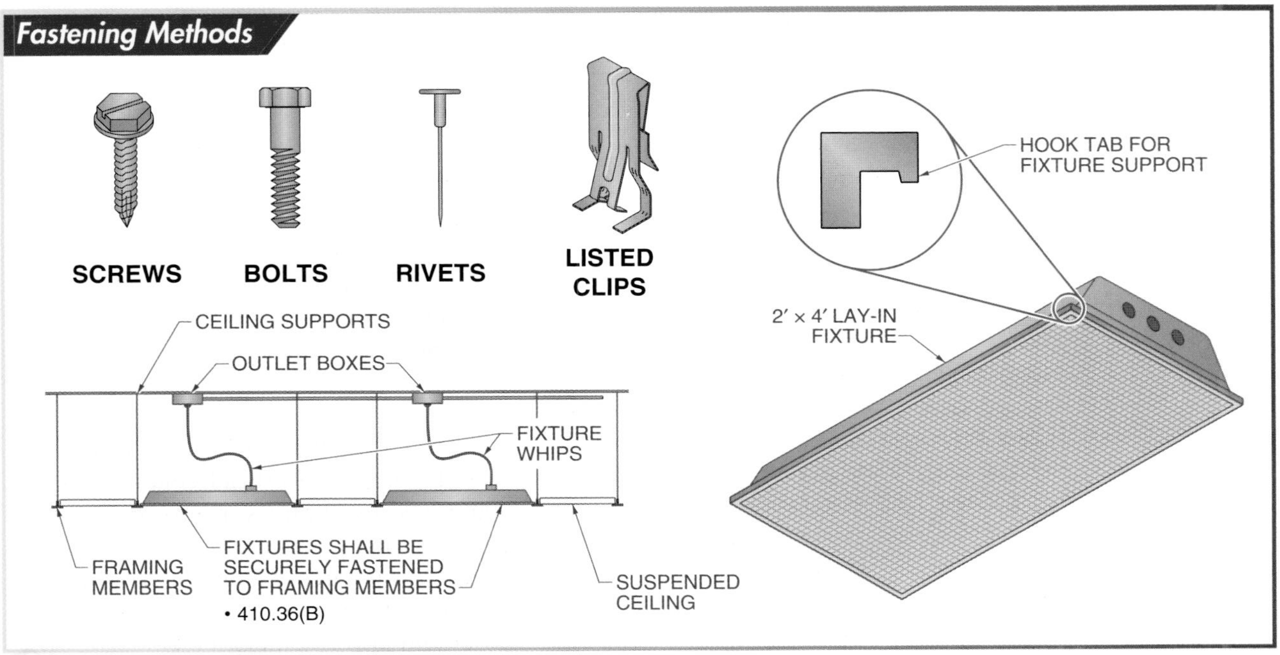

Figure 10-11. Lay-in type fixtures shall be securely fastened to the ceiling framing members by suitable mechanical means.

Lay-In Fixtures

Section 410.36(B) has led to much confusion in the field regarding the support of lay-in fixtures. Note that the requirement for attachment to the building structure only applies to the framing members of the suspended ceiling system. Many job specifications require, however, that the fixtures be independently supported by the building structure. In these cases, separate support wires are typically installed from the corners of the fixture to the actual building structure. While this practice is not required by the NEC®, it does provide a better electrical installation. Always check the job specifications before deciding how individual fixtures are to be supported.

The methods used for grounding luminaires must comply with the equipment grounding provisions of 250.118. By their design, luminaires that are constructed of metal parts must include provisions for the attachment of an equipment grounding conductor. The equipment grounding conductor provides the low-impedance return path for any ground fault that could occur within the luminaire. Section 410.44 requires that luminaires with exposed metal parts be provided with a means for connecting the EGC as specified in 250.118 and sized per 250.122.

Fixture Wiring – Article 410, Part VI

Fixture wiring shall be adequately protected from physical damage. The conductors selected for fixture wiring shall be carefully chosen to ensure that their insulation type, size, and ampacity are suitable for the specific installation. When conductors are installed within the fixture assembly, fixtures are not permitted to be used as raceways for circuit conductors. There are, however, exceptions to this general rule which are helpful when designing or installing conductors in luminaires. Article 402 contains the provisions for the construction specifications of fixture wires. In general, fixture wires are not permitted to be used as branch-circuit conductors. Fixture wires are for use within luminaires and for connecting luminaires to the branch-circuit supply conductors. See 402.10 and 402.11. They shall not be used as the branch-circuit conductors.

Physical Damage – 410.48. Section 410.48 requires that fixtures shall be wired in a neat manner. The accumulation of excess wiring shall be avoided. Installing fixture wires in a neat manner ensures that the conductors are not subject to physical damage. For example, if excess wiring is not removed, a conductor could be pinched when placing the cover on the luminaire. In addition, the conductors shall be installed within the fixture in a manner that does not subject them to excessive temperatures which could lead to damage of the conductors. Many fixtures contain ballasts or other electronic components that can generate substantial heat. Conductors shall be selected and arranged so that their temperature rating is not exceeded.

Conductor Insulation – 410.52. Ensuring that fixture wiring is done with conductors suitable for the environmental conditions which are present is closely related to protecting conductors from physical damage. Section 410.52 requires that the conductor insulation be suitable for the environmental conditions, current, voltage, and temperature to which the conductors are subjected. Section 410.68 requires that any feeder or branch-circuit conductor located within 3″ of a ballast , LED driver, power supply, or transformer shall have an insulation rating not lower than 90°C (194°F) unless the luminaire it supplies is marked as suitable for a different insulation temperature. See Table 310.104(A) for special permitted use of 75°C (167°F) insulation such as THW. Although this is a 75°C (167°F) insulation, it has been evaluated and listed for this specific use.

Size and Ampacity – 410.52, Informational Note. In general, fixture conductors shall not be smaller than 18 AWG. See 402.6.

Fixture wires are sized per Article 402. Fixture wires shall be of a type listed in Table 402.3, which provides information on insulation type, maximum operating temperature, and applicable application provisions. In general, fixture wires shall not be installed in a manner which could cause the operating temperature to exceed the maximum temperature rating listed in Table 402.3. The ampacity of fixture wires is determined per 402.5. Table 402.5 list the following ampacities for fixture wires:

- 18 AWG – 6 A
- 16 AWG – 8 A
- 14 AWG – 17 A
- 12 AWG – 23 A
- 10 AWG – 28 A

Cord-Connected – 410.62. Cord-connected lampholders and fixtures are covered in 410.62. **See Figure 10-13.** Metal lampholders, when supplied by a flexible cord, shall be equipped with an insulating bushing at the inlet per 410.62(A). If the inlet is threaded, the bushing shall provide at least a ⅜″ nominal pipe trade size opening. Otherwise, the opening shall be sized appropriately for the cord.

Fixtures that adjust or can be aimed are not required per 410.62(B) to be equipped with an attachment plug or cord connector if the portion of the cord which is exposed is of the hard or extra-hard usage type and is not larger than required for the maximum adjustment. The installation of the cord should not be subject to physical damage.

Per 410.52, luminaires shall be wired with conductors having insulation suitable for the environmental conditions, current, voltage, and temperature to which the conductors will be subjected.

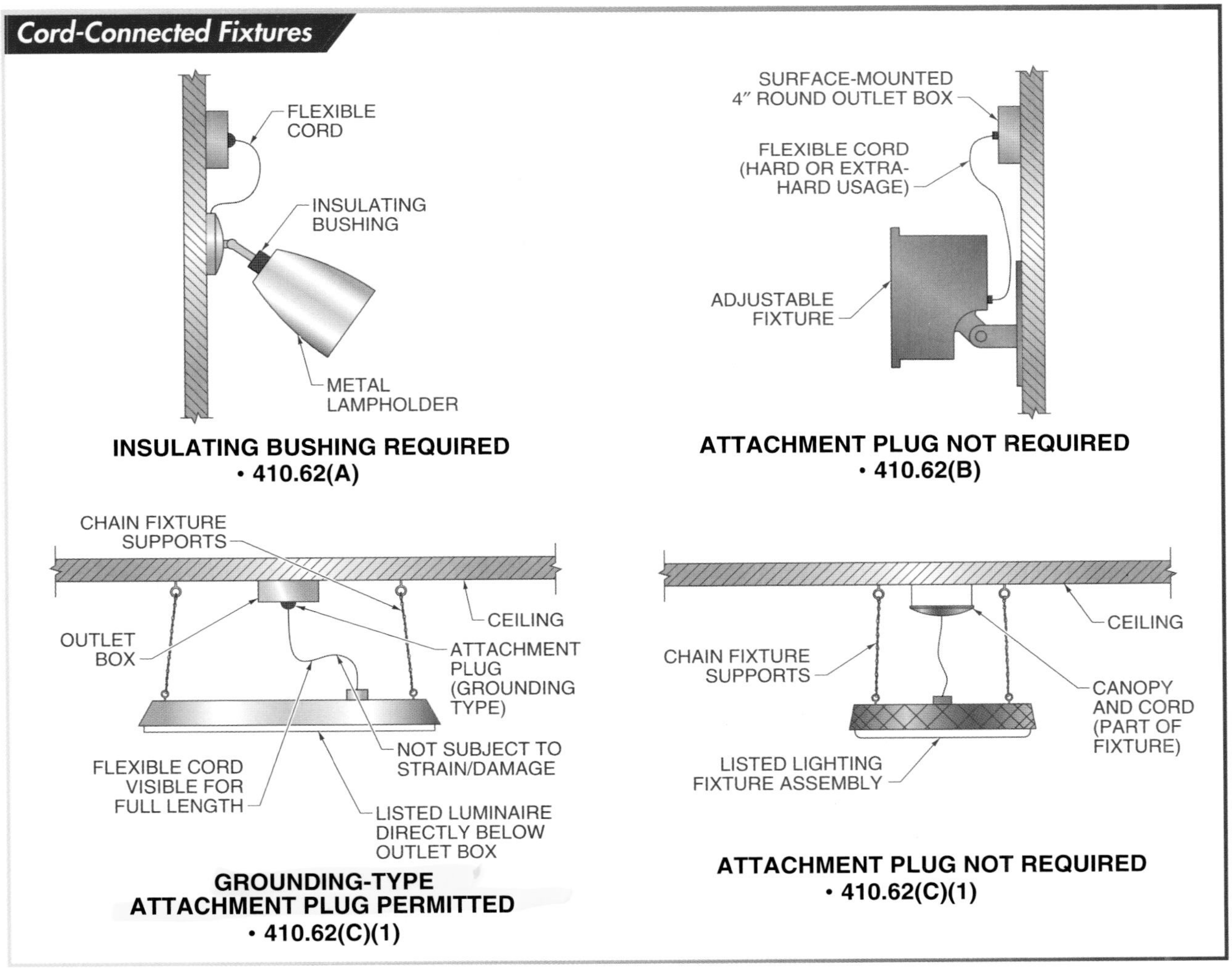

Figure 10-13. Lampholders and fixtures may be connected with flexible cord.

A listed electric-discharge fixture or listed electric-discharge fixture assembly may be supplied by a cord-connected wiring method per 410.62(C)(1)(2) provided the listed fixture or listed fixture assembly is located directly below the outlet box which supplies the fixture. In addition, the flexible cord shall be continually visible for its entire length, shall not be subject to physical damage or strain, and shall terminate at the outer end of the cord in a grounding-type attachment plug. A listed fixture or listed fixture assembly is permitted to be installed without an attachment plug per 410.62(C)(1)(2)(c) provided the fixture or assembly incorporates strain relief and canopy.

Fixtures as Raceways – 410.64. The general rule of 410.64 does not permit fixtures to be used as raceways for circuit conductors. These conductors are subject to excessive heat which may be generated within the fixture or they may be subject to physical damage. There are, however, applications in which the fixtures can be used as raceways. In order to use a luminaire as a raceway, one of the three conditions specified in 410.64 must be met. The first provision of 410.64 requires that the fixtures be listed and marked for use as a raceway. Fixtures suitable for use as raceways are typically marked or labeled to indicate such suitability.

Recessed incandescent fixtures are designed with an internal thermal protective device to sense excessive operating temperature and open the circuit to the fixture.

The second condition under which luminaires can be used as a raceway is defined in 410.64(B). This is for luminaires that are identified for through-wiring as permitted in 410.21. This section permits branch-circuit wiring to be passed through an outlet box that is an integral part of a luminaire, provided the luminaire is identified for through-wiring. The last condition under which luminaires can be used as a raceway is found in 410.64(C) for luminaires that are connected together.

Where fixtures are designed to be connected together in an end-to-end fashion, or where the fixtures are connected together with a recognized wiring method to form a continuous assembly, conductors supplying the fixtures can pass through other fixtures. The conductors shall be part of a two-wire branch circuit or a multiwire branch circuit and supply connected fixtures only. Section 410.68 requires that if the branch circuit conductors pass within 3″ of a ballast, LED driver, power supply, or transformer, they shall have an insulation rating of at least 90°C unless the fixture is listed and marked for a different insulation temperature. See Table 310.104(A) for special permitted use of 75°C (167°F) insulation such as THW.

Flush and Recessed Fixtures – Article 410, Part X

Flush or recessed fixtures are fixtures which are placed in recessed cavities of walls or ceilings. Because the fixture assembly is located above the ceiling or within the wall, special measures are required to ensure that the fixtures operate safely. Temperature is the greatest concern with this type of installation. Heat, associated with the supply conductors and the operation of the fixture, can pose a real threat to combustible materials which may be closely associated with the fixtures. The supply wiring to these fixtures shall be suitable for the temperatures that are likely to be encountered.

Incandescent Fixtures – 410.115(C). Typically, incandescent fixtures operate at a much higher temperature than fluorescent fixtures. Therefore, recessed incandescent fixtures are required by 410.115(C) to be thermally protected. A *thermally protected fixture* is a fixture designed with an internal thermal protective device which senses excessive operating temperatures and opens the supply circuit to the fixture. Excessive temperatures can result from overlamping. *Overlamping* is installing a lamp of a higher wattage than that for which the fixture is designed. Poor heat dissipation from the fixture, which occurs, for example, when insulation surrounds the fixture assembly, can also cause excessive temperatures.

Recessed incandescent fixtures are permitted to be installed without thermal protection by 410.115(C), Ex. 1 when they are identified for and used in poured concrete. This is permitted because the poured concrete surrounding the fixture is not combustible.

Recessed incandescent fixtures which are constructed and designed with equivalent thermal performance characteristics of thermally protected fixtures are permitted to be installed per 410.115(C), Ex. 2 without thermal protection. An example of one type of fixture that meets the requirements of this exception is a fixture designed to prevent overlamping. These fixtures will not accept lamps of a higher wattage than the fixture is designed for, thereby providing a level of protection equivalent to thermal protection. Fixtures of this type must be identified as inherently protected.

Recessed Lighting

The installation of thermal insulation around recessed luminaires is a concern to electricians and electrical contractors. Quite often, the electrician installs the recessed fixture and wires it long before insulation is installed. It can be difficult, under these circumstances, to ensure that the 3″ clearance is maintained. Additionally, spray insulation has become more popular, and maintaining clearances can be more difficult. Therefore, it is good design and installation practice to specify and use fixtures identified for direct thermal insulation contact whenever recessed fixtures are installed in a space that is to be insulated. Although these fixtures are more expensive than conventional fixtures, they do provide protection against excessive temperatures that could result in a fire.

Combustible Materials – 410.115(A). Fixtures shall be installed in a manner that does not subject adjacent combustible materials to temperatures in excess of 90°C (194°F) per 410.115(A). Often, recessed fixtures are installed in direct contact with combustible building materials such as wood framing members. If the fixture type is incandescent, significant heat can be generated and pose a potential for fire if not properly dissipated. If the fixture type is fluorescent, the lamp operates at a lower temperature, but the fixture ballast can generate significant heat. If a recessed fixture is installed in buildings of fire-resistant construction, 410.115(B) permits a temperature higher than 90°C (194°F) but not in excess of 150°C (302°F) if the fixture is plainly marked for that application.

Clearance and Installation – 410.116. One of the greatest concerns when installing recessed fixtures is ensuring that thermal insulation is not installed too close to the fixture. Often, recessed fixtures are installed in spaces that are provided with thermal insulation. If the insulation is installed too close to the fixture, heat can be trapped and will not properly dissipate. These high heat spots are a potential for fire if adjacent combustible materials reach their fire point.

Section 410.116 provides strict requirements for locating recessed fixtures near combustible materials and thermal insulation. A minimum of ½″ clearance from combustible materials shall be provided for recessed portions of luminaires which are not identified for contact with insulation from all combustible materials per 410.116(A)(1). This ½″ clearance does not, however, apply to the support points of the fixture and trim.

Section 410.116(B) requires a recessed fixture to be installed such that there is no thermal insulation above the fixture or within 3″ of the fixture's enclosure wiring compartment or ballast, LED driver, power supply, or transformer unless the fixture is identified for contact with insulation. **See Figure 10-14.**

Wiring – 410.117. All wiring for recessed fixtures shall be with conductors that have an insulation type suitable for the temperature that may be encountered per 410.117(A). If heat is not properly dissipated from the wiring compartment, the insulation rating of the conductors could be exceeded leading to conductor insulation degradation and possible failure. Branch-circuit wiring that supplies the fixture is permitted per 410.117(B) to terminate within the fixture if the conductors are provided with an insulation type which is suitable for the temperature encountered.

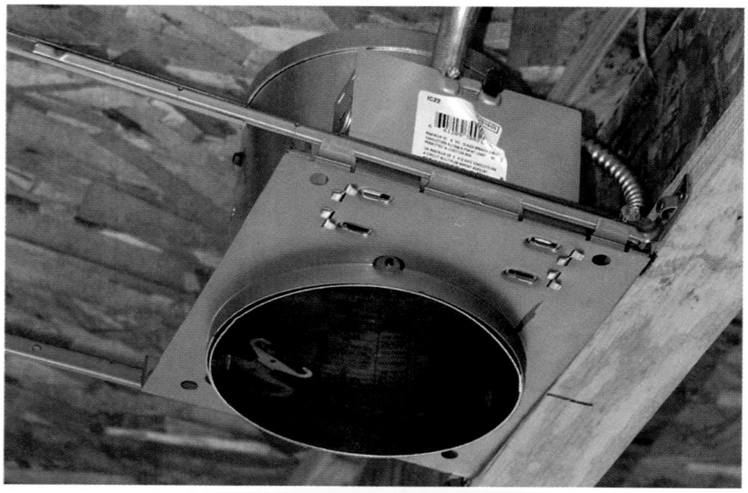

Per 410.116(B), recessed lighting fixtures should be installed so that there is no thermal insulation within 3″ of the fixture's enclosure wiring compartment, unless otherwise listed for direct contact with insulation.

Tap conductors which are run to recessed fixtures are covered in 410.117(C). If the branch-circuit conductors are not suitable for the temperature encountered in a fixture wiring compartment, an outlet box can be installed at a point which is at least 1′ from the fixture. Tap conductors are permitted to be run from the outlet box to the fixture. The tap conductors shall be suitable for the temperature encountered and at least 18″ and not more than 6′ in length. Tap conductors are permitted in a suitable raceway or they can be Type AC or MC cable. The 1′ spacing of the outlet box ensures that heat generated in the fixture does not affect the wiring in the outlet box. The tap conductor length requirement permits movement of the fixture for maintenance or repair purposes. **See Figure 10-15.**

Markings – 410.120. Recessed incandescent luminaires shall be marked with the maximum allowable wattage of lamps for use in the fixture. The marking shall be completed with permanent letters at least ¼″ high in a location that is visible during relamping. The *Underwriters Laboratories Fixture Marking Guide* identifies some other common markings found on recessed incandescent luminaires.

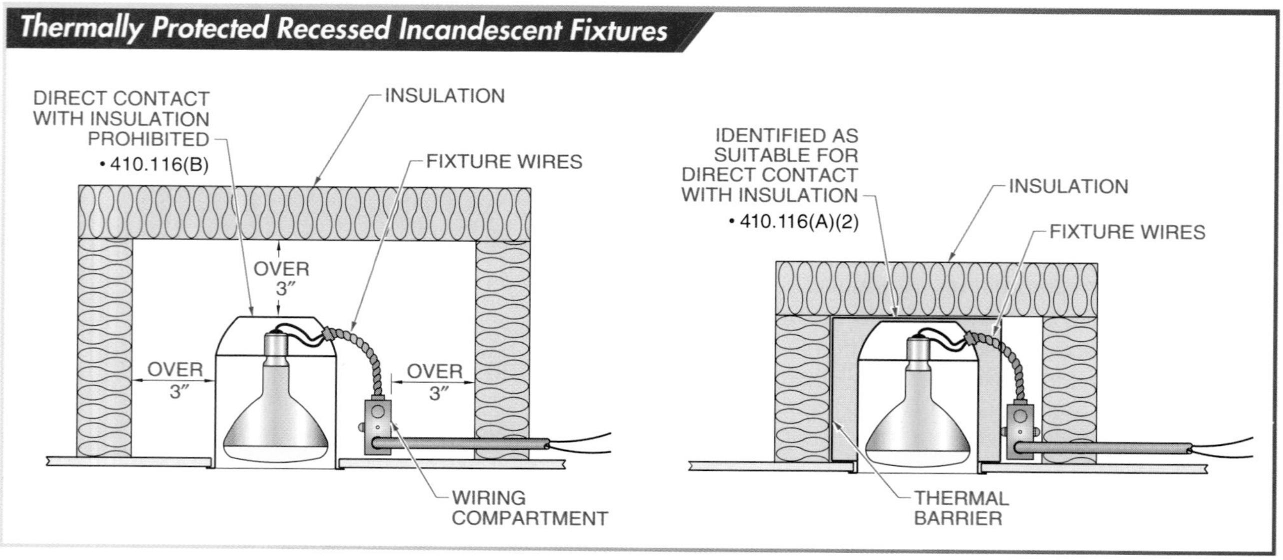

Figure 10-14. Thermally protected fixtures provide additional safety measures by ensuring that overheated fixtures do not provide a source of ignition for combustible materials.

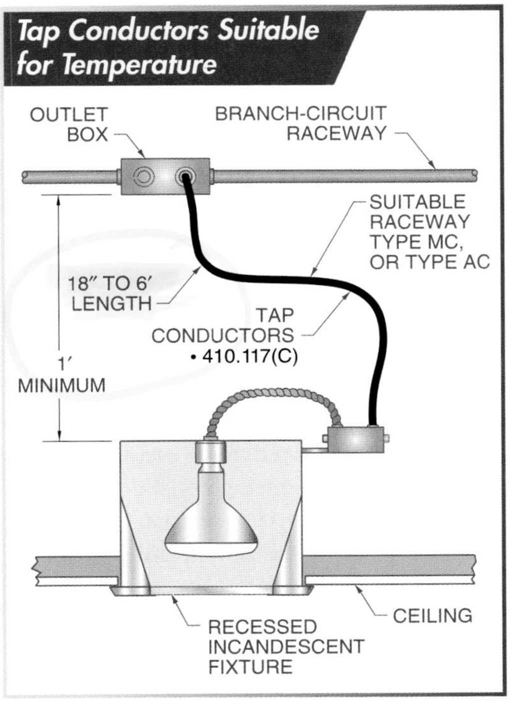

Figure 10-15. Tap conductors are used to connect branch-circuit wiring to the fixture wiring.

Fixtures which are suitable for use in direct contact with thermal insulation shall be marked "Type IC," or "Inherently Protected." If a fixture is not marked "Type IC," it shall be installed with a minimum of 3″ clearance from thermal insulation on all sides and with no insulation above the fixture. Such fixtures are typically marked "WARNING – RISK OF FIRE. Do not install insulation within 3″ of fixture sides or wiring compartment nor above fixture in such a manner as to entrap heat."

Fixtures suitable for installation only in poured concrete shall be marked "For Installation Only In Poured Concrete." Another type of marking that is common on recessed luminaires is suitability for use only in environmental air handling spaces. Fixtures that are identified for such use are marked "Suitable Only for Installation in Environmental Air Handling Spaces."

Electric-Discharge Equipment 1000 Volts or Less – Article 410, Part XII

An *electric-discharge luminaire* is a luminaire that utilizes a ballast for the operation of the lamp. The ballast provides a high starting voltage which is induced for striking the arc to light the lamp. Electric-discharge lighting fixtures are commonly used throughout the electrical industry for all types of applications. Provisions for electric-discharge lighting systems which operate at 1000 V or less are given in Article 410, Part XII.

Electric-discharge lamps operate at a much cooler temperature than incandescent lamps. This makes electric-discharge fixtures a better choice for installation in clothes closets than incandescent fixtures. See 410.16(A). Installers of electric-discharge luminaires should, however, be aware of temperatures around the ballast compartment. Some fixture assemblies incorporate remote ballasts to give the designer and installer greater latitude in specifying and selecting the ballast location.

Thermal Protection Ballasts – 410.130(E). All ballasts used in fluorescent luminaires installed indoors shall be thermally protected per 410.130(E)(1). This thermal protection is integral to the ballast and operates much like thermally protected, recessed luminaires. The ballast thermal protection device opens when the operating temperature of the ballast rises too high, interrupting the supply current to the luminaire. The thermal protection must be integral with the ballast and must be provided with replacement ballasts.

Excessive operating temperatures can be caused by improper dissipation of heat away from the ballast, burned-out fixture lamps, or the wrong ballast for the application. Replacement ballasts shall be the thermally protected type. Ballasts that meet the requirements for thermal protection in 410.130(E) are listed and identified by UL as Class P ballasts.

High-Intensity Discharge – 410.130(F). A *high-intensity discharge (HID) luminaire* is a luminaire that generates light from an arc lamp contained within an outer tube. The sustained arc produces a constant light source. The three most common HID lamps in use are mercury-vapor, sodium-vapor, and metal-halide. HID fixtures are used throughout the electrical industry in all types of applications. Some of the more common applications are for outdoor lighting and high bay areas in industrial establishments.

All recessed HID fixtures shall be thermally protected per 410.130(F)(1). Where the HID fixture and its ballast are installed in separate locations, the ballast is also required to be thermally protected. Thermally protected HID fixtures shall be identified. **See Figure 10-16.**

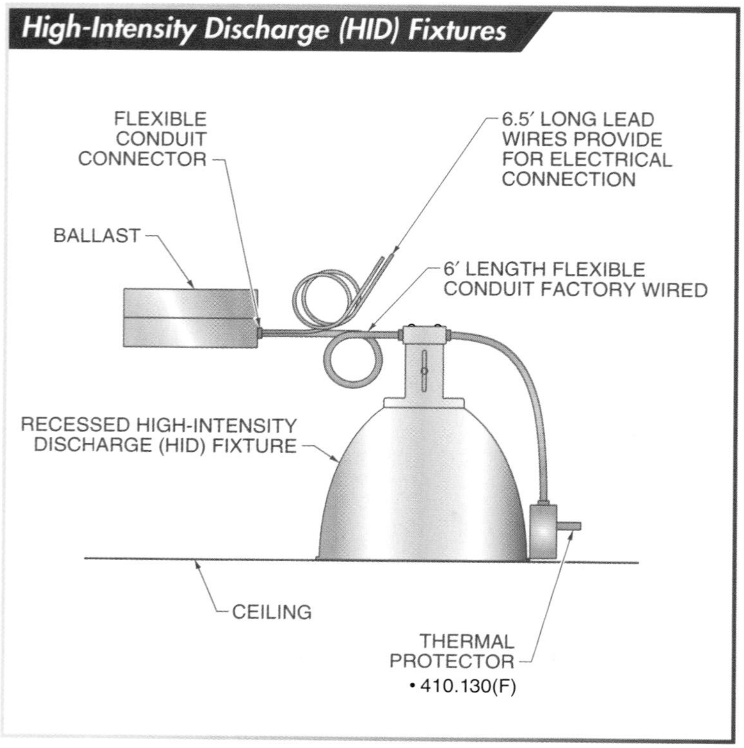

Figure 10-16. All recessed high-intensity discharge (HID) fixtures shall be thermally protected.

Lighting Track – Article 410, Part XIV

Lighting track is an assembly consisting of an energized metal track and luminaire heads which can be positioned in any location along the track. The track also supports the luminaire heads. The NEC® does not set maximum or minimum lengths for lighting track, but 220.43(B) requires that each 2′ of lighting track be considered 150 VA for the purposes of determining load calculations in other than dwelling units or guest rooms or guest suites of hotels or motels. **See Figure 10-17.**

The 150 VA requirement allows the maximum footage of lighting track on a branch circuit to be calculated. For example, a 20 A, 120 V branch circuit is permitted to supply a total load of 2400 VA (120 V × 20 A). In most nonresidential cases, the lighting is considered continuous, reducing the total allowable load on the 20 A circuit to 1920 VA. Using the 150 VA per 2′ requirement, the 20 A circuit would be permitted to supply a maximum of 25.6′ of lighting track.

Installation – 410.151. All lighting track shall be permanently installed and permanently connected to a branch circuit per 410.151(A). The use of lighting track with temporary wiring, or in an otherwise nonpermanent manner, is prohibited. In addition, the components for use on lighting track, such as the fittings, shall be identified for such use.

The load on the lighting track shall be limited to that for which the lighting track is rated per 410.151(B). Installing excess luminaires on the lighting track can subject the track to a larger load than that for which it is designed or rated. Some lighting track systems are designed to be cut in the field to a specific length. Lighting track that is identified for field cutting includes instructions on the proper methods of cutting.

Prohibited Uses – 410.151(C). Nine limitations are placed on the use of lighting track by 410.151(C). As with most electrical equipment, lighting track shall not be used where it is subject to physical damage or in wet or damp locations. Lighting track is not permitted above bathtubs in a zone which extends 3′ horizontally and 8′ vertically from the bathtub rim. See 410.10(D). Lighting track is also not permitted to be installed in storage battery rooms, hazardous (classified) locations, or other areas subject to corrosive vapors. Lighting track is not permitted to be installed in concealed locations and is not permitted to extend through walls or partitions. It is not permitted for use at less than 5′ above the finish floor level unless it operates at less than 30 V RMS and is protected from physical damage.

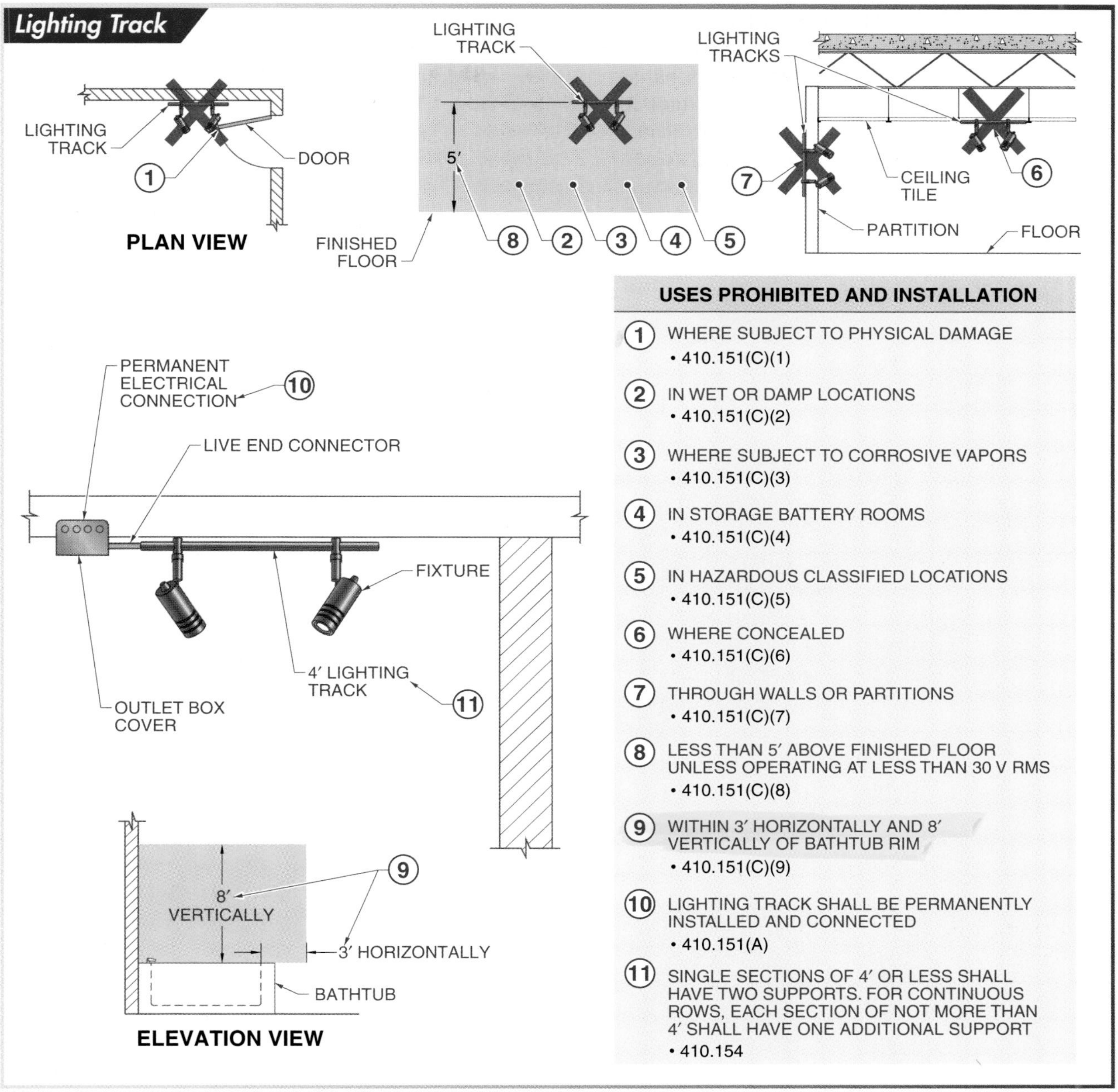

Figure 10-17. Lighting track consists of a length of energized track and luminaires that can be positioned anywhere on the track.

Support Requirements – 410.151(D). The fittings installed on lighting track shall be designed specifically for the track on which they are to be installed. The lighting track provides the support for the luminaires and the fixtures shall be securely attached to the lighting track. Fixtures which are not designed for use on lighting track cannot be modified for use on lighting track. The *UL Fixture Marking Guide* indicates that listed lighting track will be marked "CAUTION – To reduce the risk of fire and electrical shock, use only fixture assemblies marked for use with __ track." The name of the manufacturer is placed in the blank space. The luminaires are marked with a similar label to ensure that the luminaires are used with the correct lighting track.

Ruud Lighting, Inc.
Per 410.130(F), high-intensity discharge (HID) lighting is permitted to be installed in industrial building applications.

Mounting – 410.154. Because the lighting track supports the weight of the luminaires, minimum fastening requirements are necessary to ensure a safe installation. A minimum of two lighting track supports shall be installed for individual pieces of lighting track 4′ or less in length. If the lighting track is installed in a continuous row, each individual section which is 4′ or less in length shall be provided with one additional support per section. The *UL Fixture Marking Guide* requires that lighting track systems designed to have field-drilled mounting holes shall be provided with installation instructions detailing the location of the mounting holes.

RECEPTACLES, CORD CONNECTORS, AND ATTACHMENT PLUGS – ARTICLE 406

A *receptacle* is a contact device installed at outlets for the connection of cord-connected electrical equipment. A device is any unit of an electrical system that carries or controls electric energy as its principal function. In most cases, devices do not consume electric energy. However, a 2008 change to the Article 100 definition of device allows electric energy to be consumed by a device for incidental functions. For example, it is now clear that a wall switch with internal illumination is a device. The NEC® contains installation provisions for several types of receptacles used throughout the electrical industry. Article 406 is concerned primarily with receptacle types, ratings, and installation requirements of receptacles, cord connectors, and cord caps.

Receptacles are available in many voltage and current ratings and in many configurations. All receptacles, however, can be classified as either standard receptacles or special-use receptacles. **See Figure 10-18.** Standard receptacles consist of the grounding and nongrounding types. Grounding receptacles are used for new installations and nongrounding receptacles are limited for use in replacement applications. Specialty receptacles are receptacles designed for a specific electrical application. Isolated-ground receptacles, hospital-grade receptacles, and ground-fault circuit interrupter receptacles are examples of specialty receptacles.

Hubbell Incorporated (Delaware), Wiring Devices-Kellems
Receptacles are installed at outlets and are available in many voltage and current ratings.

Grounding Receptacles

Since the 1968 NEC®, the Code has required that all receptacles installed on 15 A or 20 A branch circuits be the grounding type per 406.4(A). A *grounding receptacle* is a receptacle that includes a grounding terminal connected to a grounding slot in the receptacle configuration. The grounding terminal is connected to an equipment grounding conductor which connects non-current-carrying metal parts of equipment together and to the system grounded conductor or grounding electrode conductor, or both, at the service. In general, these receptacles shall be installed only on circuits of the voltage class and current for which they are rated.

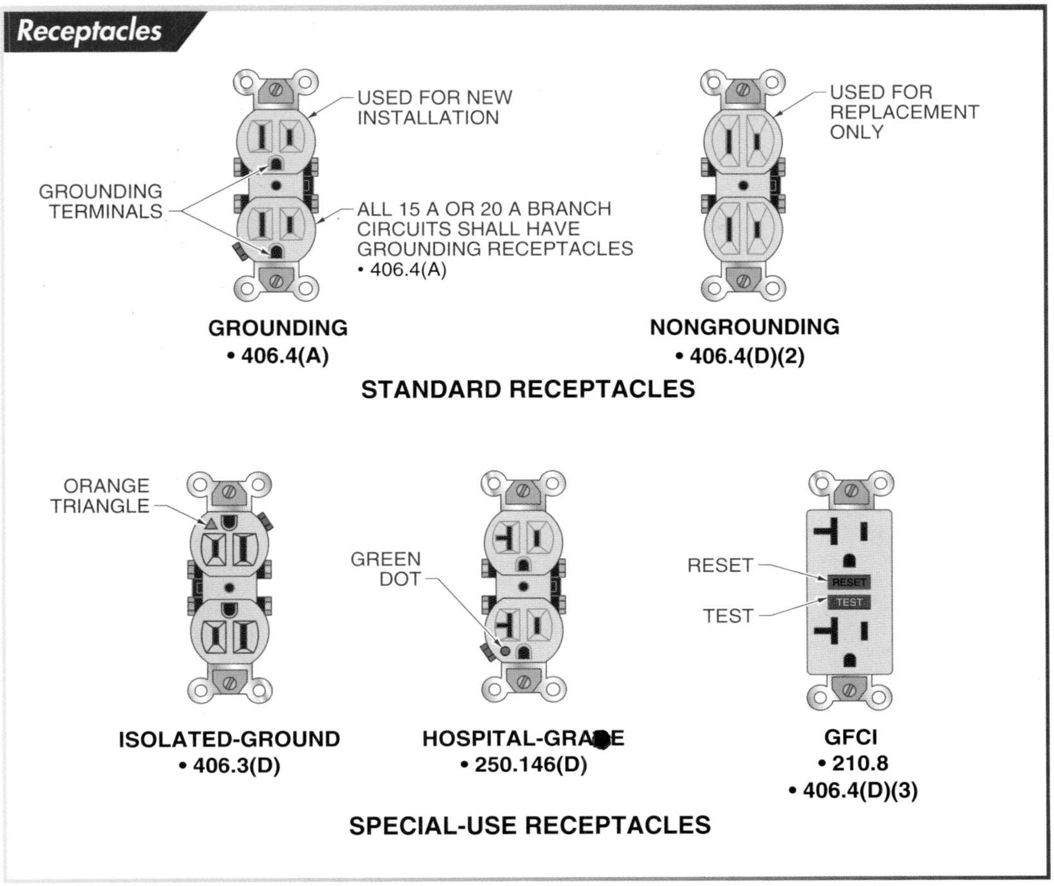

Figure 10-18. All receptacles are classified as standard receptacles or special-use receptacles.

Section 406.4(B) requires receptacles and cord connectors having grounding contacts to have those contacts connected to an EGC. Section 406.4(B), Ex. 1 does not require this on receptacles mounted on portable and vehicle-mounted generators. Per 406.4(C), the equipment grounding conductor contacts of receptacles and cord connectors shall be grounded to the EGC of the circuit that supplies the receptacle or cord connector. The branch-circuit wiring method shall provide an EGC for this purpose.

Another type of grounding receptacle is the self-grounding receptacle. A *self-grounding receptacle* is a grounding-type receptacle that utilizes a pressure clip around the 6 – 32 mounting screw to ensure good electrical contact between the receptacle yoke and the outlet box. Self-grounding receptacles are permitted to be installed without a bonding jumper between the outlet box and the receptacle per 250.146(B), if the contact devices or yokes are listed for the purpose.

Nongrounding Receptacles

A *nongrounding receptacle* is a receptacle with two wiring slots for branch-circuit wiring systems that does not provide an equipment grounding conductor. These electrical systems were commonly installed until the late 1960s. The NEC® permits nongrounding receptacles to be used for replacement of existing nongrounding receptacles only. Existing nongrounding receptacle outlets that do not contain an equipment grounding conductor can be replaced with nongrounding receptacles. See 406.4(D)(2)(a–c).

Isolated-Ground Receptacles

An *isolated-ground receptacle* is a receptacle in which the grounding terminal is isolated from the device yoke or strap. **See Figure 10-19.** The isolation ensures a clean equipment ground for electronic equipment which may be adversely affected by noise in the equipment grounding path. Isolated-ground receptacles installed for the purposes of reducing electromagnetic interference shall be identified with an orange triangle on the face of the receptacle per 406.3(D). In addition, the isolated-ground receptacles shall be used only with isolated equipment-grounding conductors.

Receptacle Position

The ground slot on the receptacle may be installed in the up position or down position, depending upon the choice of the electrician or designer. The NEC® and Article 406 do not state the position in which the receptacle shall be mounted. A good case can be made, however, for the installation of ground slots in the up position. If the plug is not fully engaged in the receptacle slots, the ground pin offers some protection from falling objects that might contact the energized receptacle prongs. For new receptacles in existing installations, the common practice is to install new receptacles in the same position as existing receptacles. Always check with the AHJ before installing receptacles.

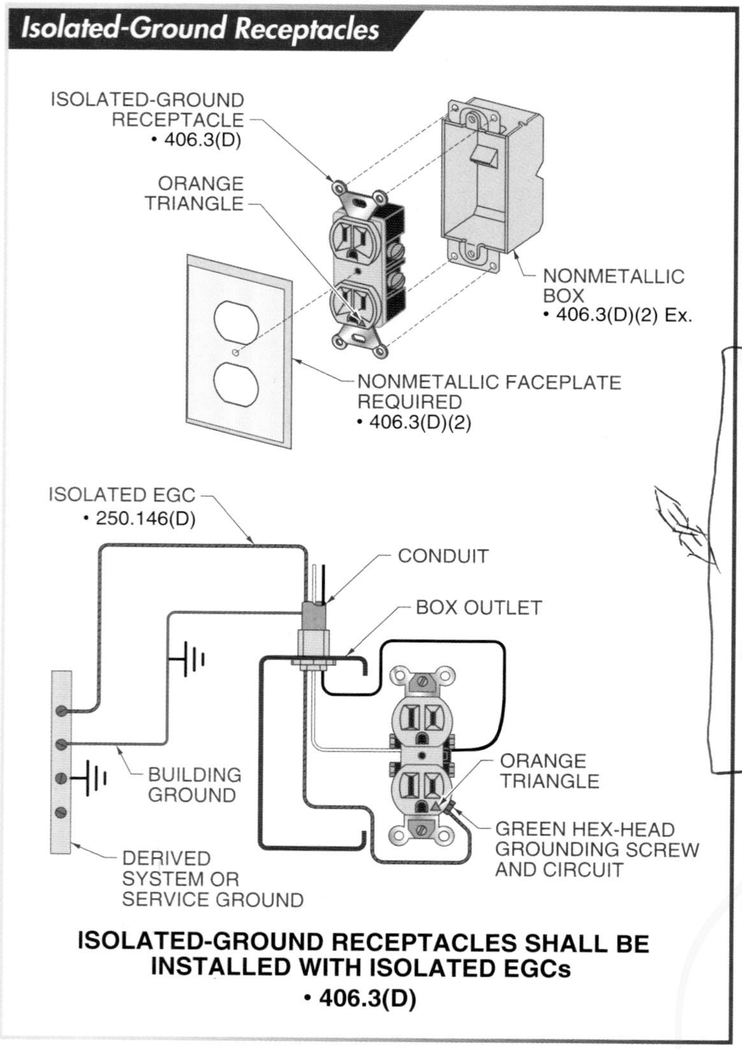

Figure 10-19. Isolated-ground receptacles shall be installed with isolated equipment grounding conductors only.

Isolated ground receptacles are not permitted to be used with the EGC that provides the normal grounding for the electrical system. A separate EGC that is isolated as permitted in 250.146(D) shall be used for receptacles that are identified as isolated ground receptacles with the orange triangle.

Hospital-Grade Receptacles

A *hospital-grade receptacle* is the highest grade receptacle manufactured for the electrical industry. These receptacles undergo the most stringent testing and quality control of any receptacle. They are designed for use in health care facilities and other areas, such as schools and plants, where they are subject to excessive abuse.

The NEC® requires that all patient bed location receptacles be listed and identified as "hospital grade" per 517.18(B). Hospital-grade receptacles are identified by a green dot on the face of the receptacle.

GFCI-Type Receptacles

A *ground-fault circuit interrupter receptacle* is a device that interrupts the flow of current to the load when a ground fault occurs that exceeds a predetermined value of current. Class A GFCIs are commonly used for personnel protection and are designed to operate when the ground fault current flow has a value between 4 mA and 6 mA. See 210.8. Article 100, Definition of GFCI.

Replacing Receptacles – 406.4(D)

Per 406.4(D)(1), a grounding type receptacle shall be used where a grounding means exists in the receptacle enclosure or an EGC is installed per Article 250. A GFCI-protected receptacle shall be used, per 406.4(D)(3), where replacements are made at receptacle outlets that are required to be GFCI-protected elsewhere in the NEC®. **See Figure 10-20.** For example, if a non-GFCI-protected receptacle in a bathroom becomes defective, it shall be replaced with a GFCI-protected receptacle. This is one of the few retroactive rules found in the NEC®.

Three methods are permitted for the installation of replacement receptacles when grounding does not exist. Per 406.4(D)(2)(a), a nongrounding receptacle shall be permitted to be replaced with another nongrounding receptacle. Per 406.4(D)(2)(b), a nongrounding receptacle shall be permitted to be replaced with a GFCI receptacle. The GFCI receptacle shall be marked "No Equipment Ground." These stickers are shipped with the GFCI receptacles. An EGC shall not be connected from the GFCI receptacle to any outlet supplied from the GFCI receptacle.

Per 406.4(D)(2)(c), a nongrounding receptacle shall be permitted to be replaced with a grounding receptacle provided it is supplied through a GFCI device. This device may be either a GFCI receptacle or a GFCI circuit breaker. The replacement grounding receptacle shall be marked "GFCI Protected" and "No Equipment Ground." An EGC shall not be connected between the grounding receptacles.

Replacement Receptacles

EXISTING NONGROUNDING RECEPTACLE	*REPLACEMENT RECEPTACLE*	*NEC®*	*REPLACEMENT PERMITTED*
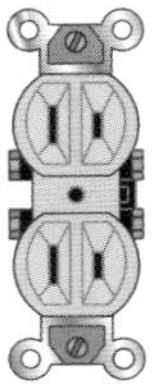	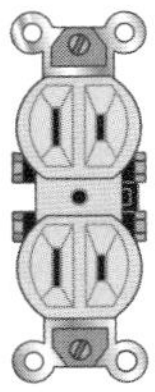	• 406.4(D)(2)(a)	REPLACE A NONGROUNDING-TYPE RECEPTACLE WITH ANOTHER NONGROUNDING-TYPE RECEPTACLE IF NO GROUNDING MEANS EXIST
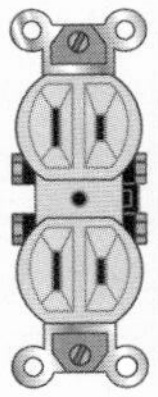		• 406.4(D)(2)(b)	REPLACE A NONGROUNDING-TYPE RECEPTACLE WITH GFCI-TYPE RECEPTACLE MARKED "NO EQUIPMENT GROUND"
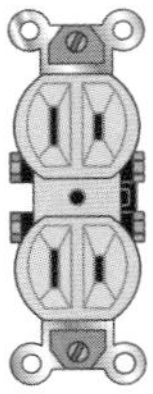	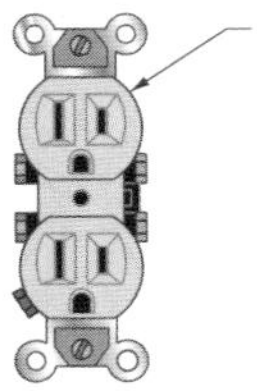 SUPPLIED BY GFCI PROTECTIVE DEVICE	• 406.4(D)(2)(c)	REPLACE A NONGROUNDING-TYPE RECEPTACLE WITH GROUNDING-TYPE RECEPTACLE WHEN SUPPLIED BY A GFCI PROTECTIVE DEVICE. REPLACEMENT GROUNDING RECEPTACLE SHALL BE MARKED "GFCI PROTECTED" AND "NO EQUIPMENT GROUND"

Figure 10-20. Three methods are permitted for the installation of replacement receptacles when grounding does not exist.

Weather-resistant receptacles include an approved weather-protective cover.

Arc-Fault Circuit Interrupter Receptacles

Beginning in the 2005 NEC®, a new form of protection was recognized in the Code. Arc-fault circuit interrupters (AFCIs) are devices that are intended to provide protection from the effects of arc faults by de-energizing the circuit when a damaging arc fault is detected. There are many types of arc faults, and many normally occur during the operation of electrical equipment and devices. AFCIs are able to distinguish the many different types of arc faults and respond to those that are potentially damaging.

Article 210 contains provisions for providing arc-fault circuit interrupter protection for 120 V, single-phase, 15 A and 20 A branch circuits supplying specified outlets in dwelling units. See Section 210.12. Where a receptacle outlet is supplied from a branch circuit that requires AFCI protection, Section 406.4(D)(4) will require, beginning January 1, 2014, that a replacement receptacle at one of these outlets comply with one of three conditions. First, it is permissible to make the replacement with a listed outlet type branch-circuit AFCI. Second, it is permissible to install a receptacle that is protected by a listed outlet branch-circuit type AFCI, and third, it is permissible to install a receptacle that is protected by a listed combination-type AFCI-type circuit breaker.

Tamper-Resistant Receptacles

Tamper-resistant receptacles are a special type of safety receptacle designed with spring-loaded shutters that close off the contact openings, or slots, of the receptacles. The design requires that both springs be compressed at the same time. Protection against insertion into the device by a child is provided because the shutters do not open when a child attempts to insert an object into only one contact opening.

Because of the enhanced safety features of these types of receptacles, the NEC® is specifying their use in several special installations. Section 406.12, for example, requires the use of listed tamper-resistant receptacles in all areas specified in 210.52 for all 15 A and 20 A, 125 V, non-locking-type receptacles. Section 406.14 has a similar requirement for receptacles in child care facilities.

Section 406.4(D)(5) requires that when receptacle replacements are made in outlets that are required to be tamper-resistant elsewhere in the Code, the replacements must be made with listed tamper-resistant receptacles.

Weather-Resistant Receptacles

Weather-resistant receptacles are designed to offer protection against the elements, such as rain, snow, ice, moisture, humidity and dust. They are designed and built with corrosion-resistant materials and plastics. This protection is provided by combining the weather-resistant receptacle and an approved weather-protective cover. Section 406.4(D)(6) requires that when replacements are made at receptacle outlets that are required to be protected against the elements, weather-resistant receptacles shall be provided at the time of replacement.

Installation – 406.5; 406.7

Because receptacle outlets are the most common point of connection to the electrical system, several safety measures concerning their installation and use are included in the NEC®. The NEC® contains provisions regarding the installation of the receptacle outlet and faceplates and the use of receptacles in damp and wet locations. In addition, 406.7(A) requires that attachment plugs and connectors be constructed in a manner that does not permit any exposed current-carrying parts, with the exception of the prongs, blades, or pins. Such attachment plug or connectors shall be constructed with dead fronts. A *dead front* is a cover required to isolate live parts of a plug or connector.

Mounting – 406.5. Section 406.5 contains provisions for the installation of receptacles mounted in boxes or assemblies designed for the purpose. If the box is set back from the wall opening, the receptacle mounting yoke or strap shall be firmly seated against the wall surface. If the box is mounted flush with the wall or projects beyond the wall surface, the receptacle yoke or strap shall be firmly seated against the box. The rigid support for receptacles required by 406.5 is dependent on the proper rigid installation of the outlet box as required in 314.23. These provisions ensure good electrical contact between the box and the receptacle for maintaining grounding continuity. The faceplates shall be installed to completely cover the entire wall opening. See 406.6.

Metal faceplates are required by 406.6(B) to be grounded. This can be accomplished by using No. 6 – 32 metal screws that attach the faceplate to a receptacle yoke that is grounded.

Noninterchangeability – 406.8. Receptacles, attachment plugs, and cord connectors shall be designed and constructed so that they cannot be interchanged with other receptacles, attachment plugs, and cord connectors of a different voltage or current rating. This ensures that the equipment will be properly protected and the personnel who might be operating the equipment can do so in a safe manner. For example, nongrounding receptacles shall be constructed so that grounding-type attachment plugs cannot be inserted into them. However, 406.8 permits T-slot 20 A receptacles to be constructed so that 15 A attachment plugs can be inserted.

Covers – 406.5(C). Receptacles which are installed in covers shall be rigidly attached to the cover by more than a single screw. **See Figure 10-21.** For many years the receptacle was attached to the cover by means of a single No. 6 – 32 screw which was provided to attach the receptacle faceplate. If the receptacle screw loosened, the grounding continuity between the receptacle and the cover could be interrupted. Covers are now designed to provide at least two points of connection between the receptacle and the cover. However, 406.5(C) permits listed devices, assemblies, or box covers to be installed with a single screw if they are identified for such use.

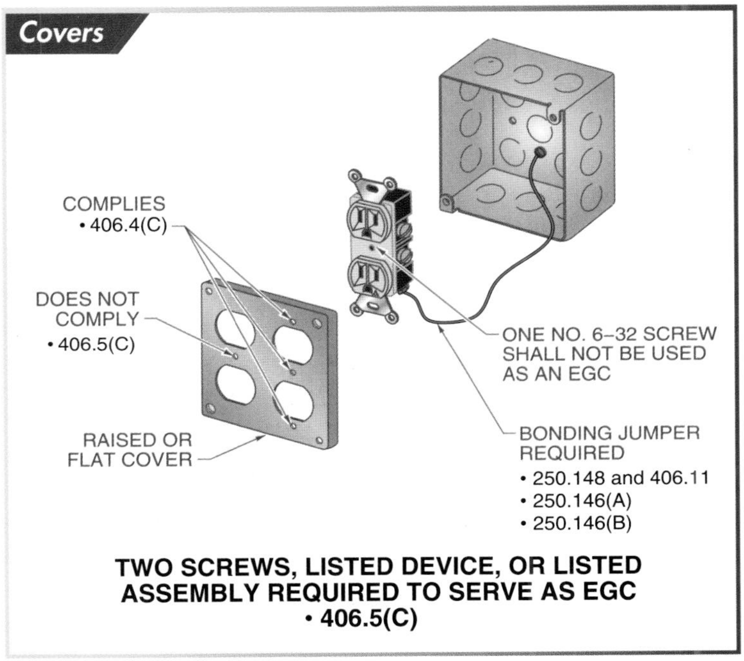

Figure 10-21. Single screws are not permitted to secure receptacles to raised covers.

Wet and Damp Locations – 406.9. Receptacles installed in damp or wet locations present additional hazards to those who might be using cord-and-plug-connected equipment. Because of the presence of moisture, the NEC® requires additional safety provisions for these receptacles. **See Figure 10-22.** Any receptacle installed in a damp location, either indoors or outdoors, shall be provided with an enclosure that is weatherproof when the receptacle is covered. This would occur when an attachment cord cap is not plugged in and the covers are in the closed position.

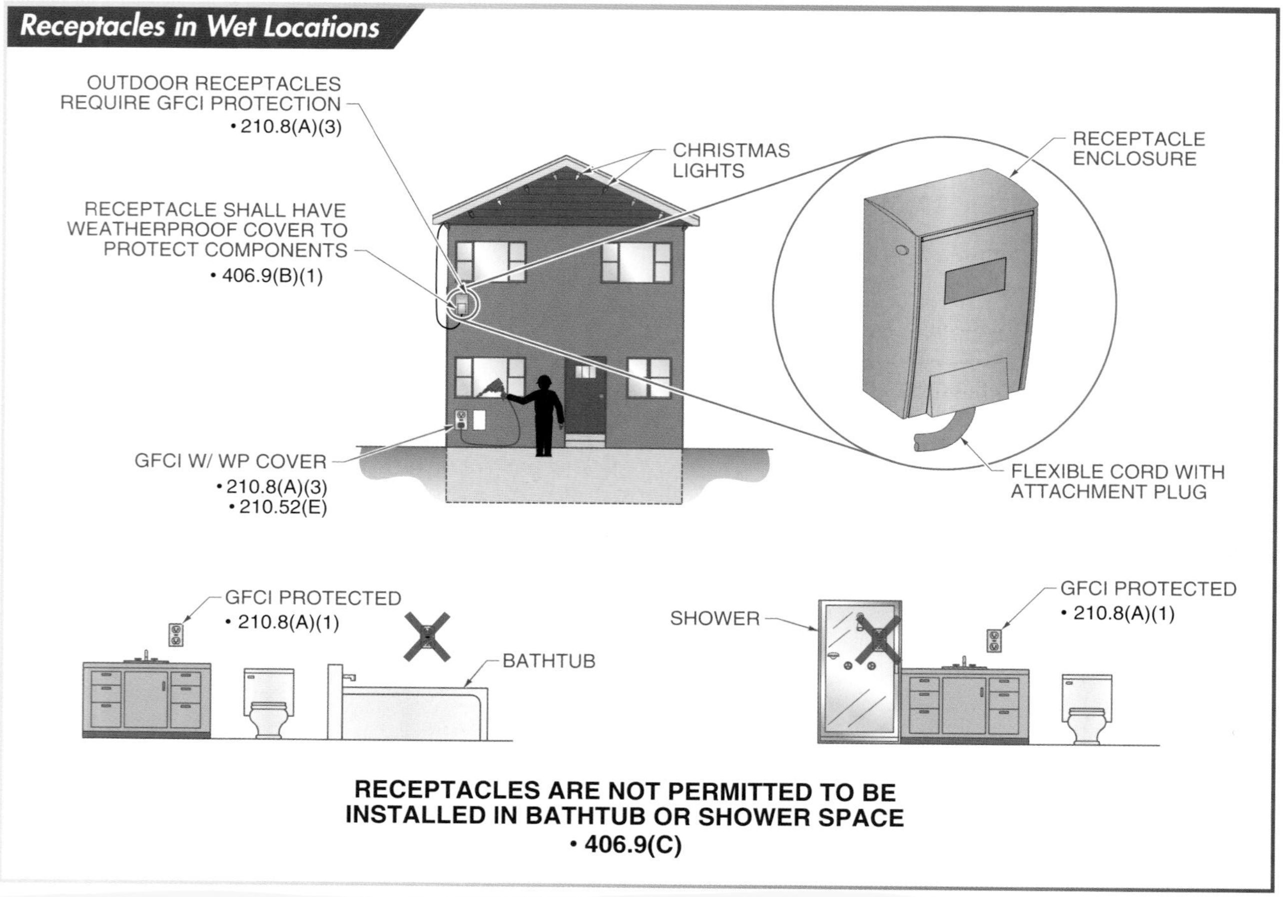

Figure 10-22. Receptacles in wet locations require additional installation provisions to protect personnel.

> **Receptacle Outlet Locations**
>
> *The location in which receptacle outlets are installed is significant to electrical designers and installers. The type of receptacle cover required depends upon whether the location is classified as a damp or wet location.*
>
> *All outdoor locations are not necessarily wet locations. Locations outdoors under open porches or other partially protected areas can be classified as damp locations. The significance of 406.9(A) and 406.9(B)(1)(2) is that spring-type receptacle covers are permitted in damp locations but not in wet locations, unless the receptacle is other than a 15 A or 20 A, 125 or 250 V receptacle.*

Per 406.9(B)(1), receptacles installed in wet locations shall have an enclosure which is weatherproof while the attachment plug cap is inserted or removed. Per 406.9(B)(2)(b), other receptacles installed in wet locations and intended for use with equipment that is attended while in use shall be installed in an enclosure that is weatherproof only when the attachment plug cap is removed. Examples of these types of equipment are portable tools and other portable equipment.

This GFCI receptacle is installed in a horizontal position.

Another wet location which presents additional hazards to personnel operating equipment is bathrooms and shower spaces. Therefore, 406.9(C) does not permit a receptacle outlet to be installed within or directly over a bathtub or shower space.

Tamper-Resistant Receptacle Installations – 406.12

A significant revision regarding types of receptacles required for certain areas of dwelling units occurred in the 2008 NEC®. Section 406.12 was added and requires that only listed tamper-resistant receptacles be installed in all areas covered in 210.52. This requirement applies to all 125 V, 15 A and 20 A nonlocking-type receptacles installed in these areas.

In the 2011 NEC®, a comprehensive exception was added to the requirement for tamper-resistant receptacles. The exception removes the requirements in four specific types of applications. The first is for receptacles that are located more than 5½′ above the floor. These are generally inaccessible to children and thus not in need of being tamper-resistant. Similarly, receptacles that are part of a lighting fixture or an appliance are exempt as well. The third condition exempts both single and duplex receptacles that are installed within dedicated space for an appliance provided the appliance is normally in use and not easily moved from one location to another. Finally, where nongrounding receptacles are used for replacements as permitted by 406.4(D)(2)(a), the replacement receptacle does not need to be tamper-resistant.

APPLIANCES – ARTICLE 422

Article 422 of the NEC® contains installation provisions for appliances, including their control, protection, and disconnecting means.

An appliance is any utilization equipment which performs one or more functions, such as clothes washing, air conditioning, cooking, etc. Utilization equipment uses electrical energy to do some useful work, such as lighting, heating, chemical, etc. While Article 422 states that the article covers appliances in all types of occupancies, typically the term appliance does not include industrial equipment, but only equipment that is used in residential or commercial occupancies.

Installation Requirements – 422.4; 422.10; 422.11; 422.12

All appliances shall be installed in an approved manner (acceptable to the AHJ). Section 110.3(B) also requires that listed or labeled equipment shall be installed, used, or both in accordance with any instructions included in the listing or labeling. This requirement is particularly important with appliances. The listing or labeling instructions for some appliances, such as ceiling fans, may include additional installation requirements that the NEC® does not address.

Per 422.4, all appliances shall be installed so that they have no live parts normally exposed to contact. However, exposed live parts necessitated by the design of the appliance are permitted. Examples of these types of appliances are toasters and grills.

Ceiling fans, because of their location and dynamics of their load, can pose a great danger to personnel if not properly installed with outlet boxes listed for the purpose.

Section 422.16(A) provides two general permissions for the use of flexible cords with appliances. The first allows the use of flexible cords to facilitate frequent interchange of appliances or to prevent the transmission of noise or vibration. The second permission is for fixed appliances that have fastening means and mechanical connections specifically designed to allow ready removal of the appliance for maintenance or repair. This permission can only be applied to fixed appliances that are intended or identified for flexible cord connection. Section 422.16(B) provides rules for specific appliances such as kitchen waste disposers, dishwashers, trash compactors, wall-mounted ovens, counter-mounted cooking units, and range hoods. Some of the rules place restrictions on the length of the cords used to connect the appliances.

Branch-Circuit Ratings – 422.10. This section contains the requirements for sizing branch circuits used to supply appliances. Part II does not apply to the conductors which are an integral part of the appliance.

The rating of an individual branch circuit which supplies an appliance shall be not less than the marked rating of the appliance per 422.10(A). If the appliance has combination marked loads, the branch-circuit rating shall be suitable for the marked appliance rating and the combination load. See 422.62. Article 100 defines an individual branch circuit as a circuit which supplies only one utilization equipment. Motor-operated appliances shall be installed per Article 430. Appliances which contain a hermetic refrigerant motor compressor shall be installed per Article 440.

For a continuously loaded, nonmotor-operated appliance, 422.10(A) requires the branch-circuit rating to be not less than 125% of the marked rating, unless the branch-circuit device and its assembly are listed for continuous loading at 100%. Household cooking appliance branch circuits shall be per Table 220.55, Note 4 and sized per 210.19(A)(3). Per 422.10(B), the rating for branch circuits supplying combination loads consisting of appliances and other loads shall be determined by 210.23.

Overcurrent Protection – 422.11. Branch circuits supplying appliances are required to be protected against overcurrents by 240.4. In general, the conductors used to supply appliances shall be protected against overcurrents in accordance with their ampacities as listed in Table 310.16. Section 240.4 does, however, contain seven subparts which are, in essence, exceptions to this general rule.

If a protective device rating is marked on an appliance, the branch-circuit device rating shall not exceed that marking per 422.11(A). Some appliances, such as air conditioners, are typically marked to include maximum branch-circuit rating information on their nameplates.

Central Heating – 422.12. Since the 1990 NEC®, Article 422 has required that an individual branch circuit supply all central heating equipment. This does not apply to fixed electric space-heating equipment, which is covered by Article 424. This rule removes the possibility that an overcurrent on another piece of equipment, connected to the same branch circuit as the central heating equipment, could result in the loss of the heating equipment when the OCPD opens. However, 422.12, Ex. 1 does permit auxiliary and associated equipment, such as power-actuated valves, pumps, humidifiers, etc. to be connected to the same branch circuit that supplies the central heater. In addition, 422.12, Ex. 2 permits permanently connected A/C to be connected to the same branch circuit.

Flexible Cords – 422.16. Specific appliances, such as kitchen waste disposers, built-in dishwashers, and trash compactors, can be cord-and-plug-connected. **See Figure 10-23.** Kitchen waste disposers are permitted to be permanently wired or cord-and-plug-connected per 422.16(B)(1). All kitchen foodwaste disposers that are cord-and-plug-connected shall be provided with flexible cord which is identified for the purpose. In addition, the following conditions shall be met:

(1) The attachment plug in which the flexible cord terminates shall be the grounding type.

(2) The cord shall be at least 18″ but no longer than 36″.

(3) The receptacle shall be located so that it does not subject the flexible cord to physical damage.

(4) The receptacle shall be installed in an accessible location.

If the kitchen waste disposer is permanently wired, the installation shall comply with the type of wiring method selected. If a kitchen foodwaste disposer is protected by a system of double insulation, and is distinctly marked to indicate such a system, the kitchen waste disposer is not required to be grounded per 422.16(B)(1)(1), Ex.

Specific Appliances

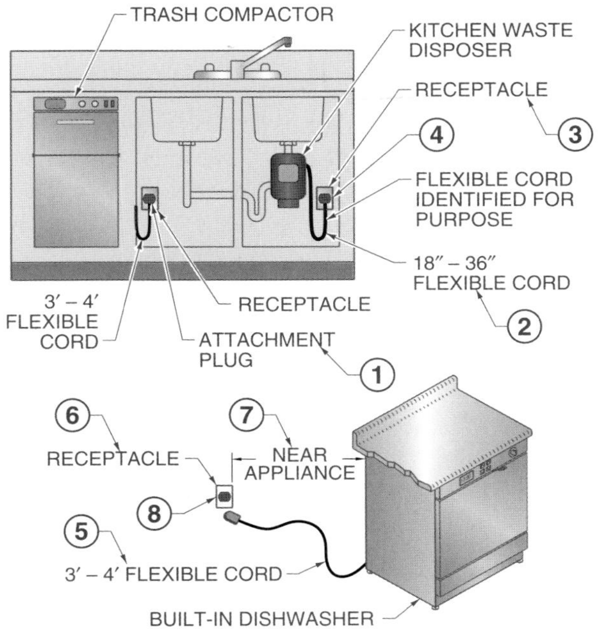

KITCHEN WASTE DISPOSERS, TRASH COMPACTORS, AND DISHWASHERS ARE PERMITTED TO BE CORD-AND-PLUG- CONNECTED IF ALL CONDITIONS ARE MET
• 422.16(B)(1)
• 422.16(B)(2)

(1) FLEXIBLE CORD TERMINATES IN GROUNDING TYPE ATTACHMENT PLUG
- 422.16(B)(1)(1)
- 422.16(B)(2)(1)

(2) FLEXIBLE CORD SHALL BE 18″ TO 36″
- 422.16(B)(1)(2)

(3) RECEPTACLES SHALL BE LOCATED TO AVOID DAMAGE TO CORD
- 422.16(B)(1)(3)

(4) RECEPTACLES SHALL BE ACCESSIBLE
- 422.16(B)(1)(4)

(5) FLEXIBLE CORD SHALL BE 3′ TO 4′
- 422.16(B)(2)(2)

(6) RECEPTACLE SHALL BE LOCATED TO AVOID DAMAGE TO FLEXIBLE CORD
- 422.16(B)(2)(3)

(7) RECEPTACLES SHALL BE NEAR APPLIANCES
- 422.16(B)(2)(4)

(8) RECEPTACLES SHALL BE ACCESSIBLE
- 422.16(B)(2)(5)

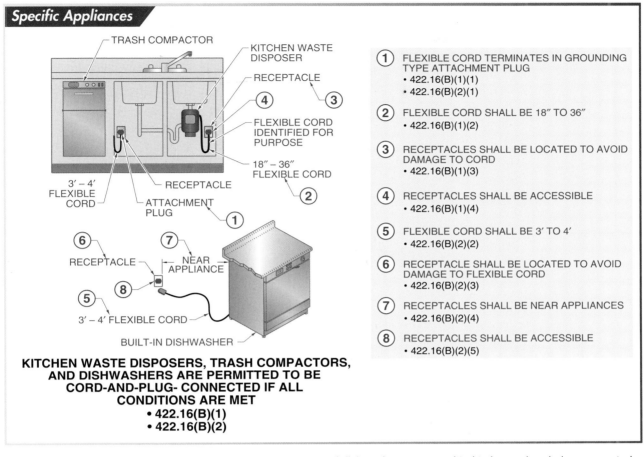

Figure 10-23. Kitchen foodwaste disposers, trash compactors, and dishwashers are permitted to be cord-and-plug-connected.

Trash compactors and built-in dishwashers are permitted be hard-wired or cord-and-plug-connected per 422.16(B)(2). Trash compactors and built-in dishwashers that are cord-and-plug-connected shall be provided with flexible cord which is identified for the purpose. In addition, the following conditions shall be met:

(1) The attachment plug in which the flexible cord terminates shall be the grounding type.

(2) The cord shall be 3′ to 4′ in length measured from the face of the attachment plug to the plane of the rear of the appliance.

(3) The receptacle shall be located so that it does not subject the flexible cord to physical damage.

(4) The receptacle shall be located in the space occupied by the appliance or adjacent thereto.

(5) The receptacle shall be installed in an accessible location.

Per 422.16(B)(2), dishwashers and trash compactors can be hardwired or cord-and-plug-connected.

If the trash compactor or built-in dishwasher is permanently wired, the installation shall comply with the type of wiring method selected. If the trash compactor or built-in dishwasher is protected by a system of double insulation, and is distinctly marked to indicate such a system, the trash compactor or dishwasher is not required to be grounded per 422.16(B)(2)(1), Ex.

Water Heaters – 422.13; 422.47. Each storage- or instantaneous-type water heater shall be provided with a temperature-limiting device to disconnect all ungrounded supply conductors per 422.47. This high-limit device is in addition to any thermostat that may be used for control purposes. The high-limit device shall be capable of sensing water temperature and be either trip-free or contain a replacement element.

The branch circuit supplying the water heater is sized by 422.10 and 422.13. For storage-type water heaters of not more than 120 gal., the branch-circuit rating shall be at least 125% of the nameplate rating of the water heater. The Informational Note to 422.13 references 422.10, which contains requirements for sizing continuously loaded branch circuits. The water heater is treated like a continuously operated appliance and is therefore, by definition, a continuous load.

Wall and Counter-Mounted Ovens – 422.16(B)(3). Wall and counter-mounted ovens can also be permanently wired or cord-and-plug-connected for ease of servicing. If a separable plug or connector and receptacle combination is used for wall or counter-mounted ovens, it must be approved for the temperature of the space in which it is located.

Ceiling Fans – 422.18. Section 422.18 provides three methods of support for ceiling fans. They shall be supported independently from the box, supported by a listed outlet box, or supported by outlet box systems identified for the use and installation per 314.27(C).

Section 314.27(C) requires that any outlet box or outlet box system that is used as the sole support of a ceiling fan must be listed and marked by the manufacturer as being suitable for that purpose. The use of these listed boxes or support systems is only permissible for ceiling fans that do not exceed 70 lb. If the ceiling fan and its accessories weigh more than 70 lb, an independent means of support must be used. **See Figure 10-24.**

If the outlet box or outlet box system is designed for the support of ceiling fans that weigh more than 35 lb, the box must be marked with the maximum weight to be supported.

Per 422.16(B)(3) a wall-oven shall be permitted to be permanently connected or cord-and-plug connected for ease on servicing.

Disconnecting Means – Article 422, Part III

The disconnecting means provides a way to remove the supply of electricity from the appliance. This is a critical safety requirement for workers who may be attempting to service or repair the appliance.

Article 422, Part III contains the requirements for providing a means of disconnecting each appliance from the source of supply. In general, every appliance shall be provided with a disconnecting means that disconnects each ungrounded conductor from the supply source. In the event that an appliance is supplied by more than one source, such as where power conductors and control conductors are both used, each source shall have a separate disconnecting means and the disconnecting means shall be grouped and properly identified.

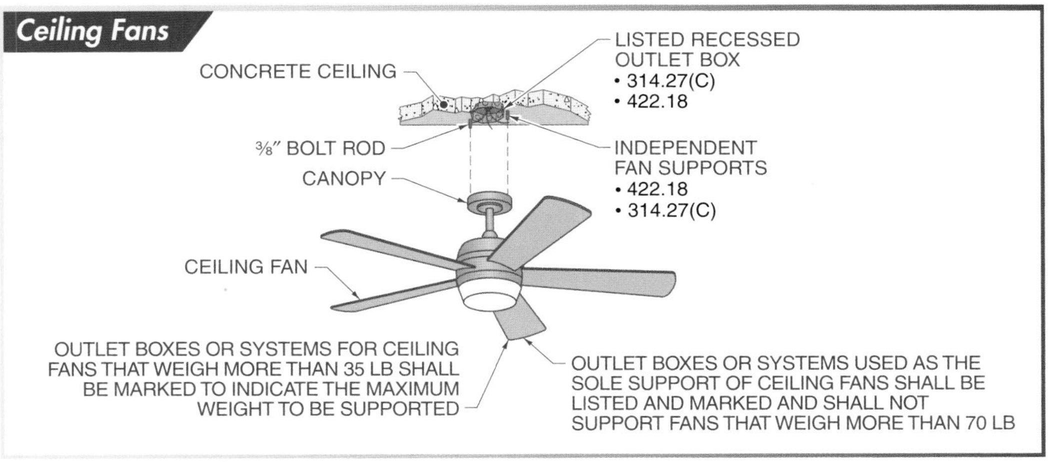
Ceiling Fans
CONCRETE CEILING
LISTED RECESSED OUTLET BOX
• 314.27(C)
• 422.18
⅜″ BOLT ROD
CANOPY
INDEPENDENT FAN SUPPORTS
• 422.18
• 314.27(C)
CEILING FAN
OUTLET BOXES OR SYSTEMS FOR CEILING FANS THAT WEIGH MORE THAN 35 LB SHALL BE MARKED TO INDICATE THE MAXIMUM WEIGHT TO BE SUPPORTED
OUTLET BOXES OR SYSTEMS USED AS THE SOLE SUPPORT OF CEILING FANS SHALL BE LISTED AND MARKED AND SHALL NOT SUPPORT FANS THAT WEIGH MORE THAN 70 LB

Figure 10-24. Ceiling fans weighing over 35 lb shall not be supported by the outlet box.

Permanently Connected Appliances – 422.31. A *permanently connected appliance* is a hard-wired appliance that is not cord-and-plug-connected. Section 422.31 divides permanently connected appliances into three groups. Appliances rated at not over 300 VA or ⅛ HP are covered in 422.31(A). Appliances rated over 300 VA are covered in 422.31(B), and motor-operated appliances rated over ⅛ HP are covered in 422.31(C).

For the smaller appliances discussed in 422.31(A), the disconnecting means for the appliance is permitted to be the branch-circuit overcurrent device. For example, the disconnecting means for a 250 VA, permanently installed appliance could be a circuit breaker.

For the larger appliances discussed in 422.31(B), the disconnecting means is permitted to be a branch-circuit switch or circuit breaker. The switch or circuit breaker shall be within sight from the appliance or it shall be capable of being individually locked in the open position. *Within sight* is visible and not more than 50′ away.

Lastly, for motor-operated appliances that are permanently connected and are rated over ⅛ HP, the branch-circuit switch or circuit breaker is permitted to serve as the disconnecting means provided it is located in sight from the appliance. The disconnecting means must comply with all of the provisions of 430.109 and 430.110. The exception allows a unit switch that complies with 422.34(A–D) that is included in a motor-driven appliance of more than ⅛ HP to serve as the in sight disconnect. The switch of a circuit breaker serving as the other disconnecting means is permitted to be installed out of sight from the motor controller.

Cord-and-Plug-Connected Appliances – 422.33. The provisions for disconnecting cord-and-plug-connected appliances are listed in 422.33. The general rule of 422.33(A) permits a separate attachment plug or connector and a receptacle to serve as the disconnecting means for the appliance if the plug and receptacle are accessible. If the plug and receptacle are not accessible, as is the case if they are located behind a fixed appliance, another disconnecting means meeting the requirements of 422.31 shall be provided.

422.16(4) – Range Hoods

In addition to dishwashers and trash compactors, range hoods are permitted to be connected by a flexible cord provided that the receptacles are connected to a flexible cord between 18″ and 36″ long, accessible, located in such a manner as to avoid physical damage to the flexible cord, and supplied by an individual branch circuit. Listed range hoods protected by double insulation or equivalent shall not be required to be terminated with a grounding-type attachment plug.

If the cord and plug is provided at the rear of the range, and is accessible from the front of the range by removal of the oven drawer, the cord and plug is permitted to serve as the disconnecting means per 422.33(B). Attachment plugs and receptacles installed to serve as the disconnecting means shall have a rating which is at least equal to the rating of the connected appliance per 422.33(C).

Unit Switches – 422.34. A *unit switch* is a switch with a marked OFF position that is part of an appliance. Section 422.34 permits unit switches to serve as the disconnecting means if the unit switch opens all ungrounded conductors and has a marked OFF position designated on the switch. In addition to these two requirements, other means for disconnection shall also be provided depending upon the type of occupancy in which the appliance is installed. **See Figure 10-25.**

For multifamily dwellings, the other disconnecting means shall be located within the dwelling unit or on the same floor as the dwelling unit per 422.34(A). The disconnecting means can control lamps and other appliances such as would be the case if a branch-circuit overcurrent device in a panelboard were used.

For two-family dwellings, 422.34(B) permits the other disconnecting means to be located either inside or outside the dwelling unit that contains the connected appliance. The disconnecting means can control lamps and other appliances such as would be the case if a branch-circuit overcurrent device in a panelboard were used.

For one-family dwellings, 422.34(C) permits the other disconnecting means to be the service disconnecting means. The service disconnecting shall be located in accordance with 230.70.

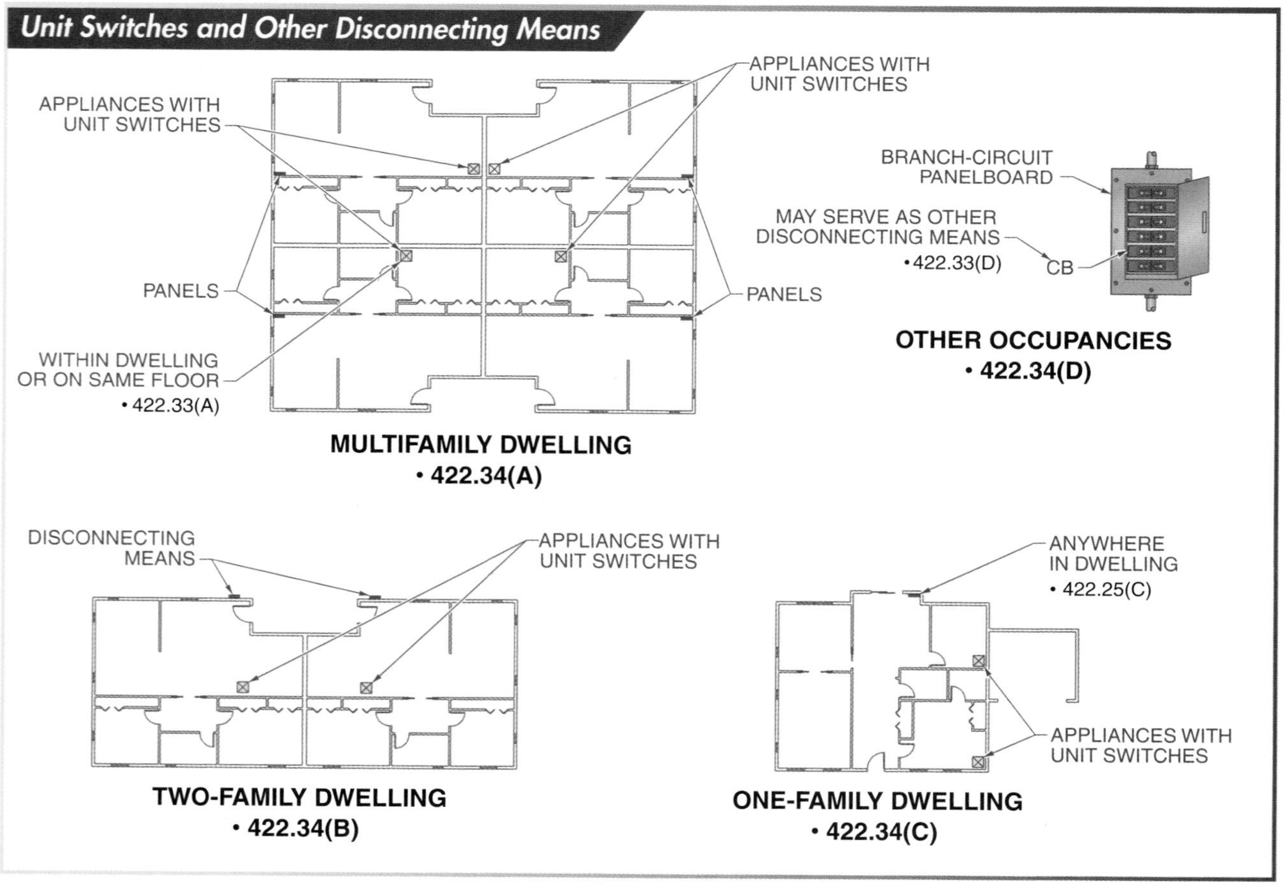

Figure 10-25. Unit switches shall have an additional or other disconnecting means that act a backup when the unit switch is used as a disconnecting means.

For all other occupancies, 422.34(D) permits the branch-circuit switch or circuit breaker to serve as the other disconnecting means. The branch-circuit switch or circuit breaker shall be readily accessible.

Safety Provisions – 422.40; 422.41

Several safety provisions are designed to protect personnel who use the appliances covered by Article 422. The first safety provision is contained in 422.40, which covers proper polarity for cord-and-plug-connected appliances. The second provision is contained in 422.41, which covers immersion protection for personnel. While both of these sections deal with design considerations, they are important safety requirements for the installation of cord-and-plug-connected appliances.

Cord-and-Plug Polarity – 422.40. Section 422.40 requires that appliances with a manually operated, line-connected, single-pole control switch, an Edison-base lampholder, or a 15 A or 20 A receptacle shall be equipped with an attachment plug that is polarized or of the grounding type. A polarized attachment plug is designed so that the ungrounded and grounded prongs cannot be inserted into a receptacle in a manner that causes an opposite or reverse polarity connection. This provision is designed to protect unqualified persons who might attempt to service or relamp an appliance equipped with an Edison-base lampholder. An attachment plug that is polarized or of the grounding type ensures that a reverse polarity condition cannot occur which could increase the risk of electrical shock.

IDCI – 422.41. An *immersion detection circuit interrupter (IDCI)* is a circuit interrupter designed to provide protection against shock when appliances fall into a sink or bathtub. An IDCI disconnects the source of supply regardless of whether the appliance is in the ON or OFF position when it is immersed in water. IDCIs are a relatively new form of personnel protection.

The NEC® requires that portable, freestanding hydromassage units and hand-held hair dryers be equipped with IDCIs when they are manufactured. IDCIs offer protection for persons who are using these types of appliances in occupancies without ground-fault circuit protection.

Markings – Article 422, Part V

The general rule of 422.60(A) requires each electric appliance to be supplied with a nameplate that contains the identifying name and rating in volts and amperes, or in volts and watts, for the appliance. The appliance shall be marked if it is designed for use on a specific frequency or if it requires external overload protection.

All markings shall be in a location that is visible or at least easily accessible after the installation of the appliance per 422.60(B). If the appliance contains field-replaceable heating elements rated over 1 A, they shall be legibly marked with either the manufacturer's part number or the rating in volts and amperes or in volts and watts. See 422.61.

Appliances shall be installed so that they have no live parts normally exposed to contact, unless it is necessitated by the design of the appliance.

Where the horsepower rating is included on the nameplate of a motor-operated appliance, the rating shall not be less than the horsepower rating on the motor nameplate. Where the appliance has multiple motors or one or more motors with other loads, the appliance shall not be less than the equivalent horsepower of the combined loads, calculated per 430.110(C)(1).

QUICK QUIZ®

Refer to Chapter 10 Quick Quiz® on CD-ROM.

Additional information is available at ATPeResources.com

Per 422.62(A)(B)(1), in addition to the markings required by 422.60, the appliance shall be marked with the minimum supply circuit conductor ampacity and the maximum rating of the circuit overcurrent protective device. The provisions of this section do not apply if the appliance is factory-equipped with cords and attachment plugs and with nameplates that are in compliance with 422.60.

An alternate marking method is provided in 422.62(B)(2). This method consists of the rating of the largest motor in volts and amperes and the other load's rating in volts and amperes or volts and watts, plus the required markings listed in 422.60. The alternate marking method does not apply if the appliance is factory-equipped with cords and attachment plugs that comply with 422.60. The ampere rating shall be permitted to be omitted from the required markings for appliances which contain a motor with a ⅛ HP or less rating or a nonmotor load of 1 A or less.

chapter 10

Equipment for General Use

REVIEW QUESTIONS

Name ______________________ **Date** ______________

______ **1.** The major difference between flexible cords and flexible cables is the number of insulated ___.

______ **2.** A(n) ___ is a complete luminaire consisting of a light source such as a lamp or lamps, reflector, lamp guards, and lamp power supply.

T F **3.** Generally, flexible cords and cables shall be installed in continuous lengths without splices or taps.

T F **4.** Lampholders shall not support lamps weighing over 6 lb.

T F **5.** Trees may be used to support outdoor luminaires.

______ **6.** An outlet box for a recessed incandescent fixture shall be at least ___′ from the fixture if the branch-circuit conductors are not suitable for the temperature.

A. ½ C. 2
B. 1 D. 4

______ **7.** An electric-discharge luminaire utilizes a(n) ___ for the operation of the lamp.

______ **8.** Lighting track fittings shall not be equipped with general-purpose ___.

hospital **9.** ___-grade receptacles are the highest grade receptacles manufactured by the electrical industry.

______ **10.** The flexible cord for a kitchen foodwaste disposer shall be from ___″ to ___″ in length.

A. 12; 24 C. 18; 36
B. 18; 30 D. 24; 36

70 **11.** Outlet boxes used as the sole support of a ceiling suspended fan shall be listed and marked as being suitable and shall not support fans that weigh more than ___ lb.

______ **12.** A(n) ___ switch is a switch designed to control a specific unit load.

______ **13.** Flexible cords not smaller than ___ AWG are considered protected by the overcurrent device.

20 **14.** Handholes are not required in ___′ or less metal poles which support luminaires if the base is hinged.

______ **15.** Any branch-circuit conductor located within ___″ of a ballast, within a ballast compartment, shall have an insulation rating not lower than 90°C.

3 **16.** A minimum of ___″ clearance shall be provided for recessed portions of luminaires from all combustible materials.

(T) F **17.** All recessed HID fixtures shall be thermally protected.

T (F) **18.** Class A GFCIs are designed to operate when the ground fault current flow is between 10 mA and 15 mA.

___ **19.** A(n) ___ is any piece of utilization equipment which performs one or more functions.

50 **20.** Within sight is visible and no more than ___′ from the object.

3 **21.** Electric fixtures are not permitted within ___′ of the horizontal bathtub rim zone.

8 **22.** Electric fixtures are not permitted within ___′ of the vertical bathtub rim zone.

identified **23.** Ceiling fans shall be supported independently of an outlet box or by a listed outlet box or outlet box system ___ for the use.

T F **24.** All lighting track shall be permanently installed and permanently connected to a branch circuit.

(T) F **25.** Metal faceplates are required to be grounded.

Clothes Closets

8″ **1.** The minimum distance at A is ___″.

12″ **2.** The minimum distance at B is ___″.

6′ **3.** The minimum depth at C is ___′.

24″ **4.** The minimum width at D is ___″.

12″ **5.** The minimum width at E is ___″.

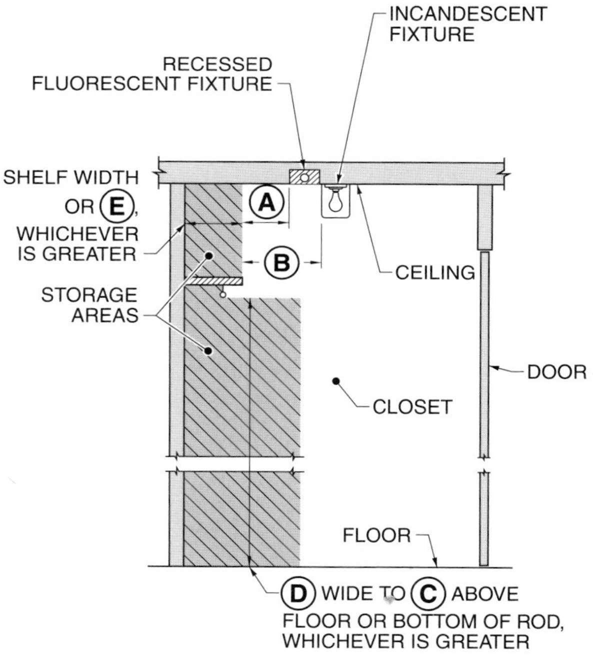

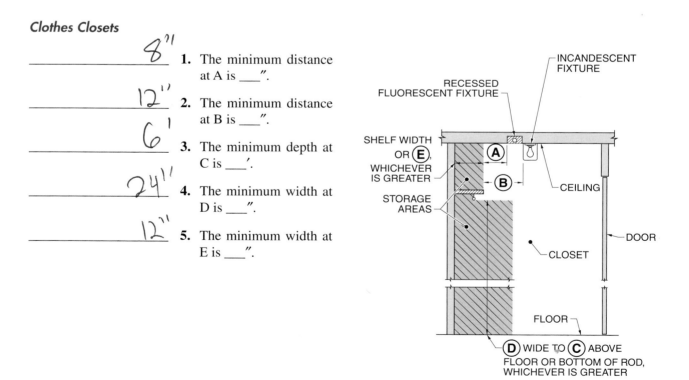

chapter 10 Equipment for General Use

TRADE COMPETENCY TEST

Name ______________________ **Date** ______________

NEC®	Answer	
________	________	**1.** Determine the ampacity of each conductor of a 16/3 AWG multiconductor Type SPT-2 flexible cord used to supply utilization equipment which is fastened-in-place. Two of the conductors are current-carrying conductors.
________	________	**2.** Determine the ampacity of each conductor of a 18/4 AWG multiconductor Type SRDT flexible cord used to supply fastened-in-place utilization equipment. All of the conductors are current-carrying conductors.
________	________	**3.** Determine the ampacity of each THW conductor of a 8/3 AWG multiconductor Type W flexible cable used to supply fastened-in-place utilization equipment. Two of the conductors are current-carrying conductors.
________	________	**4.** Determine the ampacity of each TW conductor of a 4/4 AWG multiconductor Type G flexible cable used to supply fastened-in-place utilization equipment. All of the conductors are current-carrying conductors.
________	________	**5.** Luminaires are installed outdoors under an open porch of a one-family dwelling. The fixtures are marked "Suitable for Damp Locations." Does this installation violate the provisions of Article 410 of the NEC®?
________	________	**6.** *See Figure 1.* Does this installation violate Article 410 of the NEC®?

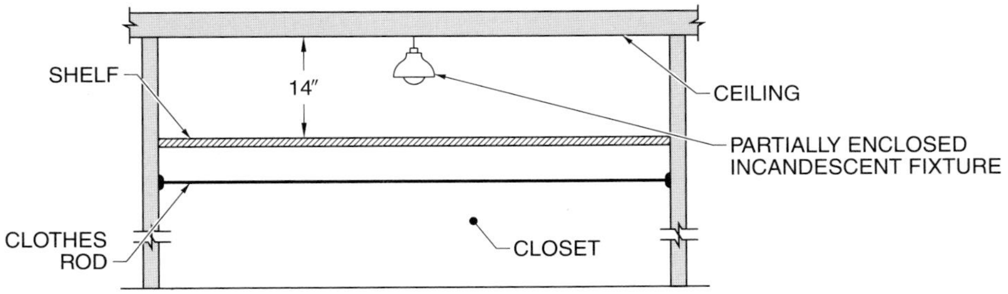

FIGURE 1

__________ __________ **7.** An electrical installation requires that recessed HID fixtures are to be installed in a ceiling of poured concrete decks. The HID fixtures are not provided with thermal protection. Does this installation violate Article 410 of the NEC®?

__________ __________ **8.** A 12′ length of lighting track is installed in the master bedroom of a one-family dwelling. How much VA shall be allocated for the lighting track when determining the calculated load?

__________ __________ **9.** Cord-and-plug-connected vending machines manufactured after January 1, 2005, shall include a(n) ___ as an integral part of the attached plug or located in the power supply cord.

__________ __________ **10.** Determine the minimum branch-circuit rating for an individual branch circuit which is used to supply a fastened-in-place appliance. The marked rating on the appliance is 35 A. The appliance is not continuously loaded.

__________ __________ **11.** Determine the minimum branch-circuit rating for an individual branch circuit which is used to supply a fastened-in-place appliance. The marked rating on the appliance is 24 A. The appliance is continuously loaded.

__________ __________ **12.** *See Figure 2.* Does this installation violate the provisions of Article 422 of the NEC®?

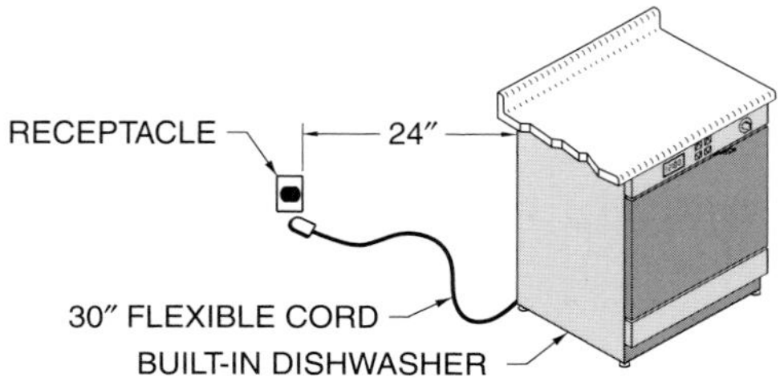

FIGURE 2

__________ __________ **13.** *See Figure 3.* Does this installation violate the provisions of Article 422 of the NEC®?

__________ __________ **14.** *See Figure 4.* Does this installation violate the provisions of Article 410 of the NEC®?

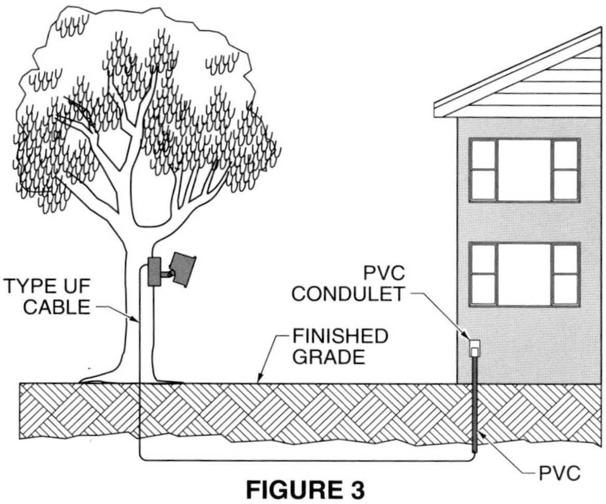

FIGURE 3

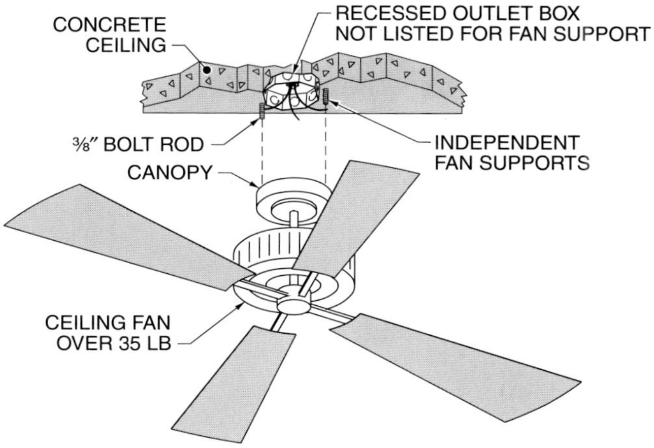

FIGURE 4

Learning the Code

- One of the most difficult Code articles to understand is Article 430, Motors, Motor Circuits, and Controllers. The article is large, complex, and, like grounding in Article 250, composed of text, terms, and vernacular that may escape the average person. The best approach to studying this article is to stick to the basics. Article 430 is full of exceptions and conditions for special motors and special equipment. Apprentices and students of the Code are advised to think in terms of the general installation first. With time, special motors, equipment, and installations can be addressed on a case-by-case basis.
- Of all the tables in Article 430, none are more important to apprentices and students of the Code than Tables 430.52, 430.248, and 430.250. Students must carefully consider the various types of protection that Article 430 seeks to provide. For example, Part II covers the motor overload protection requirements. Part IV covers the motor feeder short-circuit and ground-fault protection requirements. These protections are very different because of the different characteristics that accompany motor circuits. During the motor startup, the starting currents can approach 6 or 7 times the normal running current of the motor. The standard overcurrent protection of Article 240 would never permit these motors to start. At the same time, protection must be provided for both the motor-circuit conductors and the motor itself. Apprentices and students of the Code should continually refer to Figure 430.1 to get a picture in their mind of the point in the motor circuit that they are studying.

- Part IX of Article 430 covers the requirements for disconnecting means. As mentioned in Chapter 10 of *Electrical Systems*, the provisions for disconnecting means are critical for personnel safety. This is especially important with motors because, in addition to the electrical hazards, there are mechanical hazards from the rotating parts. Note that there are disconnecting requirements for both the motor controller and the motor itself, and there are conditions under which the same disconnecting means can be used to accomplish both sets of requirements. See 430.102(A) and (B).

chapter 11

Motors, Generators, A/C and Refrigeration, and Fire Pumps

The full load current (FLC) ratings of motors are determined to calculate the ampacity of conductors, ampere rating of switches, and branch-circuit, short-circuit, and ground-fault protection. A/C and refrigeration equipment contains motor-compressors and other motor-driven loads.

New in the 2011 NEC®

- *Additional requirements for determining the appropriate current values for valve actuator motor assemblies – 430.6(D)*
- *New provisions for 16 AWG and 18 AWG copper conductors in specific applications – 430.22(G)*
- *The addition of a new Informational Note to clarify what a "motor short-circuit protector" is by definition – 430.52(C)(7)*
- *Revised regulations for several motors or loads on a branch circuit – 430.53(C)*

MOTORS – ARTICLE 430

Article 430 covers motors, motor branch-circuit and feeder conductors and their protection, motor overload protection, motor control circuits, motor controllers, and motor control centers. The one line drawing, figure 430.1, is a guide for the designer and installer. This line drawing should be consulted to size the conductors and fuses or circuit breakers for feeders or branch circuits of a motor circuit, and to size the overload protection for the motor itself. Section 430.5 requires that motors and controllers shall comply with the applicable provisions of other articles and sections listed in Table 430.5. Three things cause problems when electricians deal with the calculations in Article 430:

(1) AC inductive loads are rated in W or kW. Because the windings of AC inductive loads offer inductive reactance to the circuit, watts are not the simple calculation of $W = V \times A$. The correct formula to use is $W = V \times A \times PF$. PF is the power factor, which is expressed as a percentage. If the load is resistive, and does not contain a significant amount of inductance, then $W = V \times A$ may be used.

(2) When finding the current rating of motors, 1 HP is often considered to be 746 W. However, this can only be true if the motor is 100% efficient, and no motor is 100% efficient. The correct formula is 1 HP = 746 W × *Eff*. Eff is the motor efficiency, which is expressed as a percentage. The closer a motor runs to its HP rating, the more efficient it becomes. If efficiency is not considered when determining the current ratings of motors, a good rule of thumb is to use 1 HP = 900 W.

(3) Always read 430.6 before any work covered in Article 430 is started. Section 430.6 addresses the proper method of determining the current rating of a motor.

Ampacity and Motor Ratings – 430.6

The most common mistake in applying the requirements of Article 430 is made when determining the motor full-load current (FLC) rating. This is an important determination since all of the calculations are based on this number. The required ampacity and motor ratings shall be determined by applying 430.6(A)(B)(C)(D).

General Motor Applications – 430.6(A). Section 430.6(A) covers all motor installations except for torque motors and AC adjustable voltage motors. The values given in the following FLC tables are used to determine the ampacity of conductors supplying the motor, the ampere rating of switches, and the branch-circuit, short-circuit, and ground-fault protection (fuse or circuit breaker).

Table	Motor
430.247	DC
430.248	1ϕ
430.249	2ϕ
430.250	3ϕ

The actual current rating marked on the motor nameplate shall not be used to determine conductor size, switch size, or the size of the branch-circuit overcurrent protective device.

Motor Full-Load Current (FLC)

The current rating on a motor nameplate is exactly what a motor pulls at full load. The NEC® does not permit the use of the nameplate current when calculating the hard-wired portion (wire, conduit, fuse, or switch) of the motor circuit. Section 430.6 requires that the current rating of a particular motor be taken from one of the FLC tables. This is because not all 3ϕ, 50 HP motors pull the same current when supplied at a particular voltage. For example, in Table 430.250, the listed current for a 3ϕ, 50 HP motor operating at 230 V is 130 A. However, the nameplate ratings of several 3ϕ, 50 HP motors operating at 230 V may range from 128 A to as low as 118 A. The range depends on the different characteristics (speed, efficiency, etc.) for which the motor was designed.

In the early 1920s, the NEC® required that the motor conductors and fuses be selected using the nameplate current. The problem was in the replacement of a defective motor. Unless all characteristics of the replacement motor matched the original motor, larger wire sizes, raceway sizes, OCPD sizes, etc. could be required. Therefore, the NEC® provides the FLC tables for motors. The current shown on a general-type motor nameplate never exceeds the current listed on FLC Tables 430.247 through 430.250.

Torque Motors – 430.6(B). The locked-rotor current (LRC) on the motor nameplate shall be the rated current in determining the ampacity of the branch-circuit conductors, the ampere rating of the motor overload protection, and the ampere rating of the motor branch-circuit short-circuit and ground-fault protection. Torque motors represent less than 10% of motor applications.

AC Adjustable Motors – 430.6(C). The maximum operating current on the motor or control nameplate shall be used in determining the ampacity of the conductors, ampere rating of switches, and branch-circuit, short-circuit, and ground-fault protection of AC adjustable voltage motors. If the maximum operating current does not appear on the nameplate, the ampacity shall be based on 150% of the values given in Tables 430.249 and 430.250. AC adjustable voltage motors represent a small percentage of motor applications but are being used in more applications each year.

Valve Actuator Motor Assemblies – 430.6(D). The rated current for valve actuator motor assemblies is the nameplate full-load current. This current is to be used to determine the maximum rating or setting of the motor branch-circuit short-circuit and ground-fault protective device, and the conductor ampacity.

Marking on Motors and Multimotor Equipment – 430.7

The motor nameplate shall contain the following 15 pieces of information, if applicable:

(1) Manufacturer's name.

(2) Rated volts and full-load amps.

(3) Rated frequency and number of phases.

(4) Rated full-load speed.

(5) Rated temperature rise or insulation system class and rated ambient temperature.

(6) Time rating (5, 15, 30, 60 minutes, or continuous).

(7) Rated horsepower if ⅛ HP or more.

(8) Code letter or locked-rotor amps if AC motor rated ½ HP or more.

(9) Design letter for B, C, or D motors.

(10) Secondary volts and full-load amps if wound-rotor induction motor.

(11) Field current and voltage for DC-excited synchronous motors.

(12) Winding type of DC motor.

(13) Motors provided with thermal protection shall be marked "Thermally Protected" or "TP."

(14) Small motors rated 100 W or less shall be marked "Impedance Protected."

(15) Motors with powered condensation protection shall include rated heater voltage, phases, and rated power.

Multimotor and Combination-Load Equipment – 430.7(D). The nameplate on multimotor and combination-load factory-wired equipment shall contain the following six items:

(1) Manufacturer's name.

(2) Voltage rating.

(3) Frequency.

(4) Number of phases.

(5) Minimum ampacity for supply conductors.

(6) Maximum amps rating of the OCPD that provides short-circuit and ground-fault protection.

Baldor Electric Co.

A motor nameplate must include motor information per 430.7.

Marking on Controllers – 430.8

The controller in a motor circuit is any switch or device that is normally used to start and stop a motor by making and breaking the motor circuit current. A controller may be a relay, contactor, starter, or combination starter. Controllers shall be marked with the following information:

(1) Manufacturer's name.
(2) Rated voltage.
(3) Rated current or HP.
(4) Short-circuit current rating.
(5) Any other information to indicate suitability.

There are four exceptions to this rule that permit the short-circuit rating to be omitted from the controller marking. See 430.8 Ex. 1–4.

Motor Nameplates

The amperes listed on the motor nameplate are the number of amps the motor pulls at full load. At no load, the motor pulls from 50% to 65% of the nameplate rating.

The rated full-load speed is the rpm the motor produces at full load. The rpm is higher when the motor is underloaded and lower when the motor is overloaded.

The frame (FR) number refers to the physical size of the motor. Shaft size, mounting hole dimensions, etc., can be obtained from manufacturer's literature and charts.

Branch Circuit – Single Motor – 430.22

The *motor branch circuit* is that point from the last fuse or circuit breaker in the motor circuit out to the motor. The minimum elements required within a motor branch circuit are the motor branch-circuit conductors, the motor OL protective device, and the motor branch-circuit, short-circuit, and ground-fault protective device (fuse or circuit breaker).

Sizing Conductors. Section 430.22 requires that the branch-circuit conductors supplying a single motor have a minimum ampacity of 125% of the FLC rating from the appropriate FLC table based on the HP rating of the motor. For example, Table 430.250 indicates the appropriate FLC for a 3ϕ motor. The appropriate FLC multiplied by 1.25 results in the ampacity required for the branch-circuit conductors per 430.22. Table 310.16 is used to select the proper size conductor. **See Figure 11-1.**

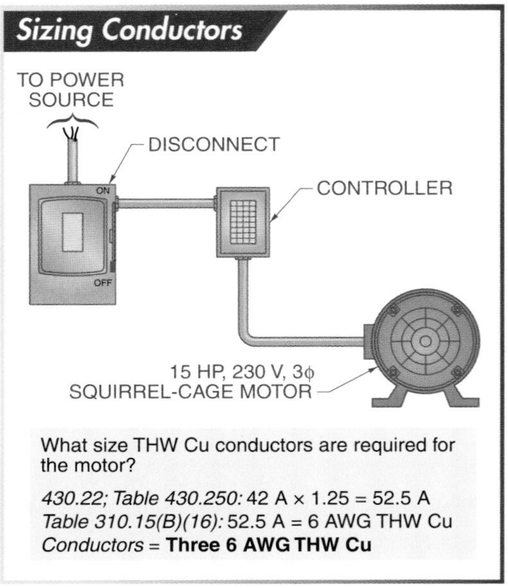
Sizing Conductors
TO POWER SOURCE
DISCONNECT
ON
OFF
CONTROLLER
15 HP, 230 V, 3ϕ
SQUIRREL-CAGE MOTOR
What size THW Cu conductors are required for the motor?
430.22; Table 430.250: 42 A × 1.25 = 52.5 A
Table 310.15(B)(16): 52.5 A = 6 AWG THW Cu
Conductors = Three 6 AWG THW Cu

Figure 11-1. The appropriate FLC multiplied by 1.25 gives the ampacity required for branch-circuit conductors.

Sizing Raceways. Once the conductors are properly sized, the appropriate size raceway can be selected. For a metallic raceway that is an approved EGC, all of the conductors are of the same size. The tables in Chapter 9, Informative Annex C may be used to select the proper size raceway that is being installed. **See Figure 11-2.**

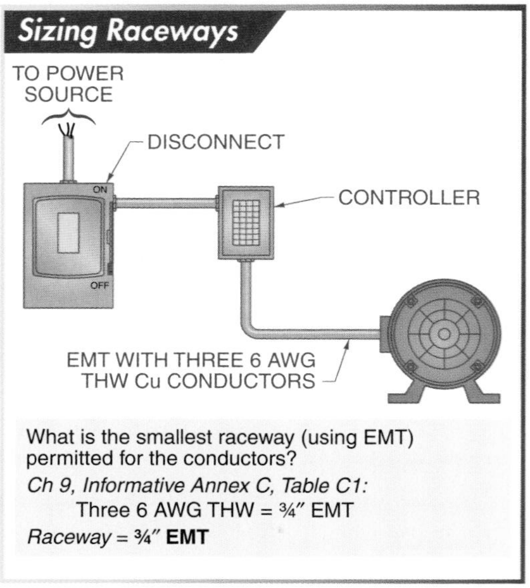
Sizing Raceways
TO POWER SOURCE
DISCONNECT
ON
OFF
CONTROLLER
EMT WITH THREE 6 AWG THW Cu CONDUCTORS
What is the smallest raceway (using EMT) permitted for the conductors?
Ch 9, Informative Annex C, Table C1:
Three 6 AWG THW = ¾″ EMT
Raceway = ¾″ EMT

Figure 11-2. Raceways for motor conductors are sized by the tables in Chapter 9, Informative Annex C.

Sizing Overload Protection. The three forms of protection that all electrical circuits require are overload protection, short-circuit protection, and ground-fault protection. In most circuits, all three types of protection are provided by the fuse or circuit breaker. In the motor circuit, the fuse or circuit breaker has to be large enough to allow the motor starting current to flow in the circuit without tripping the overcurrent protective device. As a result, it is not capable of protecting the motor when an overload occurs. In most motor circuits, the fuse or circuit breaker is referred to as the short-circuit, ground-fault protector. It only opens if there is a short or a ground fault in the motor circuit. Therefore, there is a need for another device in a motor circuit to provide overload protection. The purpose of the motor overload protector is to provide the motor and the motor circuit with this overload protection. **See Figure 11-3.**

Relays, Contactors, and Starters

The difference between a relay and a contactor is the size of the load. Relays are rated for small loads (up to 15 A or 20 A) while contactors are rated for 40 A or more. The difference between a contactor and a starter is that contactors do not provide overload (OL) protection. Starters provide OL protection within the same enclosure as the contactor. A combination starter contains a disconnect switch, fuse or circuit breaker, contactor, and OL protective device within the same enclosure.

Overload currents are any currents exceeding the rated current design of the motor. If an overload persists for a long enough period of time, it can cause damage to the insulation on the motor windings and the conductors supplying the motor. Section 430.6 requires that the nameplate current be used to determine the size of the motor overload protection.

Section 430.32(A)(1) requires that the overload protective device for a motor more than 1 HP which is rated for continuous duty be selected by multiplying the nameplate current by the following percentages:

- Nameplate service factor not less than 1.15 – 125%
- Nameplate temperature rise not more than 40°C – 125%
- All other motors – 115%

The percentage used is dependent upon the motor nameplate rating for service factor or temperature rise. Over 90% of the motors installed today use the 125% factor. If the service factor and temperature rise are not marked on the nameplate, multiply the nameplate current by 115%. **See Figure 11-4.**

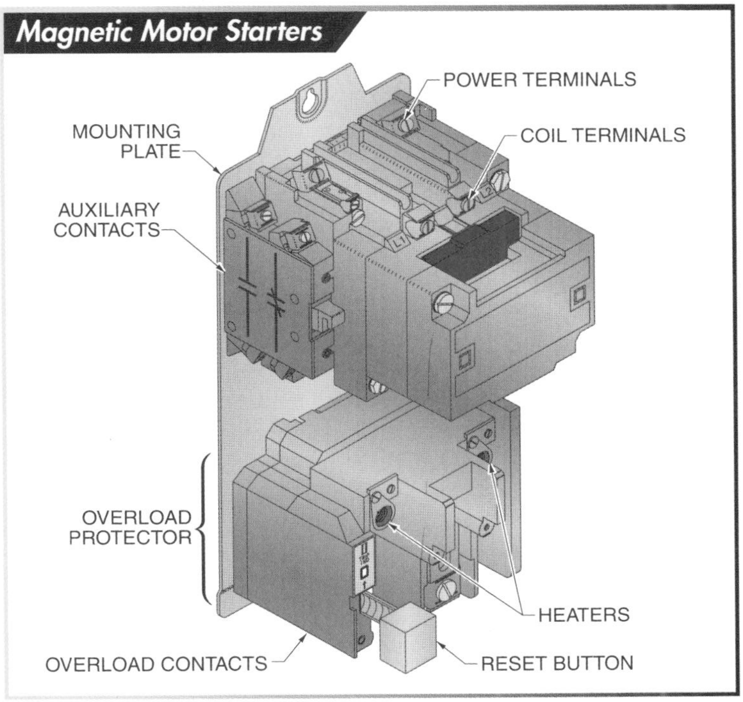

Figure 11-3. The motor overload protector provides the motor and the motor circuit with overload protection.

Section 430.32(A)(2) permits a thermal protector that is an integral part of the motor (the thermal sensor is inside the motor) to protect continuous-duty motors rated more than 1 HP. The nameplate shall indicate that the motor is thermally protected. The motor FLC used to size thermal protectors shall come from the FLC tables and not the nameplate. The table current must be increased by one of the following percentages:

Table FLC Rating	% of Increase
Not exceeding 9 A	170
9.1 A through 20 A	156
More than 20 A	140

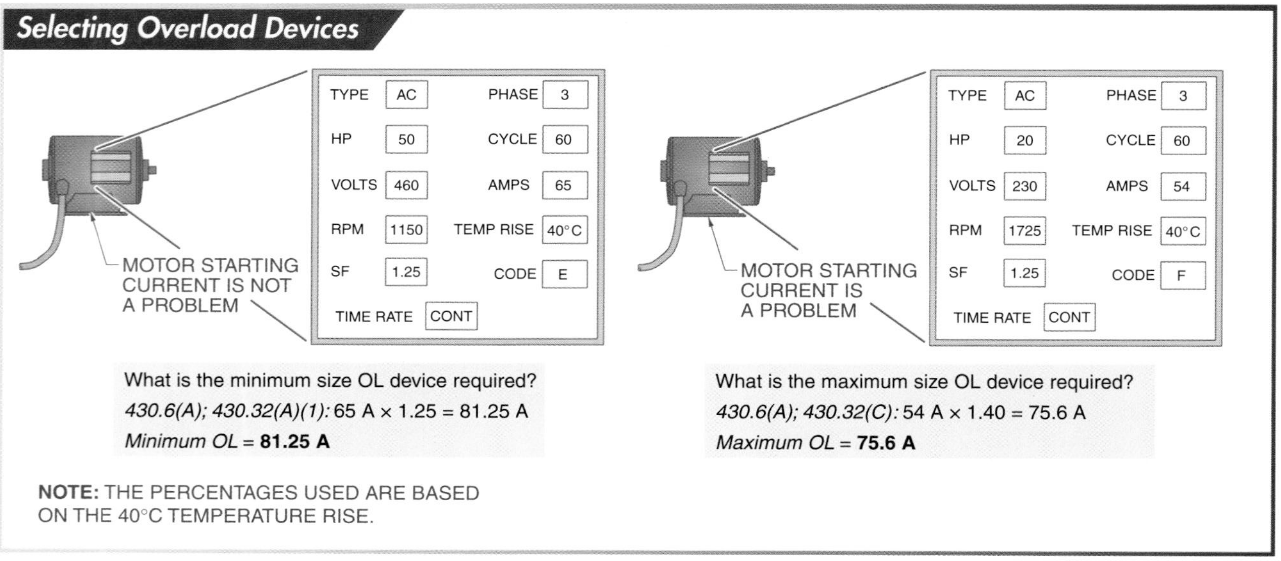

Figure 11-4. The motor nameplate current shall be used to determine the overload protection.

The larger a motor's full-load table current, the less the percentage of the increase in sizing the thermal protector that is an integral part of the motor. Section 430.32(C) allows the next higher size overload relay to be used if the one selected per 430.32(A)(1) or 430.32(B)(1) does not permit the motor to start or carry its load. In going to the next larger size, the following percentage of motor nameplate FLC ratings shall not be exceeded:

- Motors with service factor not less than 1.15% –140%
- Motors with temperature rise not more than 40°C –140%
- All other motors – 130%

The class ratings of overload relays are Classes 10, 20, and 30 per 430.32(C), Informational Note. An overload of 40% requires a certain period of time to generate the necessary heat to melt the eutectic solder and open the OL contacts. An overload of 90% requires less time to generate the same amount of heat, so it responds faster. The more severe the overload, the less time it takes for the OL contacts to open. A Class 20 OL relay takes longer for its solder to melt than a Class 10 OL relay for the same amount of current. A Class 30 OL relay carries the same current for a longer period of time than a Class 20 OL relay.

Sizing Fuses and Circuit Breakers. Section 430.52(B) requires that the motor branch-circuit, short-circuit, and ground-fault protection (fuse or circuit breaker) be large enough to carry the motor starting current. **See Figure 11-5.** Section 430.52(C)(1) requires this protective device have a rating or setting not exceeding the value calculated using the percentage of full-load current from Table 430.52. Note that two factors must be known to apply Table 430.52. The first is the type of motor and the second is the type of overcurrent device providing the protection.

Table 430.52 provides the "Maximum Rating or Setting of Motor Branch-Circuit Short-Circuit and Ground-Fault Protective Device." The general rule from 430.52(C)(1) requires a protective device with a rating or setting not exceeding the value calculated per the values in Table 430.52 to be used. If, when selecting the motor branch-circuit fuse or circuit breaker, the calculation made per Table 430.52 does not result in a standard size, the next lower size shall be selected. All numbers shall be rounded down in size, not up. Section 430.52(C)(1), Ex. 1 states that, if the values of the fuse or circuit breaker determined by using Table 430.52 do not correspond to standard sizes listed in 240.6, the next higher standard size shall be permitted. *Note:* The NEC® suggests rounding down, but permits rounding up. Note that the permitted percentage of full-load current is a function of the response time of the overcurrent device. Faster operating devices require a larger percentage for the motor to start.

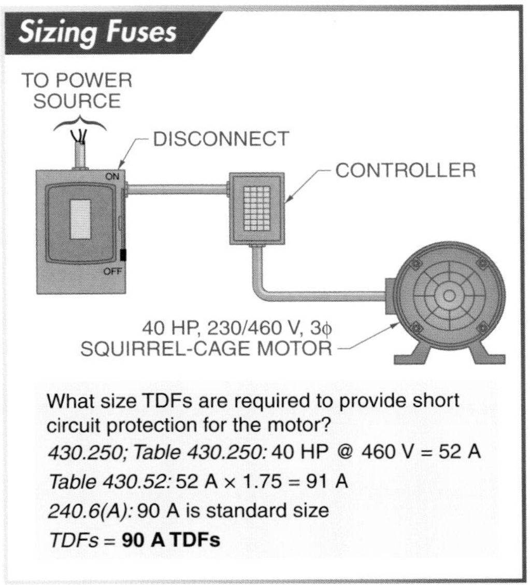
Sizing Fuses
TO POWER SOURCE
DISCONNECT
CONTROLLER
ON
OFF
40 HP, 230/460 V, 3ϕ
SQUIRREL-CAGE MOTOR
What size TDFs are required to provide short circuit protection for the motor?
430.250; Table 430.250: 40 HP @ 460 V = 52 A
Table 430.52: 52 A × 1.75 = 91 A
240.6(A): 90 A is standard size
TDFs = 90 A TDFs

Figure 11-5. The motor protection shall be large enough to carry the motor starting current.

Sizing Disconnects. The ampere rating of the fusible disconnect is determined by the size of fuse being used in the motor branch circuit. Section 430.110(A) requires that motor disconnects installed in motor circuits rated 600 V or less have an ampere rating of at least 115% of the motor's FLC from the applicable FLC table.

Sizing Feeder Conductors with More than One Motor – 430.24. When there is a common feeder supplying several motor branch circuits, the feeder conductor shall be sized per 430.24, and the short-circuit, ground-fault protection of the feeder shall be sized according to 430.62. **See Figure 11-6.**

Conductors supplying several motors or other loads have four conditions that must be added together to determine the final ampacity.

(1) 125% of the full-load current rating of the highest rated motor

(2) The sum of the full-load current ratings of all of the other motors in the group

(3) 100% of the noncontinuous non-motor loads

(4) 125% of the continuous non-motor load

The FLC ratings shall come from the appropriate FLC table. Similar to the branch circuit for a metallic raceway that is an approved EGC, all circuit conductors are the same size. Thus, Chapter 9, Informative Annex C may be used to determine the appropriate size raceway. See 430.24, Ex. 1–3 for examples of highly specialized group motor applications that permit alternative motor conductor sizing.

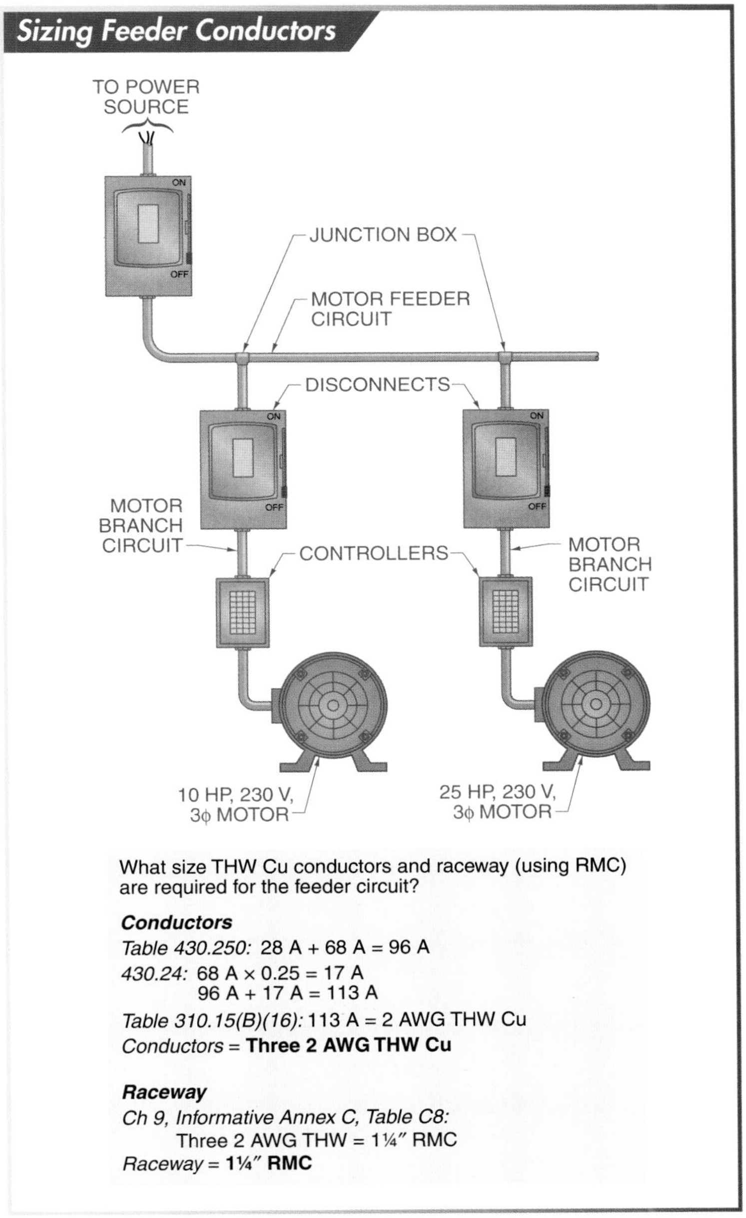
Sizing Feeder Conductors
TO POWER SOURCE
ON
OFF
JUNCTION BOX
MOTOR FEEDER CIRCUIT
DISCONNECTS
MOTOR BRANCH CIRCUIT
CONTROLLERS
MOTOR BRANCH CIRCUIT
10 HP, 230 V, 3ϕ MOTOR
25 HP, 230 V, 3ϕ MOTOR
What size THW Cu conductors and raceway (using RMC) are required for the feeder circuit?
Conductors
Table 430.250: 28 A + 68 A = 96 A
430.24: 68 A × 0.25 = 17 A
96 A + 17 A = 113 A
Table 310.15(B)(16): 113 A = 2 AWG THW Cu
Conductors = Three 2 AWG THW Cu
Raceway
Ch 9, Informative Annex C, Table C8:
Three 2 AWG THW = 1¼″ RMC
Raceway = 1¼″ RMC

Figure 11-6. Feeder conductors supplying several motors shall have an ampacity based on 430.24(1–4).

The motor branch circuit, short-circuit, and ground-fault protection (fuse or circuit breaker) shall be large enough to carry the motor starting circuit.

Sizing Feeder Fuses or Circuit Breakers – 430.62(A). The feeder shall be provided with an overcurrent protective device with a rating or setting that does not exceed the value calculated by adding the ampere rating or setting of the largest branch-circuit overcurrent device supplied by the feeder to the sum of the full load currents of the other motors of the group. **See Figure 11-7.** Where two or more branch-circuit overcurrent devices are the largest rated devices, only one device shall be used in the calculation. The remaining devices will have their associated motors included in the sum of the full load currents part of the calculation. Section 430.62(A) does not have an exception that allows the size of the OCPD to exceed the calculated value where the calculation does not correspond to a standard rating found in 240.6.

Motor Control Circuits – Article 430, Part VI

A *motor control circuit* is the circuit of a control apparatus or system which carries electric signals directing the performance of the controller, but does not carry the main power current. See 430.71. The motor control circuit is the circuit that controls the operation of the magnetic coil within the controller itself. Within this circuit are pilot devices and indicating devices. A pilot device is a sensing device that controls the motor controller. An indicating device is a pilot light, buzzer, horn, or other type of alarm. Often, the wiring of a motor control circuit is complex and requires 10 times the amount of wiring as the motor power circuit.

Overcurrent Protection – 430.72. The motor control circuit includes all of the pilot devices, indicating equipment, and magnetic coils within the circuit. The main concern of Article 430 regarding the control circuit is the protection of these items, the conductors, and the control circuit transformer.

Motor control circuits are permitted to receive their power from one of two sources: (1) the control circuit may be tapped from the load side of the motor branch-circuit fuse or circuit breaker, or (2) the control circuit may be supplied from a separate source. When the control circuit is supplied from a separate source, the circuits shall be protected against overcurrents per 725.43 or the notes to Table 11(A) and Table 11(B) in Chapter 9, as applicable. Section 430.72(B) requires that the overcurrent device for conductors shall not exceed the values shown in Table 430.72(B). Table 430.72(B) consists of three columns that address specific overcurrent protection applications.

The NEC® does not require that the control circuit conductors have an ampacity large enough to carry the current required of a motor circuit. This current is too small to be significant as most magnetic starter coils pull less than ½ A. Section 430.72 does, however, require the control circuit conductors to be protected in accordance with their ampacity. In some cases, the overcurrent device rating may be higher than the ampacity of the conductor.

When a control circuit is tapped from the load side as permitted by 430.72(A), it is not considered a branch circuit and is considered protected by either a supplementary or branch-circuit overcurrent device. Other motor control circuits may be derived from a panelboard or a control transformer. In this case, the control conductors shall be protected per 725.43.

Separate overcurrent protection is required by 430.72(B)(1) where the motor branch-circuit OCPD does not provide protection in accordance with 430.72(B)(2). The value of the separate overcurrent protection shall not exceed the values specified in Column A of Table 430.72(B).

Section 430.72(B)(2) provides two cases where the motor branch-circuit OCPD can be used to protect control circuit conductors. In the first case, where the conductors do not extend beyond the motor control equipment enclosure, the rating of the OCPD shall not exceed the value specified in Column B of Table 430.72(B). The second case requires that the rating of the OCPD not exceed the value found in Column C of Table 430.72(C) where the control circuit conductors extend beyond the motor control equipment enclosure.

Section 430.72(B), Ex. 2 permits control circuit conductors supplied from a single-phase transformer having a two-wire, single-voltage secondary to be protected by the OCPD on the primary side of the transformer. In order to use this permission, the value of the primary OCPD cannot exceed the value determined by multiplying the appropriate maximum rating of the OCPD for the secondary conductor from Table 430.72(B) by the secondary-to-primary voltage ratio. **See Figure 11-8.**

Section 430.72(B), Ex. 1 requires short-circuit, ground-fault protection for control circuit conductors that regulate circuits which would cause a hazard if they disconnected on overload, e.g., the control circuit for a fire pump.

Section 430.72(C) permits five methods of providing overcurrent protection for motor control circuit transformers. The first method is for Class 1 power-limited circuits or Class 2 or Class 3 remote-control circuits that are installed in accordance with the provisions of Article 725. Section 430.72(C)(2) permits overcurrent protection per 450.3. This is accomplished where secondary protection is not provided by a primary-side protective device rated 125%, 167%, or 300% of the primary current.

Section 430.72(C)(3) eliminates the requirement of protecting the control transformer if it is rated 50 VA or less, provided it is part of the motor controller and located within the controller enclosure. The intent is that the protection is provided by the primary overcurrent devices.

Section 430.72(C)(4) permits a control transformer with a rated primary current of 2 A or less to be considered protected at up to 500% of the rated primary current. The last option for motor control circuit transformer protection is for any protection scheme that is approved.

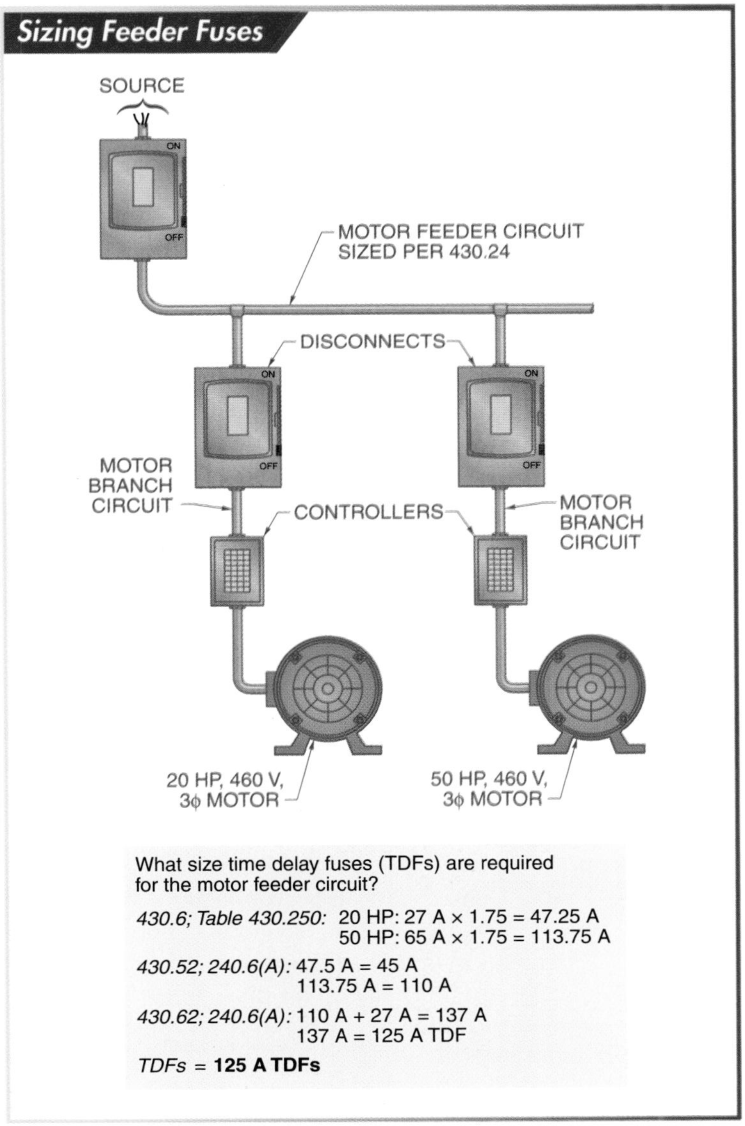

Figure 11-7. Fuses for feeder conductors supplying several motors are sized by adding the largest fuse of the branch circuit and the FLC of the other motors.

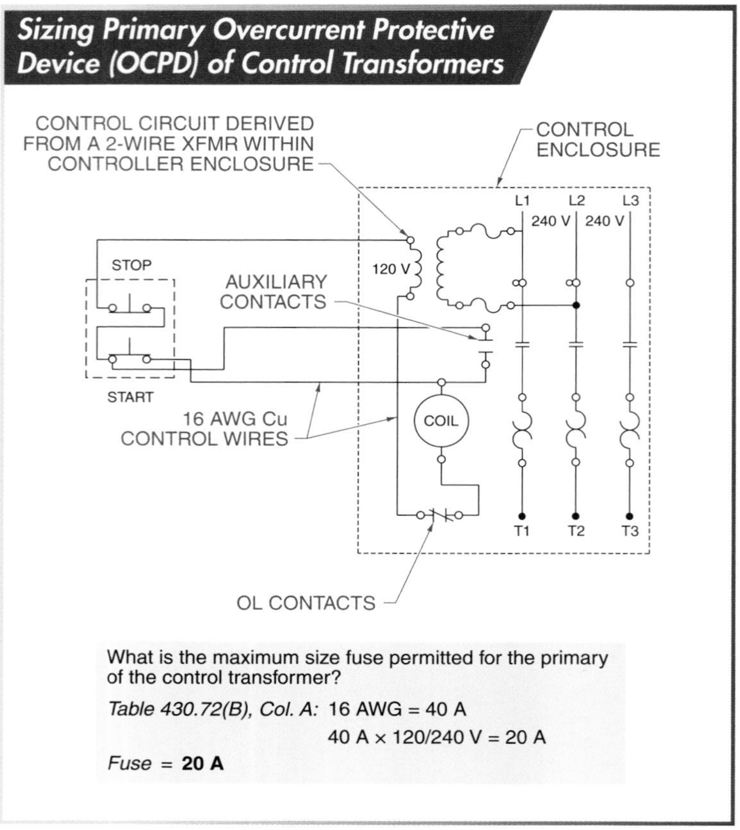

Figure 11-8. The primary OCPD of control transformers is sized per Table 430.72(B), Column A.

Motor Control Circuit Arrangement – 430.74. To ensure that personnel are adequately protected against accidental motor startup, 430.74 requires that, where one conductor of a motor control circuit is grounded, the motor control circuit must be arranged such that an accidental ground in the control circuit does not result in the motor starting and does not bypass any automatic or manual safety devices designed to shut down the motor.

Control Circuit Disconnects – 430.75. When the disconnect to the motor supply is opened, the supply to the motor control circuit shall also be opened. This disconnecting means is permitted to consist of two or more separate devices to accomplish the disconnecting of the motor power supply and the supply to the control circuit. Where separate devices are installed, they shall be located immediately adjacent to one another. There are two exceptions to this general rule.

Section 430.75(A), Ex. 1 pertains to motor control circuits that contain more than 12 motor control circuit conductors which are required to be disconnected. This is usually found in industrial locations with complex control circuits for multimotor machinery. The control circuit disconnect means are permitted at locations that are not immediately adjacent to each other provided the following two conditions are met: (A) Access to the energized parts is limited to qualified persons only. (B) A warning sign is permanently located on each equipment enclosure door or cover that permits access to the live parts of the control circuit. This sign shall specify the location and identification of each of the disconnects.

Section 430.75(A), Ex. 2 states that if the opening of the control disconnect means could result in an unsafe condition for personnel or property, then the control circuit disconnecting means may be located where they are not immediately adjacent to each other provided the conditions of (a) and (b) of Ex. 1 are met. When the control circuit transformer is mounted within the controller enclosure, it shall be connected to the load side of the disconnecting means for the motor control circuit.

Motor Controllers – Article 430, Part VII

The *controller* is the device in a motor circuit which turns the motor ON or OFF. Controllers are also known as motor starters. See Article 100 and 430.81(A). The two basic types of controllers are manual and magnetic. A manual controller requires that someone physically opens the contacts by pressing a pushbutton or throwing a switch. A magnetic controller operates when the magnetic coil is energized. The sensing devices placed into the control circuit that controls the magnetic coil circuits are the pilot devices. Some sensing devices, such as a float switch, can act as a pilot device or a controller, depending on their function in the circuit. If the contacts within the float switch control the magnetic coil of the controller, the float switch is considered a pilot device. **See Figure 11-9.** If the contacts of the float switch turn the motor ON and OFF, the float switch is considered a controller.

Part VII of Article 430 requires a suitable controller for all motors. In the case of a stationary motor that is ⅛ HP or less and its winding impedance is high enough to prevent damage to the motor when the rotor is in a locked position, 430.81(A) permits the branch-circuit disconnecting means to act as the controller. Section 430.81(B) permits the attachment plug and receptacle or cord connector to act as the controller for portable motors rated ⅓ HP or less. **See Figure 11-10.**

A manual controller requires that someone physically opens or closes the contacts.

Ratings – 430.83. The motor controller shall be rated according to the motor's application voltage, horsepower size, and type of installation per 430.83. **See Figure 11-11.**

Section 430.83(A) provides three general rules for controller ratings. Section 430.83(A)(1) requires all controllers other than inverse-time CBs and molded case switches to have HP ratings at the application voltage not less than the HP rating of the motor. Section 430.83(A)(2) permits a branch-circuit inverse-time circuit breaker, rated in amperes, to serve as a controller for all motors.

Section 430.83(C) allows a general-use switch with an ampere rating twice that of the motor FLC rating to be used as a controller provided it controls a stationary motor rated 2 HP or less and 300 V or less.

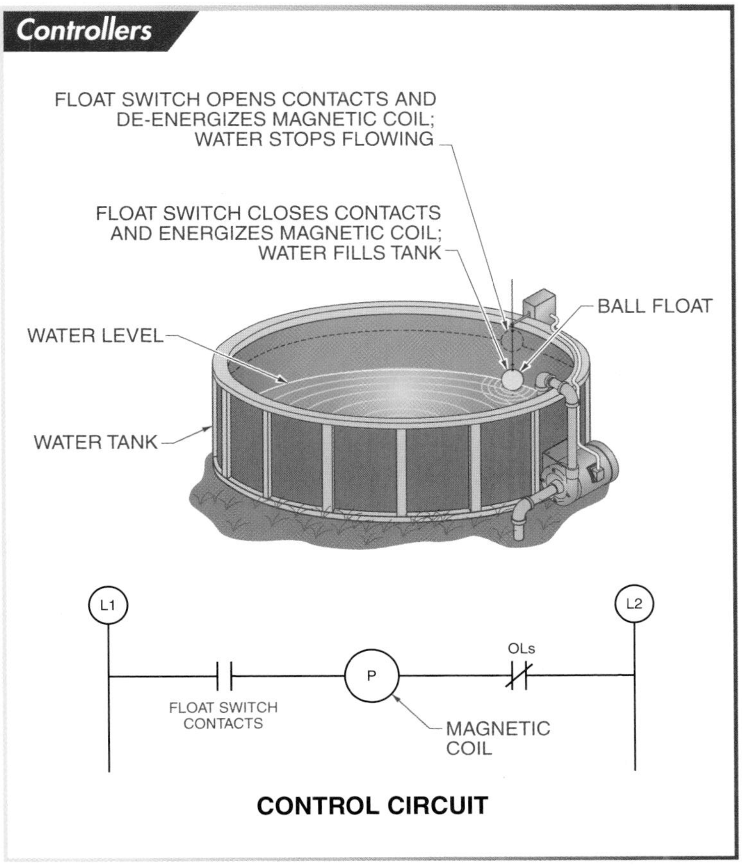

Figure 11-9. The controller is the device in a motor circuit that turns the motor ON or OFF.

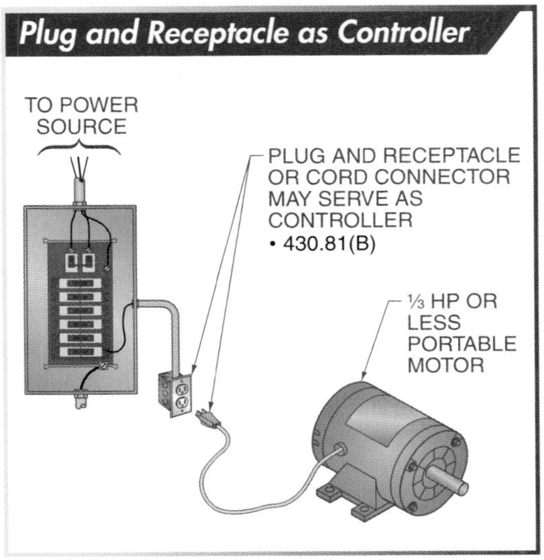

Figure 11-10. A plug and receptacle or cord connector may serve as a controller for portable motors up to ⅓ HP.

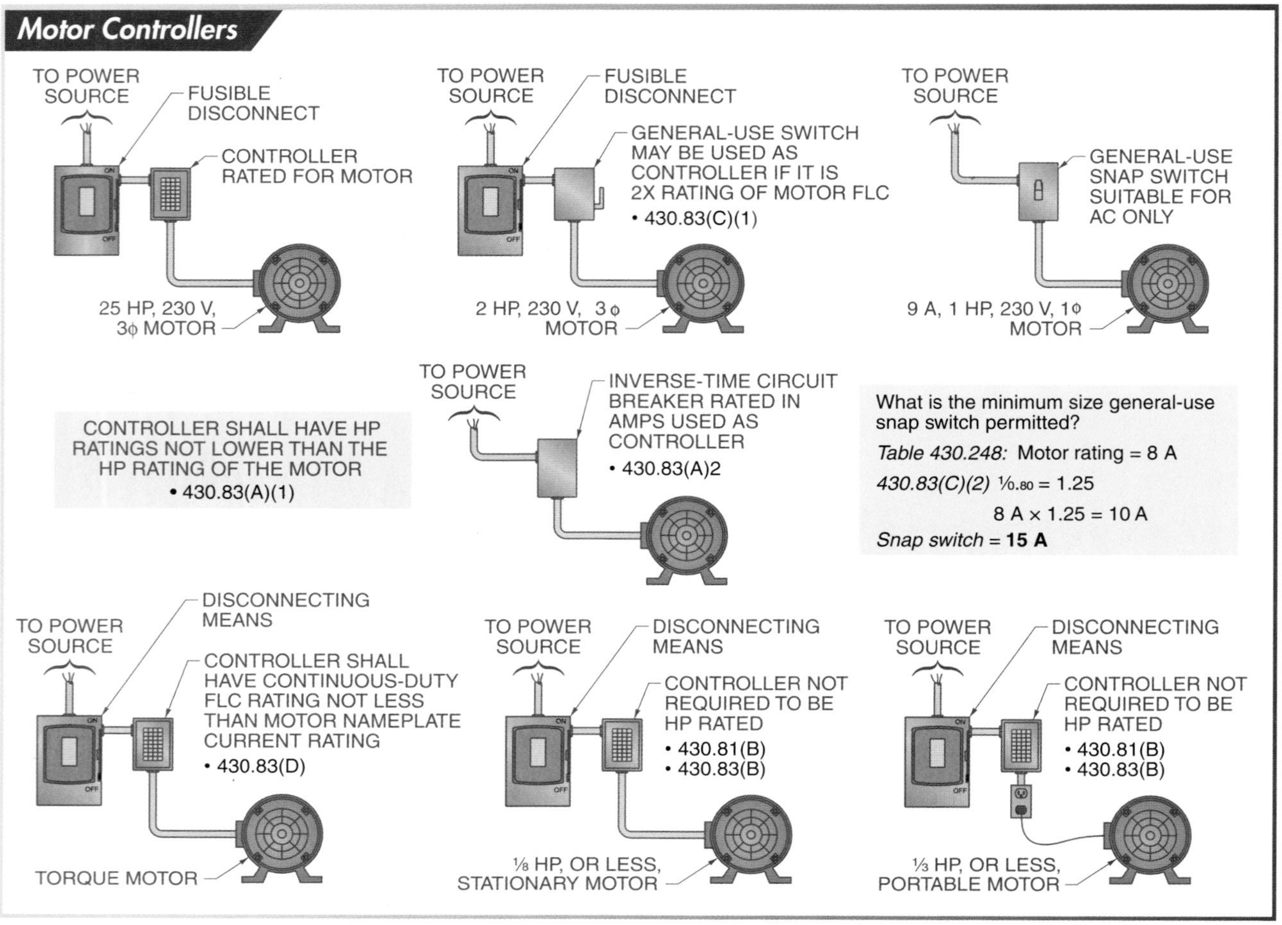

Figure 11-11. The motor controller shall be rated at the same HP rating, or greater HP rating than the motor itself.

On AC circuits, a general-use snap switch suitable for AC only shall be permitted to control a motor whose rating is 2 HP or less and 300 V or less, provided the current rating of the motor is not more than 80% of the ampere rating of the switch.

Section 430.83(D) requires that a torque motor controller shall have a continuous-duty, FLC rating not less than the nameplate current rating on the motor.

Section 430.83(B) allows devices permitted by 430.81(A) for stationary motors of ⅛ HP or less, and 430.81(B) for portable motors of ⅓ HP or less, to serve as controllers.

Section 430.83(E) requires that controllers identified with a slash voltage rating only be used on electrical systems in which the voltage-to-ground does not exceed the lower voltage rating marked on the controller and that the line voltage (phase-to-phase voltage) of the system is not greater than the higher voltage rating marked on the controller.

Opening of Conductors – 430.84. The controller is not required to open all conductors supplying power to the motor, provided the motor is stopped. If the controller is also serving as the disconnecting means, it shall open all of the ungrounded conductors to the motor.

Number of Motors Served by Each Controller – 430.87. Each motor shall be provided with its own controller. The exception to this rule permits a single controller to serve several motors provided the motors are rated 600 V or less and the controller has an equivalent horsepower rating as determined per 430.110(C)(1) for all of the motors in the group. This exception is permitted for any one of the following three conditions: (1) where all of the motors are part of a single machine or piece of apparatus, (2) where all of the motors are protected by one overcurrent device, and (3) where the groups of motors are located in a single room within sight from the controller location. Per 430.87, Ex. No. 2, if the controller is the branch-circuit disconnecting means, as permitted in 430.81(A), the branch-circuit disconnecting means (controller) shall be permitted to serve more than one motor.

Motor Control Centers – Article 430, Part VIII

A *motor control center (MCC)* is an assembly of one or more enclosed sections with a common power bus and primarily containing motor control units. See the definition of Motor Control Center in Article 100. An MCC is a large enclosure containing many sections or compartments. Within these enclosures are disconnect switches, starters, relays, contactors, fuses, circuit breakers, and any other devices found in a motor or motor control circuit. Section 430.1, Informational Note No. 1 explains that MCCs must comply with 110.26(E). Section 110.26(E) ensures that a dedicated equipment space is provided for all indoor installations. The space must extend 6′ above or to the structural ceiling to permit access to the MCC.

MCC Overcurrent Protection – 430.94. An MCC shall have its overcurrent protection based on the rating of the common power bus within the MCC. This overcurrent protection shall be provided in one of the following two locations: (1) ahead of the MCC, (2) a main overcurrent device mounted within the MCC.

MCC as Service Entrance Equipment – 430.95. An MCC is permitted to be used as service equipment provided it is equipped with a single main disconnecting means that is capable of disconnecting all ungrounded service conductors. The exception to this rule permits a second service disconnect to supply additional loads and equipment. Where a grounded conductor is provided, the MCC shall be equipped with a main bonding jumper within one of the sections. The purpose of this bonding jumper is to connect the grounded conductor on its supply side to the control center equipment ground bus. This main bonding jumper shall be sized per 250.28(D).

Grounding of MCC – 430.96. MCCs consisting of more than one section shall be bonded together with an EGC or equivalent equipment grounding bus sized per Table 250.122. The EGC shall terminate on this grounding bus or, in the case of a single-section MCC, to the grounding termination it provides.

Busbars and Conductors within MCC – 430.97. The busbars within an MCC shall be protected from physical damage and be secured firmly in place. Except for control wiring and any required interconnections, the only conductors permitted for termination in a vertical section are those conductors intended for use in that section. The exception to this rule permits conductors to travel horizontally through vertical sections provided they are isolated from the busbars by a suitable barrier. When viewing the busbars from the front of the MCC, 430.97(B) requires the busbars to be arranged A, B, C, from front to back, top to bottom, and left to right.

Marking of MCC – 430.98. MCCs shall be marked with the manufacturer's name, trademark, or other identification which indicates the organization responsible for the product. The markings shall include voltage rating, common power bus current rating, and MCC short-circuit rating. Section 110.21 provides the general guidelines for marking of all electrical equipment.

Saftronics, Inc.

Motor control centers contain enclosures that house devices such as disconnect switches, starters, relays, contactors, fuses, and circuit breakers.

Disconnecting Means – Article 430, Part IX

The general intent of Article 430 regarding the motor disconnecting means is that it be rated in HP corresponding to the motor it serves. The ideal location for the disconnecting means is readily accessible and in sight from the controller, the motor, and the driven machinery location.

Location of Disconnecting Means – 430.102. The disconnecting means shall be installed within sight from the controller location. The specified equipment therefore is required to be visible and not more than 50′ apart.

Section 430.102(A), Ex. 1 permits the disconnecting means for a controller of a motor circuit over 600 V to be located out of sight of the controller. The disconnecting means shall be capable of being locked in the open position. The controller shall be marked with a warning label indicating the location of the disconnecting means.

Per 430.102(A), Ex. 2, a single disconnecting means is permitted for a group of coordinated controllers provided the controllers drive a single machine or piece of equipment and the disconnect is located within sight of the machine or piece of equipment controlled.

The disconnecting means for a motor shall be installed within sight from the motor and the driven machinery per 430.102(B)(1). Section 430.102(B)(2) permits the controller disconnecting means required by 430.102(A) to serve as the disconnecting means for the motor, provided it is in sight of the motor and the driven machinery location.

Section 430.102(B), Ex. permits the required motor disconnect to be installed out of sight of the motor and drive machinery under the following two conditions: (1) where the location of the disconnect is impractical or introduces additional/increased hazards, or (2) in industrial installations with written safety procedures, where conditions of maintenance and supervision ensure that only qualified persons will service the equipment. The disconnecting means must be individually capable of being locked in the open position and the provisions for adding a lock must be installed on or at the switch.

Operation of Disconnecting Means – 430.103. The disconnecting means shall open all of the ungrounded conductors supplying the motor. No pole on the disconnecting means shall operate independently. Section 430.103 permits the disconnecting means to be located in the same enclosure with the controller but it cannot be capable of closing automatically.

Per 430.102(B)(1), a disconnecting means for a motor shall be installed within sight from the motor and the driven machinery.

Type of Disconnecting Means – 430.109. Part IX of Article 430 addresses motor disconnecting means. The motor disconnecting means is any device or group of devices that can isolate or remove the conductors of a motor circuit from their source of supply. Section 430.108 requires that all disconnecting means in motor circuits comply with the provisions of Sections 430.109 and 430.110.

Section 430.109 lists the permissible types of devices that may be used as disconnecting means in motor circuits. Section 430.109(A)(1) permits a motor circuit switch to be used, provided it is listed and rated in HP. Similarly, listed molded-case circuit breakers and mode case switches are also permitted by 430.109(A)(2) and (3). Sections 430.109(A)(4–7) list other specialized permissible types of motor circuit disconnecting means including instantaneous trip circuit breakers, listed self-protected combination controllers, manual motor controllers, and other system isolation equipment. **See Figure 11-12.**

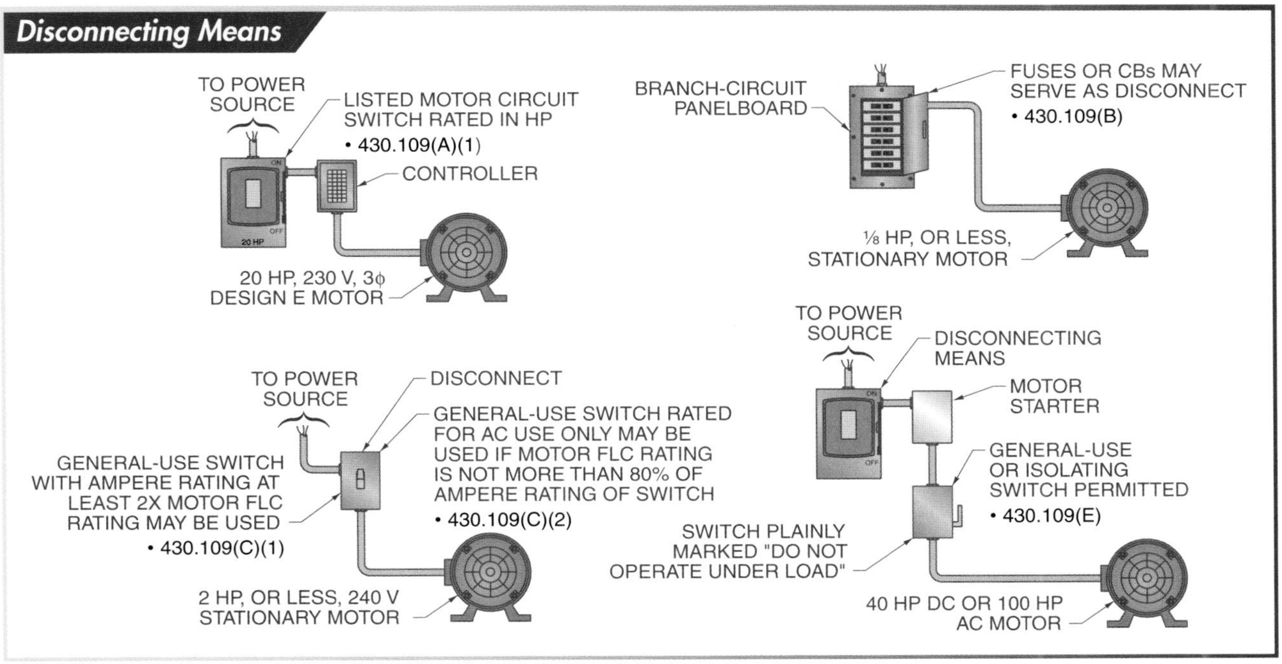

Figure 11-12. The motor disconnecting means shall be of a type listed in 430.109(A) unless the type of motor and/or conditions of use meet the requirements of 430.109(B) through (G).

Section 430.109(C) describes three permissible types of disconnecting means for stationary motors rated 2 HP or less and 300 V or less. These disconnecting means include an ampere-rated general-use switch, provided the ampere rating is at least twice the full-load current rating of the motor; a general-use switch for AC circuits only where the motor full-load current rating is less than 80% of the ampere rating of the switch; and a listed manual motor controller that has a HP rating not less than the rating of the motor. The manual motor controller must be marked and identified as being "Suitable as Motor Disconnect."

Section 430.109(D) states that for motors over 2 HP up to and including 100 HP, the disconnecting means required for a motor with an autotransformer-type controller shall be permitted to be a general-use switch if all of the following three provisions are met:

(1) The motor drives a generator which is provided with overload protection.

(2) The controller is capable of interrupting the LRC of the motor and is provided with no-voltage release (known in the field as 2-wire control) and has running overload protection not exceeding 125% of the motor FLC rating.

(3) Separate fuses or an inverse-time circuit breaker rated or set at not more than 150% of the motor FLC rating are provided in the motor branch circuit.

Section 430.109(E) permits DC stationary motors rated more than 40 HP or AC motors rated more than 100 HP to use a general-use or isolating switch where the switch is plainly marked, "Do not operate under load." An isolating switch has no interrupting rating and is not intended to be operated while the circuit is live.

For cord-and-plug-connected motors, 430.109(F) permits an HP-rated attachment plug, flanged surface inlet and receptacle, or cord connector with a rating equal to or greater than the motor rating to serve as the motor disconnecting means. The attachment plug, flanged surface inlet and receptacle, or cord connector are not required to be HP rated for cord-and-plug-connected appliances per 422.33, room A/C units per 440.63, or for portable motors rated ⅓ HP or less.

Energy-Efficient Motors

Energy-efficient motors use less energy than conventional motors because they are manufactured with higher-quality materials and techniques. Energy-efficient motors have high service factors and bearing lives and less waste-heat output and vibration than conventional motors. This reliability is often reflected by longer manufacturer warranties.

The Energy Policy Act (EP Act) of 1992 requires most general-purpose motors between 1 HP and 200 HP for sale in the United States to meet National Electrical Manufacturers Association (NEMA) standards. To be considered energy efficient, motor performance must equal or exceed the nominal full-load efficiency values provided by NEMA standard MG-1. The EP Act also provided grandfather protection to existing motors.

In 2001, NEMA created the NEMA Premium® designation to distinguish motors that are more efficient than those required by EP Act standards. The NEMA Premium® designation applies when motors meet the following criteria:

- single-speed
- polyphase
- 1 HP to 500 HP
- 2-, 4-, and 6-pole (3600 rpm, 1800 rpm, and 1200 rpm) squirrel-cage induction
- NEMA designs A or B
- 600 V or less (5 kV or less for medium voltage)
- continuous rated

The Consortium for Energy Efficiency (CEE), a nonprofit organization that includes many electric utilities among its members, recognizes NEMA Premium® motors up to 200 HP as meeting their criteria for possible energy-efficiency rebates.

Additional information is available at www.nema.com.

Section 430.107 requires that at least one of the motor disconnecting means be readily accessible.

A general-use switch is permitted as a disconnecting means for torque motors per 430.109(G). An ITCB which is part of a listed combination starter is permitted to serve as the disconnecting means per 430.109(A)(4).

Ampere Rating and Interrupting Capacity of Disconnecting Means – 430.110. If a disconnect is rated in HP, the size of the fuse is used to determine the size of disconnect to be used. The general rule found in 430.110(A) requires the disconnecting means for motor circuits to have an ampere rating of at least 115% of the FLC rating of the motor for circuits rated 600 V or less. **See Figure 11-13.**

The disconnecting means for torque motors shall have a rating of at least 115% of the motor nameplate current per 430.110(B). Section 430.110(C) requires that where two or more motors are used together, or one or more motors and other loads are served by the same disconnect, the ampere and HP ratings of the combined load shall be determined by one of the following three methods:

(1) The rating of the disconnect is determined from the sum of all currents at the full-load and at locked rotor conditions.

(2) The ampere rating of the disconnect shall not be less than 115% of all currents at the full-load condition.

(3) For small motors not shown in Tables 430.247 through 430.250, the LRC shall be assumed to be six times the nameplate FLC.

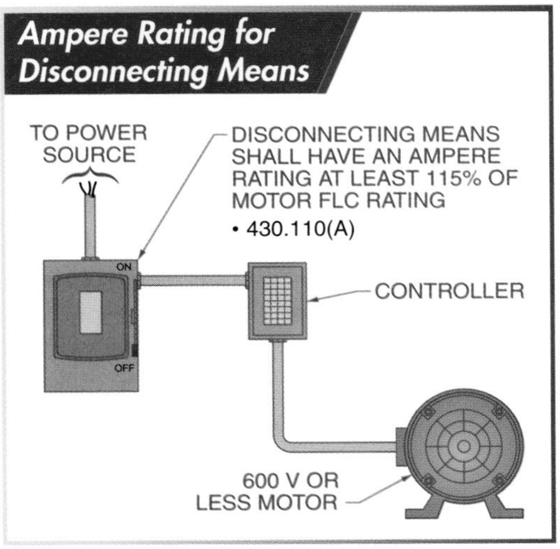

Figure 11-13. The disconnecting means for motor circuits shall have an ampere rating of at least 115% of the motor FLC rating.

Switch or Circuit Breaker as Controller and Disconnecting Means – 430.111(B)(1–3). The controller is permitted to be the disconnect means where it consists of a manually operable air-break switch, ITCB, or an oil switch with a rating which does not exceed 100 A or 600 V. The switch or circuit breaker used as a controller shall be protected by branch-circuit, short-circuit, and ground-fault protective devices which open all ungrounded conductors.

Controller Ratings

Always check controller ratings carefully when using with dual-voltage motors. HP ratings on electrical equipment vary depending on the application voltage. For example, a 230/460 V, dual-voltage motor pulls twice the current at the lower voltage. Therefore, it has a smaller HP rating at the lower voltage. A controller might be rated for 10 HP at 460 V but only 7½ HP at 230 V.

Motors Served by a Single Disconnecting Means – 430.112. Each motor shall have its own disconnecting means. The exception to this general rule permits one disconnecting means to serve more than one motor. **See Figure 11-14.** The single disconnecting means must be rated per 430.110(C).

The exception has three conditions:

(1) A number of motors are used on a single machine or piece of apparatus.

(2) More than one motor is protected by one set of branch-circuit protective devices per 430.53(A).

(3) More than one motor is in the same room within sight from the disconnecting means.

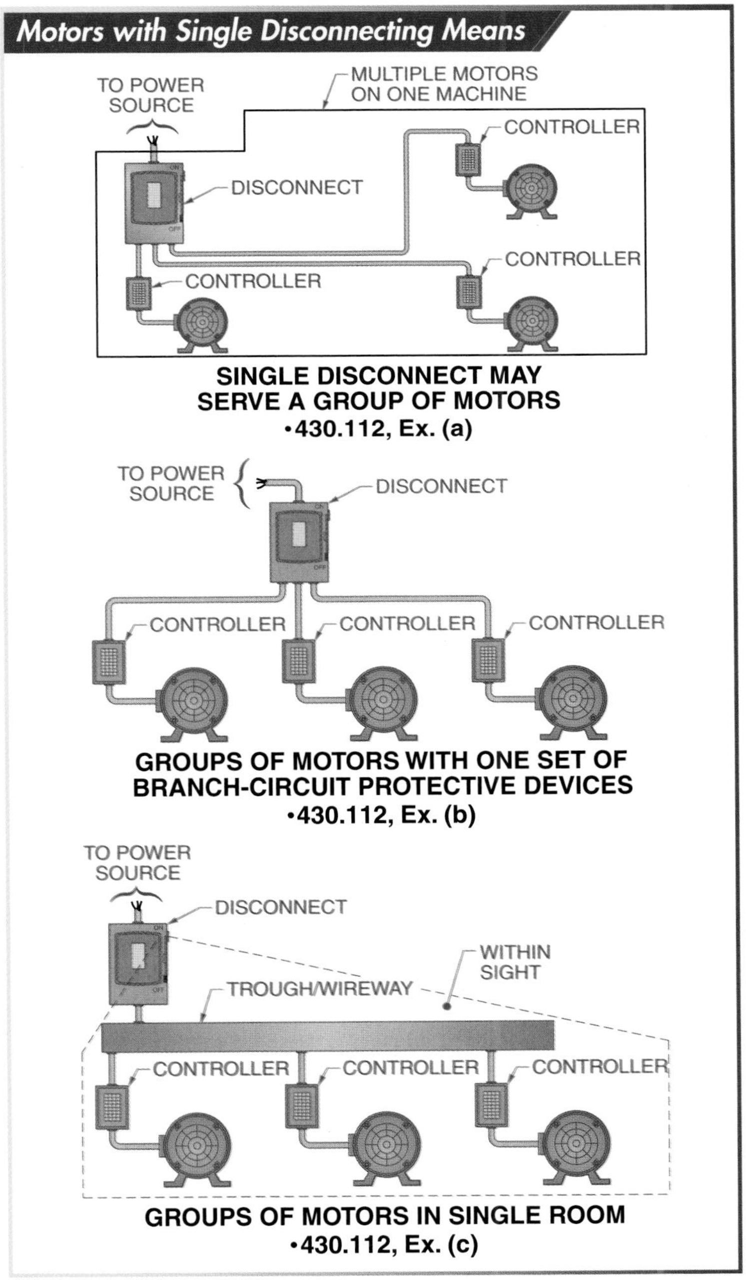

Figure 11-14. One disconnecting means may serve more than one motor.

Disconnecting Means for Energy from More than One Source – 430.113. A motor and motor-operated equipment with more than one source for receiving electrical power shall have a disconnecting means for each source of power adjacent to the equipment being served. There are two exceptions to this general rule.

Per 430.113, Ex. 1, when a motor receives power from more than one source, the disconnecting means for the main power supply to the motor is not required to be immediately adjacent to the motor, provided the disconnecting means for the controller is capable of being locked in the open position. Per 430.113, Ex. 2, a separate disconnecting means is not required for a remote control circuit conforming with Article 725 rated not more than 30 V which is isolated and ungrounded.

Other Motors

Over 90% of the motors in use today are single speed, AC induction motors. Multispeed motors, duty-cycle motors, wound rotor motors, and synchronous motors are commonly used in industry.

Multispeed Motors – 430.22(B). The branch-circuit conductors on the line side of the controller for a multispeed motor shall be sized at 125% of the highest of the FLC ratings shown on the motor nameplate. The branch-circuit conductors on the load side of the controller shall be sized at 125% of the current rating of the winding that these conductors supply. **See Figure 11-15.**

Multispeed Motors

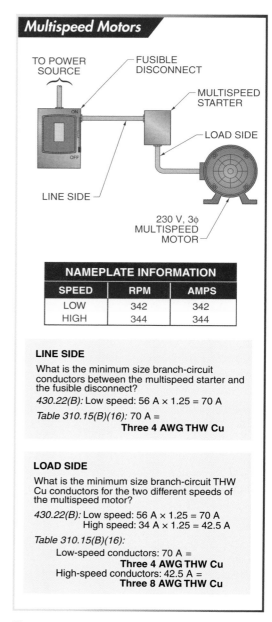

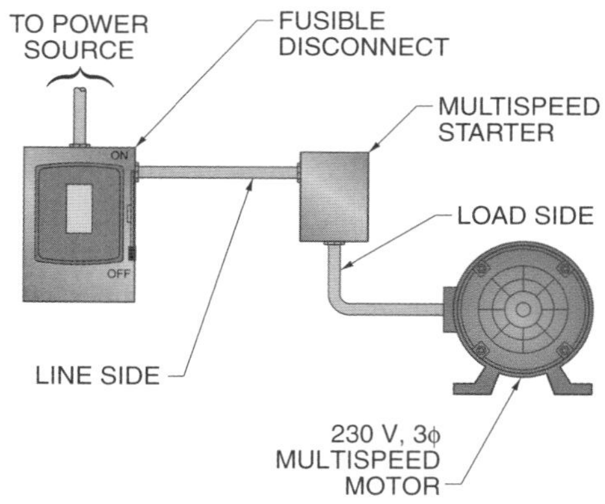

NAMEPLATE INFORMATION		
SPEED	**RPM**	**AMPS**
LOW	342	342
HIGH	344	344

LINE SIDE

What is the minimum size branch-circuit conductors between the multispeed starter and the fusible disconnect?

430.22(B): Low speed: 56 A × 1.25 = 70 A

Table 310.15(B)(16): 70 A =
Three 4 AWG THW Cu

LOAD SIDE

What is the minimum size branch-circuit THW Cu conductors for the two different speeds of the multispeed motor?

430.22(B): Low speed: 56 A × 1.25 = 70 A
High speed: 34 A × 1.25 = 42.5 A

Table 310.15(B)(16):
Low-speed conductors: 70 A =
Three 4 AWG THW Cu
High-speed conductors: 42.5 A =
Three 8 AWG THW Cu

Figure 11-15. Line side branch-circuit conductors are sized at 125% of the FLC rating and load side branch-circuit conductors are sized at 125% of the current rating of the winding.

Multispeed Motors

Multispeed motors are controlled by multispeed controllers. The most popular form of multispeed motors are two-speed and three-speed motors. A two-speed, multispeed motor is wound as two different motors. Each speed has its own set of "T" leads. It may be wound as a six-pole motor for low speed or as a four-pole motor for high speed. The two-speed controller for this motor has two controllers inside the one enclosure. A single set of supply conductors on the line side of the controller has the ampacity to safely supply the motor for each speed. On the load side of the controller, two sets of conductors go to the motor, one set for each speed.

Duty-Cycle Motors – 430.22(E). Table 430.22(E) is used to size branch-circuit conductors for duty-cycle motors. In applying this table, select the classification of service and find the percentage of nameplate current according to the motor's duty-cycle rating which is on the nameplate. **See Figure 11-16.**

Duty-Cycle Motors

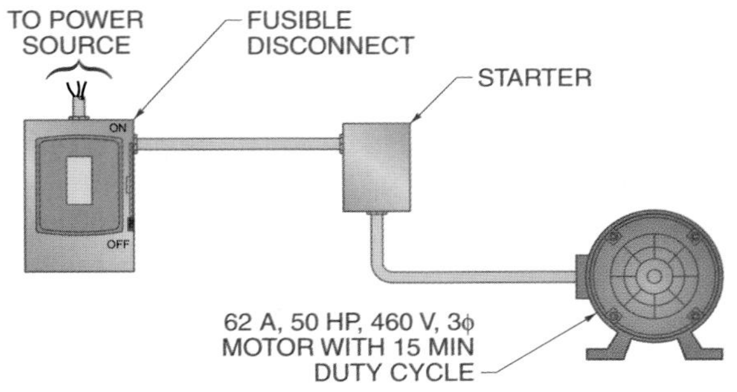

What is the minimum size THW Cu conductors required to supply this intermittent-duty motor?

430.22(E); Table 430.22(E):

15 minute rated motor = 85% of nameplate FLC
62 A × 0.85 = 52.7 A

Table 310.15(B)(16): 65 A = 6 AWG THW Cu

Conductors = **Three 6 AWG THW Cu**

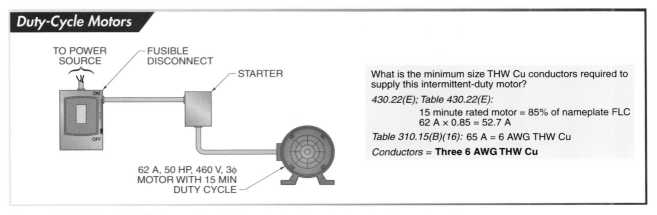

Figure 11-16. Table 430.22(E) is used to size branch-circuit conductors for duty-cycle motors.

Wound-Rotor Motors – 430.23. A wound-rotor motor is similar to a squirrel cage motor. The main difference is that there is a bank of resistors connected to the rotor through slip rings. Upon starting, all of the resistance is connected to the rotor for high-starting torque. Once the motor comes up to speed, the amount of resistance is controlled, which varies the torque and speed. These motors are seldom found in commercial installations. They are used in sewage treatment plants as lift pumps and in some crane applications. The full-load secondary current of a wound-rotor motor is found on the motor nameplate.

Per 430.23(A), for continuous-duty operation, the secondary leads between the motor and the controller shall be sized at least 125% of the secondary FLC of the motor. This rating is found on the motor nameplate. Per 430.23(B), for other than continuous duty operation, the secondary conductors may be sized less than 125% of the secondary current. These conductors shall have an ampacity, in percent of full-load secondary current, not less than that specified in Table 430.22(E). Section 430.23(C) requires the ampacity of the conductors between the controller and the resistor banks to be not less than the allowable percentages of Table 430.23(C), depending on the resistor duty classification.

Synchronous Motors. Large, 3ϕ synchronous motors start like a squirrel-cage motor. The magnetic field on the rotor is a result of the induced current flow in the rotor caused by the cutting of the rotor bars by the rotating magnetic field in the stator. When the rotor reaches 90% to 95% of its rated speed, there is little slip. As a result, the magnetic field on the rotor is too weak to overcome the bearing and windage losses to catch up with the rotating magnetic field on the stator. At this point an outside source of DC current, which is independent of induction, is sent through the rotor bars. The rotor current increases, thus strengthening its magnetic field, and the rotor locks in step with the rotating magnetic field on the stator and spins at synchronous speed. In small 1ϕ synchronous motors, the rotor is a permanent magnet.

A/C AND REFRIGERATION EQUIPMENT – ARTICLE 440

Article 440 applies to electric motor-driven air-conditioning and refrigeration equipment. It also includes the branch circuits and control equipment of these circuits. There is a great deal of similarity between Article 440 and Article 430. The main difference is the special considerations that Article 440 has for circuits supplying hermetic refrigerant motor-compressors. A *hermetic refrigerant motor-compressor* is a combination of a compressor and motor enclosed in the same housing, having no external shaft or shaft seals, with the motor operating in the refrigerant. Section 440.2 contains special definitions for terms used within Article 440. These should be reviewed carefully to ensure the provisions of Article 440 are applied correctly.

The branch-circuit selection current is marked on the equipment nameplate. This is the current used in determining the ampacity of the branch-circuit conductors, disconnecting means, and size of controller. Air-conditioning and refrigeration equipment not only contain a motor-compressor, but often a fan motor is also included. The branch-circuit selection current is always equal to or greater than the load current.

The calculations performed in Article 430 always use motor FLC. In Article 440, the calculations are based on load current or branch-circuit selection current. When the branch-circuit selection current is marked on the nameplate of a piece of A/C or refrigeration equipment, it shall be used in selecting the size of the disconnecting means, controller, branch-circuit conductors, and overcurrent protective devices.

Section 440.3 requires that the provisions of 422 (appliances) apply to such equipment as room air conditioners, household refrigerators and freezers, water coolers, and beverage dispensers. If air-conditioning and refrigeration equipment is installed in hazardous locations, motion picture and TV studios, or other similar areas, these articles shall also be followed.

Marking on Hermetic Refrigerant Motor-Compressor Nameplate – 440.4

The nameplate of a hermetic refrigerant motor-compressor shall be marked with the manufacturer's name, phase, voltage, and frequency. The rated load current of the motor-compressor shall be marked on the motor nameplate or the nameplate of the equipment in which the compressor is used. Where the motor-compressor is internally protected with a thermal device, the equipment nameplate shall be marked with the words "Thermally-Protected System." Other rated-load current markings must also be provided by the equipment manufacturer to ensure that proper branch-circuit ampacity selections can be determined in the field.

For multimotor and combination-load equipment, 440.4(B) requires that the equipment nameplate provide the manufacturer's name, voltage, frequency, number of phases, minimum supply-circuit conductor ampacity, and the maximum rating of the branch-circuit, short-circuit, and ground-fault protection device.

Marking on Controllers – 440.5

A controller suitable for a motor-compressor shall be marked with the manufacturer's name, voltage, phase, FLC and LRC or HP rating, and other data to indicate the motor-compressor for which it is suited.

Ampacity and Rating of Equipment and Compressor Motor – 440.6(A)

The FLC rating on the motor-compressor is to be used in determining the size of the branch-circuit conductor ampacity, disconnecting means, short-circuit, ground-fault protection, controller, and the motor overload protector for hermetic refrigerant motor-compressors.

If the branch-circuit selection current is available on the equipment nameplate, it shall be used in lieu of the FLC. The overload protector shall be sized from the FLC rating. See 440.6(A) Ex. 1. If the equipment nameplate does not indicate the FLC rating, it shall be taken from the motor-compressor nameplate.

Highest-Rated Motor – 440.7

To determine compliance with Article 440 and with 430.24, 430.53(B), 430.53(C) and 430.62(A), the highest-rated motor shall be the motor that has the highest-rated load current. If two or more motors have the highest rating, only one of them is considered to be the highest-rated motor. **See Figure 11-17.**

For motors other than hermetic refrigerant motor-compressors and fan or blower motors per 440.6(B), Table 430.248, Table 430.249, or Table 430.250 full-load currents shall be used to determine the highest-rated motor. The exception to 430.7 allows the branch-circuit selection current to be used to determine the highest-rated motor-compressor. **See Figure 11-18.**

Single Machine – 440.8

Air-conditioning and refrigeration equipment shall be considered as a single machine, even though the system may contain several motors. The exceptions to Sections 430.87 and 430.112 may be applied to determine how many disconnecting means are required.

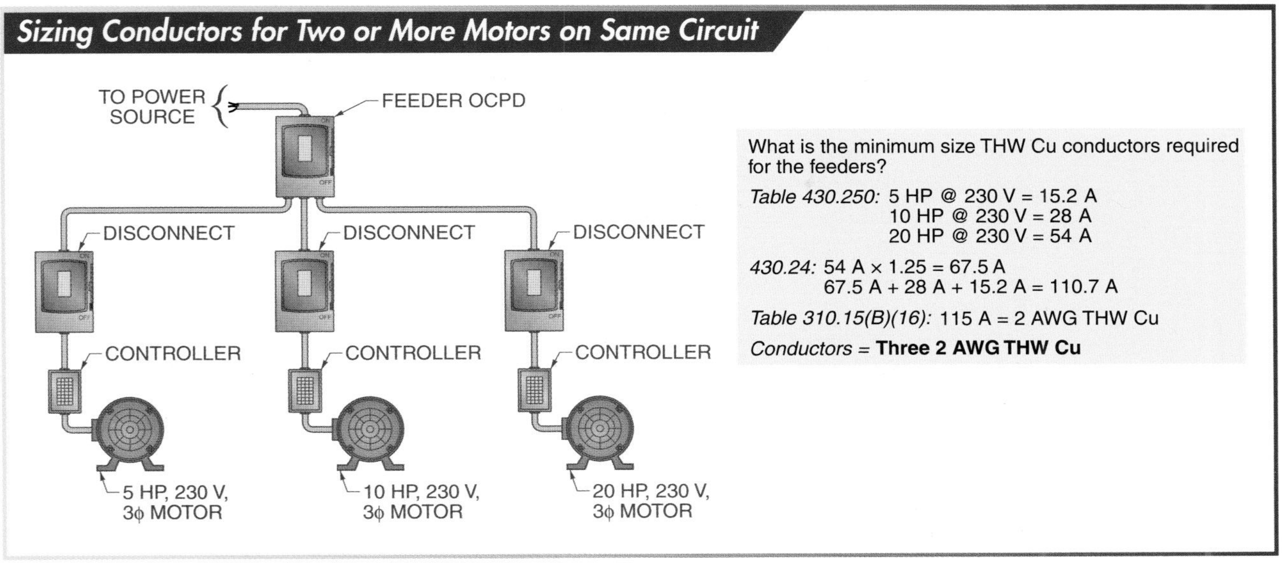

Figure 11-17. Conductors for two or more motors on the same circuit shall have an ampacity equal to 125% of the largest motor plus the sum of the FLC rating of the other motors in the group.

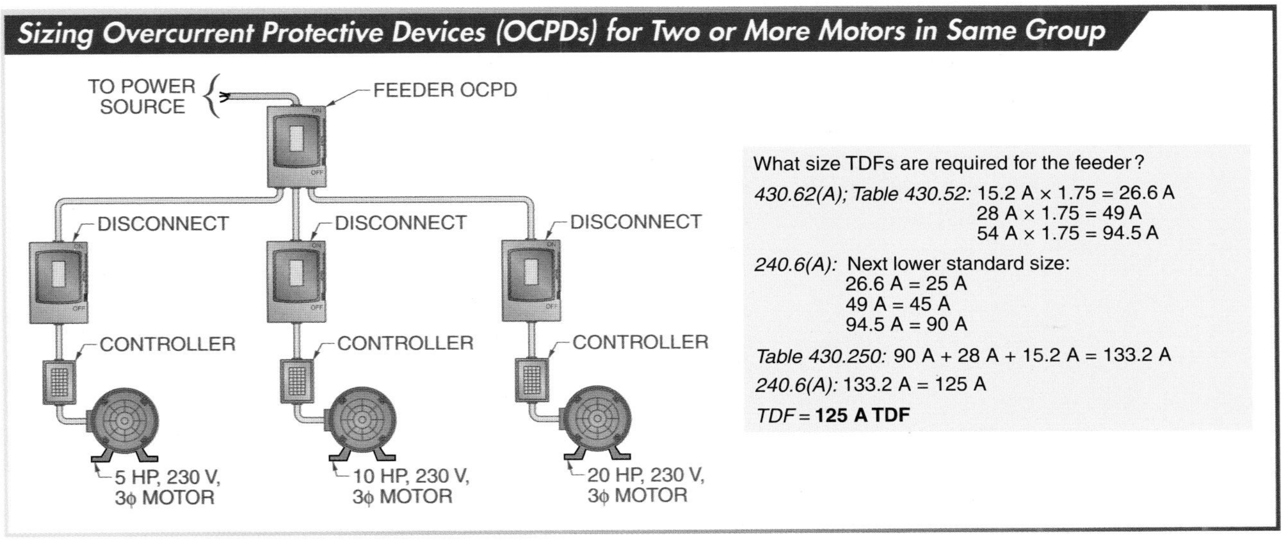

Figure 11-18. OCPDs for two or more motors on the same circuit are calculated per 430.62(A).

Disconnecting Means – 440.11

The disconnecting means for air-conditioning and refrigeration equipment shall be capable of disconnecting all of the equipment including the motor-compressor and controller from the supply circuit.

Rating and Interrupting Capacity – 440.12(A)(1). The disconnecting means for a refrigerant motor-compressor shall be at least 115% of the nameplate rated FLC or the branch-circuit selection current, whichever is greater. The disconnecting means may be a HP-rated switch, a circuit breaker, or other type switch per 430.109.

Hermetic Refrigerant Motor-Compressor – 440.12(A)(2). If the nameplate rating on the motor-compressor or the equipment nameplate rating is expressed in amperage and not HP, the following method may be used to determine the HP rating of the disconnect:

(1) Refer to the appropriate FLC Tables 430.248 through 430.250 and select the HP that corresponds to the rated-load current or the branch-circuit selection current, whichever is greater.

(2) If the LRC is marked on the nameplate, refer to Table 430.251(A) for 1ϕ motors or Table 430.251(B) for polyphase motors, and select the HP that corresponds to this LRC.

(3) Use the larger of the two HP ratings. **See Figure 11-19.**

If a circuit breaker is used as the disconnect for a hermetic motor-compressor, it shall be rated at least 115% of the nameplate FLC or the branch-circuit selection current, whichever is greater. **See Figure 11-20.**

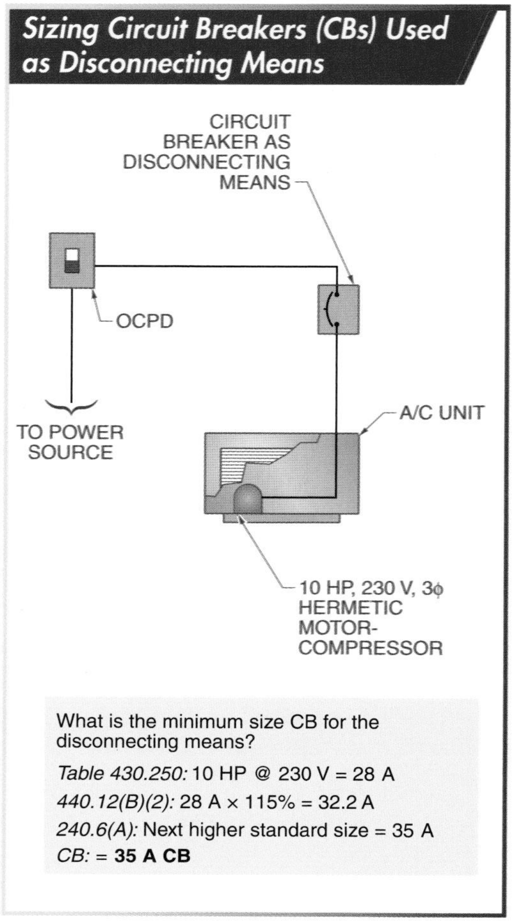

Figure 11-20. A circuit breaker (CB) used as a disconnecting means for a motor-compressor shall be rated at least 115% of the nameplate FLC or branch-circuit selection current, whichever is greater.

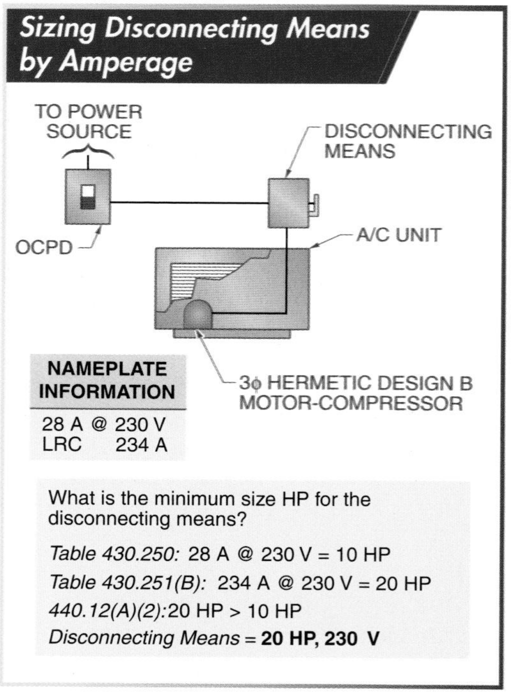

Figure 11-19. The disconnecting means is selected by the amperage if the nameplate rating is in amperage and not in HP.

The disconnecting means is the device by which circuit conductors are disconnected from their source of supply.

If a disconnecting means serves two or more hermetic motor-compressors or is used in combination with other motors or other loads, and the combined load may be on at the same time, the disconnect shall be sized as follows:

(1) Add each hermetic compressor load.

(2) Add each standard motor load.

(3) Add all other loads (if present).

The disconnecting means is then sized at 115% of the total HP.

Disconnecting Means for Cord-Connected Equipment – 440.13. Room air conditioners, household refrigerators, freezers, water coolers, and beverage dispensers that are cord-connected may use the attachment plug and receptacle as the disconnecting means. Section 440.63 allows 1ϕ, 250 V or less room A/Cs to use the cord and plug as a disconnecting means provided the manual controls for the unit are readily accessible and located within 6′ of the floor, or an approved manual switch is readily accessible and within sight of the A/C unit. **See Figure 11-21.**

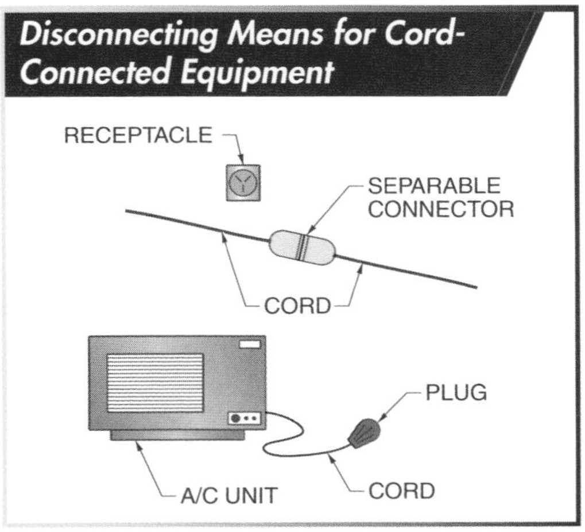

Figure 11-21. A cord and plug, or a separable connector, may serve as the disconnecting means for a room A/C unit.

Location of Disconnecting Means – 440.14. The disconnecting means for fixed-wired A/C and refrigeration equipment shall be installed within sight of the equipment and readily accessible to the user. When the equipment is not within sight of the disconnecting means, another switch shall be provided at the equipment. **See Figure 11-22.** It is important to ensure that, where the disconnected means is mounted on the equipment, it not be located on panels that are designed to be removed to allow access to the equipment.

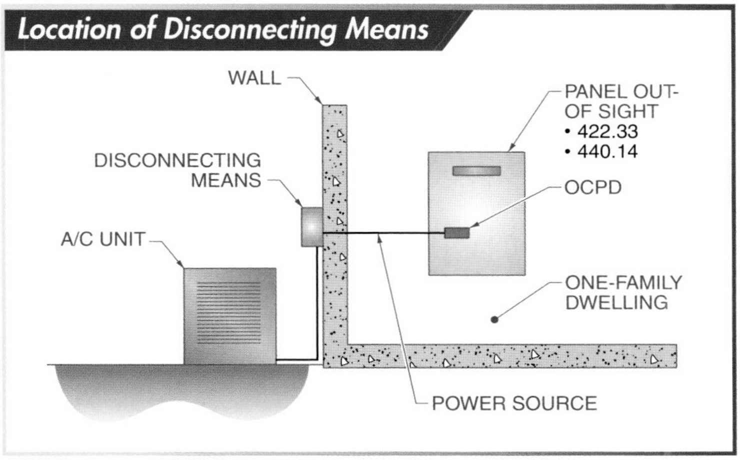

Figure 11-22. If the service panel is not within sight of the A/C unit, an additional disconnecting means shall be installed within sight.

Exception 1 to this general rule permits the disconnect to be out of sight (50′ or more) of the unit provided the following conditions exist:

(1) Disconnect is capable of being locked open.

(2) Equipment is essential to an industrial process.

(3) Installation is a facility where conditions of maintenance and supervision ensure that only qualified persons service the equipment.

(4) The means to lock the disconnecting means is permanently installed on or at the switch or circuit breaker.

Exception 2 does not require a disconnect to be readily accessible for cord-connected equipment. See 440.13.

Branch-Circuit Fuses or Circuit Breakers

When the listed A/C equipment nameplate states a certain maximum fuse size, it means the equipment was tested and listed for fuse protection only. The listing does not cover the equipment if circuit breakers are used. If the equipment is tested and listed for fuses or circuit breakers of the HACR type, it is indicated on the equipment nameplate. Listed circuit breakers which have been found suitable for HACR installations are marked "Listed HACR Type." Most molded-case circuit breakers today include the HACR rating as part of their listing.

Individual Motor-Compressor – 440.22(A). The rating or setting of the overcurrent protection shall not be greater than 175% of the nameplate FLC rating or the branch-circuit selection current, whichever is greater. If the overcurrent protection does not permit the motor to start, a larger rating may be used provided it does not exceed 225% of the nameplate FLC or the branch-circuit selection current, whichever is greater. **See Figure 11-23.**

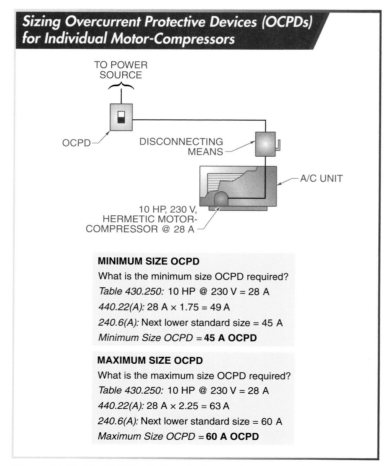

Sizing Overcurrent Protective Devices (OCPDs) for Individual Motor-Compressors

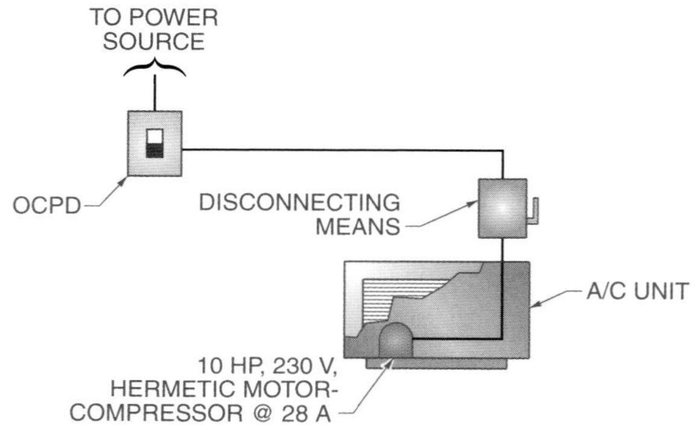

MINIMUM SIZE OCPD
What is the minimum size OCPD required?
Table 430.250: 10 HP @ 230 V = 28 A
440.22(A): 28 A × 1.75 = 49 A
240.6(A): Next lower standard size = 45 A
Minimum Size OCPD = **45 A OCPD**

MAXIMUM SIZE OCPD
What is the maximum size OCPD required?
Table 430.250: 10 HP @ 230 V = 28 A
440.22(A): 28 A × 2.25 = 63 A
240.6(A): Next lower standard size = 60 A
Maximum Size OCPD = **60 A OCPD**

Figure 11-23. The OCPD shall not be larger than 175% of the nameplate FLC or branch-circuit rating, whichever is larger, per 440.22(A).

Several Motors – 440.22(B). The rating or setting of the overcurrent protection shall be selected according to the number of hermetic motor-compressors, or the combination of hermetic motor-compressors and regular motors installed on the circuit. **See Figure 11-24.** Section 440.22(B)(1) requires that when two or more hermetic compressor motors are installed on the same circuit, the overcurrent protective device shall be selected according to the following:

(1) Take 175% of the largest motor's nameplate FLC or branch-circuit selection current rating, whichever is greater.

(2) Add to this the sum of the other motor-compressor rated-load currents or branch-circuit selection currents, whichever are greater and the ratings of other loads supplied.

If this does not permit the compressor motors to start, increase the largest motor percentage from 175% to 225% plus the sum of the other motor nameplate FLC.

(3) Round ampacity down to next standard size OCPD.

Section 440.22(C) requires that manufacturer's value for the overload protection device (heater) when marked on the equipment nameplate shall never be exceeded.

Branch-Circuit Conductors – Article 440, Part IV

Section 440.32 requires that the branch-circuit conductors supplying an A/C unit containing an individual motor-compressor shall not be sized according to 430.22 as other motor loads are. On A/C units, 125% of the marked rated load current or the marked branch-circuit selection current, whichever is greater, shall be used to determine the size of the branch-circuit conductors.

Compressor and Other Loads – 440.33. Conductors supplying one or more compressors with or without additional equipment such as fan motors shall have an ampacity of at least the sum of three items. **See Figure 11-25.** The three items are:

(1) Rated-load or branch-circuit selection current ratings, whichever is greater, of the motor-compressor(s).

(2) FLC ratings of other motors.

(3) 25% of largest motor or motor-compressor rating in the group.

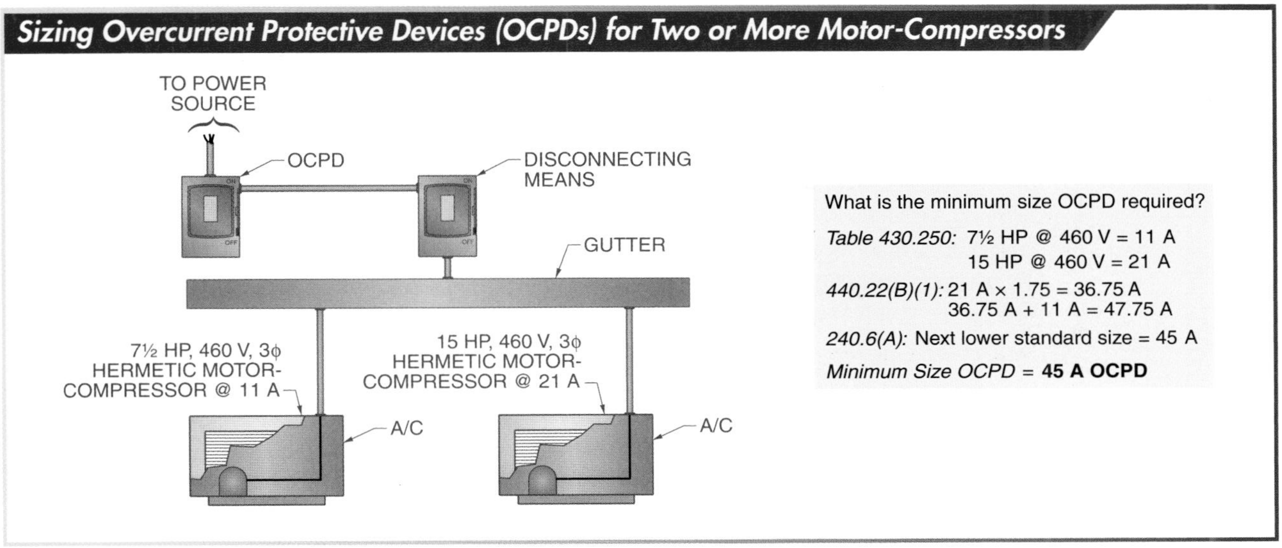

Figure 11-24. The OCPD for two or more hermetic motor-compressors on the same circuit is sized per 440.22(B).

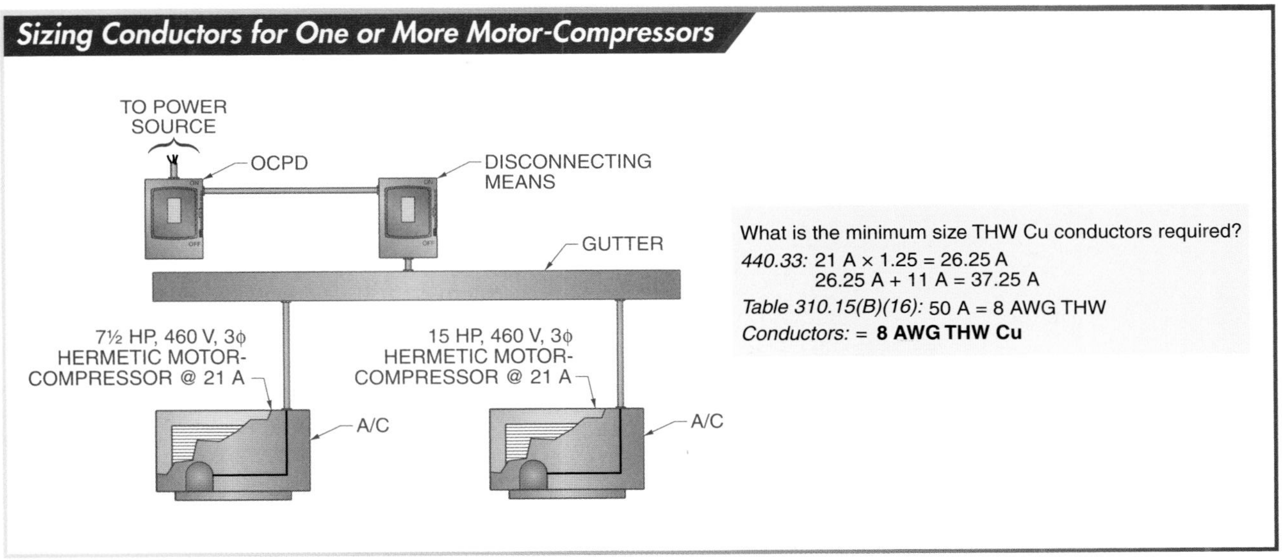

Figure 11-25. Conductors for one or more hermetic motor-compressors are sized per 440.33.

Controllers for Compressor Motors – 440.41, Part V

Controllers which are used with motor-compressors shall have both a continuous-duty full-load current rating and a locked-rotor current rating not less than the compressor nameplate FLA rating or branch-circuit selection current, whichever is larger, and the LRC rating of the motor-compressor, respectively.

In selecting a controller that is rated in HP only, the following shall be done to convert the compressor current ratings into HP:

(1) Refer to Tables 430.248 through 430.250 and select the corresponding HP rating as compared to the motor-compressor nameplate current at the proper voltage.

(2) Refer to Table 430.251(A) or 430.251(B) and select the corresponding HP rating as

compared to the motor-compressor nameplate LRC. In both of these selections, if the nameplate current does not match the table current, the next higher value shall be used in determining the equivalent HP rating. **See Figure 11-26.**

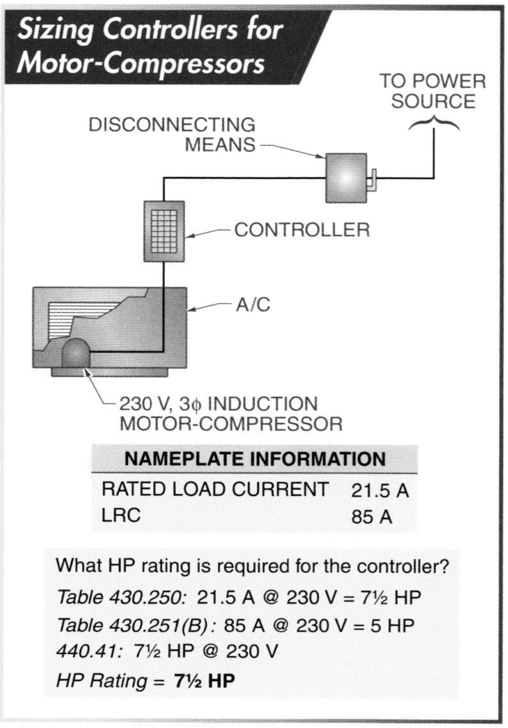

Figure 11-26. The controller for motor-compressors is sized per 440.41.

Branch-Circuit Overload Protection – Article 440, Part VI

Overload protection is required for the motor-compressor, branch-circuit conductors, and the motor control apparatus to protect them from excessive heating due to motor overloads. An overload is the operation of equipment in excess of normal, full-load current rating, or a conductor subject to a current in excess of the rated ampacity. An overload does not include short circuits or ground faults.

The general rule of 440.52(A)(1) is that the overload protection shall be rated at not more than 140% of the rated-load current of the motor-compressor. **See Figure 11-27.** Section 440.52(A)(2) permits the overload protection to be an integral part of the motor-compressor.

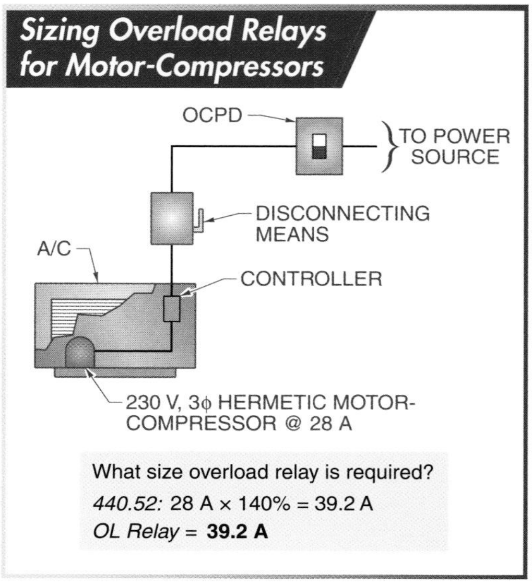

Figure 11-27. Overload protection for motor-compressors is sized per 440.52.

Fusible Disconnects

Fusible disconnects are available in standard sizes of 30 A, 60 A, 100 A, 200 A, 400 A, 600 A, 800 A, and 1200 A.

Per 440.52(A)(3), a fuse or ITCB rated at not more than 125% of the motor-compressor rated-load current used to provide overload protection shall also be permitted to serve as the branch-circuit OCPD. Section 440.52(A)(4) permits other protective systems, provided they are approved, to serve as the means of overload protection.

Overload Relays – 440.53. Overload relays do not provide short-circuit or ground-fault protection for the motor-compressor or the equipment associated with it. The overload relay requires time to open the motor circuit. The more severe the overload, the less time is required for the overload relay to open the control circuit. A short-circuit, ground-fault protector shall respond, if not instantly, within cycles of the fault.

Compressors and Equipment of 15 A or 20 A Circuits Not Cord-and-Plug-Connected – 440.54. Hard-wired equipment and motor-compressors that are rated for 15 A or 20 A,

120 V, 1ϕ or 15 A, 208 V or 240 V, 1ϕ branch circuits shall be provided with overload protection. Any of the four means listed in 440.52(A) is permitted. A separate overload protector shall not exceed 140% of the FLC rating of the hermetic motor. The fuse or circuit breaker providing short-circuit protection shall have sufficient time delay to permit the compressor and other motors to start and bring their loads up to speed.

Compressors and Equipment of 15 A or 20 A Circuits that are Cord-and-Plug-Connected – 440.55. Cord-and-plug-connected compressors and equipment shall be rated no greater than 20 A on 120 V circuits. They shall be rated no higher than 15 A on 208 V or 240 V circuits.

Room A/Cs – Article 440, Part VII

Section 440.60 recognizes room A/C units which are window-mounted, console, or the in-wall type as being A/C appliances. Units that operate at 250 V, 1ϕ or less may be cord-and-plug-connected or hard-wired. Units over 250 V, 1ϕ and all 3ϕ units shall be hard-wired.

Grounding – 440.61. Cord-and-plug-connected room A/C units shall be grounded according to 250.114. The EGC shall be run within the supply cord of each unit. Hard-wired systems shall be grounded according to 250.110, 250.112, and 250.114.

Branch-Circuit Requirements – 440.62. A room A/C is considered as a single motor unit when the following four conditions are met:

(1) It is cord-and-plug-connected.

(2) It is rated not more than 40 A, 250 V, 1ϕ.

(3) The total rated load current is marked on the unit nameplate.

(4) The rating of the fuse or circuit breaker does not exceed the branch-circuit conductors or the receptacle rating, whichever is less.

Section 440.62(B) requires that the rating of the A/C unit shall not exceed 80% of the branch-circuit conductor's ampacity where no other loads are supplied. **See Figure 11-28.** Section 440.62(C) requires that the rating of the A/C unit shall not exceed 50% of the branch-circuit conductor's ampacity where lighting or other appliances are on the same branch circuit. **See Figure 11-29.**

Duty-Cycle Ratings

Motors are rated with 5 minute, 15 minute, 30 minute, 60 minute, and continuous duty-cycle ratings. A motor with a 5 minute rating is designed to operate at its rated HP for periods not to exceed 5 minutes. After this, a cooling or nonoperating period is required before the motor is operated again. The branch-circuit conductors also take advantage of this cooling-off period and, therefore, do not have to be as large as those for a motor with a larger duty cycle.

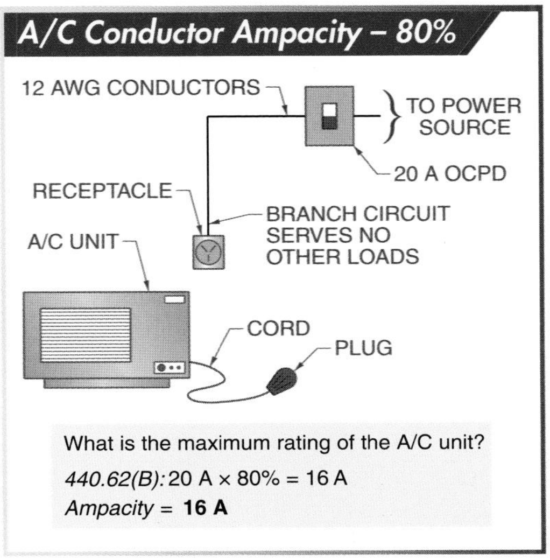

Figure 11-28. The rating of the A/C unit shall not exceed 80% of the branch-circuit conductor's ampacity.

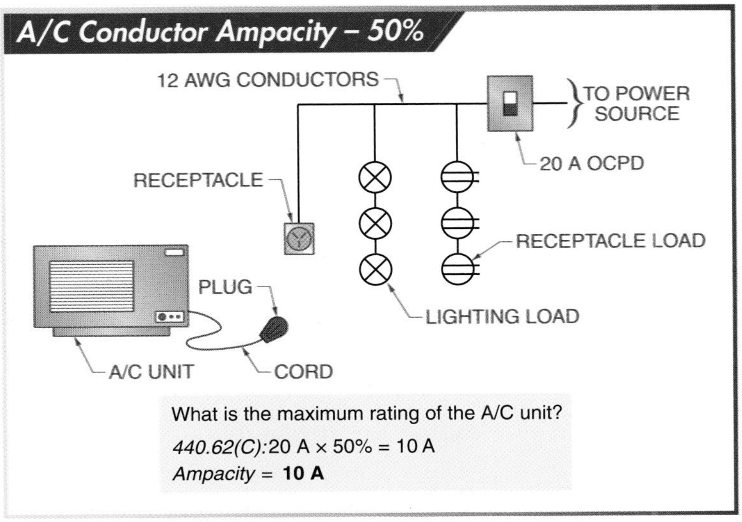

Figure 11-29. The rating of the A/C unit shall not exceed 50% of the branch-circuit conductor's ampacity with lighting or other appliances on the same circuit.

Disconnecting Means – 440.63 and 440.64. A separate disconnect is required for each A/C. The attachment plug and receptacle or cord connector shall be permitted to serve as the disconnecting means if the unit is 1ϕ and not over 250 V, provided one of the following two provisions is met:

(1) The manual controls on the unit are readily accessible and located within 6′ of the floor.

(2) An approved manual switch is installed within sight from the unit and is readily accessible.

The length of flexible cords used to supply A/C units shall not exceed 10′ for 120 V units or 6′ for 208 V or 240 V units.

Leakage Current Detection and Interruption (LCDI) Protection for Room Air Conditioners – 440.65

During the 2002 NEC® cycle, an important product safety provision was added to Part VII of Article 440 that covers electrically energized room air conditioners that control temperature and humidity. Section 440.65 was added to require all single-phase cord-and-plug-connected room air conditioners to be provided with a leakage current detection and interruption (LCDI) device. An LCDI device works similarly to an arc-fault circuit interrupter by sensing leakage currents to ground that may have been caused by physical damage to the cord set. This provision was added based on analysis of data gathered by the Consumer Products Safety Commission on over 40,000 residential fires. The provision was revised in the 2005 NEC® to permit either LCDI protection or AFCI protection provided it is an integral part of the attachment plug or located in the power supply cord within 12″ of the attachment plug.

Generators – Article 445

Article 445 covers the installation of generators. Generators can be fixed, portable, or vehicle-mounted. See 250.34. Section 445.10 requires that generators shall be suitable for the location where they are installed. Section 445.11 requires that the generator nameplate shall provide the following information:

(1) Manufacturer's name.
(2) Rated frequency.
(3) Power factor.
(4) Number of phases if an AC generator.
(5) Subtransient and transient impedances
(6) Rating in kW or kVA.
(7) Normal volts and amperes that correspond to the rating.
(8) Rated rpm.
(9) Insulation system class and rated ambient temperature or rated temperature rise.
(10) Time rating.

In-wall A/C units may be cord-and-plug-connected or hardwired.

Overcurrent Protection – 445.12. Most AC generators are designed so that during periods of time when there is an excessive overload, the voltage diminishes enough to limit the current and power output to levels that do not damage the generator. Section 445.12(A) requires that constant-voltage generators shall be protected from overloads by inherent design, circuit breakers, fuses, protective relays, or other identified OCPDs suitable for the use. It is common practice not to protect the AC generator exciter, so as not to shut down the generator due to an inadvertent opening of the exciter overcurrent protection.

Section 445.12(B) permits overcurrent protection in only one conductor of a 2-wire DC generator. This overcurrent device shall be actuated by the entire generated current other than the current in the shunt field. The shunt field shall not be opened by the overcurrent device. If the shunt field were to open, a high voltage could be induced which would damage the field winding and the generator.

Section 445.12(C) does not require overcurrent protection for generators operating at 65 V or less that are part of a motor-generator set provided the motor protection device opens when the generator delivers 150% of its rated FLC.

Section 445.12, Ex. advises that the AHJ may prefer to have the generator operate until it fails, rather than have it automatically shut down and permit a greater hazard to personnel. An overload-sensing device shall be permitted to be connected to an annunciator or other indicating device which would permit authorized personnel an opportunity to have an orderly shutdown of load-side equipment.

Ampacity of Conductors – 445.13. The phase conductors between the generator terminals and the first overcurrent device shall have an ampacity not less than 115% of the generator nameplate current rating. Neutral conductors are permitted to be sized according to 220.61. Conductors that carry ground-fault currents shall be sized according to 250.30(A). Grounded neutral conductors of DC generators shall not be smaller than the minimum size required for the largest conductor if they must carry ground-fault current.

Protection of Live Parts – 445.14. All live parts of generators that operate at over 50 V to ground shall not be exposed to unqualified persons.

FIRE PUMPS – ARTICLE 695

The NEC® has jurisdiction for electrical construction and installation of fire pumps. Article 695 covers the installation of electric power sources, interconnecting circuits, and switching and control equipment dedicated to fire pump drivers. The NEC® does not cover jockey pumps, internal wiring of fire pump system components, performance testing, maintenance testing, or acceptance testing. Jockey pumps shall be installed per Article 430. Performance, maintenance, and acceptance testing are the jurisdiction of NFPA 20-2007, *Standard for the Installation of Stationary Pumps for Fire Protection.*

Power Source to Electric-Motor Driven Fire Pumps – 695.3

Per 695.3, the source of electric supply for fire pumps shall be reliable. Recognized individual sources for supplying fire pumps are listed in 695.3(A). Where capable of carrying the required load, and where the AHJ has recognized them as reliable, a fire pump may be supplied from an electric utility service connection or an on-site power production facility.

Where a separate utility service is provided, the connection to the fire pump must be ahead of and not within the same cabinet, enclosure, or vertical switchboard section as the service disconnecting means. Taps ahead of the service disconnecting means must comply with Section 230.82(5). The location of the service equipment must comply with 230.72(B) and must be labeled per 230.2.

Disconnecting means for fire pumps are installed per 695.4.

An on-site power production facility is not a stand-by generator. The AHJ, in most cases, will not recognize the service as being reliable. Generally, utility companies will not guarantee uninterrupted service connections for fire pump supply. On-site power production facilities which have the ability to generate power for an indefinite period of time are not common in most commercial installations.

In the 2011 NEC® a new option was added to the list of permissible individual sources for supplying electric motor-driven fire pumps. 695.3(A)(3) now permits a dedicated feeder where it is derived from a service connection as described in 695.3(A).

Multiple Sources – 695.3(B)

Because of the importance of fire pumps, the NEC® requires that they have a reliable source of power to ensure that the supply of power to them is not interrupted. 695.3(A) covers the list of permissible individual sources of power for the supply of fire pumps. In the event that reliable power cannot be obtained from one of the sources listed in 695.3(A), Section 695.3(B) contains the list of permissible alternatives to the acceptable source of power.

Electric motors that drive fire pumps must have a reliable source of power to ensure that power is not interrupted.

The first option permitted is an approved combination of two or more sources listed in 695.3(A). The second option is to utilize an approved combination of one of more sources in 695.3(A) and an on-site standby generator that complies with 695.3(D).

Section 695.3(C) covers the complex requirements for multibuilding campus-style complexes. Because of the complex nature of these types of installations, they must be approved by the AHJ. If approved, two or more feeders are permitted as more than one power source if the feeders are connected to, or derived from, separate utility services.

Additionally, a feeder is permitted as a normal source of power if an alternative source of power independent from the feeder is provided. Selective coordination, which ensures the proper sequencing of overcurrent devices, must be provided for the feeders. Another option that is permissible as an alternative source is the use of an on-site standby generator that complies with provisions of 695.3(D)(1–3).

Lastly, all of the power supplies must be arranged and located such that they are protected against damage by fire from both within the premises and other exposing hazards. It is critical that where multiple power sources are provided, they be arranged so that a fire at one source does not cause an interruption at the other source.

Continuity of Power – 695.4

The general rule for supply conductors for fire pumps is that they have a direct connection to the fire pump controller. This direct connection is to eliminate the possibility of power to the fire pump being interrupted by firefighters, or others, as they attempt to disconnect the service before applying water to the fire.

Per 695.4(B), a single disconnecting means and associated overcurrent protective devices shall be permitted between the supply conductors and a listed fire pump controller described in one of the five subdivisions of this section.

Per 695.4(B)(2), the overcurrent protective devices shall be selected to carry indefinitely the sum of the locked rotor currents of the fire pump motor(s) and pressure maintenance pumps along with the full load current rating of all fire pump accessory equipment connected to the supply conductors.

Per 695.4(B)(3)(a)(1), the disconnecting means shall be suitable for use as service equipment, lockable in the closed position, and located remotely from the building disconnecting means. The location requirement is to put distance between the service disconnecting means and the fire pump disconnecting means to prevent the fire pump supply from being disconnected in a fire situation.

Per 695.4(B)(3)(b), the disconnecting means shall be visibly marked with letters at least one inch in height as "FIRE PUMP DISCONNECTING MEANS."

Per 695.4(B)(3)(c), a placard shall be installed at the fire pump controller stating the location of the single disconnecting means permitted and the location of a key if the disconnecting means is locked.

Per 695.4(B)(3)(d), the disconnecting means shall be supervised by one or more of the following methods: a signaling device to a remote or central station, an audible alarm near a constantly attended location, a lock on a closed disconnect, and sealing of the disconnect with weekly inspections where approved by the AHJ.

Transformers – 695.5

Where the utilization voltage of a fire pump motor is different from the service voltage, a transformer protected by a single disconnect and overcurrent protective devices per 695.5(A – C) shall be permitted.

Transformers shall supply only loads associated with the fire pump system. Only when applying 695.3(B)(2), which is limited to feeders in multi-building, campus-style complexes, are transformers permitted to serve other loads along with the fire pump system.

Transformers shall be sized at 125% of the sum of the full load current ratings of the fire pump and pressure maintenance motors along with 100% of all other fire pump equipment supplied. Overcurrent protection shall be selected to carry indefinitely the sum of the locked rotor current ratings of the fire pump and pressure maintenance motors along with 100% of all other fire pump equipment supplied. Overcurrent protection on the secondary of the transformer is not permitted.

Power Wiring – 695.6

Service conductors and conductors supplied by on-site power production facilities shall be installed as service entrance conductors in accordance with the provisions of Article 230. Where a single disconnecting means is installed as permitted in 695.4(B), conductors on the load side shall be kept independent of all other wiring. The load side conductors shall also be enclosed in 2″ of concrete, or be enclosed in a structure with a minimum 2 hr fire rating, or use a listed electrical circuit protective system such as type MI cable with a minimum 2 hr fire resistance.

QUICK QUIZ®

Refer to Chapter 11 Quick Quiz® on CD-ROM.

Additional information is available at ATPeResources.com

Motors, Generators, A/C and Refrigeration, and Fire Pumps

chapter 11

REVIEW QUESTIONS

Name ______________________ **Date** ______________________

______________ **1.** Motor nameplates shall include the rated horsepower if ___ HP or more.

______________ **2.** The ___ branch circuit is that point from the last fuse or circuit breaker in the motor circuit out to the motor.

______________ **3.** Branch-circuit conductors supplying a single motor shall have an ampacity of ___% of the FLC rating as indicated on the appropriate FLC table.

______________ **4.** The ___ is the device in a motor circuit which turns the motor ON or OFF.

______________ **5.** A(n) ___ is an assembly of one or more sections with a common power bus and primarily containing motor control units.

A. ACM
B. CCM
C. MCC
D. MCA

______________ **6.** A(n) ___ refrigerant motor-compressor is a combination compressor and motor in the same housing, having no external shaft or shaft seals, with the motor operating in the refrigerant.

______________ **7.** A(n) ___ is a small-magnitude overcurrent which, over a period of time, leads to an overcurrent that may operate the fuse or circuit breaker.

______________ **8.** Cord-and-plug-connected compressors and equipment shall be rated no greater than ___ A on 120 V circuits.

______________ **9.** The length of flexible cords used to supply A/C units shall not exceed ___′ for 120 V units.

______________ **10.** Table ___ is used to determine the current ratings of 3ϕ motors.

______________ **11.** The code letter or locked-rotor amps shall be listed on the nameplates of motors rated ___ HP or more.

T F **12.** Small motors rated 150 W or less shall be marked "Impedance Protected."

T F **13.** The larger a motor's full-load table current, the less the percentage of the increase in sizing the thermal protector that is an integral part of the motor.

T F **14.** A plug and receptacle may serve as a controller for stationary motors up to 2 HP.

______________ **15.** Feeder conductors supplying several 3ϕ motors shall have an ampacity equal to or larger than the sum of all FLC ratings plus ___% of the highest rated motor in the group.

______________ **16.** Over ___% of the motors in use today are single speed, AC induction motors.

______________ **17.** A circuit breaker used as a disconnecting means for a motor-compressor shall be rated at least ___% of the nameplate FLA or branch-circuit selection current, whichever is greater.

A. 100
B. 115
C. 125
D. 140

______________ **18.** A/C units that operate at ___ or less may be cord-and-plug-connected.

A. 480 V, 3ϕ
B. 208 V, 3ϕ
C. 250 V, 1ϕ
D. none of the above

T F **19.** The winding type shall be given on the nameplate of a DC motor.

T F **20.** The rated current or HP shall be given on the nameplate of a motor controller.

______________ **21.** A(n) ___ device is a pilot light, buzzer, horn, or other type of alarm.

T F **22.** The overcurrent protection for an MCC may be located ahead or within the MCC.

T F **23.** A 50 A fusible disconnect is a standard size.

T F **24.** A/C and refrigeration equipment with several motors is considered as a single machine.

______________ **25.** Equipment shall be located within ___′ and be visible to be considered within sight.

TRADE COMPETENCY TEST

Name ______________________ **Date** ______________

NEC®	Answer	
________	________	**1.** *See Figure 1.* What size THW Cu conductors are required for the motor?
________	________	**2.** *See Figure 2.* What is the smallest raceway (using EMT) permitted for the conductors?

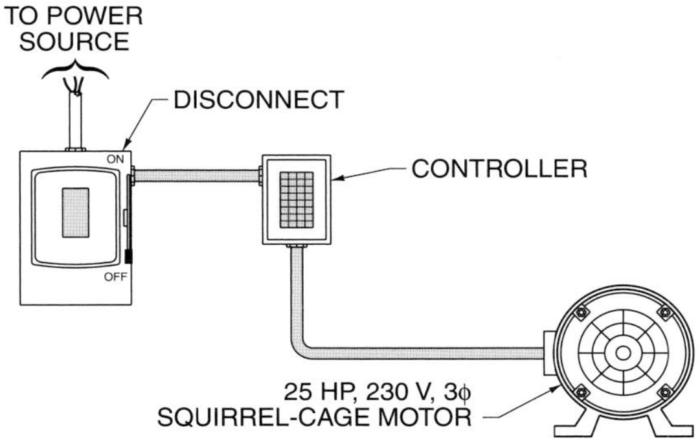

FIGURE 1

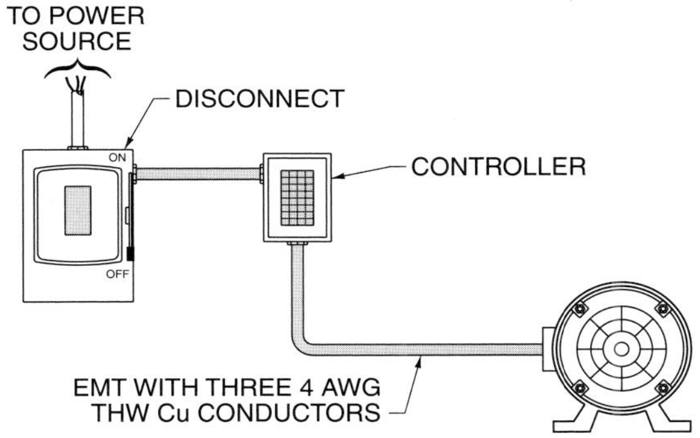

FIGURE 2

NEC®	Answer	
________	________	**3.** *See Figure 3.* What size OL device is required?
________	________	**4.** *See Figure 4.* What size OL device is required?
________	________	**5.** *See Figure 5.* What size TDFs are required to provide short-circuit protection for the motor?

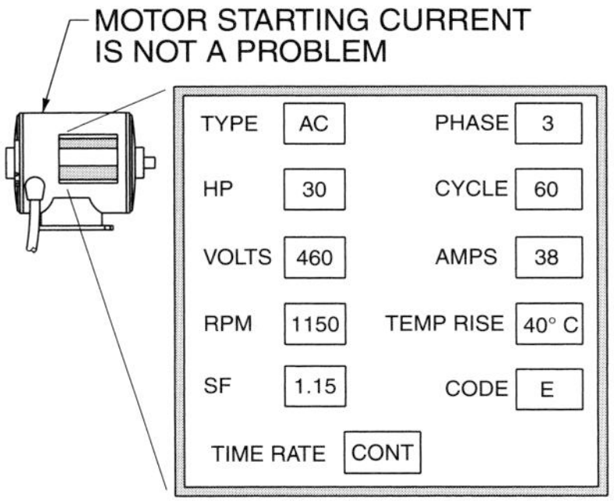

FIGURE 3

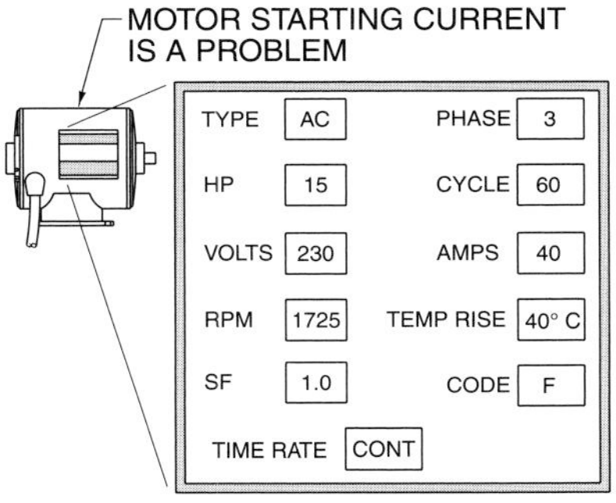

FIGURE 4

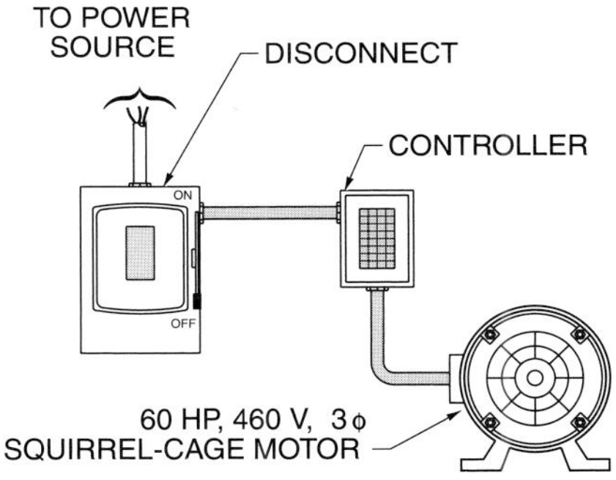

FIGURE 5

______________ ______________ **6.** *See Figure 6.* What size THW conductors and raceway (using EMT) are required for the feeder circuit?

______________ ______________ **7.** *See Figure 7.* What size TDFs are required for the motor feeder circuit?

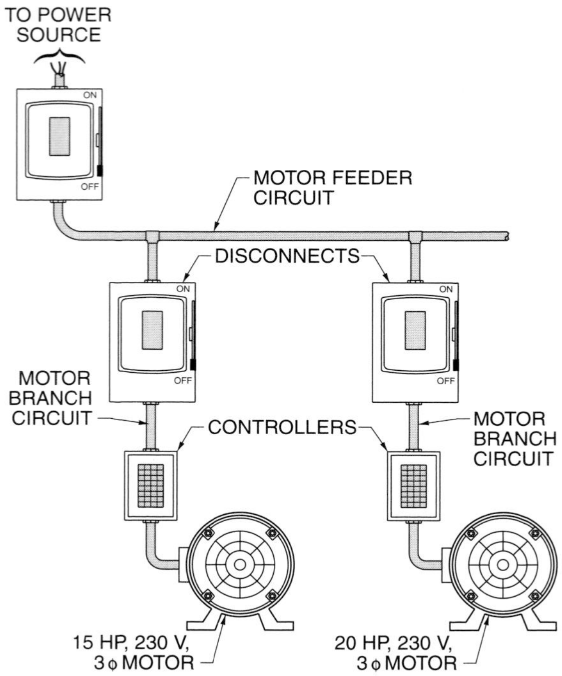

FIGURE 6

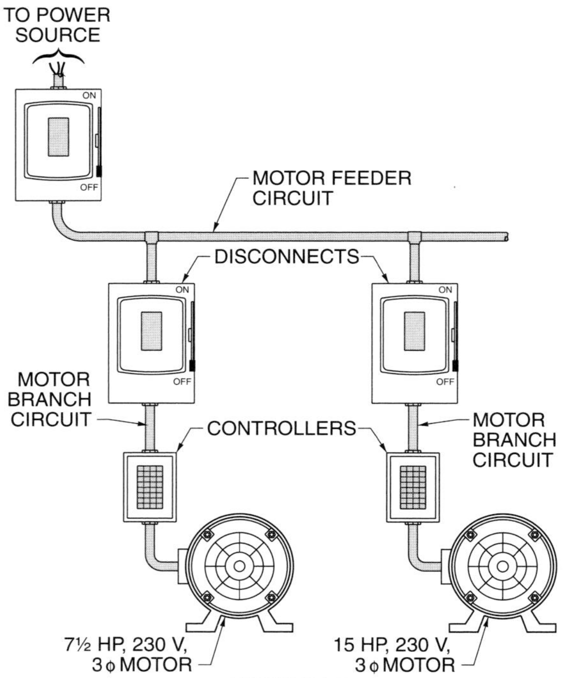

FIGURE 7

______________ ______________ **8.** *See Figure 8.* What is the minimum size disconnect required for the motor?

______________ ______________ **9.** *See Figure 9.* What is the minimum size of branch-circuit THW copper conductors between the multispeed starter and the fusible disconnect?

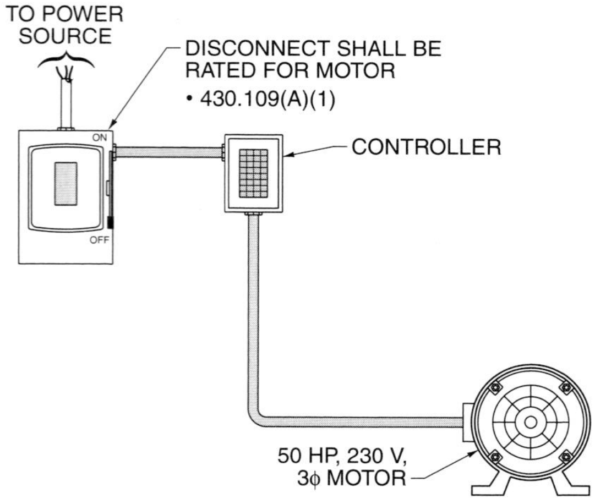

FIGURE 8

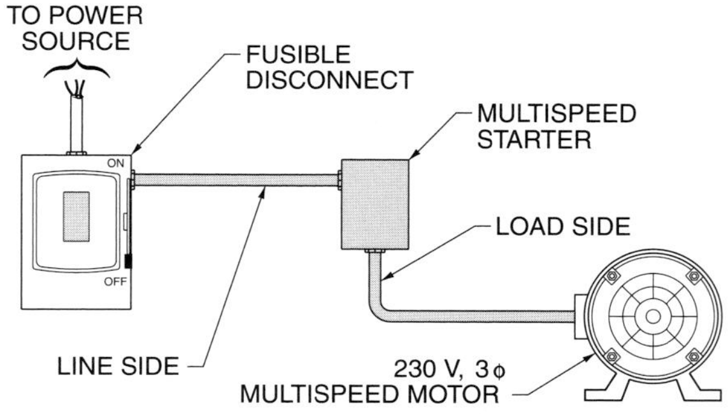

NAMEPLATE INFORMATION

SPEED	RPM	AMPS
LOW	1155	52
HIGH	1740	29

FIGURE 9

______________ ______________ **10.** *See Figure 10.* What is the minimum size of THW Cu conductors required to supply this intermittent-duty motor?

______________ ______________ **11.** *See Figure 11.* What is the minimum size of THW Cu conductors required for the feeders?

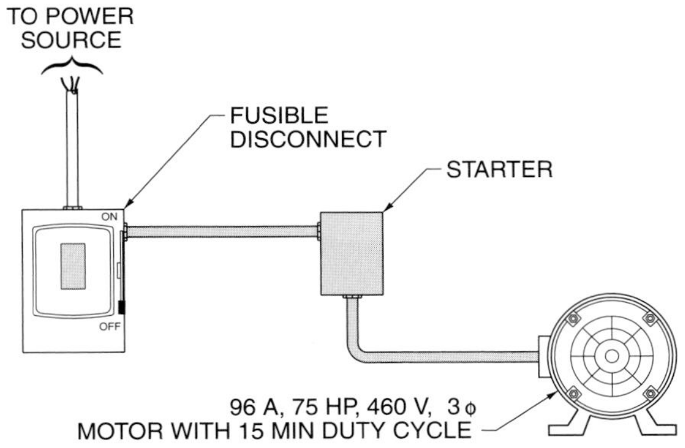

FIGURE 10

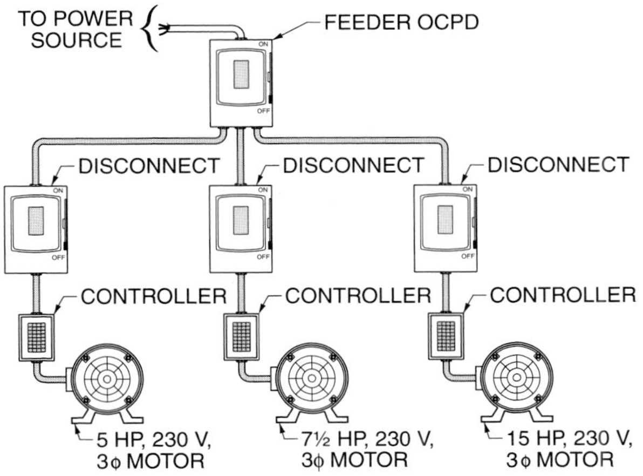

FIGURE 11

______________ ______________ **12.** *See Figure 12.* What is the minimum size HP for the disconnecting means?

______________ ______________ **13.** *See Figure 13.* What is the minimum size circuit breaker for the disconnecting means?

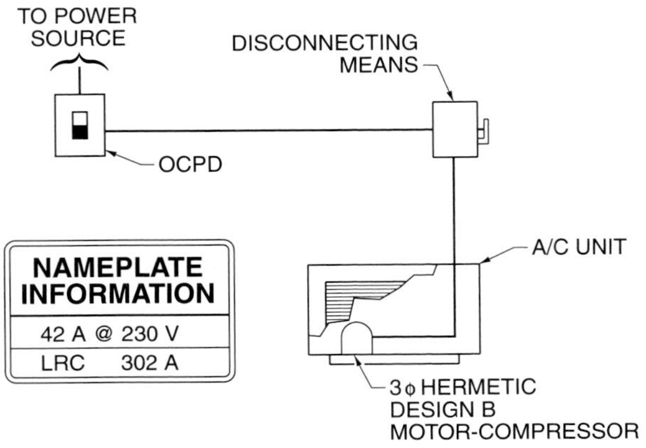

FIGURE 12

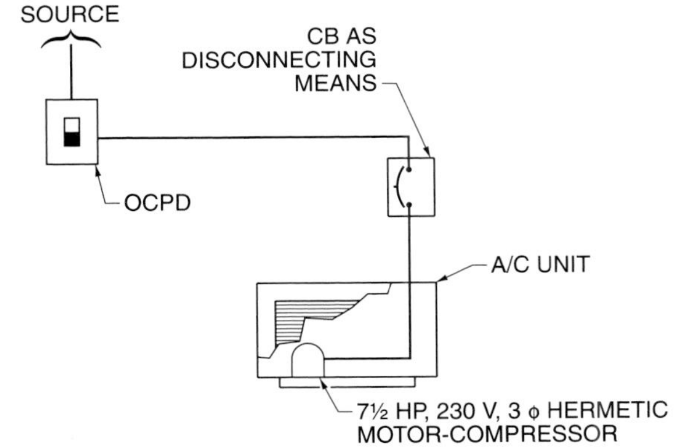

FIGURE 13

______________________ ______________________ **14.** *See Figure 14.* What size OCPD is required?

______________________ ______________________ **15.** *See Figure 15.* What are the minimum sizes of THW Cu feeder conductors required?

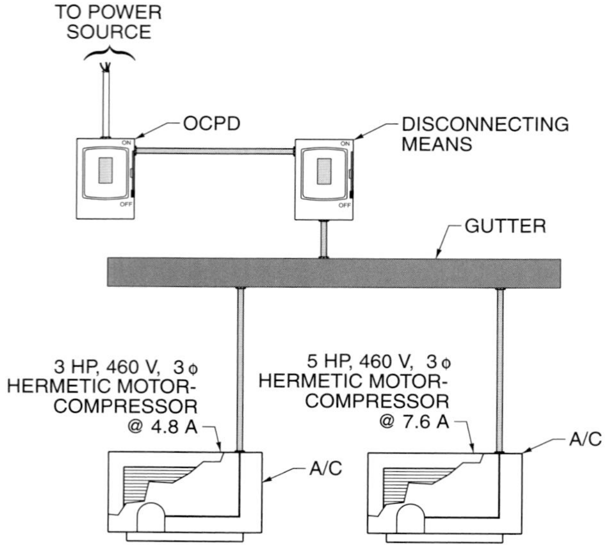

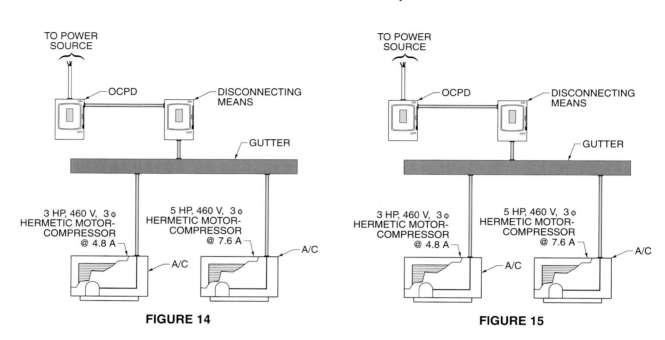

FIGURE 14

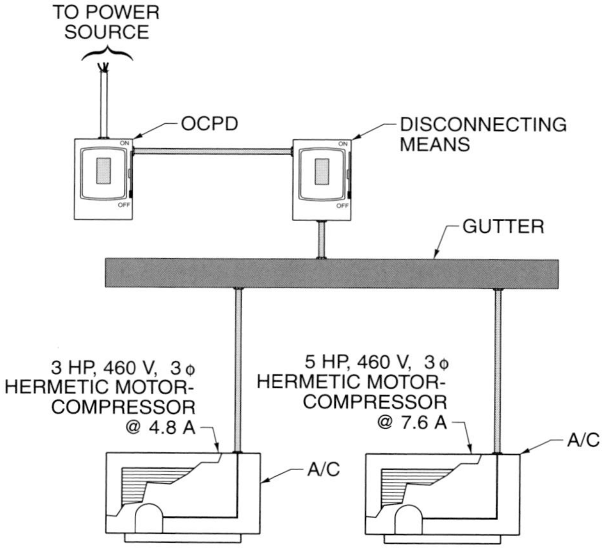

FIGURE 15

Learning the Code

- Transformers are a regular component of the electrical distribution system. Article 450 is a relatively small article that contains basic requirements for transformers and transformer vaults. Like Article 430 for motors, it is important that apprentices and students of the Code clearly understand the type of transformer that they are studying. Article 450 has separate provisions for dry-type, less-flammable liquid-insulated, nonflammable fluid-insulated, askarel-insulated, and oil-insulated transformers, and autotransformers. Be certain to first identify the transformer type and then apply the correct rules and requirements.

- Only two tables are contained in Article 450 but both are very important. Table 450.3(A) lists the maximum rating or setting of overcurrent devices for protecting transformers rated over 600 V. Table 450.3(B) covers the maximum rating or setting of overcurrent devices for protecting transformers rated 600 V and less. Once the proper table is selected, the next step is to determine whether primary only or primary and secondary protection for the transformer will be provided. Apprentices and students of the Code must remember that in the electrical distribution system there are sometimes multiple and overlapping protection requirements that must be met. For example, consider the case of a separately derived system from a transformer that supplies a panelboard. The conductors, the transformer and the panelboard all need to be protected. Tables 450.3(A) and (B) and their applicable notes, along with the overcurrent protection requirements of Article 240, all need to be properly considered.
- Part III of Article 450 covers the requirements for constructing a transformer vault. Typically, vaults are required only for dry-type transformers rated over 35,000 V. With some exceptions, all oil-insulated transformers installed indoors shall also be installed in a transformer vault.

chapter 12

Transformers

Transformers convert electrical power at one voltage and current rating to another voltage and current rating. Transformers are classified by their number of phases, method of cooling, method of mounting, current or voltage, and usage.

New in the 2011 NEC®

- *New provisions added for requirements for disconnecting means for transformers – 450.14*

TRANSFORMER CONSTRUCTION AND TYPES

Transformers are one of the most common components of electrical distribution systems. Transformers are electrical devices that contain no moving parts and are used primarily to convert electrical power at one voltage and current rating to another voltage and current rating.

Transformers operate on the principle of magnetic induction. Two separate coils or windings are wound around a steel or iron core. When a voltage is impressed upon one of the windings, a strong magnetic field is created by the magnetizing of the core. This magnetic field, in turn, induces a voltage in the second winding. The wire in the first (primary) winding, which is connected to the power supply, usually consists of smaller, heavily insulated wire to accommodate the applied voltage. The wire in the second (secondary) winding, which delivers the power to the load, is constructed from heavier gauge wire which can accommodate the increased currents.

The arrangement of the windings determines whether the transformer is classified as a step-up transformer or a step-down transformer. A *step-up transformer* is a transformer with more windings in the secondary winding, which results in a load voltage that is greater than the applied voltage. A *step-down transformer* is a transformer with more windings in the primary winding, which results in a load voltage that is less than the applied voltage. **See Figure 12-1.** Either the primary winding or the secondary winding can be the high-voltage winding, depending upon the transformer application.

Transformers can be classified in many ways. Common classifications include number of phases, method of cooling, method of mounting, current or voltage, and usage.

Dry-Type Transformers

Because of the close concentration of the conductors in the windings, heat is a factor that cannot be ignored in transformer installation and design. If excessive heat is permitted to remain in or around the transformer, the insulation on the conductors that comprise the windings would be subject to damage and perhaps failure.

Dry-type transformers are air-cooled and typically include a fan or a system of fins or radiators to help dissipate heat.

Transformers are often classified by the type of coolant or method of cooling that is employed. The NEC® classifies transformers that do not rely upon any type of liquid for cooling purposes as dry-type transformers. A *dry-type transformer* is a transformer which provides air circulation based on the principle of heat transfer. These air-cooled transformers are generally smaller in size. Larger transformers of this type are available, and they may incorporate a fan or a system of fins or radiators to aid in the dissipation of heat. Dry-type transformers are constructed as one of three major types: epoxy encapsulated, ventilated, or totally enclosed. **See Figure 12-2.**

There are four recognized NEMA insulation classifications for dry-type transformers. These insulation classifications are Class 105, Class 150, Class 180, and Class 220. Each classification has a maximum permitted insulation rating. The type of insulation used and the design of the transformer contribute to the transformer classification. Another factor used in determining the insulation class is the hot spot allowance. The value of the hot spot is approximated to equal the difference between the highest temperature inside the coil of the transformer and the average temperature of the coil.

Transformers

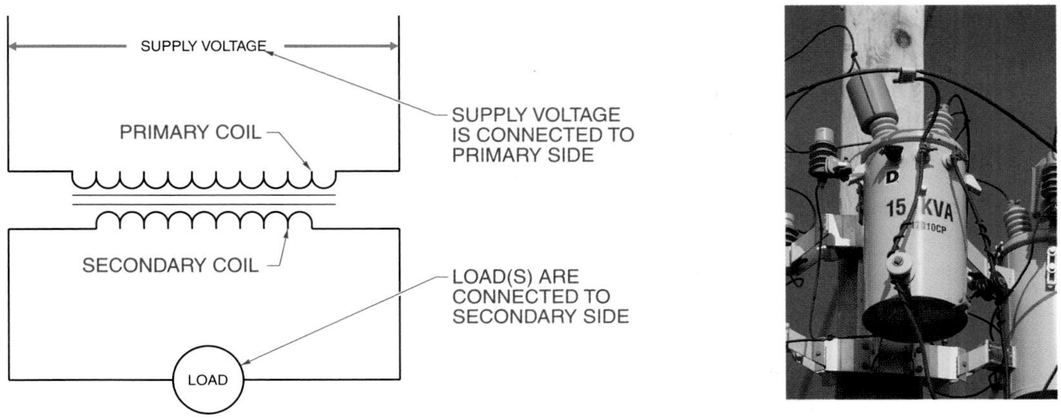

TRANSFORMER	SECONDARY			
	VOLTAGE	AVAILABLE CURRENT	POWER	TERMINAL MARKINGS
STEP-UP	↑	↓	—	PRIMARY = X1, X2, ETC. SECONDARY = H1, H2, ETC.
STEP-DOWN	↓	↑	—	PRIMARY = H1, H2, ETC. SECONDARY = X1, X2, ETC.

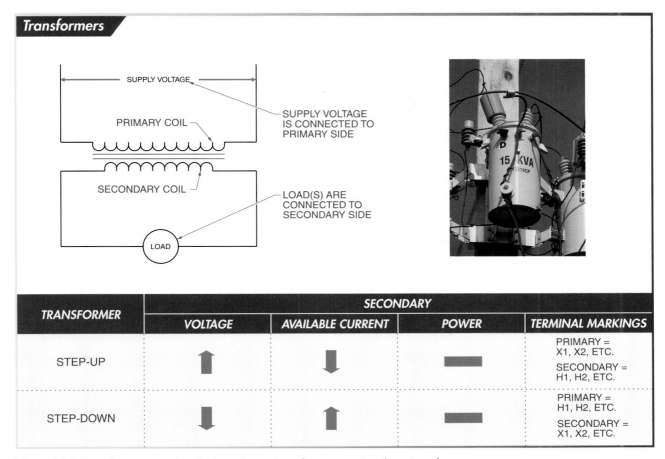

Figure 12-1. Transformers are classified as step-up transformers or step-down transformers.

Class 105 insulation is permitted to have a temperature rise of 55°C. The maximum ambient temperature is 40°C. The hot spot allowance for this class is 10°C. The class number is a value calculated by adding the temperature rise, the maximum ambient temperature, and the hot spot allowance together (55°C + 40°C + 10°C = 105°C).

Class 105 insulation was at one time termed Class A insulation. Recent standard changes replaced letter designations such as A, B, F, and H with the values of the maximum temperature limits of the transformer.

Class 150 transformer insulation operates at not more than an 80°C temperature rise on the windings provided the transformer is not overloaded and the ambient temperature does not exceed 40°C. Class 150 transformers are commonly used for distribution purposes.

Dry-Type Transformers			
Class	Temperature Rise – °C	Ambient Temperature °C	Hot Spot °C
105	55	40	10
150	80	40	30
180	115	40	25
220	150	40	30

Class 180 transformers operate at not more than a 115°C temperature rise on the windings if the transformer is not overloaded and the ambient temperature does not exceed 40°C. Class 180 transformers are rapidly replacing Class 150 transformers for distribution applications because they are smaller in size and have a higher insulation rating.

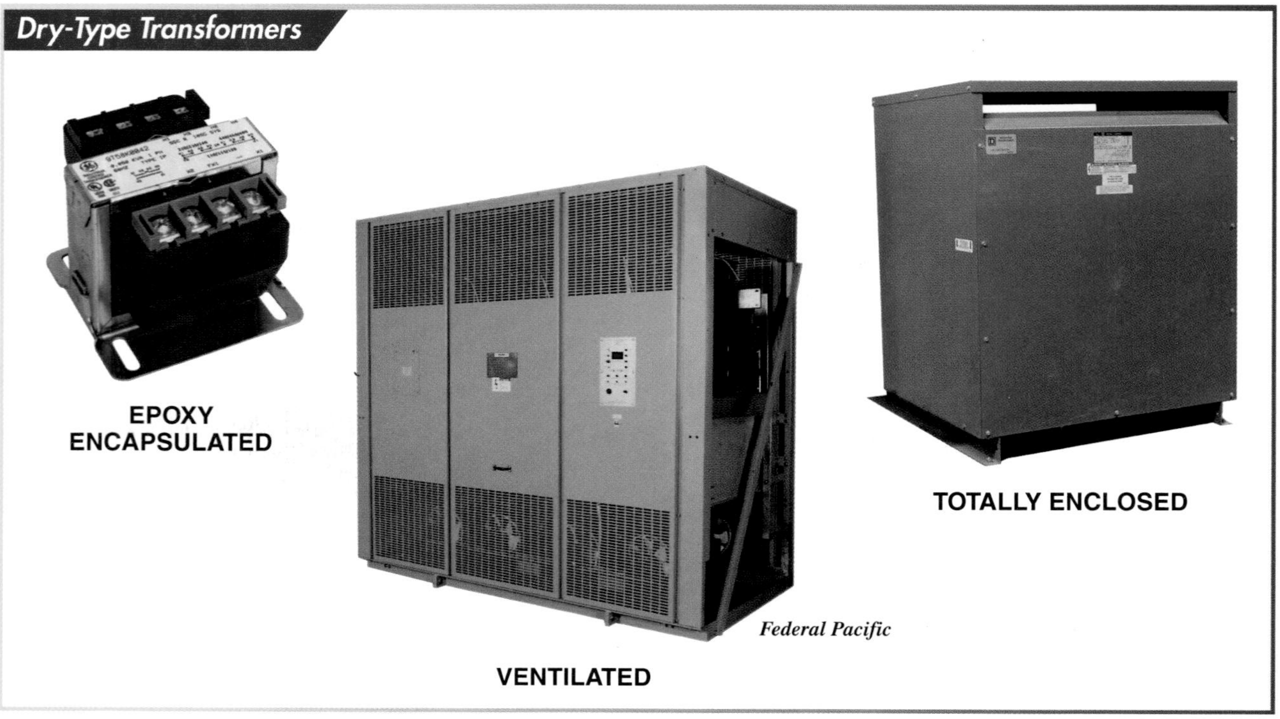

Figure 12-2. Dry-type transformers are constructed as one of three major types: epoxy encapsulated, ventilated, or totally enclosed.

Class 220 transformers operate at not more than a 150°C temperature rise on the windings if the transformer is not overloaded and the ambient temperature does not exceed 40°C. Class 220 transformer insulation is used primarily with larger distribution transformers.

Dry-type transformers are available in either single- or poly-phase configurations. Single-phase transformers consist of a single core with a primary and secondary winding around the core. Poly-phase transformers have a separate core, each with its own primary and secondary winding, for each of the connected phases. **See Figure 12-3.**

Transformer Insulating Liquids

Transformers often use liquids as insulating materials. These liquids come in different classes and have different requirements for maintenance. Routine maintenance checks of an insulating liquid are necessary because the liquid deteriorates over time. A transformer containing PCB compounds (such as askarels) as an insulating liquid should not be opened when the transformer is hot because the fumes are toxic. PCBs were declared hazardous in the mid-1970s. However, they may still be found in equipment manufactured before the mid-1970s.

Single-phase transformers are often used to construct poly-phase transformers. They are used in control circuit applications which are commonly found in motor control circuits. Control power transformers are used extensively as a source of supply for motor control circuits. These control transformers are frequently enclosed within combination starters and use the motor's supply voltage to derive a lower voltage level for the control circuit. Multitap control transformers provide multiple transformer tap locations to permit their use with a variety of supply voltages. This offers added protection against short circuits or other overcurrents that may occur on the secondary of the transformer. High-efficiency control transformers are designed to be used when voltage fluctuation must be held to a minimum and proper voltage regulation is required. These transformers operate at very low temperatures with very little energy loss. **See Figure 12-4.**

Poly-phase transformers are commonly used in electrical distribution systems because the poly-phase (most often 3ϕ) transformer is generally lighter and more efficient than a constructed transformer bank. See Appendix.

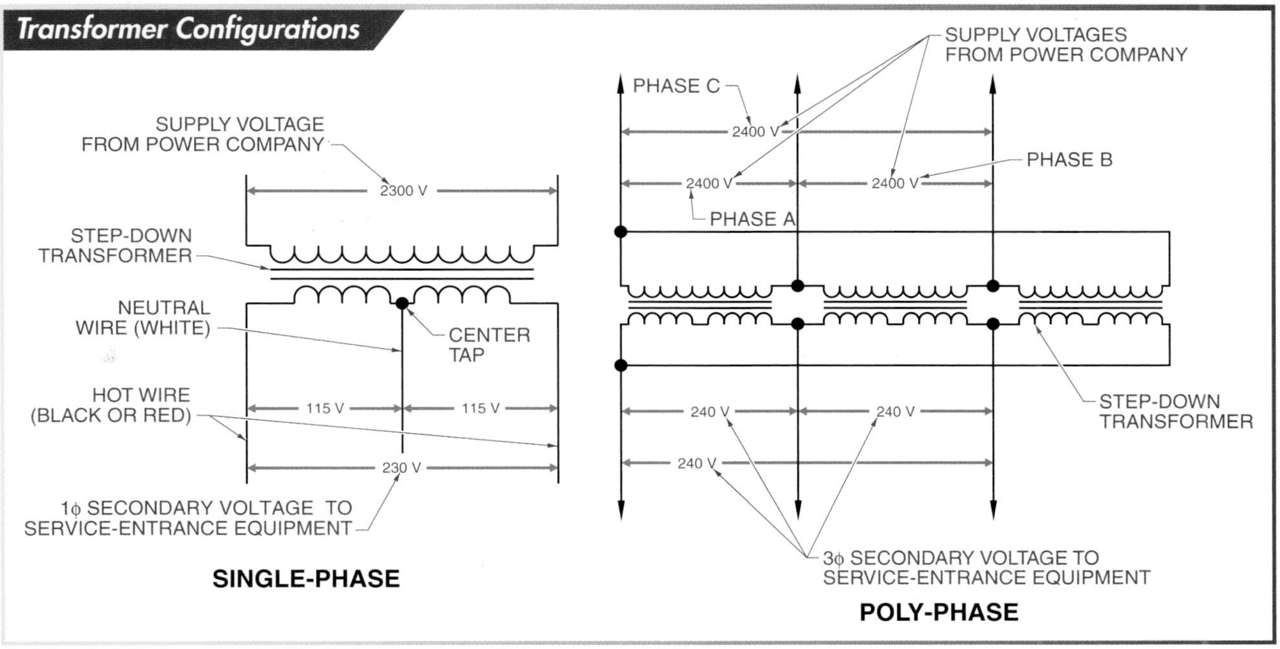

Figure 12-3. Dry-type transformers are available in either single- or poly-phase configurations.

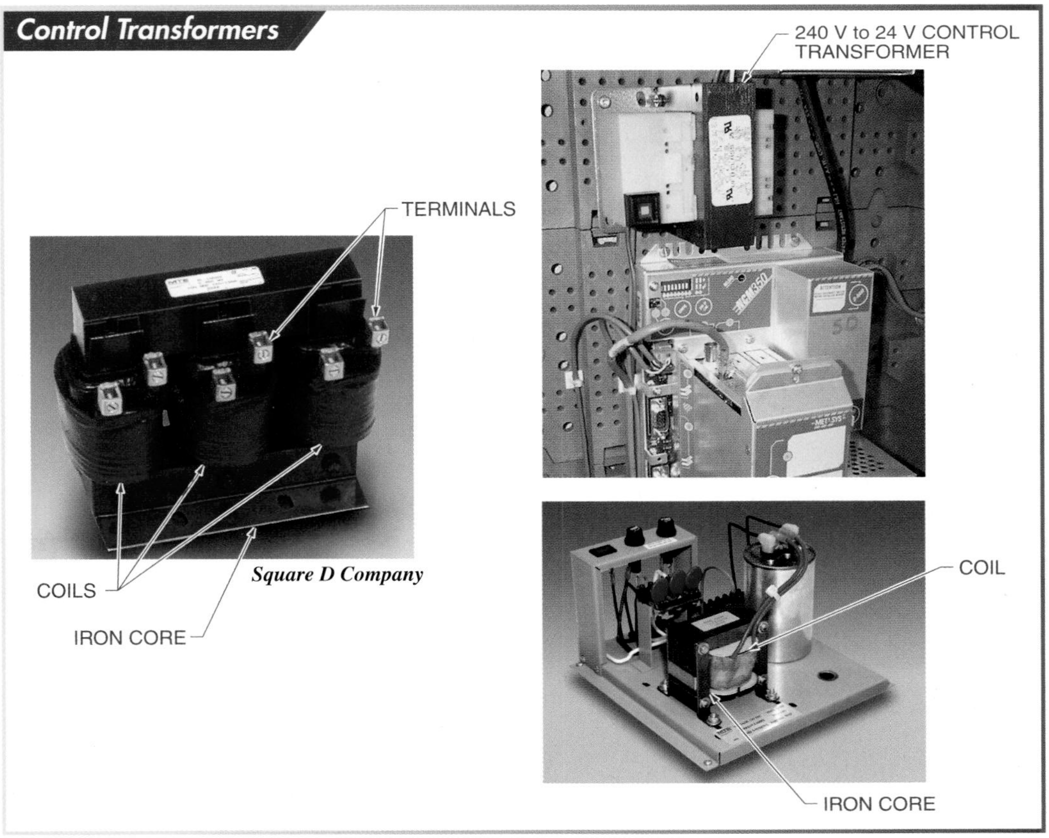

Figure 12-4. Control transformers can be used to provide voltage to operate contactors and relays in various motor control applications.

Liquid-Filled Transformers

As the size of the electrical distribution system increases, the size of the transformers for the system also increases. As the transformers increase in size, natural convection and air-cooled transformers are not capable of dissipating excessive temperature in the transformer windings. Therefore, many larger distribution transformers are cooled by a liquid.

A *liquid-filled transformer* is a transformer that utilizes some form of insulating liquid for immersion of the core and windings of the transformer. Insulating liquid around the core and windings aids in the removal of heat generated by the transformer windings. Liquid-filled transformers commonly have kVA ratings from 150 kVA to 3000 kVA and primary voltages from 2400 V to 13,800 V. The four types of liquid-filled transformers are oil-insulated, askarel, less-flammable liquid-insulated, and nonflammable fluid-insulated. **See Figure 12-5.**

The windings and core of liquid-cooled transformers are immersed in liquid. Some common liquids used are mineral oil, insulating oil, askarel, and less-flammable and nonflammable liquids. *Mineral oil* is a chemically untreated insulating oil that is distilled from petroleum. *Askarel* is a group of nonflammable synthetic chlorinated hydrocarbons that were once used where nonflammable insulating oils were required. A *less-flammable liquid* is an insulating oil which is flammable but has reduced flammable characteristics and a higher fire point. A *nonflammable liquid* is a liquid that is noncombustible and does not burn when exposed to air and a source of ignition. Nonflammable liquids do not have a flash or fire point.

Liquid is used in the transformer to insulate the transformer windings and core from each other and for cooling purposes. The liquid provides a medium which enhances heat transfer and dissipation. Liquid-cooled transformers can operate at higher voltage and current levels without damage to, and deterioration of, conductor insulations.

ABB Power T&D Company, Inc.

Figure 12-5. The core and windings of a liquid-filled transformer are immersed in an insulating liquid that aids in the removal of the heat generated by the windings.

Transformers contain no moving parts and operate on the principle of magnetic induction.

Oil-Insulated. When liquid-filled transformers were introduced, the most common insulation used was mineral oil. This oil is thin enough to circulate freely through the transformer. It does a good job of providing the necessary insulation between the transformer windings and the core. Mineral oil, however, is subject to oxidation and if any moisture enters the oil, the insulating value is dramatically reduced. *Oxidation* is the process by which oxygen mixes with other elements and forms a type of rust-like material.

Another problem associated with the use of mineral oil is that it is flammable. Therefore, oil-insulated transformers should not be located near combustible materials indoors or outdoors.

Askarel. Askarel has been banned by the Environmental Protection Agency and its use as a transformer coolant is being phased out. Askarel transformers, however, are still found throughout the electrical industry.

Askarel is a nonflammable liquid that was used for many years for liquid-filled transformers which were designed to be installed indoors. Because the liquid is nonflammable, askarel transformers were permitted to be installed outside of transformer vaults. Unfortunately, one of the chief components of askarel is polychlorinated biphenyl (PCB). So although the liquid has excellent characteristics for use in transformers, the environmental and health concerns associated with the use of PCBs have led to the banning of askarel for use in liquid-filled transformers. In fact, many askarel transformers are presently being retrofitted with newer, less-flammable liquids. The NEC® still contains provisions for installing askarel transformers.

Less-Flammable Liquids. Because of the problems associated with the use of askarel, liquids for use in transformers have been designed which are flammable but do offer a higher fire point. A *fire point* is the lowest temperature at which a material can give off vapors fast enough to support continuous combustion. The NEC® recognizes transformers that contain these liquids as less-flammable liquid-insulated transformers. Less-flammable liquids are defined in 450.23 as those liquids which are listed and have a fire point of not less than 300°C.

Liquid-Filled Transformers

Type	Liquids Used
Oil-Insulated	Chemically untreated insulating oil
Askarel	Nonflammable insulating oil
Less-Flammable Liquid-Insulated	Reduced flammable insulating oil
Nonflammable Fluid-Insulated	Noncombustible liquid

Listed liquids are those liquids which have been suitably evaluated and meet appropriate designated standards. These liquids do not contain any PCBs but they can burn. Therefore, the NEC® has special requirements for less-flammable liquid-insulated transformers. The installation requirements for these transformers depend on whether the transformer is located inside or outside.

Nonflammable Liquids. Because of the added benefit of having nonflammable liquid-filled transformers, the search continues for liquids that have the nonflammability of askarel without the use of PCBs. The NEC®, in 450.24, recognizes nonflammable fluid-insulated transformers. These transformers are filled with a liquid that has no flash or fire point and does not burn in air.

Autotransformers

An *autotransformer* is a single-winding transformer that shares a common winding between the primary and secondary circuits. They also work on the principle of magnetic induction. These transformers are often used to "buck" or "boost" the supply voltage. **See Figure 12-6.** See Appendix.

The ratio of transformer windings is used to determine if the autotransformer is a step-up (boost) or step-down (buck) transformer. The voltage is increased by a boost transformer and decreased by a buck transformer. Autotransformers are used in industry for a wide variety of applications ranging from doorbell power supplies to deriving a 3ϕ, 4-wire, grounded distribution system from a 3ϕ, 3-wire, ungrounded distribution system.

Mounting and Use

Transformers are often classified according to the method by which they are mounted. Some common transformer mountings are pole, pad, platform, and vault. Outside electrical distribution systems commonly use pole-mounted transformers, which are quite suitable for overhead connections. Indoor electrical distribution systems use floor- or platform-mounted transformers. Wall-mounted or ceiling-hung transformers are permitted in some installations. Some large capacity, high-voltage transformers are designed to be installed in transformer vaults. The NEC® sets installation requirements for transformers in vaults and the construction of the vault.

Finally, transformers can be classified by their usage. There are many special-application transformers in use throughout the electrical industry. These transformers include instrument, potential, current, and isolation transformers. **See Figure 12-7.**

An *instrument transformer* is a transformer used to reduce higher voltage and current ratings to safer and more suitable levels for the purposes of control and measurement. A *potential transformer* is a transformer which steps down higher voltages while allowing the voltage of the secondary to remain fairly constant from no-load to full-load conditions.

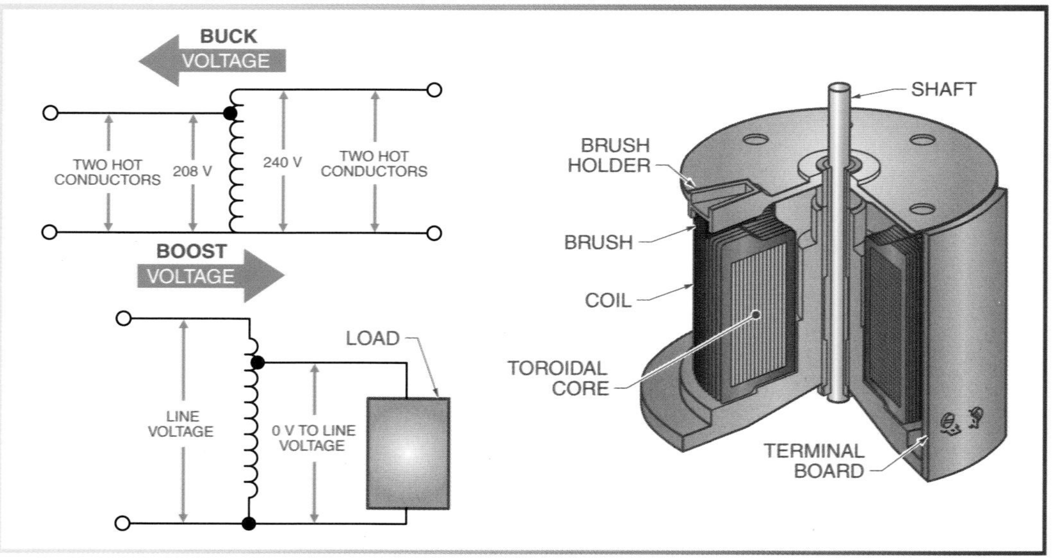

Figure 12-6. Autotransformers are single-winding transformers that share a common winding between the primary and secondary circuits.

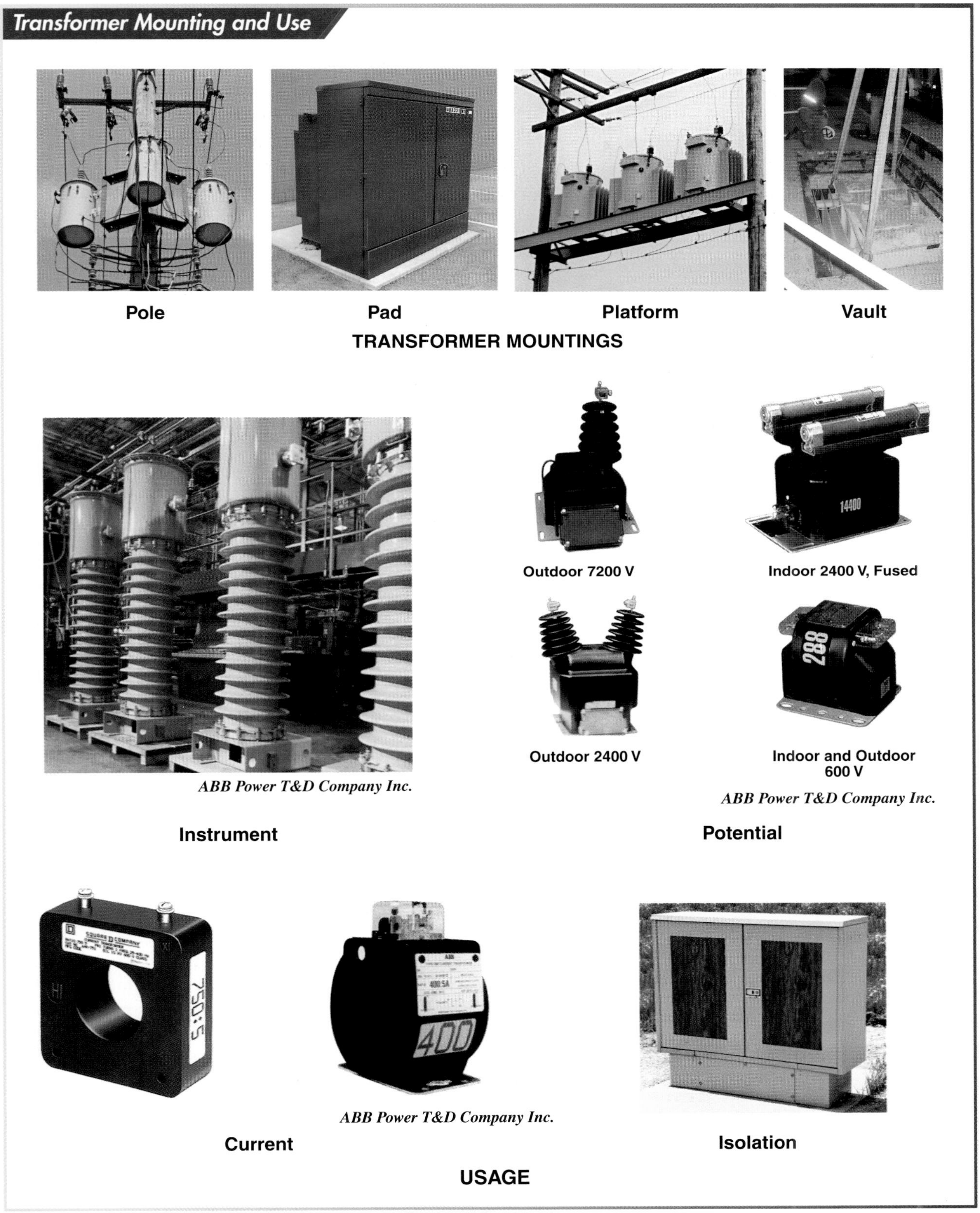

Figure 12-7. Electrical distribution and control applications require transformers of various types and configurations.

Transformer Installation

Due to severe space constraints, many electrical designers are requiring that transformers be hung from ceilings. The support system should be adequate for the weight of the transformer. Be particularly careful of modifications to transformers that are not designed to be hung. In addition, access to the space is required. In general, transformers should be installed in a readily accessible location. Although such installations are not readily accessible, 450.13(B) permits installation of transformers 50 kVA or less in hollow spaces of buildings not permanently closed in by structure, such as ceilings, with some considerations, such as ventilation and separation requirements.

Floor-mounted transformers have similar installation requirements that need to be considered when designing and installing electrical distribution systems. Perhaps the biggest pitfall occurs when installers place transformers below panelboards which are either supplied by or from the transformers. Check the work space clearance requirements and dedicated equipment space provisions of 110.26 to ensure that floor-mounted transformers do not interfere with the working space needed for other electrical equipment.

A *current transformer* is a transformer that creates a constant ratio of primary to secondary current instead of attempting to maintain a constant ratio of primary to secondary voltage. Current transformers are used extensively throughout the industry to measure large values of current that would normally be very difficult to measure.

An *isolation transformer* is a transformer that utilizes a shield between the primary and secondary windings and a transformer ratio of 1:1 to ensure that the load is separated from the power source. Isolation transformers are constructed with a shield between the primary and secondary windings. Isolation transformers can be used to provide "clean" power for information technology system equipment.

TRANSFORMER INSTALLATION

As with most electrical equipment, proper consideration must be given to several important factors before deciding on the best installation procedures for transformers. Since the transformer is an integral part of the electrical distribution system, planning for factors like physical protection, ventilation, grounding, and location helps to ensure the reliability of the electrical system. Transformers are widely used electrical devices which for the most part are extremely reliable. Transformers can create a problem for the electrical distribution system when proper installation procedures are not followed or required maintenance practices are neglected.

Autotransformers – 450.4, 450.5

An autotransformer is a transformer that utilizes one winding as both the primary and secondary of the transformer. Autotransformers have many applications in the field, due in large part to their flexibility and high-efficiency. They are most commonly used to adjust or otherwise "buck or boost" voltage. The major disadvantage is that because the autotransformer utilizes the same winding for both the primary and the secondary, there is no isolation between the load and the source.

Section 450.4(A) contains the overcurrent provisions for autotransformers rated 600 V or less. A general rule is that overcurrent protection for autotransformers must not exceed 125% of the rated full-load input current of the autotransformer. An exception to this general rule permits an increase to 167% where the rated input current of the autotransformer is less than 9 A. An overcurrent device is prohibited by 450.4(A) from being installed in series with the shunt winding of an autotransformer. The shunt winding is the winding common to both the primary and secondary winding. If a transformer is field-connected to operate as an autotransformer, 450.4(B) requires that the transformer be identified for use at the elevated voltage.

Section 450.5 covers the provisions for grounding autotransformers. Section 450.5(A) lists four general provisions for grounding autotransformers used to create 3ϕ, 4-wire from 3ϕ, 3-wire distribution systems. Section 450.5(A)(1–4) covers the provisions for point of connection, overcurrent protection, transformer fault sensing and rating. A 2005 NEC® revision clarified that zig-zag connected transformers are not permitted to be installed on the load side of any system grounding connection.

Guarding – 450.8

Physical protection of transformers can be provided by covering, shielding, fencing, enclosing, or otherwise prohibiting the approach or contact by persons or objects. Transformers are required to be protected from physical damage by 450.8(A). Dry-type transformers shall be enclosed in a case or enclosure to protect against the insertion of foreign objects per 450.8(B).

In general, equipment such as switches shall not be installed within a transformer enclosure unless the equipment operates at 600 V, nominal, or less and only qualified persons have access to such equipment. In addition, 450.8(C) requires that all energized parts within the enclosure be suitably guarded. Guarding can be accomplished by limiting access to the equipment, by use of partitions or barriers, or by elevation to restrict access to the energized parts. See 110.27 and 110.34. In the event that exposed live parts are present, 450.8(D) requires that the operating voltage of the live parts be clearly indicated by visible markings either on the equipment or on adjacent structures.

Energized parts of transformers shall be guarded per 110.27 and 110.34. Guarding transformers accomplishes two objectives. First, it protects the transformer from objects that could cause physical damage and threaten the reliability of the electrical system. Next, it protects workers and personnel in the vicinity of the transformer from inadvertent contact that could result in physical harm. All transformers are required to be guarded and marked by signs or other markings on the equipment indicating the operating voltage of the transformer. **See Figure 12-8.**

Ventilation – 450.9

Whenever electrical equipment or conductors are installed, care must be taken to protect them from the effects of heat. Heat is a natural by-product of current flow. Excessive heat, or any heat above the temperature rating of the conductor or equipment, can lead to deterioration of the conductors' insulation. Excessive heat can damage the insulation of transformer windings, leading to transformer failure. Adequate ventilation shall be provided for all transformer installations so that heat cannot lead to an excessive temperature rise per 450.9. Transformers, like motors, are rated with a maximum temperature rise. *Temperature rise* is the amount of heat that an electrical component produces above the ambient temperature. **See Figure 12-9.**

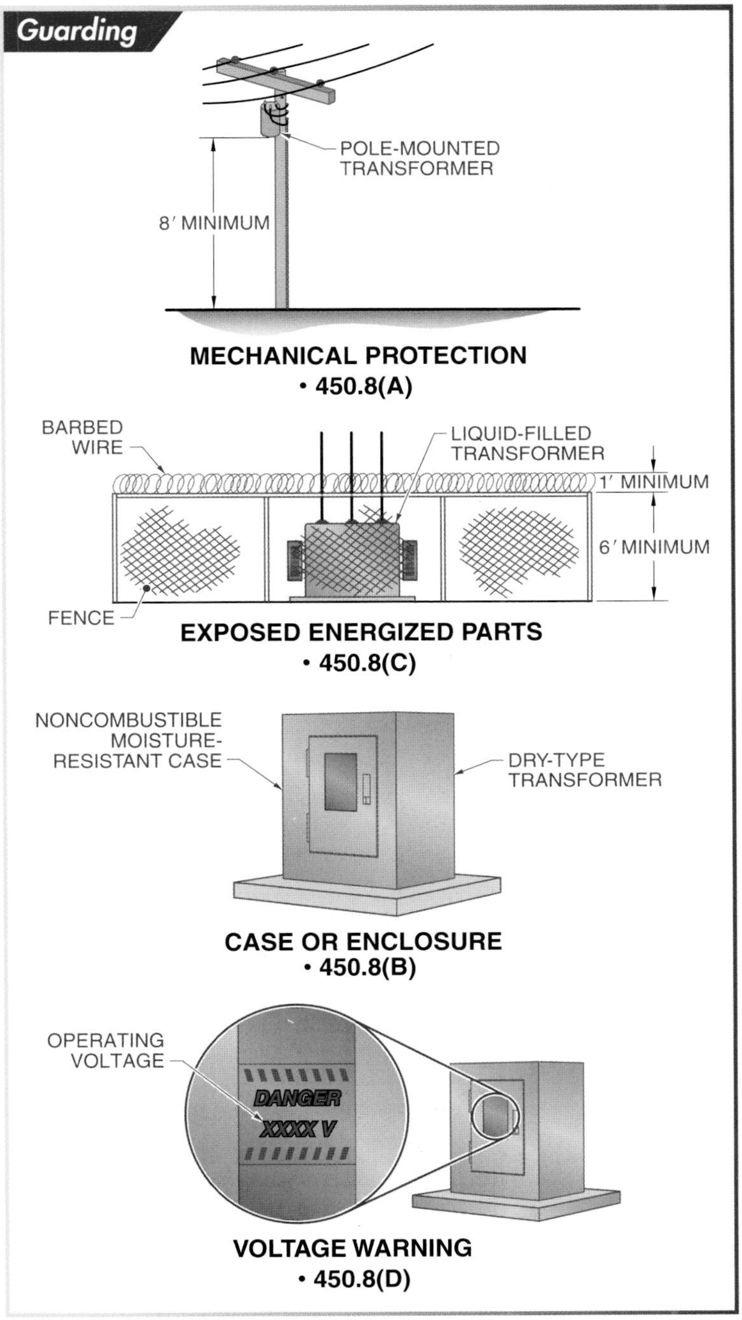

Figure 12-8. All transformers are required to be guarded.

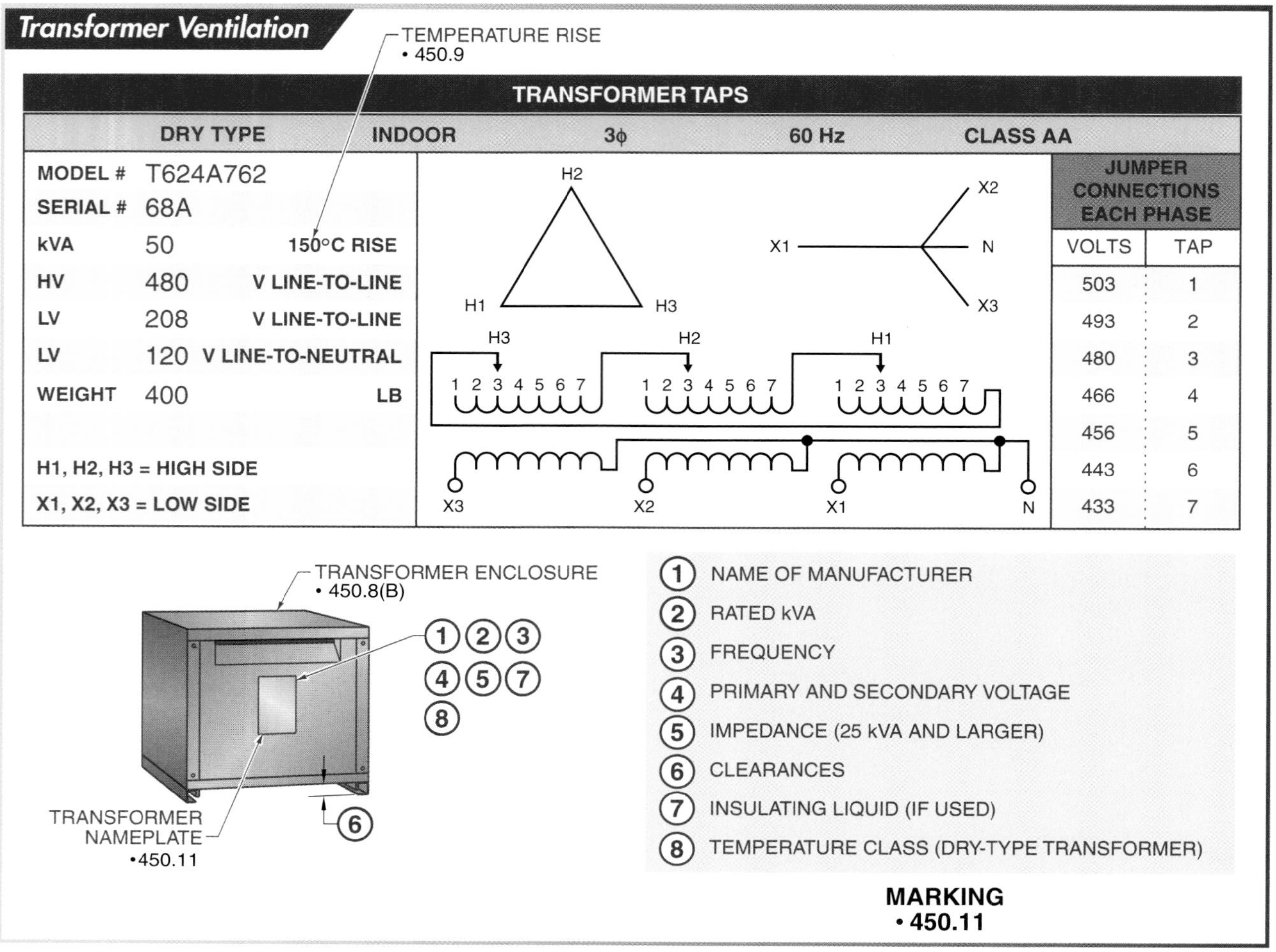

Figure 12-9. All transformers are required to be ventilated.

In the 1990 NEC®, a further requirement was added to the ventilation requirements for transformers. Transformers shall be installed in a manner which does not block or obstruct openings which are designed for cooling purposes. In addition, transformers are now required to be marked with a minimum distance or clearance from walls or other obstructions to facilitate the dissipation of heat.

Grounding – 450.10

Transformers are required to be grounded per Article 250. All metal, non-current-carrying parts of the transformer and surrounding metal objects, like fences and other guarding means, shall also be grounded and bonded. When the transformer is used as a source of a separately derived system, it shall be grounded per 250.30.

Location – 450.13

Transformers can be installed either indoors or outdoors. Because of the potential hazards associated with some types of transformers, special installation requirements apply when transformers are installed indoors. Some transformers, such as the less-flammable liquid-insulated (LFLI) type, also have special requirements when installed outdoors. In general, transformers and transformer vaults shall be installed in locations which are readily accessible to qualified personnel. Transformers require some maintenance and may even need to be replaced. For this reason, they cannot be installed where they are blocked or otherwise obstructed in a manner that quick access is denied.

All transformers shall be installed in a location per 450.13 that is readily accessible to qualified persons. Dry-type transformers 600 V, nominal, or less are not required to be installed in a readily accessible space per 450.13(B). This section permits installation in spaces above hung ceilings, etc., provided the transformers are rated 50 kVA or less, adequate ventilation is provided, and proper clearances from combustible materials are maintained. **See Figure 12-10.**

Delta Star, Inc.

Mobile transformer substations are used as backup to help provide emergency, temporary, or new service.

Disconnecting Means – 450.14. In the 2011 NEC®, new provisions were added to Article 450 requiring that all transformers have a disconnecting means located in sight of the transformer or in a remote location. If the disconnecting means is located in a remote location, 450.14 requires that the disconnecting means be lockable. In addition, the location of the disconnecting means is required to be field marked on the transformer. This requirement for a disconnecting means does not apply to Class 2 or Class 3 transformers.

Dry-Type – 450.21; 450.22. Indoor requirements for dry-type transformers depend on the size of the transformer. Dry-type transformers installed indoors, and not over 112½ kVA, shall have a clearance from combustible materials of at least 12″ per 450.21(A). **See Figure 12-11.** The exception to this section does permit the separation requirement to be omitted provided the transformer is rated 600 V or less and is completely enclosed with or without ventilation openings.

Additionally, 450.21(A) allows a separation of less than 12″ from combustible materials when a fire-resistant, heat-insulating barrier is installed between the transformer and the combustible material. For the purposes of this section, fire-resistant means the equivalent of a 1-hour minimum fire rating. **See Figure 12-12.**

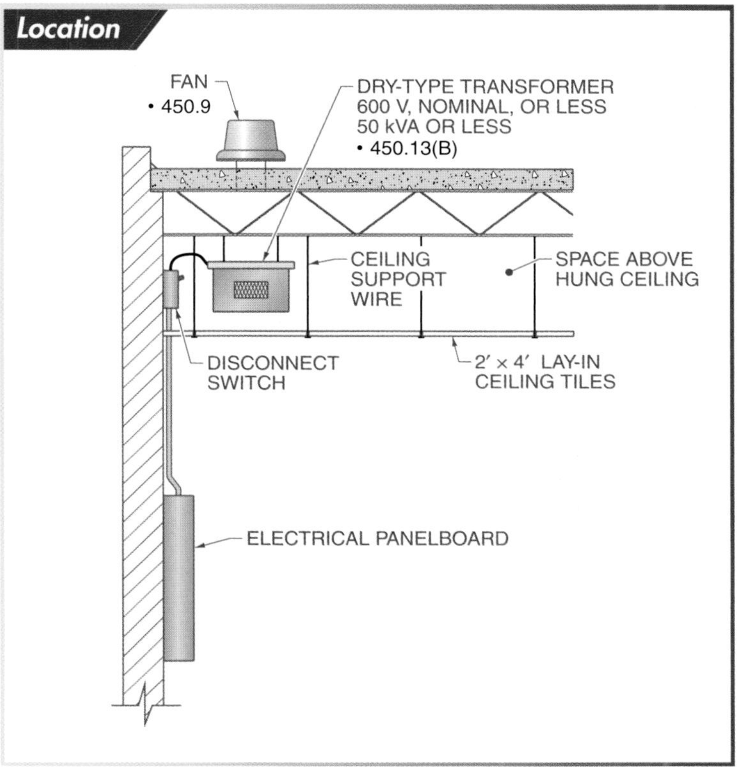

Figure 12-10. Transformers are permitted to be installed in spaces above hung ceilings provided adequate ventilation is provided.

Separately Derived Systems

Designers and installers of transformers should consider the requirements of 250.30 when installing electrical distribution systems. Specific components shall be used when grounding separately derived systems. If possible, try to install the transformers in locations that make meeting the requirements of 250.30 easier. The closest of two options shall be selected as the grounding electrode of the separately derived system per 250.30(A)(4). The first option is the metal water pipe grounding electrode per 250.52(A)(1). The second option is the structural metal grounding electrode per 250.52(A)(2). Where neither of the two applications is available, designers and installers must use one of the electrodes specified in Section 250.52(A).

Additionally, there are bonding requirements for separately derived systems. Be certain to review Article 250, especially 250.104(D), to ensure the required bonding is performed. Separately derived systems are common components of many electrical distribution systems. Careful consideration should be given to the location of such systems within the building or structure.

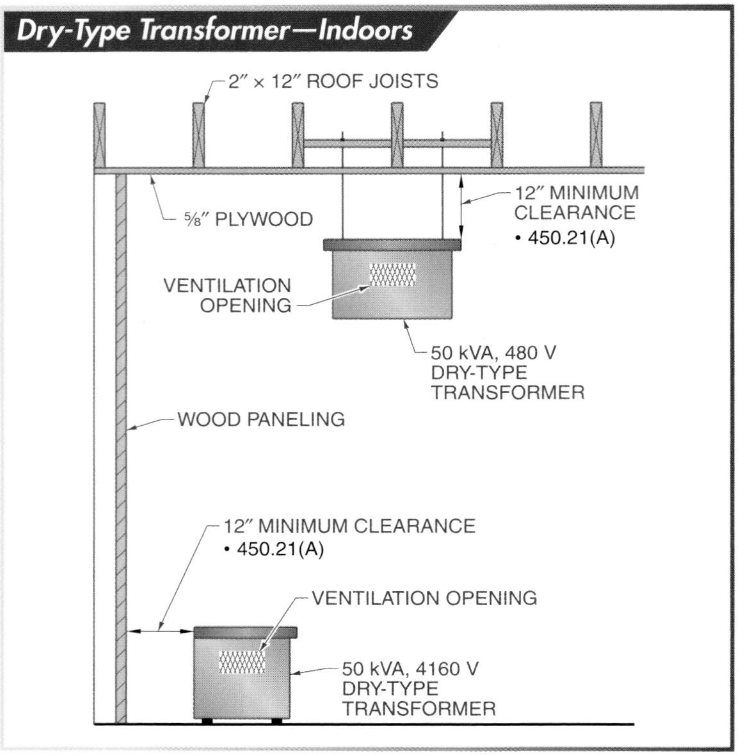

Figure 12-11. Dry-type transformers less than 112½ kVA installed indoors shall have a minimum clearance of 12″ from combustible materials.

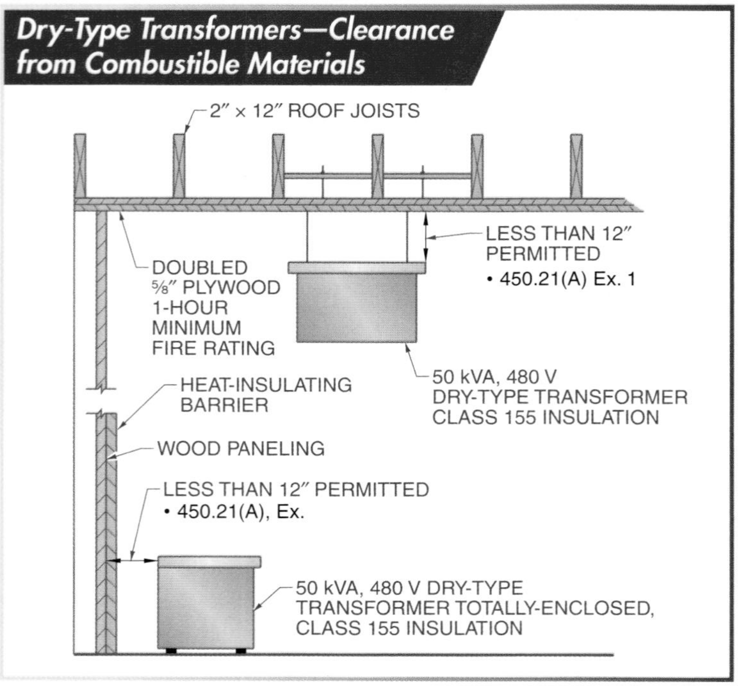

Figure 12-12. Clearance from combustible materials is not required when a fire-resistant barrier is provided or the transformer is the totally enclosed type.

In general, transformers rated over 112½ kVA are required to be installed in a fire-resistant transformer room per 450.21(B). Any construction methods or materials that provide at least a 1-hour minimum fire rating are suitable.

Two exceptions permit transformers rated over 112½ kVA to be installed in non-fire-resistant transformer rooms. Per 450.21(B), Ex. 1, transformers with Class 155 or higher insulation systems which are separated from combustible materials by a fire-resistant, heat-insulating barrier need not be installed in a transformer room constructed of fire-resistant materials. In addition, these transformers need not be in a transformer room if they are separated from combustible materials by a distance of not less than 6′ horizontally and 12′ vertically. Per 450.21(B), Ex. 2, transformers with Class 155 or higher insulation systems which are totally enclosed, other than ventilation openings, may be installed in transformer rooms constructed of non-fire-resistant materials. The insulation system temperature class is required by 450.11 to be marked on the nameplate of each dry-type transformer. **See Figure 12-13.**

Transformers Above Ceilings

Several factors should be considered when transformers are installed above ceilings. Most importantly, there shall be adequate ventilation for the transformer. For this reason, the maximum rated transformer that is permitted to be installed in these spaces is 50 kVA. The ventilation requirements of 450.9 shall also be maintained. This can be accomplished by the use of exhaust fans and/or HVAC equipment for the space.

Secondly, be sure to provide adequate access to service the transformers. Notice that 450.13(B) permits these transformers in spaces that are not permanently closed in by the structure. This permits installation above ceilings with tiles designed to be removed, but prohibits the installation above drywall ceilings that are permanently closed in.

Per 450.21(C), dry-type transformers rated over 35,000 V shall be installed in transformer vaults to reduce the hazard to building personnel and occupants. The transformer vault shall comply with Article 450, Part III.

Dry-type transformers installed outdoors shall meet the installation requirements of 450.22. In general, the transformer enclosure for transformers installed outdoors shall be weatherproof. The enclosure shall be constructed so that exposure to weather does not interfere with the transformer operation. Additionally, if the transformer rating exceeds 112½ kVA, the transformer is not permitted to be installed within 12″ of any combustible materials used in the construction of the building. However, if the transformer has Class 155 or higher insulation systems and is completely enclosed, other than the ventilation openings, the 12″ separation is not required. **See Figure 12-14.**

Less-Flammable Liquid-Insulated (LFLI) Transformers – 450.23. LFLI transformers are filled with a liquid that has reduced flammability characteristics but is not completely nonflammable. LFLI transformers installed indoors shall meet one of three general requirements. **See Figure 12-15.**

Per 450.23(A)(1), LFLI transformers are permitted to be installed indoors in Type I or Type II buildings when the transformer is rated at 35,000 V or less, the liquid is listed and installed per the listing, a liquid confinement area is provided for leaks or spills, and no combustible materials are stored in the area. Type I and II buildings generally are constructed of noncombustible materials or methods. See *ANSI/NFPA 220-2006, Types of Building Construction.*

Askarels

Askarels are nonflammable liquids used in transformers. Askarels have been banned by the Environmental Protection Agency (EPA) because of concerns with polychlorinated biphenyls (PCBs), a class of poisonous chemical present in askarels. The NEC® requires that askarel-insulated transformers installed indoors and rated over 25 kVA shall be furnished with a pressure-relief vent. Because of special handling and disposal requirements, EPA Regulation 40 CFR 761 requires labeling of transformers and other equipment that contains PCBs. The EPA office should be consulted for proper PCB disposal information.

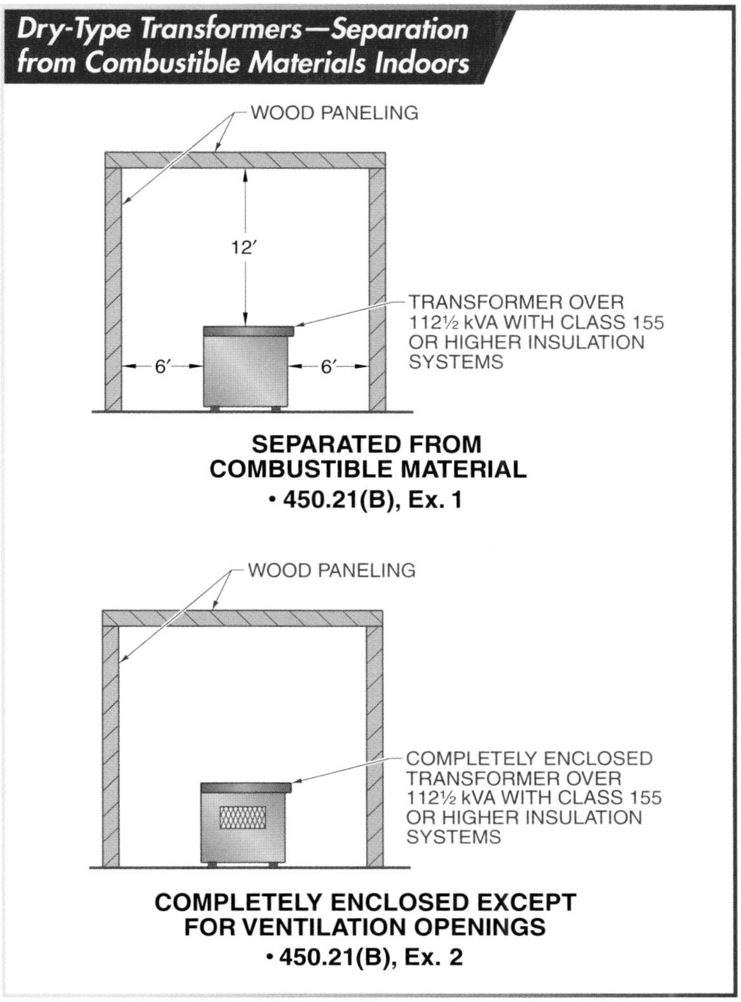

Figure 12-13. Transformers larger than 112½ kVA, with Class 155 or higher insulation systems, are permitted to be installed indoors if separated from combustible materials or completely enclosed.

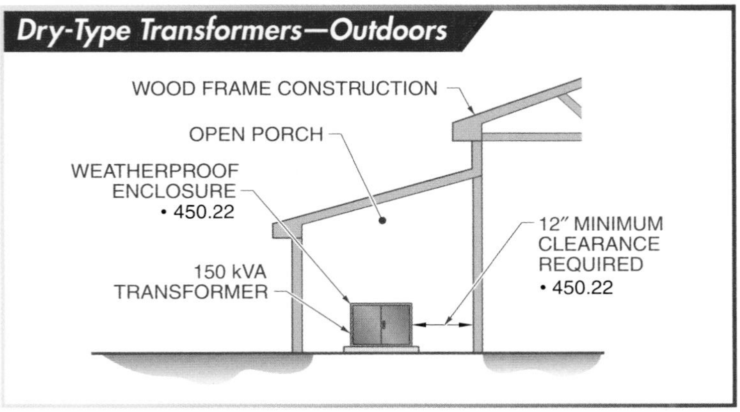

Figure 12-14. Transformers larger than 112½ kVA installed outdoors shall be located at least 12″ from combustible materials unless the transformer has a Class 155 or higher insulation system and is totally enclosed.

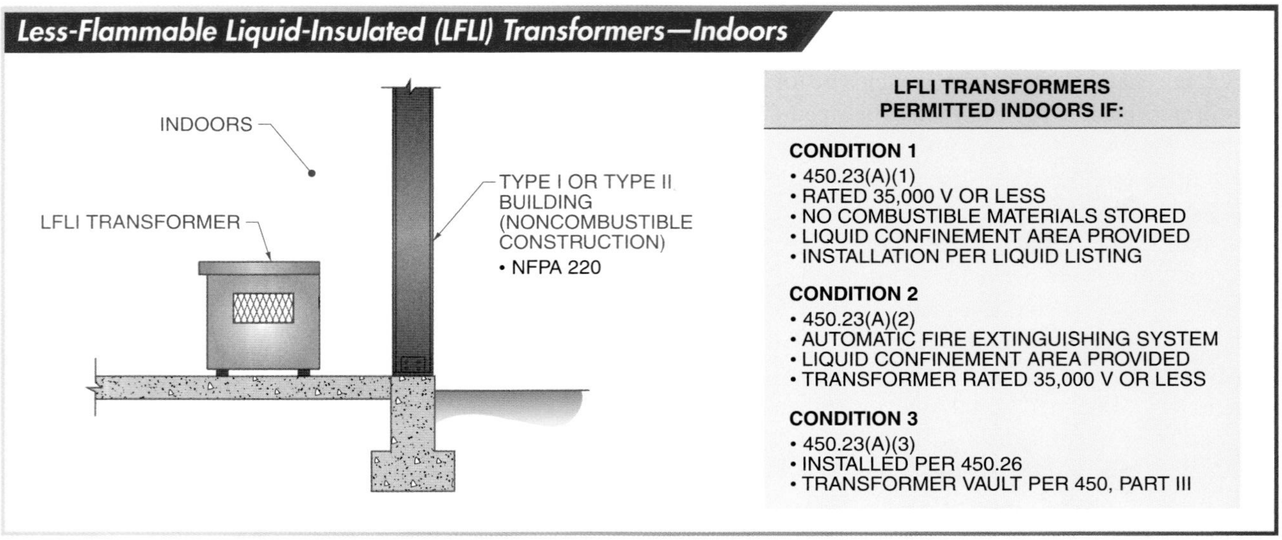
Less-Flammable Liquid-Insulated (LFLI) Transformers—Indoors
INDOORS
LFLI TRANSFORMER
TYPE I OR TYPE II BUILDING (NONCOMBUSTIBLE CONSTRUCTION)
• NFPA 220
LFLI TRANSFORMERS PERMITTED INDOORS IF:
CONDITION 1
• 450.23(A)(1)
• RATED 35,000 V OR LESS
• NO COMBUSTIBLE MATERIALS STORED
• LIQUID CONFINEMENT AREA PROVIDED
• INSTALLATION PER LIQUID LISTING
CONDITION 2
• 450.23(A)(2)
• AUTOMATIC FIRE EXTINGUISHING SYSTEM
• LIQUID CONFINEMENT AREA PROVIDED
• TRANSFORMER RATED 35,000 V OR LESS
CONDITION 3
• 450.23(A)(3)
• INSTALLED PER 450.26
• TRANSFORMER VAULT PER 450, PART III

Figure 12-15. LFLI transformers installed indoors shall meet one of three general requirements.

A *Type I building* is a building in which all structural members (walls, columns, beams, girders, trusses, arches, floors, and roofs) are constructed of approved noncombustible or limited-combustible materials. Type I buildings have fire resistance ratings that range from 0 to 4 hours.

A *Type II building* is a building that does not qualify as Type I construction in which the structural members (walls, columns, beams, girders, trusses, arches, floors, roofs, etc.) are constructed of approved noncombustible or limited-combustible materials. Type II buildings have fire resistance ratings that range from 0 to 2 hours.

Building Type Fire Rating

Type	Fire Resistance Rating
I	0–4 hours
II	0–2 hours

Per 450.23(A)(2), LFLI transformers are permitted to be installed indoors when the transformer is rated at 35,000 VA or less, a liquid confinement area is provided, and the area in which the transformer is located is protected with an automatic fire extinguishing system.

Per 450.23(A)(3), LFLI transformers are permitted to be installed indoors in accordance with the provisions for oil-insulated transformers. See 450.26. In general, the LFLI transformer has to be installed in a transformer vault meeting the requirements of Article 450, Part III.

LFLI transformers are permitted to be installed outdoors per 450.23(B). In general, these transformers can be installed outdoors where attached to, adjacent to, or even located on the building roof if one of two conditions is met. **See Figure 12-16.** Per 450.23(B)(1), LFLI transformers are permitted to be installed outdoors if the building is classified as a Type I or Type II building and the installation complies with all of the instructions included in the listing of the liquid.

Additional safeguards may be required if the transformer is installed adjacent to window and door openings, fire escapes, and other combustible materials per 450.23(B)(1), Informational Note. While Informational Notes are not mandatory, in this case, additional consideration should be given because of the potential hazards involved.

Per 450.23(B)(2), the second condition under which LFLI transformers can be installed outdoors is when they are installed in accordance with the requirements for oil-insulated transformers installed outdoors. See 450.27. Basically, LFLI transformers can be installed outdoors where one or more of the following safeguards is applied: space separations, fire-resistant barriers, automatic fire suppression systems, or enclosures.

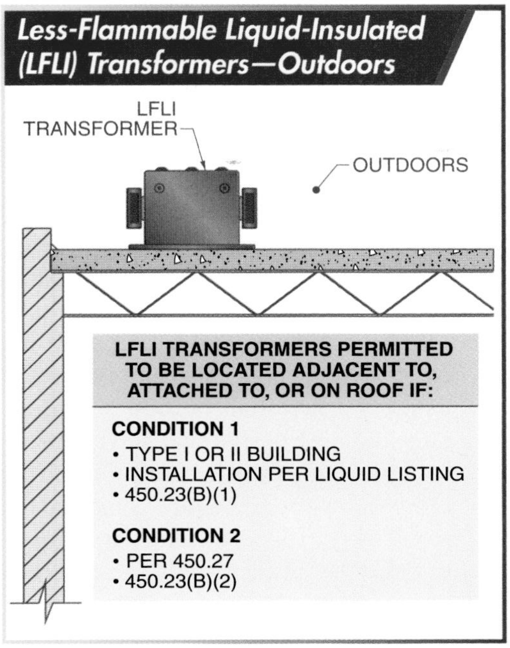

Figure 12-16. LFLI transformers installed outdoors shall meet one of two general requirements.

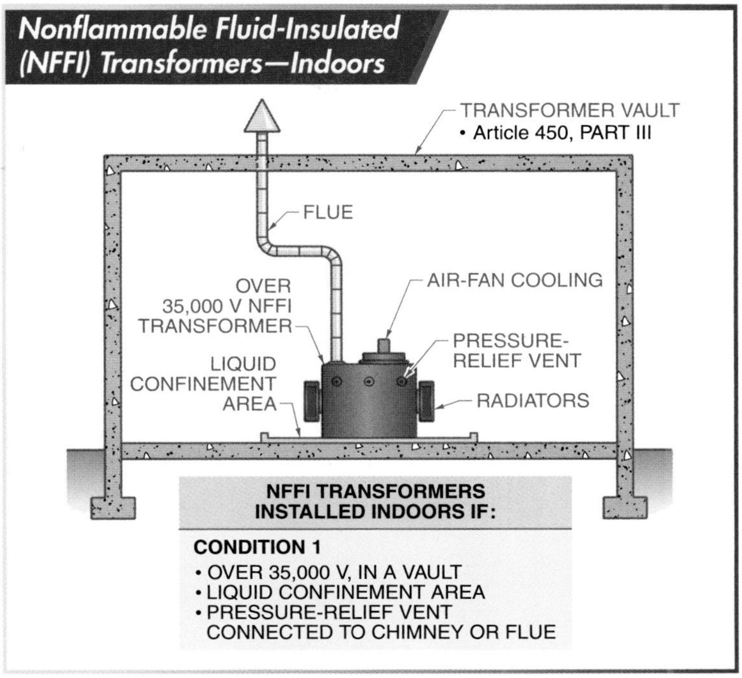

Figure 12-17. Nonflammable fluid-insulated transformers over 35,000 V and installed indoors shall be installed in a transformer vault.

Nonflammable Fluid-Insulated (NFFI) Transformers – 450.24. NFFI transformers contain a liquid that is nonflammable. Such a liquid does not have a flash or fire point; therefore, NFFI transformers can be installed in either indoor or outdoor locations with very few limitations.

Limitations only apply to NFFI transformers rated over 35,000 V. These transformers, when installed indoors, shall be installed in a vault, a liquid confinement area shall be provided, and the transformer shall be provided with a pressure-relief vent. Gases can be a byproduct of these types of transformers, so the transformer shall provide means for absorbing any of these gases, or the pressure-relief vent shall be connected to a chimney or flue for removal purposes. **See Figure 12-17.**

Askarel-Insulated Transformers – 450.25. Askarel-insulated transformers can be installed indoors per 450.25. Askarel is a nonflammable liquid that contains PCBs which are harmful to both humans and the environment. Great care and proper disposal procedures should be followed when working with askarel-insulated transformers.

Askarel-insulated transformers that are rated over 25 kVA and installed indoors shall be equipped with a pressure-relief vent to relieve gases which may be generated from arcing inside the transformer. If the area in which the transformer is installed is poorly ventilated, then the pressure-relief vent shall be connected to a chimney or flue to remove the gases from the building. Askarel-insulated transformers installed indoors and rated over 35,000 V shall be installed in a transformer vault. **See Figure 12-18.**

Oil-Insulated Transformers – 450.26; 450.27. Oil-insulated transformers contain mineral oil. They shall be installed in a transformer vault when installed indoors. There are, however, several installation methods that do not require these transformers to be installed in a transformer vault.

Per 450.26, Ex. 1, oil-insulated transformers rated up to and including 112½ kVA are permitted to be installed indoors in a vault constructed of reinforced concrete not less than 4″ thick. Vaults constructed under Article 450, Part III generally require a larger wall thickness than 4″. **See Figure 12-19.**

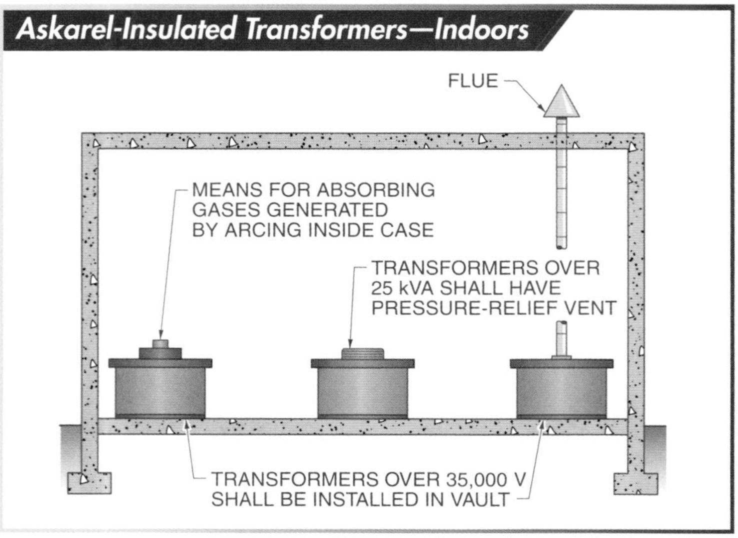

Figure 12-18. Askarel-insulated transformers installed indoors and rated over 35,000 V shall be installed in a transformer vault.

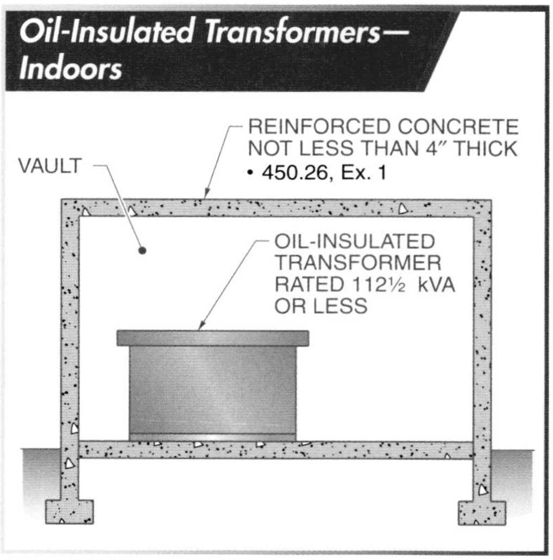

Figure 12-19. Oil-insulated transformers less than 112½ kVA can be installed indoors in a transformer vault constructed of reinforced concrete not less than 4″ thick.

Per 450.26, Ex. 2, oil-insulated transformers rated 600 V, nominal, or less are not required to be installed in a transformer vault if the installation is designed so that a fire in the transformer oil will not ignite other materials. The total capacity in one location cannot exceed 10 kVA in areas of the building classified as combustible, or 75 kVA in areas of the building classified as fire-resistant.

Per 450.26, Ex. 3, electric-furnace transformers with ratings of 75 kVA or less are allowed to be installed outside of a transformer vault if measures are taken to prevent a transformer oil fire from spreading to other materials.

Per 450.26, Ex. 4, a transformer that is an integral part of charged particle accelerating equipment need not be installed in a transformer vault if three conditions are met. First, the transformer total rating shall not exceed 75 kVA. Second, the supply voltage to the transformer shall be 600 V or less. Third, the building design shall have suitable means to prevent a transformer oil fire from spreading to other combustible materials.

Per 450.26, Ex. 5, oil-insulated transformers are permitted to be installed in detached buildings not constructed under the requirements for transformer vaults. The detached building shall not present a threat of fire to any other building or structure. Such a detached building shall be used for the purposes of electrical service or supply and it shall be accessible to qualified persons only.

Per 450.26, Ex. 6, oil-insulated transformers are permitted to be installed indoors without a transformer vault if they are used in portable or mobile mining equipment. Leaking fluid shall be drained to the ground, personnel shall have safe egress, and a minimum ¼″ steel barrier shall be provided for protection of personnel.

Oil-insulated transformers installed outdoors shall meet the installation requirements of 450.27. Care shall be taken to safeguard combustible materials, combustible buildings, parts of buildings, and other life safety apparatus, such as fire escapes, door openings, and window openings from the hazards associated with the transformer installation. At least one safeguard shall be provided to reduce the hazard when the transformer installation presents a risk of fire. The safeguard selected shall be designed to correspond to the degree of the hazard. Safeguards recognized by 450.27 include space separation, fire-resistant barriers, automatic fire suppression systems, and confinement enclosures. **See Figure 12-20.**

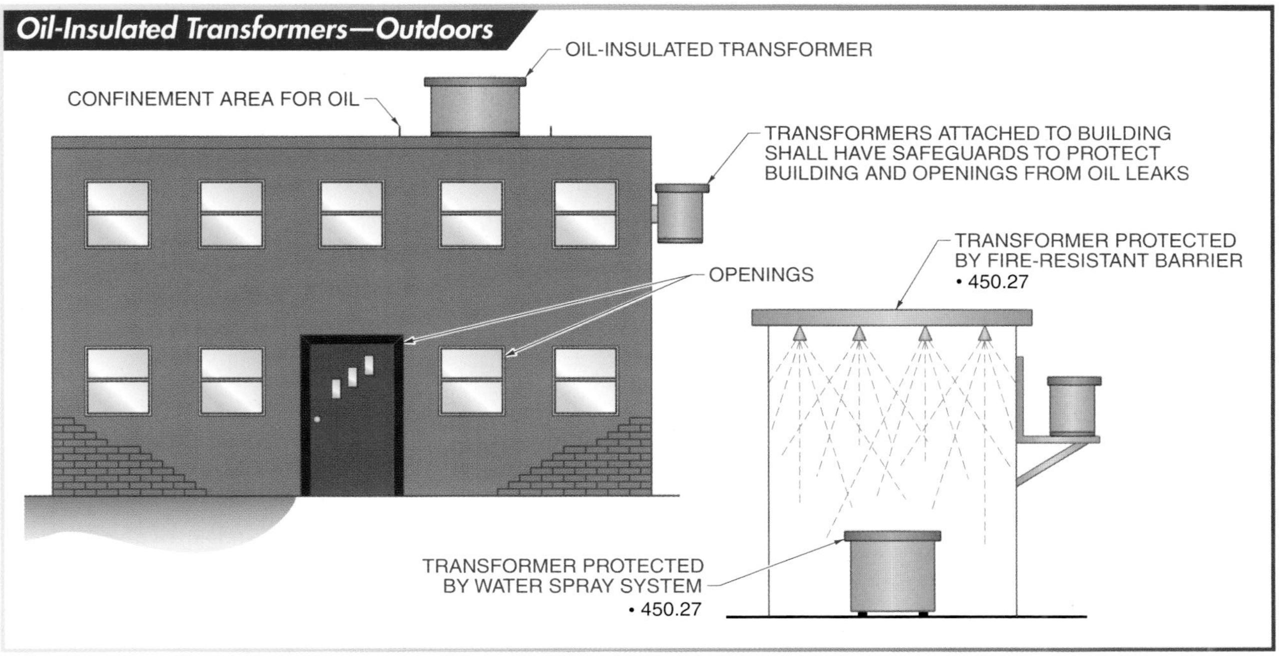

Figure 12-20. To reduce the risk of fire, additional safeguards are required when installing oil-insulated transformers outdoors.

Equipment Rated Over 600 V – Article 490

In the 1999 NEC® a new article 490 was added to address the special requirements for electrical equipment that operates at over 600 V, nominal. The article includes new material and relocated material that was formerly found in Articles 110, 240, 300, and 710. Part I of Article 490 contains the general provisions, Part II contains specific equipment provisions, and Part III covers metal-enclosed power switchgear and industrial control assemblies. Parts IV and V contain special requirements for mobile and portable equipment and electrode-type boilers. Although this book focuses on the under 600 V requirements for installing electrical distribution systems and equipment, Article 490 is an important article that should be carefully reviewed when equipment operating at over 600 V is installed.

TRANSFORMER VAULTS

The requirements for constructing a transformer vault are covered by Article 450, Part III. Transformer vaults serve two purposes. First, they provide a means to isolate potentially hazardous electrical components from unqualified personnel. Second, transformer vaults are, by design, intended to contain, or at least slow down, any fire or combustion that may occur as the result of a transformer malfunction. In order to accomplish these objectives, three factors should be considered in designing transformer vaults. The first concern is with the actual location of the transformer vault. The second concern involves the actual construction of the vault. The third concern is somewhat related, and ensures that adequate ventilation is provided for the transformer vault. **See Figure 12-21.**

Explosion Vents

High pressures are possible when an electrical fault occurs under oil. The pressure could burst the steel tank of a transformer. An explosion vent is a pipe, 4″ in diameter or greater, which extends a few feet above the cover of a transformer and is curved toward the ground at the outlet end of the pipe. A diaphragm fitted at the curved end breaks at a relatively low pressure to release the forces from within the transformer. The diaphragm may be glass or thin phenolic sheeting (0.16″).

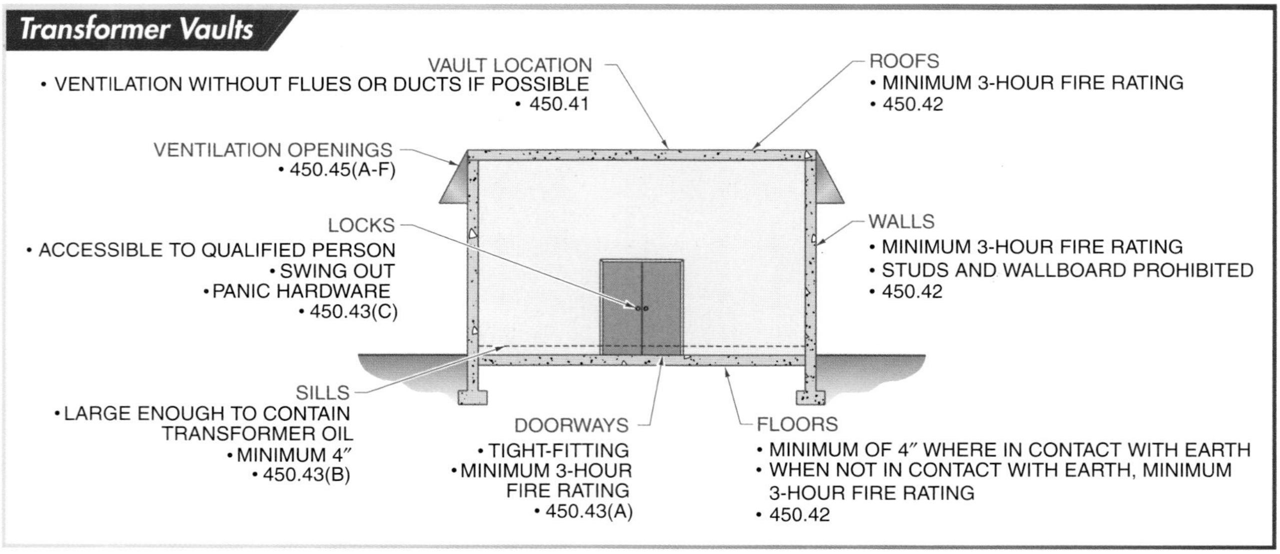

Figure 12-21. Transformer vaults help to ensure that hazards from explosion or fire are not transmitted to adjacent buildings or structures.

Location – 450.41

Where possible, transformer vaults should be located where they can be easily vented in accordance with their ventilation requirements. The use of flues and ducts should be avoided wherever practical. Another important consideration, which is not required by the NEC®, is the placement of the vault within the building. Due to the increase of acts of terrorism and the severity of storms, tornadoes, hurricanes, etc., careful consideration should be given to the placement of transformer vaults. Although no location is immune from these occurrences, every possible precaution should be taken to minimize the potential damage.

Walls, Roofs, and Floors – 450.42

Because a primary objective is to contain any products of combustion within the transformer vault, construction materials which ensure a minimum fire resistance of three hours for walls, roofs, and floors shall be used per 450.42. Standard framing members and drywall are not permitted to be used to achieve the 3-hour fire rating. Where concrete masonry units (CMU) are used to construct a 3-hour rated wall, the cells, in most cases, must be filled with an approved loose fill material. Unfilled cells are not permitted.

If the vault floor is in direct contact with the earth, the floor shall be constructed of at least 4″ of concrete. If the vault floor is not in direct contact with the earth, it shall be constructed for the load imposed on it and must have a minimum fire resistance rating of 3 hours.

Doorways – 450.43

Doors for transformer vaults shall have a minimum 3-hour fire rating unless the vault is protected with an automatic sprinkler system, water spray, carbon dioxide, or halon system. If such a system is provided, the doors are only required to have a 1-hour fire rating. A door sill or curb large enough to contain all of the oil from the largest transformer, shall be included in the vault design. The minimum height permitted for the sill or curb is 4″.

Another design consideration for transformer vaults is the door construction. For safety considerations, doors shall prohibit the entrance to unqualified personnel. Doors in transformer vaults are required to swing out and shall include panic hardware. *Panic hardware* is door hardware designed to open quickly and easily with pressure. In an emergency, a panic bar may simply be pushed to open the door quickly.

> ### Transformer Vault Doors
>
> *The 1990 NEC® added 450.43(C), which requires that doors in transformer vaults swing out and be fitted with panic hardware. After debating the issue for the past four Code cycles, CMP-1 accepted two proposals in the 2002 NEC® which required personnel doors for rooms containing equipment rated 1200 A or more and that contains overcurrent devices, switching devices, or control devices to open in the direction of egress and be equipped with panic-type hardware. For the purposes of these requirements, large equipment is defined as equipment rated 1200 A or more and over 6′ wide that contains overcurrent devices, switching devices, or control devices. See 110.26 (C)(3) and 110.33 (A)(3). These revisions are long overdue and should greatly enhance the safety of all workers who may be required to exit these spaces in the event of an electrical incident.*

Ventilation – 450.45

Ventilation facilitates the dissipation of heat from the transformer vault. Ventilation openings shall be located as far as possible from door openings, window openings, fire escapes, and other combustible material per 450.45(A). This helps ensure that if a fire occurs in the transformer vault, it is not spread to other areas by way of the ventilation openings.

Per 450.45(B), if the means of ventilation is natural circulation, ventilation openings are permitted to be installed so that one-half of the required opening area is provided at or near the floor level, and the other half is provided in or near the roof. As an alternative, all of the required ventilation openings can be provided in one or more openings in the roof of the transformer vault.

The minimum size ventilation opening is 1 sq ft for all transformers less than 50 kVA per 450.45(C). Ventilation openings for all other transformers are calculated at 3 sq in. per kVA of transformer capacity in service less the area occupied by grates, screens, etc. For example, the ventilation openings of a transformer vault housing a 75 kVA transformer shall be no less than 225 sq in. ($75 \times 3 = 225$) after deducting the area for grates, screens, etc.

Per 450.45(D), ventilation openings shall be covered or otherwise protected with a durable covering, such as grates, screens, or louvers, to prevent unsafe conditions. All ventilation openings to indoors shall be provided with automatic closing smoke or fire dampers that will respond to a fire in the vault per 450.45(E). The dampers shall have a minimum fire rating of 1½ hours. All ventilating ducts are required to be constructed of fire-resistant material per 450.45(F).

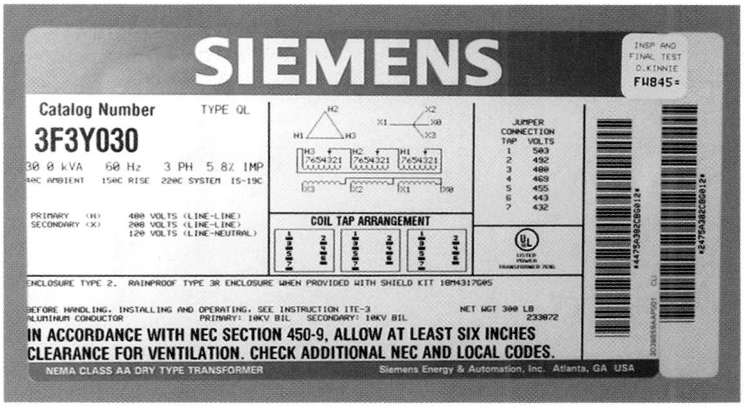

Transformer nameplate marking is required per 450.11.

TRANSFORMER OVERCURRENT PROTECTION

Transformer overcurrent protection is required to protect the primary windings from short-circuits and overloads and the secondary windings from overloads. Section 450.3 contains the requirements for overcurrent protection of transformers.

Overcurrent protection requirements depend upon several factors. First is the voltage at which the transformer operates. Section 450.3(A) contains the rules for transformers rated over 600 V, nominal. Section 450.3(B) contains the rules for transformers operating at 600 V, nominal, or less. The second factor is the location of the overcurrent device. Requirements vary depending on whether only the primary winding is protected, or both the primary and the secondary windings are protected. The last factor applies to transformers rated over 600 V only. If conditions of maintenance ensure that only qualified personnel will work on the transformers, Table 450.3(A) permits the use of primary protection without secondary protection for these supervised installations. See Table

450.3(A), Note 3 for the definition of what constitutes a supervised location.

The rules of these sections are intended to provide protection for the transformers only. Conductors on both the primary and secondary side must be protected in accordance with the provisions of Article 240.

Over 600 V, Nominal – 450.3(A)

The requirements for transformer overcurrent protection for transformers rated over 600 V depend upon whether the installation is supervised. A *supervised installation* is an electrical installation in which the conditions of maintenance are such that only qualified persons monitor or service the electrical equipment. See Note 3 to Table 450.3(A).

Primary and Secondary – Table 450.3(A). For nonsupervised installations, each transformer shall have primary and secondary OCPDs in accordance with Table 450.3(A). The Table is organized around the permissible percentages for fuses and circuit breaker ratings on both the primary and secondary sides and the impedance of the transformer. For the purposes of this section, electronically actuated fuses are rated in accordance with the circuit breaker percentages. Per Table 450.3(A), Note 1, the next higher standard rating of a fuse or circuit breaker is permitted when the calculated value does not correspond to a standard rating of a fuse or circuit breaker. **See Figure 12-22.**

Supervised Locations – 450.3(A). Supervised locations are those locations where conditions of maintenance and supervision ensure that only qualified persons will service the installation. For supervised installations, there are two options for providing the required overcurrent protection. The first is based on primary protection only. In these installations, each transformer shall be protected with a primary overcurrent device that does not exceed 250% of the rated primary current of the transformer for fuses or 300% of the rated primary current of the transformer for circuit breakers. If the calculated value does not correspond to a standard rating of a fuse or circuit breaker, the next higher standard size shall be permitted. **See Figure 12-23.** As with transformer installations of 600 V or less, individual overcurrent protection is not required at the transformer if the primary circuit OCPD provides the necessary protection.

> **Transformer Ventilation**
>
> *Ventilation shall dispose of a transformer's full-load losses without creating a temperature rise in excess of the transformer's rating per 450.9. Ventilation openings shall be provided per 450.45. Location, arrangement, size, covering, dampers, and ducts shall be considered when designing ventilation openings for transformers.*

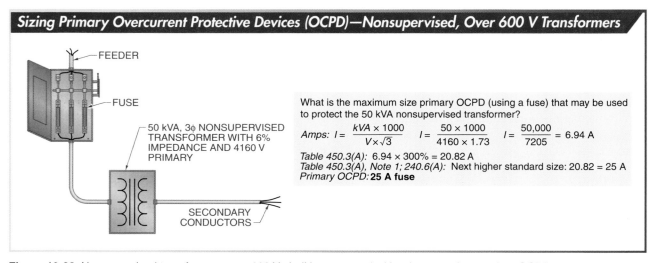

Sizing Primary Overcurrent Protective Devices (OCPD)—Nonsupervised, Over 600 V Transformers

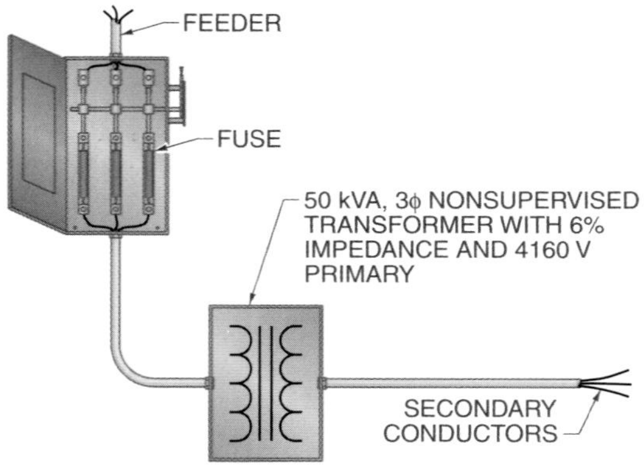

What is the maximum size primary OCPD (using a fuse) that may be used to protect the 50 kVA nonsupervised transformer?

Amps: $I = \frac{kVA \times 1000}{V \times \sqrt{3}}$ $I = \frac{50 \times 1000}{4160 \times 1.73}$ $I = \frac{50{,}000}{7205} = 6.94$ A

Table 450.3(A): 6.94 × 300% = 20.82 A

Table 450.3(A), Note 1; 240.6(A): Next higher standard size: 20.82 = 25 A

Primary OCPD: **25 A fuse**

Figure 12-22. Nonsupervised transformers over 600 V shall be protected with primary and secondary OCPDs.

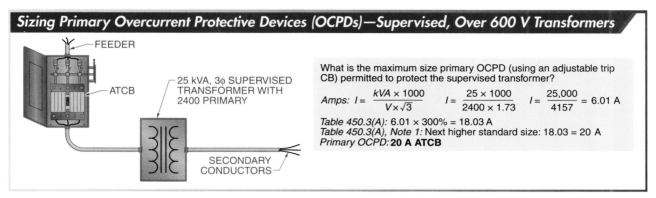

Sizing Primary Overcurrent Protective Devices (OCPDs)—Supervised, Over 600 V Transformers

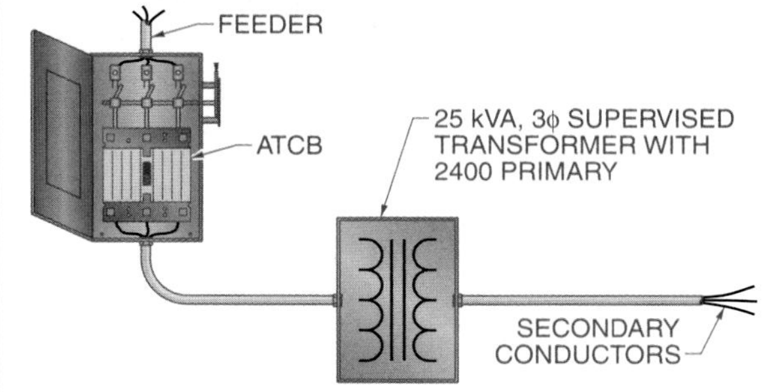

What is the maximum size primary OCPD (using an adjustable trip CB) permitted to protect the supervised transformer?

Amps: $I = \frac{kVA \times 1000}{V \times \sqrt{3}}$ $I = \frac{25 \times 1000}{2400 \times 1.73}$ $I = \frac{25{,}000}{4157} = 6.01$ A

Table 450.3(A): 6.01 × 300% = 18.03 A
Table 450.3(A), Note 1: Next higher standard size: 18.03 = 20 A
Primary OCPD: **20 A ATCB**

Figure 12-23. Supervised transformers over 600 V shall be provided with primary OCPDs not exceeding 250% for fuses and 300% for circuit breakers.

The second option for transformers rated over 600 V is to provide both primary and secondary protection. Table 450.3(A) provides the permissible percentages for fuses and circuit breaker ratings on both the primary and secondary side based upon the impedance of the transformer. For transformers with a rated impedance of not more than 6%, the maximum rating or setting of the primary OCPD shall be 600% of the primary rated current where circuit breakers are used and 300% of the primary rated current for fuse protection. The percentages used for the secondary protection are determined by the secondary voltage rating.

Transformers with a rated impedance of not more than 6% and a secondary voltage over 600 V must have the secondary OCPD sized at a maximum of 300% of the secondary rated current where circuit breakers provide the protection and at a maximum of 250% of the rated current where fuse protection is provided. Where the transformer rated impedance is more than 6% and not more than 10%, the primary percentages are 400% for circuit breakers and 300% for fuses. The secondary percentages for secondaries over 600 V are 250% for circuit breakers and 225% for fuses. The maximum secondary OCPD for all transformers in supervised locations rated 600 volts or less is based on 250% of the rated secondary current, regardless of the rated transformer impedance. This percentage is used for circuit breakers and fuses. Notice that Note 1 to Table 450.3, allowing the use of the next higher standard OCPD, is not applied to the primary and secondary OCPDs where both primary and secondary protection is provided in supervised locations. **See Figure 12-24.**

Sizing Primary and Secondary Overcurrent Protective Devices (OCPDs)—Supervised, Over 600 V Transformers

What size primary and secondary OCPDs, using CBs, are needed to protect the supervised transformer?

Primary

Amps: $I = \frac{kVA \times 1000}{V \times \sqrt{3}}$ $I = \frac{25 \times 1000}{4160 \times 1.73}$ $I = \frac{25{,}000}{7205} = 3.47$ A

Table 450.3(A): 3.47 × 600% = 20.82 A
Table 450.3(A), 240.6(A): Next lower standard size: 20.82 = 20 A
Primary OCPD: **20 A CB**

Secondary

Amps: $I = \frac{kVA \times 1000}{V \times \sqrt{3}}$ $I = \frac{25 \times 1000}{2400 \times 1.73}$ $I = \frac{25{,}000}{4157} = 6.01$ A

Table 450.3(A): 6.01 × 300% = 18.03 A
Table 450.3(A), Note 1; 240.6(A): Next lower standard size: 18.03 = 15 A
Secondary OCPD: **15 A CB**

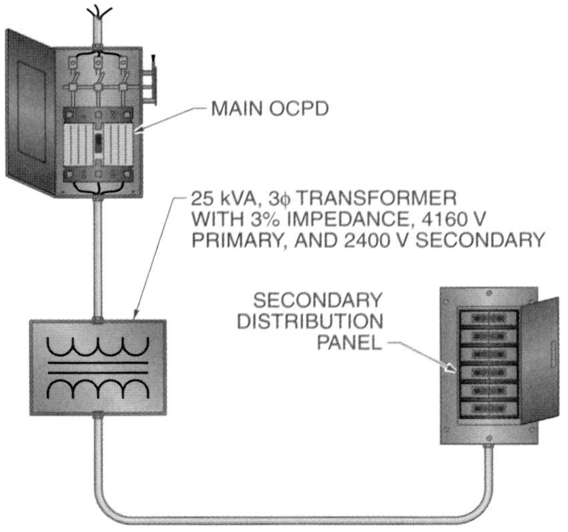

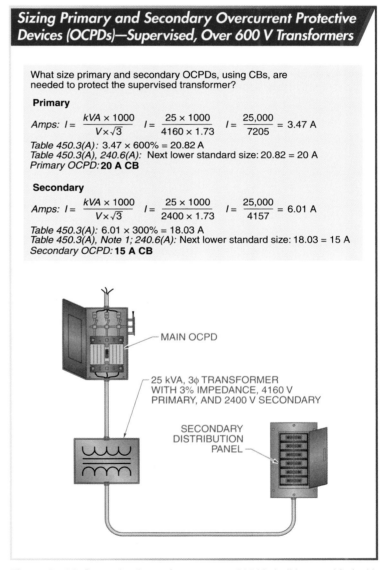

Figure 12-24. Supervised transformers over 600 V shall be provided with primary and secondary OCPDs per Table 450.3(A).

K-Rated Transformers

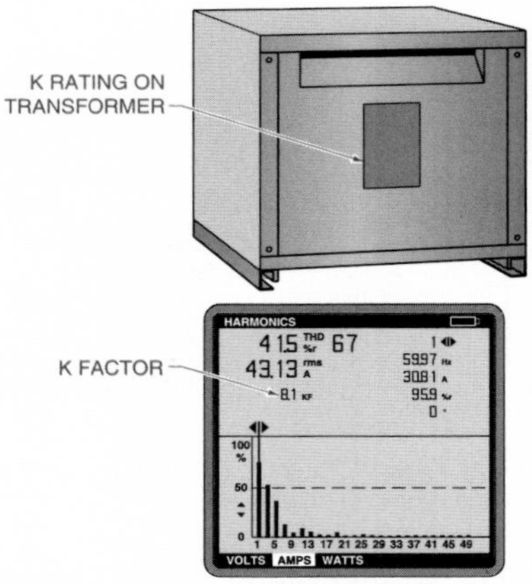

Standard transformers are designed and rated for linear loads operating at 60 Hz. Transformer manufacturers also produce K-rated transformers for supplying power to nonlinear loads as harmonics caused by nonlinear loads produce additional heating. A K rating is a listed rating on a transformer that represents its ability to operate properly with certain nonlinear loads. Typical K ratings that may be listed on a transformer include K-4, K-13, and K-20.

K factor is the measure of the additional heating effect caused by harmonics. The larger the K factor, the greater the amount of harmonics present. Linear loads without any harmonics, such as motors, incandescent lamps, and heating elements, have a factor of K-1. Nonlinear loads with some harmonics, such as welders, induction heating units, and solid-state controls, have a factor of K-4. Branch circuits that include a mix of linear and nonlinear loads, such as uninterruptible power systems and telecommunication equipment found in offices, have a factor of K-13. Circuits that include mostly nonlinear loads, such as desktop computers and variable-frequency motor drives, have a factor of K-20. The heating effect caused by eddy currents and other losses of a K-4 factor are four times that of K-1. The heating effects of a K-20 rating are twenty times that of K-1.

K-rated (nonlinear-rated) transformers are designed to reduce the effects of harmonics and extra heat. K-rated transformers have a larger-size neutral conductor and special windings to reduce eddy currents and skin effect losses. If a transformer does not have a K rating listed, the transformer is assumed to be a standard transformer with a K rating of K-1.

To determine a transformer K rating, a K factor measurement is taken with a power quality meter when the system is fully loaded. The K factor measurement will probably not equal the exact K rating of available transformers. Always round up to the transformer K rating that ensures proper transformer operation under full load.

Additional information is available at www.copper.org.

600 V, Nominal, or Less – Table 450.3(B)

The requirements for overcurrent protection for transformers rated at 600 V, nominal, or less are given in Table 450.3(B) and the accompanying notes. The protection is necessary to protect the windings of the transformer and is independent of the overcurrent protection required for conductors. The secondary overcurrent device is permitted, as it is with services, to consist of not more than six fuses or circuit breakers grouped in one location. The total rating of the six OCPDs cannot exceed that of the required value for a single OCPD.

Primary – Table 450.3(B). The general rule for transformers rated 600 V, nominal, or less is to protect the primary windings of the transformer at not more than 125% of the rated primary current of the transformer. Where this calculated value does not correspond to a standard rating for a fuse or a circuit breaker, per 240.6, and the rated primary current of the transformer is 9 A or more, Table 450.3(B), Note 1 permits the next higher standard rating to be used. **See Figure 12-25.**

Dry-type transformers 600 V or less located in open areas are not required to be readily accessible.

If the rated primary current of the transformer is less than 9 A and not less than 2 A, the rating of the OCPD is permitted to be set at not more than 167% of the rated primary current. For transformer installations where the rated primary current is less than 2 A, the rating of the OCPD shall be permitted to be set at not more than 300% of the primary current. With the lower primary current values, the impedance of the transformer will act to limit potential fault current and the rating of the primary OCPD can be increased.

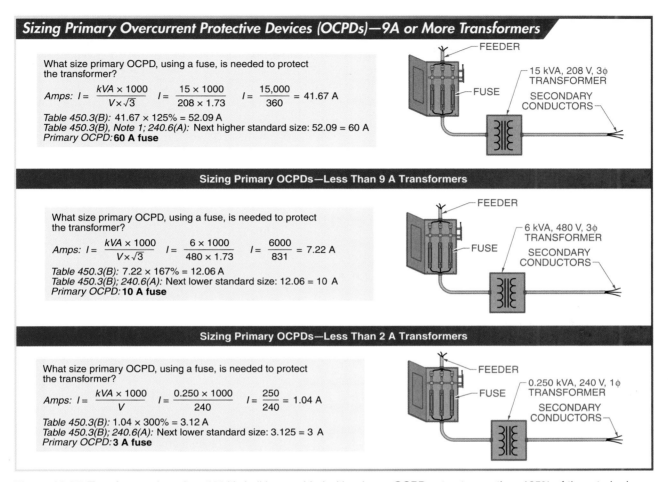

Sizing Primary Overcurrent Protective Devices (OCPDs)—9A or More Transformers

What size primary OCPD, using a fuse, is needed to protect the transformer?

Amps: $I = \dfrac{kVA \times 1000}{V \times \sqrt{3}}$ $\quad I = \dfrac{15 \times 1000}{208 \times 1.73}$ $\quad I = \dfrac{15{,}000}{360} = 41.67\text{ A}$

Table 450.3(B): 41.67 × 125% = 52.09 A
Table 450.3(B), Note 1; 240.6(A): Next higher standard size: 52.09 = 60 A
Primary OCPD: **60 A fuse**

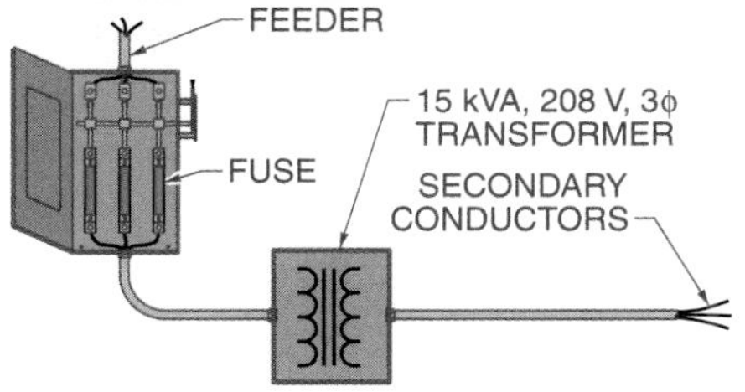

Sizing Primary OCPDs—Less Than 9 A Transformers

What size primary OCPD, using a fuse, is needed to protect the transformer?

Amps: $I = \dfrac{kVA \times 1000}{V \times \sqrt{3}}$ $\quad I = \dfrac{6 \times 1000}{480 \times 1.73}$ $\quad I = \dfrac{6000}{831} = 7.22\text{ A}$

Table 450.3(B): 7.22 × 167% = 12.06 A
Table 450.3(B); 240.6(A): Next lower standard size: 12.06 = 10 A
Primary OCPD: **10 A fuse**

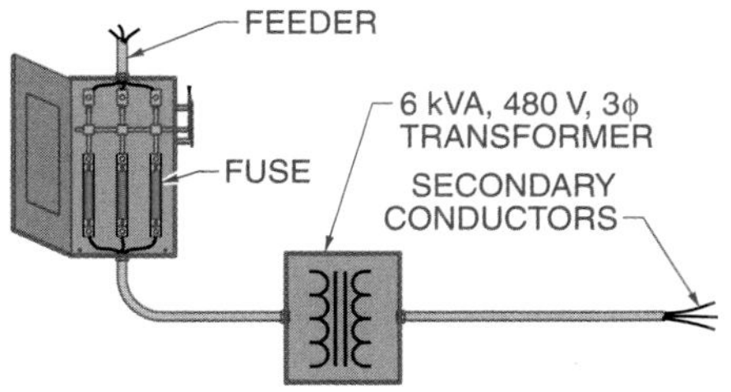

Sizing Primary OCPDs—Less Than 2 A Transformers

What size primary OCPD, using a fuse, is needed to protect the transformer?

Amps: $I = \dfrac{kVA \times 1000}{V}$ $\quad I = \dfrac{0.250 \times 1000}{240}$ $\quad I = \dfrac{250}{240} = 1.04\text{ A}$

Table 450.3(B): 1.04 × 300% = 3.12 A
Table 450.3(B); 240.6(A): Next lower standard size: 3.125 = 3 A
Primary OCPD: **3 A fuse**

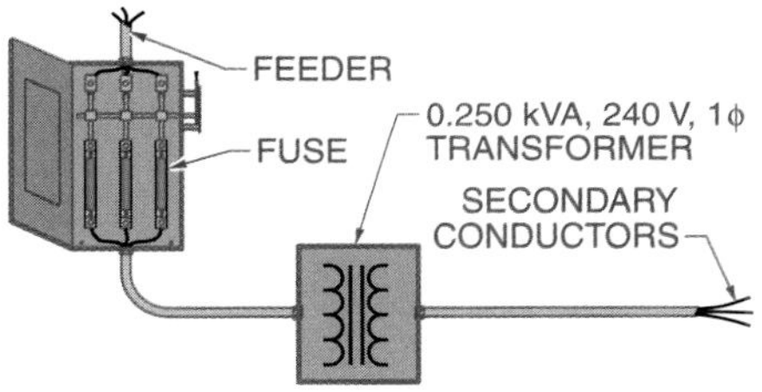

Figure 12-25. Transformers less than 600 V shall be provided with primary OCPDs at not more than 125% of the rated primary current, with exceptions.

Table 450.3(B) establishes maximum settings for transformer overcurrent protection. A separate OCPD could be provided ahead of the transformer which complies with the maximum percentages of Table 450.3(B). If, however, the circuit overcurrent device is sized so that it also provides the necessary overcurrent protection for the transformer, a separate OCPD may not be required. **See Figure 12-26.**

Transformer Loading

Each transformer has a permissible loading rate. The rate is based on several factors including the temperature of the cooling medium, its temperature rise at rated load, duration of the overloads, and altitude above sea level. A set of guides has been published by NEMA to help the operator safely load oil-immersed power transformers.

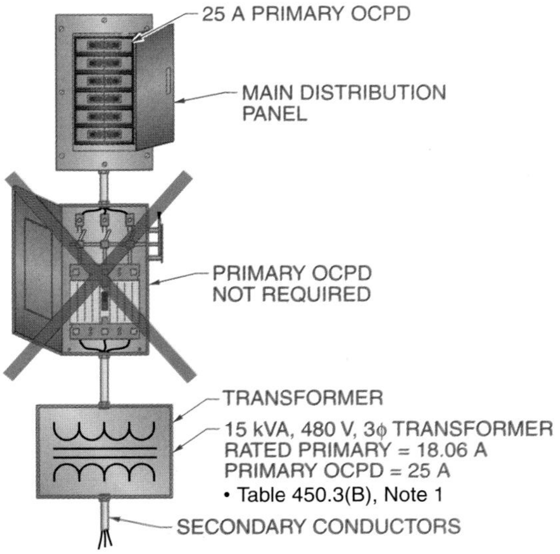

Figure 12-26. Transformers less than 600 V are permitted to be installed without individual primary OCPDs when the circuit OCPDs provide this protection.

Motor-control circuit transformers installed per 430.72(C)(1–5) are permitted to be installed without primary overcurrent protection. These transformers are generally under 50 VA and have low or limited primary current rating. **See Figure 12-27.**

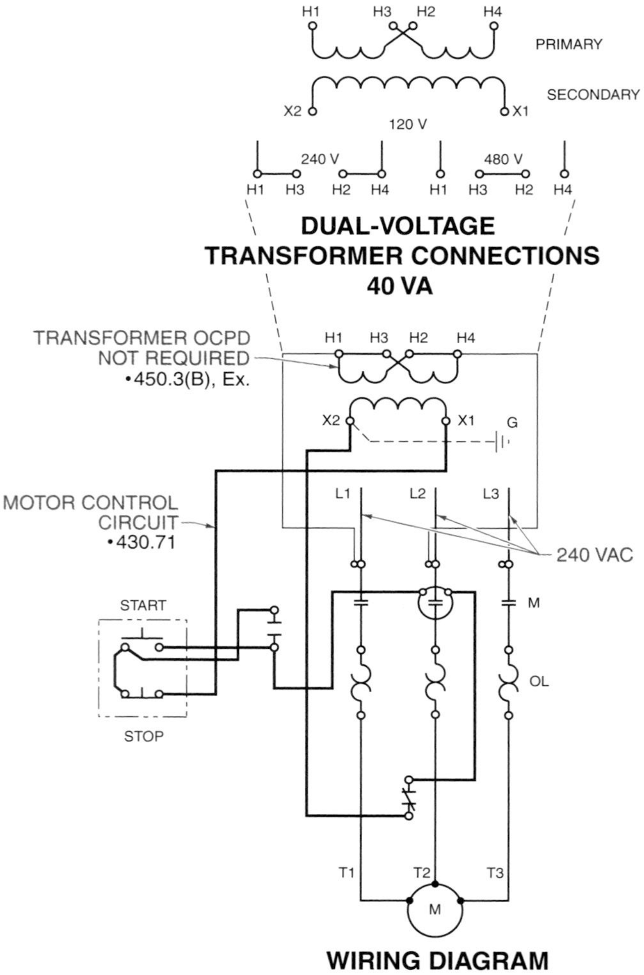

Figure 12-27. Certain motor-control circuit transformers are permitted to be installed without primary overcurrent protection.

Primary and Secondary – Table 450.3(B). Another option in providing overcurrent protection for transformers rated 600 V, nominal, or less is to protect both the primary and the secondary of the transformer. In these cases, Table 450.3(B) permits the primary feeder OCPD, set at not more than 250% of the rated primary current of the transformer, to protect the transformer primary, provided the secondary OCPD is set at a value which does not exceed 125% of the rated secondary current of the transformer. As with the primary protection only requirements, if the rated secondary current of the transformer is 9 A or more and the calculated value does not correspond to a standard rating of a fuse or circuit breaker from 240.6, the next higher standard rated OCPD is permitted. **See Figure 12-28.**

Unlike the requirements for primary only protection, there is no permission to increase the rating of the primary OCPD when it does not correspond to a standard rating of a fuse or circuit breaker. If the calculated value at 250% does not correspond to a standard rating, the next lower standard rating shall be used. If the rated secondary is less than 9 A, the rating of the secondary OCPD is permitted to be increased to a maximum value of 167% of the rated secondary current of the transformer. This rule is useful for applications in which several transformers are supplied from a single primary OCPD. Sizing at 250% prevents nuisance tripping due to the transformer current in-rush characteristics.

High-voltage electricity is transmitted from the power plant through power lines to transmission substations.

Sizing Primary and Secondary Overcurrent Protective Devices (OCPDs)—Secondary 9 A or More

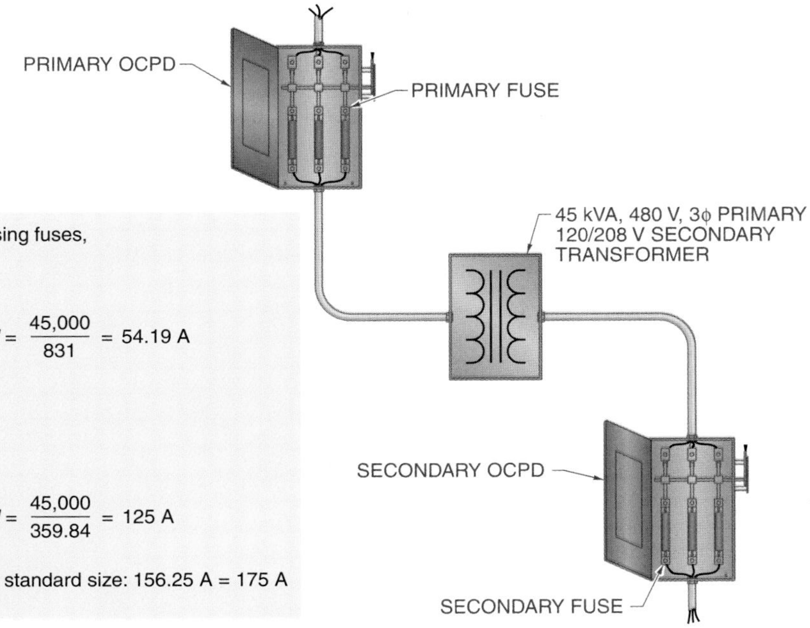

What size primary and secondary OCPDs, using fuses, are needed to protect the transformer?

Primary

$$\textit{Amps: } I = \frac{kVA \times 1000}{V \times \sqrt{3}} \quad I = \frac{45 \times 1000}{480 \times 1.73} \quad I = \frac{45{,}000}{831} = 54.19\text{ A}$$

Table 450.3(B): 54.19 × 250% = 135.47 A
240.6(A): 135.47 = 125 A
Primary OCPD: **125 A fuse**

Secondary

$$\textit{Amps: } I = \frac{kVA \times 1000}{V \times \sqrt{3}} \quad I = \frac{45 \times 1000}{208 \times 1.73} \quad I = \frac{45{,}000}{359.84} = 125\text{ A}$$

Table 450.3(B): 125 × 125% = 156.25 A
Table 450.3(B), Note 1; 240.6(A): Next higher standard size: 156.25 A = 175 A
Secondary OCPD: **175 A fuse**

Sizing Primary and Secondary OCPDs—Secondary 9 A or Less

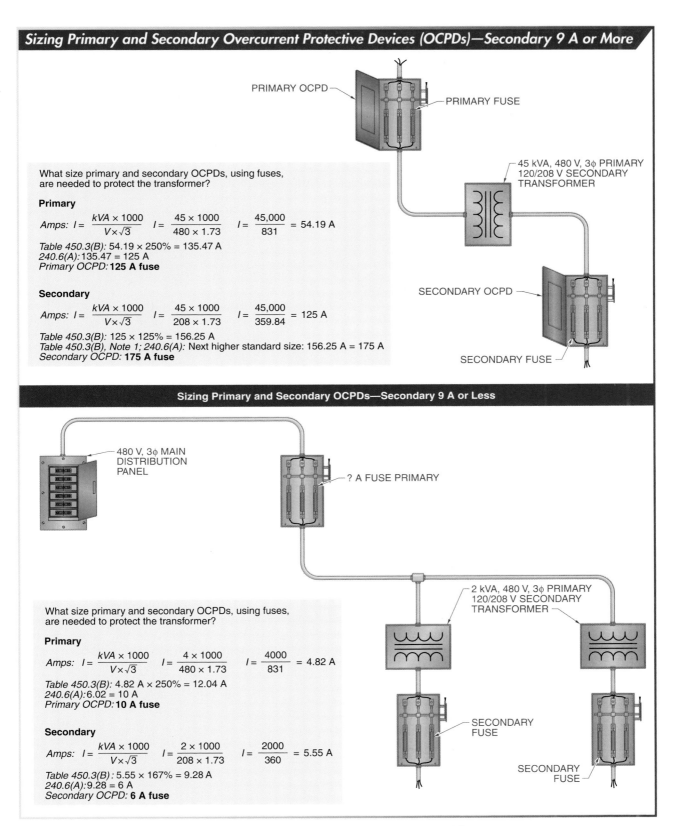

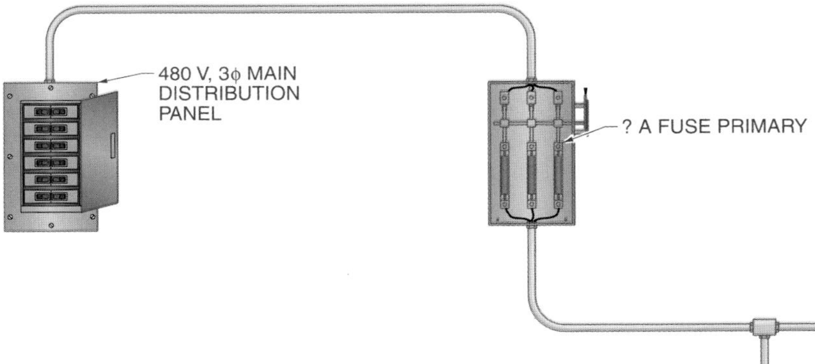

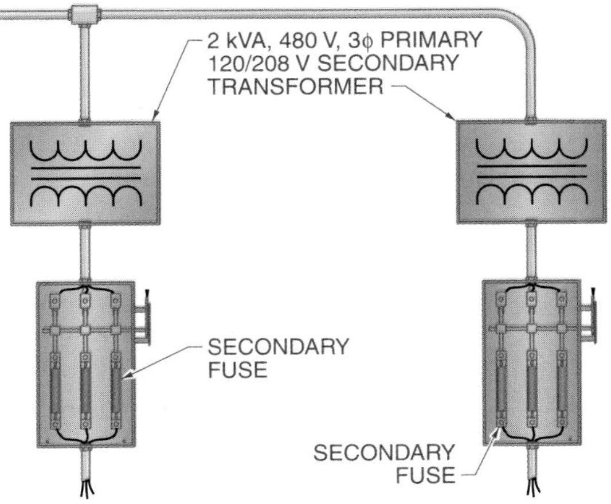

What size primary and secondary OCPDs, using fuses, are needed to protect the transformer?

Primary

$$\textit{Amps: } I = \frac{kVA \times 1000}{V \times \sqrt{3}} \quad I = \frac{4 \times 1000}{480 \times 1.73} \quad I = \frac{4000}{831} = 4.82\text{ A}$$

Table 450.3(B): 4.82 A × 250% = 12.04 A
240.6(A): 6.02 = 10 A
Primary OCPD: **10 A fuse**

Secondary

$$\textit{Amps: } I = \frac{kVA \times 1000}{V \times \sqrt{3}} \quad I = \frac{2 \times 1000}{208 \times 1.73} \quad I = \frac{2000}{360} = 5.55\text{ A}$$

Table 450.3(B): 5.55 × 167% = 9.28 A
240.6(A): 9.28 = 6 A
Secondary OCPD: **6 A fuse**

Figure 12-28. Transformers under 600 V may be protected by primary and secondary OCPDs.

Transformer and Feeder Taps – Article 240

Closely related to the application of transformer overcurrent protection are the provisions of Article 240 covering conductor overcurrent protection. Electrical installations are frequently designed in which taps from the secondary of the transformer are made to supply panelboards or other equipment. See 240.21.

The selection of the appropriate tap rule depends largely upon the length of the tap. For example, 240.21(B)(1) contains the requirements for 10′ tap feeders. The length of the tap conductors shall not exceed 10′. In addition, all of the other conditions of the tap rule shall be met. This is a common application where the requirements of 450.3 overlap with the requirements of Article 240.

Power Quality Problems

Electrical equipment, circuits, and systems often have power quality problems. The source of power quality problems can be the utility company, the in-house power distribution system, and/or nonlinear loads connected to the system. Electrical test equipment such as DMMs, power quality meters, and power monitoring equipment can be used to determine the source and magnitude of the problem. A severe power quality problem can cause a malfunction in the system or equipment. Replacing equipment that is malfunctioning can be a short-term solution to the problem. Basic troubleshooting steps include a sequence of tasks in which the system is inspected, measurements are made, and corrective action is taken.

QUICK QUIZ®

Refer to Chapter 12 Quick Quiz® on CD-ROM.

Additional information is available at ATPeResources.com

Section 240.21(B) covers the requirements for feeder taps as discussed above with the 10′ tap rule. Section 240.21(C) is an important section containing provisions for several types of transformer secondary taps. Several tap rules and general provisions are included, such as the 10′ transformer secondary tap rule (240.21(C)(2)), the 25′ transformer secondary industrial tap rule (240.21(C)(3)), the outdoor secondary conductor tap rule (240.21(C)(4)), and the 25′ secondary conductor tap rule (240.21(C)(6)). Regardless of which tap rule is used, it is important to note that the transformer overcurrent protection provisions of Article 450 are not removed or amended by the Article 240 tap rules.

Pad-mounted transformers are placed on a concrete pad and provide lateral service to buildings.

Transformer Connections

Transformers may be connected in a delta-to-wye or wye-to-delta connection. The connection used depends on the incoming supply voltage, the requirements of the loads, and the practices of the local power company. A delta-to-wye transformer connection is a common connection and delivers the same voltage output as a wye-to-wye transformer connection. The difference is that the primary is supplied from a delta system. A wye-to-wye transformer connection delivers the same voltage output as the delta-to-delta transformer connection. The difference is that the primary is supplied from a wye system.

chapter 12

Transformers

REVIEW QUESTIONS

Name ______________________________ **Date** ____________________

______________ **1.** Transformers are electrical devices that contain no ___ parts.

______________ **2.** A(n) ___ transformer has more windings in the primary than in the secondary.

______________ **3.** The NEMA insulation classifications for dry-type transformers are Classes ___.
- A. A, B, C, and D
- B. A, B, C, and F
- C. A, B, F, and H
- D. none of the above

______________ **4.** ___ is the process by which oxygen mixes with other elements and forms a type of rust-like material.

T F **5.** A less-flammable liquid has a fire point of not less than 300°C.

T F **6.** An autotransformer may be used to buck or boost the supply voltage.

______________ **7.** Temperature ___ is the amount of heat that an electrical component produces above the ambient temperature.

______________ **8.** Transformer vaults shall be constructed of materials with a minimum fire resistance of ___ hours.

______________ **9.** ___ hardware is door hardware that is designed to open easily in an emergency situation.

______________ **10.** Liquid-filled transformers are commonly used in ___.
- A. 150 kVA to 3000 kVA ratings
- B. 2400 V to 13,800 V voltages
- C. all of the above
- D. none of the above

______________ **11.** A ___ point is the temperature at which liquids give off vapor sufficient to form an ignitable mixture with the air near the surface of the liquid.
- A. flash
- B. fire
- C. either A or B
- D. none of the above

T F **12.** Transformers are required to be marked with a minimum distance or clearance from walls or other obstructions to facilitate the dissipation of heat.

T F **13.** Individual dry-type transformers rated over 112½ kVA shall be enclosed in a fire-resistant room with materials having a 3-hour minimum fire rating.

______________ **14.** Type ___ buildings have fire resistance ratings from 0 to 4 hours.

______________ **15.** A vault floor in direct contact with the earth shall be constructed of at least ___″ of concrete.

__________________ **16.** The minimum size of transformer vault ventilation openings for transformers rated less than 50 kVA is ___ sq ft.

__________________ **17.** Transformer vault ventilation openings for transformers rated over 50 kVA is calculated at ___ sq in. per kVA less the area for grates, etc.

__________________ **18.** A(n) ___ installation has conditions of maintenance which allow only qualified persons to monitor or service the electrical equipment.

__________________ **19.** The general rule for transformers rated 600 V, nominal, or less is to protect the primary windings at not more than ___% of the rated primary current.

T F **20.** Askarel has been banned by the Environmental Protection Agency and its use as a transformer coolant is being phased out.

T F **21.** All transformers are required to be insulated.

T F **22.** Transformers are required to be installed outdoors only.

T F **23.** The required clearances for proper transformer ventilation shall be clearly marked on the transformer.

__________________ **24.** Type ___ buildings have fire resistance ratings from 0 to 2 hours.

T F **25.** Doors for transformer vaults shall swing out.

__________________ **26.** Dry-type transformer insulation is based on an ambient temperature of ___°C.

__________________ **27.** Larger distribution transformers are commonly cooled by a(n) ___.

__________________ **28.** Common transformer mountings include ___.

A. pole or vault
B. floor or platform
C. all of the above
D. none of the above

T F **29.** In general, transformers are required to be grounded.

T F **30.** Askarel is a flammable liquid that contains PCBs.

Guarding

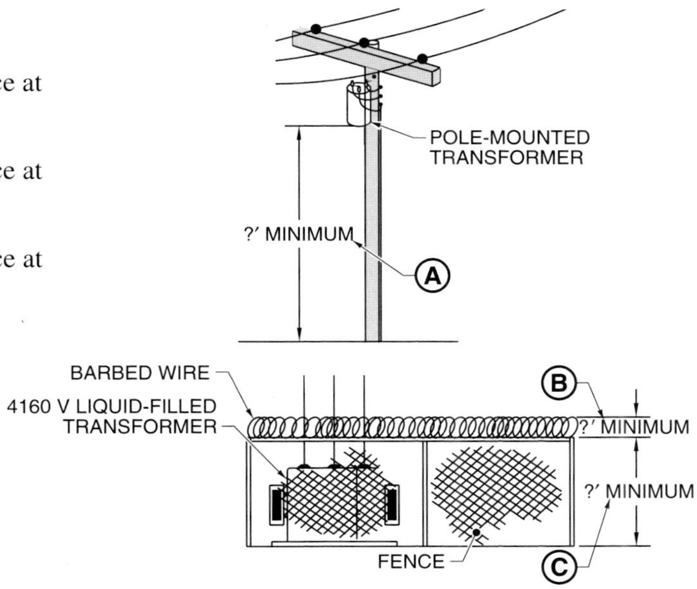

_______________ **1.** The minimum distance at A is ___′.

_______________ **2.** The minimum distance at B is ___′.

_______________ **3.** The minimum distance at C is ___′.

Transformer Vaults

_______________ **1.** The minimum fire rating at A is ___ hours.

_______________ **2.** The minimum fire rating at B is ___ hours.

_______________ **3.** The minimum concrete thickness (when in contact with earth) at C is ___″.

_______________ **4.** The minimum fire rating (when not in contact with earth) at D is ___ hours.

_______________ **5.** The minimum fire rating at E is ___ hours.

_______________ **6.** The minimum height at F is ___″.

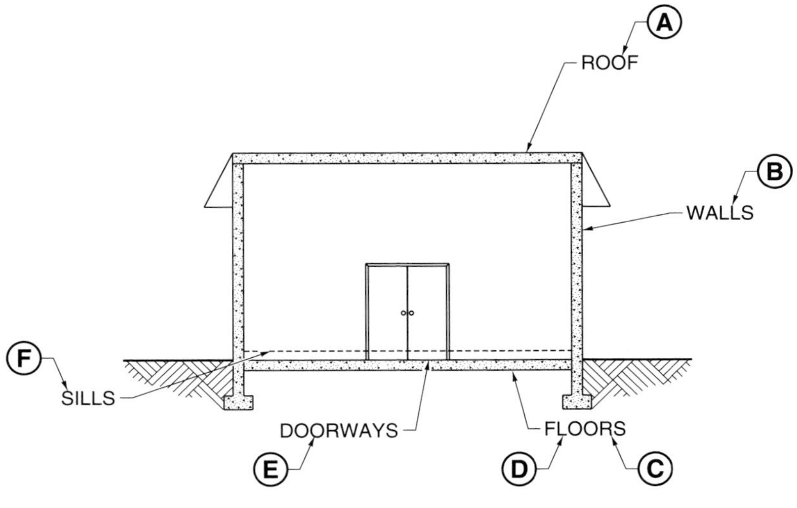

Dry-Type Transformers

______________________ **1.** The minimum clearance at A is ___″.

______________________ **2.** The minimum clearance at B is ___″.

______________________ **3.** The temperature rise at C is ___°C.

______________________ **4.** The temperature rise at D is ___°C.

______________________ **5.** The temperature rise at E is ___°C.

______________________ **6.** The temperature rise at F is ___°C.

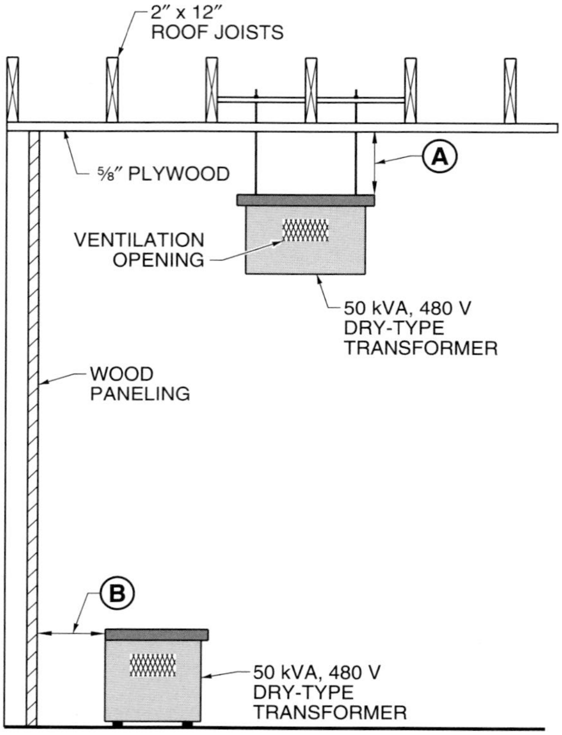

DRY-TYPE TRANSFORMERS			
CLASS	**TEMPERATURE RISE – °C**	**AMBIENT TEMPERATURE –°C**	**COMMON USE**
A	Ⓒ	40	Control
B	Ⓓ	40	Distribution
F	Ⓔ	40	Distribution
H	Ⓕ	40	Large distribution

TRADE COMPETENCY TEST

Name ______________________ **Date** ______________

NEC®	Answer	
________	________	**1.** Determine the maximum size fuse permitted to protect the primary of a 480 V, 1ϕ, 15 kVA dry-type transformer. Secondary overcurrent protection is not provided.
________	________	**2.** Determine the maximum size OCPD fuse permitted to protect the primary of a 75 kVA, 3ϕ nonsupervised transformer with 6% impedance and a 4160 V primary.
________	________	**3.** Determine the maximum size ATCB permitted to protect a 50 kVA, 3ϕ supervised transformer with a 2400 V primary.
________	________	**4.** *See Figure 1.* Determine the maximum size OCPD permitted to protect the primary of the supervised transformer.

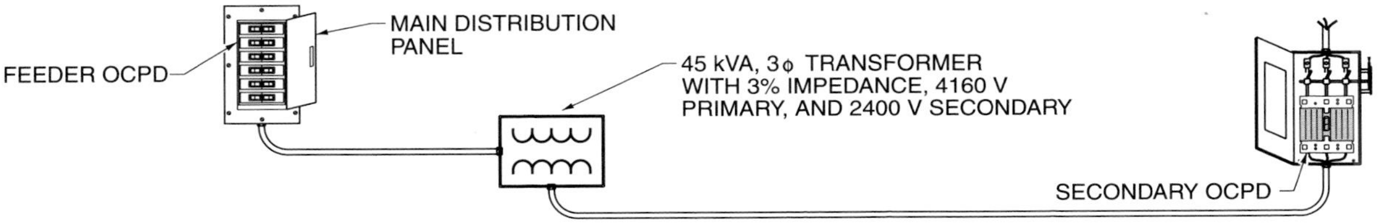

FIGURE 1

NEC®	Answer	
________	________	**5.** The rated primary current of a 480 V transformer is 8.75 A. Determine the maximum size OCPD permitted to protect the primary of the transformer. Secondary overcurrent protection is not provided.
________	________	**6.** *See Figure 2.* Determine the maximum size OCPD permitted to protect the primary of the transformer.
________	________	**7.** *See Figure 3.* Determine the maximum size OCPD permitted to protect the secondary of the transformer.

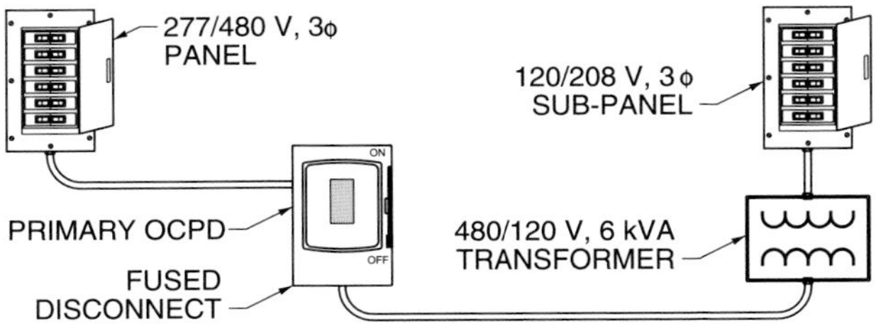

FIGURE 2

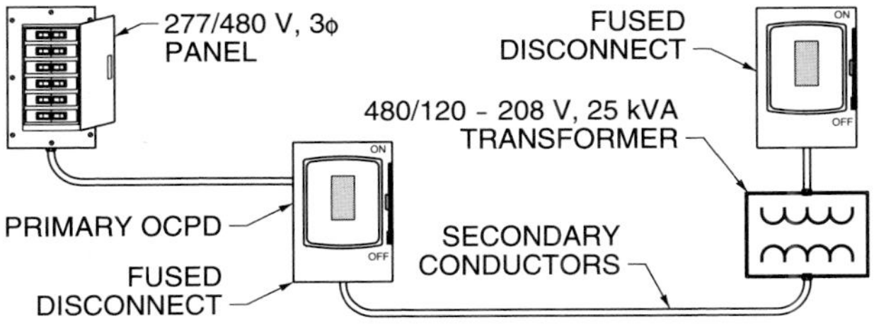

FIGURE 3

______ ______ **8.** *See Figure 4.* Determine the maximum size circuit breaker for the primary of the transformer.

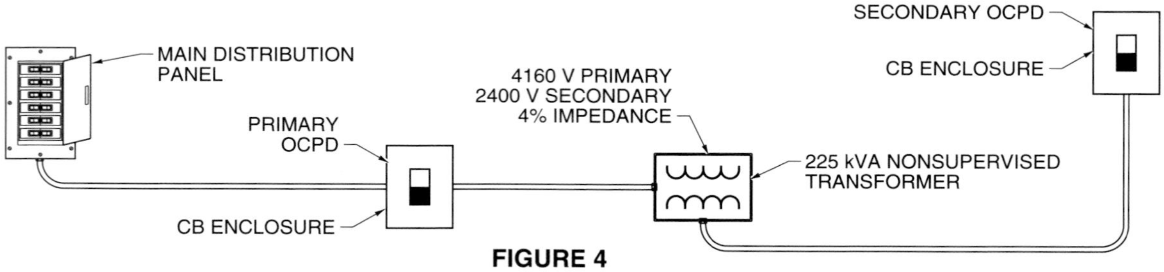

FIGURE 4

______ ______ **9.** *See Figure 5.* Determine the maximum size fuses permitted for the secondary of the transformer.

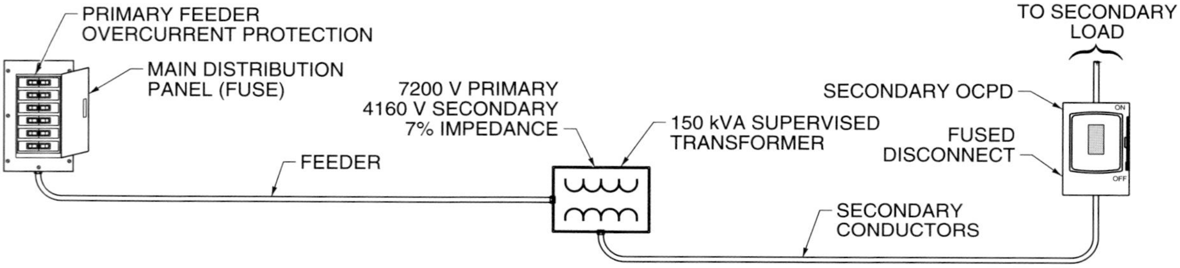

FIGURE 5

______ ______ **10.** A 480 V transformer is installed without secondary overcurrent protection. The rated primary current of the transformer is 1.875 A. Determine the maximum rating of the OCPD permitted to protect the primary.

______ ______ **11.** A 30 kVA, 3ϕ transformer operates at 480 V. Secondary overcurrent protection for the transformer is provided at not more than 125% of the rated secondary current of the transformer. Determine the maximum size OCPD permitted to protect the primary.

______ ______ **12.** A 4160 V, 50 kVA, 3ϕ transformer is installed in a supervised location. Determine the maximum size CB permitted to protect the primary. Secondary overcurrent protection is not provided.

______ ______ **13.** What size primary OCPD, using a fuse, is required to protect a 30 kVA, 208 V, 3ϕ transformer?

14. What size primary OCPD, using a fuse, is required to protect a 15 kVA, 480 V, 3ϕ transformer?

15. What size primary OCPD, using a fuse, is required to protect a 0.500 kVA, 240 V, 1ϕ transformer?

Learning the Code

- Chapter 5 of the NEC® covers the rules and requirements for special occupancies. Because these occupancies are not as commonly encountered, the best approach for apprentices and students of the Code is to become familiar with the location and existence of the various articles and not worry too much about the specific provisions. In other words, it's more important to know that there is an article that covers aircraft hangars (Article 513) than to try to memorize all of the provisions for installing electrical systems within these spaces. Concentrate on gaining a solid understanding of all of the special occupancies that are covered in Chapter 5.

- A few of the articles in Chapter 5 are worthy of special attention. If an apprentice or student of the Code is going to be exposed to a lot of hazardous location work, then Articles 500–516 are of importance. Likewise, Article 517 covers health care facilities, which are increasingly common. Note that, by definition, the term "health care facilities" covers much more than merely hospitals. See 517.2.
- Article 518 is an important article covering the requirements for places of assembly. By definition, these are assembly occupancies designed or intended for gathering together 100 or more people. Apprentices and students of the Code are often confused about the special wiring method requirements for these occupancies. Section 518.4 covers the permissible wiring methods. Because there are a lot of people present, there are greater concerns for life safety and egress from the occupancy. Additionally, there are concerns about the use of PVC products, which may produce toxic fumes if they burn. For this reason, the general wiring methods are metal raceways or nonmetallic raceways encased in concrete. It is important to note that the actual place of assembly can be a building or a portion of a building. This means that certain wiring methods may be permitted in the general building, but a portion of the building, such as an auditorium, may have restricted wiring method choices.

chapter 13

Special Locations

Special locations include hazardous locations that are subject to an increased risk of fire or explosion due to the presence of flammable gases, vapors, liquids, combustible dust, or easily ignitable fibers or flyings. These areas have requirements designed to reduce the risk of fire or explosion. Hazardous locations are classified by classes, divisions, and groups.

New in the 2011 NEC®

- New definition of combustible dust added to Article 500 – 500.2
- Clarification of termination requirements for Type MI cable – 501.10(A)(1)
- Revised methods for process seals in Class I locations – 501.17
- New requirements for wire-type equipment bonding jumpers for classified locations – 501.30(B), 502.30(B), and 503.30(B)
- New provisions for zone suitability markings for use in Class II, Division 1 or 2 locations – 502.6
- RTRC conduit added to permissible wiring methods in Class III, Division 1 locations – 503.10(A)

HAZARDOUS LOCATIONS

Special locations are grouped in Chapter 5 because they present unique installation and design characteristics which increase the risk of hazards due to the work which is performed within the occupancy or from materials which are present within the occupancy. A *hazardous location* is a location where there is an increased risk of fire or explosion due to the presence of flammable liquid-produced vapors, combustible liquid-produced vapors, combustible dusts, or ignitable fibers or flyings.

If any of these are present in the atmosphere or there is the likelihood that they could become present, special electrical installation requirements apply. This is because electrical energy can provide the source of ignition for these hazardous substances and cause an explosion or fire if sufficient quantities of the hazardous substance and oxygen are present. Not all hazardous substances, however, pose the same risk for fire or hazard. Each hazardous substance has different characteristics, which can affect the likelihood of a fire or explosion occurring in a specific location.

Hazardous Locations – Article 500

The NEC® contains a classification system for hazardous substances that groups them according to their type and their potential hazard. This classification system is the basis for determining the installation requirements when hazardous substances are present.

A hazardous location containing a fuel oil tank is classified as a Class I location.

The standard classification system used by the NEC® incorporates the use of a Class, Division, and Group system to classify the hazardous locations. **See Figure 13-1.** The three classes that are used, depending upon the properties of the hazardous material which may be present, are Class I, Class II, and Class III locations.

A *Class I location* is a hazardous location in which flammable gases, flammable liquid-produced vapors, or combustible liquid-produced vapors are present in sufficient quantities to cause an explosion or ignite the hazardous materials. A *Class II location* is a hazardous location that is hazardous due to the presence of combustible dust. A *Class III location* is a hazardous location that is hazardous due to the presence of easily ignitable fibers or flyings or where materials that produce combustible flyings are handled, manufactured, or used. The fibers or flyings may not be in suspension in the air in quantities sufficient to produce ignitable mixtures.

Each of the three classes also contains two divisions. A *division* is the classification assigned to each class based upon the likelihood of the presence of the hazardous substance in the atmosphere. A *Division 1 location* is a hazardous location in which the hazardous substance is normally present in the air in sufficient quantities to cause an explosion or ignite the hazardous materials. A *Division 2 location* is a hazardous location in which the hazardous substance is not normally present in the air in sufficient quantities to cause an explosion or ignite the hazardous materials.

Section 500.5(B)(1) provides the specific rules for assigning the Division 1 designation to Class I locations, while 500.5(B)(2) establishes the rules for assigning the Division 2 designation to Class I locations. Section 500.5(C) defines Class II locations and defines the conditions under which the location is designated as Division 1 or Division 2. Section 500.5(D) defines Division 1 and Division 2 designations for Class III locations.

Per 500.5(A), each room section, or area within an occupancy shall be considered separately for the purposes of determining area classification. Whenever possible, design personnel and installers of electrical systems should select locations that result in the least amount of equipment to be located in hazardous locations. See 500.5(A) Informational Note.

Class I, Division 1 Locations – 500.5(B)(1). A Class I, Division 1 location is any location in which sufficient quantities of flammable gases, flammable liquid-produced vapors, or combustible liquid-produced vapors are present in the air, under normal conditions, to pose the threat of explosion or ignition of the hazardous substances. If sufficient quantities do not exist during normal operations, but service or maintenance operations are performed which could cause ignitable concentrations of the flammable gases, flammable liquid-produced vapors, or combustible liquid-produced vapors to exist frequently, the area is also classified as Class I, Division 1. Additionally, an area is classified as Class I, Division 1 if a process or equipment failure could release hazardous substances in sufficient quantities for an ignition or explosion to occur.

Locations where volatile flammable liquids or liquefied flammable gases are transferred from one container to another are classified as Class I, Division 1 locations. Additionally, per 500.5(B)(1), Informational Note 1, the following occupancies are also classified as Class I, Division 1 locations:

- Transfer areas for volatile, flammable liquids
- Spray booth interiors
- Areas adjacent to spraying or painting operations using volatile flammable solvents
- Open tanks or vats of volatile flammable liquids
- Drying or evaporation rooms for flammable solvents
- Areas where fats and oil extraction equipment using flammable solvents is operated
- Cleaning and dyeing plant rooms which use flammable liquids
- Gas generator rooms
- Pump rooms for flammable gases or volatile flammable liquids which do not contain adequate ventilation
- Refrigeration or freezer interiors which store flammable materials
- All other locations where sufficient ignitable quantities of flammable gases or vapors are likely to occur during routine operations

Hazardous Locations		
Classes	**Hazardous Materials Present**	
I	Flammable gases, flammable liquid-produced vapors, combustible liquid-produced vapors	
II	Combustible dust	
III	Easily ignitable fibers; materials producing combustible flyings are handled, manufactured, or used	
Divisions	**Hazard Level Encountered**	
1	A location where the hazardous materials are present or may exist under normal operating conditions in sufficient quantities to create the possibility of fire or explosion	
2	A location where the hazardous materials are not normally present in sufficient quantities to create the possibility of fire or explosion	
Groups	**Hazardous Material Characteristics**	
Class I	**Class II**	**Class III**
A B C D	E F G	none

Figure 13-1. The standard classification system used by the NEC® incorporates the use of a Class, Division, and Group system to determine hazardous locations.

Some of these Class I, Division 1 locations have sufficient quantities of hazardous substances present continuously or for long periods of time per 500.5(B)(1) Informational Note 2. Designers and installers of electrical systems should avoid placing electrical equipment in these locations. In areas where electrical equipment is essential to the process, only equipment that is identified for the specific location and intrinsically safe systems per 504 should be used.

Determining Classifications of Hazardous Locations

Designers and installers of electrical systems in hazardous locations have a responsibility to ensure that the installation of the electrical system does not jeopardize the safety of personnel or the public at large. Often, the determination of the classification for a particular area can be a very complex issue with many factors involved. Designers of the electrical system in a hazardous location are generally responsible for the proper classification and designation of each area. If sufficient information is not available to determine the correct classification of the area, the owners of the building or structure where the classified area is located, their engineering support staff, and the AHJ should be consulted before designing or installing electrical systems for these areas.

Class I, Division 2 Locations – 500.5(B)(2). A Class I, Division 2 location is any location in which flammable gases are handled, processed, or used, but which poses no threat of explosion or ignition of the hazardous substances under normal conditions. The flammable gases or vapors are confined to containers or closed systems. The only way that the hazardous substances could be introduced into the atmosphere is through accidental rupture or breakdown of the containers.

An area is also classified as Class I, Division 2 if ignitable concentrations of flammable gases, flammable liquid-produced vapors, or combustible liquid-produced vapors are prevented by the use of positive mechanical ventilation. In these locations, the area could become hazardous if the ventilation system failed to operate. An area could also be classified as Class I, Division 2 if it is adjacent to a Class I location and the possibility of ignitable concentrations migrating from the Class I location exists.

Per 500.5(B)(2), Informational Note 1, if flammable gases or vapors are used and, in the judgment of the AHJ, their use does not pose a threat of explosion or hazard unless an accident or failure of a ventilating system occurred, the area is classified as Class I, Division 2. Similarly, if flammable liquids, gases, or vapors are stored in sealed containers that are unlikely to be opened, unless by accident, the area is classified as Class I, Division 2. Additionally, areas with piping which contain flammable gases, flammable liquid-produced vapors, or combustible liquid-produced vapors are classified as Class I, Division 2 if the piping contains any valves, checks, meters, or similar devices. If the piping contains these types of devices, the likelihood of leaks is increased. Piping that does not contain one of these devices does not normally introduce a hazard. See 500.5(B)(2), Informational Note 2.

Class II, Division 1 Locations – 500.5(C)(1). A Class II, Division 1 location is any area in which sufficient quantities of combustible dust is present, under normal operating conditions, to produce explosive or ignitable mixtures. Areas are also classified as Class II, Division 1 if mechanical failure of equipment or processes could cause explosive mixtures of combustible dusts to be produced.

Additionally, if any of the electrically conductive dusts of Group E are present in sufficient quantities to present a hazard, the area is classified as Class II, Division 1. Electrically conductive dusts, such as aluminum and magnesium, pose a greater threat for ignition or explosion because these dusts can conduct electricity if they gather on live parts of electrical equipment. Per 500.5(C)(1), Informational Note, some electrically conductive dusts, such as magnesium or aluminum, have additional hazards associated with them because of their low ignition temperatures.

Some of the combustible dusts that are nonconductive include:

- Wood
- Grain and grain products
- Pulverized sugar and cocoa
- Dried egg and milk powders
- Pulverized spices
- Starch and pastes
- Potato and woodflour
- Oil meal from beans and seeds
- Dried hay
- Any other organic materials that may produce combustible dusts during their use or handling

Class II, Division 2 Locations – 500.5(C)(2). A Class II, Division 2 location is any location where combustible dusts may be present, but not in quantities sufficient to produce explosive mixtures or interfere with the successful operation of electrical equipment or apparatus. Class II, Division 2 locations occur where combustible dust is suspended in the air as a result of occasional malfunctioning of equipment which handles or processes the material which produces the dust. Per 500.5(C)(2), Informational Note 1, a key factor in the classification of these locations is the quantity of dusts which may be present and the adequacy of the dust removal systems in these locations. Per 500.5(C)(2), Informational Note 2, if the quantity of dust is limited by the manner in which it is handled, the area may not require classification.

Class III, Division 1 Locations – 500.5(D)(1). A Class III, Division 1 location is any location in which easily ignitable fibers or materials producing combustible flyings are handled, manufactured, or used. The fibers or flyings may be present in the air, but not in quantities which would be sufficient to produce ignitable mixtures. Per 500.5(D)(1), Informational Note 1, some locations that are classified as Class III, Division 1 include:

- Portions of rayon, cotton, or other textile mills
- Manufacturing and processing plants for combustible fibers, cotton gins, and cotton seed mills
- Flax-processing plants
- Clothing manufacturing plants
- Woodworking plants
- Other establishments involving similar hazardous processes or conditions

Per 500.5(D)(1), Informational Note 2, some of the types of materials that are defined as easily ignitable fibers or flyings include:

- Rayon
- Cotton
- Sisal or henequen
- Istle
- Jute
- Hemp
- Tow
- Cocoa fiber
- Oakum
- Baled waste kapok
- Spanish moss
- Excelsior
- Other similar materials of similar nature

Class III, Division 2 Locations – 500.5(D)(2). A Class III, Division 2 location is any location where easily ignitable fibers or flyings are stored or handled. If easily ignitable fibers or flyings are in the process of being manufactured, the location is not a Class III, Division 2 location, but is a Class III, Division 1 location. Classification for Class III materials depends on the use of the material rather than the quantity which is present.

Group Classifications – 500.6

Different flammable gases and vapors have different degrees of risk associated with them. Since the 1937 NEC®, flammable gases and vapors have been grouped according to their level of hazard. A *group* is an atmosphere containing flammable gases or vapors or combustible dust. Different gases in the various groups have different ignition temperatures.

Class I locations are divided into four groups (A, B, C, and D) of flammable gases or vapors based on explosion and ignition characteristics. Class II locations are divided into three groups of combustible dusts that are grouped by electrical resistivity. There are no groups for Class III locations.

Cooper Crouse-Hinds

Explosionproof horn/strobe-light warning units are required in Class I, Division 1 locations where explosive liquids, vapors, or dusts are present.

The use of the group system primarily applies to the manufacturers of equipment which is to be installed in hazardous locations. The group system ensures that the electrical equipment installed in a particular hazardous location is suitable for the hazardous substance that may be present in the air. Section 500.6(A)(1–4) contains the specific flammable gases and vapors for Class I Groups A–D and 500.6(B)(1–3) contains the listing of Class II Groups E–G.

Hazardous Location Protection Techniques – 500.7

When electrical and electronic equipment is installed in a hazardous location, special consideration must be given to assure that the operation of the equipment does not in any way provide a source of ignition for hazardous materials that may be present in the atmosphere. The NEC® lists several protection techniques that may be used by installers and designers of electrical systems to provide the necessary assurance. Section 500.7 provides a list of some of those protection techniques.

Frequently, the first choice for protection is the use of explosionproof equipment. By definition, this equipment type provides protection by enclosure in a case that is capable of withstanding an explosion of a specified gas or vapor that may occur within it, and of preventing the ignition of the gas or vapor surrounding the enclosure by sparks, flashes, or explosion of the gas or vapor within. In other words, the equipment must contain an explosion that might occur and limit the temperature rise on the enclosure housing so that the equipment also does not provide a source of ignition. Section 500.7(A) permits explosionproof equipment as a protection technique for Class I, Division 1 or 2 locations.

Similarly, 500.7(B) and (C) address protection techniques for dust-ignitionproof and dusttight apparatus. The dust-ignitionproof technique is permitted for Class II, Division 1 or 2 locations, and the dusttight apparatus technique is permitted for Class II, Division 2 or Class III, Division 1 or 2 locations.

The purged and pressured technique listed in 500.7(D) is actually two processes. Purging requires supplying an enclosure with a protective gas at a sufficient flow and positive pressure to reduce the concentration of any flammable gas or vapor to an acceptable level. Pressurization is accomplished by supplying an enclosure with a gas with or without continuous flow at sufficient pressure to prevent the entrance of a flammable gas or vapor, a combustible dust, or ignitable fiber.

PPE—Class I Locations

In addition to using explosionproof apparatus and enclosures in Class I, Division 1 and Class I, Division 2 locations, personnel performing electrical work in such locations must also wear proper personal protective equipment (PPE) for head, eye, hearing, respiratory, foot, arm, fall, and hand protection.

Another protection technique that is utilized for circuits, equipment, and components is nonincendive protection. Circuits, equipment, and components that are nonincendive are designed to limit the available potential energy to such a minimal value that it is not capable of providing the necessary ignition source to cause an explosion. See 500.7(F), (G), and (H) and 500.2.

Recently, a new protection technique has been developed that utilizes a gas detection system to monitor, warn, and control electrical systems and components in hazardous locations. Section 500.7(K) has very specific and detailed limitations for such use. Most importantly, the combustible gas detection system must be listed for the specific gas or vapor it is intended to detect. Additionally, several installation provisions, such as installation locations, alarm and shutdown criteria, and calibration frequency, must be documented.

Wiring Methods

Chapter 5 of the NEC® has very strict requirements for wiring methods in hazardous locations. The installation of the electrical system in hazardous locations shall not create additional hazards. This is a major concern for designers and installers alike because electrical energy has the potential to ignite fires and create explosions in these locations.

Wiring method selection in hazardous locations closely follows the classification system for these locations. Wiring methods are selected based on the class and the division in which they are to be installed. In general, the wiring methods for Division 1 locations are more strict than the wiring methods for Division 2. Unlike area classifications, which are all contained in Article 500, wiring method selections are contained in separate articles as follows:

- Class I locations – Article 501
- Class II locations – Article 502
- Class III locations – Article 503

Class I, Division 1 Wiring Methods – 501.10(A)(1). Because of the nature of the hazardous substances, the selection of wiring methods is strictly limited. In general, only RMC, steel IMC, and Type MI cable terminated with fittings listed for the location are permitted as wiring methods in Class I, Division 1 locations. For RMC and IMC, all threaded joints shall have at least five threads fully engaged. Type MI cable shall be supported in a manner to ensure that the termination fittings are not stressed. The weight of the cable shall not be transferred to the terminations.

Cooper Crouse-Hinds

Heat detectors are typically installed in hazardous locations where high temperatures can cause an explosion.

All boxes and fittings used in Class I, Division 1 locations shall be designed for threaded connections and be explosionproof. An *explosionproof apparatus* is equipment which is enclosed in a case that is capable of withstanding any explosion that may occur within it, without permitting the ignition of flammable gases or vapors on the outside of the enclosure. See Article 100. The explosionproof enclosure shall prevent arcs, flashes, and sparks from igniting the flammable substances by adequately containing them within the enclosure.

The explosionproof apparatus shall keep the outside of the enclosure at a low enough temperature to prevent the ignition of the hazardous gases or vapors surrounding the enclosure. When equipment requires a flexible connection, such as with motors or equipment which is subject to vibration, flexible fittings listed for Class I locations are permitted to be used.

In general, RNC, like PVC, is not permitted in Class I, Division 1 locations. However, 501.10(A)(1)(a), Ex., does permit type PVC conduit and type RTRC conduit to be used where encased in a concrete envelope at least 2″ thick. The PVC and RTRC shall be buried at least 2′ under the earth and either RMC or steel IMC shall be used for the last 2′ of the underground run to the point of emergence or to the point of connection to the aboveground raceway. A separate EGC shall be included in the PVC or RTRC to ground any noncurrent-carrying parts of metal equipment.

In addition to Type MI cable, 501.10(A)(1)(c)(d) permits Type MC-HL cable and Type ITC-HL cable with a gas/vaportight continuous corrugated aluminum sheath to be installed in Class I, Division 1 locations in industrial establishments with restricted public access. The conditions of maintenance shall ensure that only qualified persons service the installation. These are very specialized and listed cables that must be selected with careful consideration of the industrial establishment. **See Figure 13-2.** Section 501.140 lists the permissible applications for Class I, Division 1 locations where flexible cords must be used because of the need for flexibility at the termination.

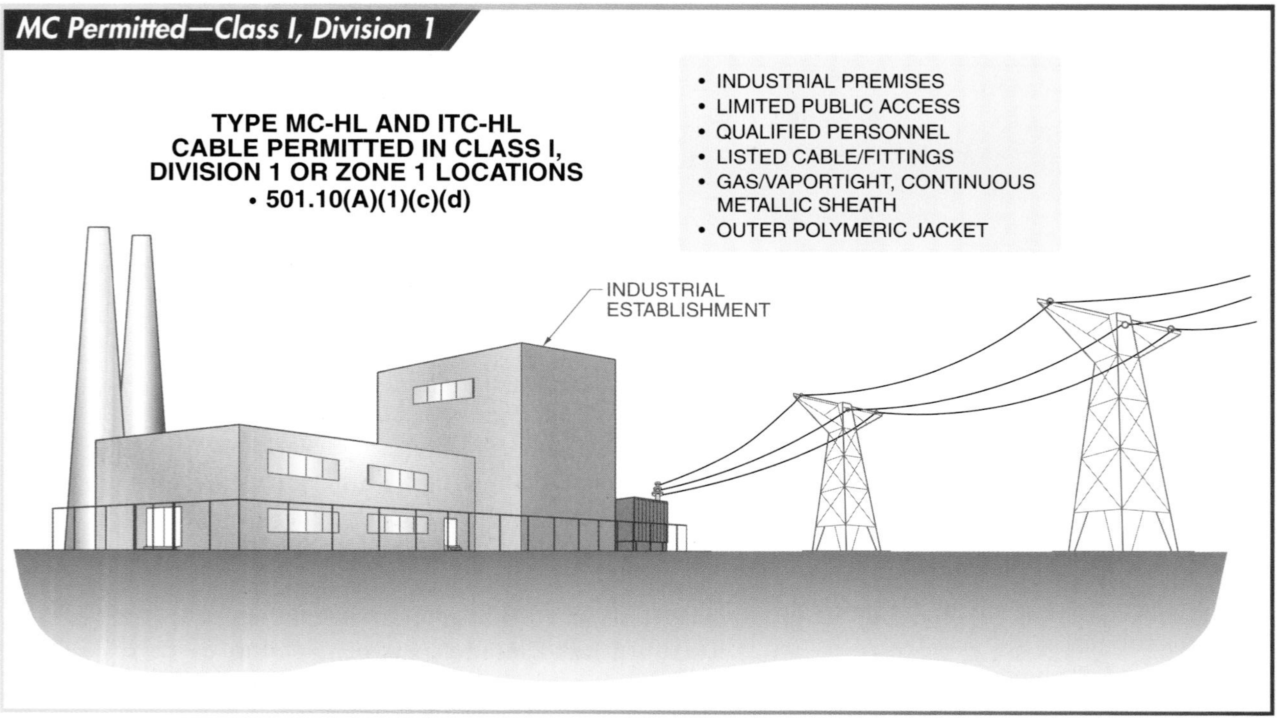
MC Permitted—Class I, Division 1
TYPE MC-HL AND ITC-HL CABLE PERMITTED IN CLASS I, DIVISION 1 OR ZONE 1 LOCATIONS
• 501.10(A)(1)(c)(d)
• INDUSTRIAL PREMISES
• LIMITED PUBLIC ACCESS
• QUALIFIED PERSONNEL
• LISTED CABLE/FITTINGS
• GAS/VAPORTIGHT, CONTINUOUS METALLIC SHEATH
• OUTER POLYMERIC JACKET
INDUSTRIAL ESTABLISHMENT

Figure 13-2. Special cable types are permitted to be used in Class I, Division 1 locations under specific conditions.

Class I, Division 2 Wiring Methods – 501.10(B). In addition to threaded RMC and threaded steel IMC, which are permitted in Class I, Division 1 locations, enclosed gasketed busways and enclosed gasketed wireways are also permitted to be used. Several different cable assemblies are permitted to be installed in Class I, Division 2 locations. These include Types MI, MC, MV, and TC cables with approved terminations.

Additionally, Type PLTC and PLTC-ER, per Article 725, and Type ITC and ITC-ER in cable trays, in raceways, supported by messenger wire, or directly buried if listed for such use are also permitted. Of these cable assemblies Types MI, MC, PLTC, and ITC shall be permitted to be installed in cable trays provided they are installed in a manner which does not impose stress or strain on the terminations. Unlike Class I, Division 1 locations, boxes, fittings, and joints are not required to be explosionproof, unless specifically required, as in 501.105(B)(1), 501.115(B)(1), and 501.150(B)(1).

Several different wiring methods, for equipment requiring flexibility, are permitted in Class I, Division 2 locations. FMC, LFMC, and LFNC are permitted with listed fittings. Additionally, extra-hard usage flexible cord terminated with listed fittings can be used, provided a separate EGC is run in the flexible cord.

Class II, Division 1 Wiring Methods – 502.10(A). The wiring methods in a Class II, Division 1 location closely follow the requirements for Class I, Division 1 locations. In general, only RMC, steel IMC, and Type MI cable are permitted as wiring methods in Class II, Division 1 locations. Fittings for Type MI cable shall be listed and the cable shall be installed in a manner which does not transmit stress and strain to the terminations. Section 502.10(A)(1)(3), contains the same provisions for using Type MC-HL cable in Class II, Division 1 locations. Listed Type MC-HL cable with a gas/vaportight continuous corrugated aluminum sheath is permitted to be installed in Class II, Division 1 locations in industrial establishments with restricted public access, provided that the conditions of maintenance ensure that only qualified persons service the installation.

Section 502.10(A)(1)(4) requires that all boxes and fittings used in Class II, Division 1 locations have threaded bosses and be equipped with close-fitting covers which contain no openings through which dust can enter. See 500.2. All fittings and boxes which contain splices, taps, terminations, etc., and which are installed in Group E locations, shall be identified as Class II. Where equipment which requires flexible connections is installed in Class II, Division 1 locations, 502.10(A)(2)(1–5) permits dusttight flexible connectors, LFMC, LFNC with listed fittings, Type MC cable, and flexible cord of the extra-hard usage type terminated with listed dusttight fittings to be used. See 502.140.

Class II, Division 2 Wiring Methods – 502.10(B). In addition to those wiring methods permitted for Class II, Division 1, the permitted wiring methods for Class II, Division 2 locations are RMC, IMC, EMT, and dusttight wireways. In addition, Type MC and MI cable with listed fittings are also permitted. Type PLTC, PLTC-ER, ITC, ITC-ER, MC, and TC are also permitted where installed in cable trays. Type MI, MC, and TC cables are permitted in ladder, ventilated trough, or ventilated channel cable trays provided they are installed in single layers, with a space not less than the larger cable diameter between adjacent cables.

Section 502.10(B)(2) permits the same wiring methods used for flexibility for Class II, Division 1 locations to be used for Class II, Division 2 locations. Wireways, fittings, and boxes installed in Class II, Division 2 locations shall be designed to minimize the entrance of dust and shall be equipped with telescoping or close-fitting covers or other means to prevent the escape of arcs or sparks. They shall contain no openings through which combustible materials might be ignited by any sparks or arcs within the enclosure. See 502.10(B)(4).

Class III, Division 1 Wiring Methods – 503.10(A). Several raceways are permitted to be used in Class III, Division 1 locations including RMC, IMC, PVC, RTRC, EMT, and dusttight wireways. In addition, Type MI and MC cable are permitted provided they are used with listed termination fittings. Additionally, Type PLTC and PLTC-ER and Type ITC and ITC-ER cables are also permitted where installed in accordance with the provisions of their respective articles: Article 725 and Article 727. All boxes and fittings used in Class III, Division 1 locations are required to be dusttight per 503.10(A)(1). *Dusttight* is construction that does not permit dust to enter the enclosing case under specified test conditions. See Article 100.

Per 503.10(A)(3), flexible wiring methods which are permitted to be used in Class III, Division 1 locations are dusttight flexible connectors, LFMC, LFNC with listed fittings, interlocked armor Type MC, and flexible cord per 503.140.

Flexible cord used in Class III, Division 1 locations shall be listed for extra-hard usage, contain a separate EGC, be terminated in an approved manner, be supported in a manner that does not put strain on the terminations, and be equipped with a suitable means to prevent the entrance of easily ignitable fibers or flyings into the box or enclosure. See 503.140.

Class III, Division 2 Wiring Methods – 503.10(B). Any of the wiring methods used for Class III, Division 1 locations are permitted to be used for Class III, Division 2 locations. In addition, 503.10(B), Ex. permits open wiring installed per Article 398 to be used in sections, compartments, or areas used solely for storage provided physical protection of the conductors is used where the conductors are not run in roof spaces and are out of reach of sources of potential physical damage. Physical protection of the conductors is required per 398.15(C) when they are run within 7′ of the floor. Physical protection can be accomplished by the use of guard strips, running boards, boxing, and insertion in other raceways such as RMC, IMC, RNC, and EMT.

Conduit Seals

A *conduit seal* is a seal provided in conduit and cable systems to minimize the passage of gases and vapors and prevent the passage of flames from one portion of the electrical installation to another through conduit. See 501.15, Informational Note 1. Conduit seals contain provisions for pouring a sealing compound into the fitting to accomplish the isolation. **See Figure 13-3.**

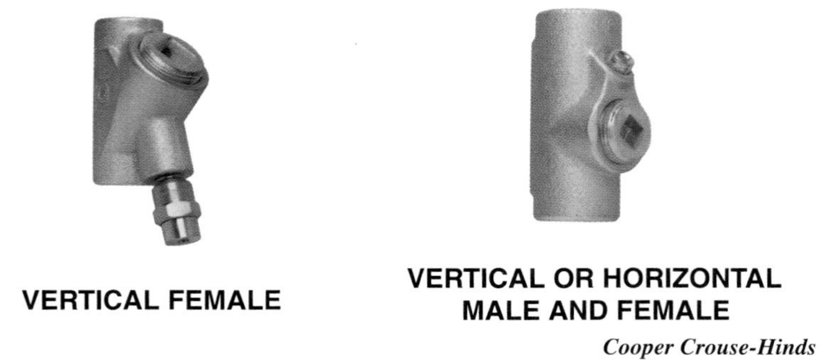

Figure 13-3. A conduit seal is a fitting which is inserted into runs of conduit to isolate certain electrical apparatus from atmospheric hazards.

> **Conduit Seals**
>
> *Conduit seals shall be installed properly to ensure that the hazardous substance does not migrate through the raceways system. Several factors are involved in selecting the proper seal fitting for the application. Seal fittings can be designed for either vertical or horizontal mounting in the conduit system and in male and female configurations. In some cases, the seal fittings may be mounted in any position.*
>
> *Conduit seals shall be installed in accordance with the manufacturer's instructions. These instructions may contain additional installation and location specifications. For example, the preferred side of a classified area boundary where the seal shall be installed may be given. Always read the manufacturer's instructions. In addition, the sealing compound used to construct the seal also contains a set of installation instructions and directions for making the seal. Always read these instructions to make a good conduit seal.*

If a run of conduit leaving a classified location and entering an unclassified area is installed without conduit seals, the flammable gases or vapors could migrate through the conduit system from the hazardous location to the unclassified area. Requirements for conduit seals are contained in the individual article which contains the requirements for the particular class and division. Installers of conduit seals should ensure that the correct type of seal is installed and that the sealing material is installed in accordance with the manufacturer's instructions included with the product listing. **See Figure 13-4.**

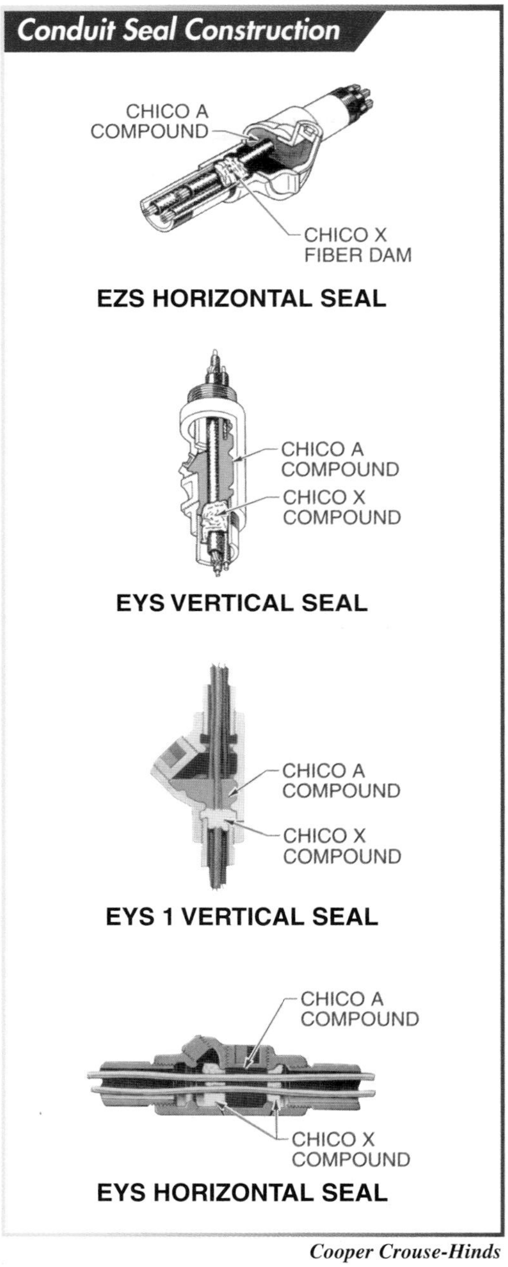

Figure 13-4. Conduit seals are constructed using sealing compound (Chico A), fiber (Chico X), and approved conduit seal fittings.

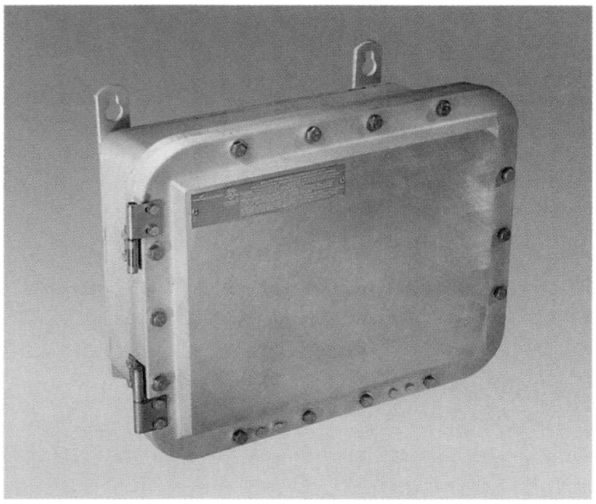

Cooper Crouse-Hinds

Junction boxes in hazardous locations must be suitable for the class and division in which they are installed.

Class I, Division 1 Conduit Seals – 501.15(A). Section 501.15(A) contains four subparts which have separate provisions for conduit seals in Class I, Division 1 locations. Per 501.15(A)(1), conduit seals shall be provided for each conduit entry into an explosionproof enclosure that contains apparatus which may produce sparks, arcs, or excessive temperatures. Examples of such electrical devices are switches, circuit breakers, relays, fuses, resistors, contactors, etc.

Excessive or high temperature is considered to be any temperature which exceeds 80% of the autoignition temperature in degrees Celsius of the gas or vapor involved. Conduit seals shall also be provided for each conduit entry into an explosionproof enclosure where the entry is trade size 2 or larger and the enclosure contains terminals, splices, or taps. The conduit seals shall be located within 18″ from the enclosure. Only explosionproof unions, couplings, reducers, elbows, capped elbows, and conduit bodies are permitted to be installed between the conduit seal and the explosionproof enclosure. **See Figure 13-5.**

Section 501.15(A)(1), Ex. permits the conduit seal to be omitted provided the current-interrupting contacts of any electrical devices contained in the enclosure are:

(1) Enclosed in a hermetically sealed chamber which prevents the entrance of gases or vapors or;

(2) Immersed in oil per 501.115(B)(1)(2) or;

(3) Enclosed within a factory-sealed, explosionproof chamber in an enclosure approved for the location and marked "factory sealed" or the equivalent unless the enclosure entry is trade size 2 or larger or;

(4) Located in nonincendive circuits.

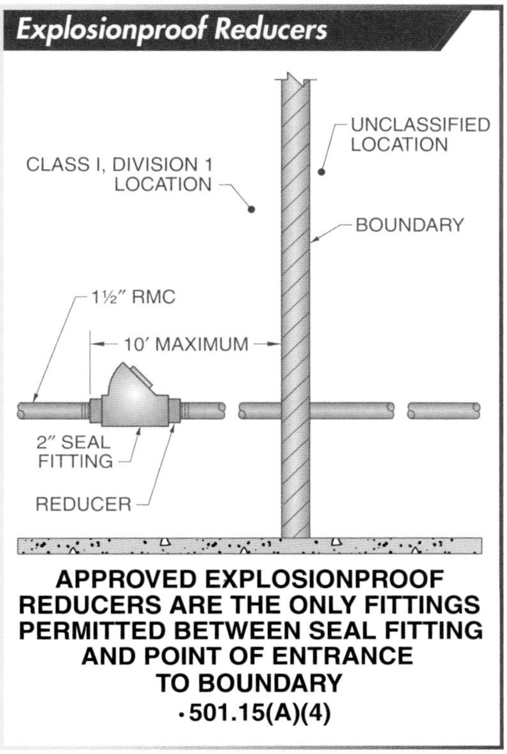

Figure 13-5. Conduit seal fittings are installed at hazardous location boundaries to prevent communication of the hazardous substance between the classified and unclassified areas.

Section 501.15(A)(2) requires that the conduit seal be located within 18″ of pressurized enclosures, unless the conduit is not pressurized as part of the protection system. Informational Note 1 states that the closer the seal is to the enclosure, the less the problems associated with purging the dead space in the pressurized conduit system.

Section 501.15(A)(3) permits a single seal to be used for two or more enclosures which are connected together by nipples or runs of conduit not exceeding 36″ in length, provided the location of the seal is not more than 18″ from either enclosure. Proper location of the conduit seal ensures that only a minimum amount of the hazardous substance can enter the conduit system. **See Figure 13-6.**

Section 501.15(A)(4) requires seals for conduit runs which leave Class I, Division 1 locations. The location of the seal is permitted on either side of the boundary provided it is within 10′ of the boundary. **See Figure 13-7.** With the exception of explosionproof reducers, no boxes or fittings are permitted to be installed between the boundary and the conduit seal. If, however, an unbroken run of metal conduit passes completely through a Class I, Division 1 location and terminates in an unclassified area, and there are no fittings less than 12″ from each boundary, seals are not required. See 501.15(A)(4), Ex. 1.

Class I, Division 2 Conduit Seals – 501.15(B). Conduit seals shall be installed for all connections to all enclosures which are required to be explosionproof. The installation of the seals shall meet the provisions of 501.15(A)(1)(1) and 501.15(A)(3). The conduit run or the nipple that is installed between the seal and the enclosure shall comply with 501.10(A).

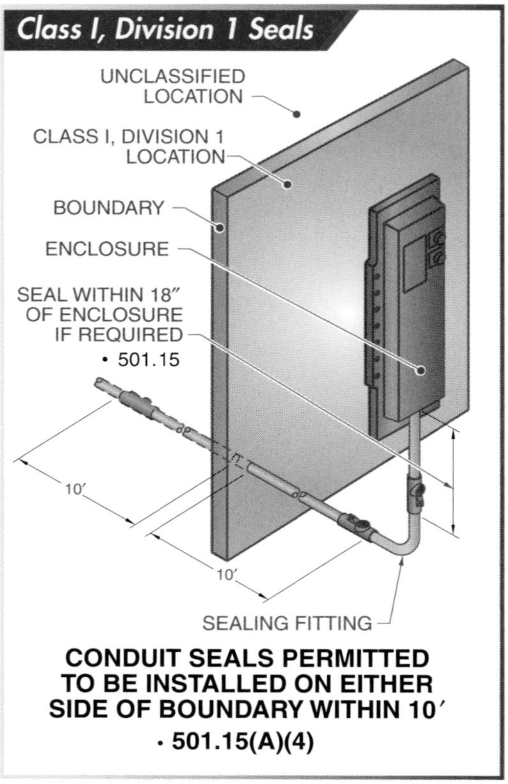

Figure 13-7. Conduit seals at hazardous locations boundaries are permitted to be installed on either side of the boundary provided they are located within 10′ of the boundary.

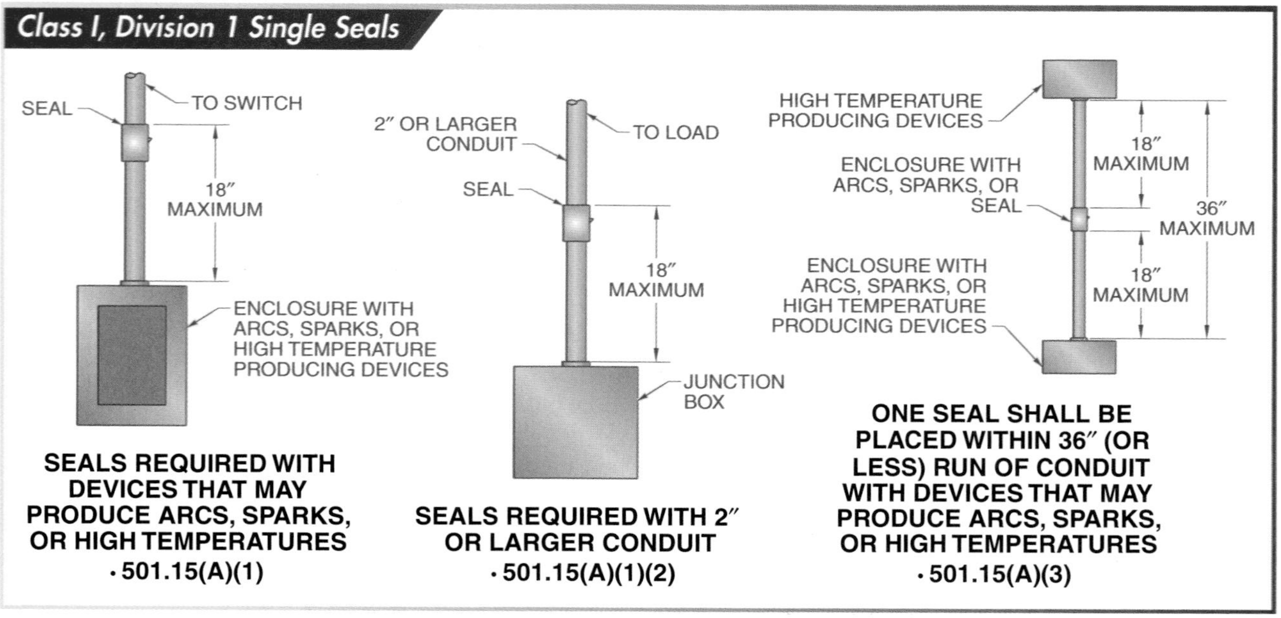

Figure 13-6. Proper location of the conduit seal ensures that only a minimum amount of the hazardous substance can enter the conduit system.

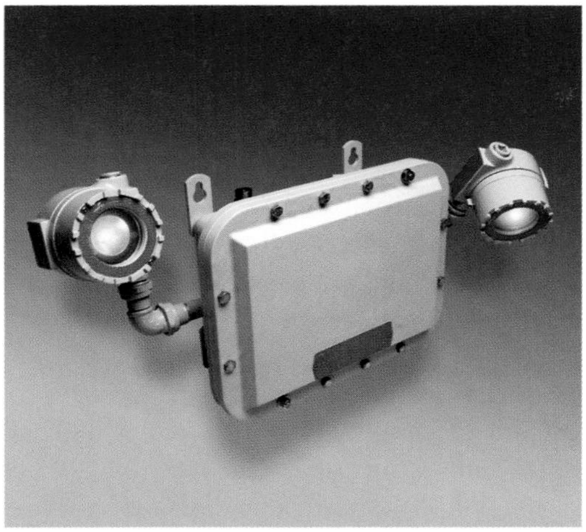

Cooper Crouse-Hinds

Explosionproof emergency lighting is installed in hazardous locations.

Section 501.5(B)(2) contains provisions for conduit runs which leave Class I, Division 2 locations. Conduit seals are required for these runs. The location of the seal is permitted on either side of the boundary within 10′ of the boundary. It should be placed in a location which minimizes the amount of gas or vapor that may have entered the conduit system within the Division 2 location being communicated to the conduit from beyond the seal. With the exception of explosionproof reducers, no boxes or fittings are permitted to be installed between the boundary and the conduit seal.

As with Class I, Division 1 locations, conduit seals are not required for an unbroken run of metal conduit which passes completely through a Class I, Division 2 location if there are no fittings less than 12″ from each boundary and if the conduits terminate in an unclassified area. See 501.5(B)(2), Ex. 1.

For outdoor locations, or locations in which the conduit system is contained within a single room, conduit seals shall not be required where a transition is made to cable tray, cablebus, ventilated busway, Type MI cable, or open wiring. The conduit system shall terminate at an unclassified location. It is not permitted to terminate at any enclosure which contains a source of ignition during normal operation. See 501.15(B)(2), Ex. 2.

Conduit seals can also be omitted from installations in which the conduit system enters a Class I, Division 2 location from an unclassified area that is unclassified because of pressurization. Pressurization requirements are contained in NFPA 496-2008, *Standard for Purged and Pressurized Enclosures for Electrical Equipment.*

Section 501.15(B)(2), Ex. 4 contains five conditions, which if present, do not warrant the inclusion of conduit seals where segments of aboveground conduit systems pass from a Class I, Division 2 location into an unclassified location.

Cable Seals

Sections 501.15(D) and 501.15(E) contain provisions for cable seals in Class I, Divisions 1 and 2 locations. While cables are used less frequently than conduit in these locations, some cables are permitted to be used under certain conditions.

Cables used in Class I locations shall be sealed for the same reasons that conduits are sealed. The provisions for installing cable seals for multiconductor cables, twisted pair, and shielded cables are included in this section.

Class I, Divisions 1 and 2 General Seal Requirements – 501.15(C)(1–6). Whether installed in a Division 1 or Division 2 location, conduit seals installed in Class I locations shall meet the provisions of 501.15(C)(1). Seals shall be an integral part of the enclosure or a sealing fitting listed for the location where it will be used. Sealing fittings shall also be listed for use with one or more specific compounds and must be accessible. The compound used for accomplishing the seal shall prevent the passage of gas or vapors. The compound shall have a melting point of at least 200°F and shall not be affected by the surrounding atmosphere or liquids. See 501.15(C)(2).

When the seal is completed, it shall have a minimum thickness equivalent to the trade size of the sealing fitting. In no case shall this be less than 5/8″ thick. Fittings used only for sealing purposes shall not contain splices or taps. Likewise, other fittings containing splices and taps shall not be filled with compound.

An excessive number of conductors in a conduit seal could result in an ineffective seal and permit transmission of the hazardous substance through the conduit system. The cross-sectional area of the conductors contained within a sealing fitting are not permitted to exceed 25% of the cross-sectional area of RMC of the same trade size unless it is approved for a higher percentage fill. **See Figure 13-8.** This provision ensures that the sealing compound can surround and fill the seal in a manner which effectively prevents the passage of gases and vapors.

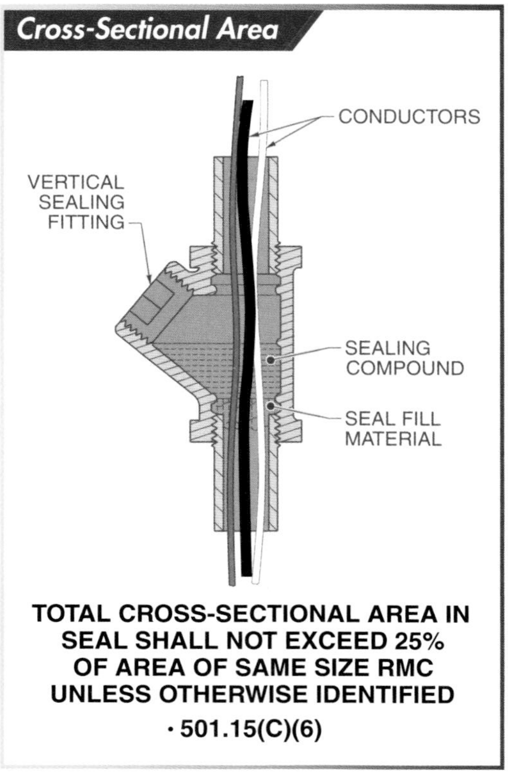

Figure 13-8. The cross-sectional area of the conductors contained within a sealing fitting is not permitted to exceed 25% of the cross-sectional area of RMC of the same trade size unless it is identified for a higher percentage fill.

Class II, Divisions 1 and 2 Conduit Seals – 502.15. Because of the nature of Class II locations, conduit seals are used to prevent the entrance of dust into a dust-ignitionproof enclosure through the raceway. *Dust-ignitionproof* is enclosed in a manner which prevents the entrance of dusts and does not permit arcs, sparks, or excessive temperature to cause ignition of exterior accumulations of specified dust. See 500.2. Where a raceway is installed between an enclosure that is required to be dust-ignitionproof and one that is not, one of the following four installations may be used to prevent the entrance of dust into the dust-ignitionproof enclosure:

(1) A permanent and effective seal may be installed. The seals are not required to be explosionproof. However, 502.10(A)(1)(4) and 502.10(B)(4) require fittings to be dusttight. Sealing putties such as electrical sealing putty are an acceptable method of providing the seal. All sealing fittings shall be accessible.

(2) A horizontal raceway of not less than 10′ in length may be installed. Seals are not required for raceways where 10′ or more of the raceway between terminations is installed horizontally.

(3) A vertical raceway of not less than 5′ in length may be installed. It shall extend downward from the dust-ignitionproof enclosure.

(4) A horizontal or vertical raceway installed in a manner equivalent to (2) or (3) that extends horizontally and downward from the dust-ignitionproof enclosure.

Class III, Divisions 1 and 2 Conduit Seals. Class III locations contain easily ignitable fibers and flyings. Contamination of these fibers and flyings is not likely to occur through conduit systems, therefore there are no conduit seal requirements for Class III, Divisions 1 and 2 locations.

Switches, Circuit Breakers, Motor Controllers, and Fuses

When electrical equipment is installed in hazardous locations, care shall be given to ensure that such equipment does not provide an ignition source for the hazardous substances. Electrical equipment such as switches, circuit breakers, motor controllers, and fuses are especially problematic because their operation frequently involves sparking and arcing which can provide the necessary ignition source. When such electrical equipment is installed in explosionproof enclosures, ignition is prevented by isolating the spark or arc from the hazardous atmospheres. Additionally, the explosionproof enclosure is designed to withstand an internal explosion without permitting ignition of the external hazardous atmosphere.

Class I locations involve flammable gases or vapors; therefore, the most stringent requirements for electrical equipment installed in hazardous locations are for Class I locations. The requirements for switches, circuit breakers, motor controllers, and fuses installed in Class II and III locations can be relaxed somewhat because the nature of the hazardous substance is not as volatile.

Equipment identified as being suitable for Division 1 locations is permitted to be used in Division 2 locations of the same class. Equipment that is marked Division 2, however, is only permitted to be installed in Division 2 locations. Equipment which is not marked to show a division is suitable for both Division 1 and 2 locations.

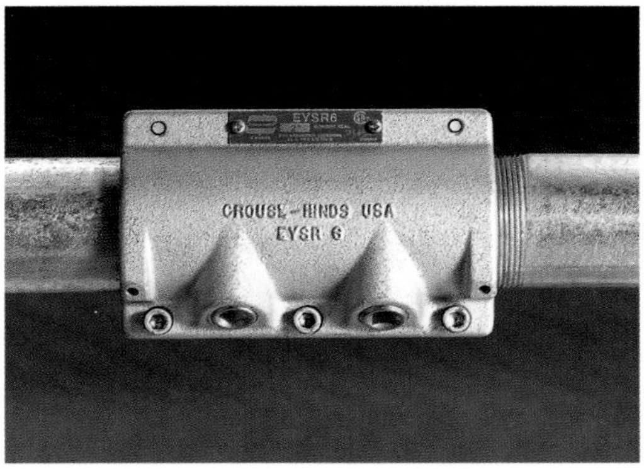

Cooper Crouse-Hinds

Sealing fittings for Class II, Division 2 locations are available in hub sizes from ¾″ to 4″.

Class I, Division 1 Electrical Equipment – Article 501 Part III. When switches, circuit breakers, motor controllers, or fuses are installed in Class I, Division 1 locations, 501.115(A) requires that they be installed in enclosures which shall be identified as a complete assembly for use in Class I, Division 1 locations. Additionally, all electrical apparatus installed in Class I, Division 1 locations shall be identified for use in Class I, Division 1 locations. **See Figure 13-9.**

Class I, Division 2 Electrical Equipment – 501.115(B). The general rule for switches, circuit breakers, motor controllers, or fuses located in Class I, Division 2 locations requires that the electrical equipment be installed in enclosures identified for Class I, Division 1 locations. There are, however, certain installation provisions which permit the electrical equipment to be installed in general-purpose enclosures. General-purpose enclosures are not permitted for Class I locations but switches, circuit breakers, motor controllers, or fuses can be contained in them if one of the following four conditions is met:

(1) The current-interrupting components of the equipment are contained within a hermetic chamber, sealed against the entrance of gas or vapor.

(2) The current make-and-break components of the equipment are oil-immersed and are of the general-purpose type, and the immersion is a minimum of 2″ for power contacts and 1″ for control contacts.

(3) The current interruption occurs within a factory-sealed explosionproof chamber approved for the location.

(4) All switching is done with solid-state components, without contacts, and where the surface temperature of the enclosure does not exceed 80% of the ignition temperature, in °C, of the gas or vapor involved.

When fused or unfused isolating switches are installed for transformer or capacitor banks which are not required to open circuits in their normal performance, general-purpose enclosures are permitted in place of enclosures approved for Class I, Division 2 locations. When fuses are installed for the protection of motors, appliances, or lamps but not for overcurrent protection per 501.115(B)(3), the fuses are permitted to be either the standard plug or cartridge type.

Where used, fuses shall be installed within enclosures identified for the location. General-purpose enclosures are permitted for fuses in which the operating element is immersed in oil or other approved liquid, or installations in which the operating element of the fuse is enclosed in a hermetically sealed chamber. Additionally, nonindicating, silver-sand, current-limiting fuses can be installed in general-purpose enclosures. Per 501.115(B)(4), listed cartridge fuses are permitted as supplementary protection within luminaires (lighting fixtures).

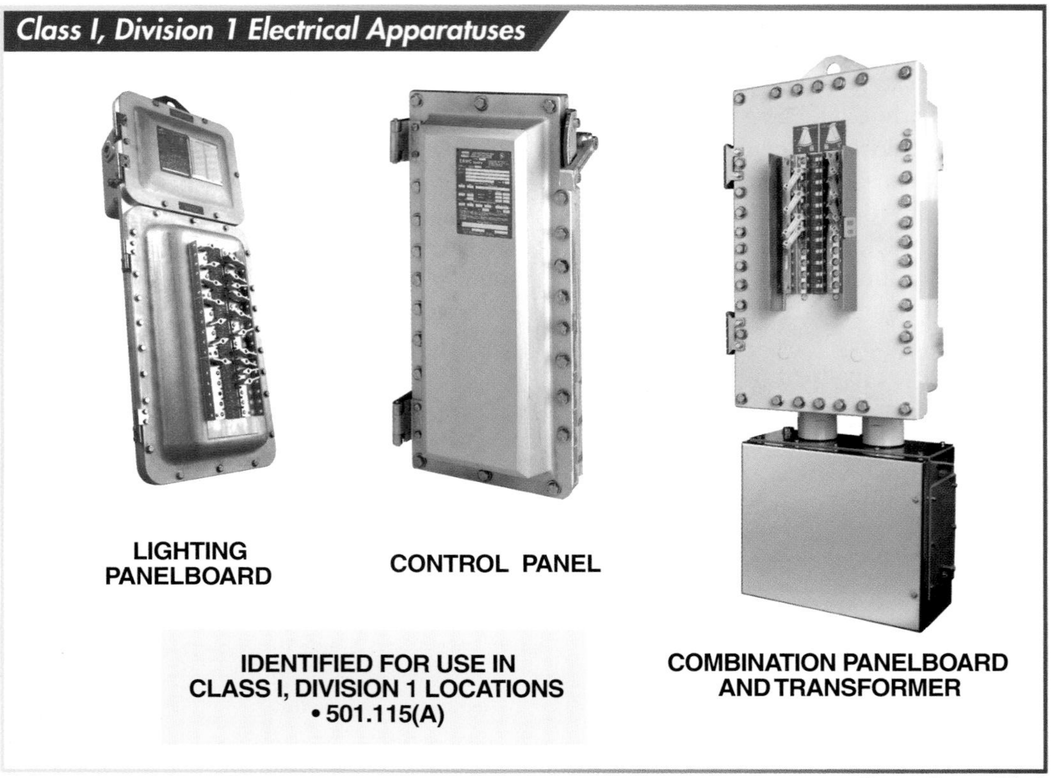

Cooper Crouse-Hinds

Figure 13-9. All electrical apparatus installed in Class I, Division 1 locations shall be identified for use in Class I, Division 1 locations.

Cooper Crouse-Hinds

Explosionproof boxes protect conductors connected within hazardous locations.

Class II, Division 1 Electrical Equipment – 502.115(A). Switches, circuit breakers, motor controllers, and fuses installed in Class II, Division 1 locations shall be installed in identified enclosures suitable for the location. Motor control devices such as pushbuttons, relays, and similar devices are included in this requirement.

Class II, Division 2 Electrical Equipment 502.115(B). When switches, circuit breakers, motor controllers, or fuses are installed in Class II, Division 2 locations, they shall be installed in enclosures which are dusttight or otherwise identified for the location.

Class III, Divisions 1 and 2 Electrical Equipment – 503.115. When switches, circuit breakers, motor controllers, or fuses are installed in Class III, Division 1 or 2 locations, they shall be installed in enclosures which are identified as dusttight.

Motors and Generators

The decision to install motors in hazardous locations should be weighed very carefully. By their nature, motors and generators tend to operate at very high temperatures which could possibly provide a source of ignition in Class I and some Class II locations. Additionally, arcing and sparking can be a normal occurrence in the operation of some motors and generators. For this reason, motors and generators should be installed outside of hazardous locations, particularly Class I and II locations, whenever practicable.

Class I, Division 1 Motors and Generators – 501.125(A). Section 501.125(A) provides four options, one of which must be used to allow motors, generators, or other electrical rotating machinery to be installed in a Class I, Division 1 location. The first option is to use motors or generators that are identified for Class I, Division 1 locations. Motors and generators approved for Class I, Division 1 are typically available and widely used for Group D and some Group C locations. Motors and generators for Group A (acetylene) and Group B (hydrogen, fuel, and combustible gases with at least 30% hydrogen or equivalent gases or vapors) locations are not typically available and one of the other three options should be used.

The second option is to use a totally enclosed motor or generator. These motors and generators must be supplied with positive-pressure ventilation from a source of clean air. The discharge of the motor or generator is to an unclassified or otherwise safe area. The control of the motor or the generator shall be such that it cannot be energized until ventilation has been established and the enclosure has been purged with at least 10 volumes of air. The control circuit shall also de-energize the motor or generator whenever the air supply is lost.

Baldor Electric Co.

Motors used in Class I and Class II locations must be of a totally enclosed type or identified for use in such locations.

The third option is to select a totally enclosed, inert gas-filled motor or generator. This is the same design as the second option, except an inert gas is used instead of clean air to purge the enclosure. The same control requirements for de-energizing the motor or generator shall be met for this option as well. Totally enclosed motors and generators are designed to prevent external surface temperatures which exceed 80% of the ignition temperature of the gas or vapor which is involved. They include safety devices that de-energize the equipment if the 80% temperature limitation is exceeded.

The fourth option relies on the use of a motor or generator of a type that is submerged in a liquid which is only flammable when it is vaporized and mixed with air. The motor or generator could also be "submerged" in a gas or vapor which has a pressure greater than atmospheric pressure and is only flammable when mixed with air. Either method shall include a control measure which prevents the motor or generator from being energized until the enclosure has been adequately purged and de-energizes the motor or generator upon a loss of the liquid, gas, or vapor which surrounds the motor or generator.

Class I, Division 2 Motors and Generators – 501.125(B). The requirements for motors, generators and other electrical rotating machinery which is installed in Class I, Division 2 locations depend on the type of motor or generator selected for the location. If the motor or generator has sliding contacts, centrifugal or other types of switches, or other integral resistance devices it shall be identified for Class I, Division 1 locations. If any of these devices, however, are contained in separate enclosures identified for Class I, Division 2 locations, then the motors or generators need not be approved for Class I, Division 1 locations.

Motors, generators, and other rotating electrical machinery not identified for Class I, Division 2 locations are permitted to be installed in the location if they are open or non-explosionproof enclosed motors such as squirrel-cage induction motors. These motors must not have brushes, switching mechanisms, or similar arc-producing devices.

Designers and installers should be aware, however, as 501.125, Informational Note 1 indicates, that the internal and external temperatures of the motors or generators shall be considered, even though the enclosures are not required to be explosionproof. This is because if the surface motor temperature is raised excessively during motor operation, it could become a possible source of ignition for some hazardous atmospheres.

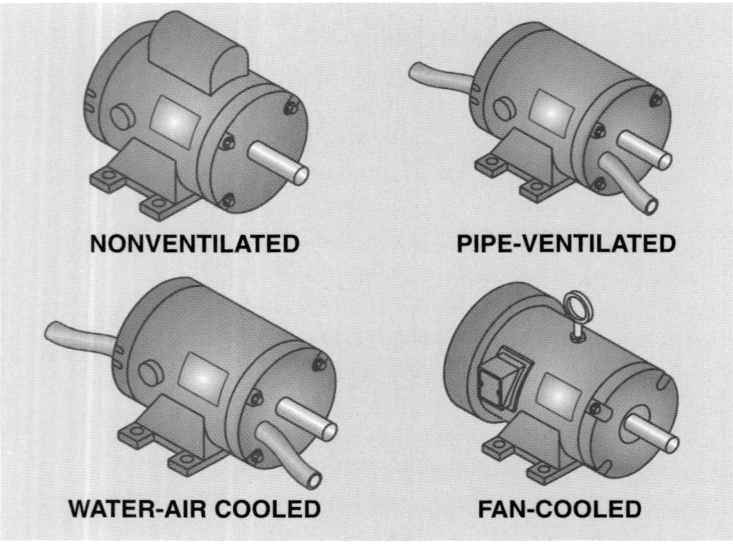

Motors or generators used for Class II, Division 2 locations shall be nonventilated, pipe-ventilated, water-air cooled, or fan-cooled.

Class II, Division 1 Motors and Generators – 502.125(A). Only two options are permitted for motors, generators, or other electrical rotating machinery installed in a Class II, Division 1 location. The first option is to install motors or generators which are identified for use in Class II, Division 1 locations. Motors or generators approved for Class II, Division 1 locations are identified as dust-ignitionproof.

The second option is to use totally enclosed motors or generators. These motors or generators shall be pipe-ventilated and designed so that dust does not enter the piping. Additionally, motors or generators installed in Class II, Division 1 locations shall be designed so that their maximum surface temperatures do not cause the dust to carbonize or become so dry that spontaneous ignition could occur. These motors and generators must also comply with the temperature provisions of 502.5.

Class II, Division 2 Motors and Generators – 502.125(B). Motors or generators used for Class II, Division 2 locations shall be one of the following totally enclosed types:

- Nonventilated
- Pipe-ventilated
- Water-air cooled
- Fan-cooled

Dust-ignitionproof motors or generators are also permitted provided they have no external openings and their maximum full-load external surface temperature is in accordance with 500.8(D)(2) for normal operation when operating in free air (not dust blanketed). Table 500.8(D)(2) establishes ignition temperatures for which the equipment was approved based on the specific Class II Group which is present.

Section 502.125(B), Ex. permits additional installation options if the AHJ determines that the likely accumulations of nonconductive, nonabrasive dusts are moderate and the motors or generators are easily accessible for maintenance and cleaning. The first option permits standard open or nonenclosed motors or generators to be installed in Class II, Division 2 locations if the motors or generators do not contain integral resistance devices, sliding contacts, centrifugal switches, or other switching mechanisms such as motor overcurrents, overloading, and overtemperature devices.

The second option permits standard open-type machines with resistance devices, contacts, and centrifugal or other switching mechanisms to be installed in Class II, Division 2 locations if the devices are enclosed within a dusttight housing which does not contain any openings or means of ventilation. The last option permits self-cleaning textile motors to be used provided the motors are of the squirrel-cage type.

Class III, Divisions 1 and 2 Motors and Generators – 503.125. Motors or generators used for Class III, Divisions 1 and 2 locations shall be one of the following totally enclosed types:

- Nonventilated
- Pipe-ventilated
- Fan-cooled

Section 503.125, Ex. permits three additional options when the AHJ determines that the likely accumulations of nonconductive, nonabrasive dusts will be moderate and the motors or generators are easily accessible for maintenance and cleaning. The three options include:

1. Self-cleaning textile motors are permitted to be used provided the motors are of the squirrel-cage type.
2. Standard open-type machines are permitted to be installed in Class II, Division 2 locations if the motors or generators do not contain sliding contacts or centrifugal or other switching mechanisms including motor overload devices.
3. Standard open-type machines with resistance devices, switching mechanisms, or contacts are permitted to be installed in Class II, Division 2 locations if the contacts, switching mechanisms, or resistance devices are enclosed within a tight housing which does not contain any openings or means of ventilation.

Luminaires

The need for luminaires is apparent in all types of hazardous locations. Even if it is desirable from a design point of view to keep luminaires out of hazardous locations, local building codes often require that lighting be provided for the area. There are, however, several concerns when installing luminaires in Class I, II, or III locations. The primary concern is to ensure that the luminaire does not provide the source of ignition for the hazardous materials which may be present in the atmosphere. While this is more of a concern for Class I and II locations, excessive temperature of the luminaire should be considered before selecting a luminaire for a particular class and division.

Another consideration is that the luminaires are installed in areas where they could be subjected to physical damage. This adds the possibility that a breakage of a lamp could provide the necessary source of ignition for the hazardous substance.

Portable lighting equipment is another concern when selecting luminaires for hazardous locations. Personnel working in hazardous classified areas must ensure that any portable lighting equipment which they are using in the area is approved for the class and division in which it is being used.

Nonsparking Hand Tools

Nonsparking hand tools such as chisels, pliers, hammers, and wire brushes are made from beryllium copper, brass, or bronze alloys and are required by OSHA to be used in locations where flammable or combustible liquids, vapors, or dusts are present.

Class I, Division 1 Luminaires – 501.130(A). Section 501.130(A)(1) requires that all luminaires used in Class I, Division 1 locations shall be identified as an assembly for use in Class I, Division 1 locations. These fixtures shall be clearly marked to indicate the maximum wattage of the lamps for which they are identified. If the fixture is designed for portable use, it shall be specifically listed, as a complete assembly, for that use. Temporary luminaires constructed of individual lighting components assembled on the job are not permitted to be used in Class I, Division 1 locations.

Section 501.130(A)(2) requires that all fixed and portable luminaires used in Class I, Division 1 locations shall be protected against physical damage. This can be accomplished by the use of a fixture or lamp guard or by the specific location of the fixture. **See Figure 13-10.**

Section 501.130(A)(3) covers the use of pendant fixtures in Class I, Division 1 locations. Pendant fixtures are required to be supported by, and supplied through, either threaded RMC stems or threaded steel IMC stems. Loosening of any threaded joints shall be prevented by providing a set-screw or other equally effective method. If conduit stems are longer than 12″, the stem shall be provided with permanent and effective bracing to guard against excessive lateral movement. The bracing shall be installed at a level which is not more than 12″ above the lower end of the stem.

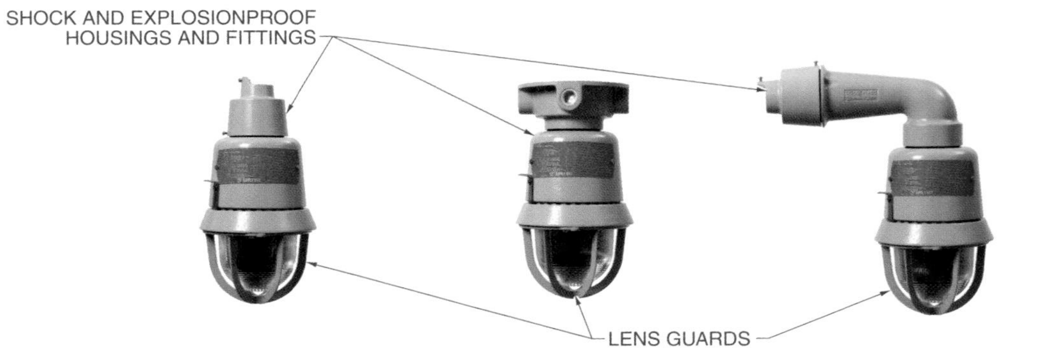

Figure 13-10. All fixed and portable luminaires used in Class I, Division 1 locations shall be protected against physical damage.

In lieu of bracing, a fitting or flexible connector which is identified for Class I, Division 1 locations may be used on conduit stems longer than 12″. The location of the fitting or flexible connector shall not be more than 12″ from the point of attachment to the box which supports the luminaire or fitting.

Section 501.130(A)(4), requires that all boxes, fittings, or box assemblies used to support luminaires in Class I, Division 1 locations shall be identified for Class I locations.

Class I, Division 2 Luminaires – 501.130(B). Portable lighting equipment is required by 501.130(B)(4) to be identified as a complete assembly for Class I, Division 1 locations. Section 501.130(B)(2) requires all luminaires installed in Class I, Division 2 locations to be protected against physical damage by fixture guards or by location. If the fixture utilizes lamps or bulbs that could cause the surface temperature of the fixture to exceed 80% of the ignition temperature of the gas or vapor involved, the fixtures shall be approved for Class I, Division 1 locations. **See Figure 13-11.** One last option would be to install a luminaire that has been tested and is marked for the appropriate temperature class of the gas or vapor.

Per 501.130(B)(3), pendant fixtures shall be suspended by either threaded RMC stems, threaded steel IMC stems, or by other approved means. If conduit stems are longer than 12″, the stem shall be provided with effective bracing to guard against excessive lateral movement. The bracing shall be installed at a level which is not more than 12″ above the lower end of the stem.

The support requirements for luminaires (lighting fixtures) in hazardous locations are intended to ensure that damage to the luminaire does not result in the development of a potential source of ignition for flammable gases or vapors. Section 501.130(B)(2) requires that suitable guards be utilized to protect the luminaire from physical damage. The luminaire location should also be considered to ensure it provides a degree of protection as well. Section 500.130(B)(1) addresses the lamps used in luminaires that are installed in Class I, Division 2 locations.

In lieu of bracing, an approved fitting or flexible connector may be used on conduit stems longer than 12″. The location of the fitting or flexible connector shall not be more than 12″ from the point of attachment to the box or fitting which supports the luminaire.

Per 501.130(B)(5), switches which are part of a fixture assembly or part of an individual lampholder in Class I, Division 2 locations shall be installed per 501.115(B)(1).

FIXED LIGHTING
CLASS I, DIVISION 1 LOCATIONS
• 501.130(A)(1)(3)

PORTABLE LIGHTING
CLASS I, DIVISION 1 LOCATIONS
• 501.130(A)(1)

Cooper Crouse-Hinds

Figure 13-11. Fixtures for hazardous locations shall be constructed so that their external surface temperatures cannot ignite the surrounding gas or vapor which may be present in the atmosphere.

Luminaire Operating Temperatures

To ensure that safe operating temperatures of luminaires are not exceeded in Class I and Class II locations, personnel should always use the proper lamp as specified by the luminaire manufacturer. This information is located on the luminaire nameplate.

The last requirement for luminaires in Class I, Division 2 locations covers starting and control equipment for electric-discharge lamps. Per 501.130(B)(6), in general, such equipment shall comply with 501.120(B).

Class II, Division 1 Luminaires – 502.130(A). Section 502.130(A)(1) requires that all luminaires used in Class II, Division 1 locations shall be identified for use in Class II locations. These fixtures shall be marked to indicate the maximum wattage of the lamps for which they are approved. Fixtures installed in locations where electrically conductive dusts such as magnesium, aluminum, aluminum bronze powders, etc. may be present shall be identified for the specific location.

Per 502.130(A)(2), all fixtures used in Class II, Division 1 locations shall be protected against physical damage. This can be accomplished by the use of a fixture or lamp guard or by the specific location of the fixture.

Per 502.130(A)(3), pendants or hanging fixtures in Class II, Division 1 locations are required to be suspended by threaded RMC stems, threaded steel IMC stems, by chains with approved fittings, or by other approved means. Loosening of any threaded joints shall be prevented by providing a set-screw or other equally effective method. If conduit stems are longer than 12″, the stem shall be provided with effective bracing to guard against excessive lateral movement. The bracing shall be installed at a level which is not more than 12″ above the lower end of the stem.

In lieu of bracing, a fitting or flexible connector which is listed for the location may be used on conduit stems longer than 12″. The location of the flexible fitting or connector shall not be more than 12″ from the point of attachment to the box or fitting which supports the luminaire. **See Figure 13-12.** Flexible cord is permitted between an outlet box or fitting and a pendant fixture provided it is of the hard-usage type and the cord is not used to support the luminaire. Flexible cord must be used in accordance with 502.10(A)(2)(5). Suitable seals must be provided where the cord enters the luminaire and the outlet box or fitting.

Per 502.130(A)(4), all boxes, fittings, or box assemblies that are used to support luminaires in Class II, Division 1 locations shall be identified for Class II locations.

Cooper Crouse-Hinds

Figure 13-12. Pendant fixtures in Class II, Division 1 locations shall be suspended by threaded RMC or steel IMC, by chains with approved fittings, or by other approved means.

Federal Signal Corporation

Explosionproof strobe lights are listed for use in Class I, Divisions 1 and 2, Groups C and D; Class II, Divisions 1 and 2, Groups E, F, and G; and Class III, Divisions 1 and 2 locations.

Class II, Division 2 Luminaires – 502.130(B). Per 502.130(B)(1), portable lighting equipment is required to be identified for the location. Fixtures for Class II, Division 2 locations shall be marked to clearly indicate the maximum wattage of lamps for which they are approved. Per 502.130(B)(2), if fixed luminaires are not identified for Class II, Division 2 locations, then they shall be provided with enclosures that are dusttight or otherwise identified for the location.

Per 502.130(B)(3), all luminaires installed in Class II, Division 2 locations shall be protected against physical damage by fixture guards or by location. Per 502.130(B)(4), pendant fixtures shall be suspended by either threaded RMC stems, threaded steel IMC stems, chains with approved fittings, or by other approved means. If conduit stems are longer than 12″, the stem shall be provided with effective bracing to guard against excessive lateral movement. The bracing shall be installed at a level which is not more than 12″ above the lower end of the stem.

In lieu of bracing, an identified fitting or flexible connector may be used on conduit stems longer than 12″. The location of the fitting or flexible connector shall not be more than 12″ from the point of attachment to the box or fitting which supports the luminaire.

The last requirement for luminaires in Class II, Division 2 locations covers starting and control equipment for electric-discharge lamps. Per 502.130(B)(5), in general, such equipment shall comply with the provisions of 502.120(B).

Class III, Divisions 1 and 2 Luminaires – 503.130. Per 503.130(A), all luminaires used in Class III, Division 1 or 2 locations for fixed lighting shall include enclosures for lamps and lampholders which are designed to minimize the entrance of fibers and flyings and to prevent sparks, hot metal, or other burning material from escaping the enclosure. In addition, the fixtures shall be marked to indicate the maximum wattage permitted for the lamps in order to limit the surface temperature to 165°C.

Per 503.130(B), all luminaires installed in Class III, Division 1 or 2 locations which may be subject to physical damage shall be protected by suitable fixture guards. Per 503.130(C), pendant fixtures shall be suspended by either threaded RMC stems, threaded IMC stems, threaded metal tubing of equivalent thickness, or by chains with approved fittings. If conduit stems are required to be longer than 12″, the stem shall be provided with effective bracing to guard against excessive lateral movement. The bracing shall be installed at a level which is not more than 12″ above the lower end of the stem.

In lieu of bracing, an identified fitting or flexible connector may be used on conduit stems longer than 12″. The location of the fitting or flexible connector shall not be more than 12″ from the point of attachment to the box or fitting which supports the luminaire.

Section 503.130(D) contains the last requirement for luminaires in Class III, Division 1 or 2 locations. In general, portable luminaires shall meet the same requirements of those for fixed lighting contained in 503.130(A). Portable lighting equipment shall be equipped with handles and shall be protected with substantial guards. Exposed, noncurrent-carrying metal parts shall be grounded and exposed current-carrying parts are not permitted in the design of the fixtures. Lampholders are not permitted to include provisions for receiving attachment plugs and the lampholders may not be switched.

Explosionproof Steady-On Beacons

Explosionproof steady-on beacons are used in factories, warehouses, and mills for applications such as safety lighting, obstacle warnings, and exit or entrance lights, and for identification of safety equipment such as showers or emergency telephones.

Grounding

Grounding in hazardous locations takes on added importance because of the presence of hazardous atmospheres that could be ignited by arcing ground faults. In addition to ensuring the operation of the overcurrent device, low-impedance ground paths ensure that arcing faults do not occur and that the surface temperatures of equipment are not raised excessively by the ground faults. If metal conduits are used as the EGC, care shall be taken to ensure that all threaded connections are tight to minimize the possibility of arcing or sparking across the joints.

Class I, Divisions 1 and 2 Grounding – 501.30. In general, all grounding shall be done in accordance with the provisions of Article 250. Section 501.30(A) covers bonding requirements and requires that bonding jumpers with proper fittings or other approved means be used where bonding is required in a Class I location. Locknut-bushings or double locknuts cannot be used for bonding purposes. Bonding is required around all intervening raceways, fittings, boxes, enclosures, etc., and between Class I locations and the point of grounding for separately derived systems or the point of grounding for service equipment.

Section 501.30(B) contains special requirements for the use of FMC and LFMC. They must include an EBJ of the wire type that complies with 250.102. FMC and LFMC are not permitted to serve as the sole EGC. Bonding requirements for these raceways must comply with 250.102. The bonding jumper may be installed either inside or outside of the conduit and in accordance with 250.102(E).

Section 501.30(B), Ex. permits listed LFMC to be installed without a bonding jumper in Class I, Division 2 locations, provided the total length of the conduit does not exceed 6′. In addition, the fittings used shall be listed for grounding, overcurrent protection for the circuit shall be limited to 10 A or less, and the load served by the conductors shall not be a power utilization load.

Grounding Electrode Resistance Factors

When the resistance of the earth grounding system is within the range specified by codes and standards, no further ground resistance modifications or measurements need to be taken. However, when the earth resistance is too high, modifications to the earth grounding system must be made. The three factors that affect the amount of resistance an earth grounding system has are resistance of the grounding electrode, contact resistance between the electrode and earth, and resistance of the earth surrounding the electrode.

Class II, Divisions 1 and 2 Grounding – 502.30. In general, all grounding shall be done in accordance with the provisions of Article 250. Section 502.30(A) covers bonding requirements and requires that bonding jumpers with proper fittings or other approved means be used where bonding is required in a Class II location. Locknut-bushings or double locknuts cannot be used for bonding purposes. Bonding is required around all intervening raceways, fittings, boxes, enclosures, etc., and between Class II locations and the point of grounding for separately derived systems or the point of grounding for service equipment.

Section 502.30(B) contains special requirements for the use of FMC and LFMC. They must include an EBJ of the wire type that complies with 250.102. If either of these conduits are installed, they shall be installed with bonding jumpers in parallel with the conduit. The bonding jumper may be installed either inside or outside of the conduit and in accordance with 250.102. LFMC and FMC are not permitted to serve as the sole ground-fault current path.

Cooper Crouse-Hinds

Articles 500 through 504 require equipment construction and installation that ensure safe performance of the system.

Section 502.30(B), Ex. permits listed LFMC to be installed without a bonding jumper in Class I, Division 2 locations, provided the total length of the conduit does not exceed 6′. In addition, the fittings used shall be listed for grounding, overcurrent protection for the circuit shall be limited to 10 A or less, and the load served by the conductors shall not be a power utilization load.

Class III, Divisions 1 and 2 Grounding – 503.30. In general, all grounding shall be done in accordance with the provisions of Article 250. Section 503.30(A) covers bonding requirements and requires that bonding jumpers with proper fittings or other approved means be used where bonding is required in a Class I location. Locknut-bushings or double locknuts cannot be used for bonding purposes. Bonding is required around all intervening raceways, fittings, boxes, enclosures, etc., and between Class II locations and the point of grounding for separately derived systems or the point of grounding for service equipment.

Section 503.30(B) contains special requirements for the use of FMC and LFMC. They must include an EBJ of the wire type that complies with 250.102. If either of these conduits are installed in a manner in which they are the sole equipment grounding path, they shall be installed with bonding jumpers in parallel with the conduit. The bonding jumper may be installed either inside or outside of the conduit and in accordance with 250.102. LFMC and FMC are not permitted to serve as the sole ground-fault current path.

Section 503.30(B), Ex. permits listed LFMC to be installed without a bonding jumper in Class III, Division 1 and 2 locations, provided the total length of the conduit does not exceed 6′. In addition, the fittings used shall be listed for grounding, overcurrent protection for the circuit shall be limited to 10 A or less, and the load served by the conductors shall not be a power utilization load.

INTRINSICALLY SAFE SYSTEMS – ARTICLE 504

Article 504, Intrinsically Safe Systems, contains an alternate method for the installation of electrical systems in Class I, II, or III locations. An *intrinsically safe system* is an assembly of interconnected, intrinsically safe apparatus, associated apparatus, and interconnected cables containing circuits in which any spark or thermal effect is incapable of causing ignition. See 504.2. The concept behind intrinsic safety is that the energy level of the circuit or equipment, which may be generated by sparking, arcs, or other short circuits or ground faults, is insufficient to cause ignition of a flammable or otherwise hazardous substance. Some of the factors involved in determining the intrinsic safety of a circuit or equipment include the voltage rating of the circuit, the current levels present, the surface temperature of the equipment, and the associated apparatus.

In general, 504.4 requires that all intrinsically safe apparatus and associated apparatus be listed. Installers of intrinsically safe systems should take care to ensure that the installation of the system does not jeopardize the integrity of the intrinsically safe system. Per 504.20, intrinsically safe systems are permitted to be installed with any of the wiring methods suitable for unclassified locations.

Sealing requirements to prevent the passage of flammable gases and vapors from classified to unclassified areas are also required for intrinsically safe systems. Section 504.70 requires that, in general, conduits or cables which are required to be sealed in Class I and II locations shall be sealed.

Section 504.30(A) contains the provisions for maintaining proper separation (1) in raceways, cable trays, and cables, (2) within enclosures, and (3) in other locations. Section 504.30(B) also requires that intrinsically safe circuits be properly separated from other intrinsically safe circuits by either (1) installation in a grounded metal shield or (2) by minimum insulation thickness.

Unless otherwise permitted by the control devices, the clearance between two terminals for the connection of field wires of different intrinsically safe circuits shall be at least 0.25″. Such severe separation requirements ensure that intrinsically safe systems stay isolated from circuits and equipment which could, under fault conditions or otherwise abnormal conditions, introduce energy levels into the intrinsically safe system and ignite flammable or other hazardous materials which may be present. Installation of intrinsically safe systems should only be done under the direct supervision of qualified personnel.

ZONE CLASSIFICATION SYSTEM

In 1996, an alternate method for classifying hazardous locations was included in the NEC®. Article 505, Class I, Zone 0, 1, and 2 Locations, incorporates a classification system which has been widely used in Europe. The zone classification system is based on the International Electrotechnical Commission (IEC) standards for area classification. Since the inclusion of this system is new, there are no NFPA or ANSI standards available for guidance in using the zone classification system.

In addition, there are many unanswered questions as to how the two different classification systems can coexist. At this time, designers and installers of electrical systems should know that this optional method is in the NEC® and they should closely follow the development of this system in future editions of the NEC®. If Article 505 is used for area classification wiring selection or equipment selection, it should be done under the supervision of a qualified person. The standard and optional hazardous location systems shall not be commingled. **See Figure 13-13.**

> ***Intrinsically Safe Equipment Applications***
>
> *Intrinsically safe equipment is primarily limited to process control instrumentation because these electrical systems have low energy requirements and are incapable of releasing significant electrical or thermal energy under normal conditions.*

The 2005 NEC® contained a new article that included the first major development in the Zone classification system since its introduction to the NEC® in 1996. Article 506, Zone 20, 21, and 22 Locations for Combustible Dusts, Fibers, and Flyings, provides an alternative method for treating electrical and electronic equipment installed in hazardous locations where combustible dusts and fibers are present. Article 506 does not cover combustible metallic dusts. Article 506 is intended to provide an alternate method to that described in Articles 500, 502, and 503. The Zone 20, 21 and 22 system is based on the modified IEC area classification system used in Europe and other parts of the world. Section 506.1 defines the scope of the article. Section 506.5 covers the classification criteria for Zone 20, 21, and 22 locations, and 506.15 lists the permissible wiring methods for the appropriate zone location.

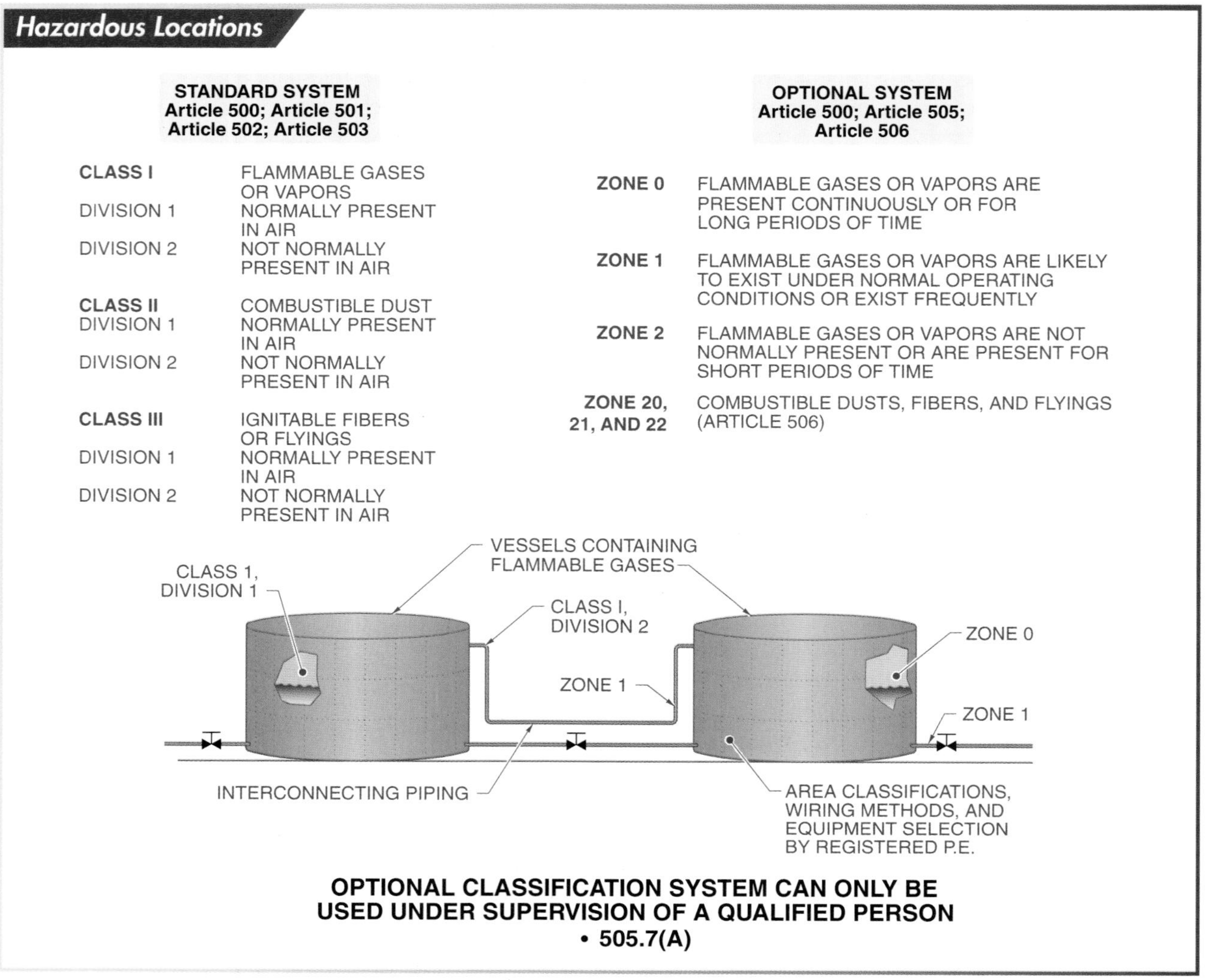
Hazardous Locations
STANDARD SYSTEM
Article 500; Article 501;
Article 502; Article 503
CLASS I
FLAMMABLE GASES OR VAPORS
DIVISION 1
NORMALLY PRESENT IN AIR
DIVISION 2
NOT NORMALLY PRESENT IN AIR
CLASS II
COMBUSTIBLE DUST
DIVISION 1
NORMALLY PRESENT IN AIR
DIVISION 2
NOT NORMALLY PRESENT IN AIR
CLASS III
IGNITABLE FIBERS OR FLYINGS
DIVISION 1
NORMALLY PRESENT IN AIR
DIVISION 2
NOT NORMALLY PRESENT IN AIR
OPTIONAL SYSTEM
Article 500; Article 505;
Article 506
ZONE 0
FLAMMABLE GASES OR VAPORS ARE PRESENT CONTINUOUSLY OR FOR LONG PERIODS OF TIME
ZONE 1
FLAMMABLE GASES OR VAPORS ARE LIKELY TO EXIST UNDER NORMAL OPERATING CONDITIONS OR EXIST FREQUENTLY
ZONE 2
FLAMMABLE GASES OR VAPORS ARE NOT NORMALLY PRESENT OR ARE PRESENT FOR SHORT PERIODS OF TIME
ZONE 20, 21, AND 22
COMBUSTIBLE DUSTS, FIBERS, AND FLYINGS (ARTICLE 506)
VESSELS CONTAINING FLAMMABLE GASES
CLASS 1, DIVISION 1
CLASS I, DIVISION 2
ZONE 0
ZONE 1
ZONE 1
INTERCONNECTING PIPING
AREA CLASSIFICATIONS, WIRING METHODS, AND EQUIPMENT SELECTION BY REGISTERED P.E.
OPTIONAL CLASSIFICATION SYSTEM CAN ONLY BE USED UNDER SUPERVISION OF A QUALIFIED PERSON
• 505.7(A)

Figure 13-13. The standard and optional hazardous location systems must not be commingled.

COMMERCIAL GARAGES – ARTICLE 511

Article 511 contains provisions for installing electrical systems in commercial garages used for repair of self-propelled vehicles and storage of volatile flammable liquids used for fuel or power. The presence of volatile flammable liquids requires areas of these garages to be classified as Class I hazardous locations and the methods for installation of electrical systems to be closely controlled. In addition to Article 511, other articles may also need to be considered when designing or installing electrical systems in commercial garages. If, for example, the commercial garage includes an area in which flammable fuel is transferred to vehicle fuel tanks, then the requirements of 514 shall also be followed.

Definitions and Classifications – 511.2, 511.3

During the 2008 NEC® cycle, the requirements for area classifications in commercial garages, repair, and storage were completely revised and restructured. The first step in determining area classifications for these types of garages is to determine the proper classification of the garage. Section 511.2 contains two definitions that are key to proper area classification.

A *major repair garage* is a commercial garage where a building or portion of a building is used for major repairs. Examples of major repairs include engine overhauls, painting, body and fender work, and repairs that require the draining of the vehicle fuel tank. This space includes the floor space used for offices, parking, and showrooms.

A *minor repair garage* is a commercial garage where a building or a portion of a building is used for lubrication, inspection, and minor vehicle repair. Examples of minor vehicle repair include engine tuneups, replacement of engine parts, fluid changes, brake-system repair, tire rotation, and similar routine maintenance work.

Most of the requirements for classifying the spaces in commercial garages have been extracted from NFPA 30A, *Code for Motor Fuel Dispensing Facilities and Repair Garages*. This should help to provide users of the Code and Article 511 with a comprehensive approach to the classification of areas within a commercial garage.

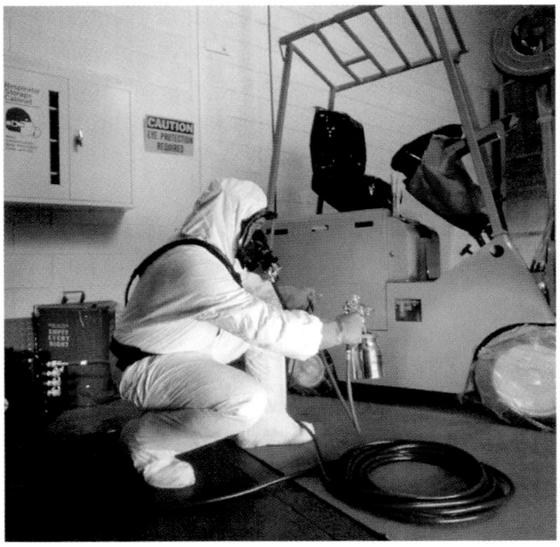

W. W. Grainger, Inc.
An industrial equipment repair and maintenance facility is an example of a major repair garage.

Section 511.3(A) states that parking garages, which are used simply for the parking or storage of vehicles, are not required to be classified. Wiring methods for these types of parking garages can be selected from the Chapter 3 wiring methods, provided the permitted uses for the respective wiring method are met.

Section 511.3(B) covers the requirements for repair garages that include motor fuel dispensing. Generally, either minor or major repair garages that contain provisions for dispensing flammable liquids with a flash point below 38°C (gasoline, hydrogen, LPG) must be classified in accordance with Table 514.3(B)(1).

Section 511.3(C) permits major repair garages that do not have fuel-dispensing provisions, other than those required for major repair activities, to have the area classification performed in accordance with 511.3(C)(1-3). These sections define classification requirements for (1) floor areas, (2) ceiling areas, and (3) pit areas in lubrication or service rooms.

Section 511.3(D) contains the provisions for the classification of minor repair garages that do not have provisions for fuel dispensing. As with the provisions for major repair garages, the area classification for minor repair garages must be performed in accordance with 511.3(D)(1-3). These sections define classification requirements for (1) floor areas, (2) ceiling areas, and (3) pit areas in lubrication or service rooms.

Section 511.3(E) covers modifications to classifications for both minor and major repair garages. There are two listed modifications that are addressed by this section. The first permits areas adjacent to classified locations to be unclassified provided that flammable vapors are not likely to be released and that mechanical ventilation is provided that ensures a rate of four or more air changes per hour, or the area is designed with positive air pressure, or where the area is effectively cut off by walls or partitions. The second listed modification to area classifications deals with the area used for the storage, handling or dispensing of alcohol-based windshield washer fluids into motor vehicles. These areas are permitted to be unclassified unless another requirement in Section 511.3 requires the area to be classified.

Wiring Methods – 511.4

Specific wiring methods are required to minimize the potential hazards of installing electrical systems in commercial garages. The selection of the wiring method depends upon the classification of the area in which the wiring is to be installed. As with all classifications, the more likely the presence of the hazardous substance, the stricter the requirements in respect to the wiring method employed.

Class I Locations – 511.4(A). All wiring installed in any area within a commercial garage which is classified as Class I per 511.3 shall comply with the applicable provisions of 501. Wiring methods in Class I, Division 1 locations shall be selected from 501.10(A) and wiring methods for Class I, Division 2 locations shall be selected from 501.10(B). Section 501.10(A)(1), Ex. permits RNC (PVC) to be installed in Class I locations, provided it is buried under not less than 2′ of cover and either threaded RMC or threaded steel IMC is used for the last 2′ of the underground run up to the point of emergence or point of connection to the aboveground raceway. **See Figure 13-14.**

Above Class I Locations – 511.7. The area above Class I locations often contains electrical equipment which is used in commercial garages. The equipment and wiring for the equipment shall not provide a source of ignition for the hazardous substances which may be present below. Section 511.7(A)(1) requires that all fixed wiring in these locations shall be installed in metal raceways, RNMC, ENT, FMC, LFMC, or LFNC. In addition, Type MI, MC, PLTC, and TC cables and manufactured wiring systems are permitted. Ceiling outlets or extensions to the floor below are permitted to be installed with cellular metal or cellular concrete floor raceways, provided such raceways have no connections leading through or into any Class I locations above the floor.

Equipment – 511.10, 511.12

Article 511 has several specific requirements for the various types of equipment used in commercial garages. The purpose of the requirements for portable lighting, fixed lighting, battery and vehicle charging equipment, and GFCI protection is two-fold.

First, the requirements ensure that the equipment which may be used in classified areas is designed and installed in a manner which does not increase the hazards associated with the flammable gases or vapors which may be present. Secondly, the requirements help to ensure that all workers within commercial garages are protected from the hazards associated with work in these types of spaces.

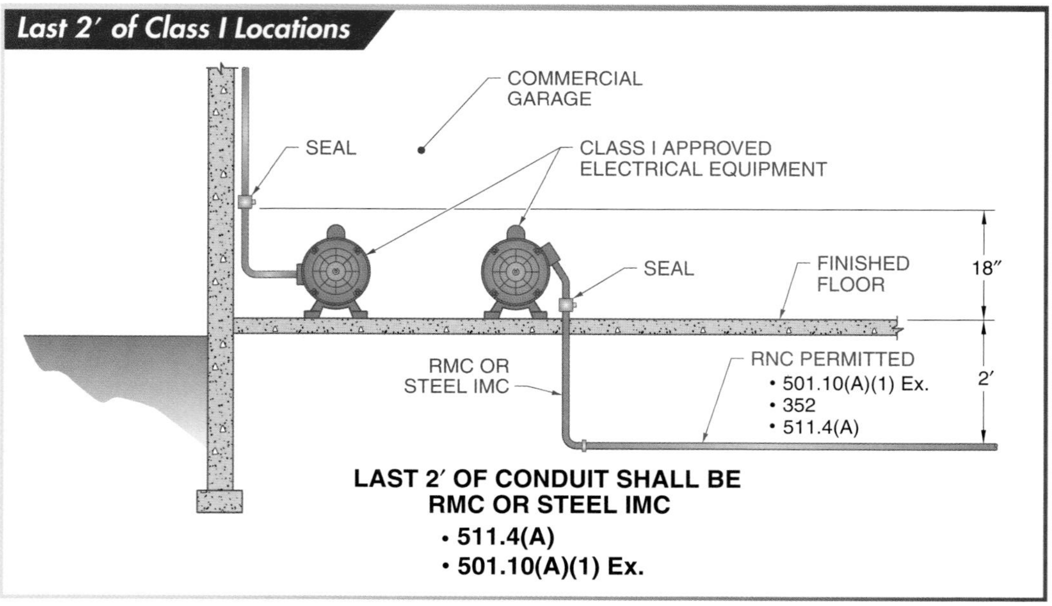

Figure 13-14. Where suitably protected and installed per Article 352, PVC is a permitted wiring method for Class I, Division 1 locations in commercial garages.

Portable Lighting – 511.4(B)(2). Portable lighting equipment shall be equipped with a handle, hook, and lampholder. It shall be protected with substantial guards. Exposed surfaces shall be constructed with nonconducting material or shall be covered with insulation.

Lampholders are not permitted to include provisions for receiving attachment plugs and the lampholders shall not be switched. The outer shell of the luminaire is required to be constructed of molded composition or other suitable materials. All portable luminaires shall be identified for Class I, Division 1 locations unless they are installed in a manner which prevents their usage in the locations that are classified as Class I, Division 1 or 2. **See Figure 13-15.**

Arcing Equipment – 511.7(B)(1)(a). If electrical equipment which may be installed above the floor level in commercial garages is capable of producing arcs or sparks or releasing particles of hot metal, and is located less than 12′ above the floor, it shall be totally enclosed or designed so that it prevents the release of arcs or sparks, or particles of hot metal. Examples of the types of equipment to which 511.7(B)(1)(a) applies include cutouts, switches, charging panels, motors, generators, and other equipment which incorporates make-and-break or sliding contacts. This section does not apply to equipment such as receptacles, lamps, or lampholders.

Fixed Lighting – 511.7(B)(1)(b). Unless the luminaires are totally enclosed or constructed so that the escape of arcs, sparks, or hot metal particles is prevented, all lamps and lampholders for fixed lighting shall be installed at a location which is not less than 12′ above the floor level if the fixtures are subject to physical damage. This includes lanes which commonly contain vehicle traffic.

PPE for arc blast protection must be worn at all times when working with energized electrical equipment.

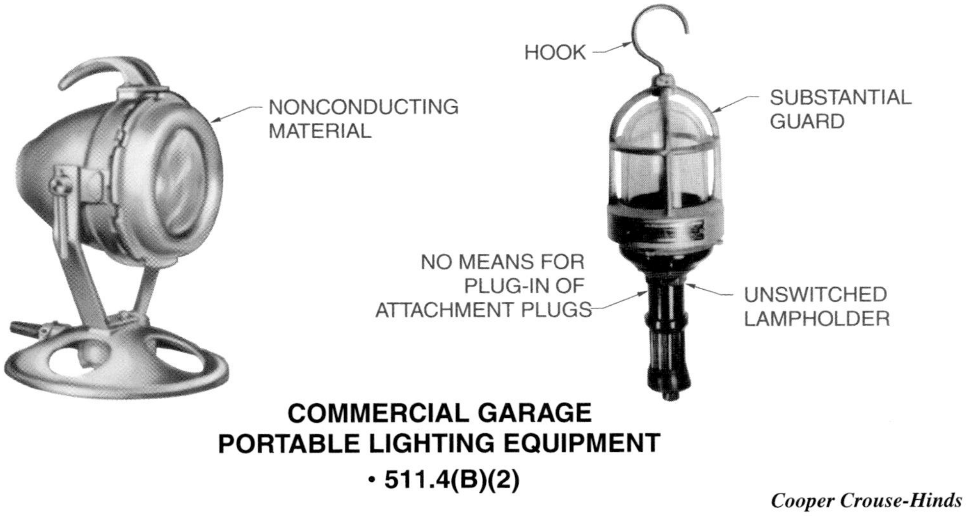

Figure 13-15. Portable lighting equipment used in Class I, Division 1 locations in commercial garages shall be identified for the location.

Battery Charging Equipment – 511.10(A). Battery charging equipment is frequently found in commercial garages. Such equipment usually consists of the battery charger, control equipment, and the batteries themselves. The installation of any of this battery charging equipment is prohibited in any of the Class I, Division 1 or 2 locations covered in 511.3.

Vehicle Charging Equipment – 511.10(B). Electrical vehicle charging equipment is covered by Article 625. Such equipment which is installed in commercial garages shall be installed per Article 625. Connectors and the point of connection between chargers and vehicles shall not occur within the Class I, Division 1 or 2 locations listed in 511.3.

GFCI Protection – 511.12. To provide protection to personnel who may be working in commercial garages, 511.12 requires that all 15 A or 20 A, 125 V, 1ϕ receptacles installed for the connection of electrical diagnostic equipment, electrical hand tools, or portable lighting equipment shall be provided with GFCI protection. This provision ensures that personnel who might be working with equipment when water or other liquids are present are protected against the hazards of electrical shock. **See Figure 13-16.**

GFCI Protection

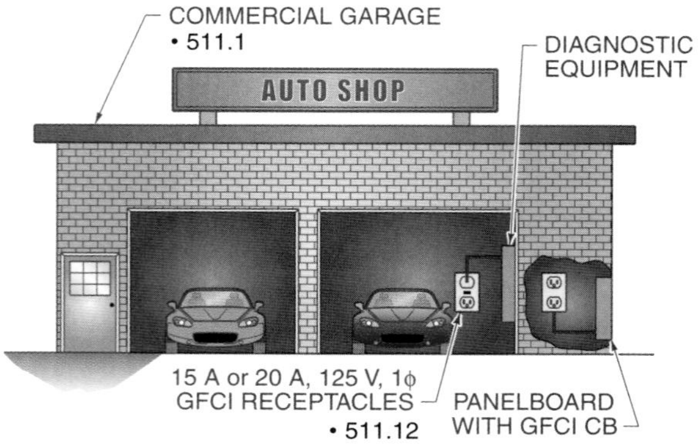

ALL 15 A OR 20 A, 125 V, 1ϕ RECEPTACLES FOR DIAGNOSTIC EQUIPMENT, HAND TOOLS, OR PORTABLE LIGHTING SHALL HAVE GFCI PROTECTION • 511.12

Figure 13-16. Where GFCI protection is required for personnel in commercial garages, it can be provided by either GFCI receptacles or from receptacles protected by a GFCI circuit breaker.

MOTOR FUEL DISPENSING FACILITIES – ARTICLE 514

Motor fuel dispensing facilities present many of the same problems for designers and installers of electrical systems that are encountered with commercial garages. Volatile flammable liquids and liquefied flammable gases may be present and are frequently transferred to the fuel tanks of many different kinds of self-propelled vehicles or other approved containers. Table 514.3(B)(1) is used when Class I liquids are stored, handled, or dispensed to determine proper area classification and delineation.

Motor fuel dispensing facilities are locations where gasoline, other volatile flammable liquids, or liquefied flammable gases are transferred to vehicle fuel tanks or other approved storage containers. Article 514 applies only to these specific areas of a motor fuel dispensing facility. Other areas in the facility include service rooms, repair rooms, offices, sales rooms, etc. These areas are not covered by Article 514. They are subject to the provisions of Article 510 or Article 511. If the AHJ determines that flammable liquids, such as gasoline, which have a flash point below 38°C (100°F) will not be handled, the area is permitted to be unclassified.

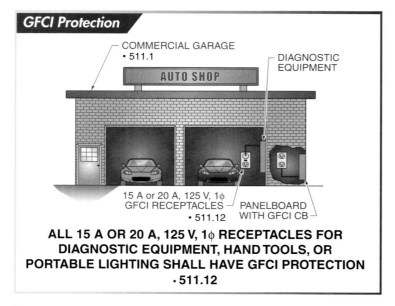

Quick-Lube Facilities

The recent development of "quick-lube" facilities has led to some interesting classifications of these types of service stations. Table 514.3(B)(1) was revised during the 1996 NEC® cycle to allow for these types of service stations which do not contain provisions for dispensing fuel, but have the potential for the presence of flammable gases or vapors.

Table 514.3(B)(1) now permits pits or depressions in quick-lube facilities to be unclassified if exhaust ventilation is provided at a rate of not less than 1 cfm/sq ft of floor area at all times that the building is occupied. This provision also extends to periods when vehicles may be parked over or near the pit or depression. The exhausted air shall be taken from a point which is within 12″ of the floor of the pit or depression. Designers and installers should check with the AHJ for proper area classification whenever electrical work is to be performed in these types of facilities.

Installation Requirements

Because volatile flammable liquids and gases are present in motor fuel dispensing facilities, several installation requirements shall be considered before designing or installing electrical systems in these occupancies. These considerations include:

- The areas of the occupancies shall be carefully classified to ensure that the proper wiring method and installation requirements for the particular class and division are followed. Table 514.3(B)(1) is designed to assist in making that determination based upon the location and extent of the classified area.
- Special disconnecting means are required for circuits supplying dispensing equipment. Additionally, other dispensing equipment controls may be required depending on whether the service station is attended or unattended.
- Sealing may be required to minimize the passage or communication of hazardous substances from classified to unclassified areas.
- Special grounding provisions may be required for classified areas within the gasoline or service station.
- Wiring methods are based upon the location within the service station and the classification of the area.

Class I Locations – Table 514.3(B)(1). Table 514.3(B)(1) lists the Class I and Division locations which may be found within motor fuel dispensing facilities. Some of the locations listed in the Table are areas around underground tanks, dispensing devices, remote pumps, lubrication rooms with and without dispensing, sales and storage rooms, equipment enclosures, and other vapor-processing equipment locations. Class I locations shall not extend beyond solid partitions, walls, or roofs. Any equipment installed within the Class I locations listed in Table 514.3(B)(1) shall be wired per the Class I, Division 1 or 2 provisions contained in Article 501.

Wiring Above Class I Locations – Table 514.7. All installations of equipment and wiring in areas above locations which are Class I locations shall comply with the provisions of 511.7.

Cooper Crouse-Hinds

Explosionproof fire alarm call points are required for hazardous locations.

Disconnecting Means – 514.11. For safety reasons, it is important to be able to quickly disconnect all circuit conductors which supply or pass through gasoline dispensing equipment. This includes all associated power circuits, communications, data and video circuits, and equipment for remote pumping systems. Section 514.11(A) requires that such disconnection be provided in the form of clearly identifiable and readily accessible switches or other acceptable means. The switch shall be located away from the dispensing equipment and shall simultaneously open all of the circuit conductors, including the grounded conductor from the supply source. Single-pole circuit breakers which use handle ties to accomplish the simultaneous opening of the circuit conductors are not permitted. **See Figure 13-17.**

For attended motor fuel dispensing facilities, 514.11(B) requires that the disconnecting means be installed not more than 100′ from the dispensers in a location which is acceptable to the AHJ. For unattended motor fuel dispensing facilities, 514.11(C) requires that the disconnecting means to be installed in a location acceptable to the AHJ that is more than 20′ but less than 100′ from the dispensers. In addition, additional disconnecting means shall be provided for each group of dispensers or the outdoor equipment which is used to control the dispensers.

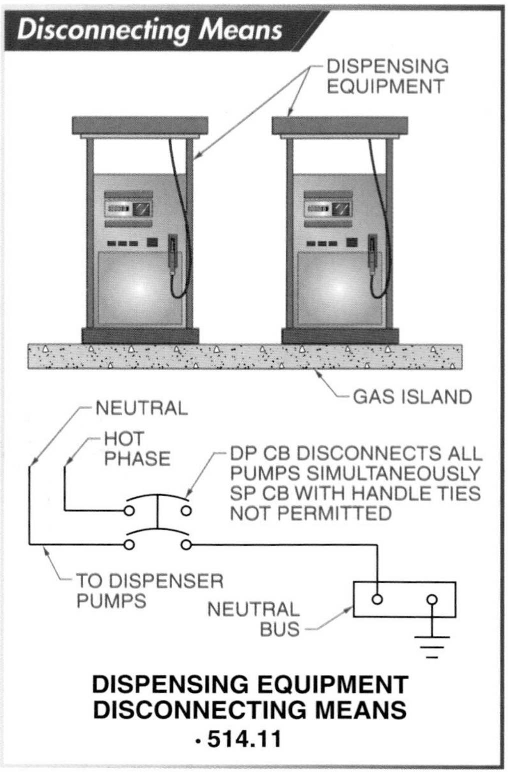

Figure 13-17. In addition to the circuit ungrounded conductors, the grounded conductor (neutral) shall be disconnected from the source of supply if it supplies or passes through dispensing equipment.

Conduit Seals – 514.9. Two specific areas within motor fuel dispensing facilities are required to have conduit seals installed to prevent the communication of hazardous substances from a classified area to an unclassified one. Per 514.9(A), a listed seal shall be installed for each conduit entering or leaving a fuel dispenser. The seal fitting shall be installed so that it is the first fitting after the conduit emerges from the earth or concrete. **See Figure 13-18.**

Per 514.9(B), all Class I, Division 1 or 2 conduit seal requirements listed in 501.15 shall be applied to gasoline and service stations. The requirements in 501.15(A)(4). and 501.15(B)(2) shall apply to both horizontal and vertical boundaries of the Class I locations. In any Class I, Division 1 or 2 location in a gasoline or service station, both vertical and horizontal runs of conduit which leave a classified area shall have a conduit seal installed within 10′ of the boundary.

Grounding – 514.16. All metal raceways, the metal armor or metallic sheath on cables, and all other metal, non-current-carrying parts of equipment shall be grounded and bonded per Article 250 regardless of voltage. Grounding in Class I locations shall also meet the requirements of 501.30. **See Figure 13-19.**

Underground Wiring – 514.8. The two general wiring methods permitted for underground wiring in gasoline and service stations are threaded RMC and threaded steel IMC. In addition, two exceptions permit alternate wiring methods to be used as well. Per 514.8, Ex. 1, MI cable may be used when installed in accordance with Article 332. MI cable for use in underground runs shall be suitably protected against both corrosive conditions and physical damage.

Per 514.8, Ex. 2, PVC conduit and Type RTRC conduit are permitted to be installed provided they are buried under at least 2′ of cover and either threaded RMC or threaded steel IMC is used for the last 2′ of the run to emerge from the ground or connection to the aboveground raceway. A separate EGC shall be installed to maintain electrical continuity of the raceway.

Mineral-Insulated (MI) Cable

Type MI cable consists of one to 17 copper conductors properly spaced and encased in tightly compressed magnesium oxide, clad in an overall copper sheath. When properly installed, Type MI cable is suitable for all Class I and Class II locations.

HEALTH CARE FACILITIES – ARTICLE 517

A *health care facility* is a location, either a building or a portion of a building, which contains occupancies such as hospitals, nursing homes, limited or supervisory care facilities, clinics, medical and dental offices, and either movable or permanent ambulatory facilities. See 517.2. *Note:* This list is not all inclusive. There may be other similar facilities that are not listed that meet the criteria for health care facilities. The definition specifically addresses portions of buildings. Section 517.2 contains a list of definitions which should be reviewed prior to attempting to apply the provisions of the article.

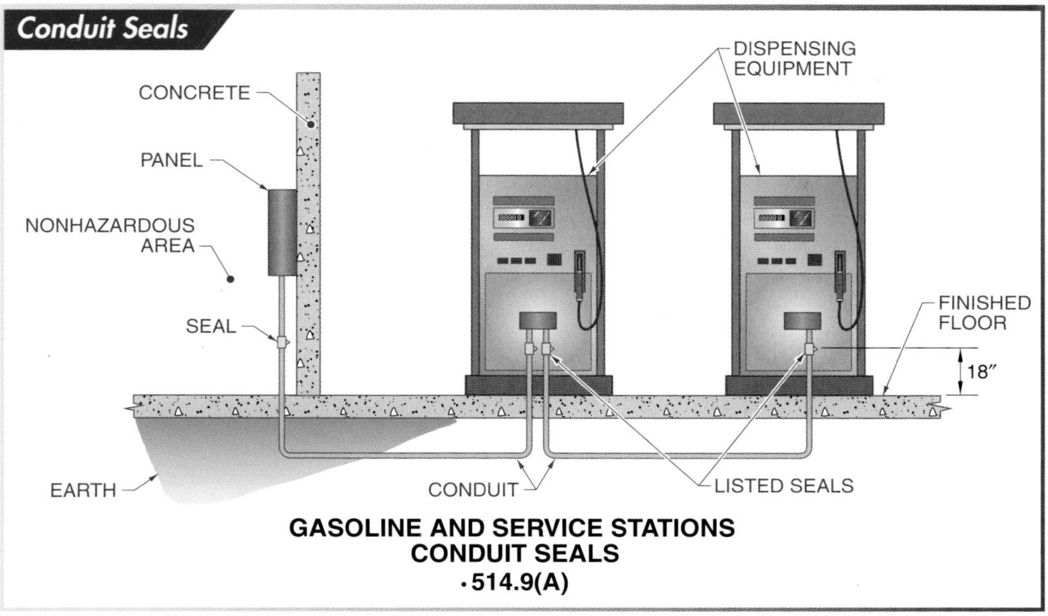

Figure 13-18. Conduit seals for dispensing equipment shall be the first fitting installed after the conduit leaves the concrete or the earth.

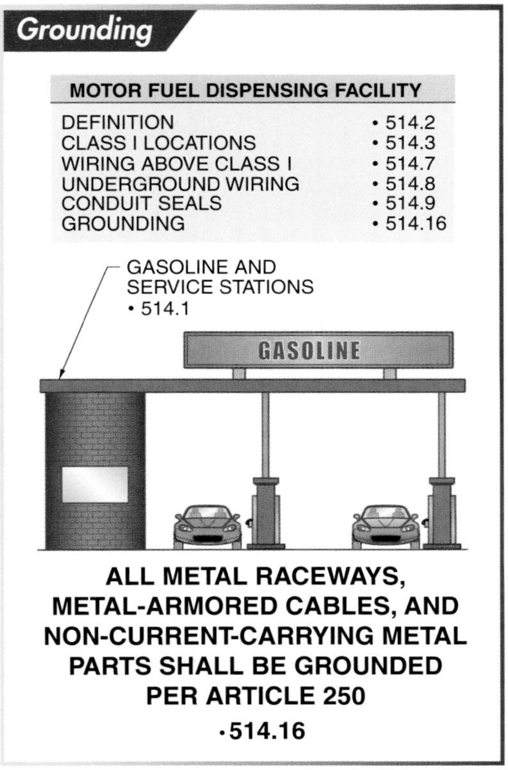

Figure 13-19. Grounding in Class I, Division 1 locations of gasoline and service stations shall be in accordance with 501.16.

Unlike the normal grounding requirements in Article 250, receptacles and equipment in patient care areas shall be grounded by an insulated copper conductor that is required to be installed in a metal raceway, which qualifies as an equipment grounding return path in accordance with 250.118. This "redundant grounding" is to ensure the safety of patients who are very vulnerable to electrical shock hazards. Section 517.13(A) does permit the use of Type MI, MC, or AC cables in these areas, but only if the outer metal armor of these cables provides an acceptable grounding return path in accordance with 250.118. Type MC cable with interlocking metal tape armor does not meet this requirement. **See Figure 13-20.**

Section 517.18(C) requires all 15 A and 20 A, 125 V receptacles installed within the rooms, bathrooms, playrooms, activity rooms, and patient care areas of designated pediatric locations to be listed tamper-resistant. A tamper-resistant receptacle is designed to limit improper access to its energized contacts. A listed tamper-resistant receptacle cover is permitted to be used in place of a tamper-resistant receptacle. **See Figure 13-21.**

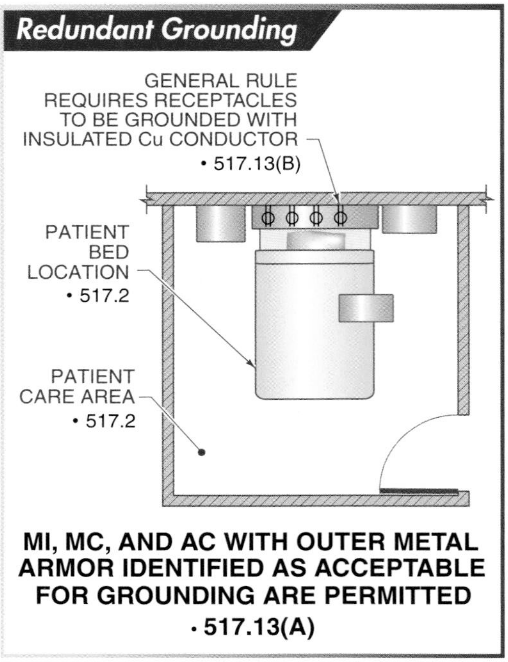

Figure 13-20. Type MI, MC, or AC cables installed in patient care areas of health care facilities shall be listed and shall be constructed with an outer metal armor which provides an acceptable grounding return path.

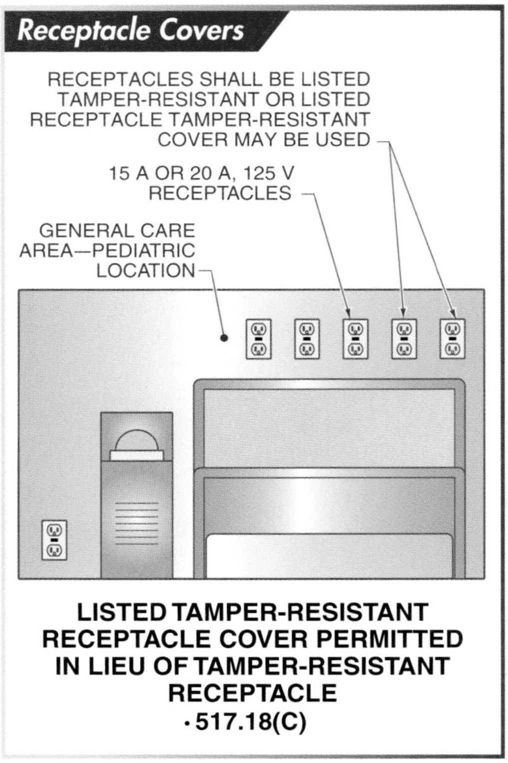

Figure 13-21. Listed tamper-resistant receptacle covers, which limit access to the energized parts of a receptacle, are permitted in place of tamper-resistant receptacles in pediatric locations of health care facilities.

Wet procedure locations in health care facilities pose additional hazards for patients. Section 517.20 requires that all receptacles and fixed equipment installed in these types of locations be protected against electric shock by one of two means. See 517.20(A)(1–2). If the interruption of power cannot be tolerated, an isolated power system shall be installed. Per 517.21, bathroom receptacles, in critical care areas where the toilet and basin are installed within patients' rooms need not be provided with GFCI protection for personnel. **See Figure 13-22.**

Health Care Facility Special Requirements – Article 517, Part III

By their nature, health care facilities have critical loads that must not be interrupted or shut down. Failure of the electrical distribution system or equipment in these facilities could seriously impact the lives of patients. For this reason, Article 517 contains special provisions for power and equipment in health care facilities.

There are several different electrical systems that exist within a health care facility electrical distribution system. By definition, the essential electrical system consists of both normal and emergency or alternative power sources. The essential electrical system is intended to ensure continuity of electrical power to designated areas and functions within the facility.

Section 517.25 defines the scope of essential systems for both lighting and power that are essential for life safety and the orderly shutdown of processes. Sections 517.30 through 517.35 contain the specific provisions for installing essential electrical systems in hospitals. However, by definition, a health care facility includes clinics, medical offices, dental offices, out-patient care facilities, nursing homes, limited care facilities and hospitals. See 517.2 and 517.25.

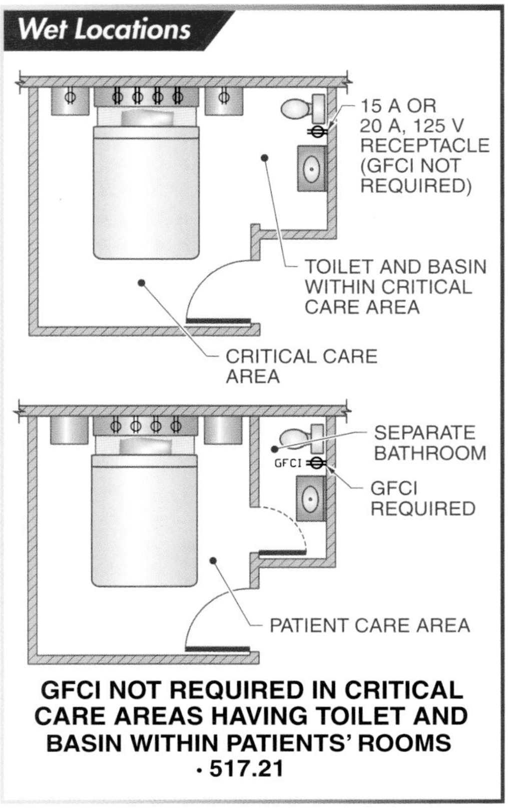

Figure 13-22. Bathroom receptacles in critical care areas shall not be provided with GFCI protection where the toilet and basin are installed within patients' rooms.

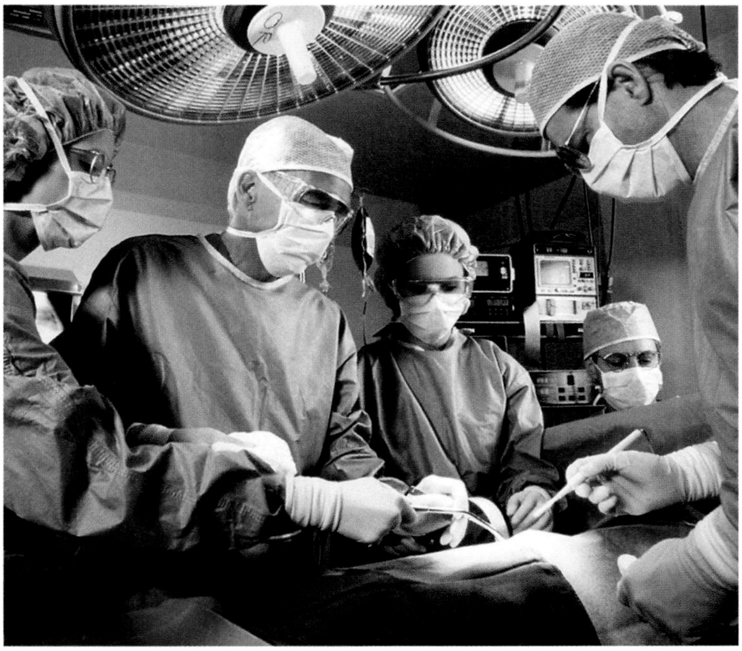

Steel Tube Institute

Article 517 covers electrical installations in special locations such as hospital operating rooms.

The provisions for installing the essential electrical system in nursing homes and limited care facilities are covered in Sections 517.40 through 517.44. Section 517.45 covers the installation requirements for the essential electrical system for other types of health care facilities.

Essential Electrical Systems – 517.30. The essential electrical system consists of two separate and distinct systems: the emergency system and the equipment system. Section 517.30(B)(1) limits the emergency system to circuits that are essential for life safety and critical patient care. Informational Note Figures 517.30 No. 1 and No. 2 provide a good single-line view of how these systems are interrelated within a hospital. The emergency system consists of two separate branches, a life safety branch and a critical branch.

The life safety branch includes all feeders and branch circuits that are intended to ensure safety to patients and personnel in the event of an interruption of normal power. This is accomplished by the use of a transfer switch that provides automatic connection to an alternative or stand-by power source. The alternate power supply is required by 517.31 to automatically restore power to operation within 10 seconds of the interruption of the normal power. Additional provisions for the installation of the life safety branch are included in Section 517.32.

The critical branch includes all feeders and branch circuits that supply lighting and special power needs during an interruption in the normal power system. This includes power for task illumination, patient care receptacles, and special power circuits that cannot be interrupted. Additional provisions for the installation of the critical branch are included in Section 517.33.

Section 517.30(B)(3) covers the second component of the essential electrical system, the equipment system. The equipment system must supply any major electrical equipment that is deemed necessary for the care of patients and normal hospital operation. Transfer switches must be provided for both the emergency system and the equipment system to ensure proper and timely switching of the power sources in the event of a power failure. Section 517.30(C) contains important wiring requirements for the emergency system in a hospital.

Wiring Requirements – 517.30 (C). Section 517.30(C)(1) requires that wiring and equipment of the life safety branch and critical branch be kept entirely separate from other wiring and equipment. There are four installations where the separation provisions are not mandated:

(1) In transfer equipment enclosures.

(2) In exit or emergency luminaires supplied from two sources.

(3) In common junction boxes attached to emergency luminaires supplied from two sources.

(4) For two or more emergency circuits supplied from the same branch circuit and same transfer switch.

There are no separation provisions required for the wiring of the equipment system.

Receptacles installed in hospitals must be of the type intended for use in health care facilities.

Healthcare Facilities—Uninterruptible Power System (UPS)

Health care facilities cannot afford interruptions to their power supplies. Health care providers, patients, and their families expect health care facilities to provide a safe and reliable supply of power.

Critical medical equipment and computers are no more reliable than the power supplied. For proper load operation, power must be supplied within a given voltage range. If voltage is interrupted, the load stops operating. When voltage is restored, the load may or may not start or operate properly again.

The effect of a power interruption on a load depends on the load and the application. For example, when a lamp load loses power, there is no light. The loss of light may produce problems, but the lamp is not damaged. However, when critical medical equipment and computers lose power, there can be a loss of life, data errors, and/or other problems.

If a power interruption could cause equipment problems that are not acceptable, an uninterruptible power system can be used. An uninterruptible power system (UPS) is a power supply that provides constant on-line power when the primary power supply is interrupted. For long-term power interruption protection, a generator UPS is used. A generator UPS is powered by a diesel, gasoline, natural gas, or propane engine connected to an electrical generator. In health care facilities, a generator is used to supply power to critical loads.

Additional information is available at www.schneider-electric.com.

Section 517.30(C)(3) includes the permissible wiring methods for the emergency system in a hospital. In general, the wiring methods of the emergency system must be protected against mechanical damage. Metal raceways are most commonly used for emergency system wiring, but Type MI cable and Schedule 80 rigid nonmetallic wiring can be used for certain applications. Nonmetallic raceways are not permitted for branch circuits that supply patient care areas. See 517.2 for the definition of patient care area. Likewise, Section 517.30(C)(3)(2–6) permits several other wiring methods, including listed MC cable with specific performance characteristics, schedule 40 PVC, FNMC, jacket metal raceways, encased in 2″ of concrete, and other listed raceways, cable assemblies, and cords.

Essential Electrical System Capacity Provisions –517.30(D). It is important that designers of essential electrical systems ensure that the system has sufficient capacity to handle the emergency critical branch and emergency branch loads. Feeders must be sized in accordance with Article 215, and while the generator sizing requirements of 700.5 and 701.6 do not apply to hospital generators, the generator must have sufficient capacity to handle the intended load. Good design practice will include consideration of historical data on the hospital demand factors, actual connected load calculations, and feeder calculations.

Health Care Facilities – Special Applications, Article 517, Parts III–VII

Article 517 is a complex and specialized article that requires a great deal of study and experience to apply correctly in the field. Many of the provisions of Article 517, for example, are derived or extracted from NFPA 99-2005, *Standard for Health Care Facilities*. A good working knowledge of this standard will greatly aid in the application of many of the Article 517 provisions. Article 517 also contains several specialized chapters or parts. Part IV covers installation provisions for anesthetizing locations. Part V covers special provisions for X-ray equipment installations. Part VI covers special signaling and communication provisions of less than 120 V, and Part VII covers isolated power systems that consist of an isolating transformer or device, a line isolation monitor, and ungrounded circuit conductors.

PLACES OF ASSEMBLY

A *place of assembly* is a building, structure, or portion of a building designed or intended for use by 100 or more persons. Because of the frequent presence of large groups of people, special installation requirements apply to places of assembly. The intent of these provisions is to ensure that the installation of the electrical system does not jeopardize the safety of the inhabitants.

Places of Assembly – Article 518

While the requirements contained in Article 518 apply to buildings or portions of buildings which are intended for use by 100 or more persons, they do not contain information on how to determine the maximum number of persons for which the occupancy is designed. Per 518.2, Informational Note, this determination may be made by the local building codes or the NFPA Life Safety Code in the jurisdiction in which the occupancies are located. See NFPA101-2006, *Life Safety Code*.

Article 518 covers requirements for places of assembly.

Wiring Methods – 518.4. In general, all permanent wiring shall be installed in metal raceways. Nonmetallic raceways are permitted if they are encased in not less than 2″ of concrete. Type MI and MC cables are also permitted to be used in places of assembly. Type AC cable is also permitted where it contains an insulated equipment grounding conductor sized in accordance with Table 250.122. All wiring methods used in a place of assembly must qualify as an EGC per 250.118 or shall contain an insulated EGC sized per table 250.122. For those buildings or portions of the building that are not required by local building codes to be of fire-rated construction, NM and AC cables, ENT, and RNC are also permitted wiring methods. See 518.4(B).

Per 518.3(B), temporary wiring in exhibition halls is permitted to be laid directly on the floor if it is located so it is protected from contact with the general public. The flexible cords or cables of temporary wiring laid directly on the floor shall be approved for hard or extra-hard usage. **See Figure 13-23.**

Temporary Cable Support Materials

For temporary installations, cables, cable assemblies, and flexible cords must be supported at intervals that ensure protection from physical damage. To provide support, staples, cable ties, straps, or other similar products that are designed to not damage the cable or cord assembly must be used. Trees or other vegetation cannot be used for support of overhead spans of branch circuits or feeders.

Per 518.3(B), the GFCI requirements for temporary wiring in 590.6 do not apply to places of assembly. **See Figure 13-24.** In addition, 518.4(C) permits ENT and RNC to be used in restaurants, conference and meeting rooms in hotels or motels, dining facilities, and church chapels, provided that either the raceways are concealed within walls, floors, or ceilings which have at least a 15-minute finish rating or the raceways are installed above suspended ceilings which provide a thermal barrier constructed of a material which also has a 15-minute finish rating.

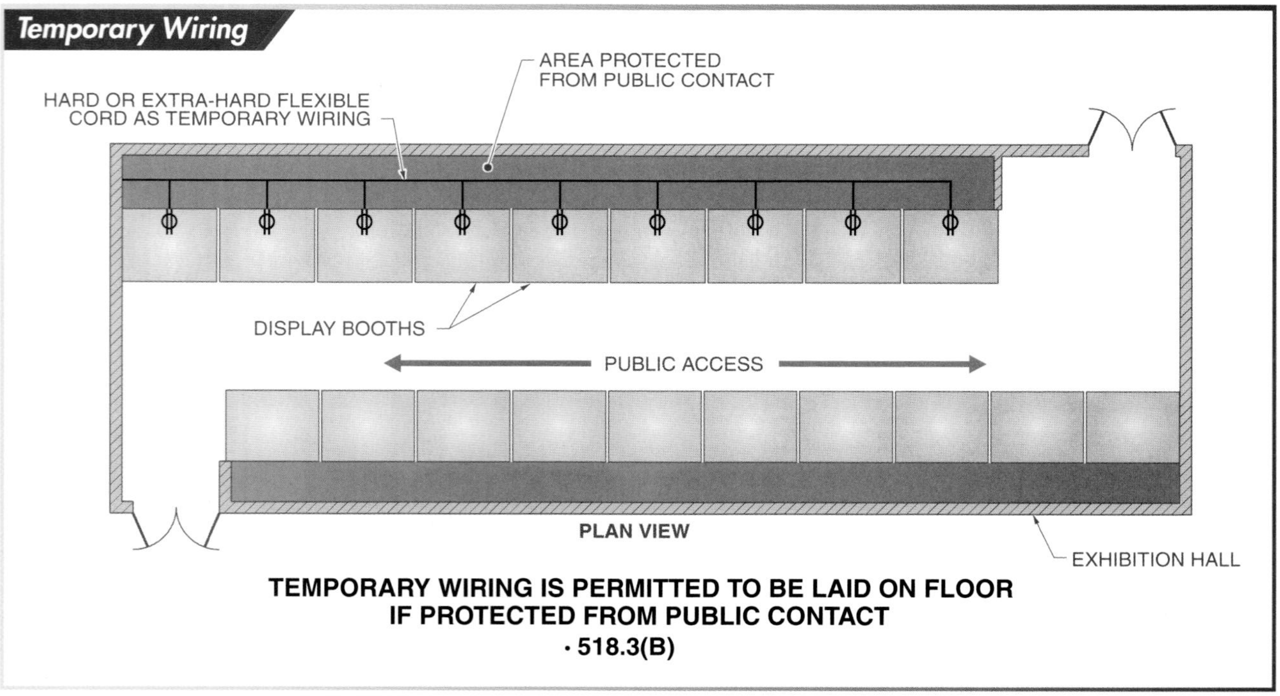

Figure 13-23. Temporary wiring is permitted to be laid directly on the floor if it is protected from public contact.

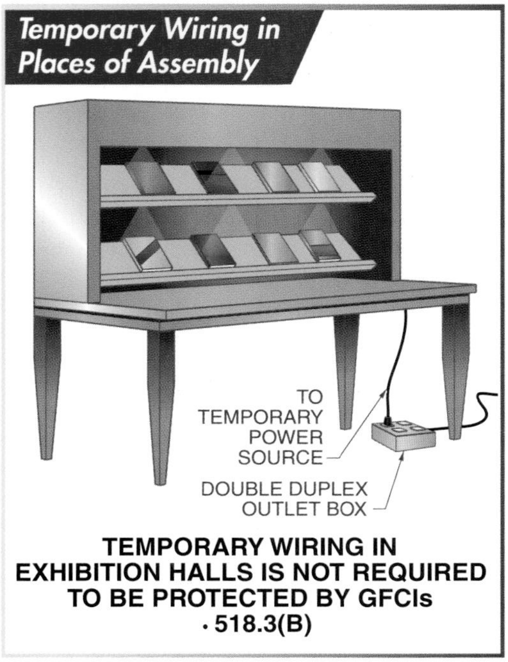

Figure 13-24. The GFCI requirements for temporary wiring in 590.6 do not apply to places of assembly.

Power Distribution – 518.5. Portable power distribution equipment such as panelboards and switchboards used in places of assembly shall be supplied only from listed power outlets. The power outlets shall have sufficient voltage and current ratings for the load supplied and they shall be protected by overcurrent devices. The overcurrent devices shall be installed in a location which is not accessible to the general public.

Theater, Motion Picture and TV Studios, and Similar Locations – Article 520

The NEC® covers requirements concerning places of assembly and locations where the public or large numbers of persons gather. These locations present special challenges for electrical safety and fire protection. Article 520 covers comprehensive requirements for electrical systems and equipment that are installed in buildings or structures that house motion picture or TV studios, dramatic, musical or other types of presentations, and theaters. Special wiring method requirements are included in 520.5 to help ensure that the electrical wiring methods do not add additional risk of fire or electrical shock to the public.

Article 520 also addresses the installation of temporary electrical equipment and wiring in these types of occupancies. For example, Part II of the article covers fixed stage switchboard provisions, and Part IV covers portable switchboards on stage. Likewise, Part V covers other types of portable stage equipment.

Carnivals, Circuses, and Fairs – Article 525

Several Code cycles ago, Article 525 was added to the NEC® to ensure that personnel and the public at large would be provided adequate protection when working at or visiting carnivals, circuses, or fairs. Because these types of events are often temporary in nature, the electrical distribution system has special needs. Typically, the wiring is less than what would be encountered in a permanent installation. While that is understandable, the hazards associated with the electrical distribution system and equipment is nonetheless equally present in these installations. The wiring methods selected must conform with the provisions of Section 525.20. Where flexible cords and cables are installed and are accessible to the public, they must be protected from physical damage and provided with nonconductive matting or other means to prevent a tripping hazard.

Wacker Corporation

Portable, trailer-mounted generators are sometimes used to provide power to carnivals, circuses, and fairs.

Section 525.23 details the provisions for GFCI protection at carnivals, fairs and circuses. Generally, all 125 V, 15 A and 20 A non-locking type receptacles are required to have GFCI protection. In addition, equipment that is supplied from 125 V, 15 A or 20 A branch circuits is required to be provided with GFCI protection if the equipment is readily accessible to the public. Egress lighting is exempted from the GFCI requirement because its failure could create a life safety issue.

QUICK QUIZ®

Refer to Chapter 13 Quick Quiz® on CD-ROM.

Additional information is available at ATPeResources.com

Agricultural Buildings – Article 547

Agricultural buildings present special challenges to designers and installers of electrical distribution systems. In many cases, severe environmental conditions are present, including excessive dust, moisture and corrosive atmospheres. In addition, animals are especially susceptible to stray voltages and low-level current flow. This necessitates the special requirements for grounding and bonding that are included in Article 547.

Section 547.5 lists the permissible wiring methods for agricultural buildings. Types UF, NMC, copper SE cables, jacketed MC cable, RNC, LFNC, and other cables and raceways that are suitable for the location are permitted. Section 547.5(C)(2) covers wiring and equipment in wet or damp locations and 547.5(C)(3) covers wiring and equipment in corrosive atmospheres.

Section 547.10 requires the installation of an equipotential plane in all concrete floor containment areas in livestock buildings and in all outdoor containment areas. By definition, an equipotential plane is an area where wire mesh or other conductive elements are embedded in the concrete, bonded to all metal structures and fixed electrical equipment, and connected to the electrical grounding system. The purpose is to remove any difference of potential within the plane that could result in a voltage drop across the animals.

chapter

Special Locations 13

REVIEW QUESTIONS

Name ______________________________ **Date** ____________________

______________ **1.** A(n) ___ is a location where there is an increased risk of fire or explosion due to the presence of flammable gases, vapors, liquids, combustible dusts, or easily ignitable fibers or flyings.

______________ **2.** The standard classification system used by the NEC to determine the type of hazardous location is the ___ system.

A. Zone
B. Class, Division, and Group
C. Zone, Class, and Division
D. none of the above

______________ **3.** A(n) ___ is a fitting which is inserted into runs of conduit to isolate certain electrical apparatus from atmospheric hazards.

T F **4.** Intrinsically safe systems may be used in Class I, II, or III locations to supply equipment.

______________ **5.** A place of assembly is a building, structure, or portion of a building designed or intended for use by ___ or more persons.

A. 12
B. 48
C. 50
D. 100

______________ **6.** The optional classification system contains ___ zones.

T F **7.** Each room or area within an occupancy shall be considered separately for the purpose of determining area classification.

T F **8.** A Class I, Division 2 location is any location in which flammable gases pose a threat of explosion or ignition under normal conditions.

______________ **9.** A Class III, Division ___ location is any location where easily ignitable fibers or flyings are stored or handled.

______________ **10.** Class I locations contain ___ groups of flammable gases or vapors.

______________ **11.** All boxes and fittings used in Class I, Division 1 locations shall be ___ for such use.

A. required
B. listed
C. approved
D. identified

T F **12.** Any wiring method permitted for use in Class III, Division 1 locations is also permitted for use in Class III, Division 2 locations.

______________ **13.** The most stringent requirements for electrical equipment installed in hazardous locations are for Class ___ locations.

______________ **14.** Motors or generators used in Class ___, Division 2 locations shall be totally enclosed, nonventilated, pipe-ventilated, water-air cooled, fan-cooled, or dust-ignition proof.

__________ **15.** Fixed electrical equipment which may produce arcs or sparks shall be totally enclosed if located less than ___′ above the floor in commercial garages.

__________ **16.** The disconnecting means for gasoline dispensers in attended service stations shall be no more than ___′ from the dispensers.

T F **17.** All 15 A and 20 A, 125 V receptacles installed in patient care areas of pediatric wards, rooms, or other areas shall be listed tamper-resistant or shall employ a listed tamper-resistant cover.

__________ **18.** A spray booth interior where volatile flammable liquids are applied is a Class I, Division ___ location.

__________ **19.** Enclosures constructed so that dust will not enter under specified test conditions are ___.

T F **20.** Portable lighting equipment used in Class I, Division 1 locations is permitted to include provisions for receiving attachment plugs.

Hazardous Locations

__________ **1.** Class I, Division 1

__________ **2.** Class I, Division 2

__________ **3.** Class II, Division 1

__________ **4.** Class II, Division 2

__________ **5.** Class III, Division 1

__________ **6.** Class III, Division 2

A. Combustible dust present in air in quantity sufficient to cause an explosion or ignite hazardous materials under normal operating conditions.

B. Possibility of combustible dust present in air in quantity sufficient to cause an explosion or ignite hazardous materials under abnormal conditions.

C. Flammable gases or vapors present in air in quantity sufficient to cause an explosion or ignite hazardous materials under normal operating conditions.

D. Possibility of flammable gases or vapors present in air in quantity sufficient to cause an explosion or ignite hazardous materials under abnormal conditions.

E. Ignitable fibers or flyings present in air in quantity sufficient to cause an explosion or ignite hazardous materials under abnormal conditions.

F. Possibility of ignitable fibers or flyings present in air in quantity sufficient to cause an explosion or ignite hazardous materials under abnormal conditions.

Class I, Division 1 Seals

__________ **1.** If a seal is required at A, it shall be within ___″ of the enclosure.

__________ **2.** A seal is required at B within ___′ of the boundary.

__________ **3.** A seal is required at C within ___′ of the boundary.

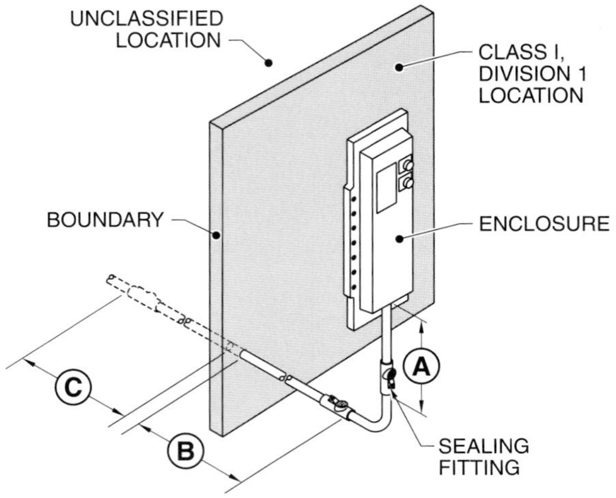

Special Locations

TRADE COMPETENCY TEST

Name ______________________ Date ______________

NEC®	Answers	
________	________	**1.** Determine the Class and Division for a location in which sufficient quantities of combustible dust are normally present in the air to cause an explosion or ignite hazardous materials.
________	________	**2.** Determine the Class and Division for a location in which sufficient quantities of easily ignitable flyings are normally present in the air during the manufacturing process, but not in sufficient quantities to cause an explosion or ignite hazardous materials.
________	________	**3.** *See Figure 1.* Determine the Class and Division for the location.

FIGURE 1

________	________	**4.** A motor is installed in a Class III, Division 1 location. The supply conductors to the motor are pulled in RMC for the entire length up to the last 3′. The last 3′ consists of FMC, which is used to provide flexibility for the motor. Does this installation violate Article 503 of the NEC®?
________	________	**5.** *See Figure 2.* Determine the maximum distance from the Class I, Division 1 boundary that the conduit seal is permitted to be installed.
________	________	**6.** *See Figure 3.* Determine the maximum distance from the electrical panelboard that the seal is permitted to be installed.

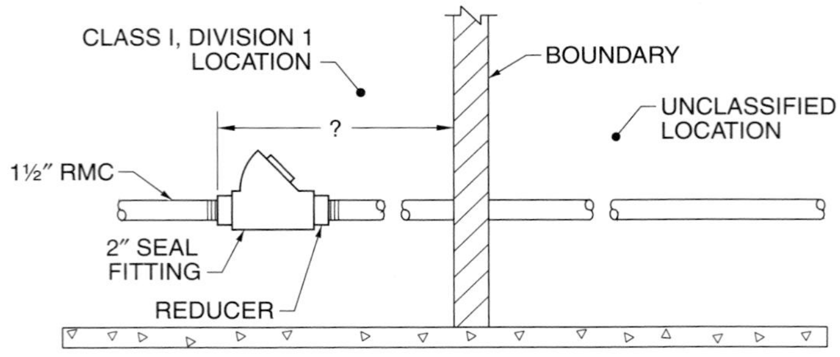

FIGURE 2

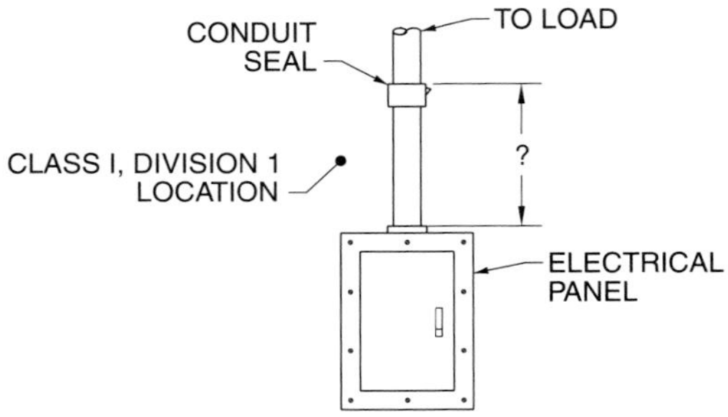

FIGURE 3

______________ ______________ **7.** *See Figure 4.* Does this installation violate Article 501 of the NEC®?

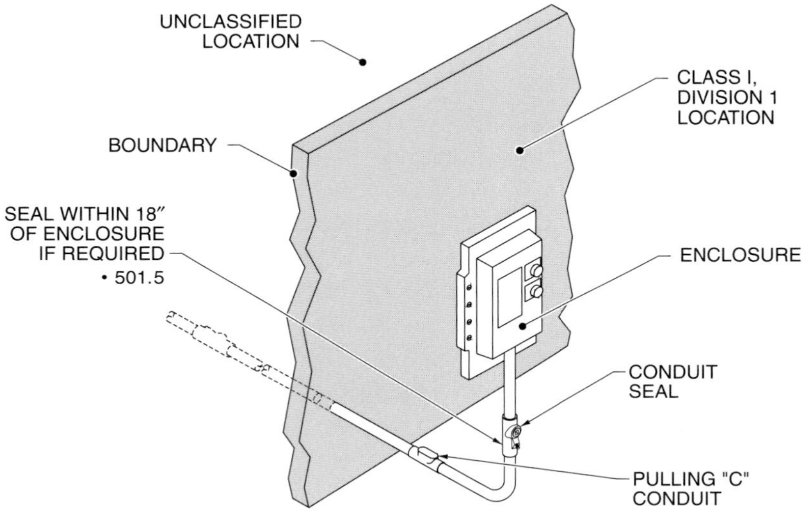

FIGURE 4

______________ ______________ **8.** *See Figure 5.* The total area of the conductors is 0.1311 sq in. The total area of the sealing fitting is listed at 0.400 sq in. Does the installation of this conduit seal violate the provisions of Article 501 of the NEC®?

______________ ______________ **9.** *See Figure 6.* Does the installation of the underground raceway violate Article 515 of the NEC®?

______________ ______________ **10.** Determine the Class and Division for a space up to 18″ above the floor and 3′ horizontally from a lubrication pit in a gasoline-dispensing service station.

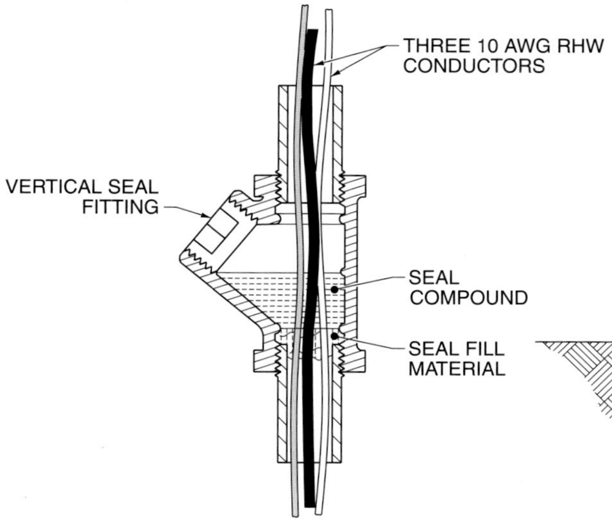
THREE 10 AWG RHW CONDUCTORS
VERTICAL SEAL FITTING
SEAL COMPOUND
SEAL FILL MATERIAL

FIGURE 5

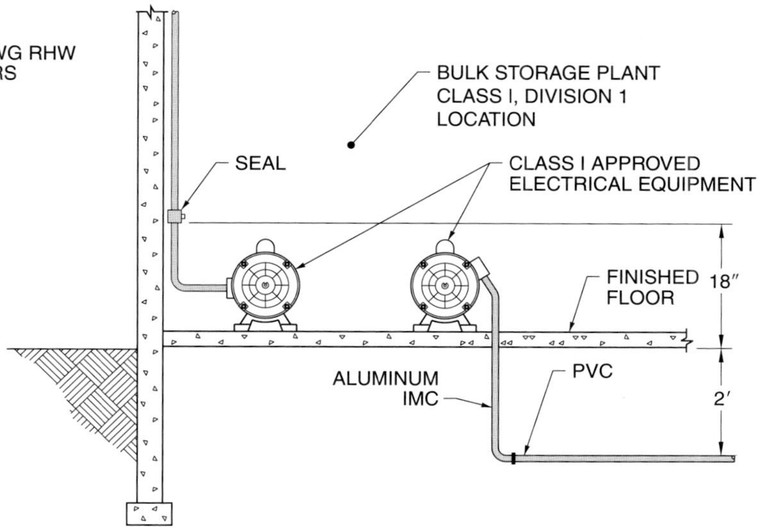
BULK STORAGE PLANT CLASS I, DIVISION 1 LOCATION
SEAL
CLASS I APPROVED ELECTRICAL EQUIPMENT
FINISHED FLOOR
18″
ALUMINUM IMC
PVC
2′

FIGURE 6

Learning the Code

- Apprentices and students of the Code are often required to demonstrate their mastery of the NEC®. Master Code examination and electrical licensing tests require a good working knowledge of the NEC®. The most important key to being successful on these types of examinations is a fundamental understanding of the layout and format of the NEC®. It takes months of study to become acquainted with and comfortable with the format and structure of the NEC®. Once an apprentice or student of the Code reaches that level, he or she must practice taking Code examinations under real-life conditions.

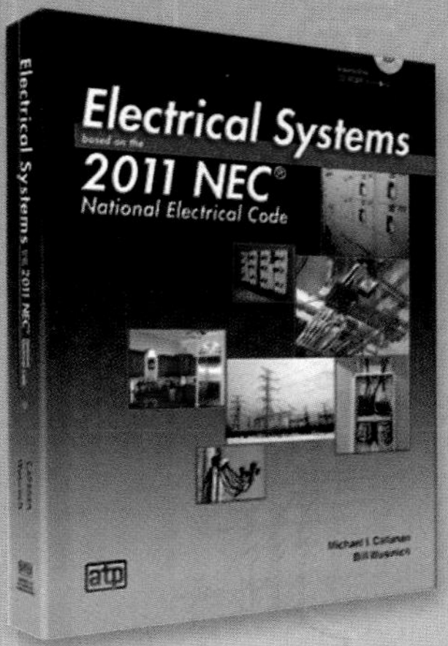

- The basic approach to taking Code examinations is no different than that for other examinations. Always read the question carefully and try to identify the key aspect of the question. In the case of the Code, this will give you a good indication as to which chapter of the NEC® most likely contains the answer. Always answer the questions that you know first, and leave the questions you are uncertain about for the end of the test.
- Most importantly, practice answering Code questions in conditions that resemble the actual test conditions. Most Code examinations have a time limit. If the test is 100 questions in a two-hour time limit, practice under a similar time constraint. For example, if you have 50 questions, give yourself one hour to complete them. Also make sure you use only the items that are permitted for the actual Code examination. For example, most Code licensing examinations allow you to bring only a copy of the NEC®. Don't practice using other Code handbooks or reference books that you will not be able to use during the examination. Finally, check about the use of calculators. If calculators are not permitted during the examination, do not practice with the use of one.
- One final point about Code examinations: when you find the section that contains what you believe to be the answer, keep reading! Frequently the actual Code answer is found within an exception that follows the general rule. Don't be tricked into thinking the answer is always found in the general rule—read the exceptions.

Final Exam

The two-fold purpose of the National Electrical Code® (NEC®) is to protect people and property from the hazards associated with the use of electricity. A strong knowledge of sound electrical principles and how to apply appropriate provisions of the NEC® is essential for those who design and install electrical systems.

FINAL EXAM

The problems in the Final Exam of *Electrical Systems Based on the 2011 NEC®* are based on Chapters 1–13 and the current edition of the National Electrical Code®. All problems are in the multiple-choice format because many testing agencies use this format for their tests.

Read the problems very carefully and select the response that most clearly answers the question. Record the letter representing that response in the blank space to the immediate left of the problem number. Record the NEC® reference that substantiates the answer in the blank space to the far left of the problem number. **See Figure 14-1.**

The option of either using or not using the NEC® to verify the answers is at the discretion of your instructor. Some instructors prefer an "open book" test while other instructors prefer a "closed book" test.

Additional information is available at ATPeResources.com

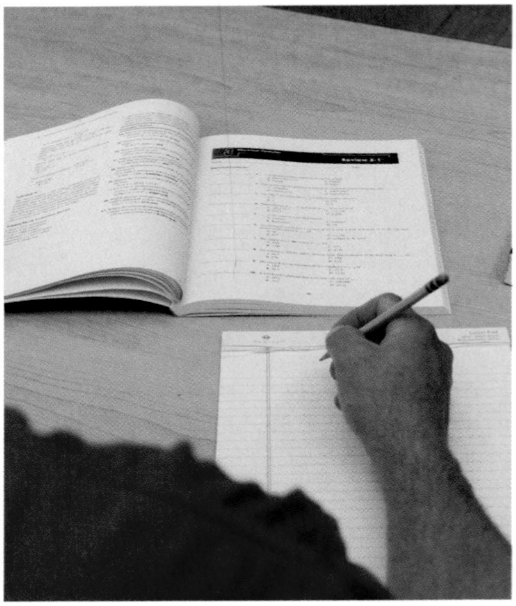

Practice answering Code questions to prepare for a Code examination.

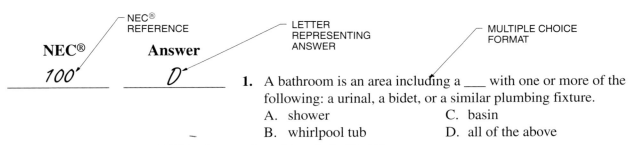

NEC® REFERENCE — LETTER REPRESENTING ANSWER — MULTIPLE CHOICE FORMAT

NEC®	**Answer**	
100	D	**1.** A bathroom is an area including a ___ with one or more of the following: a urinal, a bidet, or a similar plumbing fixture.

A. shower C. basin
B. whirlpool tub D. all of the above

Figure 14-1. Record answers and NEC® references for problems in the Final Exam.

chapter

14 Final Exam

REVIEW QUESTIONS

Name ______________________ Date ______________________

NEC®	Answer	
Art 100	B	**1.** A(n) ___ is an unintentional, electrically conducting connection between an ungrounded conductor of a circuit and the normally non-current-carrying conductors, enclosures, raceways, or ground. A. short circuit C. overload/overcurrent B. ground fault D. none of the above
		2. All branch circuits used for temporary wiring shall originate in a(n) ___ power outlet, panelboard, switchboard, or MCC. A. listed C. approved B. identified D. weatherproof
		3. ___ are not covered by Article 517, Health Care Facilities. A. Nursing homes C. Hospitals B. Dental offices D. none of the above
		4. Type ___ cable is not a permitted nonmetallic-sheathed cable for use in one-family dwellings. A. NM C. NMS B. NMC D. NMX
		5. The allowable ampacity for three 1/0 AWG THW Cu conductors installed in the same raceway in an ambient temperature of 86°F is ___ A. (assume 75°C terminations) A. 120 C 150 B. 125 D. 170
	C	**6.** Service conductors installed as open conductors without an overall outer jacket shall have a clearance of not less than ___′ from windows that are designed to be opened. A. 1 C. 3 B. 2 D. 5
	A	**7.** A(n) ___ is not a permissible grounding electrode for use on a separately derived system. A. lightning protection air terminal C. water pipe B. building steel D. ⅝″ × 8″ Cu ground rod
		8. ___ is classified as a Group E dust. A. Grain C. Wood B. Coal D. Aluminum

______________ ______________ **9.** Feeders and branch-circuit conductors within ___″ of a ballast inside a ballast, LED driver, power supply, or transformer compartment shall have an insulation rating not lower than 90°C.

A. 3 C. 9
B. 6 D. 12

______________ ______________ **10.** ___ is not a permitted wiring method for wiring from the controller of a fire pump to the pump motor.

A. RMC C. IMC
B. RNMC D. Type MI cable

______________ ______________ **11.** The maximum size of LFMC permitted to be used is the ___″ trade size.

A. 3 C. 5
B. 4 D. 6

______________ ______________ **12.** Mandatory rules of the NEC® are characterized by the use of the word ___.

A. must C. shall
B. should D. permitted

______________ ______________ **13.** Primary transformer protection, for transformers rated 600 V or less with current over 9 A, shall be provided by an individual overcurrent device on the primary side rated or set at not more than ___% of the rated primary current of the transformer, where the transformer has primary only protection.

A. 125 C. 175
B. 150 D. 250

______________ ______________ **14.** The minimum capacity required for a standard metal box that contains four 12 AWG conductors and four 8 AWG conductors is ___ cu in.

A. 16 C. 25
B. 21 D. 28

______________ ______________ **15.** Edison-base plug fuses shall be classified at not over 125 V and ___ A and below.

A. 10 C. 20
B. 15 D. 30

______________ ______________ **16.** In general, at least ___″ of free conductor shall be left at each outlet, junction, and switch point for splices or the connection of fixtures or devices.

A. 4 C. 8
B. 6 D. 12

______________ ______________ **17.** A 15 A, 125 V, 1ϕ receptacle installed ___ of a dwelling unit does not require GFCI protection.

A. in a bathroom C. in a finished basement
B. below grade in a crawl space D. at a kitchen countertop

______ ______ **18.** A(n) ___ is generally not classified as a place of assembly.

A. assembly hall
B. bowling lane
C. shopping center
D. skating rink

______ ______ **19.** The full-load table current for a 10 HP, 480 V, 3ϕ squirrel-cage motor is ___ A.

A. 14
B. 28
C. 30.8
D. 32.2

______ ______ **20.** The minimum wire-bending space required for an enclosure with 3/0 AWG conductors, one wire per terminal, where the conductors do not enter or leave the enclosure through the wall opposite the terminal is ___″.

A. 3
B. 4
C. 6
D. 8

______ ______ **21.** The minimum size THW, Cu service-entrance conductor permitted to supply a 200 A service to a one-family dwelling with a 120/240 V, 1ϕ, 3-wire service is ___ AWG.

A. 1
B. 1/0
C. 2/0
D. 4/0

______ ______ **22.** Parking garages used for parking or storage shall be permitted to be ___.

A. unclassified
B. classified Class I, Division 1
C. classified Class I, Division 2
D. classified Class II, Division 1

______ ______ **23.** The calculated feeder load for six 5.5 kW household clothes dryers installed in a multifamily dwelling is ___ VA (Standard Calculation).

A. 21,300
B. 23,100
C. 24,750
D. 33,000

______ ______ **24.** The minimum front work space width required for a 100 A, 240 V panelboard installed in a dwelling is ___″.

A. 24
B. 30
C 32
D. 36

______ ______ **25.** Branch-circuit conductors supplying a single motor shall have an ampacity not less than ___ % of the motor FLC rating.

A. 100
B. 115
C. 125
D. 140

______ ______ **26.** The panels of switchboards shall be made of moisture-resistant ___ material.

A. nonconductive
B. noncombustible
C. nonmetallic
D. nonferrous

_______________ _______________ **27.** The minimum size Cu GEC permitted for a 120/208 V, 3ϕ service which consists of two 500 kcmil Cu conductors paralleled per phase is ___ AWG.

A. 2
B. 1/0
C. 2/0
D. 3/0

_______________ _______________ **28.** The household electric range demand for the line and neutral of a 15 kW, 240 V range installed in a one-family dwelling is ___ VA.

A. 8000 line; 8000 neutral
B. 8000 line; 5600 neutral
C. 9200 line; 9200 neutral
D. 9200 line; 6440 neutral

_______________ _______________ **29.** The minimum burial depth for 600 V conductors installed in RMC, in a trench below 2″ thick concrete is ___″.

A. 4
B. 6
C. 12
D. 18

_______________ _______________ **30.** EMT shall be supported at least every ___′ and within ___″ of each outlet box, junction box, device box, conduit body, or other tubing termination.

A. 5; 12
B. 5; 36
C. 10; 12
D. 10; 36

_______________ _______________ **31.** A separate overload device is installed for a continuous-duty, 2 HP, 12 FLA motor with a marked service factor of 1.15. The motor does not have starting current problems. The maximum rating of the overload device required to protect the motor is ___ A.

A. 13.8
B. 15.0
C. 15.6
D. 16.8

_______________ _______________ **32.** The maximum length which Type AC cable, installed from an outlet box to a luminaire (lighting fixture) in an accessible ceiling, shall be installed without being supported is ___′.

A. 2
B. 4
C. 6
D. 10

_______________ _______________ **33.** The minimum ampacity for a 25′ feeder tap conductor which is tapped from a conductor protected by a 300 A overcurrent protective device is ___ A.

A. 30
B. 100
C. 150
D. 300

_______________ _______________ **34.** A 20 A, 40 A, and 70 A branch circuit are installed in the same 2″ RNC. ___ AWG is the minimum size Cu EGC required for this installation.

A. 12
B. 10
C. 8
D. 6

_______________ _______________ **35.** Equipment which is required to be in sight from other equipment shall be visible and not more than ___′ away.

A. 10
B. 25
C. 50
D. 100

______________ ______________ **36.** A conduit seal shall be installed within ___″ of the enclosure or fitting in each 2″ or larger conduit which enters an enclosure or fitting housing terminals, splices, or taps in a Class I, Division 1 location.

A. 6
B. 12
C. 18
D. 24

______________ ______________ **37.** The minimum rating for a service disconnecting means that supplies a one-family dwelling shall be ___ A, 3-wire.

A. 60
B. 100
C. 150
D. 200

______________ ______________ **38.** Receptacle outlets in dwelling unit floors shall not be counted as part of the required number of receptacle outlets unless located within ___″ of the wall.

A. 6
B. 12
C. 18
D. 24

______________ ______________ **39.** GFPE, as specified in 230.95, shall be provided for feeder disconnects rated ___ A or more in a solidly-grounded wye system with greater than 150 V to ground, but not exceeding 600 V phase-to-phase.

A. 70
B. 300
C. 600
D. 1000

______________ ______________ **40.** FMC shall be securely fastened-in-place by an approved means within ___″ of each box, cabinet, conduit body, or other conduit termination and shall be supported and secured at intervals not to exceed ___′.

A. 6; 3
B. 6; 4½
C. 12; 3
D. 12; 4½

Appendix

TEXT ABBREVIATIONS			
A	Amps	k	Kilo (1000)
A/C	Air Conditioner	kcmil	1000 Circular Mils
AC	Alternating Current	kFT	1000
AEGCP	Assured Equipment Grounding Program	kVA	Kilovolt Amps
A/H	Air Handler	kW	Kilowatt
AHCC	Ambulatory Health Care Center	kWh	Kilowatt-Hour
AHJ	Authority Having Jurisdiction	L	Line
AIR	Ampere Interrupting Rating	LFLI	Less-Flammable, Liquid-Insulated
Al	Aluminum	LP	Lighting Panel
ANSI	American National Standards Institute	LRA	Locked-Rotor Ampacity
ATCB	Adjustable-Trip Circuit Breaker	LRC	Locked-Rotor Current
AWG	American Wire Gauge	MBCSCGF	Motor Branch-Circuit, Short-Circuit, Ground-Fault
BJ	Bonding Jumper	MBJ	Main Bonding Jumper
C	Celsius	MCC	Main Control Center
CATV	Cable Antenna Television	N	Neutral
CB	Circuit Breaker	NATCB	Nonadjustable-Trip Circuit Breaker
CH	Chapter	NEC®	National Electrical Code®
CM	Circular Mils	NEMA	National Electrical Manufacturers Association
CMP	Code-making Panel	NESC	National Electrical Safety Code
CU	Copper	NFPA	National Fire Protection Association
DC	Direct Current	NO.	Number
DIA	Diameter	NTDF	Non-Time Delay Fuse
DP	Double Pole	OCPD	Overcurrent Protective Device
E	Voltage	OL	Overload(s)
EBJ	Equipment Bonding Jumper	OSHA	Occupational Safety and Health Administration
E_{ff}	Efficiency	P	Power
EGC	Equipment Grounding Conductor	PC	Personal Computer
EX.	Exception	PF	Power Factor
F	Fahrenheit	R	Resistance; Resistor
FLA	Full-Load Amps	RMS	Root Mean Square
FLC	Full-Load Current	ROC	Receipt of Comments
FPN	Fine Print Note	ROP	Receipt of Proposals
FR	Frame	SF	Service Factor
G	Ground	SP	Single-Pole
GEC	Grounding Electrode Conductor	SPCB	Single-Pole Circuit Breaker
GES	Grounding Electrode System	SQ FT	Square Foot (Feet)
GFCI	Ground Fault Circuit Interrupter	SWD	Switched Disconnect
GFPE	Ground Fault Protection of Equipment	T	Time
GR	Green	TDF	Time-Delay Fuse
HACR	Heating, Air-Conditioning, Refrigeration	TP	Thermally Protected
HP	Horsepower	UF	Underground Feeder
HRS	Hours	UL	Underwriter's Laboratory
I	Current	V	Volts
IDCI	Immersion Detection Circuit Interrupter	VA	Volt Amps
IG	Isolated Ground	VAC	Volts Alternating Current
IN.	Inch	VD	Voltage Drop
ITB	Instantaneous-Trip Circuit Breaker	W	Watts
ITCB	Inverse-Time Circuit Breaker	W	White
K	Conductor Resistivity	WP	Weatherproof

INDUSTRIAL ELECTRICAL SYMBOLS . . .

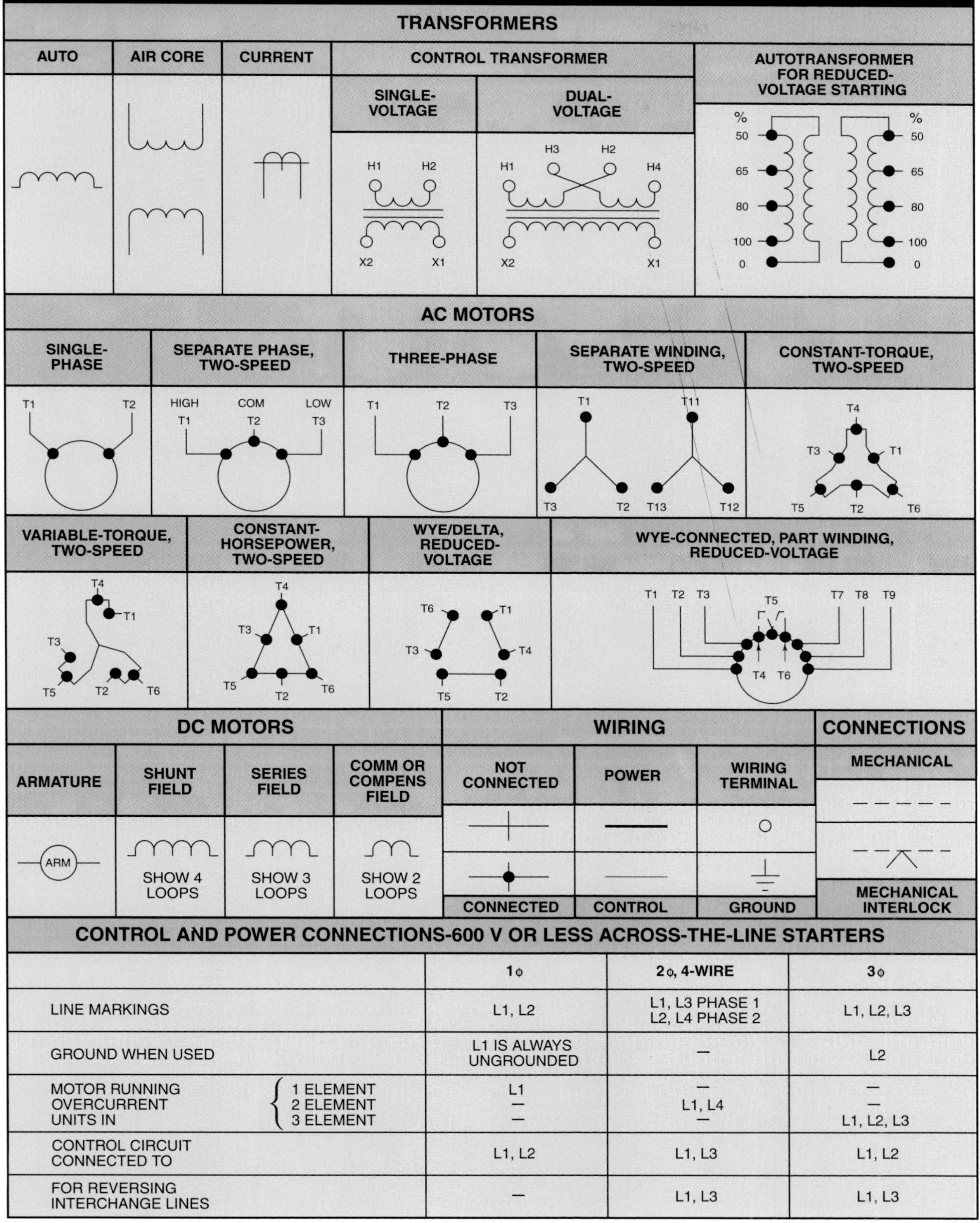

CONTROL AND POWER CONNECTIONS-600 V OR LESS ACROSS-THE-LINE STARTERS

		1φ	2φ, 4-WIRE	3φ
LINE MARKINGS		L1, L2	L1, L3 PHASE 1 L2, L4 PHASE 2	L1, L2, L3
GROUND WHEN USED		L1 IS ALWAYS UNGROUNDED	—	L2
MOTOR RUNNING OVERCURRENT UNITS IN	1 ELEMENT	L1	—	—
	2 ELEMENT	—	L1, L4	—
	3 ELEMENT	—	—	L1, L2, L3
CONTROL CIRCUIT CONNECTED TO		L1, L2	L1, L3	L1, L2
FOR REVERSING INTERCHANGE LINES		—	L1, L3	L1, L3

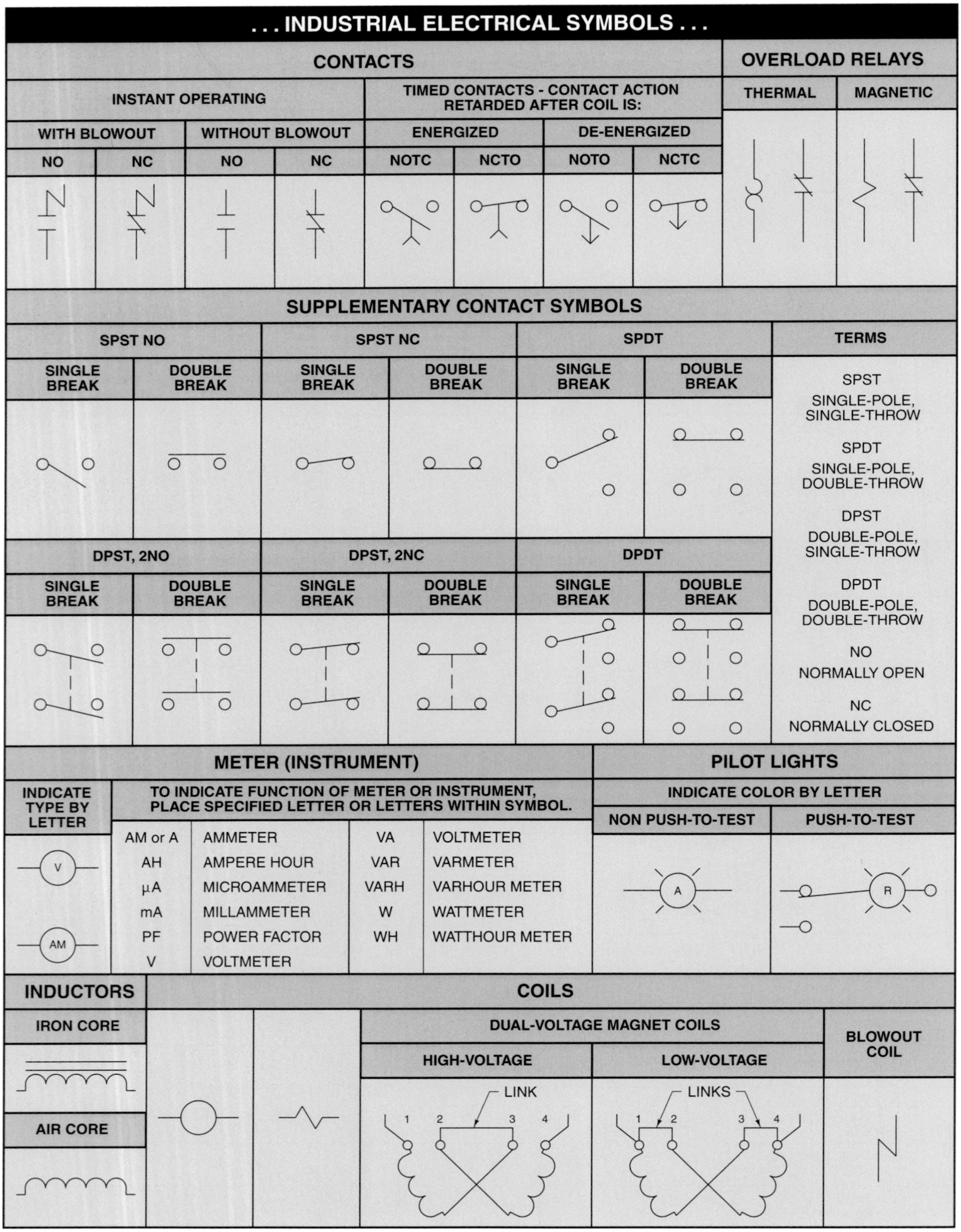
. . . INDUSTRIAL ELECTRICAL SYMBOLS . . .
CONTACTS
INSTANT OPERATING
WITH BLOWOUT
NO
NC
WITHOUT BLOWOUT
NO
NC
TIMED CONTACTS - CONTACT ACTION RETARDED AFTER COIL IS:
ENERGIZED
NOTC
NCTO
DE-ENERGIZED
NOTO
NCTC
OVERLOAD RELAYS
THERMAL
MAGNETIC
SUPPLEMENTARY CONTACT SYMBOLS
SPST NO
SINGLE BREAK
DOUBLE BREAK
SPST NC
SINGLE BREAK
DOUBLE BREAK
SPDT
SINGLE BREAK
DOUBLE BREAK
TERMS
SPST
SINGLE-POLE, SINGLE-THROW
SPDT
SINGLE-POLE, DOUBLE-THROW
DPST
DOUBLE-POLE, SINGLE-THROW
DPDT
DOUBLE-POLE, DOUBLE-THROW
NO
NORMALLY OPEN
NC
NORMALLY CLOSED
DPST, 2NO
SINGLE BREAK
DOUBLE BREAK
DPST, 2NC
SINGLE BREAK
DOUBLE BREAK
DPDT
SINGLE BREAK
DOUBLE BREAK
METER (INSTRUMENT)
INDICATE TYPE BY LETTER
V
AM
TO INDICATE FUNCTION OF METER OR INSTRUMENT, PLACE SPECIFIED LETTER OR LETTERS WITHIN SYMBOL.
AM or A AMMETER
AH AMPERE HOUR
μA MICROAMMETER
mA MILLAMMETER
PF POWER FACTOR
V VOLTMETER
VA VOLTMETER
VAR VARMETER
VARH VARHOUR METER
W WATTMETER
WH WATTHOUR METER
PILOT LIGHTS
INDICATE COLOR BY LETTER
NON PUSH-TO-TEST
A
PUSH-TO-TEST
R
INDUCTORS
IRON CORE
AIR CORE
COILS
DUAL-VOLTAGE MAGNET COILS
HIGH-VOLTAGE
LINK
1 2 3 4
LOW-VOLTAGE
LINKS
1 2 3 4
BLOWOUT COIL

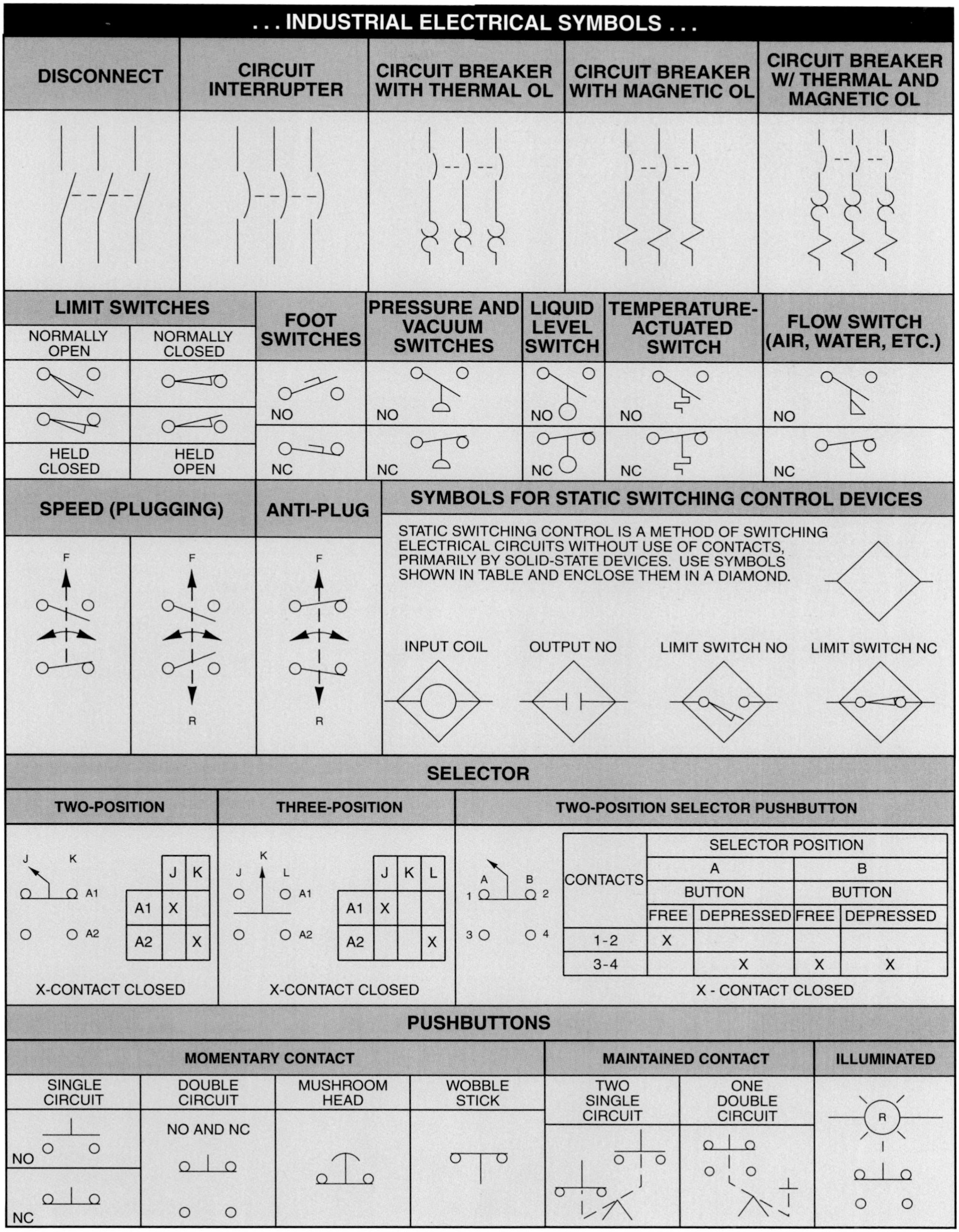

. . . INDUSTRIAL ELECTRICAL SYMBOLS . . .

DISCONNECT	CIRCUIT INTERRUPTER	CIRCUIT BREAKER WITH THERMAL OL	CIRCUIT BREAKER WITH MAGNETIC OL	CIRCUIT BREAKER W/ THERMAL AND MAGNETIC OL

LIMIT SWITCHES

NORMALLY OPEN	NORMALLY CLOSED	FOOT SWITCHES	PRESSURE AND VACUUM SWITCHES	LIQUID LEVEL SWITCH	TEMPERATURE-ACTUATED SWITCH	FLOW SWITCH (AIR, WATER, ETC.)
HELD CLOSED	HELD OPEN	NO / NC	NO / NC	NO / NC	NO / NC	NO / NC

SPEED (PLUGGING)	ANTI-PLUG	SYMBOLS FOR STATIC SWITCHING CONTROL DEVICES

STATIC SWITCHING CONTROL IS A METHOD OF SWITCHING ELECTRICAL CIRCUITS WITHOUT USE OF CONTACTS, PRIMARILY BY SOLID-STATE DEVICES. USE SYMBOLS SHOWN IN TABLE AND ENCLOSE THEM IN A DIAMOND.

INPUT COIL OUTPUT NO LIMIT SWITCH NO LIMIT SWITCH NC

SELECTOR

TWO-POSITION

	J	K
A1	X	
A2		X

X-CONTACT CLOSED

THREE-POSITION

	J	K	L
A1	X		
A2			X

X-CONTACT CLOSED

TWO-POSITION SELECTOR PUSHBUTTON

CONTACTS	SELECTOR POSITION			
	A		B	
	BUTTON		BUTTON	
	FREE	DEPRESSED	FREE	DEPRESSED
1-2	X			
3-4		X	X	X

X - CONTACT CLOSED

PUSHBUTTONS

MOMENTARY CONTACT				MAINTAINED CONTACT		ILLUMINATED
SINGLE CIRCUIT	DOUBLE CIRCUIT	MUSHROOM HEAD	WOBBLE STICK	TWO SINGLE CIRCUIT	ONE DOUBLE CIRCUIT	
NO / NC	NO AND NC					R

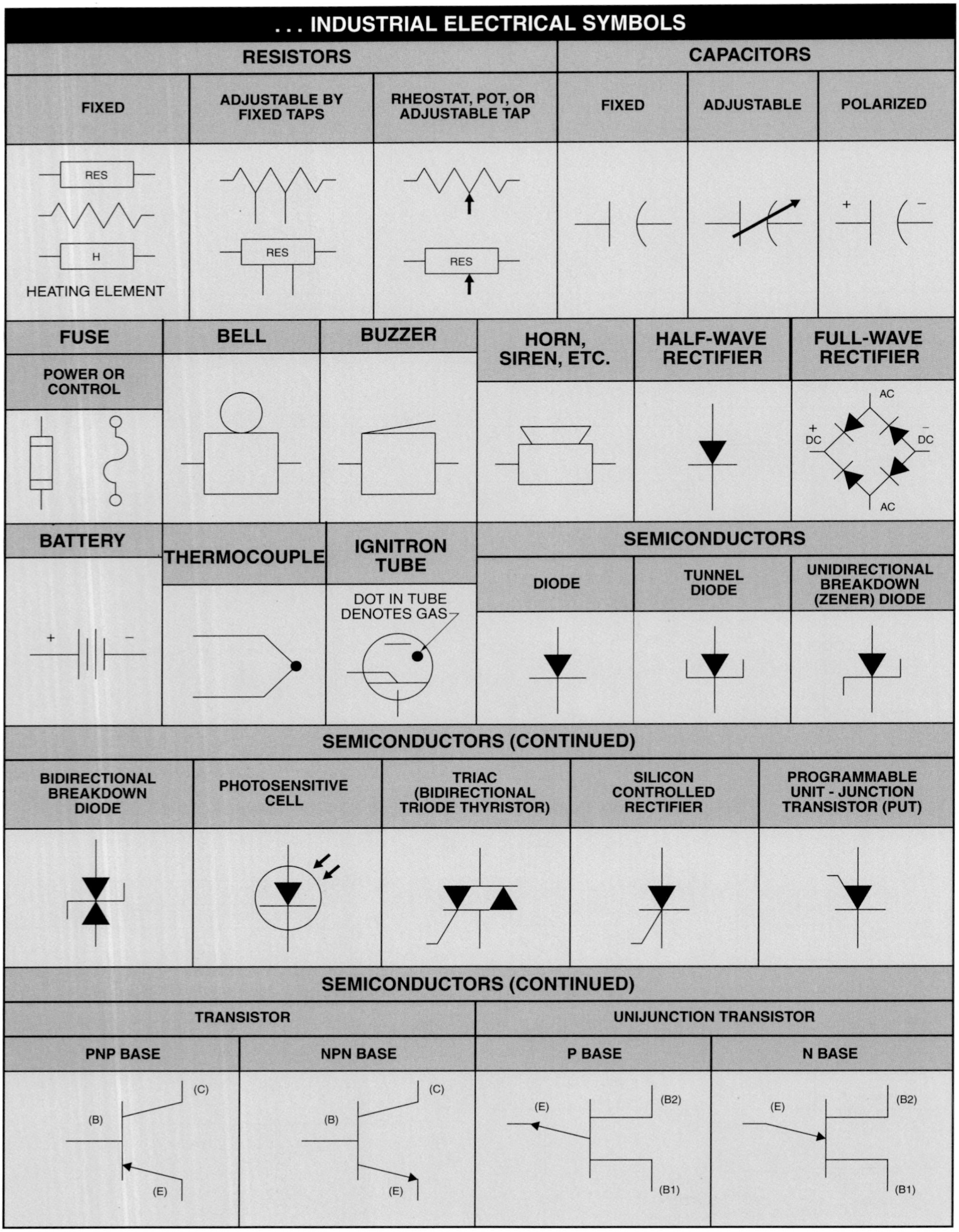
. . . INDUSTRIAL ELECTRICAL SYMBOLS
RESISTORS
CAPACITORS
FIXED
ADJUSTABLE BY FIXED TAPS
RHEOSTAT, POT, OR ADJUSTABLE TAP
FIXED
ADJUSTABLE
POLARIZED
RES
H
HEATING ELEMENT
+
−
FUSE
POWER OR CONTROL
BELL
BUZZER
HORN, SIREN, ETC.
HALF-WAVE RECTIFIER
FULL-WAVE RECTIFIER
AC
+ DC
− DC
BATTERY
THERMOCOUPLE
IGNITRON TUBE
DOT IN TUBE DENOTES GAS
SEMICONDUCTORS
DIODE
TUNNEL DIODE
UNIDIRECTIONAL BREAKDOWN (ZENER) DIODE
SEMICONDUCTORS (CONTINUED)
BIDIRECTIONAL BREAKDOWN DIODE
PHOTOSENSITIVE CELL
TRIAC (BIDIRECTIONAL TRIODE THYRISTOR)
SILICON CONTROLLED RECTIFIER
PROGRAMMABLE UNIT - JUNCTION TRANSISTOR (PUT)
SEMICONDUCTORS (CONTINUED)
TRANSISTOR
UNIJUNCTION TRANSISTOR
PNP BASE
NPN BASE
P BASE
N BASE
(B)
(C)
(E)
(B1)
(B2)

RESIDENTIAL ELECTRICAL SYMBOLS . . .

LIGHTING OUTLETS

OUTLET BOX AND INCANDESCENT LIGHTING FIXTURE	CEILING WALL
INCANDESCENT TRACK LIGHTING	
BLANKED OUTLET	B B
DROP CORD	D
EXIT LIGHT AND OUTLET BOX. SHADED AREAS DENOTE FACES.	
OUTDOOR POLE-MOUNTED FIXTURES	
JUNCTION BOX	J J
LAMPHOLDER WITH PULL SWITCH	L_{PS} L_{PS}
MULTIPLE FLOODLIGHT ASSEMBLY	
EMERGENCY BATTERY PACK WITH CHARGER	B
INDIVIDUAL FLUORESCENT FIXTURE	
OUTLET BOX AND FLUORESCENT LIGHTING TRACK FIXTURE	
CONTINUOUS FLUORESCENT FIXTURE	
SURFACE-MOUNTED FLUORESCENT FIXTURE	

PANELBOARDS

FLUSH-MOUNTED PANELBOARD AND CABINET	
SURFACE-MOUNTED PANELBOARD AND CABINET	

CONVENIENCE OUTLETS

SINGLE RECEPTACLE OUTLET	
DUPLEX RECEPTACLE OUTLET	
TRIPLEX RECEPTACLE OUTLET	
SPLIT-WIRED DUPLEX RECEPTACLE OUTLET	
SPLIT-WIRED TRIPLEX RECEPTACLE OUTLET	
SINGLE SPECIAL-PURPOSE RECEPTACLE OUTLET	
DUPLEX SPECIAL-PURPOSE RECEPTACLE OUTLET	
RANGE OUTLET	R
SPECIAL-PURPOSE CONNECTION	DW
CLOSED-CIRCUIT TELEVISION CAMERA	
CLOCK HANGER RECEPTACLE	C
FAN HANGER RECEPTACLE	F
FLOOR SINGLE RECEPTACLE OUTLET	
FLOOR DUPLEX RECEPTACLE OUTLET	
FLOOR SPECIAL-PURPOSE OUTLET	
UNDERFLOOR DUCT AND JUNCTION BOX FOR TRIPLE, DOUBLE, OR SINGLE DUCT SYSTEM AS INDICATED BY NUMBER OF PARALLEL LINES	

BUSDUCTS AND WIREWAYS

SERVICE, FEEDER, OR PLUG-IN BUSWAY	B B B
CABLE THROUGH LADDER OR CHANNEL	C C C
WIREWAY	W W W

SWITCH OUTLETS

SINGLE-POLE SWITCH	S
DOUBLE-POLE SWITCH	S_2
THREE-WAY SWITCH	S_3
FOUR-WAY SWITCH	S_4
AUTOMATIC DOOR SWITCH	S_D
KEY-OPERATED SWITCH	S_K
CIRCUIT BREAKER	S_{CB}
WEATHERPROOF CIRCUIT BREAKER	S_{WCB}
DIMMER	S_{DM}
REMOTE CONTROL SWITCH	S_{RC}
WEATHERPROOF SWITCH	S_{WP}
FUSED SWITCH	S_F
WEATHERPROOF FUSED SWITCH	S_{WF}
TIME SWITCH	S_T
CEILING PULL SWITCH	S
SWITCH AND SINGLE RECEPTACLE	S
SWITCH AND DOUBLE RECEPTACLE	S
A STANDARD SYMBOL WITH AN ADDED LOWERCASE SUBSCRIPT LETTER IS USED TO DESIGNATE A VARIATION IN STANDARD EQUIPMENT	a.b / a.b / $S_{a.b}$

. . . RESIDENTIAL ELECTRICAL SYMBOLS

COMMERCIAL AND INDUSTRIAL SYSTEMS

- PAGING SYSTEM DEVICE
- FIRE ALARM SYSTEM DEVICE
- COMPUTER DATA SYSTEM DEVICE
- PRIVATE TELEPHONE SYSTEM DEVICE
- SOUND SYSTEM
- FIRE ALARM CONTROL PANEL — FACP

SIGNALING SYSTEM OUTLETS FOR RESIDENTIAL SYSTEMS

- PUSHBUTTON
- BUZZER
- BELL
- BELL AND BUZZER COMBINATION
- COMPUTER DATA OUTLET
- BELL RINGING TRANSFORMER — BT
- ELECTRIC DOOR OPENER — D
- CHIME — CH
- TELEVISION OUTLET — TV
- THERMOSTAT — T

UNDERGROUND ELECTRICAL DISTRIBUTION OR ELECTRICAL LIGHTING SYSTEMS

- MANHOLE — M
- HANDHOLE — H
- TRANSFORMER-MANHOLE OR VAULT — TM
- TRANSFORMER PAD — TP
- UNDERGROUND DIRECT BURIAL CABLE
- UNDERGROUND DUCT LINE
- STREET LIGHT STANDARD FED FROM UNDERGROUND CIRCUIT

ABOVE-GROUND ELECTRICAL DISTRIBUTION OR LIGHTING SYSTEMS

- POLE
- STREET LIGHT AND BRACKET
- PRIMARY CIRCUIT
- SECONDARY CIRCUIT
- DOWN GUY
- HEAD GUY
- SIDEWALK GUY
- SERVICE WEATHERHEAD

PANEL CIRCUITS AND MISCELLANEOUS

- LIGHTING PANEL
- POWER PANEL
- WIRING – CONCEALED IN CEILING OR WALL
- WIRING – CONCEALED IN FLOOR
- WIRING EXPOSED
- HOME RUN TO PANEL BOARD
  Indicate number of circuits by number of arrows. Any circuit without such designation indicates a two-wire circuit. For a greater number of wires indicate as follows: (3 wires) (4 wires), etc.
- FEEDERS
  Use heavy lines and designate by number corresponding to listing in feeder schedule
- WIRING TURNED UP
- WIRING TURNED DOWN
- GENERATOR — G
- MOTOR — M
- INSTRUMENT (SPECIFY) — I
- TRANSFORMER — T
- CONTROLLER
- EXTERNALLY-OPERATED DISCONNECT SWITCH
- PULL BOX

ARCHITECTURAL SYMBOLS . . .

Material	Elevation	Plan	Section
EARTH			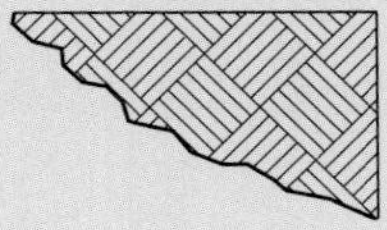
BRICK	 WITH NOTE INDICATING TYPE OF BRICK (COMMON, FACE, ETC.)	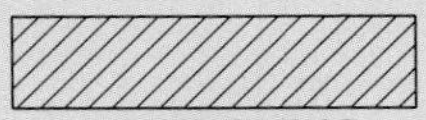 COMMON OR FACE 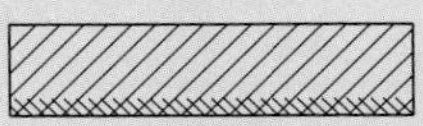 FIREBRICK	SAME AS PLAN VIEWS
CONCRETE	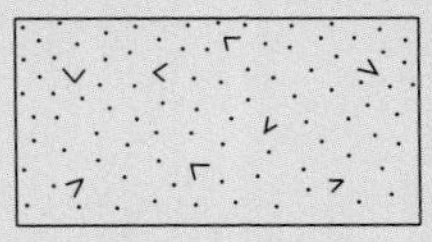	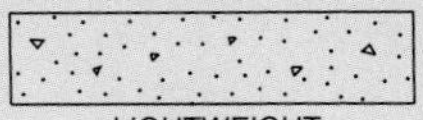 LIGHTWEIGHT 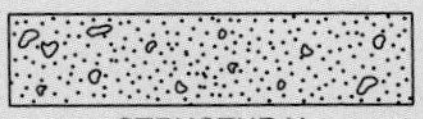 STRUCTURAL	SAME AS PLAN VIEWS
CONCRETE BLOCK	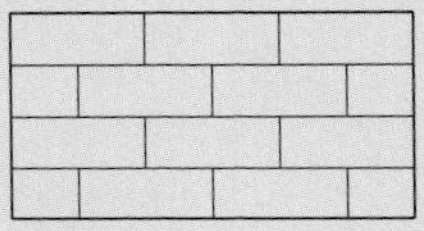	 OR 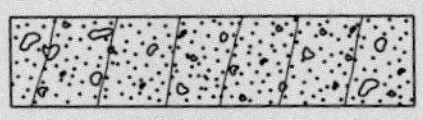	 OR 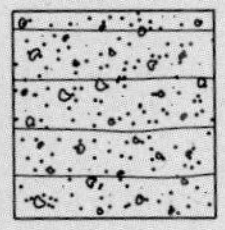
STONE	 CUT STONE 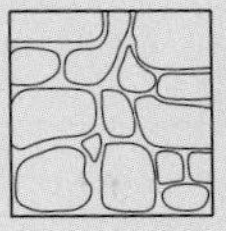 RUBBLE	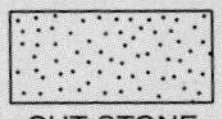 CUT STONE 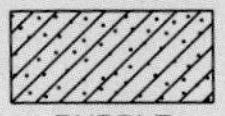 RUBBLE  CAST STONE (CONCRETE)	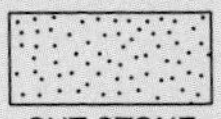 CUT STONE  CAST STONE (CONCRETE) 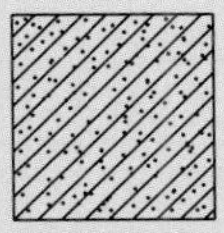 RUBBLE OR CUT STONE
WOOD	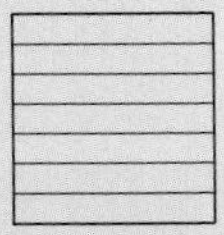 SIDING 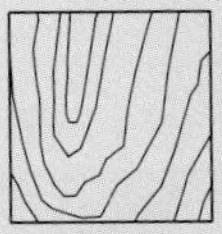 PANEL	 WOOD STUD  REMODELING  DISPLAY	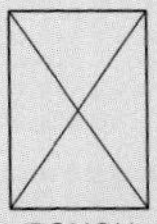 ROUGH MEMBERS  FINISHED MEMBERS
PLASTER	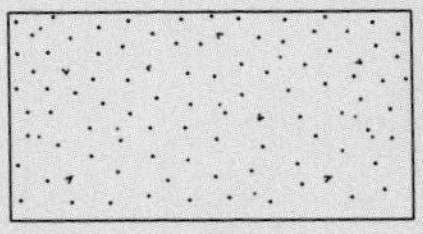	 WOOD STUD, LATH, AND PLASTER  METAL LATH AND PLASTER  SOLID PLASTER	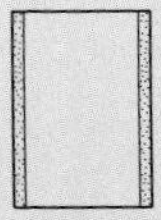 LATH AND PLASTER
ROOFING	 SHINGLES	SAME AS ELEVATION VIEW	
GLASS	 OR   GLASS BLOCK	 GLASS  GLASS BLOCK	 SMALL SCALE  LARGE SCALE

. . . ARCHITECTURAL SYMBOLS

Material	Elevation	Plan	Section
FACING TILE	CERAMIC TILE	FLOOR TILE	CERAMIC TILE LARGE SCALE CERAMIC TILE SMALL SCALE
STRUCTURAL CLAY TILE			SAME AS PLAN VIEW
INSULATION		LOOSE FILL OR BATTS RIGID SPRAY FOAM	SAME AS PLAN VIEWS
SHEET METAL FLASHING		OCCASIONALLY INDICATED BY NOTE	
METALS OTHER THAN FLASHING	INDICATED BY NOTE OR DRAWN TO SCALE	SAME AS ELEVATION	SMALL SCALE STEEL CAST IRON ALUMINUM BRONZE OR BRASS
STRUCTURAL STEEL	INDICATED BY NOTE OR DRAWN TO SCALE	OR	REBARS LI SMALL SCALE LARGE SCALE L-ANGLES, S-BEAMS, ETC.

PLOT PLAN SYMBOLS

N — NORTH	FIRE HYDRANT	WALK	E OR — ELECTRIC SERVICE
POINT OF BEGINNING (POB)	MAILBOX	IMPROVED ROAD	G OR — NATURAL GAS LINE
UTILITY METER OR VALVE	MANHOLE	UNIMPROVED ROAD	W OR — WATER LINE
POWER POLE AND GUY	TREE	BUILDING LINE	T OR — TELEPHONE LINE
LIGHT STANDARD	BUSH	PROPERTY LINE	NATURAL GRADE
TRAFFIC SIGNAL	HEDGE ROW	PROPERTY LINE	FINISH GRADE
STREET SIGN	FENCE	TOWNSHIP LINE	+ XX.00′ — EXISTING ELEVATION

THREE-PHASE VOLTAGE VALUES
For 208 V × 1.732, use 360
For 230 V × 1.732, use 398
For 240 V × 1.732, use 416
For 440 V × 1.732, use 762
For 460 V × 1.732, use 797
For 480 V × 1.732, use 831
For 2400 V × 1.732, use 4157
For 4160 V × 1.732, use 7205

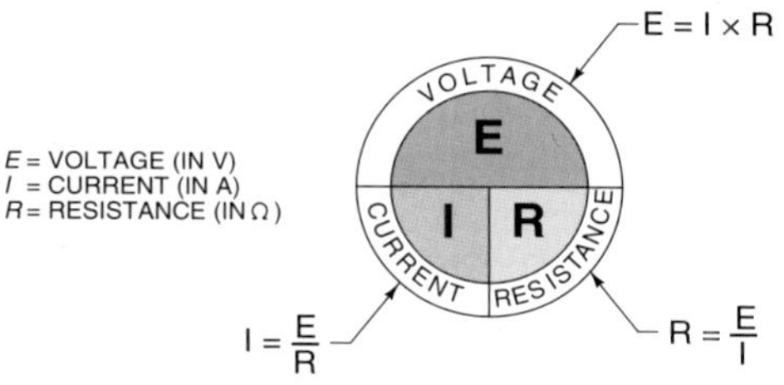

OHM'S LAW

Ohm's Law

Ohm's law is the relationship between the voltage, current, and resistance in an electrical circuit. Ohm's law states that current in a circuit is proportional to the voltage and inversely proportional to the resistance. It is written $I = E \div R$, $R = E \div I$, and $E = R \times I$.

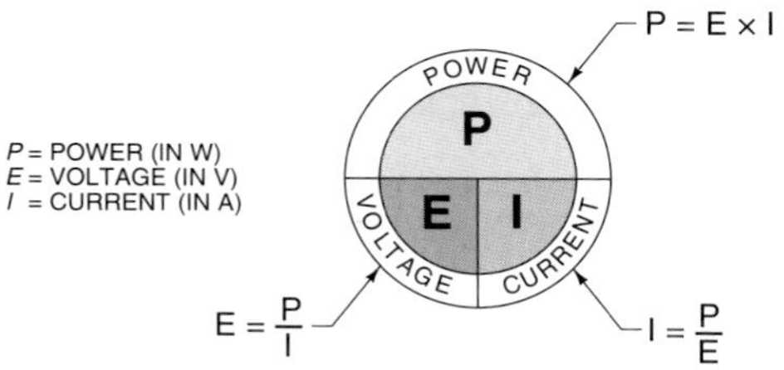

POWER FORMULA

Power Formula

The *power formula* is the relationship between the voltage, current, and power in an electrical circuit. The power formula states that the power in a circuit is equal to the I, and $I = P \div E$. Any value in these relationships is found using Ohm's Law and Power Formula.

POWER FORMULAS — 1φ, 3φ

Phase	To Find	Use Formula	Example		
			Given	Find	Solution
1φ	I	$I = \frac{VA}{V}$	32,000 VA, 240 V	I	$I = \frac{VA}{V}$ $I = \frac{32{,}000\ VA}{240\ V}$ I = **133 A**
1φ	VA	$VA = I \times V$	100 A, 240 V	VA	$VA = I \times V$ VA = 100 A × 240 V VA = **24,000 VA**
1φ	V	$V = \frac{VA}{I}$	42,000 VA, 350 A	V	$V = \frac{VA}{I}$ $V = \frac{42{,}000\ VA}{350\ A}$ V = **120 V**
3φ	I	$I = \frac{VA}{V \times \sqrt{3}}$	72,000 VA, 208 V	I	$I = \frac{VA}{V \times \sqrt{3}}$ $I = \frac{72{,}000\ VA}{360\ V}$ I = **200 A**
3φ	VA	$VA = I \times V \times \sqrt{3}$	2 A, 240 V	VA	$VA = I \times V \times \sqrt{3}$ $VA = 2 \times 416$ VA = **832 VA**

AC/DC FORMULAS				
To Find	DC	AC		
		1ϕ, 115 or 220 V	1ϕ, 208, 230, or 240 V	3ϕ—All Voltages
I, HP known	$\frac{HP \times 746}{E \times E_{ff}}$	$\frac{HP \times 746}{E \times E_{ff} \times PF}$	$\frac{HP \times 746}{E \times E_{ff} \times PF}$	$\frac{HP \times 746}{1.73 \times E \times E_{ff} \times PF}$
I, kW known	$\frac{kW \times 1000}{E}$	$\frac{kW \times 1000}{E \times PF}$	$\frac{kW \times 1000}{E \times PF}$	$\frac{kW \times 1000}{1.73 \times E \times PF}$
I, kVA known		$\frac{kVA \times 1000}{E}$	$\frac{kVA \times 1000}{E}$	$\frac{kVA \times 1000}{1.763 \times E}$
kW	$\frac{I \times E}{1000}$	$\frac{I \times E \times PF}{1000}$	$\frac{I \times E \times PF}{1000}$	$\frac{I \times E \times 1.73 \times PF}{1000}$
kVA		$\frac{I \times E}{1000}$	$\frac{I \times E}{1000}$	$\frac{I \times E \times 1.73}{1000}$
HP (output)	$\frac{I \times E \times E_{ff}}{746}$	$\frac{I \times E \times E_{ff} \times PF}{746}$	$\frac{I \times E \times E_{ff} \times PF}{746}$	$\frac{I \times E \times 1.73 \times E_{ff} \times PF}{746}$

E_{ff} = efficiency

HORSEPOWER FORMULAS				
To Find	Use Formula	Example		
		Given	Find	Solution
HP	$HP = \frac{I \times E \times E_{ff}}{746}$	240 V, 20 A, 85% E_{ff}	HP	$HP = \frac{I \times E \times E_{ff}}{746}$ $HP = \frac{20\ A \times 240\ V \times 85\%}{746}$ $HP = \mathbf{5.5}$
I	$I = \frac{HP \times 746}{E \times E_{ff} \times PF}$	10 HP, 240 V, 90% E_{ff}, 88% PF	I	$I = \frac{HP \times 746}{E \times E_{ff} \times PF}$ $I = \frac{10\ HP \times 746}{240\ V \times 90\% \times 88\%}$ $I = \mathbf{39\ A}$

VOLTAGE DROP FORMULAS – 1ϕ, 3ϕ					
Phase	To Find	Use Formula	Example		
			Given	Find	Solution
1ϕ	VD	$VD = \frac{2 \times R \times L \times I}{1000}$	240 V, 40 A, 60 L, .764 R	VD	$VD = \frac{2 \times R \times L \times I}{1000}$ $VD = \frac{2 \times .764 \times 60 \times 40}{1000}$ $VD = \mathbf{3.67\ V}$
3ϕ	VD	$VD = \frac{2 \times R \times L \times I}{1000} \times .866$	208 V, 110 A, 75 L, .194 R, .866 multiplier	VD	$VD = \frac{2 \times R \times L \times I}{1000} \times .866$ $VD = \frac{2 \times .194 \times 75 \times 110}{1000} \times .866$ $VD = \mathbf{2.77\ V}$

LOCKED ROTOR CURRENT

Apparent, 1ϕ	Apparent, 3ϕ	True, 1ϕ	True, 3ϕ
$LRC = \frac{1000 \times HP \times kVA/HP}{V}$ where LRC = locked rotor current (in amps) 1000 = multiplier for kilo HP = horsepower kVA/HP = kilovolt amps per horsepower V = volts	$LRC = \frac{1000 \times HP \times kVA/HP}{V \times \sqrt{3}}$ where LRC = locked rotor current (in amps) 1000 = multiplier for kilo HP = horsepower kVA/HP = kilovolt amps per horsepower V = volts $\sqrt{3} = 1.73$	$LRC = \frac{1000 \times HP \times kVA/HP}{V \times PF \times E_{ff}}$ where LRC = locked rotor current (in amps) 1000 = multiplier for kilo HP = horsepower kVA/HP = kilovolt amps per horsepower V = volts PF = power factor E_{ff} = motor efficiency	$LRC = \frac{1000 \times HP \times kVA/HP}{V \times \sqrt{3} \times PF \times E_{ff}}$ where LRC = locked rotor current (in amps) 1000 = multiplier for kilo HP = horsepower kVA/HP = kilovolt amps per horsepower V = volts $\sqrt{3} = 1.73$

MAXIMUM OVERCURRENT PROTECTIVE DEVICE (OCPD)

$OCPD = FLC \times R_M$

where

FLC = full-load current (from motor nameplate or NEC® Table 430.250)

R_M = maximum rating of OCPD

Motor Type	Code Letter	FLC (%)				
		Motor Size	TDF	NTDF	ITB	ITCB
AC*	—	—	175	300	150	700
AC*	A	—	150	150	150	700
AC*	B–E	—	175	250	200	700
AC*	F–V	—	175	300	250	700
DC	—	1/8 to 50 HP	150	150	150	250
DC	—	Over 50 HP	150	150	150	175

EFFICIENCY

Input and Output Power Known	Horsepower and Power Loss Known
$E_{ff} = \frac{P_{out}}{P_{in}}$ where E_{ff} = efficiency (%) P_{out} = output power (W) P_{in} = input power (W)	$E_{ff} = \frac{746 \times HP}{746 \times HP + W_l}$ where E_{ff} = efficiency (%) 746 = constant HP = horsepower W_l = watts lost

VOLTAGE UNBALANCE

$$V_u = \frac{V_d}{V_a} \times 100$$

where

V_u = voltage unbalance (%)

V_d = voltage deviation (V)

V_a = voltage average (V)

100 = constant

POWER

Power Consumed	Operating Cost	Annual Savings
$P = \frac{HP \times 746}{E_{ff}}$ where P = power consumed (W) HP = horsepower 746 = constant E_{ff} = efficiency (%)	$C_{/hr} = \frac{P_{/hr} \times C_{/kWh}}{1000}$ where $C_{/hr}$ = operating cost per hour $P_{/hr}$ = power consumed per hour $C_{/kWh}$ = cost per kilowatt hour 1000 = constant to remove kilo	$S_{Ann} = C_{Ann\ Std} - C_{Ann\ Eff}$ S_{Ann} = annual cost savings $C_{Ann\ Std}$ = annual operating cost for standard motor $C_{Ann\ Eff}$ = annual operating cost for energy-efficient motor

HORSEPOWER	
Current and Voltage Known	**Speed and Torque Known**
$HP = \frac{I \times E \times E_{ff}}{746}$ where HP = horsepower I = current (amps) E = voltage (volts) E_{ff} = efficiency	$HP = \frac{rpm \times T}{5252}$ where HP = horsepower rpm = revolutions per minute T = torque (lb-ft)

TEMPERATURE CONVERSIONS	
Convert °C to °F	**Convert °F to °C**
$°F = (1.8 \times °C) + 32$	$°C = \frac{(°F - 32)}{1.8}$

AC MOTOR CHARACTERISTICS

Motor Type	Typical Voltage	Starting Ability (Torque)	Size (HP)	Speed Range (rpm)	Cost*	Typical Uses
1φ						
Shaded-pole	115 V, 230 V	Very low 50% to 100% of full load	Fractional ½ HP to ⅓ HP	Fixed 900, 1200, 1800, 3600	Very low 75% to 85%	Light-duty applications such as small fans, hair dryers, blowers, and computers
Split-phase	115 V, 230 V	Low 75% to 200% of full load	Fractional ⅓ HP or less	Fixed 900, 1200, 1800, 3600	Low 85% to 95%	Low-torque applications such as pumps, blowers, fans, and machine tools
Capacitor-start	115 V, 230 V	High 200% to 350% of full load	Fractional to 3 HP	Fixed 900, 1200, 1800	Low 90% to 110%	Hard-to-start loads such as refrigerators, air compressors, and power tools
Capacitor-run	115 V, 230 V	Very low 50% to 100% of full load	Fractional to 5 HP	Fixed 900, 1200, 1800	Low 90% to 110%	Applications that require a high running torque such as pumps and conveyors
Capacitor-start-and-run	115 V, 230 V	Very high 350% to 450% of full load	Fractional to 10 HP	Fixed 900, 1200, 1800	Low 100% to 115%	Applications that require both a high starting and running torque such as loaded conveyors
3φ Induction	230 V, 460 V	Low 100% to 175% of full load	Fractional to over 500 HP	Fixed 900, 1200, 3600	Low 100%	Most industrial applications
Wound rotor	230 V, 460 V	High 200% to 300% of full load	½ HP to 200 HP	Varies by changing resistance in rotor	Very high 250% to 350%	Applications that require high torque at different speeds such as cranes and elevators
Synchronous	230 V, 460 V	Very low 40% to 100% of full load	Fractional to 250 HP	Exact constant speed	High 200% to 250%	Applications that require very slow speeds and correct power factors

* based on standard 3φ induction motor

DC AND UNIVERSAL MOTOR CHARACTERISTICS

Motor Type	Typical Voltage	Starting Ability (Torque)	Size (HP)	Speed Range (rpm)	Cost*	Typical Uses
DC Series	12 V, 90 V, 120 V, 180 V	Very high 400% to 450% of full load	Fractional to 100 HP	Varies 0 to full speed	High 175% to 225%	Applications that require very high torque such as hoists and bridges
Shunt	12 V, 90 V, 120 V, 180 V	Low 125% to 250% of full load	Fractional to 100 HP	Fixed or adjustable below full speed	High 175% to 225%	Applications that require better speed control than a series motor such as woodworking machines
Compound	12 V, 90 V, 120 V, 180 V	High 300% to 400% of full load	Fractional to 100 HP	Fixed or adjustable	High 175% to 225%	Applications that require high torque and speed control such as printing presses, conveyors, and hoists
Permanent-magnet	12 V, 24 V, 36 V, 120 V	Low 100% to 200% of full load	Fractional	Varies from 0 to full speed	High 150% to 200%	Applications that require small DC-operated equipment such as automobile power windows, seats, and sun roofs
Stepping	5 V, 12 V, 24 V	Very low† 0.5 to 5000 oz/in.	Size rating is given as holding torque and number of steps	Rated in number of steps per sec (maximum)	Varies based on number of steps and rated torque	Applications that require low torque and precise control such as indexing tables and printers
AC / DC Universal	115 VAC, 230 VAC, 12 VDC, 24 VDC, 36 VDC, 120 VDC	High 300% to 400% of full load	Fractional	Varies 0 to full speed	High 175% to 225%	Most portable tools such as drills, routers, mixers, and vacuum cleaners

* based on standard 3φ induction motor
† torque is rated as holding torque

ENCLOSURE TYPES				
Type	**Use**	**Service Conditions**	**Tests**	**Comments**
1	Indoor	No unusual	Rod entry, rust resistance	
3	Outdoor	Windblown dust, rain, sleet, and ice on enclosure	Rain, external icing, dust, and rust resistance	Do not provide protection against internal condensation or internal icing
3R	Outdoor	Falling rain and ice on enclosure	Rod entry, rain, external icing, and rust resistance	Do not provide protection against dust, internal condensation, or internal icing
4	Indoor/outdoor	Windblown dust and rain, splashing water, hose-directed water, and ice on enclosure	Hosedown, external icing, and rust resistance	Do not provide protection against internal condensation or internal icing
4X	Indoor/outdoor	Corrosion, windblown dust and rain, splashing water, hose-directed water, and ice on enclosure	Hosedown, external icing, and corrosion resistance	Do not provide protection against internal condensation or internal icing
6	Indoor/outdoor	Occasional temporary submersion at a limited depth		
6P	Indoor/outdoor	Prolonged submersion at a limited depth		
7	Indoor locations classified as Class I, Groups A, B, C, or D, as defined in the NEC®	Withstand and contain an internal explosion of specified gases, contain an explosion sufficiently so an explosive gas-air mixture in the atmosphere is not ignited	Explosion, hydrostatic, and temperature	Enclosed heat-generating devices shall not cause external surfaces to reach temperatures capable of igniting explosive gas-air mixtures in the atmosphere
9	Indoor locations classified as Class II, Groups E or G, as defined in the NEC®	Dust	Dust penetration, temperature, and gasket aging	Enclosed heat-generating devices shall not cause external surfaces to reach temperatures capable of igniting explosive gas-air mixtures in the atmosphere
12	Indoor	Dust, falling dirt, and dripping noncorrosive liquids	Drip, dust, and rust resistance	Do not provide protection against internal condensation
13	Indoor	Dust, spraying water, oil, and noncorrosive coolant	Oil explosion and rust resistance	Do not provide protection against internal condensation

RACEWAYS	
EMT	Electrical Metallic Tubing
ENT	Electrical Nonmetallic Tubing
FMC	Flexible Metal Conduit
FMT	Flexible Metallic Tubing
IMC	Intermediate Metal Conduit
LFMC	Liquidtight Flexible Metal Conduit
LFNC	Liquidtight Flexible Nonmetallic Conduit
RMC	Rigid Metal Conduit
RNC	Rigid Nonmetallic Conduit

FUSES AND ITCBs

Increase	Standard Ampere Ratings
5	15, 20, 25, 30, 35, 40, 45
10	50, 60, 70, 80, 90, 100, 110
25	125, 150, 175, 200, 225
50	250, 300, 350, 400, 450
100	500, 600, 700, 800
200	1000, 1200
400	1600, 2000
500	2500
1000	3000, 4000, 5000, 6000

1 A, 3 A, 6 A, 10 A, and 601 A are additional standard ratings fa fuses.

CABLES	
AC	Armored Cable
BX	Tradename for AC
FCC	Flat Conductor Cable
IGS	Integrated Gas Spacer Cable
MC	Metal-Clad Cable
MI	Mineral-Insulated, Metal Sheathed Cable
MV	Medium Voltage
NM	Nonmetallic-Sheathed Cable (dry)
NMC	Nonmetallic-Sheathed Cable (dry or damp)
NMS	Nometallic-Sheathed Cable (dry)
SE	Service-Entrance Cable
TC	Tray Cable
UF	Underground Feeder Cable
USE	Underground Service-Entrance Cable

COMMON ELECTRICAL INSULATIONS

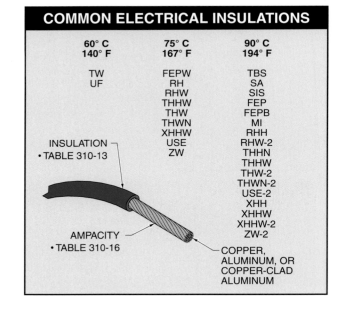

60° C 140° F	75° C 167° F	90° C 194° F
TW	FEPW	TBS
UF	RH	SA
	RHW	SIS
	THHW	FEP
	THW	FEPB
	THWN	MI
	XHHW	RHH
	USE	RHW-2
	ZW	THHN
		THHW
		THW-2
		THWN-2
		USE-2
		XHH
		XHHW
		XHHW-2
		ZW-2

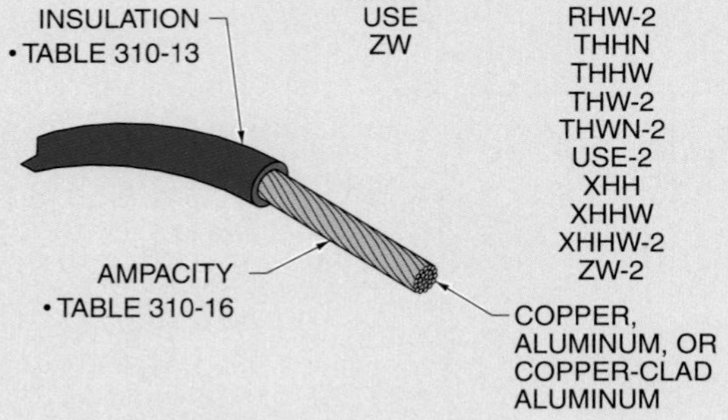

NON-LOCKING WIRING DEVICES

2-POLE, 3-WIRE

WIRING DIAGRAM	NEMA ANSI	RECEPTACLE CONFIGURATION	RATING
	5-15 C73.11		15 A 125 V
	5-20 C73.12		20 A 125 V
	5-30 C73.45		30 A 125 V
	5-50 C73.46		50 A 125 V
	6-15 C73.20		15 A 250 V
	6-20 C73.51		20 A 250 V
	6-30 C73.52		30 A 250 V
	6-50 C73.53		50 A 250 V
	7-15 C73.28		15 A 277 V
	7-20 C73.63		20 A 277 V
	7-30 C73.64		30 A 277 V
	7-50 C73.65		50 A 277 V

4-POLE, 4-WIRE

WIRING DIAGRAM	NEMA ANSI	RECEPTACLE CONFIGURATION	RATING
	18-15 C73.15		15 A 3φ Y 120/208 V
	18-20 C73.26		20 A 3φ Y 120/208 V
	18-30 C73.47		30 A 3φ Y 120/208 V
	18-50 C73.48		50 A 3φ Y 120/208 V
	18-60 C73.27		60 A 3φ Y 120/208 V

3-POLE, 3-WIRE

WIRING DIAGRAM	NEMA ANSI	RECEPTACLE CONFIGURATION	RATING
	10-20 C73.23		20 A 125/250 V
	10-30 C73.24		30 A 125/250 V
	10-50 C73.25		50 A 125/250 V
	11-15 C73.54		15 A 3φ 250 V
	11-20 C73.55		20 A 3φ 250 V
	11-30 C73.56		30 A 3φ 250 V
	11-50 C73.57		50 A 3φ 250 V

3-POLE, 4-WIRE

WIRING DIAGRAM	NEMA ANSI	RECEPTACLE CONFIGURATION	RATING
	14-15 C73.49		15 A 125/250 V
	14-20 C73.50		20 A 125/250 V
	14-30 C73.16		30 A 125/250 V
	14-50 C73.17		50 A 125/250 V
	14-60 C73.18		60 A 125/250 V
	15-15 C73.58		15 A 3φ 250 V
	15-20 C73.59		20 A 3φ 250 V
	15-30 C73.60		30 A 3φ 250 V
	15-50 C73.61		50 A 3φ 250 V
	15-60 C73.62		60 A 3φ 250 V

LOCKING WIRING DEVICES

2-POLE, 3-WIRE

WIRING DIAGRAM	NEMA ANSI	RECEPTACLE CONFIGURATION	RATING
	ML2 C73.44		15 A 125 V
	L5-15 C73.42		15 A 125 V
	L5-20 C73.72		20 A 125 V
	L6-15 C73.74		15 A 250 V
	L6-20 C73.75		20 A 250 V
	L6-30 C73.76		30 A 250 V
	L7-15 C73.43		15 A 277 V
	L7-20 C73.77		20 A 277 V
	L8-20 C73.79		20 A 480 V
	L9-20 C73.81		20 A 600 V

3-POLE, 4-WIRE

WIRING DIAGRAM	NEMA ANSI	RECEPTACLE CONFIGURATION	RATING
	L14-20 C73.83		20 A 125/250 V
	L14-30 C73.84		30 A 125/250 V
	L15-20 C73.85		20 A 3ϕ 250 V
	L15-30 C73.86		30 A 3ϕ 250 V
	L16-20 C73.87		20 A 3ϕ 480 V
	L16-30 C73.88		30 A 3ϕ 480 V
	L17-30 C73.89		30 A 3ϕ 600 V

3-POLE, 3-WIRE

WIRING DIAGRAM	NEMA ANSI	RECEPTACLE CONFIGURATION	RATING
	ML3 C73.30		15 A 125/250 V
	L10-20 C73.96		20 A 125/250 V
	L10-30 C73.97		30 A 125/250 V
	L11-15 C73.98		15 A 3ϕ 250 V
	L11-20 C73.99		20 A 3ϕ 250 V
	L12-20 C73.101		20 A 3ϕ 480 V
	L12-30 C73.102		30 A 3ϕ 480 V
	L13-30 C73.103		30 A 3ϕ 600 V

4-POLE, 4-WIRE

WIRING DIAGRAM	NEMA ANSI	RECEPTACLE CONFIGURATION	RATING
	L18-20 C73.104		20 A 3ϕY 120/208 V
	L18-30 C73.105		30 A 3ϕY 120/208 V
	L19-20 C73.106		20 A 3ϕY 277/480 V
	L20-20 C73.108		20 A 3ϕY 347/600 V

4-POLE, 5-WIRE

WIRING DIAGRAM	NEMA ANSI	RECEPTACLE CONFIGURATION	RATING
	L21-20 C73.90		20 A 3ϕY 120/208 V
	L22-20 C73.92		20 A 3ϕY 277/480 V
	L23-20 C73.94		20 A 3ϕY 347/600 V

STANDARD CALCULATION: ONE-FAMILY DWELLING

1. GENERAL LIGHTING: *Table 220.12*

_______ sq ft × 3 VA = _______ VA

Small appliances: *220.52(A)*

_______ VA × _______ circuits = _______ VA

Laundry: *220.52(B)*

_______ VA × 1 = _______ VA

_______ VA

Applying Demand Factors: *Table 220.42.*

First 3000 VA × 100% = 3000 VA

Next _______ VA × 35% = _______ VA

Remaining _______ VA × 25% = _______ VA

Total _______ VA | **PHASES** _______ VA | **NEUTRAL** _______ VA

2. FIXED APPLIANCES: *220.53*

Dishwasher = _______ VA

Disposer = _______ VA

Compactor = _______ VA

Water heater = _______ VA

_______ = _______ VA

_______ = _______ VA

_______ = _______ VA

Total _______ VA × 75% = _______ VA | _______ VA | (120 V Loads × 75%) _______ VA

3. DRYER: *220.54; Table 220.54*

_______ VA × _______ % = _______ VA | _______ VA × 70% = _______ VA

4. COOKING EQUIPMENT: *Table 220.55; Notes*

Col A _______ VA × _______ % = _______ VA

Col B _______ VA × _______ % = _______ VA

Col C _______ VA × _______ % = _______ VA

Total _______ VA | _______ VA × 70% = _______ VA

5. HEATING or A/C: *220.60*

Heating unit = _______ VA × 100% = _______ VA

A/C unit = _______ VA × 100% = _______ VA

Heat pump = _______ VA × 100% = _______ VA

Largest Load _______ VA | _______ VA | _______ VA

6. LARGEST MOTOR: *220.14(C)*

ϕ _______ VA × 25% = _______ VA | _______ VA

N _______ VA × 25% = _______ VA | _______ VA

1ϕ service: PHASES $I = \frac{\text{VA}}{\text{V}} =$ _______ A

NEUTRAL $I = \frac{\text{VA}}{\text{V}} =$ _______ A

[ ] VA | [ ] VA

220.61(B) First 200 A × 100% = 200 A

Remaining _______ A × 70% = _______ A

Total _______ A

STANDARD CALCULATION: MULTIFAMILY DWELLING

1. GENERAL LIGHTING: *Table 220.12*

_____ sq ft × 3 VA × _____ units = _____ VA
_____ sq ft × 3 VA × _____ units = _____ VA
_____ sq ft × 3 VA × _____ units = _____ VA

Small appliances: *220.52(A)*
_____ VA × _____ circuits × _____ units = _____ VA

Laundry: *220.52(B)*
_____ VA × 1 × _____ units = _____ VA
_____ VA

Applying Demand Factors: *Table 220.42*
First 3000 VA × 100% = 3000 VA
Next _____ VA × 35% = _____ VA
Remaining _____ VA × 25% = _____ VA
Total _____ VA

PHASES _____ VA **NEUTRAL** _____ VA

2. FIXED APPLIANCES: *220.53*

Dishwasher = _____ VA
Disposer = _____ VA
Compactor = _____ VA
Water heater = _____ VA
_____ = _____ VA
_____ = _____ VA
_____ = _____ VA
Total _____ VA × ___ units × 75% = _____ VA _____ VA _____ VA (120 V Loads × units × 75%)

3. DRYER: *220.54; Table 220.54*

_____ VA × _____ units × _____ % = _____ VA _____ VA × 70% = _____ VA

4. COOKING EQUIPMENT: *Table 220.55; Notes*

Col A _____ units = _____ VA × _____ % = _____ VA
Col B _____ VA × _____ units × _____ % = _____ VA
Col C _____ VA × _____ units × _____ % = _____ VA
Total _____ VA _____ VA × 70% = _____ VA

5. HEATING or A/C: *220.60*

Heating unit = _____ VA × 100% × _____ units = _____ VA
A/C unit = _____ VA × 100% × _____ units = _____ VA
Heat pump = _____ VA × 100% × _____ units = _____ VA
Largest Load _____ VA _____ VA _____ VA

6. LARGEST MOTOR: *220.14(C)*

ϕ _____ VA × 25% = _____ VA _____ VA
N _____ VA × 25% = _____ VA _____ VA

1ϕ service: PHASES $I = \frac{VA}{V}$ = _____ A

NEUTRAL $I = \frac{VA}{V}$ = _____ A

220.61(B) First 200 A × 100% = 200 A
Remaining _____ A × 70% = _____ A
_____ A

[ ] VA [ ] VA

3ϕ service: $I = \frac{VA}{V \times \sqrt{3}}$ = _____ A

$I = \frac{VA}{V \times \sqrt{3}}$ = _____ A

OPTIONAL CALCULATION: ONE-FAMILY DWELLING

1. HEATING or A/C: *220.82(C)(1 – 6)*

Heating units (3 or less) = ________ VA × 65% = ________ VA

Heating units (4 or more) = ________ VA × 40% = ________ VA

A/C unit = ________ VA × 100% = ________ VA

Heat pump = ________ VA × 100% = ________ VA

Largest Load ________ VA

Total ________ VA

PHASES

________ VA

2. GENERAL LOADS: *220.82(B)(1 – 4)*

General lighting: *220.82(B)(1)*

________ sq ft × 3 VA ________ VA

Small appliance and laundry loads: *220.82(B)(2)*

________ VA × ________ circuits = ________ VA

Special loads: *220.82(B)(3 – 4)*

Dishwasher = ________ VA

Disposer = ________ VA

Compactor = ________ VA

Water heater = ________ VA

________ = ________ VA

________ = ________ VA

________ = ________ VA

________ = ________ VA

________ = ________ VA

________ VA ________ VA

Total ________ VA

Applying Demand Factors: *220.82(B)*

First 10,000 VA × 100% = 10,000 VA

Remaining ________ VA × 40% = ________ VA

Total ________ VA ________ VA

NEUTRAL (Loads from Standard Calculation)

1. General lighting = ________ VA
2. Fixed appliances = ________ VA
3. Dryer = ________ VA
4. Cooking equipment = ________ VA
5. Heating or A/C = ________ VA
6. Largest motor = ________ VA

Total [ ] VA

1ϕ service: PHASES $I = \frac{\text{\_\_\_\_\_\_ VA}}{\text{V}} =$ ________ A

NEUTRAL $I = \frac{\text{\_\_\_\_\_\_ VA}}{\text{V}} =$ ________ A

[ ] VA

OPTIONAL CALCULATION: MULTIFAMILY DWELLING

1. HEATING or A/C: *220.84(C)(5)*

Heating unit = ________ VA × 100% × ________ units = ________ VA

A/C unit = ________ VA × 100% × ________ units = ________ VA

Heat pump = ________ VA × 100% × ________ units = ________ VA **PHASES**

Largest Load ________ VA

Total ________ VA

2. CALCULATED LOADS: *220.84(C)*

General lighting: *220.84(C)(1)*

________ sq ft × 3 VA × ________ units = ________ VA ________ VA

________ sq ft × 3 VA × ________ units = ________ VA

________ sq ft × 3 VA × ________ units = ________ VA

________ sq ft × 3 VA × ________ units = ________ VA

Small appliance and laundry loads: *220.84(C)(2)*

________ VA × ________ circuits × ________ units = ________ VA ________ VA

Special loads: *220.84(C)(3)*

Dishwasher = ________ VA

Disposer = ________ VA

Compactor = ________ VA

Water heater = ________ VA

________ = ________ VA

________ = ________ VA

________ = ________ VA

________ = ________ VA

________ = ________ VA

Total ________ VA × ________ units = ________ VA ________ VA

Total Connected Load ________ VA

Applying Demand Factors: *Table 220.84*

________ VA × ________ % = ________ VA [ ] VA

NEUTRAL (Loads from Standard Calculation)

1. General lighting = ________ VA
2. Fixed appliances = ________ VA
3. Dryer = ________ VA
4. Cooking equipment = ________ VA
5. Heating or A/C = ________ VA
6. Largest motor = ________ VA

Total [ ] VA

1ϕ service: PHASES $I = \frac{\_\_\_\_\_\_\_\_ VA}{V}$ = ________ A   3ϕ service: $I = \frac{\_\_\_\_\_\_\_\_ VA}{V \times \sqrt{3}}$ = ________ A

NEUTRAL $I = \frac{\_\_\_\_\_\_\_\_ VA}{V}$ = ________ A $I = \frac{\_\_\_\_\_\_\_\_ VA}{V \times \sqrt{3}}$ = ________ A

220.61; First 200 A × 100% = 200 A

Remaining ________ A × 70% = ________ A

________ A

Glossary

A

adjustable-trip circuit breaker (ATCB): A circuit breaker with a trip setting that can be changed by adjusting the ampere setpoint, trip time characteristics, or both, within a particular range.

ambient temperature: The temperature of air around a conductor or a piece of equipment.

ampacity: The maximum current that a conductor can carry continuously, under the conditions of use.

appliance: Any utilization equipment which performs one or more functions, such as clothes washing, air conditioning, cooking, etc.

appliance branch circuit: A branch circuit that supplies energy to one or more outlets to which appliances are to be connected.

approved: Acceptable to the authority having jurisdiction.

armored cable (AC): A factory assembly that contains the conductors within a jacket made of a spiral wrap of steel.

askarel: A group of nonflammable synthetic chlorinated hydrocarbons that were once used where nonflammable insulating oils were required.

autotransformer: A single-winding transformer that shares a common winding between the primary and secondary circuits.

auxiliary gutter: A sheet-metal or non-metallic enclosure equipped with hinged or removable covers that is used to supplement wiring space.

B

bare conductor: A conductor with no insulation or covering of any type.

bathroom: An area with a basin and one or more of the following: a toilet, tub, urinal, shower, bidet, or silmilar type of plumbing fixture.

bend: Any change in direction of a raceway.

bonding: Establishing continuity and conductivity through a connection.

box: A metallic or nonmetallic electrical enclosure used for equipment, devices, and pulling or terminating conductors.

branch circuit: The portion of the electrical circuit between the last overcurrent device (fuse or circuit breaker) and the outlets or utilization equipment.

branch-circuit rating: The ampere rating or setting of the overcurrent device protecting the conductors.

building: A stand-alone structure or a structure that is separated from adjoining structures by fire walls.

bushing: A fitting placed on the end of a conduit to protect the conductor's insulation from abrasion.

busway: A sheet metal enclosure that contains factory-assembled aluminum or copper busbars which are supported on insulators.

C

cable: A factory assembly with two or more conductors and an overall covering.

cable assembly: A flexible assembly containing multiple conductors with a protective outer sheath.

cable tray system (CTS): An assembly of sections and associated fittings which form a rigid structural system used to support cables and raceways.

cadwelding: A welding process used to make electrical connections of copper to copper or copper to steel in which no outside source of heat or power is required.

cartridge fuse: A fuse constructed of a metallic link or links which is designed to open at predetermined current levels to protect circuit conductors and equipment.

circuit breaker (CB): An overcurrent protection device with a mechanical mechanism that may manually or automatically open the circuit when an overload condition or short circuit occurs.

circular mil: A measurement used to determine the cross-sectional area of a conductor.

Class I location: A hazardous location in which flammable gases, flammable liquid-produced vapors, or combustible liquid-produced vapors are present in sufficient quantities to cause an explosion or ignite the hazardous materials.

Class II location: A hazardous location that is hazardous due to the presence of combustible dust.

Class III location: A hazardous location that is hazardous due to the presence of easily ignitable fibers or flyings or where materials that produce combustible fibers or flyings are handled, manufactured, or used.

conductor: A slender rod or wire that is used to control the flow of electrons in an electrical circuit.

conduit body: A conduit fitting that provides access to the raceway system through a removable cover at a junction or termination point.

conduit seal: Seal provided in conduit.

continuous load: A load in which the maximum current is expected to continue for three hours or more.

controller: The device in a motor circuit which turns the motor ON or OFF.

cover: The shortest distance measured between a point on the top surface of any direct-buried conductor, cable, conduit, or other raceway and the top surface of a finished grade, concrete, or similar cover.

covered conductor: A conductor not encased in a material recognized by the NEC®.

current-limiting fuse: A fuse that opens a circuit in less than one-half of a cycle to protect the circuit components from damaging short-circuit currents.

current transformer: A transformer that creates a constant ratio of primary to secondary current instead of attempting to maintain a constant ratio of primary to secondary voltage.

D

damp location: A partially protected area subject to some moisture.

dead front: A cover required for the operation of a plug or connector.

demand: The amount of electricity required at a given time.

demand factor: The ratio of the maximum demand of a system, or part of a system, to the total connected load of a system or the part of the system under consideration.

device: Any unit of an electrical system that carries or controls electricity.

device box: A box which houses an electrical device.

disconnecting means: A device or group of devices that separate or isolate the conductors of a circuit from their source of supply.

division: The classification assigned to each Class based upon the likelihood of the presence of the hazardous substance in the atmosphere.

Division 1 location: A hazardous location in which the hazardous substance is normally present in the air in sufficient quantities to cause an explosion or ignite the hazardous materials.

Division 2 location: A hazardous location in which the hazardous substance is not normally present in the air in sufficient quantities to cause an explosion or ignite the hazardous materials.

dry location: A location which is not normally damp or wet.

dry-type transformer: A transformer which provides air circulation based on the principle of heat transfer.

dust-ignitionproof: Enclosed in a manner which prevents the entrance of dusts and does not permit arcs, sparks, or excessive temperature to cause ignition of exterior accumulations of specified dust.

dusttight: Construction that does not permit dust to enter the enclosing case under specified test conditions.

dwelling: A structure that contains eating, living, and sleeping space, and permanent provisions for cooking and sanitation.

dwelling unit: A single unit for one or more persons that includes permanent provisions for living, sleeping, cooking, and sanitation.

dynamic load: A load that produces a small but constant vibration.

E

Edison-base fuse: A plug fuse that incorporates a screw configuration which is interchangeable with fuses of other ampere ratings.

electrical metallic tubing (EMT): A lightweight tubular steel raceway without threads on the ends.

electrical nonmetallic tubing (ENT): A nonmetallic corrugated raceway.

electric-discharge luminaire: A luminaire that utilizes a ballast for the operation of the lamp.

enclosure: The case or housing of equipment or other apparatus which provides protection from live or energized parts.

energized: Being electrically connected to voltage or being a source of voltage.

equipment: Any material, device, fixture, apparatus, appliance, etc. used in conjunction with electrical installations.

equipment bonding jumper (EBJ): A conductor that connects two or more parts of the equipment grounding conductor.

equipment grounding conductor (EGC): An electrical conductor that provides a low-impedance path between electrical equipment and enclosures and the system grounded conductor and grounding electrode conductor.

equipotential plane: An area in which all conductive elements are bonded or otherwise connected together in a manner which prevents a difference of potential from developing within the plane.

explosionproof apparatus: Equipment which is enclosed in a case that is capable of withstanding any explosion that may occur within it, without permitting the ignition of flammable gases or vapors on the outside of the enclosure.

exposed: As applied to wiring methods, means on a surface or behind panels which allow access.

F

feeder: All circuit conductors between the service equipment, the source of a separately derived system, or other supply source, and the final branch-circuit overcurrent device.

feeder neutral load: The maximum unbalance between any of the ungrounded conductors and the grounded conductor.

fire point: The lowest temperature at which a material can give off vapors fast enough to support continuous combustion.

fitting: An electrical system accessory that performs a mechanical function.

flexible cable: An assembly of one or more insulated conductors, with or without braids, contained within an overall outer covering and used for the connection of equipment to a power source.

flexible cord: An assembly of two or more insulated conductors, with or without braids, contained within an overall outer covering and used for the connection of equipment to a power source.

flexible metal conduit (FMC): A raceway of metal strips which are formed into a circular cross-sectional raceway.

fuse: An overcurrent protection device with a fusible link that melts and opens the circuit when an overload condition or short circuit occurs.

G

general-purpose branch circuit: A branch circuit that supplies two or more outlets for lighting and appliances.

generator: A device that is used to convert mechanical power to electrical power.

ground: The earth.

grounded: Connected to the earth or to a conductive body connected to the earth.

grounded conductor: A conductor that has been intentionally grounded.

ground fault: An unintentional connection between an ungrounded conductor and any grounded raceway, box, enclosure, fitting, etc.

ground-fault circuit interrupter receptacle: A device that interrupts the flow of current to the load when a ground fault occurs that exceeds a predetermined value of current.

grounding conductor: The conductor that connects electrical equipment or the grounded conductor to the grounding electrode.

grounding electrode conductor (GEC): The conductor that connects the grounding electrode(s) to the system grounded conductor and/or the equipment grounding conductor.

grounding receptacle: A receptacle that includes a grounding terminal connected to a grounding slot in the receptacle configuration.

group: An atmosphere containing flammable gases or vapors or combustible dust.

H

hazardous location: A location where there is an increased risk of fire or explosion due to the presence of flammable liquid-produced vapors, combustible liquid-produced vapors, combustible dusts, or ignitable fibers or flyings.

health care facility: A location, either a building or a portion of a building, which contains occupancies such as hospitals, nursing homes, limited or supervisory care facilities, clinics, medical and dental offices, and either movable or permanent ambulatory facilities.

hermetic refrigerant motor-compressor: A combination of a compressor and motor enclosed in the same housing, having no external shaft or shaft seals, with the motor operating in the refrigerant.

high density polyethylene conduit (Type HDPE conduit): Conduit constructed of high density polyethylene that is resistant to moisture and chemical atmospheres.

high-intensity discharge (HID) luminaire: A luminaire that generates light from an arc lamp contained within an outer tube.

hospital-grade receptacle: The highest grade receptacle manufactured for the electrical industry.

house load: An electrical load which is metered separately and supplies common usage areas.

I

identified: Recognized as suitable for the use, purpose, etc.

immersion detection circuit interrupter (IDCI): Circuit interrupter designed to provide protection against shock when appliances fall into a sink or bathtub.

impedance: The total opposition to the flow of current in a circuit.

individual branch circuit: A branch circuit that supplies only one piece of utilization equipment.

in sight from: Visible and not more than 50′ away.

instantaneous-trip circuit breaker (ITB): A circuit breaker with no delay between the fault or overload sensing element and the tripping action of the device.

instrument transformer: A transformer used to reduce higher voltage and current ratings to safer and more suitable levels for the purposes of control and measurement.

insulated conductor: A conductor that is encased using a material of composition and thickness that is recognized by the NEC® as an insulating material.

intermediate metal conduit (IMC): A raceway of circular cross section designed for protection and routing of conductors.

intermittent load: A load in which the maximum current does not continue for three hours.

interrupting rating: The maximum amount of current that an overcurrent protective device can clear safely.

intrinsically safe system: An assembly of interconnected, intrinsically safe apparatus, associated apparatus, and interconnected cables containing circuits in which any spark or thermal effect is capable of causing ignition.

inverse-time circuit breaker (ITCB): A circuit breaker with an intentional delay between the time when the fault or overload is sensed and the time when the circuit breaker operates.

isolated-ground receptacle: A receptacle in which the grounding terminal is isolated from the device yoke or strap.

isolation transformer: A transformer that utilizes a shield between the primary and secondary windings and a transformer ratio of 1:1 to ensure that the load is separated from the power source.

J

junction box: A box in which splices, taps, or terminations are made.

K

kick: A single bend in a raceway.

L

labeled: Equipment acceptable to the authority having jurisdiction and to which a label has been attached.

lampholder: A device designed to accommodate a lamp for the purpose of illumination.

less-flammable liquid: An insulating oil which is flammable but has reduced flammable characteristics and a higher fire point.

lighting outlet: An outlet intended for the direct connection of a lampholder, luminaire (lighting fixture), or pendant cord terminating in a lampholder.

lighting track: An assembly consisting of an energized metal track and luminaire heads which can be positioned in any location along the track.

line surge: A temporary increase in the circuit or system voltage or current that may occur as a result of fluctuations in the electrical distribution system.

liquid-filled transformer: A transformer that utilizes some form of insulating liquid for immersion of the core and windings of the transformer.

liquidtight flexible metal conduit (LFMC): A raceway of circular cross section with an outer liquidtight, nonmetallic, sunlight-resistant jacket over an inner helically wound metal strip.

liquidtight flexible nonmetallic conduit (LFNC): A raceway of circular cross section with an outer jacket and with an inner core and reinforcement that varies based on the intended use.

listed: Equipment, materials, or services that are included in a list published by an organization acceptable to the authority having jurisdiction.

luminaire: A complete lighting unit consisting of a lamp or lamps together with the parts designed to distribute the light, to position and protect the lamps and ballast (where applicable), and to connect the lamps to the power supply.

M

main bonding jumper (MBJ): The connection at the service equipment that ties together the equipment grounding conductor, the grounded conductor, and the grounding electrode conductor.

major repair garage: A commercial garage where a building or a portion of a building is used for major repairs.

medium voltage cable (Type MV cable): A single or multiple copper, aluminum, or copper-clad aluminum conductor cable that is constructed of a solid dielectric insulation rated 2001 V or higher.

metal-clad (MC) cable: A factory assembly of one or more conductors with or without fiber-optic members.

mil: 0.001″.

milliampere (mA): $\frac{1}{1000}$ of an ampere (1000 mA = 1 A).

mineral insulated, metal sheathed cable (Type MI cable): A factory assembled cable construction that consists of one or more conductors that are insulated with a highly compressed refractory mineral insulation.

mineral oil: A chemically untreated insulating oil that is distilled from petroleum.

minor repair garage: A commercial garage where a building or a portion of a building is used for lubrication, inspection, and minor vehicle repair.

motor branch circuit: The point from the last fuse or circuit breaker in the motor circuit out to the motor.

motor control center (MCC): An assembly of one or more enclosed sections with a common power bus and primarily containing motor control units.

motor control circuit: The circuit of a control apparatus or system which carries electric signals directing the performance of the controller, but does not carry the main power current.

multifamily dwelling: A dwelling with three or more dwelling units.

multioutlet assembly: A surface, flush, or freestanding raceway which contains conductors and receptacles.

multiple receptacle: A single device with two or more receptacles.

multiwire branch circuit: A branch circuit with two or more ungrounded conductors having a voltage between them, and a grounded conductor having equal voltage between it and each ungrounded conductor, and is connected to the neutral or grounded conductor of the system.

N

neutral conductor: A conductor connected to the neutral point of a system that is intended to carry current under normal conditions.

neutral point: The common point on a wye connection in a polyphase system; midpoint on a 1ϕ, 3-wire system; the midpoint on a 1ϕ portion of a 3ϕ delta system; or the midpoint of a 3-wire DC system.

nipple: A short piece of conduit or tubing that does not exceed 24″ in length.

nominal voltage: A nominal value assigned to a circuit or system for the purpose of designating its class.

nonadjustable-trip circuit breaker (NATCB): A fixed circuit breaker designed without provisions for adjusting either the ampere trip setpoint or the time-trip setpoint.

noncoincidental load: A load that is not on at the same time as another load.

nonflammable liquid: A liquid that is noncombustible and does not burn when exposed to air and a source of ignition.

nongrounding receptacle: A receptacle with two wiring slots for branch-circuit wiring systems that does not provide an equipment grounding conductor.

nonlinear load: A load where the wave shape of the steady-state current does not follow the wave shape of the applied voltage.

nonmetallic-sheathed (NM) cable: A factory assembly of two or more insulated conductors having an outer sheath of moisture-resistant, flame-retardant, nonmetallic material.

nonmetallic underground conduit with conductors (Type NUCC): A factory assembly of conductors or cables that are contained within a nonmetallic, smooth wall circular raceway.

non-time delay fuse (NTDF): A fuse that may detect an overcurrent and open the circuit almost instantly.

O

offset: A double bend in a raceway, each containing the same number of degrees.

one-family dwelling: A dwelling with one dwelling unit.

outlet: Any point in the electrical system where current supplies utilization equipment.

outlet box: A box which houses a piece of utilization equipment.

overcurrent: Any current in excess of that for which the conductor or equipment is rated.

overlamping: Installing a lamp of a higher wattage than that for which the fixture is designed.

overload: The operation of equipment in excess of normal, full-load rating, or a conductor that has an excess amount of current.

oxidation: The process by which oxygen mixes with other elements and forms a type of rust-like material.

oxide: A thin, but highly resistive coating that forms on metal when exposed to the air.

P

panelboard: A single panel or group of assembled panels with buses and overcurrent devices, which may have switches to control light, heat, or power circuits.

panic hardware: Door hardware designed to open quickly and easily with pressure.

parallel conductors: Two or more conductors that are electrically connected at both ends to form a single conductor.

pendant: A hanging luminaire that uses flexible cords to support the lampholder.

permanently connected appliance: A hard-wired appliance that is not cord-and-plug-connected.

phase-to-ground voltage: The difference of potential between a phase conductor and ground.

phase-to-phase voltage: The maximum voltage between any two phases of an electrical distribution system.

place of assembly: A building, structure, or portion of a building designed or intended for use by 100 or more persons.

plug fuse: A fuse that uses a metallic strip which melts when a predetermined amount of current flows through it.

potential transformer: A transformer which steps down higher voltages while allowing the voltage of the secondary to remain fairly constant from no-load to full-load conditions.

power and control tray cable (Type TC cable): A factory assembled cable consisting of one or more insulated copper, aluminum, or copper-clad aluminum conductors.

premises wiring: Basically all interior and exterior wiring installed on the load side of the service point or the source of a separately derived system.

pull box: A box used as a point to pull or feed electrical conductors into the raceway system.

Q

qualified person: A person who has knowledge and skills related to the construction and operation of electrical equipment and has received appropriate safety training.

R

raceway: A metal or nonmetallic enclosed channel for conductors.

raceway system: An enclosed channel of metal or nonmetallic materials used to contain the wires or cables of an electrical system.

readily accessible: Capable of being reached quickly.

receptacle: A contact device installed at outlets for the connection of cord-connected electrical equipment.

receptacle outlet: An outlet that provides power for cord-and-plug-connected equipment.

rigid metal conduit (RMC): A threadable conduit made of metal. It is the universal raceway.

rigid polyvinyl chloride (PVC) conduit: A conduit made of materials other than metal designed for the installation of electrical conductors and cables.

S

self-grounding receptacle: A grounding type receptacle that utilizes a pressure clip around the 6–32 mounting screw to ensure good electrical contact between the receptacle yoke and the outlet box.

separately derived system: A system that supplies premises with electrical power derived or taken from storage batteries, solar photovoltaic systems, generators, transformers, or converter windings.

service: The electrical supply, in the form of conductors and equipment, that provides electrical power to the building or structure.

service conductor: The conductor from the service point to the service disconnecting means.

service conductor—overhead: A conductor that connects the service equipment for the building or structure with the electrical utility supply conductors at the service point.

service drop: The overhead conductors between the service point and the utility's electrical supply system.

service-entrance (SE) cable: A single or multiconductor assembly with or without an overall covering.

service-entrance conductor—underground system: A conductor that connects the service equipment with the service lateral.

service equipment: **1.** The necessary equipment, usually consisting of a circuit breaker or switch and fuses and their accessories, connected to the load end of service conductors to a building or other structure, or an otherwise designated area, and intended to constitute the main control and cut-off of the supply. **2.** All of the equipment necessary to control the supply of electrical power to a building or structure.

service lateral: The underground service conductors that connect the utility's electrical distribution system and the service point.

service mast: An assembly consisting of a service raceway, guy wires or braces, service head, and any fittings necessary for the support of service-drop conductors.

service point: The point of connection between the local electrical utility company and the premises wiring of the building or structure.

short circuit: **1.** The unintentional connection of two ungrounded conductors that have a potential difference between them. **2.** The condition that occurs when two ungrounded conductors (hot wires), or an ungrounded and a grounded conductor of a 1ϕ circuit, come in contact with each other.

single receptacle: A single contact device with no other contact device on the same yoke.

special permission: The written approval of the authority having jurisdiction.

step-down transformer: A transformer with more windings in the primary winding, which results in a load voltage that is less than the applied voltage.

step-up transformer: A transformer with more windings in the secondary winding, which results in a load voltage that is greater than the applied voltage.

strut-type channel raceway: A surface raceway formed of moisture-resistant and corrosion-resistant metal.

supervised installation: An electrical installation in which the conditions of maintenance are such that only qualified persons monitor or service the electrical equipment.

surface raceway: An enclosed channel for conductors which is attached to a surface.

switch: A device, with a current and voltage rating, used to open or close an electrical circuit.

switchboard: A single panel or group of assembled panels with buses, overcurrent devices, and instruments.

system bonding jumper: The bonding jumper that is used to connect the grounded circuit conductor and the supply-side bonding jumper and/or the equipment grounding conductor at a separately derived system.

T

temperature rise: The amount of heat that an electrical component produces above the ambient temperature.

thermally protected fixture: A fixture designed with an internal thermal protective device which senses excessive operating temperatures and opens the supply circuit to the fixture.

time delay fuse (TDF): A fuse that may detect and remove a short circuit almost instantly, but allows small overloads to exist for a short period of time.

torque: A turning or twisting force, typically measured in foot-pounds (ft-lb).

transformer: A device that converts electrical power at one voltage or current to another voltage or current.

two-family dwelling: A dwelling with two dwelling units.

Type I building: A building in which all structural members (walls, columns, beams, girders, trusses, floors, and roofs) are constructed of approved noncombustible or limited-combustible materials; has a fire resistance rating of 0 hr to 4 hr.

Type II building: A building that does not qualify as Type I construction in which the structural members (walls, columns, beams, girders, trusses, arches, floors, roofs, etc.) are constructed of approved noncombustible or limited-combustible materials; has a fire resistance rating of 0 hr to 2 hr.

Type S fuse: A plug fuse that incorporates a screw and adapter configuration which is not interchangeable with fuses of another ampere rating.

U

underground feeder and branch-circuit cable (Type UF cable): A factory assembly of one or more insulated conductors contained within an overall covering of nonmetallic material that is suitable for direct burial applications.

unfinished basement: The portion or area of a basement which is not intended as a habitable room, but is limited to storage areas, work areas, etc.

unit switch: A switch with a marked OFF position that is part of an appliance.

utilization equipment: Equipment that utilizes electrical energy for electronic, electromechanical, chemical, heating, lighting, or similar purposes.

V

voltage-to-ground: The difference of potential between a given conductor and ground.

W

wet location: Any location in which a conductor is subject to saturation from any type of liquid or water.

wireway: A metallic or nonmetallic trough with a hinged or removable cover designed to house and protect conductors and cables.

within sight: Visible and not more than 50′ away.

Index

Page numbers in italic refer to figures.

T

U

V

W

Z

USING THE ELECTRICAL SYSTEMS INTERACTIVE CD-ROM

Before removing the interactive CD-ROM from the protective sleeve, please note that the book cannot be returned for refund or credit if the CD-ROM sleeve seal is broken.

System Requirements

To use this Windows®-compatible CD-ROM, your computer must meet the following minimum system requirements:

- Microsoft® Windows® 7, Windows Vista®, or Windows® XP operating system
- Intel® 1.3 GHz processor (or equivalent)
- 128 MB of available RAM (256 MB recommended)
- 335 MB of available hard disk space
- 1024 × 768 monitor resolution
- CD-ROM drive (or equivalent optical drive)
- Sound output capability and speakers
- Microsoft® Internet Explorer® 6.0 or Firefox® 2.0 web browser
- Active Internet connection required for Internet links

Opening Files

Insert the interactive CD-ROM into the computer CD-ROM drive. Within a few seconds, the home screen will be displayed allowing access to all features of the CD-ROM. Information about the usage of the CD-ROM can be accessed by clicking on Using This Interactive CD-ROM. The Quick Quizzes®, Illustrated Glossary, Flash Cards, Interactive Load Calculation Forms, Media Clips, and ATPeResources.com can be accessed by clicking on the appropriate button on the home screen. Clicking on the ATP web site button (www.go2atp.com) accesses information on related educational products. Unauthorized reproduction of the material on this CD-ROM is strictly prohibited.

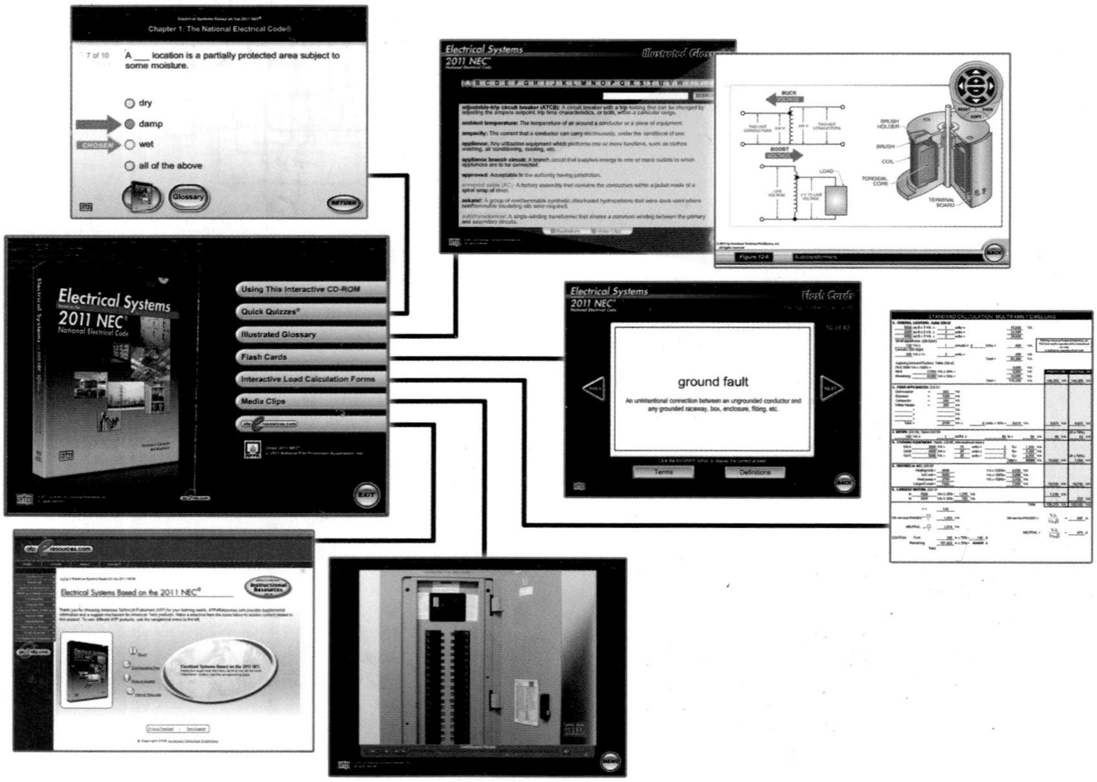